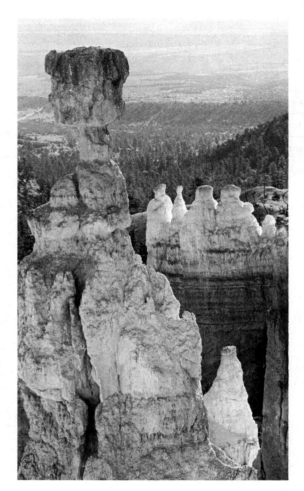

Geology of National Parks

FOURTH EDITION

Ann G. Harris
Youngstown State University

Esther Tuttle
Science Editor, Iowa City, Iowa

Sherwood D. Tuttle
Contributing Editor
University of Iowa

KENDALL/HUNT PUBLISHING COMPANY
2460 Kerper Boulevard P.O. Box 539 Dubuque, Iowa 52004-0539

Cover photo: Thor's Hammer, looking east, in mid-morning light. Processes of weathering and erosion, attacking beds of Eocene sedimentary rock, developed this striking pillar, topped by a large knob. Thor's Hammer is located in Bryce Canyon National Park, Utah. Photo by Kaz Hagiwara.

Library of Congress Catalog Card Number: 89–84799

ISBN 0–8403–4619–0

Printed in the United States of America
10   9   8   7   6   5

Contents

Foreword

To the Readers of this Book:

Welcome to *Geology of National Parks.*

Knowing something about geology will increase your enjoyment and awareness as a national park visitor and will also enhance your understanding of the natural world. National parks have a wealth of challenging and interesting illustrations of geologic concepts and are unparalleled as places to study geology. The presence of "geologic features worthy of preservation and protection" is one of the criteria for the selection and establishment of an area as a national park by Act of Congress.

This fourth edition of *Geology of National Parks,* like the previous editions, was developed for those who use the book to "get right into the parks" and learn about geology as they go along. Explanations of rocks and landscapes in the parks offer to beginners in geology vivid examples of geologic processes.

The design of the book groups 49 national parks into five parts or divisions based roughly on categories of geologic concepts. Each section of the book is introduced by a brief overview of geologic features and processes dominant in the parks making up that group; discussions of specific geologic components are given in the appropriate chapters. "Boxes" throughout the text deal with geologic topics of special interest. The five sections of the book are as follows: Part I Scenery developed by weathering and erosion on flat-lying, sedimentary rocks; Part II Caves and reefs in national parks; Part III Landscapes shaped by continental or alpine glaciation; Part IV Volcanic features and volcanic activity; and Part V Landscapes and structures in areas of complex mountains. Some arbitrary choices were made in cases of national parks having scenery and features representing several of these categories. Our general plan, however, was to proceed from simpler geologic themes to more complex ones; or, in another sense, from discussions of landscapes and features developed by the earth's *external* processes to the study of regions that display the results of the earth's *internal* processes.

Geology of National Parks presents basic elements of physical and historical geology in a nontraditional format with terms and concepts defined or explained as they are introduced. (Definitions are also presented in a glossary at the end of the book.) The chapter on each park includes discussions of significant geologic features and processes, a summary of the geologic history, information about the park's history and geographic setting, a location map, illustrations, and a simplified geologic column. To help readers grasp essential aspects of a park's geology, the important rock units and the relative times of geologic events are shown in the geologic columns, although the columns are not intended to be comprehensive or to show detailed stratigraphy.

Readers who desire more information about a particular park may find the references at the end of each chapter helpful. Most of the references are not highly technical and are easily accessible and understandable.

Users of this book may find it more convenient to learn about the national parks and their geology in a sequence different from our suggested groups. This is still practical because each chapter can stand alone and is not necessarily dependent on preceding or following chapters. Those who use *Geology of National Parks* may select or arrange topics in an order that is appropriate for their individual interests.

It is important that you pursue your interest in natural surroundings wherever you are, so that you can "see" what's around you and relate your observations to the dynamic processes that develop and continually change natural scenery. For this kind of "active" learning, the national parks offer a multitude of spectacular examples.

The National Park System

Over one century ago, in 1872, by Act of Congress and the signature of President U.S. Grant, Yellowstone National Park was established as the first national park. Fifty national parks now exist. The most recent, the National Park of American Samoa, was authorized by the 100th U.S. Congress late in 1988. This park midway between Hawaii and Australia, consists of volcanic islands and coral reefs near the eastern edge of the great Tonga Trench in the southwest Pacific. Publication schedules did not permit inclusion of a chapter on the geology of the National Park of American Samoa in this edition, but we do have a chapter on the recently established Great Basin National Park in Nevada, which came into the system in 1986.

Generally, a national park covers a large area containing features of great scenic and scientific quality and encompasses sufficient land and water for the adequate protection of the natural features and resources. A national monument, usually smaller than a national park, preserves at least one unique, significant resource but lacks the diversity of a national park. Although some national monuments are famous for remarkable geologic features

(Dinosaur National Monument in Colorado and Utah, for example), full coverage of monuments is beyond the scope of this book. We do, however, briefly mention some national monuments, particularly those in which the geologic features are related to the geology of a national park nearby.

When establishing the National Park Service, Congress stated that its fundamental purpose was to ". . . conserve the scenery and the natural and historic objects and the wild life therein and to provide for the enjoyment of the same in such manner and by such means as will leave them unimpaired for the enjoyment of future generations." Accordingly, the National Park Service has found that a strong interpretive program has been invaluable in effectively carrying out this mandate. In short, park rangers and other national park personnel have been teachers as well as caretakers.

Studying the geology of the national parks helps you to be aware of how interactions occur between rocks and landscapes in different climates, and how types of soil cover evolve. You notice how life forms—both plants and animals—select habitats controlled by geology. Fossils in the rocks tell you about ancient living things that grew in environments sometimes different from present conditions and sometimes very similar.

Traces of our more recent past are also found in areas set aside as national parks. Dwelling sites of ancient people, Indian relics, trails and wagon ruts of pioneers, abandoned mines and ranches, and much more are protected in national parks. In short, geology helps us to better understand the lives and work of those who came before us.

In recent years, increasing use of national parks and budgetary constraints have placed severe strains on the National Park Service's ability to manage and to operate its many diverse areas. National parks such as Great Smoky Mountains, Grand Canyon, Yellowstone, and Yosemite attract millions of visitors annually. Seeing the need, people in different parts of the country who are devoted to the national parks have donated many volunteer hours by helping according to their interests and abilities. Information about this program may be obtained at any national park or regional office.

Each park in the National Park System has its own beauty and uniqueness, and its own geologic story. Some of our national parks are considered so valuable that they have been designated by the United Nations as Biosphere Reserves and/or World Heritage Sites. But all are treasures of our own national heritage. They are ours to learn about, appreciate, and protect.

Sherwood D. Tuttle

Preface

National parks tend to be in regions that are on the cutting edge of developments in the earth sciences. Active tectonic plate boundaries, with seismic phenomena; subduction zones generating volcanoes; ancient plate boundaries; suspect terranes of assorted origins; hot spots at depth associated with volcanism at the surface; fossil beds that reveal clues as to the evolution of life forms; refinements in dating and geophysical techniques that move rock units higher or lower on the geologic time scale are some examples that challenge geologists to rethink, reevaluate, and update geologic histories and interpretations.

The rocks and the magnificent scenery in the national parks are still the same. But geology is a dynamic, speculative science. The challenge in preparing the fourth edition of *Geology of National Parks* has been to provide scientific explanations that are current, accurate, and understandable to readers who may not have a geologic background. We recognize the hazards of trying to synthesize incomplete or controversial geologic information, but we used what seemed to be the best explanations for each park in the light of what is known at this time. Some may fault us for not including enough detail about a significant process or feature in what may be their favorite park. Other readers may be overwhelmed by geologic terms and complex concepts. In all cases, when faced with doubts about what to include, we have based our decisions on what would help users of this book to expand their general knowledge and appreciation of geologic features in the national parks.

Seven new chapters have been prepared for the fourth edition of *Geology of National Parks;* and each chapter from the third edition has been revised and updated. Four of the new chapters are prepared by guest authors: Kenai Fjords National Park by Donald S. Follows; Wrangell-St. Elias National Park and Preserve by Monte Wilson; Glacier Bay National Park and Preserve and Great Basin National Park by Sherwood D. Tuttle. In addition, Donald F. Palmer rewrote the chapter on Yosemite National Park, combined and rewrote as one chapter Sequoia and Kings Canyon National Parks, and revised the chapter on Channel Islands National Park that he prepared for the third edition.

Ann F. Budd updated the chapter on Virgin Islands National Park, and with Katherine A. Goebel co-authored the chapter on Guadalupe Mountains National Park. Rodney M. Feldmann, the author of chapters on Theodore Roosevelt and Wind Cave National Parks, revised them for the fourth edition. Arthur N. Palmer revised his material on Mammoth Cave and Carlsbad Caverns National Parks. Lisa A. Rossbacher contributed the chapter on Redwood National Park in the third edition and updated the chapter for the fourth edition. Sherwood D. Tuttle revised his chapters from the third edition on the Petrified Forest, Yellowstone, Grand Teton, and Big Bend National Parks and also wrote new introductions to Parts I through V.

Instructors who use *Geology of National Parks* as a textbook draw on their own scientific interests and geologic experiences to meet the special needs of their students. A number of instructors who have used previous editions of the book have made suggestions for this revision, calling our attention to errors and pointing out significant items that were overlooked. We are indebted to them and hope their interest and helpfulness continues.

Other geologists who are knowledgeable about particular regions have answered questions and reviewed chapters or sections of this edition. They not only made comments, corrections, and criticisms that we could hardly have done without; but they also suggested additional sources of information, sent us unpublished materials, and lent us diagrams and photographs. Their help was invaluable. It is with grateful appreciation that we list their names below:

Richard Baker, University of Iowa
Robert R. Churchill, Middlebury College
Julie A. Dumoulin, U.S. Geological Survey
J. Thomas Dutro, Jr., U.S. Geological Survey
Arthur T. Fernald, U.S. Geological Survey (emeritus)
Brian F. Glenister, University of Iowa
Mary L. Griffitts, Boulder, Colorado
Lee Grim, Rainy River Community College
George Haselton, Clemson University
Carol A. Hill, Cave Research Foundation, Albuquerque, New Mexico
Richard A. Hoppin, University of Iowa
Patricia Humberston, Youngstown State University

Gwendolyn W. Luttrell, U.S. Geological Survey
David McGeary, University of California, Sacramento
Walter L. Manger, University of Arkansas
Willard H. Parsons, Wayne State University (emeritus)
Charles C. Plummer, University of California, Sacramento
Weldon W. Rau, Department of Natural Resources, State of Washington
Mark Reagan, University of Iowa
Bruce Reed, U.S. Geological Survey
John C. Reed, Jr., U.S. Geological Survey
Holmes A. Semken, University of Iowa
Keene Swett, University of Iowa
Rowland W. Tabor, U.S. Geological Survey
Paul Tychsen, University of Wisconsin, Superior
Robert E. Wallace, U.S. Geological Survey

We acknowledge with appreciation the continuing help and support from the Department of Geology and the Department of Geography at Youngstown State University and the Department of Geology at the University of Iowa.

Special thanks go to Kay Chambers, Alec Pacella, and Robert Tynal, who drafted maps and diagrams.

We cannot adequately express our gratitude for their assistance to National Park Service personnel and to those who staff the affiliated Natural History Associations. We especially appreciate the chapter reviews, suggestions, and corrections from national park scientists. Invariably, superintendents, naturalists, rangers, interpretive specialists, technicians, and secretaries responded efficiently and helpfully to our requests. Those who work in the national parks are special people. We admire their dedication to the philosophy and ideals of the National Park Service. As we ourselves have learned more about the irreplaceable wonders of the national parks, our own commitment to the preservation of America's parklands has been strengthened. All of us, as citizens, must help protect the national parks for future generations to enjoy.

Ann G. Harris
Esther Tuttle

PART I

Scenery Developed by Weathering and Erosion on Flat-Lying Rocks

© by G. Shimer

Figure I.1 Grand Canyon National Park, Arizona. The rocks are ancient, but the Grand Canyon and the Colorado River are relatively young. Drawing by G. Shimer from *Field Guide to Landforms in the United States*, J. A. Shimer. The Macmillan Company, © 1972 by J. A. Shimer. Used by permission.

Spectacular cliffs, canyons, arroyos, pinnacles, arches, pedestals, and dunes are some of the striking landforms produced by weathering and erosion that we see when we visit the national parks in this group. All are located in drier areas of the United States and all but three are in the Colorado Plateaus physiographic province. (Portions of the earth's surface that display a similar array of landforms and scenery are recognized as *physiographic provinces.*) The high plateaus of western Colorado, Utah, northern Arizona, and northwestern New Mexico make up the Colorado Plateaus province. Eight national parks (and more than a dozen national monuments) have been set aside in this remarkable and beautiful region. We next describe two national parks in the northern part of the Great Plains province, Badlands National Park (South Dakota) and Theodore Roosevelt National Park (North Dakota). The last national park in Part I is Kobuk Valley National Park in northwest Alaska, where some 500 square miles are covered by either active or stabilized dunefields.

Despite the dry climate of these areas, running water is the dominant agent of erosion shaping the land surface. Weathering processes, which prepare rock for erosion, operate wherever water is present, even in small amounts. Stream erosion—the most important external geologic process—sculptures the slopes and valleys. Scarcely a landscape exists anywhere that does not have erosional and depositional features produced by the action of streams.

Landforms are the individual features we see, such as a hill or a slope; *landscapes* are combinations of landforms making up an expanse of natural scenery—what the eye sees in a view. *Geologic agents* and *geologic processes* continually form and alter landforms and landscapes. Running water and wind, for example, are geologic agents; stream erosion and wind erosion are the geologic processes. *Weathering* of rock at the earth's surface, which includes both chemical decomposition and physical disintegration; *mass wasting,* the downslope movement of material under the influence of gravity; *entrainment,* the picking up of particles by running water and wind; transportation by a variety of means; and eventual deposition, either temporary or permanent, are some of the natural processes that recycle rocks and modify landscapes over time.

The generally low precipitation (rainfall and snow) in the national parks of the Colorado Plateau and the other parks in Part I supports a sparse or spotty cover of vegetation. We can see the geologic framework, the rock structures, and the dynamic interaction of erosional processes more easily and more dramatically here than in humid areas where forests tend to obscure the geology. The bedrock of the Colorado Plateaus has been elevated so that much of the present land surface stands well over a mile above sea level. The elevation gives streams so much erosive power that they have cut deeply into the rocky crust

and the resulting landforms are unmatched for size and scale in most other places in the world.

Visitors to the national parks of the Colorado Plateaus (fig. I.2) invariably are struck by the prevalence of red-colored rocks, reds of every hue, pinks light and dark, purples, orange-reds, yellows, browns. When you see them for the first time, names like Pink Cliffs, Chocolate Cliffs, Redwall Limestone, and the Painted Desert take on a vivid meaning. (The name Colorado itself means "red-colored" in Spanish.) A principal attraction for visitors to these parks is watching the reddish tints of the rocks change as sunlight and shadow move and shift from dawn to dusk. Depictions of these landscapes on film, postcards, and paintings seldom capture the grandeur of the view as seen by the human eye.

How does the sculpturing of the land take place? And how is the rock removed? Wherever moisture exists, weathering processes soften the rocks, alter the hues, and loosen the constituent particles. Water enters minute cracks, soaks into spaces between rock or mineral grains, and percolates slowly downward. Atmospheric gases and organic acids from decaying plants combine with the water to form a weak acid solution that slowly decomposes rock. When temperatures drop below freezing, trapped water freezes and expands, shattering rocks and breaking grains apart. Fragments, individual grains, clay, and soluble compounds are shed from rocks of all sizes and shapes. Loose sediment accumulates on rock surfaces, around the edges of blocks, at the base of cliffs, and at the foot of slopes. Impelled by gravity, the processes of mass wasting (sheetwash during rainstorms, creep of soil, slumping, sliding, and avalanching) move vast amounts of rock material downslope toward the valley bottoms.

Wind, a powerful geologic agent in dry climates, picks up and drops loose sediment. The size and the amount of the particles moved depend on weather conditions and how rock fragments are exposed. Wind also scours and abrades rock and piles up sand in dunes. Ancient or "fossil" dunes form prominent cliffs in Zion National Park in Utah (chapter 2), while modern, active dunes still accumulate and move in Kobuk Valley National Park in Alaska (chapter 11).

Running water, in trickles and in streams, picks up material and moves its load by rolling or jumping fragments, by carrying fine particles in *suspension,* and soluble matter in *solution.* Sediments are continuously eroded, deposited, and re-eroded. Fragments and particles become rounded, abraded, sorted, and mixed. They may lie undisturbed in a channel bottom or a river bar for many years or be spread across a valley floor during a flood. But inevitably sediments move downhill and downstream and away from their source area. Estimates of rates of erosion suggest that even high regions may be reduced to lowlands in a few million years. Erosion rates in the national

Figure I.2   Location on the Colorado Plateaus of the first eight national parks described in Part I. The Colorado Plateaus, a large physiographic province, is about 130,000 square miles in area. The region's distinctive characteristics are 1) flat-lying sedimentary rocks, with red the dominant color; 2) relatively high elevations, between 5000 and 11,000 feet; 3) deep canyons dissecting plateau surfaces; 4) retreating escarpments produced by erosion of rock layers of varying resistance; 5) arches, natural bridges, pinnacles, fins, and many other erosional features; 6) semiarid climate.

A

B

Figure I.3   Two views of erosional features in Canyonlands National Park, Utah. *A.* Canyon wall rising above the Colorado River. The beds here are Lower Triassic in age. Logs of petrified wood at the base of the cliff weathered out from the Chinle Formation. *B.* A hummock of resistant rock, an erosional remnant, sticks up from the flat mesa top, about 2000 feet above the Colorado River. Photos by S. D. Tuttle.

parks in Part I are comparatively high because downcutting goes rapidly in regions of sparse vegetation and high elevation. The Grand Canyon, deep as it is, was eroded only within the last one or two million years.

The Geologic View of Time

Geology draws on other sciences, both physical and biological, for many of its concepts; but as a science, its unique intellectual aspect is its historical approach and its view of time. Geologists think in terms of sequences of events and develop explanatory patterns that closely follow cause and effect within a framework that's called "geologic time." A geologist observes running water gullying a slope and carrying off soil and postulates that in the past running water eroded slopes, transported sediment, and deposited it in a similar way. This fundamental principle is called *uniformitarianism,* which means that natural processes have operated in the past (recent or ancient) in much the same way and at about the same rate as now. "The present is the key to the past" is another way of expressing this idea.

Therefore, the time ordering, or sequencing, of events in the geologic history of a feature or region becomes very important in studying and understanding modern landscapes. By going farther back, a geologist attempts to fit the events into the overall time framework of earth history. For example, we look at a glaciated mountain range and assume that glaciation occurred after the range was uplifted. Streams dissecting the glacial deposits began after (or while) the glaciers melted.

By various techniques, the rocks in the mountain range might be dated Jurassic and Cretaceous in age with the glaciation occurring late in Pleistocene time. These are *relative* time terms (shown in the Geologic Time Scale, inside front cover). Relative time units are like days of the week or cultural and historical eras of human history, except for their much greater expanse of earth time. Relative dating techniques, which are based on paleontology (study of ancient life forms) and stratigraphy (interpretation of rock layers), are mostly logical applications of careful observations and tests in order to build a sequence of geologic history. Prior to the discovery of radioactivity, these relative dating methods, which are still necessary and important, were the only means of determining when rocks formed.

The determination of *absolute ages* of rocks became possible with radiometric techniques that measured the rates of radioactive decay, or disintegration. *Radioactive decay* is the spontaneous nuclear disintegration of certain isotopes of radioactive elements, a process that begins when some types of rock are formed and one that continues at a constant rate. The "radioactive clock" in such a rock can be read by laboratory techniques that compare the amount of radioactive elements with the amount of radioactive decay products. This determines how much time has elapsed since the rock was formed because radioactive elements in rocks have characteristic and constant rates of decay. The time units are expressed as "years before the present," or y.b.p. An absolute age, for example, might be 265 million years, plus or minus 30,000 years. Note that absolute age is not the same as exact age.

The Flat-Lying Rocks

In the national parks discussed in Part I, most of the solid bedrock, whether exposed or underlying the soil cover, is relatively undeformed, flat-lying sedimentary rock. In many places in the Colorado Plateaus, a sequence of sedimentary rock layers, one to two miles thick overall, is exposed. You can stand on the floor of a canyon or before the face of a cliff and count the beds (*strata*) from bottom to top. (This is called "layer-cake geology.") Many varieties of sedimentary rock, formed in different periods of geologic time and by different ways are found in these national parks, but all resulted from the consolidation of loose sediment (particles, grains, shells, etc.) that accumulated in layers. Sand grains, compacted and cemented, become *sandstone*. Finer quartz grains, clays, and muds may lithify (form rock) as *shales*. *Limestones* may precipitate from seawater (an inorganic process) or accumulate from the skeletal remains of marine organisms.

By careful examination of the layered rocks, geologists produce data and make interpretations about the compositions of the rocks, their ages, the environments in which the sediments accumulated, where the sediments came from, and the remains of ancient life (fossils) preserved in the rocks. The remarkable colors of the rocks, due in part to the presence of minor amounts of iron compounds, traces of carbon, manganese, copper, and other elements, provide clues to the origins of rocks, how they were lithified, and under what conditions.

With this sort of information, geologists can reasonably deduce what it was like in a park area, say 400 million years ago. Of course, this much earlier time was long before the canyons were cut and the slopes eroded back. The present topography (the landscapes we see) was formed recently and with dramatic speed, if compared with the earlier geologic history of the region.

We describe the sedimentary rocks in the Colorado Plateaus and the Dakota badlands as "undeformed" and "flat-lying," but this is true only in a general sense. Some of the layered rocks have been subjected to stress and in places have yielded to bending (*folding*) or breaking (*faulting*). As a park visitor, you may recognize broad open folds, or you may see faults where rock has moved several hundred feet. However, in comparison to deformation that rocks have undergone elsewhere, mild bends and breaks in layers can be regarded as reflections of local circumstances rather than as a regional characteristic.

To summarize, the scenery of the national parks in Part I—except for Kobuk Valley National Park—developed because of deep dissection of brightly colored sedimentary rocks in areas of low rainfall where vegetation does not blanket the landscape. The difference in the Kobuk Valley, which is also dry and sparsely vegetated, is that the sediments are comparatively young and are unconsolidated. But in all these national parks the interactions of complex weathering, downslope movements, and wind and stream erosion have produced magnificent landscapes.

National Parks in Part I

1

The incomparable Grand Canyon of the Colorado River is ten miles wide and over a mile deep. The national park encompasses 277 miles of the river, plus adjacent uplands.

Weathering and erosion have produced unparalleled scenery of colorful cliffs, slopes, spires, buttes, and mesas. Exposed in these features are an immense variety of rock formations that illustrate a vast span of geologic history. Two sequences of Precambrian rocks record very early mountain-building episodes. A classic Paleozoic section includes carbonates, shales, and sandstones deposited in five geologic periods.

Grand Canyon National Park

Northern Arizona

Figure 1.1 The north wall of the Grand Canyon as depicted in John Wesley Powell's report on his 1869 and 1871 expeditions. Rising from the waterline are ancient crystalline rocks of the Precambrian Vishnu Group. The dipping beds next above are rocks of the Precambrian Unkar Group. A distinct angular unconformity separates the Precambrian rocks of the Inner Gorge from the flat-lying Paleozoic beds of the canyon wall. From *The Colorado River Region, John Wesley Powell.* 1969. U.S. Geological Survey Professional Paper 669.

Local History

The Indians of the Southwest had legends to explain the origin of the Grand Canyon. These legends indicate that the Indians were as much in awe of the Grand Canyon as the white men who arrived later. The stories reveal an awareness of natural causes and effects. The Navajos, for example, believed that rain for many days and nights caused a great flood that covered the land with water that rose higher and higher. Finally an outlet formed, and as the rushing waters drained away, the Grand Canyon was cut deeply into the earth.

Nomadic hunters camped in the Grand Canyon region between 3000 and 10,000 years ago. Twigs shaped into forms resembling animals, found in caves in the canyon walls, may have been left by the early hunters. More advanced cultures developed as new groups settled in the Southwest. More than 500 Indian sites in the park, both on the rim and within the canyon, indicate that as recently as 600 years ago large numbers of Indians lived in the Grand Canyon vicinity. Havasupai Indians still live in a branch canyon watered by Havasu Creek in the western section of the park.

In 1540, a band of Coronado's conquistadores were the first white men to see the Grand Canyon, after being guided by Hopi Indians to the rim. They tried to descend to the Colorado River but managed to get only a third of the way down. Discouraged, they gave up and left.

More than 300 years later, Lt. Joseph C. Ives, a U.S. Army surveyor, came upriver from the southwest and reached the lower Grand Canyon. After an unsuccessful attempt to reach the rim, he concluded that surveying the Grand Canyon of the Colorado River was impractical.

John S. Newberry, a member of the Ives party and a geologist, thought otherwise. He convinced Major John Wesley Powell, a fellow geologist (who had lost a forearm in the Civil War), that an expedition by boat down the Colorado River through the Grand Canyon would be worth the risk in order to complete the survey. With scientific help from the Smithsonian Institution, the Powell expedition of four boats and nine men left Green River, Wyoming, in May 1869, and began their hazardous journey downstream. One boat was smashed; scientific instruments and food were lost. Most of the time the party was wet, tired and hungry, and convinced that they would never get out of the canyon alive. Three of the men climbed out of the canyon but were killed on the plateau by hostile Indians. The men who stayed with Powell survived and reached the settlements near the western end of the Grand Canyon. Two years later, Powell went down the Colorado River again in order to collect more data for his geographical and geological report on the region.

Theodore Roosevelt, after a trip to the Grand Canyon in 1903, was determined that it should be preserved. He was able to have part of the area placed under government protection in 1908, but opponents blocked the establishment of a national park until 1919. The park was enlarged in 1975 by combining Marble Canyon National Monument and Grand Canyon National Monument, plus additional lands. Grand Canyon National Park was designated a World Heritage Site in 1979.

The Colorado River carries tons of sediment past any one point each day. The rate of erosion is not constant throughout the course of the river, but it is estimated that every thousand years the drainage basin of the Colorado River system is lowered by an average of 6½ inches. It was his observation of the sediment load that convinced Powell that he would not be washed over high falls (as some people had predicted) on his initial expedition. He reasoned correctly that a river carrying so much sediment would have scoured its bed down to a more or less even grade, or slope.

The Colorado River no longer rampages every spring or falls to lowest low-water stage in summer. The Colorado has been a regulated stream since 1966, when Glen Canyon Dam, just upstream from Lees Ferry, impounded river water to fill Lake Powell in Glen Canyon National Recreation Area. The tremendous sediment loads that used to scour the canyon during spring floods now settle to the bottom of Lake Powell, and the pre-1966 average loads of half a million tons per day have been greatly reduced. The river below the Grand Canyon has been controlled since 1935 by Hoover Dam, which backed up water in Lake Mead to Grand Wash, where the river leaves the national park.

Grand Canyon National Park extends from the mouth of the Paria River, which joins the Colorado River just below Glen Canyon Dam, to the eastern boundary of Lake Mead National Recreation Area, a river distance of 277 miles. The Grand Canyon is over a mile deep, with the distance from rim to rim ranging from 9 to 18 miles. The elevation of the North Rim (8900 feet) averages some 1200 feet higher than the South Rim (6900 feet); and the North Rim is about twice as far from the river as the South Rim.

Dimensions of such magnitude have biological as well as geological significance. Life forms on the two canyon rims have evolved differently because of physical separation over time. At the higher altitude of the North Rim, under conditions of cooler average temperatures and greater annual precipitation, plants and animals characteristic of Canadian life zones became established. Trees and shrubs such as blue spruce, ponderosa pine, fir, and aspen, that are darker in color, took over. A typical North Rim animal is the Kaibab squirrel, which has a black body, black stomach, and white tail.

On the South Rim, vegetation that is lighter in color and typical of the Upper Sonoran life zone (junipers, pinon, etc.) predominates. The Abert squirrel is found here, with its grizzled gray body, white stomach, and gray tail. The

GRAND CANYON NATIONAL PARK

Figure 1.2 Grand Canyon National Park, Arizona.

common ancestor of these two modern species was a black and white squirrel that lived on the plateau before the Grand Canyon was incised. When the squirrels were no longer able to migrate back and forth across the physical barrier of the canyon, and as differences in climate became pronounced between the North Rim and South Rim, new squirrel species developed due to evolutionary adaptation—the darker Kaibab squirrel on the North Rim and the lighter Abert on the South Rim.

Geologic Features

In terms of geologic time, the Grand Canyon of the Colorado River is surprisingly young. The river accomplished the major part of the downcutting during the last two million years. It is the rocks that are ancient, not the canyon itself. Exposed by this downcutting is one of the most nearly complete geologic columns on earth, encompassing some two billion years of geologic history. The layers of sedimentary rock seen in the Grand Canyon are now, as when they were deposited, essentially flat-lying; many of these horizontal beds extend throughout the Colorado Plateaus region.

How Does a River System Evolve?

The Colorado River is a major through-flowing stream with a total length of 1450 miles from the Rocky Mountains to the Gulf of California. Like other major streams, the Colorado developed in response to environmental variables over time; but in the Colorado's case, its history has been more complicated. For over a hundred years, geologists have argued about possible courses and directions of flow of tributaries and "ancestral" streams that may have had some part in the Colorado's evolution. So far no hypothesis has met with overall acceptance. The details of the problem are fairly technical, but some general principles that apply to stream development are outlined below.

Base Level. The limiting surface to which a stream can cut down is called *base level*. Sea level is the ultimate base level for through-going streams like the Colorado, but sometimes local base level, such as a lake, a reservoir, or an interior basin without an outlet, controls a stream's downcutting ability. The Colorado's base level has been affected by uplift and also by sea level fluctuations, especially during the Pleistocene Ice Ages. Moreover, segments of the river, probably before it became through-flowing, may have emptied into inland lakes that functioned as temporary base levels.

Gradient. A stream whose *gradient* (rate of descent, or slope) is low does not have much erosive power. But if an area is uplifted, the gradient steepens, increasing the velocity of stream flow. When this happens, a stream downcuts rapidly, especially if its volume of water also in-

Figure 1.3   Natural caves and a flowing spring (left) in Redwall Limestone in the Grand Canyon. Ground water, seeping down through rock layers, over time dissolves parts of the limestone and enlarges openings in the rock. Prehistoric animals and humans found shelter in caves in the canyon walls. From D. Nations and E. Stump, *Geology of Arizona.* © 1981 Kendall/Hunt Publishing Company. Used by permission.

creases. Sometimes, during uplift, a stream is able to maintain its original course and cut a narrow trench down through bedrock.

Headward Erosion and Stream Capture. While a trunk stream is downcutting or entrenching itself, its tributaries lengthen themselves by *headward erosion,* which means that they cut back into a plateau or upland at the head of each valley. A drainage system that is favorably situated, perhaps flowing on weaker rock, for example, is able to erode headward faster and eventually intersect and *capture* the headwaters of a neighboring stream or drainage system. Geologists are sure that stream capture occurred numerous times while the Colorado was evolving into a major stream; but the details of how, when, and where these events happened is not clear.

The key factors in the Colorado River's evolution are probably these:

1. The river's source area in the Rocky Mountains and also the Colorado Plateaus have undergone a long series of relatively rapid uplifts since the end of Mesozoic time (65 million years ago). Overall, the Colorado Plateaus rose 5000 to 10,000 feet. Stream gradients became steeper, increasing stream velocity and erosion capability.

2. More rain and snow fell in the Rocky Mountains as their elevation increased. In the colder Pleistocene climates, glaciers formed and grew. As more melt-water and runoff became available, streams gained volume. The streams that drained the Rocky Mountains were powerful eroders as they poured out on the plateaus.

3. The climate of the plateaus was (and is) drier than the mountain climate. Therefore, the high-volume, swift rivers cut narrow, deep canyons. Because rainfall is low on the uplifted plateaus, runoff is not ample enough to widen the valley side slopes.

4. When the Gulf of California opened approximately 6 million years ago creating a depression, this lowered base level for the Grand Canyon region and caused rapid headward erosion. The modern course of the through-flowing Colorado River was probably established during this time. Previously to the opening of this arm of the sea, the Colorado—or an ancestral stream that drained the Rockies—flowed into the ocean at a different location.

5. The east–west Grand Canyon segment of the Colorado's course across the Kaibab Plateau is asymmetrical; that is, the river flows closer to the south side due to a slight southward tilting that occurred during the last uplift. The fact that the North Rim receives more moisture and sheds more sediment also tends to push the river toward the south side of the canyon. The rate of erosion on the Kaibab Plateau is greater because all of the precipitation falling on the north side of the river drains to the south and into Grand Canyon. On the South Rim, rain and meltwater also drain to the south; but since the runoff goes away from the canyon, it has less erosional effect.

6. How the Colorado River is regulated will determine how it evolves in human time in this century and the next. Lake Powell became filled to capacity in 1980. How long will it be before silt accumulating in the lake bottom will cause the river to run over the dam? If such overflow is allowed to happen, the dam will inevitably wash out.

The controlled average release of water by Glen Canyon Dam is 12,200 cubic feet per second. Contrast this with the range of flow before dams were built above the Grand Canyon: a low of 1000 cubic feet per second during the driest time of year and a recorded flood high of 325,000 cubic feet per second. The Paria River, which joins the Colorado just below Glen Canyon Dam, brings some sediment, seasonal high-water flow, and occasional flash-flooding to the Grand Canyon. To a lesser extent, the Little Colorado River, with its confluence just above the big bend where the main river turns westward, does the same. But these contributions, added to sediment-free water released by the dam, are not powerful enough to keep the canyon bottom free of debris. Will the inevitable slumping of weak zones in the walls and accumulation of debris from the tributary canyons create obstructions in the canyon?

Restoring the river's normal flow is not a practical answer because the Colorado River is the only significant source of surface water in the entire Southwest. More of its water is legally allocated to users than the river normally carries. How national parks and recreational areas are affected by the regulation of the Colorado River is only one aspect of a difficult problem of crucial importance to millions of people.

Igneous and Metamorphic Rocks in the Grand Canyon

As the Colorado River sliced its way deeper into the up-warped rocks of the plateau, it let itself down onto ancient crystalline rocks and structures beneath the flat-lying Paleozoic beds. These igneous and metamorphic rocks, exposed in the Inner Gorge of the Grand Canyon, are Precambrian, which means that they are over 600 million years old. The oldest is the Vishnu Schist at the very bottom of the canyon. The rock originally accumulated as silts, muds, and volcanic material and lithified about 2 billion years ago. During mountain-building episodes, the layered rocks were highly deformed and metamorphosed. On dark-colored, contorted Vishnu outcrops alongside the river, little, rounded garnets stick out from the rock surface like a scattering of seeds.

A body of granitic magma, the Zoroaster Granite, intruded the metamorphosed Vishnu rocks deep underground and gradually cooled. The granite shows up as light-colored, irregular bands and blotches in the Vishnu Schist.

Later in Precambrian time, many layers of shales, sandstones, and other sedimentary rocks—interbedded with countless lava flows—covered the region. These rocks, which comprise the Grand Canyon Supergroup, were also metamorphosed by heat and pressure during tectonic activity. Some of the lavas show up as dark ribbons or bands between or cutting through the old sedimentary beds. (Geologists call such features *dikes,* when they cut through older rocks, and *sills,* if they are squeezed between rock layers.) A long period of erosion removed a great portion of the Precambrian rocks.

In comparatively recent geologic time, around a million years ago, dark (basaltic) lava erupted from fissures and flowed down into the Grand Canyon in the western section of the park. Volcanic ash, thrown out from small cinder cones on the canyon rim, fell on the land. These flows and ash falls made the youngest rocks in the park.

Box 1.1
How Rocks Are Classified

How a rock originated determines its category. The three main divisions are *igneous, sedimentary,* and *metamorphic.* Igneous rocks are (1) those that cool from molten rock material, called magma, deep in the earth; or (2) those that cool from *lava,* derived from magma that rises to the earth's surface. Sedimentary and metamorphic rocks are formed from preexisting rocks. Loose particles of rock (sediment) are consolidated, or *lithified,* to make up sedimentary rocks. Metamorphic rocks, which originally may have been igneous, sedimentary, or other metamorphic rocks, have been so altered by burial, heat, chemical fluids, and pressure that they acquire a new rock identity.

When you look at a hand specimen of a rock, you may or may not be able to tell how the rock originated; but you can probably see grains or particles. Either with the naked eye or with the aid of a hand lens, you may be able to recognize some of the mineral constituents by their physical characteristics.

Rocks are made up of grains of one or more varieties of minerals. In the geological sense, a *mineral* is a naturally occurring, inorganic, crystalline solid with a definite chemical composition and characteristic physical properties that are either uniform or variable within definite limits. Therefore, because the grains in rock represent the constituent minerals and because each mineral has a specific crystalline structure, the mineral grains, their size, and how they combine provide the basis for identifying and classifying rocks.

Rock texture, a means of describing the grain size of component minerals (or sediments, in the case of sedimentary rocks), ranges from extremely fine to extremely coarse. For example, most *extrusive igneous rocks* (those erupted at the earth's surface) have a fine-grained texture with crystals too small to be seen by an unaided eye. This is because at surface temperatures and pressures, lava cools too fast for crystals to attain much size while the rock hardens. A rock that cools so quickly that *no* crystals have time to form becomes a natural glass, called *obsidian.* (In the plateau country, weathered pieces of obsidian are called "Apache tears.")

Intrusive igneous rocks (which harden far below the earth's surface) tend to have a coarse-grained texture because magma cools at great depth under confining pressure over thousands, and perhaps millions, of years. As the mineral crystals separate from the melt, they grow, interlocking and filling all the available space.

When igneous rocks and any other rocks are exposed at the earth's surface, they are subjected to processes of *weathering* that bring about chemical changes (*decomposition*) and physical changes (*disintegration*). The rock particles produced by weathering are usually transported and then deposited, commonly in layers, and may eventually become sedimentary rock; that is, rock made up of consolidated sediment.

Sediment is a collective term for loose, solid particles of rock that can originate in several ways. *Clastic* sediment refers to any accumulation of unconsolidated fragments freed by weathering and erosion of preexisting rock. Sizes of fragments range from large boulders to microscopic clay particles. During transportation by running water, glaciers, or other agents, these rock particles become rounded and broken further by processes of *abrasion,* which is the wearing away of rock by friction.

Decomposition of rock involves chemical alteration by the addition or subtraction of various substances. Some minerals, for example, go into *solution* when exposed to water and are transported in dissolved form. Later, the minerals may be *precipitated out* by evaporation or by organic secretion (by marine organisms, etc.). Sediment formed by chemical precipitation or by organic secretion is referred to as *nonclastic.*

A sedimentary rock made up of grains or fragments of clastic sediment is described as having *clastic texture.* A *conglomerate,* for example, is a coarse-textured rock made up of consolidated rounded gravel. *Shale* is a fine-grained rock made of silt or wet mud that has been lithified by compaction and cementation.

continued

Box 1.1 continued

Table 1.1 Classification of Common Sedimentary Rocks		
Source	Sediments	Sedimentary rocks
coarse-textured	boulders cobbles pebbles granules	conglomerate (rounded fragments) or breccia (angular fragments)
Clastic	quartz grains; other mineral and rock particles of sand size	sandstone, (graywacke, arkose)
fine-textured	particles of silt and/or clay size	shale (thin-bedded), mudstone, siltstone
Chemical precipitation	calcium carbonate $CaCO_3$ (calcite)	limestone
	calcium magnesium carbonate $CaMg(CO_3)_2$ (dolomite)	dolomite/dolostone
and/or	silicon dioxide SiO_2 (silica)	chert
evaporation	sodium chloride NaCl (halite)	rock salt
	calcium sulfate $CaSO_4$ (gypsum)	rock gypsum
Organic remains plus biochemical secretions, etc.	shells, shell fragments, skeletal material (bones, teeth, etc.) calcareous or silicious ooze	limestone, some chert
	buried vegetable matter	peat, coal

Grade sizes of clastic sediments:

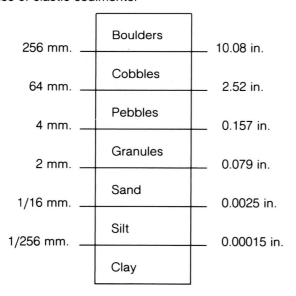

256 mm.	Boulders	10.08 in.
64 mm.	Cobbles	2.52 in.
4 mm.	Pebbles	0.157 in.
2 mm.	Granules	0.079 in.
1/16 mm.	Sand	0.0025 in.
1/256 mm.	Silt	0.00015 in.
	Clay	

Cementation is the process by which clastic sediments are converted into sedimentary rock by precipitation of a mineral cement (such as silica or calcite) between the sediment grains. As mineral cement binds the sediment grains together, it becomes an integral part of the rock.

The most common sedimentary rock having a *nonclastic texture* (i.e., made up of nonclastic sediment) is *limestone*, which is composed mainly of interlocking crystals of calcite (calcium carbonate) that have been precipitated by inorganic or organic means.

Table 1.1 shows how the more common sedimentary rocks are classified. Exposures of these rocks in cliffs, buttes, and mesas form the distinctive landscapes of the Colorado Plateaus.

The third major rock category, metamorphic rocks, is the least abundant group in the Colorado Plateaus; but examples are found in some localities, such as in the deeper parts of the Grand Canyon. Generally (but not always), a metamorphic rock has a texture and composition different from those properties in its "parent" rock. For example, an erodible sandstone may be changed into highly resistant *quartzite;* a soft shale into hard *slate;* and a dull limestone into gleaming *marble.* If metamorphism persists or intensifies, slate may change into *phyllite,* which, in turn, may become a *schist,* and then a *gneiss.* During metamorphism, if the melting point of a rock is exceeded, the rock becomes molten. The resulting magma may eventually be "recycled" as an igneous rock.

This continual reworking of the earth's crust over geologic time can be depicted graphically by a theoretical *rock cycle* (fig. 1.4). A rock cycle model shows the interrelationships among weathering, erosion, sedimentation, deposition, burial, tectonic forces, melting, volcanism, and the major rock types. With the passage of time, new rocks are formed from old rocks under many different conditions, and minerals combine and recombine to produce the infinite variety of rocks found in nature.

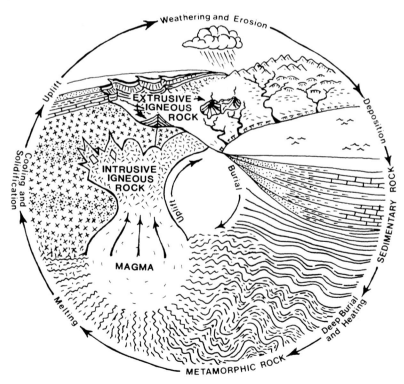

Figure 1.4   Rock cycle. From D. Nations and E. Stump, *Geology of Arizona.*
© 1981 Kendall/Hunt Publishing Company. Used by permission.

Figure 1.5   Geology students in dories looking at Vishnu Schist and Zoroaster Granite in the Inner Gorge. Photo by K. Swett.

Geologic Structures

Unconformities. Before erosion planed off the igneous and metamorphic rocks of the Inner Gorge, the rock units had been broken into blocks and tilted about 15° by mountain-building. But the surface was relatively flat by the time an ocean covered the region and the sedimentary beds of Paleozoic time began to accumulate. The surface or contact between Paleozoic and Precambrian rocks, which is called an *unconformity,* represents a gap in the geologic record. In this case, it is an *angular unconformity* because the bedding planes of the rock units above and below the erosion surface are *not* parallel; specifically, the younger overlying sediments rest upon the eroded surface of tilted or folded older rocks (fig. 1.6).

What causes gaps in the rock record? The marine sedimentary rocks in the walls of the Grand Canyon cover a large portion of geologic history, but some geologic periods are not represented. For example, a major erosional unconformity exists between horizontal Cambrian and Devonian strata in the canyon sequence. Were the marine beds of the missing geologic periods completely eroded away? Or did marine sedimentation not occur during those periods?

Tectonic forces generated within the earth have resulted in uplift, movement, and deformation of portions of the earth's crust. Continents have stood at times higher and at times lower during geologic history. Ocean levels have fluctuated gradually over millions of years.

When sea level rises, the oceans *transgress* onto the land; and when sea level falls, they *regress* from the land. During the long and slow process of transgression or regression, streams carrying sediment down to the ocean continue to erode. If the ocean is regressing, the streams may strip and redeposit (at lower elevations) most or all of the sediment layers that were put down during an earlier period of transgression. When the seas finally come back and again cover an area that has been stripped of sediment, a gap, shown as an unconformity in a geologic column, occurs in the geologic record. The rock above an unconformity is considerably younger than the underlying rock.

Geologists who try to reconstruct ancient changes in elevation and sea level find unconformities to be useful indicators. Typically, an unconformity can be recognized by the presence of a *basal conglomerate* that rests on the old

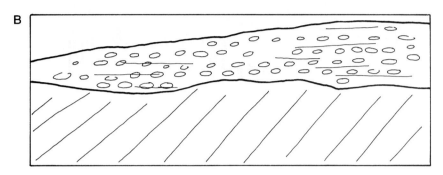

Figure 1.6 **Unconformities** represent a gap in the geologic record. They are old erosional surfaces separating younger rock units from older rocks exposed to erosion before being covered over by younger rock layers. *A.* This unconformity may be due to an interruption in sedimentation or to uplift and erosion. Evidently deformation did not occur while processes of weathering and erosion operated because beds above the unconformity are approximately parallel to layers below the contact. *B.* This example of an **angular unconformity** shows that the older rocks below the contact were folded ("deformed") and subsequently eroded. Eventually younger beds were deposited on the old erosional surface.

erosion surface and is composed of fragments of old eroded sediment incorporated with younger sediment deposited by a transgressing sea.

As a shoreline shifts landward or seaward (depending on whether the sea is transgressing or regressing), beach deposits accumulate in this narrow depositional environment. A *time-transgressive rock unit,* with the rocks becoming younger in the direction in which the sea was moving, may show the shifting location of the old shoreline. The Tapeats Sandstone, which is on the upper side of the angular unconformity between the Precambrian rocks and the Paleozoic beds, is an example of a time-transgressive rock unit (fig. 1.7).

Folds and Faults. Layered sedimentary rock units, such as those exposed in the canyon walls above the igneous and metamorphic rocks of the Inner Gorge, have other interesting clues about geologic conditions in the past. The presence of marine fossils leads to the obvious conclusion that these rock layers have been uplifted more than a mile from their original position beneath an ancient sea. Uplift must have been essentially uniform over a large area in order to keep the layers so nearly horizontal.

What happens to layered bedrock when the uplifting forces are not uniform? Keep in mind that most tectonic activity takes place at considerable depth when rocks are subjected to great heat and pressure. Under such conditions, layered rock that would fracture or break in a surface environment may become *plastic,* or malleable, and bend into *folds* rather than break. The sides, or arms, of

a fold are called *limbs.* Most folds are either *anticlines* (upfolds) or *synclines* (downfolds) (fig. 1.8). If the limbs are parallel (as with a hairpin), the fold is called an *isocline* or *isoclinal fold,* and we know it must have been formed by intensely compressive forces (i.e., squeezed together). Folds of this type are found in some of the Precambrian metamorphic rocks in the Grand Canyon.

A type of fold more characteristic of the Colorado Plateaus is the monocline, which usually shows up in flat or slightly tilted beds. *Monoclines,* or *monoclinal folds,* have one limb and a gentle bend, or flexure, connecting rock layers at one level (elevation) with the same layers at another level (fig. 1.8). The East Kaibab monocline, which outlines the eastern edge of the Kaibab Plateau, is an example.

A *fault* is the result of movement along a break or fracture in bedrock causing *displacement.* Faults are named and identified according to the type of movement or displacement that occurred. Faulting may or may not be related to the topography of an area. In some regions the surface features give no indication of the geologic structure of the bedrock below; but in the Basin and Range Province, just west of the Grand Canyon, great mountains and valleys display evidence of faulting on a grand scale.

Normal faults separate blocks that have been either raised or dropped, mainly as a result of tensional and vertical stresses affecting the earth's crust. Tectonic forces that stretch the crust cause normal faults. When a series of such faults develop (as in the Basin and Range), the

Figure 1.7   Time-transgressive rock layers may be deposited when a shoreline shifts laterally, causing the age of a formation to become progressively younger across an area. Beach deposits of the Tapeats Sandstone accumulated as an ocean advanced over Precambrian rocks; mud and silt settled in shallow water offshore and became the Bright Angel Shale; the Muav Limestone formed in deeper water. The time line in the cross section represents a theoretical equal age across the formations shown. From D. Nations and E. Stump, *Geology of Arizona.* © 1981 Kendall/Hunt Publishing Company. Used by permission.

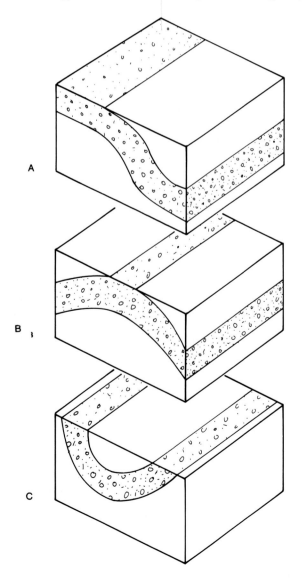

Figure 1.8   Block diagrams of simple folds. *A.* **Monocline.** Local steepening of flat-lying beds produces a bend in the rock unit. *B.* **Anticline.** Compression of the layered rocks creates an upfold, opening downward, with older beds in the center of the fold. *C.* **Syncline.** A downfold, also the result of compression, opens upward, with younger beds in the center of the fold.

uplifted blocks are called *horsts* and the downdropped blocks are called *grabens.* The normal faults in the Precambrian rocks in the Grand Canyon indicate block-faulting during ancient geologic times and are unrelated to present surface features.

Faults involve relative movement between two blocks, one of which is called the footwall block and the other the hanging wall block. If a person were able to stand on a fault surface within the earth (as if in a mine), his feet would be supported by the *footwall,* while the *hanging wall* would be the rock surface over his/her head. In a normal fault, the hanging wall has moved *down* in relation to the footwall.

If the hanging wall has moved *up* in relation to the footwall, it is called a *reverse fault.* A special type of reverse fault, in which the fault plane has a low angle (about 10° or less), is referred to as a *thrust fault* (fig. 1.9). Reverse faults are produced when tectonic forces compress crustal rocks, causing them to break vertically or at an angle. In the Grand Canyon, some of the Precambrian normal faults (produced by tensional forces) were changed millions of years later to reverse faults by compressive forces.

Sometimes compressive forces cause blocks to slide past each other, producing a *strike-slip fault* (fig. 1.9) that shows lateral displacement.

Sedimentary Environments

Studying the sizes, arrangement, and composition of sediments can often reveal a great deal about the origin of sedimentary rocks and the environmental conditions that prevailed at the time the sediments accumulated.

For example, a geologist may be able to tell what type of weathering produced the sediment that comprises a sedimentary rock. (*Weathering* refers to destructive processes that change or break down rock at or near the earth's surface.) In the Grand Canyon many interesting examples of weathering processes can be seen. *Brecciated* deposits of rock that had been broken up and crushed into angular fragments are found where a cave ceiling has fallen or in fault zones where movement along fault surfaces has crushed rock. Rock broken by faulting weathers more rapidly because more surfaces are exposed to water and the atmosphere. Tributaries of the Colorado River have tended to form along fault zones where the rock is susceptible to weathering and erosion. An example is Bright Angel Canyon.

Some of the Grand Canyon limestones have *karst* features, which are the result of slightly acidified underground water dissolving soluble rock. Because limestone consists principally of the mineral calcite (calcium carbonate), which is soluble under humid conditions, water trickling down through cracks and crevices in rocks tends to enlarge any voids into hollows or caverns. The caves in the Red Wall Limestone developed during an earlier time when the climate was more humid than now.

In the arid climate of the present, the limestones of the Colorado Plateaus tend to be resistant to weathering. The Kaibab Limestone, which forms the caprock on the rim of the Grand Canyon, is an example. However, visitors approaching the North Rim can see circular pits and hollows, some containing water, beside the highway. These are *sinkholes*, which are depressions dissolved beneath the surface in the Kaibab Limestone. The Kaibab Plateau is flat enough so that rain and meltwater tend to collect in pools that over time eat away the soluble rock.

Precipitation that falls on the Kaibab Plateau also percolates down through the sedimentary beds of the North Rim. When the water reaches an impervious rock unit, it emerges from the canyon wall as seeps or natural springs. Roaring Springs, one of the larger springs supplied in this way, gushes from a chasm high on the north wall. Roaring Springs provides water (with the aid of a pumping station) for the Park Service facilities on the South Rim.

When water containing dissolved calcium carbonate from the limestones evaporates, deposits of finely crystalline, pale *travertine* are precipitated. Along Havasu Creek, travertine deposits have encrusted the canyon walls and formed low, sinuous dams in the creek bed.

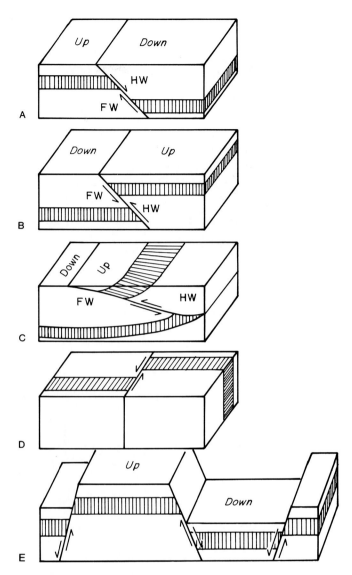

Figure 1.9 Block diagrams of common faults. Stresses that create faults may cause one block to move up, down, obliquely, horizontally, or even to rotate, in relation to another block. *A.* **Normal fault.** The hanging wall side of the fault moves *down* in relation to the footwall side. *B.* **Reverse fault.** The hanging wall side moves *up* in relation to the footwall side. *C.* **Thrust fault.** A reverse fault in which the fault plane dips at a low angle. *D.* **Strike-slip fault.** One side of the fault moves horizontally by the other side. *E.* **Horst and graben structure.** The upthrown block, bounded by normal faults is a horst. The downdropped block, also bounded by normal faults, is a graben. Normal faults are the result of extensional forces acting on the earth's crust. Compressive forces produce reverse faults.

Sandstones may be formed from old dunes, such as the *eolian* (wind-blown) deposits that make up the Coconino Sandstone. Other sandstones were formed from water-laid deposits, either in fresh-water lakes or in marine environments. The degree of sorting and rounding of sand

Box 1.2
A Brief Geologic History of the Colorado Plateaus

The national parks of the Colorado Plateaus have many individual differences and characteristics, but they share, in broad outline, a common geologic history.

1. Precambrian time
The Precambrian rock units exposed in the Grand Canyon do not crop out in the other national parks of the Colorado Plateaus, presumably because erosion and stream downcutting have not reached low enough depths. Some Precambrian rocks are exposed, however, in areas north and east of the Colorado Plateaus.

Figure 1.10   In the Grand Canyon, the processes of erosion have cut into highly resistant Precambrian rocks to form the Inner Gorge. National Park Service photo.

2. **Lower Paleozoic: Cambrian to Devonian Periods**

   The Colorado Plateaus were a tectonically stable area for millions of years, as probably the trailing, subsiding edge of a *continental plate* (a segment of continental crust). It is believed that continental rifting (the splitting up of a continent) had occurred, possibly before the end of Precambrian time. Shallow seas spread over a vast area of the continental plate that would eventually become North America. In these seas, thousands of feet of sediment were deposited.

3. **Upper Paleozoic and Mesozoic: Mississippian through Cretaceous Periods**

   A series of *orogenies* (mountain-building episodes) deformed parts of western North America surrounding the Colorado Plateaus. Instead of breaking up or pulling apart, some land masses were coming together and closing up former oceans during this time. Sediments were shed from uplifted, encircling highlands and deposited on coastal plains, deltas, and in lagoons along shores. Basins between coastal mountains rimming continental margins filled with sediments. In some places, muddy, limey bottoms stagnated in shallow seas. Elsewhere, sand dunes overspread vast plains. The abundance of quartz-rich sands that built the great Jurassic dune belts in coastal deserts were derived from the weathering and recycling of quartz grains from older rocks.

   In Cretaceous time, the last great inundation by shallow ocean waters spread over western North America. Thick units of richly fossiliferous, marine rocks were laid down. But the Mesozoic Era drew to a dramatic close as accelerated orogenic activity lifted the lands from the ocean toward the end of the Cretaceous Period. This great event, which began 65 million years ago and lasted for millions of years, drastically changed the face of western North America. It is called the Laramide orogeny.

4. **Cenozoic Era**

   Minor uplifts and adjustments, accompanied by gentle deformation and moderate outpourings of dark lava, characterized early Cenozoic time in the Colorado Plateaus. Nonmarine sediments, washed down from surrounding highlands and ranges, accumulated in basins and spread over plains.

   Toward the middle of the Cenozoic Era, tectonic activity resumed. The Colorado Plateaus region was uplifted still higher and tilted. Streams began to dissect the landscape into plateaus, benches, mesas, buttes, arches, and deep canyons. Each successive uplift increased the power of the erosional processes. Uplift was uneven, causing some faults and open folds to develop in the rocks. Upwarping and stretching of the continental crust in the Basin-and-Range region to the west of the Colorado Plateaus produced several large faults, such as Hurricane fault, between the two regions.

   In the meantime, because of the major elevation of the Sierra Nevada to the west, the Colorado Plateaus were deprived of rain, and the climate became increasingly arid. (When moisture-laden clouds from the ocean are forced to drop rain or snow when rising over a mountain range, the far side of the range is said to be in a *rain-shadow zone.*)

   The Pleistocene Epoch was a time of colder climates, increased precipitation, and intensified erosion. Mountain glaciers formed at higher elevations, and ice and frost broke up rocks everywhere on the plateaus. The results of colder temperatures, and more rain during this time show up in the Colorado Plateaus landscapes as dry valleys that once held streams, abandoned lakebeds, landforms eroded by fluvial sculpturing, and stream deposits unrelated to today's creeks and rivers.

Table 1.2 Geologic Column, Grand Canyon National Park

Time Units		Rock Units		Geologic Events
Era	Period	Group & Super-group	Formation	
Cenozoic	Quaternary Tertiary		Lavas Slump and slide deposits, terrace gravels, river deposits, travertine.	Minor lava flows Development of Grand Canyon by uplifting and tilting, accompanied by down-cutting. Uplift, erosion, minor deformation.
Mesozoic	Cretaceous Jurassic Triassic		～～～～～ Erosion ～～～～～ Moenkopi	Erosion ———— Flood-plain deposition. Withdrawal of seas.
Paleozoic	Permian		～～～ Erosion ～～～ Kaibab Limestone	Erosion ———— Transgression of shallow seas; regression.
			～～～ Erosion ～～～ Toroweap Coconino Sandstone	Erosion ———— Transgression of shallow seas, flattening sand dunes. Flood-plain deposits covered by migrating sand dunes.
			～～～ Erosion ～～～ Hermit Shale	Erosion ———— Deposition on flood plain, lagoons.
	Pennsylvanian	Supai	～～～ Erosion ～～～ Esplanade Sandstone Wescogame Manakacha Watahomigi	Erosion ———— Flood-plain deposition Retreat of seas Marine sedimentation
	Mississippian		～～～ Erosion ～～～ Surprise Canyon ～～～ Erosion ～～～ Redwall Limestone	Erosion ———— Channel deposits; uplift Erosion ———— Karst development Marine sedimentation
	Devonian		～～～ Erosion ～～～ Temple Butte Limestone	Erosion ———— Marine deposits in channels
	Cambrian	Tonto	～～～ Erosion ～～～ Muav Limestone Bright Angel Shale Tapeats Sandstone	Erosion ———— Deposition in a transgressing sea, forming a facies change in these formations
Late & Middle Precambrian time (Protero-zoic)		Grand Canyon — Chuar	～～～ Erosion ～～～ Sixty Mile ～～～ Erosion ～～～ Kwagunt Galeros	Erosion ———— Stream Deposition Erosion ———— Volcanic activity; Grand Canyon orogeny, block-faulting, thrust-faulting, folding.
		Grand Canyon — Chuar	～～～ Erosion ～～～ Nankoweap	Erosion ———— Deposition in a shallow sea.
		Grand Canyon — Unkar	～～～ Erosion ～～～ Cardenas Lava Dox Sandstone Shinumo Quartzite Hakatai Shale Bass Limestone (including Hotauta Conglomerate)	Erosion ———— Faulting, igneous intrusions, lava flows Shallow sea deposits Deposited as a marine sandstone Temporary retreat of the sea Deposition in warm, shallow sea Transgression of seas; major unconformity
			～～～ Erosion ～～～ Trinity and Elves Chasm Gneisses	Erosion ———— Block-faulting; uplift. Igneous intrusions; metamorphism
			～～～ Erosion ～～～ Zoroaster Granite	Erosion ———— Uplift of Mazatzal Mountains; orogeny, folding, faulting, intrusions, metamorphism
Early Precambrian time (Archean)			Vishnu Schist	Deposition of marine sediment; volcanic activity.

Source: Modified after Huntoon et al. 1981, 1982, 1986.

grains, the presence or absence of fossils, the type of bedding, and the kind of cementing (or matrix) holding the grains together can all provide clues as to the geologic history of an area.

Geologic History of the Grand Canyon

1. **Accumulation of sediments early in Precambrian time.**

   Around 2 billion years ago, sands, silts, muds, and volcanic ash were deposited in a shallow marine basin associated with an orogenic belt. Some lava flows may have poured out into the ancient ocean. Fossil evidence of life has not been found in these marine beds.

2. **Ancient mountain-building and igneous intrusions.**

   About 1.7 million years ago, tectonic activity, resulting from plate collision, converted the accumulation of sedimentary beds and volcanic material into the metamorphic Vishnu Schist. Then molten rock pushed into the schist, slowly cooling to form Zoroaster Granite. Some of the rock, injected under great pressure by hot, rising chemical fluids, became metamorphosed to gneiss. In Trinity Chasm near Isis Temple, light-colored gneiss (containing appreciable amounts of quartz, feldspar, and micas) shows up as veins in the ancient rocks.

3. **Uplift and erosion of the early Precambrian rocks.**

   We know little about the ancient Mazatzal Mountains that were constructed; only their roots are left. But even as the mountains rose, they were subjected to the processes of weathering and erosion. For millions of years, sediments were stripped from highlands and deposited in lowlands or oceans.

4. **Deposition of the Grand Canyon Supergroup (1.2 billion to 800 million years ago).**

   The rock units of the *supergroup* (an assemblage of vertically related formations and groups)consist mainly of sedimentary beds that were later subjected to varying degrees of metamorphism. (Since they have retained the layering and certain other characteristics of sedimentary rocks, most have been given sedimentary rock names.) The Unkar Group, the older section of the supergroup, was deposited in an offshore environment. Waves worked over and smoothed out the early Precambrian land surface. Larger fragments accumulated as gravel, creating a basal conglomerate, known as the Hotauta Member of the Bass Limestone, a formation. (A *group* consists of two or more geographically associated formations with notable common features. A *formation* is a rock unit containing one or more beds with distinctive physical and chemical characteristics. A *member* is a minor unit within a formation.) Ranging from 120 to 340 feet in thickness, the Bass Lime-

stone accumulated in a shallow sea as a mixture of limestone, sandstone, and shale.

The Hakatai Shale, the next younger formation of the Unkar Group, is composed of units of thin-bedded shales, mudstones, and sandstones. Since most of the shale is nonmarine, the formation represents a temporary retreat of the sea.

The Shinumo Quartzite, a marine sandstone before it was metamorphosed, was so resistant to erosion that it juts up as "fossil islands" into overlying Cambrian beds. Apparently cliffs of Shinumo Quartzite withstood the wave action of the Cambrian ocean until the islands were completely engulfed. The Dox Sandstone, mostly sandstone but containing some shale layers, is another accumulation of shallow ocean deposits.

The flows of dark (basaltic) Cardenas Lava, the youngest rock unit of the Unkar Group, intruded any available openings, fractures, faults, or zones of weakness in the sedimentary rock. Eventually lava flows up to 1000 feet thick erupted from fissures and spread over older rocks.

The sedimentary beds of the Nankoweap Formation, which are shallow-ocean deposits, lie unconformably over the eroded surface of Cardenas Lava and also separate the Unkar Group formations from the younger rocks of the Chuar Group. Three formations—Galeros, Kwagunt, and Sixty Mile, in order of age—make up the Chuar Group. Evidence of algae has been found in Kwagunt rocks, but fossils are rare in rocks this ancient.

5. **Block-faulting of the Grand Canyon Supergroup.**

   Remnants of the Grand Canyon Supergroup crop out in scattered sections of the Inner Gorge and the deeper tributary canyons. The pockets of rock are isolated from each other because when tilting and block-faulting occurred some rocks were elevated and some were dropped down, creating horst-and-graben structures (fig. 1.9E). Tensional tectonic forces were stretching and rifting crustal segments at that time. Movement was along fault lines, such as the Bright Angel fault, creating north–south trending fault-block mountain ranges.

6. **"The Great Unconformity."**

   During the very long erosional period that followed the uplifting and tilting of the fault-block mountains, most of the Chuar beds and some of the Unkar Group were stripped from the mountain slopes. This exposed some of the more resistant beds, such as the Shinumo Quartzite, that stood as islands in the encroaching ocean of Cambrian time. Most of what is left of the Grand Canyon Supergroup is preserved in the down-faulted blocks. Because of the length of the erosional period, the contact between the tilted Precambrian beds and the Paleozoic beds is close to the horizontal.

In 1869 John Wesley Powell recognized the angular relationship between the older tilted rocks low in the Grand Canyon and the flat-lying sedimentary formations above. He named the contact the Great Unconformity (fig. 1.1).

7. Deposition of Cambrian rock units.

About 550 million years ago, a western ocean began to spread over the eroded coastal lowlands. Three formations were deposited, each representing a different environment. The way in which the sediments were laid down illustrates the principle of time transgression (fig.1.7).

Imagine yourself standing on a sandy beach that would eventually become the Tapeats Sandstone. Pebbles, cobbles, and a few boulders are on the sand after falling from the cliffs along the shore. This is a turbulent coast. Pounding storm waves attack the cliffs; cobbles and sand are rolled back and forth in the surf and tossed up on the shore. These are the constituents of the basal conglomerate at the bottom of the Tapeats. Also preserved in the Tapeats are the shapes of ancient coastal sand dunes.

Farther out in the water, the sea bottom becomes muddy. These muds, which contain the remains of primitive brachiopods, trilobites, and worms, formed the Bright Angel Shale. Offshore, calcium carbonate precipitates from seawater and accumulates as a limy ooze on the bottom. From this ooze was formed the Muav Limestone.

Because sea level is very gradually rising, the shoreline and the beach are gradually shifting eastward, and the pattern of sediment deposition is also moving eastward without a break in continuity. Thus when the Tapeats Sandstone and the Bright Angel Shale were first laid down, it was early in Cambrian time; but by the time the last grains were being deposited some distance to the east, the time was Middle Cambrian, about 30 million years later. Therefore, these beds can be considered time-transgressive rock units. Evidence that the sea was transgressing is shown by finer-grained shale deposited over coarser-grained sandstone. Had the sea been regressing, coarser sediment would have overlain the finer.

The Tapeats Sandstone, Bright Angel Shale, and Muav Limestone belong to the Tonto Group and are the beds that make up the Tonto Platform. These beds lie in their original horizontal position, virtually undisturbed since they were deposited on top of the Precambrian erosional surface. The Tapeats Sandstone and the Muav Limestone are consistent cliff-formers in the walls of the Grand Canyon, while Bright Angel Shale is a slope-former (box 1.3).

8. Unconformity between the Cambrian and Devonian beds.

Beds from the Ordovician and Silurian periods are missing from the Grand Canyon geologic column. There is no evidence to indicate whether beds were deposited in the region during these time periods and subsequently removed by erosion, or whether deposition did not occur. However, deep channels were carved by streams (or possibly by marine scour) on top of the Muav. Devonian Temple Butte Limestone was deposited in these depressions. The channel deposits are especially well displayed in Marble Canyon in the eastern section of the park. Toward the western end of the park, near Grand Wash Cliffs, however, Temple Butte is a cliff-former, in some places a thousand feet thick. Because the limestone in Temple Butte has largely been replaced by dolomite, only a few fossil fish teeth remain in the formation. Erosion, producing an unconformity, brought the Devonian Period to a close.

9. Deposition of the Mississippian Redwall Limestone.

The visually striking cliffs of the Redwall in the Grand Canyon today are stained red by rainwater dripping down from the overlying Supai and Hermit Shale redbeds. The Redwall is actually a bluish gray limestone containing chert nodules. It was deposited in a shallow tropical sea in early to middle Mississippian time, some 330 million years ago. Thickness averages 500 feet. Fossil remains of crinoids, brachiopods, bryozoans, and other marine creatures are abundant in the formation. A few years ago, chambered nautiloid fossils were discovered in the streambed of a branch of Marble Canyon. Following the deposition of the Redwall Limestone, the region was gradually uplifted and the beds underwent erosion late in Mississippian time. Because limestone is highly susceptible to solution by ground water, some natural caves developed.

10. A geological "Surprise" in the Grand Canyon.

In very latest Mississippian time (and possibly very earliest Pennsylvanian time), a singular depositional event occurred in what is now the Grand Canyon region (Beus 1986). Recently a sedimentary rock unit, heretofore unrecognized, was discovered cropping out in several locations in western and central Grand Canyon, which are accessible only by helicopter. This "new" unit has been appropriately named the Surprise Canyon Formation. The exposures, which are in discontinuous lenses a few hundred feet thick at the top of the Redwall Limestone, record the evolution of tidal estuaries on a low coastal plain that gradually subsided, drowning the valleys. At the end, a short period of erosion was followed by deposition of the lowermost beds of the Supai Group. Fossils in the Surprise Canyon Formation—from abundant logs and plant materials mixed with coarse sands in the bottom beds to marine shells in beds of sand and lime deposits—reflect the changing environments that produced this unusual sequence.

Rim of The Grand Canyon (South Side)

Kaibab Limestone

Toroweap Formation

Coconino Sandstone

Hermit Shale

The Esplanade

Supai Group (Sandstones)

Redwall Limestone

Temple Butte Limestone

Tonto Group { Muav Limestone

Bright Angel Shale

Tapeats Sandstone

Tonto Platform

Unkar and Chuar Groups

Inner Gorge

Plutonic Complex and Gneisses

Vishnu Group

Figure 1.11   Schematic profile of rock units exposed on the south wall of the Grand Canyon. Relative thicknesses of formations are approximate, but slope angles are greatly exaggerated. In the semiarid climate, limestones tend to be cliff-formers, while shales form gentle slopes. The resistance of sandstones depends on their purity and cementation. Weathering and erosion are much less effective on the crystalline rocks at the bottom of the canyon. (Adapted from Billingsley and Breed 1980).

11. Deposition of the Supai Group during Pennsylvanian and early Permian time.

The Supai Group, a prominent ledge- and cliff-former in the canyon walls, is 600 to 700 feet thick. Mostly non-marine, red siltstones and sandstones accumulated in a swampy environment. Distinctive crossbedding may be an indication of ancient deltas and coastal dunes. Fossils include footprints of amphibians and an abundance of plant materials. The Esplanade Sandstone of early Permian age has been oxidized to a bright red color.

12. Early and middle Permian deposition, 280 to 250 million years ago.

The Hermit Shale, a red slope-former, accumulated in an environment similar to the underlying Supai Group. The beds contain ripple marks, mud cracks, footprints, plant fossils, and a few insect remains.

Migrating sand dunes that gradually buried the swampy flood plain formed the Coconino Sandstone. This formation is the whitish cliff-former near the top of Grand Canyon. The uniformly sized, perfectly rounded, frosted quartz grains, arranged in a cross-bedded pattern characteristic of dunes, testify to a wind-blown (eolian) origin. The dunes must have been fairly large because the Coconino is 400 feet thick. Reptile tracks preserved in the sandstone show that animals wandered around the dunes.

The Toroweap Formation, with its nearly 200 feet of sandstones and limestones interbedded with gypsum, records marine advances and retreats. Brachiopods, mollusks, corals, and other marine organisms lived in the ancient ocean. The Toroweap is a ledge- and cliff-former.

Forming the rim of the Grand Canyon and the edge of the Kaibab Plateau is the massive, thick-bedded Kaibab Limestone of middle Permian age. More than 300 feet thick, the Kaibab Limestone is a prominent cliff-former. Some zones in the Kaibab have sand grains and chert nodules. Fossils include marine organisms such as crinoids and brachiopods that lived in warm, shallow seawater.

13. Withdrawal of the Permian Sea; Mesozoic deposition.

As the Mesozoic Era began, the continent was rising. Streams carved wide, low valleys on the land surface. Sediment was transported from higher land to lower slopes and basins. These deposits became the Moenkopi Formation of early Triassic age, a mixture of brightly colored sandstones and shales with interbedded gypsum layers. The number and variety of fossils indicate an abundance of life forms.

Presumably the early Triassic rock units, such as those now exposed along the Colorado River above Marble Canyon, may have overspread the uplands of the Grand Canyon region. The soft, easily weathered Triassic shales, less resistant to erosion than the Permian Kaibab and Toroweap rock units, were probably stripped off long ago.

Younger Mesozoic and early Cenozoic beds, identified in nearby areas, may have been deposited and later removed by erosion.

14. Mesozoic uplift and mountain-building.

The compressive tectonic forces generated by the Laramide orogeny about 65 million years ago had profound effects on western North America at the end of Mesozoic time. Nevertheless, the horizontal beds of the Colorado Plateaus remained relatively stable and undeformed while still being elevated like a great platform. All around the Colorado Plateaus, rock units were being folded, faulted, and uplifted, and great mountain ranges were being pushed up. The East Kaibab monocline (and other monoclinal folds) developed along the edge of the Kaibab Plateau during this time. (Presumably a south-flowing segment of the Colorado River began to erode Marble Canyon along the lower flank of the monocline, long before the stream cut across the Kaibab Plateau.)

15. Middle Tertiary uplifting and faulting.

About 20 million years ago, as the Colorado Plateaus were uplifted and warped by tensional forces, local faulting—sometimes along old fault lines—occurred, along with moderate volcanic activity. The stress of crustal stretching that was going on to the west in the Basin and Range province created several major north-south trending faults between the Colorado Plateaus and the Basin and Range. The Grand Wash fault, which intersects the Colorado River at the western edge of the park, is on the boundary between the two provinces. The Hurricane fault extends north into Utah. By late in Tertiary time, the Colorado Plateaus region had attained most of its present elevation.

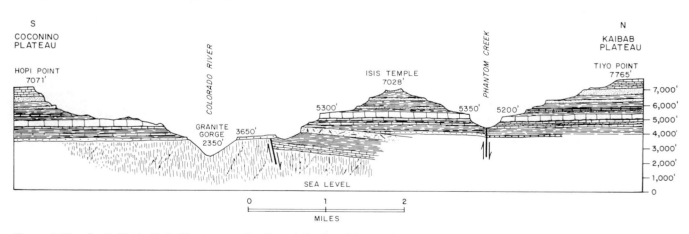

Figure 1.12   South Rim to North Rim cross section through the Grand Canyon from Hopi Point to Tiyo Point. Canyon features are shown in their relative sizes and correct proportions. Uplifting of the Colorado Plateau enabled the Colorado River, with the help of mass wasting, to remove immense amounts of rock and carve the Grand Canyon in a few million years. (Adapted from F. Matthes 1906, U.S. Geological Survey Bright Angel quadrangle).

16. The connecting of Colorado River segments about 6 million years ago and establishment of its present course.

The uplift and tilting of the plateaus accelerated the development of the drainage system. With its gradient increasing and base level lowered by the opening of the Gulf of California, the young Colorado River began to downcut rapidly. This enabled the river to capture the headwaters of other streams and to carve out the configuration of the Grand Canyon.

17. Intensification of weathering and erosion during Pleistocene glacial episodes.

The colder, wetter climates of Pleistocene time increased erosional rates in and around the Grand Canyon. However, the excavating of the canyon by the river was not evenly spread over the year. Most of the work was done by mighty floods of meltwater in the spring and by sudden flash floods from violent summer storms. This was how in the last two million years most of the downcutting of the Grand Canyon was accomplished.

18. Late Cenozoic volcanism.

Only a million years ago, flows of fluid basaltic lava covered parts of the plateau. Some poured down side canyons and dammed the Colorado River in the western section of the Grand Canyon. These lava flows are still being eroded by the Colorado River. The remains of small cinder cones on the rim mark the sites of ash eruptions.

19. Continued deepening of the Grand Canyon by downcutting; widening of the Canyon by mass wasting.

The climatic patterns of spring and summer floods, along with the age-old processes of weathering and erosion, continued into Holocene time, although at a somewhat slower pace. Since the Colorado River has become a regulated stream system from its source area to its outlet, its power to incise its channel has been reduced. However, as weathering loosens particles, mass wasting—powered by gravitational energy—moves materials downslope. In an arid climate these processes operate slowly except when quantities of loose sediment are transported rapidly during flash floods or landslides. Erosional processes, mostly very slow or very fast, deliver sediment to canyon streams that carry it away.

Box 1.3
The Terraced Walls of the Grand Canyon

Observers who look down into the Grand Canyon from either the South Rim or the North Rim are usually fascinated by the "stairstep" appearance of the canyon walls (figs. 1.11, 1.13). The upper part of the Grand Canyon is terraced while the lower part, or Inner Gorge, is narrower and V-shaped.

Four geological rock units are responsible for the most prominent cliffs or terrace risers. They

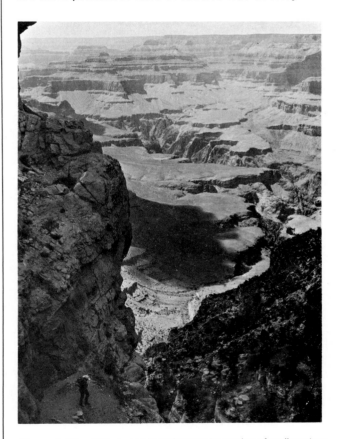

Figure 1.13   Differences in resistance to erosion of sedimentary rock layers produced the stairstep topography of the Grand Canyon. The Tonto Platform (flat surface, center of photograph) was eroded from weak Bright Angel Shale, but is upheld by the underlying, resistant Tapeats Sandstone. Muav Limestone, a resistant formation, makes up the first tier of cliffs that partially ring the Tonto Platform. National Park Service photo.

continued

Box 1.3 continued

are, from the top, the Permian Kaibab Limestone and the Supai Group, the Mississippian Redwall Limestone, and the Cambrian Tapeats Sandstone. Each cliff has at its base blankets of loose rock fragments, or *talus,* overlying gentle slopes. On top of the Tapeats Sandstone a broad terrace top, the Tonto Platform, has been created by removal of less resistant rock units. A similar platform, called the Esplanade, in the western part of the park, has developed on the resistant Esplanade Sandstone. These slope segments, formed as the Grand Canyon was sculptured downward through successive geological formations, illustrate how in the same climate some rock units are more resistant to weathering and erosion than others. This phenomenon is called *differential erosion.*

The narrow Inner Gorge is not terraced or stepped because the metamorphic and intrusive igneous rocks have fairly uniform hardness and are more resistant to weathering and mass wasting than the Paleozoic formations that form the cliffs and slopes.

A first impression of the Grand Canyon regards the immensity of its erosional downcutting. However, because the canyon is more than 10 miles wide from rim to rim, it becomes evident that the processes of valley widening and cliff recession are also significant. Everywhere in the Grand Canyon, except on the faces of the steepest cliffs, great sheets of broken rock lie on the valley sides as proof of the effectiveness of weathering, which produces the fragments, and mass wasting, which moves loose rock downslope.

North of Grand Canyon National Park, stepped topography on a larger scale is produced by the retreat of the margins of Mesozoic and Cenozoic formations in the form of the Chocolate Cliffs, Vermilion Cliffs, White Cliffs, and Pink Cliffs. These cliffs, which make up "the Grand Staircase" of the Colorado Plateaus, trend generally east–west, and step up toward the north. The most prominent features in Zion National Park (chapter 2) are formed in the White Cliffs. The Pink Cliffs are strikingly displayed in Bryce Canyon National Park (chapter 3).

Geologic Maps and Cross Sections

Billingsley, G. H., and Breed, W. J. 1980. Geologic cross section along Interstate 40, Kingman to Flagstaff, Arizona. Petrified Forest National Park: Petrified Forest Museum Association.

Breed, W. J., Stefanic, V., and Billingsley, G. H. 1986. Geologic guide to Bright Angel trail, Grand Canyon, Arizona. Tulsa, Oklahoma: American Association of Petroleum Geologists.

Huntoon, P. W., Billingsley, G. H., and Clark, M. D., 1981. Geologic map of the Hurricane fault zone and vicinity, western Grand Canyon, Arizona. Grand Canyon Natural History Association.

———— 1982. Geologic map of the lower Granite Gorge and vicinity, western Grand Canyon, Arizona. Grand Canyon Natural History Association.

Huntoon, P. W., Billingsley, G. H., Breed, W. J., Sears, J. W., Ford, T. D., Clark, M., Babcock, R. S., and Brown, E. H. 1986. Geologic map of the eastern portion of the Grand Canyon National Park, Arizona. Grand Canyon Natural History Association.

Maxson, J. H. 1966. Geologic map of Bright Angel quadrangle, Arizona. Grand Canyon Natural History Association.

Bibliography

Baars, D. L. 1983. *The Colorado Plateau, a geologic history.* Albuquerque, New Mexico: University of New Mexico Press. p. 3–48.

Beal, M. D. 1978. *Grand Canyon, the story behind the scenery.* Las Vegas, Nevada: KC Publications.

Beus, S. S. 1986. A geologic surprise in the Grand Canyon. *Fieldnotes* 16 (3): 1–4. Tucson, Arizona: Arizona Bureau of Geology and Mineral Technology, University of Arizona.

Chronic, H. 1983. *Roadside geology of Arizona.* Missoula, Montana: Mountain Press.

Collier, M. 1980. *An introduction to Grand Canyon geology.* Grand Canyon Natural History Association.

Hamblin, W. K., and Murphy, J. R. 1980 (revised edition). *Grand Canyon perspectives.* Provo, Utah: Brigham Young University Geology Studies, special publication no. 1, 48 p.

Hunt, C. B. 1969. Geologic history of the Colorado River, in *Colorado River Region: John Wesley Powell.* U.S. Geological Survey Professional Paper 669, pp. 59–130.

King, P. B. 1977 (rev. ed.) *The evolution of North America.* Princeton, New Jersey: Princeton University Press (chapter VII).

McKee, E. D. 1969. Stratified rocks of the Grand Canyon, in *Colorado River Region: John Wesley Powell.* U.S. Geological Survey Professional Paper 669, pp. 23–58.

Nations, D., and Stump, E. 1981. *Geology of Arizona.* Dubuque, Iowa: Kendall/Hunt Publishing Company.

Powell, J. W. 1981 reprint (1875 *Scribner's Monthly* articles) W. R. Jones, editor. *The canyons of the Colorado.* Golden, Colorado: Outbooks. 64 p.

Rigby, J. K. 1977. *Southern Colorado Plateau,* K/H Geology field guide series. Dubuque, Iowa: Kendall/Hunt Publishing Company.

Trimble, S. 1979. *The bright edge, a guide to the national parks of the Colorado Plateau.* Flagstaff, Arizona: Museum of Northern Arizona Press.

Grand Canyon National Park

Area: 1,218,375 acres; 1903 square miles

Proclaimed a National Monument: January 11, 1908

Established as a National Park: February 26, 1919

A separate Grand Canyon National Monument was proclaimed in 1932; Marble Canyon National Monument was proclaimed in 1969. On January 3, 1975, the three areas were combined with additional lands to form the present Grand Canyon National Park.

Designated a World Heritage Site: October 24, 1979

Address: P.O. Box 129, Grand Canyon, Arizona 86023

2

Zion National Park

Southwest Utah

Near-vertical canyon walls and great towers on a grand scale display hues of bright red, salmon pink, orange, yellow, and off-white.

Rock patterns of faults, fractures, and ancient dunes have created spectacular landforms and landscapes.

Zion Canyon, 3000 feet deep, is a comparatively young erosional feature, cut by the Virgin River through Mesozoic clastic sedimentary rocks.

Sandstone Kolob Arch, with a span of 310 feet, is the largest of its type in the world.

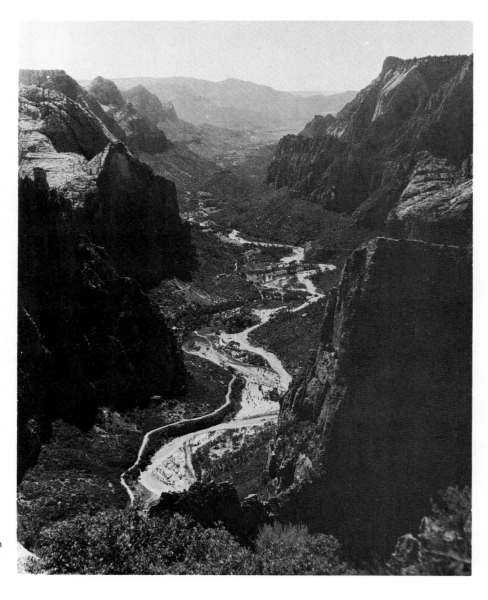

Figure 2.1 Downcutting by the North Fork of the Virgin River, weathering, and mass wasting have sculptured Zion Canyon in Zion National Park. National Park Service photo by G.A. Grant.

Local History

In 1869, when Major John Wesley Powell and his fellow explorers were in desperate straits because of their loss of boats and supplies during their first trip through the Grand Canyon, they knew that they would be safe if they reached the Mormon settlements below the canyon, near the confluence of the Colorado and Virgin Rivers. Mormons started settling the Virgin River region in 1847 with plans to farm and grow cotton. The Mormons' search for farmland led them to Zion Canyon in 1858, about 75 miles up the Virgin River from the older settlements. Within a few years, Mormon pioneers began to move into the canyon area, which they named Zion, meaning a place of refuge. They farmed on tillable lands scattered in narrow fields along the river banks and on an old lakebed in the canyon until 1909 when a national monument was first set aside.

After his successful and arduous passage through the Grand Canyon, Powell was determined to complete the task of surveying the whole Colorado River region. He was intensely interested in finding how the Colorado Plateau and the Colorado River had evolved. His exploration party included geologists Clarence Dutton and Grove Karl Gilbert, photographer Jack Hillers, and artist William H. Holmes—all famous names in the history of scientific exploration in the American West. Over the next two decades, these hardy scientists put together a picture of time and territory on a giant scale: how through periods of geologic history the Colorado Plateau was uplifted while river systems steadily cut down, layer by layer, through sedimentary rocks.

Powell and Gilbert made a geological reconnaissance of Zion Canyon in 1872 and named several features. Later Dutton carried on extensive mapping of the region. It was their glowing reports, plus Hillers' photographs and Holmes' drawings, that first created interest in Zion. In 1903, artist Frederick S. Dellenbaugh painted scenes in the canyon and exhibited his canvases at the St. Louis World's Fair. Spectators were doubtful that such strange and remarkable landscapes could exist, but the attention led to the presidential proclamation of a national monument. Congress established Zion National Park in 1919.

The Zion-Mt. Carmel highway (Utah 9), an engineering marvel of its time, was opened in 1930 to provide access to Zion. From the park's east entrance, the highway leads to the famous Pine Creek tunnel, which is over one mile long and has six large windows cut into the rock for views. From the tunnel, switchbacks go down to the scenic drive on the floor of Zion Canyon. The Kolob section of Zion National Park has a visitor center and scenic drive and is accessible from Interstate 15.

The Zion Climate

In Zion National Park, climatic zones, determined by elevation, affect weathering and erosion as well as the distribution of plant and animal life. In the high desert (Upper Sonoran) zone of the canyons (elevation, 3500 to 5000 feet), summers are hot and dry and winters are mild. On the canyon rims (elevation, 5500 to 7500) in the transition zone, snow may last all winter and summer temperatures are more moderate. On the higher parts of the plateau (around 8000 feet), in the montane (Canadian) zone, stands of fir and aspen thrive under conditions of increased moisture and colder winters.

Zone boundaries are not sharply defined. Within the desert zone, for example, protected tributary canyons and shaded slopes with springs support plants and animals characteristic of higher elevations. Wet seasons are normally in early spring and late summer. Most of the scouring in the canyons is done by spring floods or by flash floods during intense summer storms.

Geologic Features

Zion Canyon, a deep chasm with nearly vertical walls, was cut by the North Fork of the Virgin River, a stream with great erosive power whose gradient, or slope, ranges between 50 and 80 feet per mile. In addition to high velocity because of the steep gradient, the Virgin River has a higher volume during wet seasons and runs rampant when flooding. At such times, the river transports the greater part of the 3 million tons of rock and sediment it carries downstream annually.

Since the Virgin River is downcutting more rapidly than it is widening, its tributaries with smaller volumes of water are literally "left hanging" because they are unable to downcut as fast as the main stream. Their valley floors are at a higher level than the floor of the main stream, making them *hanging valleys*. Waterfalls connect the hanging valleys with the river below. The tributaries have smaller drainage basins and some of them are *intermittent;* that is, they flow only during wet seasons. Angels Landing is a typical hanging valley, as are the valleys between Twin Brothers peaks and Mountain of the Sun (located on the eastern side of the canyon, north of Pine Creek).

How much more can the Virgin River cut down before it "hits bottom," or reaches base level? Geologists estimate that the Virgin can cut down another thousand feet and still have enough energy to transport its sediment to the Colorado River.

Figure 2.2   Zion National Park, Utah.

Figure 2.3 *A.* Rapid downcutting of resistant rock in arid regions tends to create slot canyons. *B.* Weathering and mass wasting eventually widen most canyons into V-shaped valleys. From *Physical Geology,* 4th ed., C. C. Plummer and D. McGeary. © 1979, 1982, 1985, 1988 Wm. C. Brown Publishers. Used by permission.

The erosive power of sediment-laden water running at high velocity is well shown in the famous Narrows at the head of Zion Canyon. Here the stream runs through an erosion-enlarged fracture about 12 miles long. The floor is only a few feet wide but the enclosing walls rise about 2000 feet, nearly straight up. Sunlight reaches the streambed only when the sun is directly overhead. When the water level is low, visitors are permitted to hike through the Narrows. However, the canyon walls are so close to each other that the water level may rise 25 feet in 15 minutes during a heavy rain or flash flood. This explains why the rock walls show evidence of scouring high above normal stream levels (fig. 2.4).

Mass wasting in Zion Canyon is gradually widening the valley, and in many places the canyon rims are developing a scalloped look as they retreat unevenly. Sapping by springs and the undermining of sandstone cliffs as the softer underlying shales give way has had a twofold effect. First the sandstone cliffs become overhanging; and then, due to lack of support, blocks of rock break off and fall to the canyon floor. Occasionally a block weighing several tons crashes down. Sometimes a "blind arch," or *inset arch,* forms on a cliff face when the upper rock layers remain

intact after the lower part has fallen. A prominent example is Great Arch on a cliff that can be seen from tunnel windows on the way into Zion. Great Arch measures 600 feet across its base and is 400 feet high. Some arches, such as Kolob Arch in the Kolob section of the park, have become freestanding instead of inset. With a span of 310 feet, Kolob is the largest freestanding arch in the world.

Slump blocks, large masses of material that slide downslope as a coherent unit, are another example of mass wasting in Zion National Park. Mudstone layers in the Kayenta Formation, underlying the Navajo Sandstone, become slippery when saturated. Sometimes when the mudstones give way, slump blocks of the Navajo break loose along vertical joints (fractures) and slide down to the canyon floor. About 4000 years ago, a great slump block in Zion Canyon dammed the Virgin River and created a lake that backed up in the canyon as far as Weeping Rock. Lake-bottom deposits accumulated on the canyon floor until the Virgin River was able to cut through the slump and drain the lake. Eventually all traces of the former lake will be removed and the river will resume downcutting bedrock in that part of the canyon.

The great variety of landform shapes in Zion that astonished the early explorers are due to a combination of factors. Resistant and nonresistant types of bedrock are exposed to the ravages of weathering, and shales tend to be less resistant than sandstones. Providing access to the interior of the layered rocks are many bedding planes, joints, and faults, some of which are curved, others horizontal or vertical, some close together, and still others widely separated.

Frost action has a significant effect in the park, mainly at the higher elevations on the canyon rims and on the plateau where there are more freeze-thaw days and where more moisture is available. Rainwater, seepage, and tree roots are also effective agents of weathering, even in the arid to semiarid climate of southern Utah. The massive Navajo Sandstone is sufficiently porous so that water slowly percolates through it. On its downward path, some water seeps out along fractures and bedding planes and drips off the cliff face; still more water collects along the top of an impermeable rock layer and flows out in a spring. On Weeping Rock in Zion Canyon, hanging gardens of wildflowers cling to dripping crevices in the wall above a spring near the base. Weeping Rock is the loveliest of many springs that issue from the "spring line" between the Navajo and Kayenta Formations. The Kayenta is an impermeable shale.

The generally rectangular stream pattern is the result of closely spaced *joints* (fractures in bedrock without displacement) and vertical faults that permit broken or exposed rock to weather and erode more easily to form

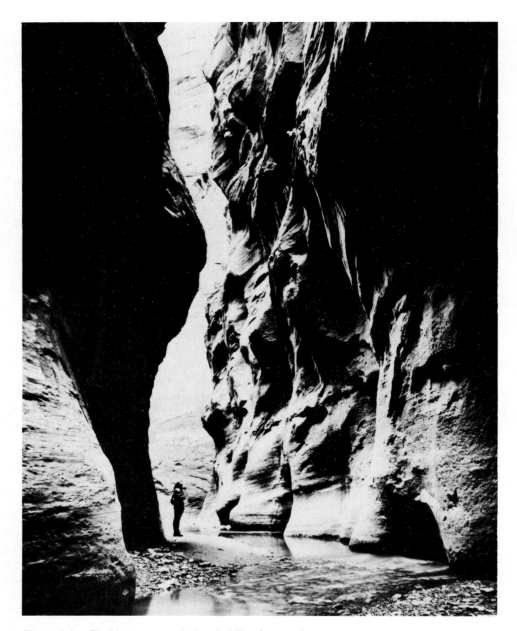

Figure 2.4   The Narrows, near the head of Zion Canyon. Turbulent waters of the North Fork of the Virgin River have cut a slot canyon through resistant Moenave beds of purplish sandstone and siltstone. The canyon walls are fluted and polished by stream abrasion. A smooth coating of calcium carbonate (tufa), precipitated out from flowing water, partly covers the rock surface. Photo by D. Pettit.

A

B

Figure 2.5   Processes of weathering and mass wasting continually attack the rocks in Zion National Park. *A.* An inset (blind) arch beside Hepworth Wash, an intermittent tributary of Pine Creek, near the tunnel on the Zion-Mt. Carmel highway. National Park Service photo by J. Fishbein. *B.* Cavernous weathering in a sandstone cliff above the North Fork of Taylor Creek in the Kolob section of park. Ground water percolating down through the cliff dissolves cementing compounds in the rock. Loosened sand grains then fall or are washed away. National Park Service photo by J. Staebler.

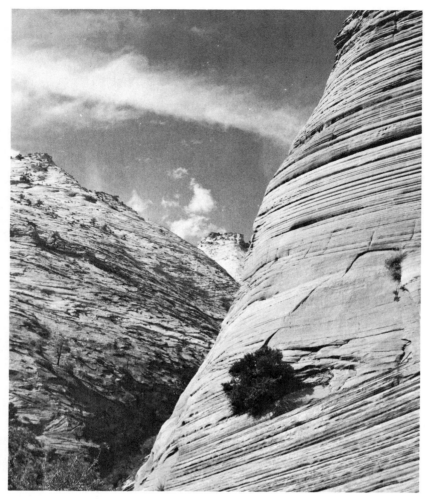

Figure 2.6   Cross-bedding in Navajo Sandstone beside the Zion-Mt. Carmel highway. National Park Service photo.

canyons. The "temples" are the result of resistant, unfractured rock standing as columns or pyramids after the highly fractured rock between has been eroded and transported away. Examples are East Temple, Checkerboard Mesa, Sentinel Mountain, Twin Brothers, Temple of Sinawava, West Temple, the Three Patriarchs, Towers of the Virgin, and the Beehives.

The rock unit that dominates Zion National Park is Navajo Sandstone. This massive formation is 1500 to 2000 feet thick and is famed for its *cross-bedding* that developed as sand dunes migrated on the shores of an ancient shallow sea (fig. 2.6). Some lower parts of the formation were deposited in shallow water with the cross-bedding caused by currents. The eolian cross-bedding formed as shore dunes were impelled by the prevailing winds. In this

process, wind picks up sand grains on the gentle upwind slope of a dune and deposits them on the (steeper) downwind side. The loose grains cascade down the face of the dune, which tends to keep the slope at about 34°, the "angle of repose." As sand underneath becomes compacted and the top of the dune becomes oversteepened, the face of the dune shifts direction slightly as more sand grains keep tumbling down. Gradually the entire dune is overridden as it moves in a downwind direction. Any shift in wind direction intensifies the cross-bedding effect (fig. 2.7). The most striking cross-bed sets are displayed in the upper parts of the Navajo, as found in Checkerboard Mesa and other cliffs along the Zion–Mt. Carmel highway.

The quartz sand grains in the Navajo are well rounded and frosted due to long exposure to abrasion by water and

Figure 2.7 Cross-bedding may develop in wind-blown sand (*A,B*) and in current-deposited sand (*C,D*). *A.* Sand accumulates in inclined beds on the downwind side of a dune. *B.* With a change in wind direction, another dune, with a different orientation, covers the first dune. *C.* Sand grains, dropped by currents, accumulate in layers in a depression on a sea floor or river bottom. *D.* Continued sedimentation may cover the first set of cross-beds with another. From *Physical Geology,* 4th ed., C. C. Plummer and D. McGeary. © 1979, 1982, 1985, 1988 Wm. C. Brown Publishers. Used by permission.

wind. The sand was washed down into the ocean from the continental interior in an earlier time. Waves and currents piled the sand onto the emerging coastal plain, and shifting westerly winds shaped the great coastal dune belt that became the Navajo Sandstone.

The Great White Throne, a natural feature sculptured by erosion of the Navajo Sandstone, is a large monolith rising approximately 2450 feet from the canyon floor. The color of the throne—the upper part of the monolith—is actually a light tan; but from a distance, particularly if the sunlight is striking it, the rock appears to be a brilliant white. The base of the throne is stained a deep red. This is the result of iron oxide leaching out from higher rock layers as they were weathered and eroded. The Great White Throne is also an example of how the lower beds of the Navajo Sandstone tend to form vertical cliffs, while the upper portion tends to weather as rounded domes.

The highest point in the park is the summit of Horse Ranch Mountain, elevation 8726 feet, which is located in the Kolob section, north of Kolob Finger Canyons. The lowest point, near South Campground, is about 3900 feet above sea level. Thus the *relief* in the park (i.e., the difference in elevation between the highest and lowest points) is a little over 4800 feet.

The colorful purple and gray Chinle Formation is exposed near Cougar Mountain and near Taylor Creek trailhead in the Kolob section. The sediments composing the Chinle are a mixture of sandstone, shale, limestone, and volcanic ash. The volcanic ash is interesting because it supplied the silica necessary for wood petrification in the Chinle Formation (chapter 8). Pieces of petrified wood weather out of Chinle beds in the Zion region and other parts of the Colorado Plateaus.

Volcanic Rocks

The only igneous rocks in the park are young basaltic lavas that broke through to the surface along north–south trending normal faults. One flow, about one half million years ago, flooded and dammed the Coalpits Wash drainageway in the southwest corner of the park. After the lava cooled, a large lake filled the wash. Crater Hill, a cinder cone, represents the last eruption in the Coalpits Wash area. Eventually, as the lake filled with sediment, a spillway developed that gullied and drained off the water. The Coalpits Wash area is now the driest, most desolate area in the park. Lava Point, near the northern edge of the park, and Horse Ranch Mountain in the Kolob section are also locations accessible by trails where lava flows can be seen.

Box 2.1
Plate Tectonics: An All-Encompassing Explanation

The spectacular scenery of western North America is largely the result of the past 65 million years of tectonic pulses and continuous erosion. We have suggested that tectonic forces generated within the earth elevated the Colorado Plateaus and built mountain ranges all around the plateaus. Moreover, we have seen that at times tectonic forces compressed the rocks of the crust and at other times have pulled segments of the crust apart. How does this happen?

A Composite, Unifying Hypothesis

Historians of science regard the plate tectonics hypothesis, including sea-floor spreading, to be the most significant idea developed in the physical sciences in the twentieth century. Plate tectonics is now the dominant geological concept in planning research strategies, framing hypotheses, and synthesizing geologic history.

The essential idea of plate tectonics states that the earth's outer shell consists of eight major crustal *plates,* or slabs, plus about a dozen smaller ones. Continents and ocean floors occupy the larger plates; some smaller plates are the size of islands. The plates range in thickness from 10 to 100 miles and they move slowly over the global surface. They interact mostly along their edges because they tend to be internally rigid. Each plate consists of three zones: a shallow upper part deforms by brittle breaking or by elastic bending; a deeper part yields plastically; and the lowest part is viscous. It is on this viscous layer that a plate slides, or "drifts."

Plates are pushed together in some places because they are being pulled apart at other places on the global surface, mainly at *spreading centers* (rifts) in the deep ocean basins along the crest of the great mid-oceanic ridge system. A nearly continuous series of submarine mountain ranges makes up the *mid-oceanic ridge.* Along its crest, upwelling magma continuously flows from the spreading center and adds new material to the earth's crust while the plates move away from the ridge in either direction.

Sometimes rifting occurs on a continental plate, breaking it apart. When this happens, an arm of the sea may move in over the spreading center. The opening of the Gulf of California (mentioned in chapter 1) is an example of continental rifting.

Since the earth is not expanding, new crust added at a spreading center, or *diverging plate boundaries,* must be consumed elsewhere. This happens in *subduction zones,* at *converging plate boundaries,* where plates collide and some crustal material plunges several hundred miles to depths where rock material is melted and recycled. Spreading centers and converging zones, where plate boundaries interact, are where the most intense tectonic activity takes place. Earthquakes are frequent; volcanic belts may develop; and mountain ranges may rise.

The plates that are composed mainly of continental crust tend to ride higher on the global surface because they are less dense than plates of oceanic crust. That is why continental plates tend to override oceanic plates when they converge. When an oceanic plate tilts downward, or is subducted, the bend creates an *oceanic trench* in the subduction zone.

Continental rocks and oceanic rocks get jammed together in a subduction zone and may get pulled down into the oceanic trench as a jumble of mashed and broken rock material. Other continental rocks are folded and faulted by the compression of colliding plates, and mountains are pushed up. Deep below the surface, rocks subjected to intense heat and pressure are metamorphosed or melted. The less dense molten rock gradually works its way toward the surface. Some of the melt collects in *magma chambers* that may eventually cool as great batholiths beneath

Figure 2.8 Schematic diagram of the North American tectonic plate from the mid-oceanic ridge in the Atlantic Ocean to the San Andreas fault in western California. The plate boundary at the mid-oceanic ridge is a spreading center where the North American and European plates are pulling apart and new oceanic crust is being formed. At the transform plate boundary along the San Andreas fault, the Pacific plate is rotating slowly in a counterclockwise direction and sliding past the westward-moving North American plate.

mountains. Some magma chambers are reservoirs from which lava erupts in volcanoes.

In a third type of plate interaction, the converging plates may scrape past each other as they collide, instead of one being subducted. This is called a *transform plate boundary.* The San Andreas fault in California marks such a boundary between the Pacific plate and the North American plate (fig. 2.8). California earthquakes are caused by shifting of crustal segments along the fault zone. Channel Islands National Park (chapter 43) is on the Pacific side of the plate boundary.

Plate Tectonic Activity and the Colorado Plateaus

Much of the geologic history of the Colorado Plateaus is based on the principle of *uniformitarianism;* that is, that geologic processes in the past operated in much the same way as they do in the present. For example, the "fossil" sand dunes that we see preserved in the cliff faces of Zion National Park probably accumulated and migrated in the same way as the modern dunes in Kobuk Valley National Park (chapter 11).

Geologists examine the ancient, contorted, crystalline rocks in the bottom of the Grand Canyon and assume that plate interactions were going on in Precambrian time that were not unlike the behavior of colliding plates and opening ocean basins that are being investigated today. Evidence suggests that for millions of years, periods of intense tectonic activity occurred involving the deformation of rocks, mountain-building, and volcanism. Interspersed with this activity were long spans of quiet erosion.

The land mass that eventually became North America was probably in the Southern Hemisphere in Precambrian time. Throughout Paleozoic time, the North American continental plate slowly drifted in a northwesterly direction, crossing several global climatic belts. Some rotation occurred, which means that the plate has had different orientations in the past. Part of the time the North American plate was joined with other plates in a supercontinent (called Pangaea). And for long periods of geologic time, vast shallow oceans covered the stable interior of the plate. Many of

continued

Box 2.1 continued

the rock layers of the Colorado Plateaus accumulated in these oceans. The sediments and structures of the rocks faithfully record times of emergence, submergence, and reemergence during the drifting of the plate over degrees of latitude and many miles of distance.

In the latter part of Mesozoic time, the lengthy and complicated period of Laramide mountain-building began when the westward-moving North American crust began to be affected by compression from the eastward-moving Pacific Ocean crust. The ocean floor slipped under the western edge of the continent and broke pieces off the continental shelf, dragging down sediments that thickened the continental crust and increased its buoyancy. The far western mountain ranges developed due to the compression and crumpling

of the parts of North America closer to the converging zone. The Colorado Plateaus were a solid, unyielding, discrete block that resisted the compression from the west. The plateaus rose, partly from pressure underneath and partly from buoyancy, and rotated slightly as the continent moved northward. The Rocky Mountain area was crumpled and pushed up due to intense compression along the eastern boundary of the plateaus (fig.2.9).

In Cenozoic time, toward the end of the Tertiary Period, the Colorado Plateaus again underwent uplift with minor warping and deformation. This time the tectonic forces were extensional as the crust in the Basin and Range to the west thinned and stretched, causing extensive block-faulting and tilting in that neighboring area.

Figure 2.9   Schematic diagram illustrating a shallow type of subduction that might account for the uplifting of the Colorado Plateaus and the development of the Rocky Mountains. Along the western continental margin, sediments accumulated in basins while volcanic arcs (volcanic chains, offshore or on continental crust near the coast) erupted quantities of pyroclastics and lava. Inland, an unyielding block of plateaus rose, buoyed up by the thrust of a huge slab of oceanic crust sliding beneath continental crust. Along the eastern edge of the plateau block, intense compression resulted in folding, thrust faulting, and the pushing up of a great mountain range.

Table 2.1 Geologic Column—Zion National Park

Era	Period		Epoch	Group	Formation	Member	Geologic Events
Time Units				Rock Units			
Cenozoic	Quaternary		Holocene Pleistocene		Basaltic lavas		Weathering, canyon-cutting, cliff retreat, slump blocks. Volcanic eruptions
	Tertiary	Neo-gene	Pliocene				Gravel deposits
			Miocene				Block faulting; uplift, erosion
		Paleo-gene	Oligocene		Lava flows		Episodes of volcanic eruption; uplift
Mesozoic	Cretaceous			San Rafael	Dakota Sandstone		Stream deposition
							Erosion ——————
	Jurassic	Upper			Carmel		Silts deposited in shallow sea
							Erosion ——————
		Middle and Lower			Temple Cap		Flooding followed by desert conditions
				Glen Canyon	Navajo Sandstone		White Cliffs Area buried by migrating sand dunes; increasing aridity.
							Erosion ——————
	?	?			Kayenta		Intermittent stream deopsits
	Triassic				Moenave	Springdale Sandstone	Stream deposits
						Whitmore Point	Lake deposition
						Dinosaur Canyon	Stream deposits
							Erosion ——————
					Chinle	Shinarump Conglom-erate	Stream, swamp deposits, plus volcanic ash; wood fossils. Basal conglomerate
							Erosion ——————
						Upper Red	Inland stream deposits
					Moenkopi	Shnabkaib Shale	Coastal plain; gradually changed from marine deposition.
						Virgin Limestone	
Paleozoic	Permian				Kaibab Limestone		Deposition in shallow sea
				Older units not exposed.			

Source: Hamilton, 1984.

Figure 2.10   Cross section of the southern part of the Colorado Plateaus "staircase" showing the relative positions of Precambrian and Paleozoic rocks in Grand Canyon National Park and younger Mesozoic formations in Zion National Park. From *Physical Geology*, 4th ed., C. C. Plummer and D. McGeary. © 1979, 1982, 1985, 1988 Wm. C. Brown Publishers. Used by permission.

Geologic History

The geologic history of Zion Canyon begins where the history of the Grand Canyon ends. The rock sequence in Grand Canyon National Park is primarily Precambrian and Paleozoic and covers about 2 billion years, whereas Zion formations are Mesozoic and represent only about 150 million years of geologic time.

1. Deposition of the Permian Kaibab Limestone.

The Kaibab, which is the same formation that caps the rim of Grand Canyon, is a yellowish gray marine limestone that was deposited in a shallow sea. In Zion National Park, the Kaibab Limestone is exposed along the Hurricane Cliffs, behind and to the north of the Kolob Visitor Center.

2. The varied sediments of the Lower Triassic Moenkopi Formation.

The shales, siltstones, sandstones, gypsum, mudstones, and limestones of the Moenkopi Formation crop out on pink, white, and chocolate-brown slopes along the scenic drive in the Kolob section of Zion National Park. When these beds were deposited, the vast Kaibab sea had withdrawn, leaving a low coastal plain that was recovered by seawater from time to time as sea levels rose or fell. The gypsum beds formed when seawater left in lagoons evaporated. The limestones and limey mudstones contain marine fossils. The three members of the Moenkopi, from oldest to youngest, are the Virgin Limestone, Shnabkaib Shale, and Upper Red Member.

3. Deposition of the Shinarump Conglomerate at the bottom of the Chinle Formation.

By the middle of Triassic time, the land had risen and the Moenkopi surface was being eroded. Rivers draining highlands picked up sands and gravels and washed down trees that became buried in stream bars during floods. Later these channel deposits were strongly cemented by silica dissolved in ground water and the wood became petrified. The resistant Shinarump is a prominent cliff-former in Zion and is up to 200 feet thick. The name *Shinarump* means "wolf's rump." Indians in the region where the Shinarump crops out thought the gray beds (which have a tendency to weather to rounded hills) looked like the back end of a wolf.

4. Deposition of the Chinle Formation in a warm, humid environment.

Fossils of amphibians, pond-dwellers, and other freshwater creatures suggest the presence of tropical swamps when Chinle beds were laid down. Fine-grained sediments, slowly accumulating on the flat plain, became the purple and gray claystones, mudstones, and thin-bedded limestones of the Chinle Formation in Zion. The volcanic ash that supplied the silica, which percolated down into the Shinarump, was wind-deposited and probably came from volcanoes some distance away.

5. Uplift in late Triassic time; deposition of the Moenave Formation.

In Zion National Park, a reddish, slope-forming siltstone called the Dinosaur Canyon Member makes up the older beds in the Moenave (mo-en-ah-vee) Formation. Ripple marks and cross-bedding suggest that the sediments were deposited in sluggish streams and ponds. The cross-bedding was due to shifting currents and the building of sandbars. The upper member, the Springdale Sandstone, was deposited by streams with greater velocity and volume. Gray mudstone and shale, which accumulated in lakes, make up the thin-bedded Whitmore Point Member between the upper and lower members of the Moenave Formation. The fossil remains of large freshwater fish similar to sturgeon have been found in Zion Canyon in the Springdale Sandstone. They lived at the end of Triassic

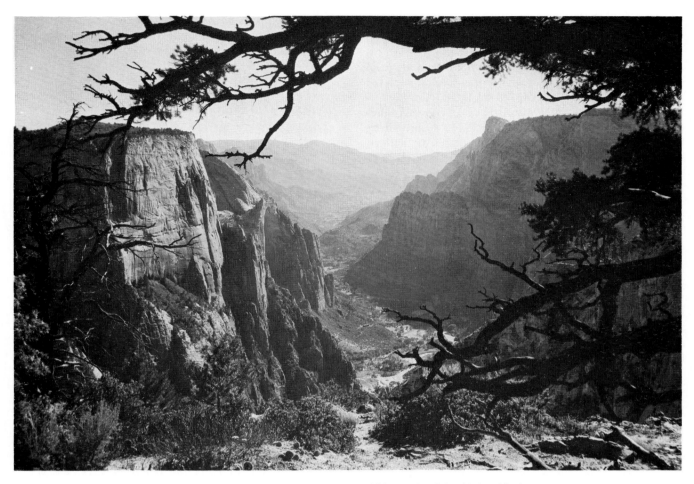

Figure 2.11   Sandstone cliffs in Zion Canyon; viewed from the east side of Observation Point. National Park Service photo by G.A. Grant.

time when the land mass that was to become North America was still drifting through equatorial latitudes in a northward direction.

6. Increasing aridity in early Jurassic time as the Kayenta Formation accumulated.

The climate became like today's northern tropical belt, with wet summers and dry winters. Streams deposited silt and sand in channels and on the flood plain. Dinosaurs coming down to the stream for water left footprints in the damp sediment along the banks. They were three-toed and walked upright, but have not been identified because their bones haven't been found in the park. The Kayenta Formation, a slope-former underlying the Navajo Sandstone, consists of red and mauve siltstones and sandstones.

7. Migrating dunes preserved in Navajo Sandstone.

A vast coastal desert, similar to the modern Sahara, spread over thousands of square miles of western North America. The thickest accumulation of the dune sands (about 2200 feet) was in the Zion region. Strong dry winds picked up sand grains and piled them in dunes that migrated because no vegetation could grow to anchor them. Some cross-bedded sands were transported by currents and accumulated in the shallow waters of the nearshore environment, but most of the sand was in coastal dunes that spread gradually inland. At first the dunes interfingered with the Kayenta stream deposits, but eventually desert sands covered the entire area. Because of differences in the type and amount of cement, the Navajo Sandstone varies in color from pale tan to red.

8. Early mid-Jurassic deposition of the Temple Cap Formation.

The caprock at the top of East Temple and West Temple records a marked change in climate at about the midpoint of Jurassic time. Streams loaded with red mud flooded the dunes and partially leveled them. After thin beds of clay and silt were deposited, desert conditions resumed, but not for long.

9. **Deposition of the Carmel Limestone in a nearshore marine environment from middle to late Jurassic time; erosional unconformity.**

About 150 million years ago a shallow, warm sea gradually spread over the area. Wave action flattened the sand dunes as calcareous silt buried them. The Carmel Limestone accumulated in massive beds of compact limestone that were one to four feet thick and separated by thin, sandy limestone beds containing fossils of shallow-water organisms. Exposures of Carmel Limestone overlie the Navajo Sandstone on Horse Ranch Mountain in the Kolob section of the park.

Late Jurassic beds are not present in Zion and were probably removed by uplift and a long period of erosion that brought the Mesozoic Era to a close in this region. Nevertheless, a small remnant of the Dakota Formation, probably deposited late in Cretaceous time, crops out at the top of Horse Ranch Mountain (Hamilton 1984). The beds consist of a basal conglomerate and a fossiliferous sandstone.

During the latter part of Mesozoic time, the layered rocks of Zion were mildly affected by the compressive effects of tectonic activity to the west. The results of folding and subsequent thrust faulting can be seen in beds in the Taylor Creek area in the Kolob section of the park.

10. **Normal faulting, uplift and erosion; lava flows.**

Late in Tertiary time (about 20 million years ago), the Zion rocks began to "feel" the tensional forces that were stretching and breaking up the crust in the region to the west that eventually became the Basin and Range. Two major north–south-trending vertical faults, which form the margins of the Markagunt Plateau, developed on either side of Zion National Park: Sevier fault on the east and Hurricane fault on the west (fig. 2.12). Several normal faults developed in the park, including the Cougar Mountain faults and the Beartrap fault. It was after the uplifting and tilting of the Markagunt Plateau that the Virgin River began downcutting.

Much later and after considerable erosion had taken place, flows of basaltic lava reached the surface by pushing up through the fractured rocks in the fault zones. The oldest dated lava flows, at Lava Point, are 1.4 million years old; the youngest, near the lower end of Cave Valley, are only a little more than 250,000 years old (Grater and Hamilton 1978). Earthquakes that occur in the region from time to time suggest that lava flows may still be working their way toward the surface.

11. **Dissection of the Markagunt Plateau by the Virgin River and its tributaries.**

As renewed faulting and tilting increased stream gradients, downcutting and mass wasting sculptured the Zion landscape. The differences in the lithology of the rock layers, their resistance or lack of resistance to weathering, and the location of the faults and joints (fractures) produced the features that we now view.

Geologic Map and Cross Section

Hamilton, W. L. 1978. Geological map of Zion National Park. Springdale, Utah: Zion Natural History Association.

Bibliography

Eardley, A. J.; and Schaack, J. W. 1971. *Zion—the story behind the scenery*. Las Vegas, Nevada: KC Publications.

Goetzmann, W. H. 1982. Explorer, mountain man, and scientist, in National Park Service Handbook 116: *Exploring the American West, 1803–1879*. p. 17–95.

Grater, R. K. and Hamilton, W. L. 1978. *Zion* (Natural history and geology) with illustrations by W. L. Chesser, M. E. Lacey, and R. Redfern. Unicorn Associates, Denver, Colorado, in cooperation with Zion Natural History Association

Gregory, H. E. 1950. Geology and geography of the Zion Park region, Utah and Arizona. U.S. Geological Survey Professional Paper 220.

———— 1956. *Geological and geographic sketches of Zion and Bryce Canyon National Parks*. Springdale, Utah: Zion-Bryce Natural History Association.

Hagood, A. 1985 (2nd edition). *This is Zion*. Zion Natural History Association and National Park Service. 73 p.

Hamilton, W. L. 1984. *The sculpturing of Zion*. Springdale, Utah: Zion Natural History Association.

Rigby, J. K. 1976. *Northern Colorado Plateau*, K/H Geology Field Guide Series. Dubuque, Iowa: Kendall/Hunt Publishing Company

Woodbury, A. M. 1950. *A history of southern Utah and its national parks*. Originally published by Utah State Historical Society (1944, v. XII, n.3,4); revised and reprinted by A. M. Woodbury, Salt Lake City.

Figure 2.12   Major plateaus and faults in Zion National Park and vicinity, southwestern Utah. Hurricane fault extends south past the western edge of Grand Canyon National Park.

Zion National Park

Area: 146,551 acres; 229 square miles

Proclaimed Mukuntuweap National Monument: July 31, 1909

Incorporated in Zion National Monument: March 18, 1918

Established as Zion National Park: November 19, 1919

Separate Zion National Monument proclaimed January 22, 1937; incorporated in national park July 11, 1956

Address: Springdale, Utah 84767

3

Innumerable highly colored, bizarre pinnacles, walls, and spires stand in horseshoe-shaped amphitheaters along the edge of the Paunsaugunt Plateau.
Bryce Canyon is not a true canyon, but an amphitheater hollowed out from the Pink Cliffs by erosion.
The unusual erosional sculptures developed in red, pink, mauve, buff, and off-white beds of the Tertiary Wasatch Formation.

Bryce Canyon National Park

Southern Utah

Figure 3.1   Queens Garden Trail in Bryce Canyon National Park. Weathering and erosion of the horizontal beds of the Wasatch Formation have produced pinnacles, gullies, the tunnel and many other fascinating and colorful features along the trail. National Park Service photo by R. Frear.

Figure 3.2   Locations of Cedar Breaks National Monument and Zion and Bryce Canyon National Parks in relation to major faults, fault blocks (the plateaus), and river valleys of southwestern Utah. In Tertiary time, as the Colorado Plateaus were uplifted, high-angle normal faults developed between crustal blocks. The plateaus that resulted are slightly higher on their western edges and tilt gently eastward. Weathering and erosion have sculptured an intricate and spectacular topography around the plateau escarpments and in the canyons, especially in the areas set aside as national parks and monuments and state parks.

Geographic Setting and Local History

Bryce Canyon National Park is only about 50 miles to the northeast in a direct line from Zion National Park, but the two parks have striking differences in geology, topography, climate, and scenery. Because Bryce Canyon is about one thousand feet higher in elevation, the climate is cooler and moister. The two canyons have undergone intense erosion for about the same length of time, but the rocks in Bryce Canyon are around a hundred million years younger than those found in Zion Canyon. When you arrive in Zion Canyon, you stand on the canyon floor and look up; but at Bryce Canyon, you come in at the top of the Paunsaugunt Plateau and look down over the rim. The cliffs and towers in Zion Canyon National Park are bold and colossal. The amphitheaters of Bryce Canyon National Park are immense, but within them the standing walls, fins, and pinnacles display intricate, delicately formed spheres, knobs, windows, and pendants in a complex array of colors.

An interesting way to go from Zion National Park to Bryce Canyon National Park is by way of Cedar Breaks National Monument, which is about 25 highway miles north of Zion's Kolob section. Cedar Breaks National Monument is on the western rim of the Markagunt Plateau at an elevation of over 10,000 feet. The rocks and erosional features of the amphitheater at Cedar Breaks are similar in age and appearance to those in Bryce Canyon National Park, located about 35 miles to the east on the Paunsaugunt Plateau. In an earlier time, one continuous, unbroken deposit covered the two areas and the region between. Now the two plateaus stand at different elevations, separated by the great, north–south-trending Sevier fault and the Sevier River valley.

The western explorers and pioneers used the term "breaks" to describe places like Cedar Breaks and Bryce Canyon, where the edge of an upland descends abruptly or breaks down by eroding to a lower level. Thus Bryce Canyon is not a real canyon at all, but a series of breaks where amphitheaters are developing by headward erosion.

Bryce Amphitheater Detail

BRYCE CANYON
NATIONAL PARK

.... Trail

———————— Highway

Park Border

0   2 mi.

Figure 3.3   Bryce Canyon National Park, Utah.

Few white men had visited the Bryce Canyon area before Major John Wesley Powell and his team of mapmakers and geologists began to survey southern Utah in the 1870s in their systematic study of the Colorado Plateaus. At that time small bands of nomadic Paiute Indians roamed through the region hunting and gathering. The surveyors retained many of the Paiute place names on their maps. For example, *Paunsaugunt,* means "home of the beaver."

A few Mormon pioneers tried to settle along the Paria River, which with its tributaries, drains the eastern side of the Paunsaugunt Plateau. Among the pioneers were Ebenezer and Mary Bryce, for whom the park was named. Overgrazing, floods, and drought forced out most of the settlers, including the Bryce family, after an attempt to divert water from the Sevier drainage to the Paria Valley had failed. Those who were left, a determined lot, dug a 10–mile ditch from the east fork of the Sevier River and this brought water from the western side of the Paunsaugunt Plateau over the rim and down into Tropic Valley. Farms on the east side then had a more dependable supply of water.

Conservationists realized, however, that the fantastic but fragile scenery of the amphitheaters was threatened by overgrazing and lumbering on the plateau. A movement was begun to protect the area. A national forest was set aside in 1905, and in 1923, Bryce Canyon National Monument was proclaimed. Utah National Park was authorized in 1924 and four years later, the name was changed to Bryce Canyon National Park. Bryce Amphitheater and the other amphitheaters along the plateau rim extend for more than 20 miles through the park from north to south. A 35–mile scenic drive on the plateau connects viewpoints in the park.

Geologic Features

On a landscape scale, it can be seen that the terraces ("steps") on the plateaus are formed on the softer rocks, such as shales. The cliff-formers are the resistant sandstones and limestones. The cliffs that rim the plateaus are associated with various formations. The Pink Cliffs of the Wasatch Formation, for example, are most strikingly exposed on the west side of the Markagunt Plateau in Cedar Breaks National Monument and on the east side of the Paunsaugunt Plateau in Bryce Canyon National Park. The White Cliffs of the Navajo Sandstone rim the southern side of the Markagunt Plateau in Zion National Park.

The Pink Cliffs developed very differently from the White Cliffs because of their lithologies; that is, the physical characteristics of the rocks. Instead of the imposing monoliths of the White Cliffs, the Pink Cliffs display barren badlands with very little soil or vegetation. However, these are not typical badlands that are entirely made up of stream-dissected slopes. W.L. Stokes (1986) describes the scenery of Bryce Canyon and Cedar Breaks as displaying "a combination of slopes and cliffs, a compromise between the erosion of material too soft to form uniform cliffs and too hard to form uniform slopes."

In Bryce Canyon National Park, the processes of weathering, *slope wash* (downward trickling or running water, not in channels), and mass wasting have been mainly responsible for the remarkable scenery that visitors see. Downcutting by streams has been less significant. Bryce Amphitheater, the largest and most spectacular scenic feature, is a horseshoe-shaped basin about 12 miles wide, 3 miles long, and 800 feet deep. The arches and windows in Bryce, its sculptured terraces, and its intricate pinnacles have all been weathered and eroded out of the pink and white Wasatch Formation—a mixture of limestone, limy siltstone, shale, sandstone, and a few remnants of conglomerate, laid down in an inland lake of vast extent. The pink hues of the beds are due to the presence of iron oxide and manganese in the sediments.

The inland lake, which existed early in Eocene time, spread over low-lying, shallow basins that covered several thousand square miles. The size and shape of the lake system varied as it expanded and contracted in response to changes in climate and environment over millions of years. The differences in the Wasatch beds reflect the different depositional environments in which they accumulated.

Subsesquent uplift and faulting formed blocks that streams carved into plateaus. The joint systems and cracks produced by uplift control the location of walls, arches, windows, and natural bridges. The pinnacles are the result of differential weathering acting along vertical joints (fractures) on beds with varying degrees of hardness. The lime-rich beds are more resistant to weathering in the Bryce climate than the softer, lime-poor beds. The difference in weathering rates produces the characteristic parallel grooving on the features; protrusions follow lime-rich layers and indentations appear along weaker, lime-poor beds (fig. 3.5).

A feature called Queen Victoria, located in the Queen's Garden area, is one of many notable examples of differential weathering. Some beds have been completely removed. For example, the conglomerate capping two buttes at Bryce Point is about all that remains of that particular layer in the entire park.

Most of the arches in the park, such as the famous Natural Bridge, are carved from sandstone beds in the Wasatch Formation. Despite their appearance of solidity,

Figure 3.4   Looking north over Bryce Amphitheater from Bryce Point. The pink and red hues of pinnacles and towers standing in the vast amphitheater appear to change from hour to hour as the sun passes overhead. National Park Service photo.

they are not permanent features. In the Fairyland area, an arch known as Oastler's Castle collapsed under its own weight in 1964. However, as old ones are destroyed, new ones are created by the unremitting processes of weathering (table 3.1).

Weathering Processes

The fragmentation or *disintegration* of rock is a *physical* process that is inextricably linked to *decomposition* of rock, which involves *chemical* processes. In other words, rock weathers physically and chemically at the same time. In Bryce Canyon National Park, the climatic zonation strongly influences how the various weathering processes interact. Moreover, the intermittent pattern of precipitation throughout the year creates cycles of wetting and drying that affect how the cliffs retreat.

At the higher elevations along the plateau (roughly 8500 to over 9000 feet) in the montane, or Canadian, life zone, precipitation averages 25 inches annually, much of it in the form of snow. Winters are severe, and late spring snow squalls are not infrequent. Spruce and fir thrive in the cool, moist conditions. Water from rain and melting snow running off the edge erodes the rim of the plateau. Moisture seeping through the soil and transporting organic acids down into bedrock increases chemical weathering.

In the Upper Sonoran (high desert) life zone at the base of the plateau (elevation around 6500 feet), the average annual precipitation is 13 inches, with summers tending to be hot and dry. Only hardy trees and shrubs can survive on the semiarid eastern slopes. Rain falls mostly in late winter and in late summer. Thunderstorms may cause rapid erosion of loose sediments on dried out or weakened slopes.

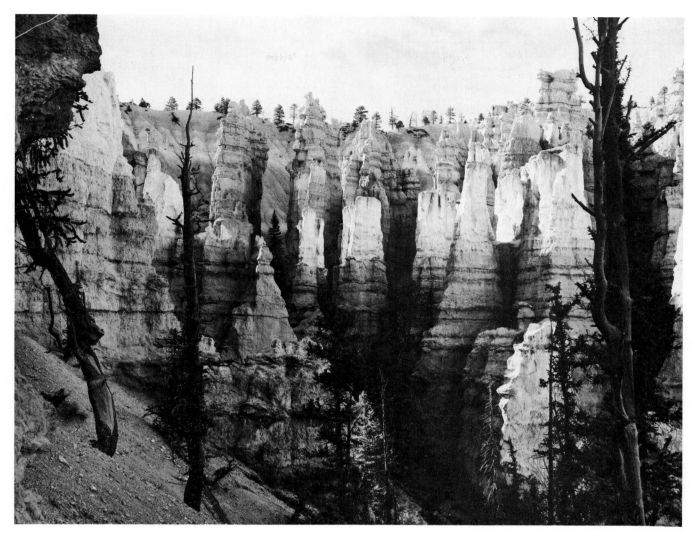

Figure 3.5 The "Silent City" area near Sunset Point; viewed from a point below the rim of Bryce Amphitheater. Turreted, vertical columns stand above gullies several hundred feet deep. Joint systems determine the locations of columns and gullies. National Park Service photo.

In the transition zone (7000 to 8000 feet), where the park headquarters is located, ponderosa pines are the dominant trees and precipitation averages 15 inches per year. Average temperatures range from lows in the 30°s (Fahrenheit) to highs in the 80°s.

On the steep, unstable slopes within the park amphitheaters, very little vegetation is available to hold soil and sediment and soak up rainwater. Moreover, the high clay content in many of the beds prevents water from penetrating deeply. That is why a sudden, intense downpour can sometimes change the appearance of features within minutes.

Zone boundaries throughout the park are not distinct. Southwest-facing slopes and erosional features, for example, are exposed more to sunlight, to greater evaporation, and to more freeze-thaw days. Sometimes when the morning sun warms a southern exposure after a freezing night, rocks come crashing down. The north-facing slopes that stay in the shadow remain covered with snow longer and have fewer freeze-thaw days. As a result, erosional processes operate a little more slowly.

Frost action is especially effective in attacking bedrock on the exposed and wind-swept promontories at higher elevations. Fairyland View, for example, is 7586 feet above sea level; Bryce Point, 8294 feet; and Rainbow Point, the highest spot on the rim, 9105 feet. Rim erosion is intensified at locations that are subject to greater extremes of temperature.

Along with freezing and thawing, extremes of temperature also tend to cause *granular disintegration* of coarse-grained rock masses such as conglomerate and coarse sandstone, both of which are found in the Wasatch Formation. The mineral grains tend to separate from one another along their natural contacts to produce coarse sand or gravel.

Table 3.1 Weathering Processes

A. Mainly physical (mechanical) weathering—disintegration

Type	Process
1. Granular disintegration	Separation of coarse mineral grains due to temperature extremes
2. Frost wedging	Grains and particles forced apart by freezing water
3. Frost heaving	Rock fragments raised up by ice pressure
4. Exfoliation	Stripping of concentric rock slabs from outer surface of a resistant rock mass (principally in igneous and metamorphic rocks; usually large-scale)
5. Spheroidal weathering	Rounding by weathering (partly chemical) from an initial blocky shape; forms boulders
6. Organic weathering	Root pry, animal burrows, human activity, etc.

B. Mainly chemical weathering—decomposition

Type	Process
7. Oxidation	Reaction of mineral grains with atmospheric oxygen
8. Carbonation	Reaction of carbonic acid (from water and carbon dioxide) with mineral grains; frequently in combination with organic acids
9. Hydration	Minerals in rock combining with water
10. Solution	Soluble minerals go into solution; insoluble mineral grains transported as sediment

Frost action attacks rock primarily by physical (mechanical) weathering. Moisture in cracks and joints freezes and expands causing *frost wedging.* Cracks enlarge and rock fragments break off. Subsurface freezing of water forces a portion of rock upward and causes more fragmentation by a process called *frost heaving.* As cracks deepen, moisture penetrates farther into the rock and enables chemical weathering to attack the minerals by loosening and altering the grains.

Spheroidal weathering is a physical and chemical process that has the effect of rounding off pinnacles and knobs in the amphitheaters. Most of the features tend to be angular as they break up along vertical joints (fig. 3.6B). A sharp corner has more surface area exposed than a smooth flat rock face, so weathering works faster on cor-

ners. Angular shapes become rounded as corners retreat. If enough moisture is present, rounded layers may spall off as concentric shells (fig. 3.6D).

A similar process on a more massive scale is called *exfoliation.* Large curved sheets or slabs of rock peel off, or *exfoliate,* from a bedrock surface. Igneous rocks and resistant sandstones that have widely spaced fractures or joints tend to be subject to this type of weathering.

Plants and animals (including humans) cause fragmentation of rock by both physical and chemical processes. As plant roots pry open cracks, plant acids react chemically with rock. Animals dig burrows, and engineers blast out rock when building roads. All of these activities expose fresh rock surfaces to moisture and organic and atmospheric acids.

The effects of chemical weathering are evident, even in the drier sections of the park where evaporation rates are high. *Oxidation,* which involves the reaction of certain mineral constituents in rock with the oxygen in the air, is the process responsible for the many shades of color characteristic of Bryce Canyon. The reds, pinks, and browns are produced by the iron oxide hematite (Fe_2O_3); the yellows by another iron oxide, limonite ($Fe_2O_3 \cdot nH_2$); and the purple hues by manganese oxide (MnO_2). Most of the white or light-colored beds were deposited as relatively pure limestones lacking the minerals that produce the colors. In a number of places in the amphitheaters, a natural stucco has given white layers an orange coating.

Hydration is a weathering reaction involving the combining of water with minerals in rock. It causes the grains to expand and the rock to be weakened. (Limonite is an oxide that has been hydrated.)

The *carbonation* reaction begins with the combination of carbon dioxide in air and soil with water to form carbonic acid, which plays an important role in the weathering of many kinds of rock. Some feldspars, for example, weather by carbonation to produce soluble calcite ($CaCO_3$) that becomes the main constituent of limestone and also forms a cementing medium in clastic sedimentary rocks.

Examples of *solution weathering* (i.e., the dissolving of the soluble minerals in rock) can be seen in such features as the Wall of Windows along Peekaboo Trail. Here again, physical and chemical processes interact as spheroidal weathering and exfoliation also aid in carving out the windows (fig. 3.7).

Many of the odd and unusual rock shapes in the park, such as the standing pillars, or *hoodoos,* are due in part to solution weathering of less resistant portions of the original bedrock. Hoodoos form where gullying has cut through resistant rock and exposed softer layers beneath. Many of the pillars are capped by remnants of resistant rock. In the Hat Shop on the Under-the-Rim Trail some hoodoos are topped by boulders that have fallen from the

A

B

C

D

Figure 3.6   Examples of weathering processes that break up rocks. *A.* **Granular disintegration.** Massive rocks (with few bedding planes or joints) tend to break into irregular blocks and fragments. As more surfaces are exposed, mineral grains gradually weather out from the disintegrating rock. *B.* **Separation along joint blocks.** Weathering penetrates fractures, causing a rock to come apart in blocks. Once loosened, blocks break into smaller and smaller pieces. *C.* **Frost wedging.** Pressure of freezing water, expanding in cracks and fractures, enlarges openings in rock. Many cycles of freezing and thawing cause slabs or blocks to split off. *D.* **Spheroidal weathering.** While processes of physical and chemical weathering operate on exposed edges and corners of outcrops, cracks tend to form just beneath a rock's exterior. Eventually concentric shells or layers loosen and spall off.

cliffs above. The boulders act like umbrellas by protecting the weak silty or clayey beds from erosion by pelting raindrops.

Frost action on jointed rocks in a region of high relief accounts for the effectiveness of mass wasting in Bryce Canyon National Park. Evidence for this can be seen in the extensive accumulations of talus that overspread the base of slopes, pile up against the cliffs, and surround the columns (fig. 3.8). Typically the steepness of a talus slope or cone is at the angle of repose (roughly 35°). Loose fragments continually roll downslope or shift position as they move downward.

The Paria River and its headwaters, which drain Bryce Canyon, carry away sediment and debris plus rock material in solution. The Paria flows southeastward and empties into the Colorado River at Lees Ferry, just above the Grand Canyon.

Geologic History

1. Deposition of the Dakota Sandstone in a shallow Cretaceous sea.

A few remnants of thin Dakota Sandstone, the oldest formation exposed in the vicinity of the park, have been found well below the plateau in the Paria Valley. The Dakota Sandstone lithified from a *blanket sand;* that is, a sedimentary layer thinly covering a large geographic area. In this region the deposit represents the beach area of a transgressing sea (fig. 3.9), the last sea to spread over the Colorado Plateaus.

2. Deposition of the Tropic Shale in a swampy coastal area.

At the lowest elevations in the park are scattered outcrops of the coal-bearing Tropic Shale that accumulated

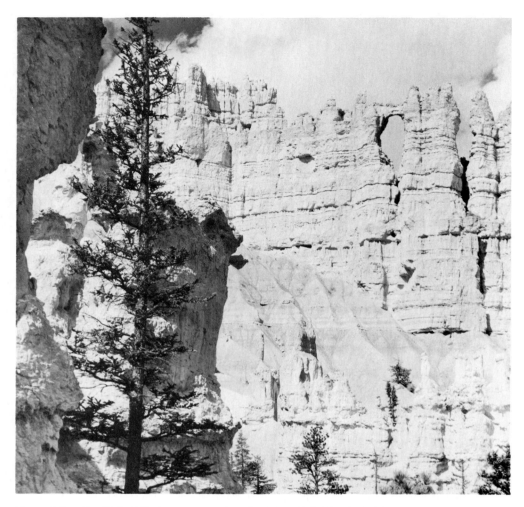

Figure 3.7   The Wall of Windows, as seen from Peekaboo Trail. Chemical and spheroidal weathering and exfoliation, attacking less resistant parts of the rock, produced windows in the wall. National Park Service photo.

in a regressing sea and in coastal swamps. Coal from the Tropic Shale is mined commercially in areas near the park.

3. Erosion of the Late Cretaceous formations, creating unconformities.

Most of the sandstones and shales that were deposited in the Bryce Canyon vicinity late in Cretaceous time were stripped away by erosion as the period came to a close. A few outcrops of the Straight Cliffs and Wahweap Sandstones and the Kaiparowitz Formation are found at low elevations in the southern portion of the park.

4. Compressive effects of plate convergence to the west late in Cretaceous and early in Paleocene time.

The Laramide orogeny, a mountain-building episode that profoundly affected western North America (box 2.1), began as the westward-moving North American plate slid over the Pacific oceanic plate. The compressive tectonic forces produced by this convergence caused low uplands

to be uplifted in what is now southern Utah. Between the uplands, low-lying basins gently subsided.

5. Deposition of the Wasatch Formation in inland lakes early in Eocene time.

A broad, shallow lake system that spread over the basins was fed by numerous rivers that sorted and transported vast quantities of clastic sediment. Over the millions of years that the lake system (called Lake Flagstaff) existed, it expanded and contracted many times; and as it did so, alternating beds built up, some thick, some thin, some coarse, some fine, some in water, and others along shorelines. The continually changing depositional environments created a formation lacking uniformity in texture; nearly pure limestones, laid down in deep water, graded into calcium-rich mudstones, which were often interspersed with lenses of sandstone and conglomerate.

Fossils are not abundant in the Wasatch beds, but some subtropical, aquatic snails, turtles, reptiles, and other species have been identified.

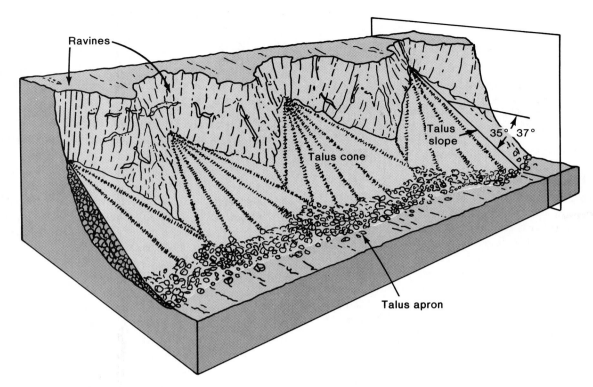

Figure 3.8 Sediment grains and fragments, loosened by frost action, chemical weathering, slope wash, and other processes, accumulate in talus slopes around a retreating escarpment. Steepness of the slopes is determined by the angle of repose of accumulating sediments. © 1964 Wards Natural Science Establishment; used by permission.

Table 3.2 Geologic Column, Bryce Canyon National Park

Era	Period	Epoch	Rock Units Formation	Geologic Events
Cenozoic	Quaternary	Holocene		Weathering and erosion of Wasatch beds to form present landscape. Block-faulting ————— Uplift and erosion —————
		Pleistocene		
	Tertiary	Eocene	Wasatch	Streams depositing in vast, inland lakes. ————— Erosion —————
Mesozoic	Upper Cretaceous		Kaiparowits	Flood-plain deposition ————— Erosion —————
			Wahweap Sandstone	Deposition along a shifting shoreline.
			Straight Cliffs Sandstone	Deposition in a shallow sea and brackish lagoons. ————— Erosion —————
			Tropic Shale	Deposition in a shallow sea and coal swamps
			Dakota Sandstone	Beach deposits as a blanket sand in a transgressing sea.
	(Older units not exposed in park)			

Source: Gregory 1956; Rigby 1976; Stokes 1986

Figure 3.9   Beach sand, accumulating on a shoreline migrating inland, may form a blanket sandstone. The rock formation shown here becomes progressively younger toward the right as the sea encroaches on the land.

6. Mid-Tertiary (Miocene) uplift and block-faulting, creating the plateaus.

From Rainbow Point (elevation 9105 feet), you can look out over a broad expanse of southern Utah and see most of Bryce Amphitheater, the Table Cliffs, the Aquarius Plateau, White Cliffs, Vermilion Cliffs, and the Henry Mountains. The culminating uplift of the high plateaus of southwest Utah that raised these features (and the whole Colorado Plateaus region) several thousand feet from near sea level occurred in Miocene time, some 16 million years ago.

During this episode, the tectonic forces were tensional rather than compressional. A portion of the western continent, especially in Nevada and southwest Utah, was being pulled apart as well as elevated. As a result, deformation was greater on the western margins of the Colorado Plateaus than toward the east. The stress and strain of the crustal disturbance created the great north-south normal faults (such as the Sevier and Paunsaugunt faults) that broke up and separated crustal blocks (fig. 3.2). Some blocks dropped, but the blocks that formed the southern Utah plateaus rose and tilted. Occasional earthquakes in the region are an indication that tectonic adjustments are still occurring far below the surface.

7. Erosion of the Paunsaugunt Plateau and development of the amphitheaters in late Tertiary and Quaternary time.

The river systems that transported the Wasatch sediments were completely disrupted by the Miocene upheavals. Moreover, as the North American plate continued to drift toward northern latitudes, while moving generally westward, the climate cooled, then became more moist. During the times of greatest cold in the Pleistocene Epoch, the low temperatures maintained snow year-round at higher elevations. For a time glaciers existed at the top of windswept Cedar Breaks to the west.

The steeper gradients of the reorganized drainages and the increased volume from rain and snowmelt made the streams of the region swift and powerful. The headwaters of the Paria River (a tributary of the Colorado River) began cutting rapidly headward into the eastern side of the Paunsagunt Plateau. The more erosion broke through protecting soil and vegetation, the more the bedrock layers were laid bare, exposed in all their colors and structures, to the endless processes of weathering.

Figure 3.10   At Bryce Amphitheater, the edge of the Paunsaugunt Plateau is retreating appproximately one foot every 50 years. National Park Service photo.

Bibliography

Bezy, J. 1980. *Bryce Canyon, the story behind the scenery.* Las Vegas, Nevada: K.C. Publications. 48 p.

Gregory, H. E. 1951. The geology and geography of the Paunsaugunt region, Utah (with geologic map). U.S. Geological Survey Professional Paper 226.

Hirschmann, F. *A natural history guide to Bryce Canyon.* Bryce Canyon Natural History Association.

Rigby, J. K. 1976. *Northern Colorado Plateau,* K/H Geology Field Guide Series. Dubuque, Iowa: Kendall/Hunt Publishing Company.

Stokes, W. L. 1986. *Geology of Utah.* Salt Lake City: Utah Museum of Natural History and Utah Geological and Mineral Survey. p. 161–163.

Yardell, M. D. et al. 1972. *National parkways guide to Bryce Canyon and Zion Canyon National Parks.* Casper, Wyoming: World-Wide Research and Publishing Company. p. 3, p. 41–64.

Bryce Canyon National Park

Area: 36,835 acres; 56 square miles

Proclaimed Bryce Canyon National Monument: June 8, 1923

Established as Utah National Park: June 7, 1924

Changed to Bryce Canyon National Park: February 25, 1928

Address: Bryce Canyon National Park, Bryce Canyon, Utah 84717

4

Capitol Reef National Park

South Central Utah

The dome-shaped, white Navajo Sandstone rimrock, uplifted and tilted by the Waterpocket Fold, gives the park its name. The Waterpocket Fold, a monocline, dips east and trends north–south through the park.

Narrow, high-walled gorges, cut through the uplifted fold, reveal vividly colored beds of Mesozoic sedimentary rocks.

". . . a maze of cliffs and terraces, red and white domes, rock platforms gashed with profound canyons, burning plains barren even of sagebrush—all glowing with bright color and flooded with blazing sunlight."

Clarence E. Dutton, 1880

Figure 4.1 View from the south end of Capitol Reef National Park, looking north along the Waterpocket Fold. Navajo Sandstone, the oldest formation exposed, forms the high ridge on the right; the Carmel Formation is a narrow band along the base. On the valley floor, going toward the left, the Entrada Sandstone, Summerville, and Morrison Formations crop out. Next, the Dakota Sandstone forms a low, sinuous ridge. On the left, a butte of Mancos Shale is capped by beds of the Mesa Verde Group, youngest rocks in the sequence. Photo by K. Swett.

Box 4.1
How Mass Wasting Operates in Arid Regions

The effects of mass wasting on landforms in the national parks of the Colorado Plateaus region are plain to see. Loose pieces or chunks of sedimentary rock that fall from cliffs accumulate in piles of *talus* and talus slopes that surround the base of a mesa or butte. Roadcuts made through bedrock may bring about *rockslides* that block a highway. On some steep slopes, scars of old slides can be seen above a block or mass of

hillside material that has separated and slipped down along a curved surface, slumping at the bottom so that the base bulges out as a hummocky toe. Sudden torrential rains, perhaps accompanied by pelting hail, may loosen, wash down, or saturate so much debris that a desert stream turns into a thick *mudflow* that continues down a canyon picking up or burying everything in its path. When the rain stops, momentum keeps

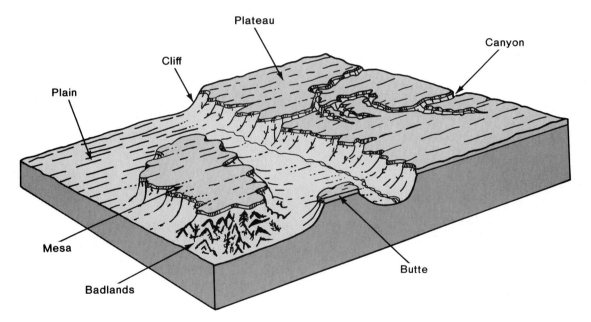

Figure 4.2   Landforms typical of uplifted, horizontal sedimentary rock layers in an arid climate. Resistant limestones and sandstones cap plateaus and form rimrock on escarpments. Less resistant shale beds, exposed by cliff retreat, erode more rapidly. Mesas and buttes are erosional remnants. Slot canyons are cut by streams during uplift. Badlands may develop in weak shales and mudstones. © 1964 Ward's Natural Science Establishment; used by permission.

continued

Geographic Setting and Local History

Capitol Reef National Park lies along a north-south-trending monocline, called the Waterpocket Fold, about halfway (in a direct line) between Bryce Canyon National Park and Canyonlands National Park. Only about 7 inches of precipitation a year falls in the Capitol Reef region, some of it in brief, intense storms that may cause flash flooding. Capitol Reef National Park is in desert country,

except for the oasis along the Fremont River, a year-round stream. The Fremont River cuts through the Waterpocket Fold, from west to east, on its way to the Dirty Devil River, a tributary of the Colorado.

Indians of the Fremont culture, who raised corn, beans, and squash by irrigation, lived along the river around a thousand years ago. They drew pictographs on canyon walls. The stone granaries they built for storing

Box 4.1 continued

the mudflow going until the water dissipates, leaving the channel or the canyon mouth choked with boulders, uprooted trees, gravel, and silt.

With the driving force of gravity supplying the necessary energy, loosened rock or debris may fall, slide, slip, slump, or flow from higher locations to lower locations. Controlling the rate and type of mass wasting processes are such factors as climate, degree of slope, relief, characteristics of the bedrock or sediments, and vegetation cover. However, climate is usually the key factor.

Relief and slope angle are important because steep slopes are more susceptible to mass wasting than gentle slopes. Slopes in areas of high relief tend to be less stable than slopes in areas of low relief. Climate influences what types of weathering occur and governs how much and what type of vegetation grows in an area. In cold climates, freezing and thawing contribute to downslope movement in both humid and arid

regions. In desert country, infrequent but heavy rain pelting on slopes unprotected by vegetation aids the processes of mass wasting. Depending upon the amount of runoff, debris may wash down or may become saturated, forming a debris slide or a mudflow.

In contrast to the more gently rounded hills and valleys typical of humid regions, the desert landscape looks angular with its steep slopes and walled canyons. This kind of topography tends to increase the effectiveness of mass wasting. Due to the thin soil and lack of moisture in the desert, one type of mass wasting characteristic of humid areas is extremely slow here. This is *creep,* the slow, continuous downslope movement of soil or any unconsolidated debris. In temperate regions, which have water in the soil with seasonal freezing and thawing, even slopes covered with thick vegetation are subject to the continual action of creep.

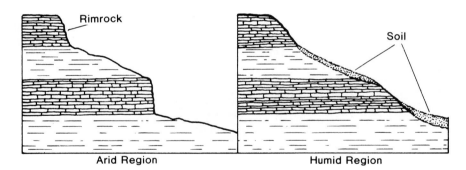

Figure 4.3   Slope profiles in arid regions tend to develop differently from those in humid regions even when the underlying bedrock is similar. Because more moisture is available for weathering processes in humid regions, slopes are typically rounded and mantled with soil. In an arid climate, steep slopes with angular rock faces are common, and soil cover is sparse. From *Landforms and Landscapes* (3rd ed.) by Sherwood D. Tuttle. © 1970, 1975, 1980 Wm. C. Brown Company Publishers, Dubuque, Iowa. Reprinted by permission.

corn were called "moki huts" by the Paiute Indians, who came into the region many years later. The Paiutes thought that a race of tiny people, or moki, lived in the huts, some of which are preserved in sheltered locations in the park.

Almon H. Thompson, one of Major J.W. Powell's surveyors, explored the formidable region to the east of Bryce Canyon and crossed the Waterpocket Fold in 1872. Subsequently, geologist Clarence Dutton spent several summers in the area engaged in field work. His findings are

reported in his classic monograph on the High Plateaus geology (1880).

Mormon settlers moved into the Fremont River valley in the 1880s. They lived in and around a community called Fruita. Their orchards and lawns are now a picnic area and campground within the national park. The old Fruita Schoolhouse and a settler's cabin have been restored by the National Park Service. Old lime kilns in Sulfur Creek and near the campgrounds on Scenic Drive are all that

Figure 4.4   Capitol Reef National Park, Utah.

remains of an early industry here. The settlers used lime from some of the limestone beds in Capitol Reef for white-wash, fertilizer, and cement. Earlier in this century several small uranium mines were worked. The last residents of Fruita moved away in the late 1960s.

The name Capitol Reef was applied to an especially scenic line of white domes and cliffs of Navajo Sandstone that extend along part of the Waterpocket Fold from the Fremont River south to Pleasant Creek. Early visitors were reminded of the white, domed buildings of Washington, D.C. The term "reef" meant any rocky barrier or escarpment that made traveling difficult. (Reef in this sense has nothing to do with the rocky masses built by marine organisms.) A national monument that was proclaimed in 1937 comprised roughly the Capitol Reef area only. Its expansion in 1971 to a national park encompassed the greater part of the Waterpocket Fold.

Utah highway 24, which crosses the park from east to west, was built along the Fremont River in 1962. Before that time, a frequently washed-out "highway" went through Capitol Gorge, about ten miles to the south. Abandoned now except for travel by foot, the old wagon road is strewn with flood boulders and hummocks of re-worked gravel. In some places, Capitol Gorge, with towering cliffs on either side, was barely wide enough for a wagon or car to pass through.

Geologic Features

The Waterpocket Fold, a large monoclinal fold dipping eastward and trending north-south for over 90 miles, forms the geologic framework of Capitol Reef National Park (fig. 4.1). A *monocline* is a fold creating a local steepening of dip in layered rocks in an area where the bedding is otherwise generally flat or very gently dipping. All but about 30 miles of this dominant structure lies within park boundaries. Along the east side of the fold, erosion sculpted the escarpment, or so-called reef, that was for many years a formidable barrier to travelers and settlers going west.

Capitol Reef National Park is an ideal place to see large-scale examples of differential erosion of harder and softer beds because of the way they are exposed in the monoclinal fold. The more resistant beds in the structure form imposing ridges, cliffs, domes, mesas, and buttes, as well as smaller features, such as pinnacles, towers, and arches.

Major cliff faces in the park are formed from Wingate Sandstone, while the Navajo Sandstone and the Carmel Formation form the summit and much of the eastern slope of Capitol Reef and the Waterpocket Fold. The eroded turrets of the Castle, for example, are carved from the orange and red Wingate Sandstone, which juts up in westward-facing rimrock along the "spine" of the fold. The Wingate Sandstone also forms the Circle Cliffs,

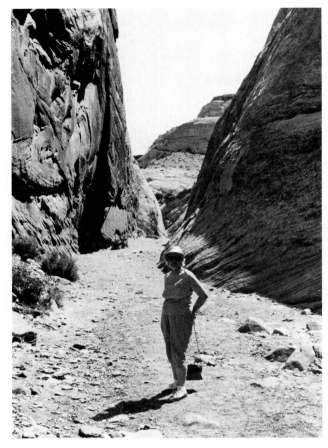

Figure 4.5   The abandoned road through Capitol Gorge is strewn with rocks left by flash floods that sometimes fill the normally dry canyon with water 10 to 20 feet deep. Photo by S. D. Tuttle.

which lie just outside the west boundary of the park's south section. The Circle Cliffs are an eroded remnant of an uplift that flanks the west side of the Waterpocket Fold.

The Hickman Natural Bridge, 133 feet wide and 125 feet above the streambed, is a true natural bridge that was formed over a watercourse as weathering and stream erosion wore a hole through a narrow "fin" of sandstone in the Kayenta Formation. Successive floods deepened the hole and the canyon, leaving the bridge high and dry most of the time. Hickman Bridge can be reached by a trail from the north side of Highway 24.

The flat-topped hills in the park, capped by resistant rock layers, are erosional remnants carved from the nearly horizontal sedimentary rock layers. These landforms, called *mesas,* have steeply sloping erosion scarps on all sides. *Buttes* are very similar but are smaller in size with less summit area than mesas. Most buttes were probably mesas originally. Examples of easily accessible buttes in the park are Chimney Rock, Twin Rocks, and Egyptian Temple. In these buttes, soft Moenkopi beds are capped by the resistant Shinarump Member of the Chinle Formation (fig. 4.6).

Figure 4.6 The Shinarump Conglomerate, which caps Chimney Rock, protects softer, underlying beds of the Moenkopi Formation. National Park Service photo.

The "waterpockets," for which the monocline is named, are really *potholes* that have been carved in bedrock by the abrasive action of fine and coarse sediment whirling in turbulent water and deepening a hole that may have started as a small depression. Some may owe their origins to the meltwater torrents of waning Pleistocene time, but many are still being actively eroded whenever flash floods fill narrow gorges and canyons in the park with rushing water. Boulders and cobbles, trapped in low places and swirled around by the water, may act like a drill boring holes in the rock. Some waterpockets are found at the foot of a cliff in the bed of an *intermittent stream;* that is, a temporary or seasonal stream. The action of falling water when the stream is in flood may create bowl-shaped potholes. The water stored in these natural pockets is an important source of drinking water for park wildlife (fig. 4.7).

Rockhounds find many examples of rock layers in the park and outside its boundaries that exhibit *primary structural features,* such as ripple marks and footprints of dinosaurs and amphibians, which were put there during deposition but before consolidation. Slabs of ripple-marked stone from red Moenkopi beds are quarried outside the park and sold as "ripple rock" for decorative stonework.

Igneous Rocks. Dark-colored basaltic magma squeezed between some sedimentary beds in Tertiary time, while they were still deeply buried. The intrusions hardened and formed sills. Subsequently, more magma pushed up and cut through both the sedimentary layers and the sills and created dikes. (By definition, *sills* are intrusions that harden parallel to bedding while *dikes* cut across structures.) Because the basalt is more resistant than the sedimentary rocks, some sills and dikes are now exposed and standing in relief at the northern end of the Waterpocket Fold.

The black basaltic boulders scattered about the park on top of Mesozoic bedrock are extrusive in origin, rather than intrusive. They came from lava flows capping Boulder and Thousand Lake Mountains west of the park and were transported by rockslides, glacial action, and floods of glacial meltwater. The Fremont people used some of the boulders for the foundations of their pit-house dwellings.

Patterns of Stream Erosion

Stream erosion has been significant in the shaping of the Capitol Reef even though in today's arid climate most of the streams are intermittent. The Fremont River and a few others are permanent streams. Sulphur Creek enters the park from the west, cuts through the Goosenecks, and joins the Fremont east of the Park Headquarters and Visitor Center. (The Goosenecks are an example of "incised meanders," which are explained in chapter 5.) Another year-round stream, Pleasant Creek, which flows northeast, crosses the park about ten miles south of the Fremont. The only other permanent stream in the park is Oak Creek, which crosses the Waterpocket Fold in the area south of the domes.

It is interesting to note the effect of structure on the stream patterns of the area. All the present year-round streams had enough erosive power to cut across the monocline from west to east while it was being uplifted. However, some canyons, such as Capitol Gorge, which were carved out by east-flowing streams, are now dry or intermittent. Probably as the climate became more arid, these streams lost out to more favorably located drainages.

Several intermittent streams follow the trend or strike of the monoclinal structure. (The *strike* is the direction, or trend, of the bedding. In this case, it's generally north–south.) A tributary that follows the strike of the beds has adjusted its course to the underlying rock structure. For example, Hall Creek in the southern part of the park flows south, parallel to the east flank of the Waterpocket Fold. The Hall follows a course of less resistance along the base of Navajo Sandstone cliffs, cutting down its channel in softer beds of the Carmel Formation and the Entrada Sandstone. However, just before Hall Creek crosses the

Figure 4.7   Sediment swirling in turbulent flood waters may abrade round, undrained depressions, locally called "waterpockets" or "tanks." National Park Service photo.

park's southern boundary, it cuts the Narrows, a three-mile defile through Navajo Sandstone, and then empties into the Colorado River. In a sense, Hall Creek "let itself down" in order to reach the Colorado's lower level.

The Fremont River flows in an easterly direction *across the strike* of the beds, following the regional slope toward the Colorado River. The regional slope coincides approximately with the tilt, or dip, of the beds. (*Dip* is defined as the vertical angle that beds form with a horizontal surface.) The Fremont River is a swift-flowing stream that rampages through its valley when augmented by even a moderate increase in volume. A drop of roughly 2000 feet between the west side of the monocline (near Torrey) and the junction with the Dirty Devil (near Hanksville) gives the Fremont River enough gradient to generate considerable erosive power. The Mormon farmers finally gave up their orchards and fields in the Fremont valley after years of trying to maintain dams and irrigation ditches on an uncontrollable river that alternated between extremes of too little flow and too much.

Alluvial fans have been built up where streams loaded with sediment come out of steep canyons onto an open, flat floor or a basin or valley. The sudden loss of velocity causes the streams to unload as water spreads out and dissipates.

Sudden flash floods that come with little warning are dangerous to persons who may become trapped in narrow, steep-walled gorges and canyons. Grand Wash and Capitol Gorge are well known examples of areas subject to flash flooding. Park visitors must be especially cautious if rain threatens.

Geologic History

1. Deposition of the Cutler Formation and Kaibab Limestone during Permian time.

Toward the end of the Paleozoic Era, the Cutler Formation and the Kaibab Limestone were deposited along the margin of a shallow, transgressing sea. Only the White Rim Sandstone, the youngest member of the Cutler Formation, is exposed in Capitol Reef National Park. The

Table 4.1 Geologic Column, Capitol Reef National Park

Era	Period	Group	Formation	Member	Geologic Events
Time Units		Rock Units			
Era	Period	Group	Formation	Member	Geologic Events
Cenozoic	Quaternary	Sands, gravel, glacial erratics, boulders			Mass wasting, stream erosion Canyon cutting Glacial climate
Cenozoic	Tertiary	Dikes, sills, mostly basalt Volcanic plugs, lavas			Extrusive and intrusive activity Uplift and deep fluvial erosion Waterpocket Fold formed during Laramide orogeny
Mesozoic	Cretaceous	Mesaverde			
Mesozoic	Cretaceous		Mancos Shale		Sandy shales deposited in oscillating sea
Mesozoic	Cretaceous		Dakota Sandstone		Nonmarine, then shore deposits in trangressing sea Erosion—
Mesozoic	Jurassic		Morrison	Brushy Basin	Gray-green and maroon clays, dinosaur bones in flood-plain deposits
Mesozoic	Jurassic		Morrison	Saltwash	Conglomerates, siltstones; contains uranium Erosion —
Mesozoic	Jurassic	San Rafael	Summerville		Sand dunes; shales on tidal flats and in shallow lakes
Mesozoic	Jurassic	San Rafael	Curtis		Green shales, tidal-flat deposits Erosion —
Mesozoic	Jurassic	San Rafael	Entrada Sandstone		Reddish brown sandstone, shallow-water deposits
Mesozoic	Jurassic	San Rafael	Carmel		Shales, limestone, gypsum deposited by advancing and retreating seas Erosion —
Mesozoic	Triassic	Glen Canyon	Navajo Sandstone		Accumulation of white, well rounded sand grains in cross-bedded dunes; cliff-former
Mesozoic	Triassic	Glen Canyon	Kayenta		Fluvial, interbedded red shale and sandstone; slope-former
Mesozoic	Triassic	Glen Canyon	Wingate Sandstone		Bright-red sandstone, cross-bedded dunes; cliff-former Erosion —
Mesozoic	Triassic		Chinle		Shale, sandstone, siltstone; slope-former; contains petrified wood
Mesozoic	Triassic		Chinle	Shinarump	Conglomerate; contains uranium Erosion —
Mesozoic	Triassic		Moenkopi		Brownish red shales, gypsum, sandstones in tidal-flat deposits; "ripple rock"
Paleozoic	Permian		Kaibab Limestone		Dolomitic limestone, siltstone
Paleozoic	Permian		Cutler	White Rim Sandstone	Marine sandstone
Older rocks not exposed in park					

Sources: Baars 1985; Billinglsey, Breed, Huntoon 1981; Collier 1987; Stokes 1986

White Rim, a cross-bedded dune deposit, crops out in the bottom of Sulphur Creek and in the center of the Circle Cliffs outside the western edge of the park. However, the White Rim Sandstone and the other members of the Cutler Formation are prominently displayed in Canyonlands National Park (chapter 5), which is roughly 60 miles in a direct line to the east. The Kaibab Limestone is well known as the rim of Grand Canyon (chapter 1). In the Capitol Reef area, where the Kaibab Limestone is exposed in the deeper parts of some canyons, the formation consists not of limestone but dolomite with interbedded sandstone. (In dolomitic rocks, magnesium has replaced much of the calcium of the original limestone.)

2. Deposition of the Triassic Moenkopi and Chinle Formations.

Mesozoic rocks (about 10,000 feet of sedimentary layers) dominate the landscape in Capitol Reef National Park. First laid down were the reddish brown, Triassic Moenkopi beds of siltstone. Many of them bear ripple marks that indicate their deposition by streams flowing toward a marine embayment to the north.

After uplift, erosion of the Moenkopi Formation produced an unconformity, on which discontinuous beds of basal conglomerate of the Shinarump Member of the Chinle Formation were deposited. Uranium salts that accumulated in this sedimentary unit were taken out as ore from the Oyler Mine in Grand Wash. The first claim was filed in 1904, many years before the uranium boom of the 1950s.

Sediments accumulating in the Chinle beds above the Shinarump tended to be finer as the rivers slowed and deposited irregular layers of sand and mud over the subsiding basin. Logs washed down, were buried, and eventually petrified, probably aided in this process by volcanic ash blown in from distant eruptions. The soft, red and pink shales of the Chinle Formation tend to weather into rounded hills and slopes.

3. Accumulation of the Glen Canyon Group of sandstones in late Triassic and early Jurassic time.

The most visible and prominent rocks in Capitol Reef National Park are the three formations of the Glen Canyon Group that make up the spine of the Waterpocket Fold; namely, the Wingate Sandstone, the Kayenta Formation, and the Navajo Sandstone. Actually, these three formations, and especially the Navajo, "star" in the majority of well known landmarks of the national parks of the Colorado Plateaus.

The Triassic Period ended in a time of increasing aridity. Migrating sand dunes on the shore of an ancient sea advanced over the Chinle surface and buried it with 350 feet of a cross-bedded, orange sandstone that became the cliff-forming Wingate Sandstone.

As the Triassic Period came to a close and the Jurassic Period began, sluggish streams deposited the limey, thin-bedded sandstones of the Kayenta Formation in channels and on low-lying plains. At places where dinosaurs and tritylodonts (crocodilelike animals) came to drink on the banks of streams, their footprints have been preserved in the rock. The Kayenta tends to form ledgy slopes where it crops out between the massive Wingate Sandstone below and the Navajo Sandstone above.

Again desert sands overspread the land, encompassing a vast area for millions of years. Migrating dunes accumulated to a thickness of 1000 feet. These remarkably clean cross-bedded sands became the white Navajo Sandstone that is famous for its beautiful canyons, gleaming domes, smooth, unbroken cliffs, and striking monoliths. Examples are the Golden Throne in Capitol Reef National Park and the Great White Throne in Zion National Park (chapter 2). The cross-bedding of the Navajo Sandstone causes it to weather into curving canyons and rounded domes.

4. Deposition of the San Rafael Group unconformably over the Glen Canyon Group.

The four formations of the San Rafael Group—Carmel Formation, Entrada Sandstone, Curtis Formation, and Summerville Formation—are exposed on the dipping east side of the Waterpocket Fold. Around the middle of Jurassic time, after an erosional period had planed off the Navajo dunes, the limey silts, sands, and gypsum beds of the Carmel Formation accumulated in what may have been a down-faulted basin that was periodically invaded by seawater. At the north end of the park, in Cathedral Valley, some of the Carmel gypsum dissolved, causing a surface area to collapse, which formed a sinkhole. A more resistant part of the Carmel Formation caps the Golden Throne, located south of Grand Wash.

The environment in which the overlying Entrada Sandstone accumulated consisted of tidal flats, low shores, offshore bars and barrier islands. The reddish brown color of the Entrada Sandstone is the same here as in Arches National Park (chapter 6), but since the joint systems in the rock are different, arches did not form in the Capitol Reef exposures of the Entrada. The oddly shaped monuments and "goblins" (hoodoos) that stand in Cathedral Valley are typical of the weathering of Entrada Sandstone in this park.

The Curtis Formation, which is light gray and only 50 feet thick, is resistant to erosion and has a basal conglomerate overlain by sandstone and shale beds. The presence of glauconite, a green iron potassium silicate mineral, indicates that the beds were deposited in a shallow sea. The formation crops out only in the northern section of the park where numerous features in Cathedral Valley are capped by Curtis bedrock.

Figure 4.8   Looking north over the Waterpocket Fold from the vicinity of Circle Cliffs. Here the Navajo Sandstone, which crops out at the extreme right, caps the slope-forming Kayenta Formation. Next is the Wingate Sandstone, dipping toward the right and overlying Chinle beds. The lower slope, extending into the valley, exposes the Moenkopi Formation. National Park Service photo.

As the Curtis sea retreated, the reddish brown mudstones and white sandstones of the Summerville Formation accumulated on tidal flats. The Summerville tends to form ledgy cliffs.

5. Deposition of the nonmarine Morrison Formation, bringing the Jurassic Period to a close.

The Morrison Formation, overlying the San Rafael Group, consists of stream-deposited sandstones and mudstones laid down in lakes on swampy plains. The lower Salt Wash Member was the source of most of the uranium that was mined in the park area in the 1950s. The upper Brushy Basin Member is best known for its dinosaur remains. Because the bones are scattered, however, the dinosaurs have been more difficult to assemble and identify than in Morrison exposures like those in Dinosaur National Monument in northeastern Utah and elsewhere.

6. Cretaceous deposition of Dakota Sandstone and Mancos Shale.

The lower units of the Dakota Sandstone are nonmarine and resemble the Jurassic beds on which they were deposited. But eventually a vast sea overspread the Colorado Plateaus and much of the region beyond. Exposures of Dakota Sandstone, some containing fossil oyster shells, are found in the southern section of Capitol Reef National Park.

Units of Mancos Shale were deposited during times of advancing and retreating shorelines. Within the park's southern section, some of the mesas and buttes are eroded from Mancos Shale beds. A few mesas of Mancos Shale on the park's eastern boundary are capped by beds of the more resistant Mesaverde Group, the youngest sedimentary rocks in the park. East of the park, some soft Mancos units have formed badlands.

A

B

C

Figure 4.9 How the Waterpocket Fold developed. *A.* Sandstones, shales, and limestones, both marine and nonmarine, accumulated for millions of years in flat-lying beds. *B.* Compressive forces, generated during the Laramide orogeny, produced a great monocline, the Waterpocket Fold. *C.* As the region was uplifted, streams began to dissect the monocline, cutting narrow canyons through the Navajo Sandstone and wider canyons in the Chinle Formation. From M. Collier, *The Geology of Capitol Reef National Park.* © 1987 Capitol Reef Natural History Association. Used by permission.

Figure 4.10 Magma from a small pluton intrudes sedimentary rock beds. Dikes cut across rock layers, while a sill squeezes between bedding planes.

7. Folding and uplift of the monocline at the end of Cretaceous time.

Like the other monoclines of the Colorado Plateaus, the Waterpocket Fold has a north-south alignment with the steep side facing east. The structure is the result of the intense horizontal compression exerted on the Colorado Plateaus during the Laramide orogeny about 65 million years ago. Perhaps the fold formed here because a weak zone may have resulted from a major basement fault in Precambrian rocks far below the surface. Minor earthquakes in 1979 that centered below the Waterpocket Fold may indicate shifting of a basement fault.

The uplifting of the Capitol Reef region was part of the overall elevation of the Colorado Plateaus as the land was raised several thousand feet from being near sea level at the beginning of Cretaceous time.

8. Tertiary igneous activity.

The compression, uplifting, and fracturing that accompanied the Laramide orogeny were followed by relatively gentle stretching and adjusting of the crust. Magma began to rise and pushed into weakened crustal zones. In the South Desert and Cathedral Valley, dikes and sills were intruded along joints and bedding planes in the sedimentary beds. Some magma reached the surface and erupted as basaltic lava through fissures. Along with dikes and sills, volcanic plugs have been stripped of their sedimentary covers and now stand exposed above the desert sands. The igneous activity probably began in Oligocene time and continued sporadically. A relatively young dike that cuts the Curtis Formation has been dated at just under 4 million years B.P., or Pliocene time (Collier 1987).

9. Development of the Fremont River and other Colorado River tributaries; effects of Pleistocene glaciation.

The uplifting of the Waterpocket Fold caused rearrangement of the drainage in the area. While the Fremont River and several other streams continued to carve their courses across the fold during uplift, streams such as Sandy Creek and Halls Creek began to flow parallel to the strike, or trend, of the monocline, downcutting their channels along less resistant beds. Some streams carved steep-sided canyons and later abandoned these courses. The Fremont River and its tributaries continue to shape the landscape in the park, but the processes of weathering and mass wasting play a significant part as well.

Erosional processes were more intensive during the Pleistocene Epoch and during moister climatic cycles of the Holocene. Pleistocene mountain glaciers (outside the park area) and the accompanying cooler, wetter climate kept streams well supplied with water. Glacial ice plucked basaltic boulders from Boulder Mountain; meltwater streams washed them downslope and distributed them over much of the park area.

The present landscape is remarkably varied with dry canyons, plateaus, mesas, buttes, domed cliffs, arches, and colorful rocks—a place of magnificent scenery and few people.

Geologic Map and Cross Section

Billingsley, G. H., Breed, W. J., and Huntoon, P. W. 1987. Geologic map of Capitol Reef National Park and vicinity, Utah (4 sheets). Utah Geological Survey.

Billingsley, G. H. and Breed, W. J. 1985. Geologic cross section from Cedar Breaks National Monument through Bryce Canyon National Park to Escalante, Capitol Reef National Park, and Canyonlands National Park, Utah. Capitol Reef Natural History Association.

Bibliography

Baars, D. L. 1983. *The Colorado Plateau, a geologic history.* Albuquerque: University of New Mexico Press.

Collier, M. 1987. *The geology of Capitol Reef National Park.* Capitol Reef Natural History Association.

Davidson, E. S. 1967. *Geology of the Circle Cliffs area, Garfield and Kane Counties, Utah.* U.S. Geological Survey Bulletin 1229.

Dutton, C. E. 1880. *Geology of the High Plateaus of Utah.* U.S. Geographical and Geological Survey of the Rocky Mountain Region.

Olson, V. J.; and Olson H. 1972 (1979 printing). *Capitol Reef—the story behind the scenery.* Las Vegas, Nevada: KC Publications.

Rice, L. 1982. Journeying into the Fold. *National Parks* 56(9–10): 4–8, September/October.

Rigby, J. K. 1976. *Northern Colorado Plateau,* K/H Geology Field Guide Series. Dubuque, Iowa: Kendall/Hunt Publishing

Stokes, W. L. 1986. *Geology of Utah.* Salt Lake City: Utah Museum of Natural History and Utah Geological and Mineral Survey. p. 111–112.

Trimble, S. 1978. *Rock glow, sky shine; the spirit of Capitol Reef.* Capitol Reef Natural History Association.

Capitol Reef National Park

Area: 241,904 acres; 378 square miles
Proclaimed a National Monument: August 2, 1937
Established as a National Park: December 18, 1971
Address: Torrey, Utah 84775

5

Utah's largest national park is a geological wonderland of spires and mesas rising to more than 7800 feet.

A large triangular plateau, called Island in the Sky, stands high above the confluence of the Green and Colorado Rivers.

The two rivers come together through deep, sheer-walled, meandering canyons; below their confluence is Cataract Canyon, a 14-mile-long stretch of white-water rapids that challenge the most daring river-runners in the world.

The Maze, an almost inaccessible jumble of canyons, lies west of the rivers; beyond are the red towers and walls of the Land of Standing Rocks.

An eroded salt dome, Upheaval Dome, exposes rings of colorful Mesozoic rocks. On the Colorado River's southeast side are the strange, sculpted landforms of the Needles district, a natural exhibit area of arches, fins, spires, grabens, canyons, potholes, and Indian ruins.

Canyonlands National Park

Southeast Utah

Figure 5.1   Church Rock, beside the entrance road to the Needles section of Canyonlands National Park. This huge sandstone monolith looks as though it had been "turned on a lathe." Weathering and erosion by water and wind produced the impressive landform. Photo by J. W. Kusiak.

CANYONLANDS
NATIONAL PARK

CANE CREEK ANTICLINE

GREEN RIVER

Upheaval Canyon

DEAD HORSE POINT STATE PARK

Dead Horse Point

The Neck

Shafer

Road

Trail

Park

Shafer Dome

ANTICLINE OVERLOOK

Upheaval Dome

ISLAND IN THE SKY

COLORADO RIVER

HATCH POINT

WHITE RIM

White Rim Trail

Anderson Bottom

GLEN CANYON NATIONAL
RECREATION AREA

Stillwater Canyon

Grand View Point

Junction Butte

Monument Basin

WHITE RIM

Lockhart Basin

ELATERITE BASIN

The Loop

Gibson Dome

THE MAZE

Land of Standing Rocks

Canyon

Lens Canyon

Elephant Canyon

Silver Stairs

Squaw Spring

Cave Spring

Elephant Hill

Squaw Flat

Spanish Bottom

Twin Canyon

Red Lake Canyon

Cyclone Canyon

Devils Lane

Devils Pocket

Wooden Shoe

Paul Bunyan's Potty

North Sixshooter Peak

Cataract

SOB Hill

THE

Tower Ruin

Chesler Park

THE GRABENS

Virginia Park

Druid Arch

NEEDLES

Horse Canyon

South Sixshooter Peak

COLORADO R.

Cross Canyon

Bobby's Hole

Angel Arch

All American Man

Natural Arch

Fisheye Arch

Kirk Arch

The Jump

N

Park boundary

0   6 mi.

Tynac

Figure 5.3   The confluence of the Colorado and Green Rivers, where Major Powell and his men camped in 1869. In this view, looking downstream (south), the Green River comes in from the right. The mesa top in the right foreground is the south end of Island in the Sky. Photo by M. Agnew.

Local History

Canyonlands National Park is a wild and primitive desert region between Capitol Reef and Arches National Parks in the heart of the Colorado Plateaus. The Green and Colorado Rivers, which join in the park, and much of the surrounding wilderness are little changed from when Major John W. Powell first explored the canyons in 1869. He entered what is now the park from the north, through the Green River's Labyrinth Canyon. He described the landscape between Labyrinth Canyon and the confluence of the Green and Colorado Rivers as

". . . naked rock with giant forms carved on it, cathedral-shaped buttes towering thousands of feet, cliffs that cannot be scaled, and canyon walls that make the river shrink into significance, with vast hollow domes, tall pinnacles, and shafts set on the verge overhead, and all the rocks, tinted with buff, gray, red, brown, and chocolate, never lichened, never moss-covered, but bare, and sometimes even polished." (Powell 1981 reprint)

After resting several days at the confluence in order to dry out gear and provisions, the intrepid explorers started out again through a stretch of white water that Powell named "Cataract Canyon," because of "bad rapids in close succession." (In parts of the canyon, huge blocks of rock that have fallen from the cliffs impede the water, producing waves 30 feet high.) One of Powell's boats swamped, three oars were lost, and all the boats were damaged and leaked badly from banging against rocks. After a narrow escape from being trapped in the gorge near the foot of Cataract Canyon by a flash flood, the party emerged safely in an open area of the river beside deserted Indian dwellings.

Being as interested in archaeology as geology, Powell carefully described the ruins, flint chips, arrowheads, broken pottery, and what he called "etchings" (pictographs) on the cliffs. He was excited to find stairsteps cut into a cliff leading up to a watchtower. "I stood," he wrote, "where a lost people had lived centuries ago, and looked over the same strange country."

Powell's "lost people" may have been Indians of the Fremont Culture, who lived in this area until A.D. 900. Or they could have been a later group, the Anasazis, who arrived around A.D. 1075 and left during the twelfth century. The pictographs and camp sites of these early people can be seen in numerous places in Canyonlands National Park and the surrounding area. The Indians discovered that they could make pictures on the rock by scratching through the coating of desert varnish to the lighter-colored, unweathered rock beneath. *Desert varnish* is a thin dark film of iron and manganese oxides that with long exposure forms on the surface of desert rocks.

Since this is desert country (only 5 to 9 inches of precipitation per year), water is precious and water holes are few and far between. Settlers were not attracted to the region, but in the late 1800s and early 1900s, cowboys and sheepherders passed through the area from time to time. Butch Cassidy and his "Wild Bunch" knew the region well and used their knowledge to evade posses of lawmen on numerous occasions. The area was ideal for outlaws and cattle rustlers who knew the location of water holes and could come and go without getting lost in the confusing jumble of canyons, cliffs, and exceedingly rough terrain.

Canyonlands National Park was established in 1964 in order to preserve its outstanding scenic, scientific, and archeological resources. As the Colorado River leaves Cataract Canyon and crosses the park's southern boundary, the river flows into the head of Lake Powell at the northern end of Glen Canyon National Recreational Area. A strip of Glen Canyon National Recreational Area also extends northward along the entire western boundary of Canyonlands National Park.

Geologic Features

Island in the Sky

Island in the Sky is the best observation tower from which to gain perspective on the geological processes that have created the "architecture" of Canyonlands. This broad, level, roughly triangular mesa is wedged between the Green and Colorado Rivers and stands over 2000 feet above the two streams that come together at the foot of the mesa point. Views from the overlooks encompass canyon after canyon, stretching off to mountains ringing the horizon a hundred miles distant. Within the park, the Maze is to the west and the Needles is to the south. Closer to the mesa edge, about 1200 feet down, is the White Rim Sandstone bench, the habitat of desert bighorn sheep. The rivers are another 1000 feet down (fig. 5.3).

Red sandstones are the dominant rock of the Canyonlands landscape; but a rich variety of sedimentary rocks in hues ranging from near-white through orange, brown, or near-black provides contrast and diversity to individual landforms, as well as the overall panorama. The sandstones consist mainly of uniformly sized, clear quartz grains. The red rocks are cemented with iron oxide, which gives them their color. During and after the uplifting of these sedimentary rocks, differential weathering along joints and erosion by wind and water dissected the flat layers into hundreds of brightly colored canyons, mesas, buttes, fins, arches, and spires, and thousands of smaller features.

Upheaval Dome

Upheaval Dome, located in the northwest corner of Island in the Sky, is a circular feature about three miles across. From the air it looks like a bull's-eye (fig. 5.4). The view from the top of Whale Rock (elevation 5820 feet), a smooth, whale-shaped monolith of Navajo Sandstone, gives an idea of Upheaval Dome's enormous size. From the Wingate Sandstone rim (5700 feet), you can look down into Upheaval Dome's core more than 1000 feet below. Concentric rings expose the sedimentary beds between the core and the rim. The more resistant layers jut up in ridges and the less resistant layers form valleys. Upheaval Canyon breaches the western wall and provides a flume for runoff to carry away erosional debris. The canyon drains to the Green River five miles downslope (fig. 5.5 A,B).

Geophysical studies indicate that a *salt dome* forcing its way through sedimentary beds pushed up Upheaval Dome. The salt is from the Paradox Formation (a unit of the Hermosa Group) that was deposited during the Pennsylvanian Period. The Paradox salt beds evaporated from the brine of a landlocked ancient sea, trapped by a rising mountain block (the Uncompahgre Uplift). In this particular location, a hump in the granite floor deep below the sea bottom caused a slight thinning of salt layers above the crest, creating a weak zone.

Eventually, many feet of sediment washed down from highlands, filled the basin, and covered the salt beds. The pressure of about a mile of sediment weighing down the salt caused it to become *plastic* (i.e., able to flow without rupture). Salt flowing upward over the hump as a huge plug (not unlike toothpaste oozing from a tube) arched the overlying beds into a dome, or circular anticline. As the beds were pushed up almost 1000 feet and turned almost on edge, they became cracked and severely deformed. This made them more susceptible to weathering and erosion and the top beds were stripped off. Since the beds exposed in the center of Upheaval Dome are softer than those forming the rims, the less resistant rock eroded more rapidly, creating the depression that resembles an open-pit mine.

The Maze, the Land of Standing Rocks, and the Fins

West of the Colorado and Green Rivers is the Maze, the wildest section of Canyonlands, described in the park brochure as a "30-square-mile puzzle in sandstone." Outstanding examples of prehistoric rock art are hidden in some of the deepest canyons. Beyond the jumbled canyons of the Maze is the Land of Standing Rocks, where weirdly

Figure 5.4   Upheaval Dome, in the northwest part of Canyonlands National Park, was formed by the intrusion and subsequent collapse of a salt dome. The feature is three miles across and almost 1600 feet deep. National Park Service photo.

shaped pinnacles, buttes, and mesas display eroded sandstone layers.

The Fins are narrow sandstone ridges, or walls, that are eroding along parallel joints. Cracks between closely spaced vertical joints become enlarged by chemical and physical weathering as water seeps down, dissolving iron oxide cement and loosening sand grains. Frost action, with its alternate freezing, expanding, and thawing, also breaks up the sandstone. Loose sediment is removed by wind and runoff. The joint spaces gradually enlarge into narrow valleys that separate the narrow walls, called *fins*.

The Needles, Arches, and Grabens

The east side of the Colorado River is an area of great diversity. The Needles themselves are similar to fins but have more crisscross fractures. Needles, like fins, are the result of weathering and erosion along joint lines and fault lines, some of which are quite close together, a feature characteristic of the Cedar Mesa Sandstone. The Joint Trail, which goes through a crack in this formation, is only a couple of feet wide. The distinctive red-and-white banding in the Needles (and elsewhere in the park) is due

to the interfingering of two units of the Cutler Group: the red Organ Rock Shale and the white Cedar Mesa Sandstone.

The *needles* and *pillars* develop when two intersecting joint sets are spaced closely together. The second set of joints, which is at an angle to the first set, prevents fins from developing. Instead, pointed needles or rounded pillars evolve as weathering and erosion sculpture the rock. Some pillars are capped by resistant material that makes balancing rocks. Where fins have developed, they tend to evolve into a line of pillars as weathering attacks cross joints.

Natural arches, some of the distinctive landmarks in the Needles district, may form from fins. As the joint spaces widen and deepen, beds at the base of the fins are exposed. If the lower beds are slightly softer, they are removed more rapidly than the top beds. Because the fins are narrow, a hole, or *window,* appears where the rock is worn completely through. Sand blown by wind (sandblasting) polishes and rounds off the window edges. Snow, rain, and ice continue to attack the rock, dislodge fragments, and cause rockfalls (a process of mass wasting) until a window is enlarged to form a natural arch.

Figure 5.5   *A.* Upheaval Dome in map view. The topography is directly related to the rock formations, with the more resistant formations, such as the Wingate Sandstone, forming concentric walls. *B.* Cross section of Upheaval Dome, showing the salt dome at the core and the sedimentary layers that were arched up by pressure of the rising salt plug. Adapted from 1981 ''Upheaval Dome,'' National Park Service and Canyonlands Natural History Association.

Figure 5.6   Needles and Grabens section, south central part of Canyonlands National Park, east of the Colorado River. Chesler Park (foreground) is a natural meadow, walled in by the Needles (left) and the Grabens, the parallel troughs beyond Chesler Park that run from left to right across the picture. Geologic structure controls the location and patterns of the canyons, mesas, and other erosional features. National Park Service photo.

Because of their composition, type of cement, and structure, certain rock formations have a tendency to form arches. Canyonlands has 25 arches, mostly sculpted in Permian Cutler Group sandstones, which are more likely to form standing rocks than arches. In nearby Arches National Park (chapter 6), more than 300 arches have developed in Jurassic Entrada Sandstone because they tend to form rapidly in that area.

"Potholes" form in Cedar Mesa Sandstone because the calcium carbonate that cements its sand grains is soluble in natural carbonic acid (carbon dioxide plus water). Carbonic acid attacks any weak spot or zone in the sandstone. Loosened sand grains fall, are washed, or blown away. Once a slight depression is started, more and more of the sandstone is exposed to the processes of weathering and erosion. Potholes, caves, and sometimes arches develop in this way. Paul Bunyan's Potty is a well known example.

The name Chesler Park was given to an expanse of a thousand acres of flat, grassy land surrounded by pinnacles of the Cedar Mesa Sandstone. Chesler Park is east of the Grabens.

The Grabens is an area of elongate collapse valleys parallel to Cataract Canyon. As was the case in Upheaval Dome, the underlying salt deposits of the Paradox Formation acted as an unstable platform for the millions of tons of overlying shales and sandstones. Great thicknesses of salt accumulated in subsiding, elongate, down-faulted valleys parallel to the rising, up-faulted ranges of the Uncompahgre Uplift. As the weight of sedimentary overburden (washed down from the mountains) became ever greater, the salt began to flow as a plastic mass toward the southwest, away from the source and greatest weight of sediment. When the flowing salt came against the fault-block ridges, it was deflected upward, forming salt walls and plugs that penetrated the overlying beds. In this way, the salt took the path of least resistance to reach areas where the pressure of overburden was lower.

When the Colorado Plateaus region was uplifted millions of years later, enough of the sedimentary cover was stripped off so that ground water seeped into the salt, dissolving some of the upper parts and causing the overlying sandstones and shales to collapse into the salt-valley grabens.

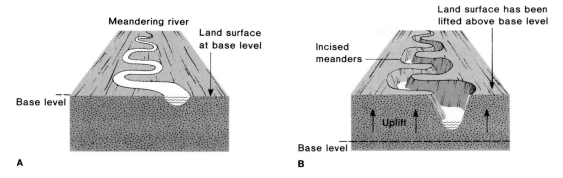

Figure 5.7 During and following regional uplift, incised meanders may develop as major streams downcut. From *Physical Geology,* 4th edition, C. C. Plummer and D. McGeary. © 1979, 1982, 1985, 1988 Wm. C. Brown Publishers. Used by permission.

The Rivers

The Colorado River flows south through Canyonlands in a spectacular, deep, narrow-walled gorge. The Colorado is joined in the park by its largest tributary, the Green River, which also has a deep, narrow canyon (fig. 5.3). Together they flow quietly for about three miles until they enter Cataract Canyon at Spanish Bottom. The turbulent water in Cataract Canyon, which rushes through at a rate of around 40 million gallons per minute, is constantly boiling up as "white water" because of changes in the bedrock of the stream channel, slumping of the lower beds in canyon walls, and blocks of rock falling from the cliffs into the stream.

The Colorado River appears to follow a course through Cataract Canyon along the edge of the block-faulted area. The large faults on the east wall of the canyon do not appear on the western wall—nor do the intricate fracture patterns in the rocks on the southeast side of the Colorado, such as the lines of needles and their intervening, north-south, "racetrack" valleys, cross the river. Presumably, the path of salt-dissolving ground water terminates where it is diverted into the Colorado River drainage.

Both the Green River and the Colorado River have extremely abrasive bedloads, a significant factor in their ability to cut down their channels to form *incised meanders* (fig. 5.7). These two rivers may have meandered back and forth on a fairly low, flat surface in the geologic past, and may have retained their meandering pattern as the region was raised. However, some geologists believe that during uplift, as base level was lowered, the streams began downcutting their flood plains so rapidly by headward erosion that they could do very little lateral eroding. In soft alluvial sediments, meanders migrate over a flood plain, but when meanders are trenched into resistant rock layers, as in Canyonlands, the meandering pattern becomes "trapped." The stream's energy, supplied by increased velocity, cuts its channel ever deeper but cannot widen it unless it encounters a zone of weak or easily erodible bedrock, as at Spanish Bottom, where gypsum beds have been eroded.

The water in the Green and Colorado Rivers is not potable because of the saline ground water that seeps into the rivers from the salt beds. People and animals in the park use water in potholes or rock depressions in the White Rim Sandstone and Cedar Mesa Sandstone or depend on spring water that seeps from canyon walls along contacts between the Organ Rock Shale and the White Rim Sandstone. The hollows where water collects—locally referred to as "tanks" or "charcos"—are formed by the combined process of solution of calcium carbonate cement and removal of loose rock grains by wind (deflation). Tanks are found by Junction Butte near Grand View Point in the central portion of the park and also in the Needles district near Chesler Park.

Geologic History

1. The development in Pennsylvanian time of the Paradox Basin and the Uncompahgre Uplift.

The oldest rocks exposed in Canyonlands National Park are beds of the Paradox Formation of Pennsylvanian age in the bottom of Cataract Canyon. These exposures are of rock gypsum and interbedded black shale that are less soluble than the halite salt beds underneath Upheaval Dome and the Needles section. A gypsum plug, similar to a salt plug but more resistant, has been exposed by erosion at the mouth of Lower Red Lake Canyon, near the head of Cataract Canyon. The plug was pushed up into the overlying Honaker Trail Formation, the uppermost unit deposited in Pennsylvanian time. Both formations are units of the Hermosa Group.

The thousands of feet of evaporites that make up the Paradox Formation began to accumulate in Middle Pennsylvanian time when block-faulting caused the Paradox

Table 5.1 Geologic Column, Canyonlands National Park

Time Units		Rock Units		Geologic Events
Era	Period	Group	Formation	
Cenozoic	Quaternary	Alluvium Dune sand Landslide deposits		Arches and needles formed Entrenching of Green and Colorado Rivers Continuing uplift and block-faulting Forming of salt valleys by solution and collapse
	Tertiary			
Mesozoic	Cretaceous			Uplift and warping due to compression during Laramide orogeny Erosion ————
	Jurassic	San Rafael	Entrada Sandstone	Scattered patches exposed in park vicinity
			Carmel	
	——?——			Erosion ————
	Triassic	Glen Canyon	Navajo Sandstone	Cross-bedded, white sands deposited by migrating dunes; forms cliffs, knobs, hummocks
			Kayenta	Deposited on flood plains and in lakes; slope-former
			Wingate Sandstone	Cross-bedded sands deposited by migrating dunes; forms red rimrock
			Chinle	Bright-colored shales; slope-former; contains petrified wood Erosion
			Moenkopi	Tidal flat deposits; shales with ripple marks, mud cracks Erosion ————
Paleozoic	Permian	Cutler	White Rim Sandstone Organ Rock Shale Cedar Mesa Sandstone Elephant Canyon Halgaito Shale	White, cross-bedded sandstone deposited in coastal dunes Reddish brown siltstones, sandy shales and sandstones deposited on oscillating shoreline and interfingered with deposits on beach dunes, in sand bars, and in arkosic alluvial fans. (Because formations thin and facies change from west to east, rocks are mapped as "Undivided Cutler" along the eastern margin.) Erosion ————
	Pennsyl-vanian	Hermosa	Honaker Trail	Shallow marine deposits
			Paradox	Gypsum and salt evaporites interbedded with shale and black limestone

Huntoon, Billingsley, and Breed, 1982

Basin to begin to subside and the Uncompahgre ranges to be uplifted. The cause of this tectonic activity is not clear, but may be related to plate collisions far to the south and east.

Water from a shallow sea that had covered the area early in the Pennsylvanian Period became landlocked in the sinking basin, and a system of interior drainage developed as mountains rose along the margin. In the *closed basin,* water escaped only by evaporation; and since the climate was hot and dry, water evaporated rapidly, concentrating salts and mineral matter in shallow lagoons.

When the salinity of a brine reaches a ratio of 10 parts salt per 1000 parts water (3 times the ratio of salt in seawater), the minerals anhydrite and gypsum begin to precipitate as crystals. If evaporation continues and no water is added to dilute the brine, the ratio of salt to water may reach 35 parts per 1000 parts water, or 10 times that of normal seawater. Under these conditions, halite (rock salt) precipitates. The heavier salt crystals tend to sink, accumulating over the bottom of the lagoon. This was what happened in the Paradox Basin.

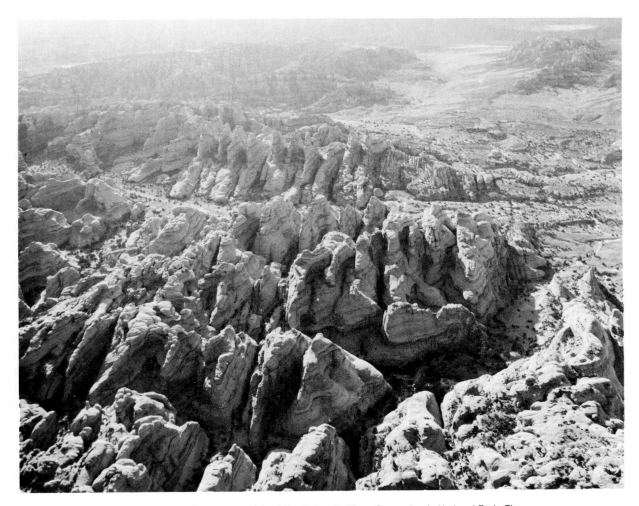

Figure 5.8   The Land of Standing Rocks, west side of the Colorado River, Canyonlands National Park. The location of the fins (closely spaced ridges) is controlled by the joint system of the Cedar Mesa Sandstone. National Park Service photo.

From time to time, floods washed sediment down from the mountains. The coarser gravels accumulated in wedges along the base of the range, and the silts and muds spread over the salt flats. Occasionally fresh seawater breached the lagoons and moved into the basin over a shallow threshold in the margin. The salty brines were not flushed out because the fresh seawater acted as a barrier that prevented the escape of the dense bottom water. Thin black shale layers between some salt beds mark these incursions of seawater. For thousands of years, the layers of salt and clastic sediment accumulated, becoming ever thicker—up to 5000 feet in some places. The gypsum layers in the bottom of Cataract Canyon represent the last phases of the evaporite deposits.

Presumably, the weight of the accumulating overburden caused the salt beds to begin flowing plastically before the close of the Pennsylvanian Period, and this process continued throughout the Permian and well into Mesozoic time. The migration of the salt beds had probably stopped by the end of the Jurassic Period.

Toward the end of Pennsylvanian time, a warm tropical sea spread over the region. About 1500 feet of fossiliferous limestones, shales, and sandstones blanketed the salt basin. These are the gray beds of the Honaker Trail Formation that crop out at lower elevations in the deep canyons of the park, especially along the Colorado River.

2. Permian deposition of the extensive, varied formations of the Cutler Group.

In some places the red, terrestrial Permian beds lie unconformably over gray, marine Pennsylvanian beds; but elsewhere the transition is more gradual. Early in Permian time, advancing seas deposited the Halgaito Shale, which grades into the Elephant Canyon Formation, a rock unit deposited in coastal lowlands. In Canyonlands, beds of these formations are exposed in Cataract Canyon, Elephant Canyon, and upriver from the confluence.

All during this time the Uncompahgre Mountains to the east were being severely eroded. Streams stripped sediment off the slopes and dropped it on enlarging alluvial fans that gradually filled the basin at the foot of the range. The material in the fans lithified to iron-rich, red, arkose sandstones that are mapped as "undivided" rock units of the Cutler Group.

Interfingering with Cutler red beds are white Cedar Mesa Sandstone beds that were deposited in submarine bars and coastal dunes. The zone of facies change between these continental and marine rock units occurs as a 4- to 5-mile-wide belt across Canyonlands National Park from south of the Needles through the Maze and into the Elaterite Basin along the park's western boundary. Many of Canyonlands' most dramatic and colorful rock sculptures are in this zone where the two types of sedimentation shifted back and forth.

Muds that oxidized and became the Organ Rock Shale were laid down over the white sands of the Cedar Mesa. The Organ Rock beds make up brightly colored red, orange, and brown slopes that separate cliff-forming units above and below. Some of the most interesting exposures of the Organ Rock Shale are in the Land of Standing Rocks (fig. 5.8).

The "type locality" of the White Rim Sandstone is the topographic bench, mentioned earlier, about 1200 feet down from the surface of Island in the Sky. (A *type locality* is the location from which a stratigraphic unit takes its name.) The White Rim Sandstone that forms striking cliffs between Cutler and Organ Rock red beds includes ancient sand dunes as well as large marine sand bars. Interesting exposures of this rock unit can be examined along the White Rim Trail.

In the Elaterite Basin along the park's western boundary, the White Rim Sandstone displays a "fossil" offshore bar that may have been a barrier island late in Permian time. This structure contains petroleum that seeps out as a dark-brown, tarry substance called "elaterite." The oil probably migrated into the sandstone from a source rock and became trapped by the impermeable red shale beds along the east side of the sandbar. Because the structure has been exhumed by erosion and the gas has escaped, the oil is not recoverable.

3. Mesozoic Moenkopi and Chinle Formations

An unconformity between Paleozoic and Mesozoic beds represents withdrawal of seas and a lengthy period of erosion when the continent was relatively high. The Triassic units in the Canyonlands area are red beds consisting of shallow-water clastic sediments deposited by streams in channels and on flood plains throughout a broad lowland that sloped gently toward a western ocean. The mudstones of the Moenkopi Formation, which accumulated on tidal flats, are exposed in the northern and western sections of the park. Mudcracks and ripple marks are plainly visible on many of the outcrops.

An unconformity separates Moenkopi beds from brightly colored, overlying Chinle beds, which are mostly slope-forming shales. A geologist describes Chinle beds as showing up on canyon walls in the park like "a brillant ring of fire" (Baars 1983, p. 167). In some places, pieces of petrified wood from the Petrified Forest Member of the Chinle Formation have weathered out of cliffs and collected around the base.

4. Sandstones of the Glen Canyon Group deposited in late Triassic and early Jurassic time.

The prominent cliffs of Wingate Sandstone, the cliffs and slopes of the Kayenta Formation, and the cliffs, knobs, and rounded monoliths of the Navajo Sandstone characterize Glen Canyon Group rock units in Canyonlands. Imposing outcrops of these strata encircle the western and northern sections of the park; and the cliffs, especially the red Wingate cliffs, form an effective barrier to travel across many parts of Canyonlands and vicinity. Much of the beauty and uniqueness of the Colorado Plateaus is due to the units of the Glen Canyon Group, which are well exposed in Canyonlands and other national parks and national monuments in the region.

As the Triassic climate became drier, sand dunes migrated across the region, burying the rivers and their flood plains. The dune sediment became the Wingate Sandstone. The Wingate cliffs rise several hundred feet throughout Canyonlands and run for hundreds of miles with very few breaks in their semicircular wall. Separating the Wingate dunes and the windblown desert deposits of the Jurassic Navajo Sandstone are the stream-deposited beds of the Kayenta Formation. Apparently a cycle of somewhat moister climate prevailed between the windy desert environments of the Wingate and Navajo. The Kayenta Formation, consisting of reddish brown to lavender sandstones with interbedded siltstones and shales, forms more gentle, ledgy slopes. The Navajo Sandstone, the top formation of the Glen Canyon Group, has well developed cross-bedding, and forms light-colored (buff to pale orange), steep, rounded cliffs, knobs, and hummocks. Some of the more striking ones resemble beehives, turbans, and temples in shape. Canyon View Point Arch and Millard Canyon Arch are also examples of Navajo Sandstone features.

5. San Rafael Group deposition, ending the Jurassic Period.

Rock units of the San Rafael Group unconformably overlie the Glen Canyon Group in the Canyonlands vicinity, mainly in scattered patches east and west of the park. In this area, the Carmel Formation beds accumulated on mud flats. The Entrada Sandstone, a massive sandstone, forms vertical cliffs and hummocky knobs that can be seen along some of the approaches to Canyonlands National Park.

Erosion stripped off most of the San Rafael beds in this area, along with any Cretaceous rock units that may have overlaid them.

6. The relationship of the Laramide orogeny to the development of Canyonlands.

The Laramide orogeny, which brought the Mesozoic Era to a close and lasted into Tertiary time, elevated the Canyonlands region thousands of feet, but left the sedimentary rock units essentially horizontal. The dip of the beds is so slight that they appear flat. However, the tectonic compressive forces that pushed up the surrounding mountains also produced joint systems and fractures that strongly influenced patterns of landform development in Canyonlands. Erosional processes, intensified by uplift, tended to follow the joint patterns while dissecting the flat-lying rocks.

The presence of the salt beds of the Paradox Formation, as explained earlier, complicated the fracturing and jointing. Consequently, they affected the ways in which landforms evolved, especially in the northwestern and southeastern sections of the park. Sometime after uplift began, and as erosion stripped off the younger beds, ground water began to reach the buried salt beds. Some of the upper salt layers were dissolved, but the less soluble gypsum was left in place. In places where salt near the surface was reduced in volume—as it was under the salt valleys and grabens—the overlying sedimentary beds collapsed.

7. Quaternary erosion and deposition.

The periods of cold and increased moisture that characterized Pleistocene climates on the Colorado Plateaus greatly increased the dissection and deepening of the canyons, especially the incising of the Green and Colorado River canyons. The volume of the two rivers was augmented by increases in runoff from glaciers in the Rocky Mountains and other highlands drained by these streams. Some of the dune deposits, landslide deposits, alluvial fans, and other unconsolidated sediments in the park accumulated—or at least began to form—in Pleistocene time. These processes continue today, although at a somewhat slower rate in the drier climate.

Geologic Map and Cross Section

Huntoon, P. W., Billingsley, G. H., and Breed, W. J. 1982. Geologic Map of Canyonlands National Park and Vicinity, Utah. Moab, Utah: Canyonlands Natural History Association.

Lohman, S. W. 1974. The geologic story of Canyonlands National Park. U.S. Geological Survey Bulletin 1327.

Bibliography

Baars, D. L. 1983. *The Colorado Plateau, a geologic history.* Albuquerque: University of New Mexico Press. 279 p.

Campbell, J. A. 1987. Geology of Cataract Canyon and vicinity, a field symposium. Guidebook, Four Corners Geological Society, 10th Field Conference.

Huntoon, P. W. 1982. The Meander anticline, Canyonlands, Utah: An unloading structure resulting from horizontal gliding on salt. *Geological Society of America Bulletin* 93:941–950, October.

Lohman, S. W. 1974. *The geologic story of Canyonlands National Park.* U.S. Geological Survey Bulletin 1327.

Powell, J. W. 1981 reprint. *The canyons of the Colorado.* Originally published in *Scribner's Monthly,* 1875. Golden, Colorado: Outbooks. 64 p.

Rigby, J. K. 1976. *Northern Colorado Plateau.* K/H Geology Field Guide Series. Dubuque, Iowa: Kendall/Hunt Publishing Company.

Canyonlands National Park

Area: 337,570 acres; 559 square miles
Established: September 12, 1964
Address: 446 S. Main Street, Moab, Utah 84532

6

The greatest density of natural arches in the world is in Arches National Park. More than 300 arches, displayed in all stages of development, are carved in the salmon-colored Entrada Sandstone, Joint sets and faults controlling the size, location, and orientation of fins and arches are related to salt beds in the Paradox Formation that forms an unstable base beneath the sandstones.

Arches National Park

Southeast Utah

"These are natural arches, holes in the rock, windows in stone, no two alike, as varied in form as dimension. They range in size from holes just big enough to walk through to openings large enough to contain the dome of the Capitol in Washington, D.C. Some resemble jug handles or flying buttresses. . . . In color they shade from off-white through buff, pink, brown, and red, tones that change with the time of day and the moods of the light, the weather, the sky."
Desert Solitaire, Edward Abbey 1968

Figure 6.1 Landscape Arch, 291 feet across, is the longest arch in Arches National Park and one of the longest natural arches in the world. National Park Service photo.

Local History

Arches National Park, near Moab, Utah, is about 25 miles northeast of Canyonlands National Park (chapter 5) by road, and approximately the same distance up the Colorado River. A few miles upstream from Canyonlands, the river's enclosing walls become farther apart, the flood plain widens, and the water has less turbulence. Just beyond Moab, the Colorado's course becomes the southern boundary of Arches National Park. The finest panorama of natural arches, fins, windows, pinnacles, and balanced rocks to be seen on the Colorado Plateaus—and probably the finest in the world—is displayed in Arches National Park. The area was set aside as a national monument in 1929 and established as a national park in 1971.

Yet, for all its strange beauty, the lack of water and the barren appearance of the terrain discouraged settlers. The climate of the park area is harsh. Temperatures range from 110° F in summer to −20° F in winter. Precipitation averages 5 to 9 inches per year, and shelter from the strong winds is hard to find.

Prehistoric Indians were attracted to the area around Delicate Arch by the presence of chalcedony, a variety of quartz that can be chipped to make a sharp edge. Probably the Indians spent winters making tools, arrowheads, and spear points at their campsites. They also left many pictographs on rock walls throughout the park.

One hardy pioneer family, named Wolfe, operated a cattle ranch around the turn of the century and managed to eke out a living for more than twenty years. Their weathered cabin and corral (now listed in the National Register of Historic Places) are in a protected hollow beside the trail to Delicate Arch.

Geologic Features

Since Arches National Park is only a short distance up the Colorado River from Canyonlands National Park, the geology of the two parks is similar in many ways; but significant differences are obvious. Canyonlands and Arches have several of the same sandstones—the Cutler sandstones of Permian age, the Triassic Wingate, and the Jurassic Navajo. Some Cutler red beds produced huge fins and some natural arches in Canyonlands but not in Arches National Park. In Arches most of the formation is missing or covered by younger beds.

Several sandstones with very minor exposures in Canyonlands are present in Arches National Park and vicinity. The Jurassic Entrada Sandstone is the most prominent and the most important because most of the arches are in this rock unit. With all of these sandstones in the same arid climate, why are most of the arches concentrated in Arches National Park? And why are natural bridges not present in Arches?

Arches and Natural Bridges

By definition, a *natural bridge,* which is chiefly the result of erosion by running water, spans a watercourse (that may now be a dry stream bed). A natural *arch,* on the other hand, is formed by weathering and processes such as wind erosion, but stream erosion is not involved. Once a natural bridge is begun, however, it is enlarged and shaped by the same forces that mature natural arches.

Two national monuments in southern Utah, south of Arches and Canyonlands National Parks, protect outstanding examples of natural bridges: Natural Bridges National Monument, with three spans in White and Armstrong Canyons, carved by tributaries (now intermittent) of the Colorado River; and Rainbow Bridge National Monument, farther downstream and also located in a tributary canyon. These features are formed in massive cross-bedded sandstones. The spans in Natural Bridges National Monument are in Permian Cedar Mesa Sandstone; Rainbow Bridge is in Jurassic Navajo Sandstone.

To find answers to why arches but not natural bridges formed from the Entrada Sandstone in Arches National Park, we look at the processes by which the features are made, the differences in the origin of the sandstones, variations in joint sets and composition, differing reactions to uplift and weathering, and—most important of all—the local geologic structure. Although stream erosion is not an important process in arch formation in Arches National Park, streams do have a clean-up function. During flash floods or times of snow melt, sediments accumulated from weathering processes are swept down dry washes by intermittent streams that drain to the Colorado River.

How the Arches Form

Like other massive cross-bedded sandstones, the Entrada is highly susceptible to weathering by *exfoliation;* that is, the peeling off from bedrock of slabs or shells usually in a series of concentric layers. (This process was discussed in the examples of weathering in Bryce Canyon National Park, chapter 3.) Massive rocks, such as quartzite, granite, and some sandstones, have a tendency to weather in this manner.

Several weathering processes help exfoliation get started. The vertical linear joints in the Entrada Sandstone, 10 to 20 feet apart, are zones of weakness through which weathering agents, such as water, ice, and frost, attack the rock. Rain water, being slightly acidic from dissolved atmospheric carbon dioxide, slowly breaks down the calcium carbonate cement, releasing the sand grains. *Fins* develop in this way. The process of exfoliation, working on both sides of the fin, eventually perforate the rock, forming windows. Water and frost enlarge these holes, and the wind, carrying fine sand, abrades and polishes them.

ARCHES NATIONAL PARK

Figure 6.3   How the arches developed. 1) Upwarping of massive beds of Entrada Sandstone caused deep vertical, parallel fractures to form, exposing rock surfaces to the elements. 2) As weathering and erosion enlarge cracks and fractures, narrow sandstone walls, or fins, develop. 3) Frost action, exfoliation, and crumbling round off the tops of the fins. Some fins are breached; blind arches and windows form in others. 4) Rockfalls and weathering gradually enlarge holes to arches. Eventually arches wear through and collapse, leaving buttresses that will, in time, succumb to erosion. Adapted from 1985 Arches National Park brochure, National Park Service.

At the base of the fins, added moisture from seepage evaporates at a slower rate, so that beds at the base weather more rapidly. This undercuts the fin, causing loosened fragments and blocks to fall, enlarging the opening.

These processes are significant in starting many of the Entrada arches because the lower rock unit, the Dewey Bridge Member, is less resistant than the overlying Slickrock Member. Thus the Dewey Bridge weathers more intensely and forms hollows or niches. Left unsupported, sections of the Slickrock Member collapse and form (or enlarge) openings. A few windows appear to have been started as potholes on the sides of cliffs where the abrasion of rushing water formed hollows. This process also enlarges the "sills" of the windows. It becomes apparent that a number of processes in various combinations have formed arches; and it would be a mistake to assume that every arch in the park has the same erosional history. What is important is that these features are impermanent, that they are constantly changing, and that new ones form as old ones are destroyed (fig. 6.3).

Most of the arches in the park are in Devils Garden, which has 64 of them in all stages of development from the faintest suggestion of a depression to the collapsed remains of those that have all but weathered away. Landscape Arch, one of the longest natural spans in the world, is 291 feet long and 105 feet high. One part of the span is only 6 feet wide, which is an indication that in terms of geologic time, the arch is in the last stage of its existence and may soon be no more (fig. 6.1).

Unique in its isolation, Delicate Arch stands alone at an elevation of 4829 feet, northeast of Wolfe Ranch. Estimations are that approximately 70,000 years were re-

quired for its sculpturing by water, wind, frost, and sun. Because of a narrowed segment on one side, this arch may also collapse in the near geologic future. Delicate Arch, 85 feet high and 65 feet wide, is formed at the contact of the Slickrock and Moab Tongue Members of the Entrada Sandstone, the Moab being the uppermost unit of the formation.

In the Devils Garden area, Skyline Arch last changed its profile in November 1940, when a large section fell from the arch to the ground and noticeably increased the size of the opening. This event is another reminder that arches continue to change until they collapse under their own weight.

Balanced Rocks

Balanced Rock, near the Windows, is a striking example of a feature shaped by wind erosion as well as by solution from seepage (fig. 6.5). A resistant block of Slickrock sandstone is balanced on top of a pedestal of crinkly Dewey Bridge beds, which, in turn, rest on a Navajo Sandstone base. (A *balanced rock* is formed by weathering and erosion in place and rests more or less precariously on its base.) Normally the rock on top is a cap rock that is more resistant than the rock below. Although the dissolving of intergranular cement by moisture plays a role in shaping the feature, the dominant process in the development of the balanced rocks in Arches National Park (and in most desert regions) is *sandblasting,* or abrasion by wind-blown sand. Wind velocity is at its greatest about 20 inches above the ground, so the most intense erosion occurs at this level. First a pedestal is formed. As it becomes thinner, the weight of the cap rock causes the top part to tilt very slowly

Figure 6.4   North Window of the Spectacles. The wavy beds visible at the lower part of the opening belong to the Dewey Bridge Member of the Entrada Formation. From a distance, North and South Windows of the Spectacles resemble a pair of eyeglasses. Photo by J. W. Kusiak.

Figure 6.5   Balanced Rock, 55 feet high, weighs 3500 tons. The huge block, eroded from the Slick Rock Member of the Entrada Sandstone, is perched on crinkly beds of the Dewey Bridge Member. The pedestal, in turn, rests on Navajo Sandstone. Weathering and sandblasting (wind abrasion) sculptured this erosional oddity. From *Northern Colorado Plateau*, J. K. Rigby. © 1976 Kendall/Hunt Publishing Company. Used by permission.

to one side, thus compressing the rock on that side and making it more difficult to erode. The opposite side, not under as much compression, weathers away slightly faster until this causes the rock to tilt back and compress *that* side, and then the process starts all over again. The rock keeps tilting very slightly back and forth until the pedestal gives way or it falls from its own weight.

The Controlling Influence of an Ancient Structure

Many feet below the surface in Arches National Park there are layers of rock salt considerably older than the rocks we see exposed. These salt beds belong to the same Pennsylvanian-age Paradox Formation responsible for so much of the unique scenery in Canyonlands National Park. Because the relationship of rocks and structure is somewhat different in the Arches area, however, the resulting expression of scenery is not the same.

Figure 6.6   Block diagram showing how the pattern of surface erosion is controlled by an underlying salt anticline so that some formations are locally thick, thin, or missing. (From Doelling 1985)

Arches National Park lies over a deeper part of the Paradox Basin, closer to the northwest-trending Uncompahgre fault and the upthrown mountain block beyond, which shed sediment in vast quantities into the basin. In a landlocked arm of the sea, held in this closed drainage basin, layers of salt and interbedded sediments accumulated for millions of years (as explained in chapter 5).

The increasing pressure of the overburden on the salt beds caused the rock salt to flow plastically, pushing upward through old fault lines, which were zones of weakness. In the Arches area, the subterranean flowage of salt created northwest-trending "salt anticlines." Much later, dissolution of the upper beds by ground water caused collapse, creating sunken valleys (like grabens), bounded by escarpments. Salt Valley and Cache Valley are examples inside the park. The Moab-Spanish Valley (with the town of Moab at its northern end) is a sunken valley mostly outside the park. Nearly all of the arches and fins in the park stand on the cliffs rising above Salt and Cache Valleys.

So much salt flowed into the salt anticlines that the thickness under Salt Valley in Arches National Park is at least 10,000 feet (Doelling 1985). Rock salt is not exposed at the surface. As fresh water seeping down through fractures dissolves the buried salt, a residue of impurities like gypsum and shale is left behind, slumped down, or contorted in masses. Gypsum exposures of this type occur in Salt Valley beside the jeep trail to Klondike Bluffs.

By the time the Entrada Sandstone was deposited on tidal flats, beaches, and coastal dunes, the flowage of salt had slowed and may have stopped. The Entrada beds do not slope over the salt anticlines as much as the older beds, which were much thinner—or missing—over the tops and thicker on the sides. The arching and flexure of beds over the anticlines produced parallel, northwest-trending fractures. When compressive tectonic forces later caused gentle warping of the rocks in the region, additional closely spaced, parallel joints developed. Several thousand feet of regional uplift, which accelerated erosion, set the stage for the development of the present landscape. The solution of salt by ground water led to the collapse of the salt valleys. Ongoing weathering and erosion along joints in the Entrada escarpments, as described above, created the remarkable arches and fins that we see today.

Products of Weathering

Most of the loose sediment that accumulates in shifting dunes or piles up in joints between fins and arches is almost entirely fine sand grains that are clean, well rounded, and very smooth. In fact, they are so spherical that they do not cling together well, even when wet. When the sand grains in stream beds become saturated with water, they may form *quicksand*. Such a deposit is found in Salt Wash in the eastern part of Arches National Park. Water flowing upward due to hydrostatic pressure beneath the surface keeps the grains separate and neutralizes their frictional strength. The soft, semiliquid mass of quicksand instantly shifts under the weight of an object, an animal, or a person and has a tendency to suck anything down. However, few of the quicksand beds are more than waist-deep to an adult. The important thing to remember if you find yourself in quicksand is not to struggle, because the sand is reluctant to release anything. The best method for extricating yourself is to move slowly in a breast-stroke motion so as to

Table 6.1 Geologic Column, Arches National Park

Time Units		Rock Units			
Era	Period	Group	Formation	Member	Geologic Events
Cen-ozoic	Quaternary		Alluvium Terrace gravels Sand dunes		Weathering and erosion forming arches, etc. Solution along faults, forming salt valleys
	Tertiary				Block faulting Uplift; Laramide orogeny
Mesozoic	Cretaceous		Mancos Shale	Upper	Marine shales
				Ferron Sand-stone	Lenticular, deltaic, and barrier island sands
				Lower	Deposited as marginal marine beds, some coal; regressing sea
			Dakota Sandstone		Beach deposits along the edge of a transgressing sea
			Cedar Mountain		Stream deposits ∿∿∿ Erosion ———
	Jurassic		Morrison	Brushy Basin	Greenish gray, maroon gray mudstones; contains dinosaur bones
				Saltwash	Channel sandstones, siltstones; contains petrified wood
				Tidwell	Thin, dark-red siltstone; contains concretions ∿∿∿ Erosion ———
		San Rafael	Entrada Sandstone	Moab Tongue	White sandstone; dune deposits
				Slickrock	Orange-red, resistant, marine sandstone; most arches eroded from this member
	—— ? ——			Dewey Bridge	Red and gray beds deposited in shallow sea; has crinkly appearance ∿∿∿ Erosion ———
		Glen Canyon	Navajo Sandstone		Yellow and gray sandstone deposited in dunes—cliff-former
			Kayenta Sandstone		Deposited by streams and in lakes; produces slabby fragments
	Triassic		Wingate Sandstone		Sandstone (accumulated in dunes) cliff-former
			Chinle		Reddish brown, silty sandstone Stream, beach, and lake deposits Erosion ———
			Moenkopi		Reddish brown sandstone and shale, accumulated on tidal flat ∿∿∿ Erosion ———
Paleo-zoic	Permian		Cutler		Red sediment deposited in fans ∿∿∿ Erosion ———
	Pennsyl-vanian		Honaker Trail Paradox		Limestone, sandstone and shale deposited in shallow sea
					Gypsum, salt and shale deposited in lagoons

Source: Doelling 1985

distribute body weight more evenly. Then gradually work your way out of the quicksand. If you have to cross a quicksand area, walk slowly and steadily without stopping or standing in one place.

The *chalcedony,* from which the Indians made stone tools, is a cryptocrystalline variety of quartz (i.e., having no visible crystals) that is sometimes found in chert or as concretions that are weathered out from sedimentary rocks. Because chalcedony can be chipped or made to break with a smooth, curved surface, the Indians used the stone for making tools and points with sharp, scalloped edges. This property or characteristic of some minerals to break along curved surfaces is called *conchoidal fracture.*

For "rockhounds," this region yields chalcedony, apache tears, almandine garnet, quartz crystals, geodes, flint, petrified wood, and other kinds of fossils. **Specimens must not be collected in the park; it is illegal to take or disturb natural objects or Indian relics in a national park.**

Geologic History

1. Deposition of Paleozoic beds; erosion.

Paleozoic sedimentary beds exposed in nearby Canyonlands were also deposited in the Arches National Park area, but underwent more extensive erosion. Although the Paradox salt beds are not exposed in the park, a few nonsalt, contorted, gypsiferous masses of the Paradox Formation crop out in Salt Valley. Nongypsiferous, marine beds of the Honaker Trail Formation, which were laid down over the Paradox beds late in Pennsylvanian time, have mostly been eroded away or covered, although some are exposed near the Arches National Park Visitor Center and Highway 191. Honaker Trail limestones in these outcrops contain fossils of corals, bryozoans, crinoid stems, and other organisms. Permian Cutler red beds, so well displayed in Canyonlands, are missing in Arches except in the cliffs bordering the southwest corner of the park.

2. Deposition of the Moenkopi and Chinle Formations and the Glen Canyon Group in Triassic/Jurassic time.

Remnants of Moenkopi and Chinle Formations are best exposed along the canyon of the Colorado River where it forms the southern boundary of Arches National Park. The Chinle also underlies the Wingate Sandstone in the cliffs along Salt Valley. The sandstones of the Glen Canyon Group—the Wingate, Kayenta, and Navajo—make up the cliffs and the plateau tops throughout much of the park. By the time these sandstones were deposited, the flowage of salt beds had slowed down so the sedimentary beds completely covered the salt anticlines, although some thinning occurred over the top. In Arches National Park, the Wingate, Kayenta, and Navajo look much as they do in the other Colorado Plateau parks. The Wingate and

Navajo are cliff-formers and the Kayenta in between forms ledgy slopes. The "petrified" sand dunes of the Navajo beside Scenic Drive in Arches look as if a slight increase in the wind's velocity might set them moving again!

3. How the Entrada Sandstone, with its three members, accumulated in Jurassic time.

When an arm of the sea moved from the north over central Utah in middle Jurassic time, the Arches region became an area of tidal flats, beaches, low islands, and coastal dunes. The sediments deposited in this environment became the three distinctive members of the Entrada Sandstone, the formation in which most of the arches are found. The bottom member is the dark, reddish brown Dewey Bridge, a fine-grained, silty sandstone. It's easy to recognize because it tends to be "crinkly" or wavy. This waviness, or contorted bedding, affects some of the lower parts of the overlying Slickrock Member. Slickrock, an orange-red, smooth, more resistant unit is the main body of the Entrada Sandstone. The gleaming red cliffs, fins, arches, and windows of Slickrock are a memorable sight as they stand against the blue desert sky (fig. 6.7).

The uppermost member of the Entrada, the Moab Tongue, is a lighter-colored (yellowish, pink, or buff) resistant sandstone that crops out on the flanks of the Salt Valley anticline and in some of the erosional features.

4. Deposition of the Morrison Formation late in Jurassic time.

Three members of the Morrison Formation are exposed in Arches National Park. The Tidwell Member (formerly mapped in this area as part of the Summerville Formation) is an interesting unit that crops out in tilted blocks in Cache Valley and Salt Valley. Tidwell is a thin, dark red siltstone that contains large, white, siliceous concretions. (This unit was probably a source of the Indians' chalcedony.) Good exposures occur beside the trail to Delicate Arch.

The middle Salt Wash Member and the upper Brushy Basin Member are interbedded channel sandstones, siltstones, mudstones, and conglomerates that were deposited under varying conditions on broad flood plains. The channel sandstones in the Salt Wash Member contain fossil wood fragments and a few dinosaur bones. The colors of the Brushy Basin slope-forming mudstones range from purplish outcrops just north of the park to greenish exposures in the southern part of the park like those found near Delicate Arch. In the park vicinity some of the thicker sandstone beds of the Salt Wash Member contain uranium, vanadium, and copper ores. The minerals have been transported as solutes in ground water and concentrated or precipitated by organic material (plant fragments, etc.) that was present in ancient stream channels, sandbars, lakeshores, and similar features.

Figure 6.7   Arches are developing in this enormous fin of Entrada Sandstone. Exposed here are wavy beds of the Dewey Bridge Member underlying massive layers of the Slick Rock Member. Photo by S. D. Tuttle.

5. Cretaceous deposits, nonmarine and marine.

During a long period of erosion, thin beds of the Cedar Mountain Formation were deposited by streams. The last terrestrial deposits were thin, discontinuous beds of Dakota Sandstone that accumulated in stream valleys in coastal regions. The Dakota Sandstone now forms a rimrock over the shaley Cedar Mountain strata in parts of Salt Valley.

In Late Cretaceous time, the last great sea spread over the Colorado Plateaus. In this region, several thousand feet of gray mud, which became the Mancos Shale, covered everything. Between upper and lower Mancos Shale members, the Ferron Sandstone was deposited. Within the park, all Mancos beds have been removed by erosion except for a few remnants that are exposed in collapse features in Salt Valley.

6. Laramide orogeny and uplift.

At the end of Cretaceous time and the beginning of Tertiary time, stress and strain accompanying mountain-building around the Colorado Plateaus and the subsequent elevation of the region caused minor warping and faulting, along with fracturing of the rock layers. Deformation during Tertiary time produced gentle anticlines and synclines that generally followed the same northwest trends that were there before. The Courthouse syncline, which follows the Courthouse Wash, is an example.

The Moab fault, which skirts the southwest corner of the Arches National Park near the Visitor Center, displays the effects of relatively recent movement that occurred about 6 million years ago. The beds on either side

Figure 6.8 Fins eroded along vertical, parallel fractures of Entrada Sandstone in Devils Garden, Arches National Park. National Park Service photo.

of U.S. Highway 191 do not match. The Navajo Sandstone and the Entrada Sandstone, which were dropped down, can be seen on the northeast side of the road. Across the highway, in the upthrust block, Paleozoic beds of the Hermosa Group and the Cutler Formation are exposed at the bottom of the canyon. Prominent in the high cliff above are beds of the Moenkopi and Chinle Formations; at the top is the Wingate Sandstone. Vertical movement along the fault amounted to 2600 feet.

7. Development of the present landscape mainly in Quaternary time.

Continuing uplift, accompanied by enlargement of the joint spaces and intensified erosion in Tertiary time, initiated the development of the unique landforms in Arches National Park. The most rapid period of development was during the colder, wetter episodes of the Pleistocene. In the desert environment of the present day, the construction and destruction of arches and other features continues at a slower rate.

Modern sediments accumulate mostly in the form of talus along the base of cliffs, as landslide deposits, in dunes and sheets of wind-blown sand, and terrace gravels along the through-flowing Colorado River, which brings sediments into the park area as well as carries them away. Between floods, rock fragments and soil accumulate in the intermittent stream beds. In the desert, such a feature is called a dry wash.

Geologic Map and Cross Section

Doelling, H. H. 1985. Geologic map of Arches National Park and vicinity, Grand County, Utah. Map 74. Utah Geological and Mineral Survey, Department of Natural Resources.

Box 6.1
Dinosaur National Monument and the Morrison Formation

About a hundred miles north of Arches National Park, the Colorado Plateaus drop off into the Uinta Basin, which lies between the plateaus and the east-west-trending Uinta Mountains. At the eastern end of the Uintas, straddling the Utah-Colorado boundary, is Dinosaur National Monument, famous as a dinosaur quarry.

Although this interesting national monument is actually in the same physiographic province as the Middle Rocky Mountains, its geologic history and many of its rock formations are closely related to the Colorado Plateaus. Paleozoic and Mesozoic formations dip southward off the Uinta Mountains, sink under the Uinta Basin, and continue southward as layers on and beneath the surface of the Colorado Plateaus.

The Jurassic Morrison Formation is the rock unit holding the concentration of dinosaur remains in Dinosaur National Monument. From this "graveyard," once a large sandbar in a powerful stream, have come hundreds of skeletons that have been assembled and shipped to museums all over the world.

The Morrison Formation, which is exposed in many localities in western North America—as well as in several national parks of the Colorado Plateaus—has been one of the most thoroughly studied formations on the continent, both for its unique fossil assemblages (dinosaurs, other animals, plants, etc.) and for its mineralization, especially uranium deposits. The entire formation consists of terrestrial deposits that accumulated in and alongside streams and in shallow lakes. Judging by the variety and abundance of life forms in the fossil record, the late Jurassic climate must have unusually benign. However, in no place is the abundance and variety of dinosaur fossils greater than in Dinosaur National Monument. Apparently season after season dead animals were washed downstream by flood waters, stranded on the sandbar, and buried by sediment and debris.

Long after the lithification and burial of the rocks of the Morrison Formation, the Laramide orogeny began to alter western North America. The monument area, which underwent more severe deformation and uplift than did the Colorado Plateaus rocks to the south, became part of the Uinta Mountains anticline. Erosion by the Green and Yampa Rivers eventually exposed the dinosaur boneyard.

Bibliography

Abbey, E. F. 1968. *Desert solitaire, a season in the wilderness.* New York: Ballantine Books. 303 p.

Baars, D. L. 1983. *The Colorado Plateau, a geologic history.* Albuquerque, New Mexico: University of New Mexico Press.

Doelling, H. H. 1985. *Geology of Arches National Park.* Utah Geological and Mineral Survey, Department of Natural Resources. 15 p.

Hagood, A. 1972. *Dinosaur, the story behind the scenery.* Las Vegas, Nevada: KC Publications. 30 p.

Hoffman, J. F. 1981. *Arches National Park, an illustrated guide and history.* San Diego, California: Western Recreational Publications.

Lohman, S. W. 1975. *The geologic story of Arches National Park.* U.S. Geological Survey Bulletin 1393.

Rigby, J. K. 1976. *Northern Colorado Plateau,* K/H Geology Field Guide Series. Dubuque, Iowa: Kendall/Hunt Publishing Company.

Stokes, W. L. 1986. *Geology of Utah.* Salt Lake City: Utah Museum of Natural History and Utah Geological and Mineral Survey. p. 120–121.

Arches National Park

Area: 73,379 acres; 115 square miles
Proclaimed a national monument: April 12, 1929
Established as a national park: November 12, 1971
Address: 446 South Main Street, Moab, Utah 84532

7

Pre-Columbian cliff dwellings, multi-room pueblos, irrigation systems, and other works of Anasazi people are the best preserved in the United States.
Natural alcoves in which the cliff houses were built were produced by weathering and erosion of Late Cretaceous sandstone and shale beds.

Mesa Verde National Park

Southwest Colorado

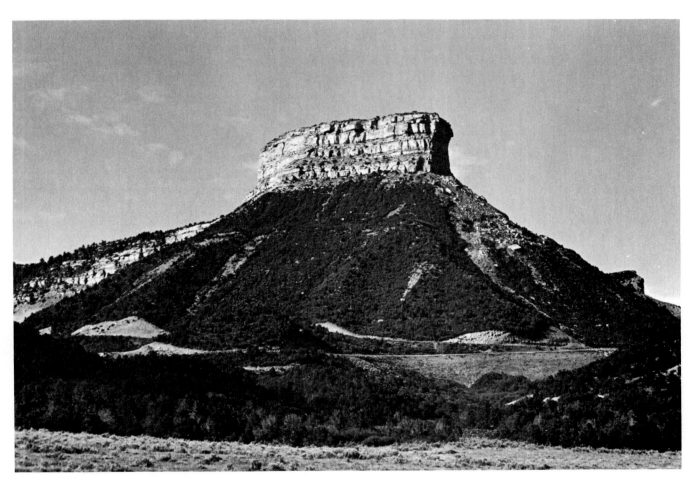

Figure 7.1 Point Lookout, Mesa Verde National Park, as seen from the Mancos Valley, nearly 1500 feet below. The park entrance road climbs from the base of the mesa to the mesa top in a series of switchbacks on the east side of the mesa (left in photo). Photo by M. L. Griffitts.

MESA VERDE NATIONAL PARK

Figure 7.2   Mesa Verde National Park, Colorado.

Local History

Mesa Verde National Park, west of Durango, is in the Colorado corner of the Four Corners area. The park lies on the top of a high mesa that rises from the canyon of the Mancos River, a tributary of the Colorado River. Early Spanish explorers, who camped along the Mancos, named the feature *La Mesa Verde,* meaning "the Green Table."

The first American to make an official report on the area was William H. Jackson, the famous photographer with the Hayden Survey, who photographed cliff dwellings in Mancos Canyon in 1874. Virginia McClurg, a New York journalist, was sent to the Four Corners area in 1882 to investigate rumors of "lost cities." Because of Indian uprisings, she was unable to do much exploring until her third visit in 1886, at which time her small expedition went into several lower canyons that drain into the Mancos River. Her most important discovery was Balcony House (near the park's southern boundary), which was one of the last cliff dwellings occupied by the Anasazi people. She was determined that the archeological resources of Mesa Verde should be protected and preserved, and for the next 20 years, she campaigned vigorously in writings and speeches to gain support for this effort.

The existence of the pueblo ruins on the mesa top and the larger, hidden cliff houses in the alcoves of inner canyons was unsuspected (by non-Indians) until 1888, when

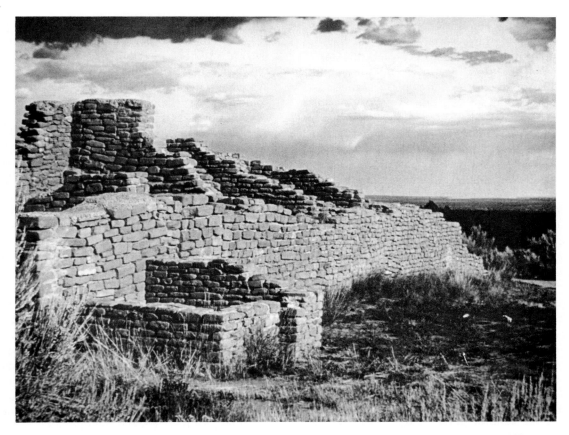

Figure 7.3   Far View Ruins, located near the center of the mesa top, were occupied by Anasazi around 800 years ago. Skilled masons built the multi-room dwellings and also constructed extensive stone-lined irrigation systems. National Park Service photo.

two ranchers, Richard Wetherill and Charles Mason, happened on Cliff Palace and Spruce Tree House when they were out searching for cattle. Fascinated by their discovery of abandoned communities, obviously occupied in the not-too-distant past, Wetherill and his brothers and Mason, their brother-in-law, set about exploring the dwellings during the winter when ranch chores were less demanding. Later, after word of the discoveries had spread, John Wetherill helped Gustaf Nordenskiold (the young son of a Swedish nobleman) to excavate several of the cliff dwellings. Nordenskiold published his findings in 1893 in a book entitled *The Cliff Dwellings of Mesa Verde*. His collection of artifacts from Mesa Verde is displayed in the Finnish National Museum in Helsinki.

In 1900 the Colorado Cliff Dwellings Association was founded by Virginia McClurg. The members rallied support to stop the removal and sale of artifacts and worked on arrangements to preserve the ruins. Due to their efforts, Mesa Verde National Park was established in 1906 in order to protect the archeological sites so that they could be scientifically excavated and studied. In 1978, the significance of the ancient Pueblo culture of Mesa Verde was further recognized when the park was designated a World Heritage Cultural Site, one of the first seven sites selected

for such an honor. Several activities that are now an integral part of the National Park Service program were first begun at Mesa Verde; namely, the development of a museum, interpretive services with guided tours, and "campfire programs" with lectures.

The Natural World of the Mesa Verde Anasazi

The Anasazi (from a Navajo word meaning "Ancient Ones") were a Stone Age people who were exceedingly skillful at making tools and furnishings from natural materials, such as stone, bone, wood, clay, feathers, skins, plant fibers, and whatever was at hand. They also learned how to use the local geology to their utmost advantage in securing shelter and dependable food and water supplies. Management of water resources for agricultural purposes and daily life was vitally important in Anasazi culture. The task demanded foresight and a high degree of community cooperation.

The climate of the region is semiarid, but more rain falls on the high plateaus and mesas than in the parched lowlands of the Montezuma and Mancos Valleys, some 2000 feet below the mesa. The heaviest snowfalls are in the area of the north rim of Mesa Verde where elevations are 8000 feet or more. Runoff from melting snow drains south due to the slight tilt of the mesa surface. Furthermore, because of prevailing storm patterns, the heaviest average annual rainfall, 18 inches per year, falls in the middle section of the mesa, where elevations are around 7000 feet. The most favorable growing conditions are, therefore, in this middle area that receives more moisture from rain and snowmelt. Not surprisingly, the villages of Far View, which had the densest population, were located near the central part of the mesa. With no permanent streams (except the Mancos), the Indians depended on springs and seeps for drinking water. Summer thunderstorms and melting snow in spring replenished the springs and provided water for crops. In addition to higher precipitation, the growing season on the mesa is longer than in the valleys at the foot of the escarpment because temperatures are more moderate in both winter and summer on the mesa.

The presence of fertile soil on the mesa top is another favorable geologic factor. While some of this soil may have been formed by weathering of underlying bedrock, most of it accumulated as wind-deposited silt (loess), blown from the southwest during the Pleistocene Epoch and early in Holocene time.

Much of what we know about the Anasazi who lived on the mesa top and in the cliff alcoves is due to excavations and restorations begun in 1916 by Jesse Walter Fewkes, a Smithsonian Institution archeologist. His work is being carried on today by Park Service archeologists and other scientists.

About A.D. 550, the first Anasazi, called by archeologists, the "Modified Basketmakers," settled in Mesa Verde. They lived in pithouses on the mesa top, cultivated crops, and made handsome baskets from split willow and shrubs. Lining the baskets with pitch made them waterproof. The Anasazi descended to the alcoves in the cliffs for shelter in winter, made crude dwellings there, and stored food in hollows in the sandstone.

During the evolving of the Pueblo culture (A.D. 750 to 1100), the Anasazi built rows of houses with poles and mud above ground on the mesa. The oblong excavations for pithouses gradually acquired the round shape of kivas, while the crude row houses developed into stone-walled pueblos. By A.D. 1000, the Anasazi had become skilled stone masons. They fashioned sandstone into rectangular building blocks and used mud and water for mortar. Some buildings were three stories high with thick, double-coursed walls, joined in units of 50 or more rooms.

Ceremonial and other activities were carried on in the underground kivas. Also, scattered over the mesa are the remains of round stone towers that may have been watch or signal towers. Perhaps the towers were for keeping track of the sun's position so that solar observers could tell the people when to plant and harvest crops.

The "Classic" or most culturally advanced period was between 1100 and 1300. Probably several thousand people lived in the stone-walled villages, and most of the mesa top was under cultivation. Irrigation ditches channeled available water into the fields, and check dams in stream courses all over the mesa collected silt and helped to prevent erosion of precious topsoil.

A water supply system developed for the Far View area (south of the Visitor Center) was remarkable. A manmade reservoir, Mummy Lake (now dry), was constructed to trap and hold runoff from rain and snow in order to supply water for crops and household use. The Anasazi hollowed out a large circular depression that was then lined with sandstone blocks and rimmed with a low retaining wall. A stone-lined intake channel, where silt could settle out, allowed clear water to go into the reservoir. A ditch above Mummy Lake may have diverted water for irrigating.

Around 1200 the Anasazi began to move back into the cliff alcoves where their ancestors had sheltered centuries before. Perhaps the soil on the mesa had become depleted from overuse, or perhaps increasing aridity and a harsher climate forced the mesa dwellers to move closer to springs in the more protected environment of the canyons. Most likely a combination of reasons led to construction of cliff dwellings in the middle decades of the thirteenth century. They range in size from one- or two-room houses built into an overhang to cliff alcoves, such as Cliff Palace, which holds over 200 rooms. Spruce Tree House, which is the third largest in the park, the best preserved, and the one easiest to get to, was constructed in an alcove measuring 216 feet wide and 89 feet deep. A natural spring at the head of the canyon supports lush vegetation and is protected from evaporation by shade most of the day. The canyon springs provided the Indians with a reliable supply of water close at hand, except in times of extreme drought. The narrow canyons provided protection from possible enemies as well as from the elements.

The cliff houses were inhabited less than one hundred years. By the beginning of the fourteenth century, after a series of droughts and crop failures, the Anasazi left Mesa Verde. They may have moved south into Arizona and New Mexico. Some of the Pueblo Indians of today are believed to be descendants of the people of Mesa Verde.

Geologic Features

Mesa Verde is a *cuesta* that slopes gently to the south. Its steep, 2000-foot high escarpment on the north side is formed by an outcropping of resistant sandstone (fig. 7.1). The dip of the sedimentary beds southward from the escarpment is slightly steeper than the slope of the mesa

Figure 7.4 Cliff Palace, the largest cliff dwelling in the park, had over 200 living rooms, 23 kivas, and many storage recesses. As many as 400 people at one time lived in the complex, which had at least 8 floor levels. National Park Service photo.

surface. The Mesa Verde cuesta is an erosional remnant of an ancient pediment that once connected Mesa Verde with the La Plata and San Juan Mountains. A *pediment* is a broad, flat (or gently sloping) bedrock erosional surface that develops at the base of a mountain front. Usually a veneer of sediment called a *bajada* covers the pediment. (A bajada is built by sediment that washes down from mountain slopes and spreads over the pediment surface.) These landforms are characteristic of arid regions.

When the Mesa Verde region was uplifted, streams incised themselves into the pediment and divided it into its separate components. McElmo Creek, which flows west outside the park, captured the headwaters of streams that used to flow south along the north and west sides of the cuesta. The *stream capture* occurred because the south-flowing streams had lower gradients even though they were at higher elevations than McElmo Creek. When headward erosion by the McElmo tributaries cut back far enough, they encountered the south-flowing streams and captured their headwaters, thus changing their direction of flow from south to west.

The Mancos River, which drains the cuesta, has cut a deep, broad valley along the mesa's eastern and southern edge. Intermittent streams, fed by rain and meltwater,

dissected the mesa by headward erosion into 15 long, parallel, steep-sided canyons, all of which drain into the Mancos. Narrow uplands separate the fingerlike canyons. The Anasazi built their dwellings in alcoves in the canyon walls and on the intervening mesa surface (fig.7.5).

How the Alcoves Formed

The shallow, cavelike alcoves that the Anasazi enlarged and adapted for cliff dwellings developed along a shale-sandstone contact where massive, cliff-forming sandstone is underlain by soft, erodible shale beds. Some of the cliff dwellings in the park are at the base of the top member (the Barker Dome Tongue) of the Cliff House Sandstone, which contains a discontinuous layer of shale (fig. 7.6).

Because the Cliff House Sandstone has many pore spaces (openings) and can hold a considerable quantity of ground water, it has high *porosity*. It is also *permeable,* which means that water can pass through it. On the mesa top, melting snow and rain soaks in and readily percolates down through sandstone until reaching shale beds, which are porous, but not permeable because the pore spaces are so small that water is trapped and cannot pass through. The ground water is forced to move laterally along the contact until it seeps out on the face of the canyon wall or at the back of an alcove, creating a spring.

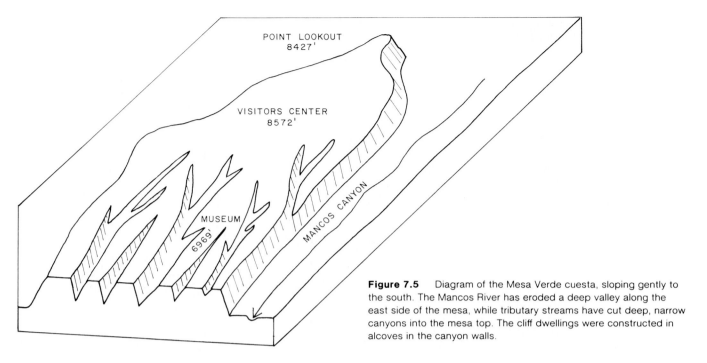

Figure 7.5 Diagram of the Mesa Verde cuesta, sloping gently to the south. The Mancos River has eroded a deep valley along the east side of the mesa, while tributary streams have cut deep, narrow canyons into the mesa top. The cliff dwellings were constructed in alcoves in the canyon walls.

Figure 7.6   Park Headquarters and Museum on top of Chapin Mesa. The massive Cliff House Sandstone is exposed in the canyon wall. National Park Service photo.

Table 7.1 Geologic Column, Mesa Verde National Park

Time Units		Rock Units			Geologic Events
Era	Period	Group	Formation	Member	
Cenozoic	Quaternary	Stream deposits, talus Loess soil			Weathering of alcoves in cliff walls Headward erosion of canyons Uplift and tilting; isolation of Mesa Verde cuesta by erosion Development of pediment Uplift of La Platas and San Juans Uplift of Colorado Plateaus during Laramide orogeny. Erosion ———————————
Cenozoic	Tertiary	Pediment gravels, cobbles			
Mesozoic	Upper Cretaceous	Mesaverde	Cliff House Sandstone	Barker Dome Tongue	Cliff-former; thick accumulations of fine, cross-bedded sand and thin shale beds on oscillating shoreline
Mesozoic	Upper Cretaceous	Mesaverde	Cliff House Sandstone	Lower Member	
Mesozoic	Upper Cretaceous	Mesaverde	Menefee		Backshore and lagoonal deposits; silts, muds, sand, coal beds
Mesozoic	Upper Cretaceous	Mesaverde	Point Lookout Sandstone		Cliff-former; massive, fine, cross-bedded sands
Mesozoic	Upper Cretaceous		Mancos Shale		Slope-former; deep-water deposits; muds, silts, fine sands
Mesozoic	Upper Cretaceous		Dakota Sandstone		Not exposed in park

Source: M. Griffitts (personal communication); Wenger, 1980.

In geological terms, the percolating ground water dissolves cement along joints and bedding planes and loosens sheets and blocks of sandstone. Seasonal freezing and thawing of the ground water breaks up the bedrock. Exfoliation, mass wasting, and trickling water remove the loosened material, enlarging hollows into alcoves.

Geologic History

1. Deposition of the Mancos Shale over the Dakota Sandstone late in Cretaceous time.

The Dakota Sandstone, which underlies most of Colorado, crops out west of Mesa Verde National Park, but is not exposed in the park. The Mancos Shale (above the Dakota) forms the long talus slope of the north escarpment of Mesa Verde. The highway up to the mesa top from the park entrance has to climb unstable slopes on weak Mancos Shale in order to reach the sandstone rimrock. Because the Mancos has a high clay content, it expands when wet and causes slumping and sliding so that the road has to be repaired frequently.

The Mancos beds in this area are fine-grained sandstones, mudstones, and shales that accumulated in deep water, far enough from shore so that coarser sediments had settled out. The members that crop out at the surface in Mesa Verde National Park are the highly fossiliferous Carlile Shales; the Niobrara beds of sandy, limey shale; and a Transition Member, containing sandy shale and shaley sandstone.

2. Shallow-water deposits of the Point Lookout Sandstone, Mesaverde Group.

The three formations comprising the Mesaverde Group—Point Lookout Sandstone, Menefee Formation, and Cliff House Sandstone—reflect the changes in depositional environment that were occurring toward the end of Cretaceous time when the area that is now the park lay along an oscillating shoreline. The type locality of the Mesaverde Group is Mesa Verde National Park.

The Point Lookout Sandstone, which forms the rim of the escarpment above the entrance to the park, is named for Point Lookout (elevation, 8427 feet) (fig. 7.1). The sediments making up the formation, which is about 400 feet thick, accumulated as the shoreline of the Cretaceous sea was shifting toward the northeast. The upper layers of the sandstone contain fossils of invertebrates. The Point Lookout Sandstone is massive (because it has no bedding planes), fine-grained, cross-bedded, and very resistant. Since the formation is virtually all marine, the cross bedding is probably the result of wave action and currents in shallow water.

Figure 7.7   Cliff Palace, in a great alcove on the east wall of Cliff Canyon, as seen from Sun Point, a half mile away on the other side of the canyon. Wetherill and Mason first saw the cliff dwelling from Sun Point in 1888. From *Northern Colorado Plateau*, J. K. Rigby. © 1976 Kendall/Hunt Publishing Company. Used by permission.

3. Nonmarine deposits of the Menefee Formation.

The middle formation of the Mesaverde Group consists of interbedded carbonaceous shales, siltstones, and sandstones that were deposited in brackish water in bays, swamps, and lagoons. In areas outside the park so much organic material accumulated that coal beds of commercial grade were produced. A few thin coal seams are exposed in canyon walls in the park. The Upper Member of the Menefee Formation interfingers with the overlying Cliff House Sandstone.

4. A shifting shoreline during deposition of the Cliff House Sandstone.

Great quantities of fine-grained sand accumulated in sand bars, beaches, and shore dunes, with clastic materials overlapping and interbedding as the shoreline kept shifting back and forth. The sediments became the massive, cross-bedded Cliff House Sandstone, the youngest bedrock in the park. The formation averages 400 feet in thickness and forms the bold, yellow-orange cliffs in the canyons. Fossil beds, containing clam shells, snail shells, fish teeth, etc. occur in lenses in the upper 100 feet of the

Cliff House. Concretions lacking nuclei that erode out of the Cliff House Sandstone appear to be the result of variations in porosity.

The shale zones in the Cliff House Sandstone determine the location of the alcoves in which the cliff dwellings were constructed (figs. 7.6, 7.7).

5. Beginning of successive uplifts at the close of the Cretaceous, continuing through Tertiary time.

Although marine sediments were deposited over the formations of the Mesaverde Group, the youngest Cretaceous beds were stripped off by erosion while the Colorado Plateaus were uplifted during the Laramide orogeny.

The uplifting of the San Juan and La Plata Mountains in Tertiary time and the accompanying erosion created the Mesa Verde pediment, which gently slopes southward. On this erosional surface, meandering streams flowed toward the southwest depositing gravels brought down from the mountains. Scattered patches of the gravels remain in the Mesa Verde area. These waterworn pebbles and cobbles, generally of greater hardness than the sandstones, were prized by the Indians for toolmaking and other uses.

6. Dissection of the pediment plateau and isolation of Mesa Verde from the La Plata and San Juan highlands.

A final uplift and southward tilting late in Tertiary time caused the streams on the pediment to downcut rapidly—removing sediment, creating canyons and broad valleys, and isolating erosional remnants, such as the Mesa Verde cuesta. A slight difference in steepness between the slope of the mesa surface and the dip of the sedimentary formations is shown by the fact that Point Lookout Sandstone forms the north rim of the mesa but is buried under younger beds of the Mesaverde Group in the center of the mesa. Only the Menefee Formation and the Cliff House Sandstone are exposed in the canyons.

Headward erosion of the canyons in the cuesta, developing a parallel drainage pattern, was especially intense during the cooler, more humid climates of Pleistocene time. Also, because the bedrock was wetter, older alcoves formed, deepened, and collapsed in the canyon walls, while new alcoves were being created. In the drier climate of the present, the erosional processes continue at a slower rate.

Geologic Maps and Cross Sections

Tweto, O. 1979. Geologic Map of Colorado. U.S. Geological Survey.

Wanek, A. A. 1959. Geology of fuel resources of the Mesa Verde area, Montezuma, and LaMata counties, Colorado. U.S. Geological Survey Bulletin 1072M (map in pocket).

Bibliography

Griffitts, M. L. (1990). *Guidebook to the geology of Mesa Verde National Park, Colorado.* Mesa Verde Museum Association, Mesa Verde National Park.

Hunt, C. B. 1969. *Geologic history of the Colorado River.* U.S. Geological Survey Professional Paper 669.

LeRoy, L. A. 1960. *Generalized composite stratigraphic sections of western Colorado.* Colorado School of Mines Department of Publications.

National Park Service. n.d. Geology information sheet, Mesa Verde National Park.

Rigby, J. K. 1977. *Southern Colorado Plateau,* K/H Geology field guide series. Dubuque, Iowa: Kendall/Hunt Publishing Company.

Wanek, A. A. 1959. *Geology of fuel resources of the Mesa Verde area, Montezuma, and LaMata counties, Colorado.* U.S. Geological Survey Bulletin 1072M.

Wenger, G. R. 1980. *The story of Mesa Verde National Park.* Mesa Verde Museum Association, Mesa Verde National Park.

Yardell, M. D. (ed.) 1972. *National Parkways, a photographic and comprehensive guide to Rocky Mountain and Mesa Verde National Parks.* Casper, Wyoming: World Wide Research and Publishing Company.

Mesa Verde National Park

Area: 52,085 acres; 81 square miles
Established: June 29, 1906
Designated: World Heritage Site, September 6, 1978
Wilderness designated, 8100 acres: October 10, 1976
Address: Mesa Verde National Park, Colorado 81330

8

A profusion of logs and parts of trees that have been changed to multicolored stone.
Indian ruins and petroglyphs.
Brightly colored badlands in the Painted Desert

Petrified Forest National Park

East Central Arizona

by Sherwood D. Tuttle
University of Iowa

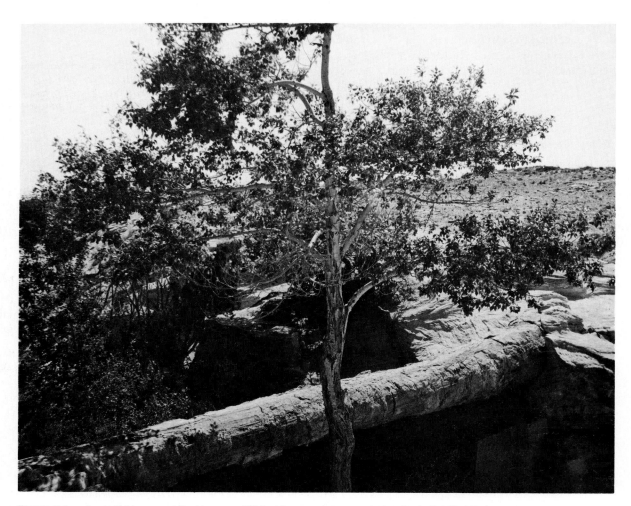

Figure 8.1   Agate Bridge, a petrified log over 100 feet long, spans an eroded ravine in Petrified Forest National Park. The National Park Service has placed supports beneath Agate Bridge to prevent its collapse. National Park Service photo.

Geographic Setting

Petrified Forest National Park is located near the southern edge of the Colorado Plateaus about 120 miles (air distance) southeast of Grand Canyon National Park. The geology of Petrified Forest is similar to other national parks of the Colorado Plateaus. Erosional landforms, characteristic of arid regions and sculpted from flat-lying rocks, dominate the scenery. However, scattered over the desert surface in Petrified Forest National Park are thousands of jewellike logs, broken branches, segments of tree trunks, and sparkling wood chips. All have been changed from wood to stone. Here is an array that is unsurpassed anywhere in the world for variety and profusion.

The petrified trees have been eroded from a sequence of Triassic rocks that are a part of the vast Painted Desert of Arizona. Although the American Southwest has many areas of brightly colored rocks referred to as "painted deserts," the geographic entity officially designated as the Painted Desert curves southwestward across Arizona from Lees Ferry, at the northern tip of Grand Canyon National Park, to Holbrook, then east across Petrified Forest National Park, and on to the New Mexico border.

The Little Colorado River, which, in a general way, skirts the southern edge of the Painted Desert, flows northwest to join the Colorado River, collecting the drainage of the area as it goes. The intermittent Puerco River, which cuts across the middle of the park in a wide, flat-bottomed arroyo, meets the Little Colorado west of the park at Holbrook.

Evidence of stream erosion by the Puerco and its tributaries is everywhere about the landscape, but water is seldom seen on the surface. The arid plateau on which the park is located (elevation roughly 5500 feet) is smooth in some places, but elsewhere the landscape is intricately dissected. Mesas, buttes, gullies, arroyos, sinkholes, and other features show that even in the desert, water is the dominant agent of erosion.

The desert climate is variable with wide fluctuations in rainfall and sudden changes in temperature. Mean annual rainfall is 9 to 10 inches. Most of it comes in brief, violent thunderstorms during the summer months. Annual temperature readings may range as much as 127 Fahrenheit degrees between average highs and lows! Occasionally snow falls in winter, but does not remain long on the ground. In times of drought, many months may pass without precipitation. In this harsh climate, rocks are well exposed, but vegetation is sparse. Sagebrush, greasewood, cactus, and yucca are some of the hardy plants that survive.

Local History

Since prehistoric times, a well traveled, east-west trade route has followed the valley of the Puerco River through this part of Arizona. Indians, and later explorers and pioneers, used the ancient trade route to get to Baja California and the Pacific Coast. A transcontinental railroad followed the same route; and later a transcontinental highway, Route 66, was built. Today Interstate 40, on its way across the country, goes through the Petrified Forest National Park and provides convenient access to its features.

The oldest Indian ruins in the park are on The Flattops, an area of sandstone-capped mesas near the southern entrance. Ancestors of the Pueblo people probably lived here in a village of pit houses that are partially dug into the earth and lined with sandstone slabs. More than 2000 years ago, the inhabitants raised crops on the flat mesa tops, when the climate was less arid than now. Nearby, south of Rainbow Forest Museum, is Agate House, a partially restored pueblo built entirely of petrified wood.

A village at Twin Buttes (west of the park) was occupied by people of a later, less primitive culture. Pit houses were excavated more deeply and had more rooms and storage chambers. The inhabitants built long, low structures of stone that may have served as windbreaks to protect crops and reduce the movement of sand over the land surface. The Hopi Indians of today still build such walls. Pottery and tools found at the site are more advanced and include artifacts made from Pacific seashells and other materials foreign to this region that were probably received in trade for articles made from petrified wood.

Droughts that began in the thirteenth century brought about changes in the Indians' way of life. They began building their villages on the flood plains where the water supply was more dependable. The Puerco Indian Ruin, which at one time may have had a population of more than 200 people, is a large, rectangular pueblo of uncemented masonry construction on a terrace overlooking the Puerco River. About 150 rooms, many of them interconnected, are laid out around a large courtyard. The Indians grew crops on the level terrace tops.

Around the beginning of the fourteenth century, precipitation increased; but instead of being evenly distributed throughout the growing season, the rain came in heavy bursts and thunderstorms. Time and time again, the river flooded, washing out fields, eroding topsoil, and ruining crops. Finally the Pueblo people could no longer maintain their preferred way of life. According to Hopi

PETRIFIED FOREST NATIONAL PARK

Figure 8.2   Petrified Forest National Park, Arizona.

Box 8.1
Silica Can Assume Many Forms

Inasmuch as silicon (Si) and oxygen (O) are the two most abundant chemical elements in the rocks of the earth's crust, it is not surprising that naturally occurring silica (silicon dioxide, SiO_2) is found in many forms, both crystalline and noncrystalline. Variations in color are caused by minute amounts of other compounds that are precipitated with silica. The silicified logs in Petrified Forest National Park display interesting and unique examples of this mineral. In the petrified logs the color tones of yellow, red, and brown were produced by iron in the ground water. Copper added tints of blue and green; manganese or carbon blended in black tones.

All of the crystalline varieties of precipitated silica are quartz, but they have different names based on their color. The most common are rock crystal (clear), amethyst (lavender or purple), smoky quartz (gray to black), milky quartz (white or cream), and citrine (yellow). If the mineral can precipitate, or grow, in an unoccupied space,

it may develop its regular, distinct crystal form as a six-sided pyramid or prism. If the quartz grows in very small or irregular spaces, it becomes crystalline grains.

When the internal crystalline arrangement of the quartz is so irregular that it has no characteristic external form, or when its internal structure cannot be detected by a petrographic microscope, the specimen is described as amorphous or *cryptocrystalline*. (Extremely fine crystallinity may, however, be demonstrated by the use of an electron microscope.) Such varieties include: *chalcedony*— white, pale blue, gray, brown, or black with a waxlike luster and uniform tint; *jasper*—red chalcedony or chert (a sedimentary rock composed primarily of silica); *carnelian*—a translucent red or orange-red variety of chalcedony; *agate*—a banded or variegated chalcedony, characterized by colors arranged in alternating stripes or bands but occurring in other forms as well.

tradition, their ancestors moved from the Puerco region many generations ago to what is now the Hopi Indian Reservation located about 80 miles to the northwest.

The first official report on the petrified wood by American explorers was made by an army officer who led an expedition down the Zuni and Colorado Rivers in 1851. He told of finding large amounts of petrified wood a few miles southeast of the present park area. Five years later, Amiel Whipple, while surveying a route for the transcontinental railroad, found petrified wood in such abundance that he named a tributary of the Puerco River Lithodendron ("stone-tree") Creek. Later geologic investigations throughout the Colorado Plateau revealed that fossilized tree remains are not uncommon in Mesozoic rocks, especially in the Chinle Formation. However, nowhere else are specimens found so well preserved and in such abundance as in Petrified Forest National Park.

When General W. T. Sherman toured the western territories in 1879, he was so impressed by the natural display of petrified wood in the Puerco region that he asked that specimens be sent to the Smithsonian Institution in Washington, D.C. An Army detail, sent to accomplish the

task, found huge buried logs that seemed suitable near Bear Springs at the head of Lithodendron Wash. They dug up two large segments and transported them by wagon to Fort Wingate, New Mexico. From there, the larger piece was shipped by rail to the Smithsonian, where it remains on display.

When the transcontinental railroad was extended through Arizona, the railroad company constructed a coaling and water stop in the Puerco River Valley that became the town of Adamana. Train passengers were encouraged to take excursions to "Chalcedony Park," as the area was then called, to see the "petrified forests." Word spread, and from then on, souvenir hunters, gem collectors, and commercial jewelers began hauling away quantities of petrified wood, quartz and amethyst crystals, and Indian relics. Some collectors, looking for crystals, used dynamite to break up the logs. Fossil wood was carted away by the wagonload. When a stamp mill was erected at Adamana to crush the logs for the manufacture of abrasives, the local people became alarmed and intensified their efforts to preserve the petrified forests. The Arizona territorial legislature petitioned the United State

Figure 8.3   Differential weathering and erosion uncovered the petrified logs, which are more resistant than the Chinle Formation beds in which they were buried. From *Geology of Arizona*, D. Nations and E. Stump. © 1981 Kendall/Hunt Publishing Company. Used by permission.

Congress to set aside Chalcedony Park as a national preserve. Lester Ward, a paleobotanist of the U.S. Geological Survey, investigated the area in 1899 and recommended that the land be withdrawn from homesteading and placed under government protection.

In 1906, after Congress passed the Antiquities Act, President Theodore Roosevelt declared the area a national monument. In 1962, after several additions, including a part of the Painted Desert, the monument was redesignated Petrified Forest National Park. From the north-south highway through the park, the petrified forests and other geologic features, as well as Indian ruins, are easily accessible. Several overlooks in the northern part of the park provide sweeping views of the Painted Desert. **Removal of petrified wood or any unauthorized materials from Petrified Forest National Park is prohibited**.

Geologic Features

The Petrified Wood

In Petrified Forest National Park, fossil logs and chunks of all sizes lie scattered on the ground or stick out of the soft, shaly bedrock. Examination of the logs shows that some have roots, but many are bare trunks without roots, limbs, or twigs. Much of the bark is scarred and abraded. The silica that infiltrated the wood made the logs heavy as well as brittle. Some are split lengthwise and some are broken clean across their width as if sawed. Probably some logs were broken after petrification during uplift, possibly by earthquakes, or perhaps when tumbling down slopes as they were uncovered and loosened by erosion (fig. 8.3).

Most of the trees appear to have been uprooted, transported, tumbled, and swirled about by raging streams and then left in a jumble of woody trash and sediment far downstream from where they originally grew. To account for so much wood being washed out by floods, transported, dumped, and then buried, we have to visualize large rivers that overspread flood plains and deltas during times of high water. Perhaps the climate was monsoonal, with winds bringing in tropical typhoons and seasonal downpours from a coast that lay to the west.

About 40 different species of fossil plants (most of them extinct) have been identified in the Chinle Formation, the principal rock unit exposed in the park. In addition to the petrified logs, careful searches have uncovered roots, leaves, seeds, and other plant debris preserved in

Box 8.2
How Does Wood Turn to Stone?

Like a number of other geochemical processes, petrifaction is not completely understood. How helpful it would be if wood could be silicified and the process studied in a laboratory under controlled conditions. So far no one has successfully accomplished this, although some investigators have tried. A human lifetime is too short for the completion of experiments to duplicate processes that operate on geologic time in nature.

The logs in Petrified Forest demonstrate that processes of petrifaction are infinitely varied. In some logs, the internal woody structure has been preserved with amazing fidelity. Others are *pseudomorphs;* that is, they are recognizable as logs, but the internal woody structure was not retained in the precipitated jasper, chalcedony, or agate. The wood has been entirely replaced by silica. Moreover, in some logs both replacement of woody tissue by silica and preservation of cellular structure have occurred in close proximity.

The process by which the details of cellular structure are preserved is called *permineralization;* it is a type of fossilization whereby mineral matter is deposited within and around the cells or pores without distorting or destroying them. Wood that has been silicified in this way can be immersed in hydrofluoric acid so that the silica is chemically removed, which leaves a soft mass of woody tissue from which slides can be made for microscopic study, as with any modern wood.

What probably helps to make this process of fossilization possible under certain natural conditions is the fact that inorganic molecules are much smaller than organic molecules. In other words, the preservation of wood structure in faithful detail by petrifaction is not accomplished

by replacement of wood cells by silica ("molecule for molecule," as was once thought to be the case), because there is considerable discrepancy in size and geometry between organic and inorganic molecules. What does happen is that the woody components are penetrated and surrounded by "microcrystalline mineral matter often without visible disturbance of either the tissue or the interpenetrating microscopic crystal patterns" (Schopf 1971, p. 523).

Natural conditions under which permineralization of wood may take place involve ample moisture as well as time. In the swampy lowlands in which the wood was deposited, the water table was close to the surface. Ground water circulated very slowly and the plant material was constantly saturated. From time to time, fine volcanic ash from volcanic eruptions to the west fell from the atmosphere or was washed down from the highlands. The ash was the main source of the silica. Ground water, somewhat acidic from decaying plants, began dissolving silica and traces of other mineral matter from the ash. Thus enriched, the ground water solutions gradually permeated the woody tissue, infiltrating the cells, filling the intercellular cavities, and coating the cell walls. We can only speculate as to what chemical or environmental conditions caused some of the precipitating silica to *replace* cells and some to *preserve* cells. We cannot know whether the processes operated concurrently or successively.

The well developed crystals of quartz and amethyst (some of gem quality) that are found in some of the petrified logs grew in hollows or open spaces that probably existed in the wood of the original tree.

fine-grained shale beds. Included in the Triassic flora are rushes and horsetails, large and small ferns, and large fernlike cycads. Such plants thrive on warm, swampy flood plains and in marshes.

Most of the logs are conifers belonging to a group named *Araucarioxylon.* Distant living relatives of these conifers are the monkey puzzle tree (native to Chile) and

the Norfolk Island pine (native to Norfolk Island in the South Pacific). Some of the petrified trees were more than 200 feet tall and 10 feet in diameter. Evidently the conifers grew on highland slopes upstream before they were buried in the lowlands by sediment-laden flood waters. Some trees were healthy specimens, perhaps growing on riverbanks undercut during high water; but other trees were dead or dying of natural causes when they toppled down or were blown down. Evidences of insect borings, fungus rot, and charring by forest fires have all been preserved in the petrified remains of the fossil trees. Burial was rapid once the logs came to rest. Not many logs rotted before being covered. The water-saturated muds and sands sealed out air and prevented decomposition.

Fossils

Although plant fossils are by far the most abundant remains in the park, teeth, scales, and bones of extinct species of fish, clams, amphibians, and dinosaurs record the presence of animal life in the Triassic swamps. Many of these fossils were petrified by permineralization or pseudomorphic replacement. In 1985, the bones of a small, very early species of dinosaur, perhaps the oldest dinosaur skeleton yet discovered, was found near Chinde Point in the northern part of the park. The dinosaur, the size of a large dog, lived early in Triassic time and is regarded as an ancestor of the giant Brontosaurs, which evolved much later in Mesozoic time.

Petroglyphs and Desert Varnish

Carving petroglyphs in desert varnish on rocks seems to have been a favorite pastime of the prehistoric people who lived in the American Southwest. Thousands of their mysterious inscriptions are in Petrified Forest National Park. Some of the petroglyphs are drawings of human and animal figures, and others are abstract, geometric patterns, similar to the designs on Indian pottery. Newspaper Rock, near the Puerco Indian Ruin, is a massive sandstone block with many interesting examples of rock art on its surface. The meaning of the petroglyphs is not clear, although some of them are clan symbols and many probably have ceremonial significance. Recently astronomers have discovered solar markers, carved on rock in Petrified Forest National Park, which correlate with the sun's position as the seasons change. Ancient sun-watchers used these petroglyphs as a calendar to keep track of the sun's progress across the sky so that they would know when to plant crops.

Archeologists and anthropologists who have studied the rock art assume that the drawings were done at different times and by individuals from different cultures, but no satisfactory method of dating the petroglyphs has been discovered. The Indians made the drawings by using a sharp tool, probably a piece of petrified wood, to scratch or peck through the dark desert varnish that coated the sandstone, thus exposing the fresh, light-colored rock underneath.

As sandstone weathers in the desert climate, a thin, dark brownish red or black patina, or *desert varnish,* develops on the rock surface. The varnish consists of iron and manganese oxides with some silica and is probably caused by the repeated evaporation of water drops from the rock surface. As the moisture dries, dissolved mineral matter exuded from the rock is left. Some authorities believe that acid-secreting lichens help to fix the varnish coating to the rock. Blowing sand polishes the varnish, giving it a gloss. Because the rate at which desert varnish accumulates is highly variable, it cannot be used for dating inscriptions.

Caliche and Other Desert Features

Tubes and nodules of caliche are exposed in ravines and on the flanks of mesas. *Caliche* is a general term for concentrations of soluble calcium salts (such as calcium carbonate) that precipitate from soil moisture at or near the surface. Under arid conditions caliche accumulates because rainfall is insufficient to wash calcium carbonate away or dissolve it. Sometimes caliche forms an impermeable "hardpan" layer beneath desert soil, as well as nodules and tubes. The latter represent roots that have been coated and eventually replaced.

Desert pavement tends to form on flat surfaces, such as on the flood plain of the Puerco River. Wind has removed soil and fine particles and left a closely packed surface of wind-polished pebbles and rock fragments, usually somewhat cemented together by calcium carbonate. Loose rocks, many of them fragments of petrified wood, have been so abraded by wind-blown sand that they have smooth, polished faces and sharp edges. These are *ventifacts.*

Badland Topography and the Painted Desert

To the north along Interstate 40 in this part of Arizona is the panorama of the Painted Desert—a landscape of brilliant, improbable colors against a deep blue sky. The multicolored rocks of the Chinle Formation are exposed in the myriad shapes of the badlands—gullies, ridges, arroyos, washes, mounds, escarpments, and mesas. Like the colors of Badlands National Park in South Dakota (chapter 9), but much more brilliant, the hues of the Painted Desert change with the hours of the day. Colors are most brilliant in the morning and late afternoon but least intense at midday. Colors change when the surface is wet; and on a partly cloudy day, shaded areas and sunlit slopes contrast beautifully. Thus the natural variations in daylight act on the sedimentary rocks to produce spectacular chromatic effects. The colors of the various shales, marls, sandstones, and limestones of the Chinle are purple, pink, gray, yellow, ash gray, lavender, rose, maroon, sienna, lilac, cream, chocolate, and various other shades of red and brown. Patches of blue, white, and even black are seen. Dominating overall are the reds, pinks, and purples.

Although annual rainfall is low, the relatively nonresistant Triassic rocks are being eroded. Landforms change with every cloudburst and thunderstorm. Much of the badlands topography, however, was formed when rain was more plentiful than in today's climate. Interbedded with the thin layers of poorly cemented layers of gravel, mud, and sand are water-laid units of volcanic ash. Weathering has altered the ash to *bentonite,* a type of soft clay that absorbs water and swells as it becomes wet. When saturated, the clay slumps or may even flow as a fine muck. After it dries, the clay hardens again. Local slopes that develop on the bentonite clay units are rounded and bulbous because of the clay's plasticity (fig. 8.4). *Piping* causes sinkholes to develop in the tops of mounds and in gully bottoms. During downpours, water drains through tunnels, or "pipes," inside the mounds and the gullies, washing out loose material and undermining slopes that later collapse as sinks and slumps. New slope profiles and sinkholes may develop in a few hours when flash floods occur.

Steeper slopes and escarpments develop on the sandstone and conglomerate beds of the Chinle. Some of the more resistant units form cap rocks on the buttes and mesas of the landscape.

Geologic History

1. Subsurface Paleozoic rocks.

Paleozoic rock units that are well exposed in many other parts of the Colorado Plateau are not seen in the region of Petrified Forest National Park. Some of the Paleozoic rocks were probably removed by erosion; the rest are buried by younger rocks.

2. Deposition of the Moenkopi Formation early in Triassic time.

At the beginning of Triassic time, the landmass that eventually became North America was still part of the supercontinent Pangea, and the Colorado Plateaus region was close to the equator. On a low coastal plain that sloped gently westward across northern Arizona, sandy and muddy beds accumulated, 300 to 600 feet in thickness. Marine and calcareous units interfingered with continental beds, indicating that at times the sea encroached on the coastal plain. In some places gypsiferous beds formed in shallow lagoons between layers of mud. Sand-filled channels trace the courses of ancient streams. The climate was alternately warm and dry or warm and humid.

Only a few beds of the Moenkopi Formation remain in the Petrified Forest area because most of the formation was eroded as the ancient plain was upwarped and the sea withdrew.

3. Deposition of the Shinarump Member of the Chinle Formation.

In a subsiding basin in north-central and northeastern Arizona, a typical basal conglomerate, containing many Moenkopi pebbles, was deposited unconformably over the eroded Moenkopi Formation. The gravelly, sandy, lower layers are indicative of high source areas and (presumably) streams that are able to transport the clastic

Figure 8.4 Badland topography in the Painted Desert, Petrified Forest National Park. Bentonite clay and shale beds of the brightly colored Chinle Formation erode into bulbous slopes and slumps. A sandstone bed capping the ridge in the foreground provides some protection to underlying clay and shale. G. Shimer drawing from *Field Guide to Landforms in the United States,* J. A. Shimer. The Macmillan Company © 1972 by J. A. Shimer. Used by permission.

material. The coarse sediment grades upward into finer material. This conglomerate, averaging 35 to 50 feet in thickness, is the Shinarump Member of the Chinle Formation.

4. Deposition of Chinle beds later in Triassic time.

The sea to the west continued to regress, leaving an immense alluvial plain in the Colorado Plateaus area. Sediment transported by the rivers became finer. The climate, still warm, became more humid, and tropical grasslands and marshes developed. Aprons of sand, mud, and silt spread over the northern Arizona lowlands, covering old flood plains, deltas, and backswamp areas. Localized units of gypsiferous clay represent evaporite deposits in lagoons or playa lakes, perhaps due to seasonal wet and dry periods.

Several hundred feet of sandy, shaly material accumulated during this time to form the Lower Petrified Forest Member and the Upper Petrified Forest Member of the Chinle Formation. These two members are separated in some places by a discontinuous unit, the Sonsela Sandstone Member, which contains more petrified wood than any of the other rock units in the park. The Owl Rock Member at the top of the section completes the Chinle Formation in Petrified Forest National Park (Billingsley and Breed 1980).

Tectonic activities toward the end of the Triassic began to affect the northern Arizona basin. In the western sea, a chain of volcanoes (about where the California-Nevada boundary is now) began erupting large quantities of ash that were carried eastward by winds and dropped on the plains, basins, and highlands of the interior. Heavy rains washed quantities of ash, mixed with mud and other debris, into the Petrified Forest area. The volcanic ash is the source of the silica in the petrified wood.

Table 8.1 Geologic Column, Petrified Forest National Park

\multicolumn{3}{Time Units}			Rock Units		Geologic Events
Era	Periods	Epochs	Formations	Members	
Cenozoic	Quaternary	Holocene	Stream, wind, slope wash, and slump deposits		Fluctuating climate, becoming more arid; development of badland topography. Intensified erosion during pluvials
		Pleistocene			
	Tertiary	Pliocene	Bidahochi		Coarse sediment interbedded with and topped by pyroclastics from explosive eruptions northwest of park area
					———— Extensive erosion ————
		Paleocene			Upwarping, faulting. Laramide orogeny
Mesozoic	Cretaceous Jurassic		(rock units eroded from park area)		Extensive marine and nonmarine sedimentation
					———— Erosion ————
	Triassic		Chinle	Owl Rock	Gypsiferous clays accumulating in playas and lagoons; ashfalls
				Upper Petrified Forest	
				Sonsela Sandstone	Contains largest accumulation of trees and plant debris
				Lower Petrified Forest	Extensive deposition in interior basin; shales, sandstones
				Shinarump	Basal conglomerate containing Moenkopi fragments
					———— Erosion ————
			Moenkopi		Shallow marine and nonmarine sedimentation
			(older rock units not exposed in park)		

Source: modified after Billingsley and Breed 1980

5. The wood in the Chinle Formation.

The Chinle Formation is famous in the western states for its petrified wood. Although the finest and most abundant specimens are in the Petrified Forest area, the same species of conifers have been found in Chinle beds in other places where the formation crops out. Obviously the forests were widespread. A late Triassic highland extending from central Colorado to northern New Mexico may have been the source of the large conifers washed into the northern Arizona basin. It was there in the swampy lowlands that the ferns and cycads flourished. The fact that their fossil remains are found together, showing little evidence of decay, suggests that the uprooting, transportation, and swamp burial of the conifers occurred in relatively short spans of time, possibly during monsoonal or seasonal floods of considerable magnitude.

Just when and for how long the processes of petrifaction and permineralization went on is not known. Perhaps the elevation of the Colorado Plateau region, beginning as the Mesozoic Era closed, terminated the silicifying of the buried wood.

6. Accumulation of Jurassic and Cretaceous sediments.

Several thousand feet of Jurassic and Cretaceous sediments were deposited on top of the Chinle Formation in the Colorado Plateaus region; but in the Petrified Forest area, these beds were subsequently eroded. The weight of the rock units helped to compact the Chinle Formation, while ground water, percolating through the sediments, probably affected the wood buried in the Chinle.

7. Tertiary uplift, warping, and erosion.

Through the last of the Mesozoic and the beginning of the Cenozoic Eras, the Laramide orogeny, with its pulses of tectonic activity, shifting from west to east, profoundly affected western North America. Although the Colorado Plateaus were uplifted and slightly tilted during Laramide mountain-building episodes, the flat-lying rocks remained relatively undeformed. Around the margins of the plateau, some broad folding and upwarping occurred. In northeastern Arizona, the upwarping (Defiance uplift) that took place resulted in the shaping of the park landscape that we see today. Erosion stripped off all the Jurassic and Cretaceous rocks and some of the Chinle beds from the Petrified Forest region. (Some of these units are preserved in downwarped basins only about 50 miles away from the park.)

Meanwhile the climate was becoming cooler and increasingly arid. The mountain barriers rising to the west cut off moisture-laden winds from the Pacific.

8. Late Cenozoic sedimentation and volcanism.

In Pliocene time, an irregular blanket (in places several hundred feet thick) of coarse sediment and volcanic debris was spread unconformably across the eroded surface of Mesozoic rocks in this part of Arizona. This is the Bidahochi Formation. Only a few remnants are left in the park, mainly as cap rock on mesas and divides in the Painted Desert. Overlying the Bidahochi sediments (and interbedded with them) are units of basaltic debris derived from volcanic vents northwest of the park. Large *volcanic bombs* (ellipsoidal fragments that were ejected in a molten state and then hardened in the air) can be seen in roadcuts in the northern part of the park. The bombs, as well as large angular boulders of basalt in the formation, show that the eruptions were nearby and highly explosive.

9. Pleistocene and Holocene erosion.

During the waxing and waning of the great Pleistocene ice sheets to the north and northeast, regions outside glaciated areas such as Petrified Forest had intervals of *pluvial climate,* that is, periods of high precipitation. In general, the pluvials correspond to transitional or glacial periods in the glaciated regions. Erosion accelerated during the pluvial intervals, and probably more rain fell throughout the year instead of being largely concentrated in summer as at the present time. The nearly complete removal of the Bidahochi sediment and volcanic debris indicates that the intense erosion of Tertiary time continued through the Pleistocene but that erosion is somewhat slower today due to the increasing aridity of the region. By Quaternary time, or before, the Little Colorado River and its tributary, the Puerco, had cut down their valleys into the soft Triassic shales of the Chinle Formation.

10. Changes in the regimen of the Puerco River early in the fourteenth century.

In the arid lands of the Southwest, slight changes in weather patterns and annual precipitation that become climatic trends or persist for more than a few years have profound effects on plant, animal, and human life. During the long thirteenth-century droughts, prehistoric people survived by abandoning their fields on the mesa tops and tilling new areas on the flood plain terraces of the Puerco River. But when rainfall increased again in the early 1300s, the added precipitation came in the form of cloudbursts and heavy summer thunderstorms that caused the river to rise out of its banks, wash out terraces, and cover the precious soil of the flood plain with coarse gravels. Because of this change in the environment, the Indians moved away. The climate in the Petrified Forest region has tended to become more, rather than less, arid since that time.

Bibliography

Ash, S. R.; and May, D. D. 1969. *Petrified Forest, the story behind the scenery.* Petrified Forest National Park, Arizona: Petrified Forest Museum Association.

Bezy, J. W.; and Trevena, A. S. 1975. *Guide to twenty geological features at Petrified Forest National Park.* Petrified Forest National Park, Arizona: Petrified Forest Museum Association.

Billingsley, G. H., and Breed, W. J. 1980. *Geologic cross section along Interstate 40 from Kingman to the New Mexico-Arizona state line.* Petrified Forest National Park, Arizona: Petrified Forest Museum Association.

Chronic, H. 1985. *Roadside geology of Arizona.* Missoula, Montana: Mountain Press. 320 p.

Gregory, H. E. 1917. *Geology of the Navajo country.* U.S. Geological Survey Professional Paper 93.

Mears, B., Jr. 1968. Piping. *In Encyclopedia of geomorphology,* ed. R. W. Fairbridge, p. 849. New York: Reinhold Book Corporation.

Nations, D. and Stump, E. 1981. *Geology of Arizona.* Dubuque, Iowa: Kendall/Hunt Publishing Company.

Rigby, J. K. 1977. *Southern Colorado Plateau,* K/H Geology Field Guide Series. Dubuque, Iowa: Kendall/Hunt Publishing Company.

Schopf, J. M. 1971. Notes on plant tissue preservation and mineralization in a Permian deposit of peat from Antarctica. *American Journal of Science* 271:522–543 (December).

———. 1975. Modes of fossil preservation. *Review of paleobotany and palynology* 20:27–53.

Wagoner, J. J. 1971. *This is Painted Desert.* Petrified Forest National Park, Arizona: Petrified Forest Museum Association.

Wilson, E. D. 1962. A resume of the geology of Arizona. Arizona Bureau of Mines Bulletin 171.

Petrified Forest National Park

Area: 93,493 acres; 146 square miles
Proclaimed a National Monument: December 8, 1906
Established as a National Park: December 9, 1962
Wilderness designated (50,260 acres): October 20, 1976
Address: Petrified Forest National Park, Arizona 86028

9

An intricately sculptured landscape created by rain, ground water, frost, and wind in soft, layered rocks, nearly barren of vegetation.
Badlands slopes, pinnacles, and buttes are eroding back into remnants of the shortgrass prairie on high tablelands. Ancient soil horizons and a rich variety of fossils, exhumed by erosion, depict a Tertiary time of verdant flood plains, grasslands, and forests, teeming with life. Also exposed by erosion are concretions, geodes, clastic dikes, fissures, and faults.

Badlands National Park

West Central South Dakota

Figure 9.1 Castle Butte, part of the Wall in Badlands National Park. A hard sandstone caprock slows the erosion of the pinnacles on the skyline. National Park Service photo.

Local History

The Badlands terrain is beautiful and desolate; fascinating, yet formidable. To the Dakota Indians the region was *Ma-koo-si-tcha*, or "bad land." The French trappers who had to cross it to get to their hunting grounds called the area *Les mauvaises terres a traverser*, meaning "a bad place to travel across." Early settlers avoided this land because it was poor for farming and a barrier to travel.

Yet, from this hostile terrain has come much of what we know about North America's "Golden Age of Mammals" in the Oligocene Epoch. The strange bones in the walls of the canyons and cliffs and along stream banks aroused the curiosity of nineteenth-century naturalists. As it turned out, this region was probably the richest storehouse of vertebrate fossils in all of North America. Among the first scientists were paleontologists Fielding B. Meek and Ferdinand V. Hayden, who were with an expedition of the Corps of Topographical Engineers sent out to explore the Upper Missouri River and the Dakota Badlands in the 1850s. Meek and Hayden brought back many specimens from the great caches of extinct mammal fossils so that these bones could be studied and identified by scholars. Dr. Joseph Leidy of Philadelphia, who found in the collection bones of tiny primitive horses, wrote a paper on the evolution of the horse, the first significant study of the subject.

Paleontologists came to the area in increasing numbers. The U.S. Geological Survey, the great natural history museums, and universities of America and Europe sent expeditions and established "digs." One can open practically any textbook on historical geology or vertebrate paleontology and find pictures of fauna from the Brule and Chadron Formations of the White River Group—famous names, these, in American geology. Probably the best collection of Badlands fossils is displayed in the Museum of Geology of South Dakota School of Mines and Technology in Rapid City.

Unfortunately, private collectors and amateurs—perhaps more enthusiastic than discriminating—began collecting in alarming numbers. Congress decided in 1929 that this record of ancient life forms, set in the most remarkable and extensive badlands of the world, should be preserved. Ten years later the national monument was proclaimed. After several additions to the protected area, primarily to conserve the shortgrass prairie, the South Dakota Badlands were redesignated a national park in 1978.

Today a 28-mile scenic highway with numerous overlooks winds through the park's North Unit. The best times to take this drive are early in the morning and late in the afternoon. At those times of day, the sun's rays slant into the deep gorges and gullies and light up hues of pink, yellow, green, maroon, cream, gray, and red on the spires and pinnacles. Buttresses and cliffs stand out in relief. After every rainstorm—the short, infrequent, heavy rains typical of a dry climate—the colors and shapes are different.

In the park's South Unit (which lies within the Pine Ridge Indian Reservation) is Stronghold Table, a large butte where "ghost dances" were performed by Oglala Sioux Indians late in the nineteenth century. They believed that the ceremony would bring back the buffalo and cause the white men to disappear from Indian lands. In December 1890, Chief Big Foot tried to elude the 7th Cavalry by leading his people through the Badlands toward the Pine Ridge Reservation. However, cavalry troops caught up with the Indians on December 28 and escorted them to Wounded Knee Creek (south of the park) where they camped for the night. In the morning, when the troops were searching and disarming the Indians, a rifle discharged. Indiscriminate shooting followed until more than 200 Sioux—men, women, and children—were dead. Twenty-five soldiers died and 39 were wounded, some of them hit by their own fire. In the bitter cold as a blizzard approached, the wounded soldiers and about 50 wounded Indians were taken in wagons to Pine Ridge. The bodies of the dead Sioux were later interred in a mass grave above the creek. This was the last major event of the Indian Wars.

Geologic Features

The Prehistoric Animals of the Badlands

During the Oligocene Epoch, between 38 and 24 million years ago, the region that is now the Badlands supported many kinds of animals. The climate was especially benign early in the Cenozoic Era when it was warmer and wetter than later in the era. Plants and animals flourished on the flood plains. The remains of successive generations of animals were covered over by relatively rapid sedimentation, and their bones and teeth were preserved in the rocks. Sometimes animals perished in seasonal floods and were quickly buried; thus whole skeletons became fossilized.

Some Oligocene species were the direct ancestors of modern horses and camels. Strangely enough, although most of the evolution of the camel and the horse took place in North America, both had become extinct on this continent before the arrival of the white men. What caused their disappearance is a mystery. Nevertheless, horses survived and multiplied on the western plains after the Spaniards reintroduced them to the New World. Clearly, the North American habitat was suitable for horses after the 1500s.

Some groups, the saber-toothed cats, for example, have died out completely. Ancient turtle shells, on the other hand, the most common fossil found in the Badlands, look just about the same as the shells of modern species. Other

Figure 9.2 Badlands National Park, South Dakota.

reptiles, such as alligators and lizards, lived in this lush, well watered land. Many groups of birds were present, including eagles, owls, gulls, and pelicans.

Most significant, however, were the mammals. They began to take over after the dinosaurs (the rulers of the Mesozoic) disappeared from the face of the earth. By the time the Oligocene Epoch began, due to the mammals' rapid expansion into habitats left by the dinosaurs, most of the modern mammalian orders had become established. Carnivores (flesh-eaters) diversified rapidly as their main source of food, the ungulates (hoofed mammals), proliferated in forests and marshes and on flood plains. Not only was this a time of great diversity among mammals; several species attained impressive size.

Largest of all mammals at that time were the *titanotheres,* an extinct order of *perissodactyls,* or odd-toed ungulates, that attained their peak in the Oligocene. *Brontotherium,* a giant titanothere, the remains of which are found in the sandstones of the Chadron Formation, stood about 12 feet high and had a pair of bony protuberances that grew hornlike above the snout. Like other ungulates, the titanotheres were herbivorous (eaters of foliage and grasses).

Most perissodactyl groups had three toes to begin with and tended to carry most of their weight on the middle toe, which evolved into the single hoof of modern horses. *Mesohippus,* an early horse that lived in the forests and ate foliage, still had three toes, but the middle one was large. *Pliohippus,* his successor and a grass-eater, had one great toe with two tiny ones on either side. The bones and teeth of the perissodactyl *Metamynodon,* an aquatic rhinoceros, have been collected from a channel sandstone in the Brule Formation. The rhinoceros, like the horse and camel, died out on this continent. A third group of odd-toed ungulates, the tapirs, were present in the Badlands. Their descendants migrated to South America during the Pleistocene and have survived to the present.

The even-toed ungulates, the *artiodactyls,* were more successful mammals than the perissodactyls in terms of abundance, variety, and survival. Most of the Oligocene artiodactyls had four toes and carried their weight on the third and fourth digits. Bones of camels (*Tylopoda*) at varying evolutionary stages have been found in the South Dakota Badlands, ranging from a four-toed animal the size of a jack rabbit to a species similar to modern camels.

Many bones of piglike animals, called *entelodonts,* are preserved in the Badlands. These animals, which became extinct in the Oligocene, were not the ancestors of domestic pigs.

Primitive flesh-eaters, called *creodonts,* are classified as "archaic carnivores." They were forerunners of more efficient hunters, the true carnivores. *Hyaenodon,* about the size of a small bear, was an Oligocene, carrion-eating

creodont. In the evolutionary development of carnivores, some became doglike and others became catlike; but they retained features in common for some time after they differentiated from their shared ancestor. For example, several of the Badlands dogs had curved, hooded, catlike claws rather than the curved cylindrical cones that are more typical of dog claws. Most of the dogs were about the size of foxes, but some were as large as black bears.

Impressive members of the Oligocene cat family in the Badlands were the large saber-toothed cats. Two species have been found in the White River beds, both belonging to an extinct group called the "stabbing cats." They had the ability to swing their lower jaw out of the way (as some snakes do) and stab their prey with their large, curved, daggerlike upper fangs. These awesome cats preyed on large, slow-moving mammals.

Insectivores are represented in the fossil record by the golden mole, hedgehogs, and shrews. Other rodent groups were ancestors of modern beavers, rabbits, squirrels, and rats. Squirrels and rabbits were the most common rodent ancestors in the Badlands.

Most of the modern ruminants (cud-chewers), the most diverse and abundant of all mammal orders today, evolved in Eurasia and Africa. However, a large group of primitive ruminants, unrelated to modern species, did thrive in the Oligocene. Fossilized bones of these animals, called *oreodonts* (fig. 9.3), are found only in North America. The cud-chewing oreodonts, about the size of foxes, roamed the plains in large herds. Many of their skeletons are found in huddled groups as if they were overcome by a natural disaster such as a blizzard. They became extinct well before the beginning of the Ice Age.

The Animals' Environment

The Great Plains are part of a vast midcontinent area overlying the North American craton. (A *craton* is a portion of the earth's crust that has been stable for a very long time.) Extending eastward from the Rocky Mountains to the Mississippi valley, the Great Plains supported grasslands from southern Canada nearly to Mexico before the settlers came.

Although they remained generally flat-lying, the sedimentary rocks of the Great Plains were strongly affected by the intense tectonic activities that took place to the west at the end of the Mesozoic. During the Laramide orogeny, the regional tilt of the Great Plains was increased from west to east as the land rose gradually.

In an early Tertiary uplift, the dome of the Black Hills bulged upward, along what is now the South Dakota-Wyoming border, only about 70 miles west of the Badlands. As erosion removed Cretaceous marine shales and sandstones from the Black Hills, and even part of the core of Precambrian crystalline rocks, rivers spread these great

Figure 9.3   Articulated oreodont skeleton displayed in the Visitors Center, Badlands National Park. Oreodonts, about the size of foxes, are extinct Oligocene mammals, unrelated to modern species. National Park Service photo.

quantities of sediment over the Badlands in Oligocene time. Late in the epoch, clouds of volcanic ash, borne by winds from eruptions far to the west, were a main source of sediment. All these materials became consolidated into the conglomerates, sandstones, mudstones, and associated *paleosols* (ancient buried soil horizons) that formed the tablelands and badlands of the Dakota Great Plains.

Interpreting Paleosols in the Badlands

Periods of uplift alternated with quieter times and resulted in alternating intervals of stream erosion and deposition. During periods of stability, well developed soils formed over wide areas. The earlier soils, which formed under warm, moist conditions, resemble soils that are presently found in southeastern United States. Although the vegetation is poorly known, it was probably a dense, canopied forest of warm-temperate and subtropical angiosperm trees (tree species that bear seeds enclosed in a seed case).

Strongly renewed erosive activity by streams early in Oligocene time, resulting from an episode of upwarp and tilting to the west, deepened the valleys temporarily in the Badlands region. However, as the valleys filled with sediment, the landscape became flatter. The paleosols of this time exhibit calcium carbonate horizons and other features that indicate a progressively drier, cooler climate. Forests became increasingly confined to flood plains. The uplands changed first to open savannas, and finally, in latest

Oligocene time, to semiarid steppes. (A *savanna* is a subtropical grassland having scattered trees and a pronounced dry season; a *steppe* is a broad, flat, treeless, dry region of scattered bushes and short-lived grasses.)

Recent careful study of Tertiary sedimentary layers in the Badlands has revealed that the entire sequence contains abundant evidence of paleosols (Rettalack 1983). At least 87 different paleosols were examined, and each one provides evidence of soil-forming factors, such as climate, organisms, parent material, and topography that prevailed at the time. Characteristic features—including fossil root traces, fossil burrows, mammal bones, mollusk shells, typical soil structures, color banding, and calcareous layers—help to identify the fossil layers and interpret their environment of formation. Many of the reddish bands that are seen throughout the Badlands represent paleosols.

How the Badlands Developed

The *badlands topography* of Badlands National Park is characterized by complex, intricately dissected steep slopes and sharp ridges. Fine drainage networks are developed on soft, erodible, flat-lying sedimentary rocks that lack a vegetation cover. Local relief is about 100 to 400 feet. Conditions favorable for the sculpturing of the Badlands landscape came about as follows: 1) erosive activity became more intense when renewed uplift occurred late in Tertiary time; 2) the climate became semiarid; 3) precipitation occurs mainly as infrequent heavy rains or

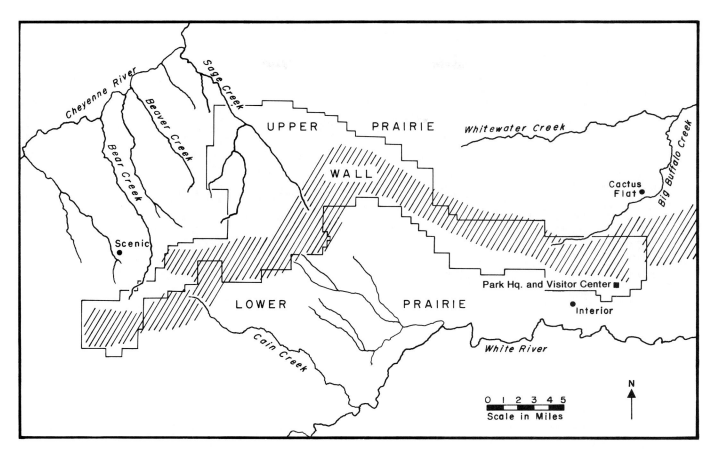

Figure 9.4   Sketch map of the Wall, an erosional escarpment that acts as a drainage divide in Badlands National Park. (Adapted from R. R. Churchill 1979).

cloudbursts; 4) vegetation did not become well established even during moister episodes in Pleistocene time; and 5) the high bentonite clay content in the sedimentary rocks causes rapid surficial erosion when they are exposed to moisture.

Bentonite is a weak sedimentary rock, containing a high proportion of clay minerals (belonging to the montmorillonite group), which are derived from weathered volcanic ash. Because the clay minerals swell in contact with moisture, bentonite can absorb large quantities of water and may increase its volume about eight times. When bentonite dries out after a soaking by rain or melting snow, it "slakes," which means that the surface material crumbles and disintegrates so that fine sediment is either blown away or washed away by the next rain.

The South Dakota Badlands are being eroded into the Wall, a south-facing, retreating escarpment, 60 miles long and 3 to 5 miles wide (figs. 9.4, 9.5). From the top of the Wall, sloping gently to the north is the Upper Prairie, a nearly flat grassland, slightly incised by north-flowing tributaries of the Bad River. In some places, the Wall juts up above the Upper Prairie and is being eroded from both sides. The deeply dissected badlands terrain of the Wall

is drained by tributaries that flow southward into the White River. The top of the Wall is, therefore, a drainage divide. The Cheyenne River, west of the park, also receives some Badlands drainage. The water of all three rivers eventually discharges into the Upper Missouri.

Below the Wall is an abrupt transition to the Lower Prairie, a partly grassy pediment area that slopes down to the flood plain of the White River. Between the top of the Wall and the distinct boundary where the Lower Prairie begins, relief averages about 150 feet but rises to more than 200 feet in some places. Incised, intermittent stream channels and "sod tables" (erosional remnants) break up the surface of the Lower Prairie. Park Headquarters is on the Lower Prairie, which is about 3 miles in width. The White River flood plain, below the Lower Prairie, is about half a mile wide.

Once much closer to the White River, the Wall is steadily eroding northward as the river widens its flood plain and the escarpment is dissected into badlands topography. The White River has cut all the way down through the Cenozoic rock units into the Mesozoic formations below. It is this downcutting and escarpment retreat that has exposed the fossils and paleosols.

Figure 9.5   Looking northward at the Wall, which is cutting back into the Upper Prairie. The Lower Prairie grassland is in the foreground. National Park Service photograph.

Erosional Features of the Badlands

Erosion rates in the Badlands are among the highest known. When photographs of Badlands landforms taken early in the century are compared with recent photographs of the same features, it is apparent that marked changes in form and height have taken place. During World War II, planes from a U.S. Army Air Force base near Rapid City sometimes discharged surplus ammunition in the Badlands on the way back to base after target runs. About thirty years ago a park ranger found a projectile in the ground, its top flush with the surface on a sloping hillside. Today the projectile sticks up more than a foot above the ground.

The rate of erosion varies with location and rock type. Spires in loosely consolidated ash may lose half a foot of height per year. Tops of mudstone mounds may be lowered about one inch per year. Resistant sandstone caprock, on the other hand, may show an erosion rate of an inch in 500 years.

The dominant geologic agent of weathering and erosion in the Badlands is running water. During cloudbursts pelting raindrops loosen grains and particles, and running water washes loose material downslope. Miniature gullies, eroding headward, enlarge rapidly. When infiltration occurs, the slope material may become saturated and move downward by mass wasting processes, slowly as creep or faster by sliding and slumping. Sometimes an entire slope face moves at one time. Cliff Shelf, a large slump area near Cedar Pass, continues to slip downward, especially during wet seasons.

The lithology of the rocks strongly influences erosional processes. The high clay content and fine-grained texture of the shale, siltstone, and ash of the Brule rocks limits infiltration of rain water, making runoff the dominant process. Thus many spires and pinnacles have formed. In the Chadron rocks, which have more sandstone and weaker cementation, a higher rate of infiltration makes mass wasting processes and downwastage more significant.

Frost action, which can be severe in the South Dakota climate, speeds the breaking up of the rocks. Repeated wetting and drying also causes cracks to form, with each tiny fissure providing more access for water.

Topographic position is a factor in slope development of tables, or buttes. The north-facing slopes, being protected to some extent from the sun's radiation, retain moisture longer than south-facing slopes. Water has more time to react with calcium carbonate cement that holds the rock together, so that chemical weathering penetrates farther into the rock. Thus creep or slump of weathered material is more likely to occur on north-facing slopes; and when rain comes, the loose, weathered material is easily washed downslope. North-facing slopes tend to have more risers or slope segments and are usually not so steep as south-facing slopes.

South-facing slopes, on the other hand, tend to have high, steep slopes and angular dropoffs. Weathering on south-facing slopes is extremely limited because they dry very quickly when storms end and the sun comes out. Once dried out, any loosened fragments or blocks fall rapidly,

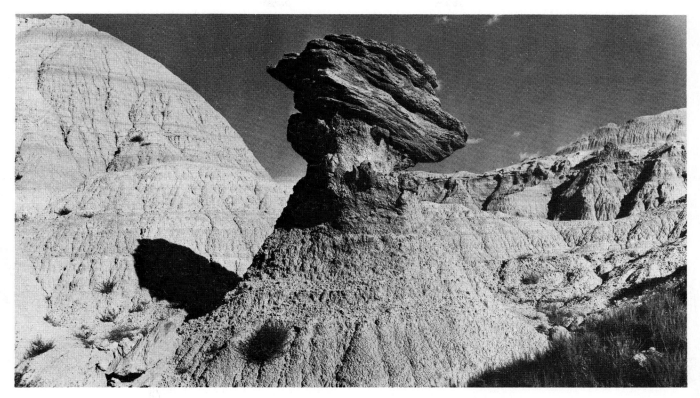

Figure 9.6   Resistant sandstone caprock forms a "mushroom" after most of the surrounding mudstones and siltstones have eroded away. In time, as undermining continues, the caprock will tumble down the slope. National Park Service photo.

collecting as debris at the bottom of the slope face (Churchill 1979).

Some of the strange-looking erosional effects in the Badlands are caused by layers or patches of resistant caprock. The rounded tops of the "toadstools" (rock pedestals or hoodoos) found here and there in the park are made up of channel sandstone boulders that have broken off and dropped down during slumping. The clay pedestals supporting the sandstone caps erode faster and become thinner and thinner until eventually the structure topples over (fig. 9.6). Arches have formed where weakly consolidated beds have washed out faster than overlying beds that are more firmly consolidated. These arches have a short life and cave in after a few years. Meanwhile, new ones are carved out by the never-ending processes of erosion.

A ground water process called *piping* (a term from engineering geology) speeds mass wasting in the Badlands, particularly in the weakly cemented sandstones of the Chadron Formation. Percolating ground water, infiltrating soil and rock and dissolving cement in sandstone, initiates the formation of little tunnels, or pipes, in sediment and rock. These conduits enlarge as rivulets drain into them or as storm waters rush into them and scour out tunnels. Eventually a valley slope, an arch, or a slump body becomes undermined by the pipes and may collapse suddenly during or right after a hard rain. Landslide masses that have temporarily dammed tributary valleys may give way or wash out after being undermined by water flowing in pipes or tunnels beneath the surface.

Clastic dikes stand out in relief against adjacent rocks that have been eroded back. These features were formed by inwash of clastic sediment (i.e., sand, silt) that filled cracks and fissures in the sedimentary rock. Chalcedony (silica) precipitated by circulating ground water cemented the sandy sediment into hard, resistant rock. Little veinlets of chalcedony also filled joints and fractures in the rock. Once the less resistant enclosing rock has been removed by erosion, the dikes and veinlets are left as thin, resistant ridges, walls, and "shingles." Some of them are a foot or more wide and others only an inch or two wide (fig. 9.7).

Why the Water Is "White"

When streams are running in the Badlands, they are always loaded with sediment. However, even when streams slow down or water stands so that material in suspension drops, the water remains cloudy and light-colored. This is because submicroscopic particles of clay are in *colloidal dispersion;* that is, the particles carry a slight charge of electricity that causes them to repel each other and prevents them from settling to the bottom. The White River was given its name by early visitors to the region because of its cloudy water.

Figure 9.7   This resistant clastic dike, jutting above softer rock, was formed by volcanic ash that washed into a fissure or fracture in sedimentary rock and then hardened. Vertical clastic dikes crisscross many areas of the Badlands. National Park Service photo.

Pioneers in the Badlands region discovered that pouring a pail of "alkali water" (taken from a waterhole in shale) into a barrel of white water would cause it to settle so that it could be used. When a railroad was built through the Badlands in 1907, providing drinking water for the teams of horses used in construction work was a serious problem until someone discovered that throwing slices of cactus pads into a horse trough or waterhole would clear the water enough so that the animals could drink it. Obtaining a supply of clear water in the Badlands region can be a problem even today; however, the National Park Service has drilled shallow wells in gravel deposits near the White River that furnish enough clear water for park facilities.

Faults

A series of linear normal faults (associated with a structural trough) crosses the Badlands area. The amount of displacement, a few feet in most cases, shows up in the alternating bands of light- and dark-colored sediment that can be seen on either side of a fault. The fault at Dillon Pass, however, shows that movement there was as much as 49 feet. The faults are related to movement in the Precambrian-Paleozoic basement complex far below the sedimentary rocks of the Badlands.

Concretions and Geodes

Like the fossils, these curious objects are to be seen but *not collected* within the park. The calcareous concretions, found in all the formations of the Badlands, are more nu-

Figure 9.8   The White River, which drains the South Dakota Badlands, has a milky look due to its high content of clay particles in suspension and colloidal dispersion that make the water unpalatable. National Park Service photo.

Box 9.1
How Are Geodes Formed?

Geodes are like snowflakes; no two are exactly alike. They may be large, small, hollow, or nearly solid. Some, called "rattle rocks," contain loose crystals; some are filled with connate water (water trapped in sediment). In the Badlands, geodes have even been found clustered together *inside* a calcareous concretion or—even more surprising—encased in the long bone of a mammal fossil.

Geode occurrences are not very common worldwide. They are found most typically in limestone, but occasionally in marine shales. The origin of geodes is not clear. No one theory seems to account satisfactorily for all varieties of geodes and their different sedimentary environments. The puzzle involves getting what was inside *out* and putting what is now inside *in*.

A widely accepted theory that appears to fit many geodes suggests that they began with the burial of a shelled animal or other object in the limy ooze of an ocean bottom. The shell or bone provided the original cavity; then silica gel accumulated in the void, surrounding and trapping a small amount of sea water. Later, after the sedimentary unit had been uplifted, ground water began circulating through the rock. Osmotic pressure between the trapped sea water inside the silica gel and fresh water outside caused the membrane of silica to expand and eventually to dry out and become the chalcedony layer. Minute cracks in the chalcedony permitted mineral ions in circulating ground water to enter the geode and gradually crystallize out of solution. Layers of clay particles may have become cemented to the exterior because they could not pass through.

Obviously, some modifications in this theory must be made in order to account for the small Badlands geodes. The only Badlands marine sequence is the Cretaceous Pierre Shale. All the overlying beds formed in a terrestrial environment. Most of the rock layers are shales, siltstones, and sandstones; there is no marine limestone, and there are only a few lenses of freshwater limestones here and there. Yet geodes are found in all formations (but not in every layer).

But what was happening during and after lithification of the sediment to make such a varied assortment of both calcareous concretions and geodes in the Badlands formations? Ground water must have been plentiful and it was probably close to the surface (high water table). But did it just percolate downward, or was it also forced upward by hydrostatic pressure that brought mineral ions from rocks farther down? Silica in soluble form must have been readily available over a long period of time. Was the highly siliceous volcanic ash that settled on and washed into the region the source? Why are Badlands geodes sometimes associated with calcareous concretions and sometimes not? Rockhounds, who regard geodes as fascinating natural objects, can doubtless add more questions!

merous and more varied in size, shape, and type than those ordinarily found in sedimentary beds. The nodular zones of the Brule Formation have the most concretions. These zones, which are probably flood-plain deposits, show up in the sedimentary sequence as red beds with a lumpy appearance. The nodules (concretions) that have weathered out tend to be rusty brown due to oxidation. Many of the concretions contain a fossil fragment as a nucleus. Around the nucleus are tightly cemented layers of clay particles.

Like the calcareous concretions, geodes are present in all of the park's formations and are most numerous in the nodular zones of the Brule.* The nodules that weather out are apt not to show by their outward appearance whether they are calcareous concretions or geodes. But geode nod-

ules that have been broken can be identified as such by small crystals in hollow interiors. By definition, *geodes* are hollow (or partly hollow), globular-shaped bodies that occur in sedimentary rock and that weather out as discrete nodules. (Geodes are a type of concretion.) A geode's outer layer is made of dense chalcedony, and crystals project from its lining into a central cavity. The crystals, most commonly quartz or calcite, are usually perfectly formed and are not the same minerals as those of the enclosing rock (box 9.1). Sometimes the chalcedony shell is further enclosed by coatings of cemented clay.

Unusual specimens of calcareous concretions and geodes, as well as fossils and minerals, are on display in the Museum of Geology at the South Dakota School of Mines and Technology in Rapid City and in the Badlands National Park Museum.

*Dr. Philip Bjork, Director, Museum of Geology, South Dakota School of Mines and Technology. Personal communication.

Table 9.1 Geologic Column, Badlands National Park

Time Units			Rock Units			Geologic Events
Era	Period	Epoch	Group	Formation	Member	
Cenozoic	Quaternary	Holocene / Pleisto-cene	Alluvium Landslide deposits Windblown sand Sod tables River gravels Older alluvium			Deposits in streambeds Slumped masses of shale, silt, and ash Stabilized dunes Erosion of badland topography during last 500,000 years Channel fills Pediment deposits
						Erosion ————————————
	Tertiary	Oligocene	White	Sharps (Arikaree)	Rockyford Ash	Volcanic ash, silt, pebbles, concretions, sands, poorly consolidated beds
				Brule	Poleslide Scenic	Arid climate, sheet flows, mudflows, flash floods; ashfalls, concretions
				Forming of red Interior paleosols		
			River	Chadron	Peanut Peak	Interfluve deposition; sandstone, clay, conglomerate
					Crazy Johnson	Climate changing from humid to semiarid,
					Ahearn	Flood-plain deposition; channel fill
						Erosion and Uplift ————————————
Mesozoic	—?—┴—?— Upper Cretaceous			Yellow Mounds paleosols formed by deep weathering in warm, wet climate		
				Pierre Shale		Black marine shale deposited in Pierre Sea

Sources: Raymond and King 1976, Retallack 1953, National Park Service 1988

Geologic History

1. Deposition of Pierre Shale, over 1200 feet in thickness.

The geologic record exposed in the White River Badlands began in Late Cretaceous time, between 80 and 65 million years ago, with the deposition of silts and muds in the shallow Pierre Sea, which stretched from the area of the Gulf of Mexico north into Canada. The skeletal remains of marine animals, including sea lizards, fish, and turtles, have been preserved as fossils in the shale beds. When the land rose, due to Laramide mountain-building in the west at the end of the Cretaceous Period, the sea retreated.

2. Development of Yellow Mounds paleosols.

Millions of years passed while sun, rain, and vegetation weathered the upper layers of the black Pierre Shale to a bright yellow and purple soil called the Yellow Mounds paleosol. Erosion has exposed these layers in the park at the Yellow Mounds Overlook and in the Sage Creek area.

3. Severe erosion and extensive deposition due to the doming of the Black Hills; Chadron Formation.

By the beginning of the Oligocene Epoch, uplift had brought the Black Hills to their maximum elevation. Rivers draining the mountains carved a rolling terrain of hills and valleys. In the Badlands region, where relief was not great, streams loaded with sand and gravel dumped the sediments as they slowed down. Layers of clastics covered the old weathered and eroded surface. These deposits became the Chadron Formation, the oldest formation of the White River Group.

Three members, Ahearn, Crazy Johnson, and Peanut Peak, make up the Chadron Formation. The Ahearn was deposited as channel fill in an ancient drainage network that antedated the drainage patterns of today. Because the river system's conditions changed frequently, cross-bedding and cut-and-fill deposits are common. The streams that deposited the Crazy Johnson and Peanut Peak Members* had lost some volume and erosive energy. Crazy Johnson Member is a mixture of mudstone and channel fill, its color ranging from green-gray conglomerate to gray-blue or green mudstone. Peanut Peak Member is similar to Crazy Johnson but thinner and lighter in color. It has more tan and orange mudstone and less of the green and gray. As rainfall decreased, vegetation changed, bringing about changes in faunal habitat. The titanotheres were approaching the culmination of their development, and many of their bones are found in the Chadron Formation, along with bones of other mammal orders. Chadron rocks tend to weather to low, rounded "haystack hills."

4. Interior paleosols.

Transitional between the Yellow Mounds paleosols and the Oligocene formations are the red, weathered clays of the Interior soils, named for the nearby town of Interior. These red paleosol layers, which overlie the Yellow Mounds, also developed on flood-plain deposits and channel fills in the Chadron and the Brule Formations. The paleosols represent stable times during which soil horizons formed.

5. Deposition of the Brule Formation.

A thin, discontinuous nonmarine limestone marks the boundary between the Chadron and Brule Formations, suggesting the existence of shallow ponds at that time. Overlying this limestone sheet is the Lower Nodular Zone that contains, in addition to concretions, many fossils of turtles and oreodonts (fig. 9.3). Although the overall trend during Brule time was toward a cooler and more arid climate, cycles of warm-wet climate with widespread flooding did occur. The flooding laid down sheets of thick, gooey mud that preserved fossil bones as it dried. Above the Lower Nodular Zone is the Upper Nodular Zone. Since both of these are continuous across the Badlands, as well as distinctive in appearance, they serve as excellent markers (fig. 9.9).

The Scenic and the Poleslide are the two members of the Brule Formation. The upper member, the Poleslide, is thicker. The Scenic Member, which includes the nodular zones, consists of interbedded clays, silts, sand, and volcanic ash, as well as channel fillings. In the Poleslide Member, increasing amounts of volcanic ash are found, some of it fluvially reworked and mixed with silts, fine sands, and clays. The Poleslide channel-fill deposits are called "Protoceras sandstones" because they contain bones from this extinct, distant relative of the camel (fig. 9.9).

*The names of these Chadron members refer to a Badlands settler, dubbed "Crazy Johnson" by his neighbors, because he tried to grow peanuts on a large, flat-topped sod table, still called "Peanut Peak."

Figure 9.9 Composite diagram of Oligocene rock units in the South Dakota Badlands. (Adapted from H. F. Osborn 1909.)

The Brule rocks, which make up the major part of the Wall, tend to form steep cliffs, knife-edge ridges, and pinnacles. On Brule slopes are many steps, overhangs, and flutings that reflect various responses to weathering processes. Some Brule beds are made up of loose clay; others of firm clay. Some are finer, some coarser. These differences suggest rapidly changing conditions of deposition and consolidation that result in variations in slope as the rock weathers and erodes. The distinctive, reddish, horizontal banding in the Brule Formation is due to the presence of beds having a high concentration of oxidized iron. Lighter bands contain less iron not so heavily oxidized.

6. Deposition of the Sharps Formation at the close of Oligocene time.

The highest and youngest rocks in the Badlands are those of the Sharps (Arikaree) Formation, which contains fluvial deposits, as well as large amounts of volcanic airborne and water-laid ash. The Rockyford Ash Member in the Sharps is a layer of nearly pure, stream-deposited, white volcanic ash. Capping the Rockyford Ash are resistant sandstone beds that form some of the ridge tops in the Badlands.

7. Development of badlands topography.

A long quiet span with little erosion or deposition followed the eventful years of the Oligocene. Then about half a million years ago, during Pleistocene time, erosion of the formations of the White River Group began. Large deposits of river gravels are associated with the general disruption of drainage that went on during the severe climates of Pleistocene time. The Badlands were not covered by continental glaciers, but were close to glacial margins. In post-Pleistocene time, high winds created sand dunes, most of which later stabilized. Even during times of greater moisture than the present, vegetation never became established well enough to control the erosion of the poorly consolidated beds of ash and clastics. As long as these conditions continue, the Badlands will continue to erode.

Animals that have adapted to the Great Plains environment thrive in the park, as did ancient species who lived in the region millions of years ago. American bison (buffalo) and bighorn sheep have been successfully reintroduced on the open plains and slopes. Prairie dogs busily dig their "towns" on the prairie. Pronghorns, mule deer, and many species of small animals find their favorite habitats in this beautiful and fascinating national park.

Geologic Map and Cross Section

Raymond, W. H., and King, R. U. 1976. Geologic map of the Badlands National Monument and vicinity, west-central South Dakota. U.S. Geological Survey, Miscellaneous Investigations Series Map I-934.

Bibliography

Churchill, R. R. 1979. Topoclimate as a controlling factor in Badlands hillslope development. Ph.D. thesis, University of Iowa.

———. 1979a. The importance of mass movement and piping in Badlands slope development. In *Proceedings, Iowa Academy of Science* 86(1):10–14.

Clark, J., Beerbower, J. R., and Kietzke, K. K. 1967. Oligocene sedimentation, stratigraphy, paleoecology, and paleoclimatology in the Big Badlands of South Dakota. *Fieldiana, Geology Memoirs*, vol. 5. Chicago Museum of Natural History.

Goetzmann, W. H. 1982. Explorer, mountain man, and scientist, in *Exploring the American West, 1803–1879*, Handbook 116, National Park Service Division of Publications. p. 81.

Graham, R. W., Semken, H. A. Jr., and Graham, M. A. (editors). 1987. *Late Quaternary mammalian biogeography and environments of the Great Plains and prairies*. Illinois State Museum Scientific Papers, v XXII, 491 p.

Harksen, J. C., and Macdonald, J. R. 1969. Type sections for the Chadron and Brule Formations of the White River Oligocene in the Big Badlands, South Dakota. South Dakota Geological Survey, Report of Investigation 99.

Hauk, J. K. 1969. *Badlands, its life and landscape*. Bulletin no. 2. Interior, South Dakota: Badlands Natural History Association.

Kurten, Bjorn. 1972. *The age of mammals*. New York: Columbia University Press.

Madson, J. 1981. South Dakota's castles in clay. *National Geographic* 159(4):524–539, April.

National Park Service. 1988. Geologic history, Badlands National Park. Information sheet.

Osborn, H. F. 1909. *Cenozoic mammal horizons of western North America*. U.S. Geological Survey Bulletin No. 361.

Retallack, G. J. 1983. Late Eocene and Oligocene paleosols from Badlands National Park, South Dakota. Special Paper 193, Geological Society of America.

Scheidegger, A. E., Schumm, S. A., and Fairbridge, R. A. 1968. Badlands. *The encyclopedia of geomorphology,* ed. R. A. Fairbridge. New York: Reinhold Book Corporation.

Schumm, S. A. 1956. The role of creep and rainwash on badland slopes. *American journal of science* 254: 693–706.

Smith, K. G. 1958. Erosional processes and landforms in Badlands National Monument, South Dakota. *Geological Society of America Bulletin* 69: 975–1008.

Wanless, H. R. 1923. The stratigraphy of the White River Badlands, South Dakota. *American Philosophical Society* 62:190–269.

Badlands National Park

Area: 243,302 acres; 380 square miles
Authorized by Congress: March 4, 1929
Proclaimed a National Monument: January 25, 1939
Wilderness designated: October 20, 1976
Redesignated a National Park: November 10, 1978
Address: P.O. Box 6, Interior, South Dakota 57750

10

Scenic badlands along the Little Missouri River.

Huge slump blocks on the sides of the valley.

Flat-lying sedimentary rocks; interbedded coal seams; burned-out lignite beds; "scoria."

Quantities of petrified wood; "cannonball" concretions.

Continental glacier boundary.

Theodore Roosevelt National Park

North Dakota

by Rodney M. Feldmann
Kent State University

Figure 10.1   Badlands along the Little Missouri River in Theodore Roosevelt National Park. Photo by R. Feldmann.

Local History

In 1883, when he was 24 years old, Theodore Roosevelt traveled west on the newly constructed Northern Pacific Railroad. This was his first visit to Dakota Territory. His purpose was to hunt buffalo, but he was so impressed with the badlands along the Little Missouri River that he purchased the Maltese Cross Ranch, located six miles south of Medora, and began a 15-year career as a rancher. He expanded his enterprise by developing the Elkhorn Ranch site in the badlands along the river 35 miles north of Medora. (These badlands are some 300 miles north of Badlands National Park, South Dakota, chapter 9.)

Although the badlands were referred to as *les mauvaises terres à traverser* ("the badlands to cross"), they possessed the characteristics necessary for sustaining successful ranching operations in the northern Great Plains. In an article written by Roosevelt and illustrated by Frederic Remington for *Century* magazine in February 1888, Roosevelt observed that the success of a ranching venture in the northern Great Plains was dependent upon the essential elements of food, water, and shelter. During the hot summer months, food and water were the two essentials and, because the Little Missouri River was one of the few perennial streams in the area, proximity to it assured a continuous, abundant water supply for cattle. Additionally, the badlands region of the Little Missouri River supported adequate vegetation to permit grazing. As Roosevelt pointed out, during the long harsh winters, the most important element for survival of cattle was shelter. Availability of water was less essential because cattle could derive adequate moisture from the snow cover. The numerous canyons and gulleys of the region, as well as the cottonwood thickets along the Little Missouri River, provided the animals with excellent protection against the winter winds that severely limited ranching enterprises in the bleak, exposed country of the Missouri Plateau.

Thus, the badlands were a paradox; they were both help and hindrance to the development of the West. Wagon travel across the area was extremely difficult. Wagons had to be disassembled and hand-carried across the more rugged parts of the terrain. On the other hand, that very ruggedness provided shelter that permitted development of a ranching industry in a climate that was otherwise too severe to permit anything but summer grazing.

Even at its best, the badlands did not support a large number of animals. By the mid-1880s, the number of cattle in the area had increased to the point where overgrazing of the range land resulted. The winter of 1886–1887 was extremely severe and dealt an almost killing blow to the ranching industry along the Little Missouri. The ranchers did survive, however, and Roosevelt maintained his ranching interests in the area until 1898.

Before the ranchers brought cattle into the northern Great Plains, herds of buffalo (bison) had thrived on the grazing lands for thousands of years. Hardier than cattle, the buffalo were able to survive periods of drought and the most severe winters. The buffalo were a mainstay of the Indians, who used the meat for food and hides for clothing and shelter. Hostilities between Indians and ranchers were the result of Indian resistance to the loss of their hunting lands and the slaughtering of the buffalo herds. Organized resistance by the Indians had ended only a few years before Theodore Roosevelt came to the region.

Upon being elected President of the United States, Theodore Roosevelt initiated and supported numerous policies to assure conservation of lands in the National Park System. His interest in protecting these areas was based in large part upon the influence of his years as a cattleman in the badlands. Therefore, it is fitting that on April 25, 1947, two large separate tracts of land (North Unit and South Unit) along the Little Missouri River were established as the Theodore Roosevelt National Memorial Park. A part of Elkhorn Ranch, between the two units, is also under park jurisdiction. In 1978, the area was redesignated Theodore Roosevelt National Park.

Geologic Features

The topographic expression of any area is a function of structure, stratigraphy, and climate. In the badlands region of Theodore Roosevelt National Park, the rocks were elevated after deposition and burial but have not been substantially deformed tectonically. Therefore, with the exception of local mass wasting features, the beds are essentially horizontal (fig. 10.1). As stream erosion dissects the area, the drainage pattern tends to be dendritic, and no obvious "grain" related to structural deformation has developed. Much of the park's land lies within the deeply dissected valley walls of the Little Missouri River, which meanders across the southwestern corner of North Dakota, and minor ephemeral tributaries that feed it. (*Ephemeral streams* flow only after heavy rains or snow melt.) The valley is bounded by an upland surface, the Missouri Plateau, which represents a stripped structural surface.

Much of the interest in this particular badlands region lies in the variety of exposed sediments. The sedimentary types include sandstone, siltstone, shale, *lignite* (brown coal intermediate in degree of coalification between peat and subbituminous coal), and occasionally limestone deposited in a rhythmic sequence during the Paleocene Epoch (table 10.1). These sediments are just a part of a thick wedge of fluvial, lacustrine, and flood-plain deposits formed during the uplift and erosion of the Rocky Mountains in Cretaceous and early Cenozoic time. Faunal and floral evidence suggests that the climate then was warm and humid so that lush vegetation covered the land. Sluggish streams meandered across the region, depositing a variety of sediments now referred to as the Fort Union

Figure 10.2 Theodore Roosevelt National Park, North Dakota.

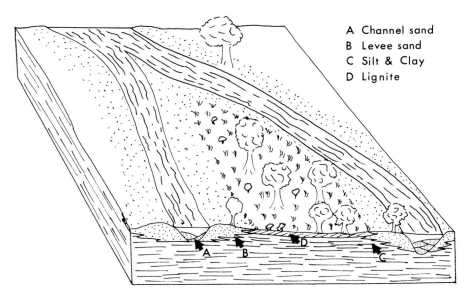

A Channel sand
B Levee sand
C Silt & Clay
D Lignite

Figure 10.3   Schematic representation of environments in which the more common Fort Union sediments were deposited. (Modified from Jacob 1976.)

Group. The streams deposited sand as channel fill and in *natural levees,* or embankments, paralleling the channels (fig. 10.3). During episodes of flooding, finer-grained material, silt and clay, accumulated as overbank deposits. Low-lying regions adjacent to the river valley were swampy, and in these areas organic material was buried, compacted, and eventually became lignite. Thin beds of nonmarine limestone accumulated in lakes. Finally, occasional bursts of volcanic activity in the mountain ranges to the west yielded sufficient volcanic ash to deposit layers several inches thick over the region. These volcanic ashes have either been mixed with fine-grained flood-plain sediments or remain as discrete layers. In either case, weathering of the material has yielded bentonitic clays with a lithology distinct from other units in the Fort Union.

Perhaps the most distinctive rock unit in the Theodore Roosevelt National Park, as contrasted to the sediments of Badlands National Park, South Dakota, is lignite. The badlands in South Dakota do not contain lignitic beds and, therefore, the general appearance of the terrain is quite different. The lignite in Theodore Roosevelt National Park is prominently exposed throughout the area and, additionally, has produced another interesting rock type, locally called "scoria." (The term *scoria* usually refers to vesicular rocks of volcanic origin only.) The North Dakota scoria is baked clay with a pottery-like consistency that has resulted from the *in situ* burning of lignite. The incinerated bright red to black clinker beds are diagnostic of the Fort Union Group (fig. 10.4).

The variation in lithology seen in the Fort Union Group is expressed as sediments with a wide range of colors and weathering characteristics so that the terrain is studded with erosional remnants capped by resistant rock

Figure 10.4   Small knolls capped by "scoria," baked clay resulting from burning lignite. National Park Service photo.

units in a myriad of forms and sizes (figs. 10.5, 10.6). The variation in lithology also produces an interesting distribution of vegetation types. In the present climate, most of the vegetation is typical of a semiarid region and dominated by short prairie grasses and sagebrush. In areas where particularly permeable rocks are exposed, however, stands of cedar are common. On lowland surfaces adjacent to the Little Missouri River, where ground water and surface water are plentiful, thick groves of cottonwood are the dominant vegetation.

The regional climate is best described as severe. Annual temperatures fluctuate over a range of 150° F; and

Figure 10.5 Erosional remnants capped by resistant sandstone. National Park Service photo.

Figure 10.6 Sandy concretions protect pedestals from being eroded away. National Park Service photo.

rainfall, which averages about 15 inches per year, typically occurs in short, violent bursts. This climatic setting results in a relatively sparse vegetal cover. The absence of dense vegetation, coupled with periodic heavy rainfalls, makes an ideal setting for development of the high drainage density of badlands topography (fig. 10.7).

Petrified wood, which is common throughout the park, is mostly preserved as kindling-sized and smaller pieces that weather out from slopes or accumulate in low spots. However, a few nearly complete stumps (fig.10.8) and logs of petrified wood can be found in place. Most of the petrified wood occurs in the silty and clayey sediments representative of the flood-plain environment during the time of deposition of the Fort Union Group.

Concretions of various sizes also weather out from the beds of the Fort Union Group. Some of them are remarkably large, especially in North Unit. Near Squaw Creek Campground alongside the Scenic Drive, large, round concretions called "cannonballs" have accumulated at the base of the bluffs.

The occurrence of lignite in the Fort Union Group has yielded another interesting combination of geological events. The lignite beds exposed throughout the region are subject to spontaneous combustion. If beds begin to burn, they not only bake the overlying sediments, but also affect the topography of the area. Coal seams tend to burn from the area of exposure back into the hillside, and this results in collapse of the overlying material and development of a hummocky terrain. Burning continues until such time as the lignite has burned so far back into the hillside that cracking and slumping do not provide access to oxygen for further combustion. A major coal seam that smoldered and burned until 1978 is in the area of Buck Hill in the South Unit of the park. The effect on the topography is still evident. Areas in which scoria has been produced lack the lignite beds that should occupy that stratigraphic position.

Although the area is structurally simple, that is, the beds typically dip at angles of less than 1°, there are several places where much greater dips, on the order of 45°,

Figure 10.7 Panoramic view of Painted Canyon, illustrating the high density drainage pattern. The Missouri Plateau is the flat surface on the horizon. Photo by R. Feldmann.

can be observed. Many of these tilted blocks are tens of meters in length and width. Invariably, however, they occur along slopes, and investigation of the rock on the slopes suggests that the dips are of local significance and the result of mass wasting processes. Huge slump blocks (fig. 10.9) have moved downslope and rotated under the influence of gravity.

Geologic History

1. Pre-Paleocene deposition of sediments in the Williston Basin.

Throughout the Paleozoic Era and most of the Mesozoic, the area comprising western North Dakota, eastern Montana, and southern Saskatchewan and Manitoba was one of continued, slow subsidence and accumulation of mostly marine sediments. Although none of the Paleozoic or Mesozoic rock sequence is exposed in the vicinity of Theodore Roosevelt National Park, the sequence is of interest. The Williston Basin is a significant oil-producing region, and the area surrounding the park is dotted with pumps and drill rigs extracting petroleum from the Paleozoic rock sequence. Until near the close of the Mesozoic Era, the entire area was either slightly below sea level or about at sea level.

2. Uplift of the Rocky Mountains and the end of marine conditions.

Beginning late in the Cretaceous Period and continuing into Paleogene time, the Laramide orogeny resulted in the development of the Rocky Mountain chain. One of the effects of the uplifting of the Rocky Mountains, hundreds of miles west of the park, was that erosion of these newly formed mountains produced hundreds of cubic miles of sediment that was carried eastward from the mountains and deposited in great clastic wedges. This event is recorded in a sequence of terrestrial sediments ranging in age from Late Cretaceous through Oligocene time in many parts of the Dakotas and Montana. The Fort Union Group represents the Paleocene element of that clastic wedge. During the time of deposition of the Fort Union Group in the badlands area, the last vestige of the midcontinental seaway in North America was rapidly being filled east of the park in central North Dakota. These rocks, comprising the Cannonball Formation, are well exposed along the Missouri River and some of its tributaries.

Scattered remnants of post-Paleocene sediments are found throughout western North Dakota in isolated remnants capping butte tops. These units, the Golden Valley Formation and the White River Formation, span Eocene and Oligocene time.

Figure 10.8   Petrified stump preserved in a standing position. National Park Service photo.

Figure 10.9   Large slump block of Sentinel Butte material in North Unit. National Park Service photo.

Table 10.1 Geologic Column, Theodore Roosevelt National Park

Era	Period	Epoch	Group	Formation	Marker Bed	Geologic Events
			Time Units		**Rock Units**	**Geologic Events**
Cenozoic	Quaternary	Holocene	Alluvial deposits, landslide features, slump features, "scoria" formation			Development of present badlands topography
		Pleistocene	Cut-and-fill structures and terraces along Little Missouri River and tributaries, glacial erratics, slump features			Glaciation reroutes Little Missouri River and Missouri River; possible early Pleistocene glaciation
			/////////////////////////////			
	Tertiary	Paleogene Paleocene	Fort Union Group	Sentinel Butte Formation	Upper sand	Grayish to brownish siltstone, clay, and sandstone deposited in habitats similar to Tongue River sediments; lignites less common; "Blue" bed represents weathered volcanic ash marker.
					Bullion Butte lignite	
					Upper "yellow" bed	
					Lower "yellow" bed	
					"Blue" bed	
					Basal sand	
				Tongue River Formation	HT lignite	Yellowish, pastel siltstone, clay, and sandstone deposited in fluvial, lacustrine, and overbank environments with numerous lignitic beds formed in swampy, flood plain setting.
					Meyer lignite	
					Garner Creek lignite	
					Harmon lignite	
					Hanson lignite	
					H lignite	
					Basal sandstone bed	

Modified from Royse 1970

3. Uplift and erosion of sediments following the Oligocene.

No rock units of any extent in western North Dakota are younger than the Oligocene White River beds. The area had apparently been uplifted and was in the process of being dissected by rivers ancestral to the modern Little Missouri River system. The only evidence of deposition subsequent to Oligocene time is in the form of gravel and sand accumulations that are difficult to date.

4. Pleistocene glaciation and development of the modern drainage system.

Pleistocene continental glaciers advanced southward from Manitoba and Saskatchewan at least as far as the North Unit of Theodore Roosevelt National Park. Evidence of glaciation may be seen at "Edge of Glacier" pullout in the North Unit.

In preglacial time, the Little Missouri River flowed north toward Hudson Bay; but with the onset of glaciation, north-flowing streams were blocked. The drainage of the Little Missouri, along with that of the Yellowstone River to the west, was diverted eastward and southward along the edge of the ice front. This ice-marginal drainage eventually developed into today's Missouri River, to which the Little Missouri is tributary. The rate of erosion and downcutting of the Little Missouri River was not constant, and as erosion and deposition continued, the stream cut a series of terraces, remnants of which can still be seen in several places in the park.

Geologic Map

Clayton, L., Moran, S. R.; Bluemle, J. P.; and Carlson, C. G. 1980. Geologic map of North Dakota: U.S. Geological Survey.

Bibliography

Bluemle, J. P., and Jacob, A. F. 1981. Auto tour guide along the South Loop Road. North Dakota Geological Survey Educational Series 4.

Chronic, H. 1984. *Pages of stone, geology of western national parks and monuments, v.1, Rocky Mountains and Western Great Plains.* Seattle, Washington: The Mountaineers, p. 128–136.

Clayton, L.; Carlson, C. G.; Moore, W. L.; Groenewold, G.; Holland, F. D., Jr.; and Moran, S. R. 1977. The Slope (Paleocene) and Bullion Creek (Paleocene) Formations of North Dakota. North Dakota Geological Survey Report of Investigation 59.

——— and Freers, T. F. 1967. *Glacial geology of the Missouri Plateau and adjacent areas* (guidebook). North Dakota Geological Survey Miscellaneous series 30.

Eliot, J. L. 1982. Roosevelt Country: T. R.'s Wilderness Legacy. *National Geographic* 162(3):340–363, September.

Holtzman, R. C. 1978. Late Paleocene mammals of the Tongue River Formation, western North Dakota. North Dakota Geological Survey Report of Investigation 65.

Jacob, A. F. 1976. Geology of the upper part of the Fort Union Group (Paleocene), Williston Basin, with reference to uranium. North Dakota Geological Survey Report of Investigation 58.

Laird, W. M. 1950. *The geology of the South Unit Theodore Roosevelt National Memorial Park.* North Dakota Geological Survey Bulletin 25.

———. 1956. *Geology of the North Unit Theodore Roosevelt National Memorial Park.* North Dakota Geological Survey Bulletin 32.

McCullough, D. 1981. *Mornings on Horseback.* New York: Simon & Schuster.

Royse, C. F., Jr. 1967. Tongue River-Sentinel Butte Contact. North Dakota Geological Survey Report of Investigation 45.

———. 1970. A sedimentologic analysis of the Tongue River-Sentinel Butte interval (Paleocene) of the Williston Basin, western North Dakota. *Sedimentary Geology* 4:19–80.

Schoch, H. A. 1974. *Theodore Roosevelt; the story behind the scenery.* Las Vegas, Nevada: KC Publications.

Ting, F. T. C., ed. 1972. Depositional environments of the lignite-bearing strata in western North Dakota. North Dakota Geological Survey Miscellaneous Series 50.

Theodore Roosevelt National Park

Size: 70,345 acres; 110 square miles

Established: April 25, 1947; redesignated: November 10, 1978

Address: Box 7, Medora, North Dakota 58645

11

Located entirely north of the Arctic Circle, the park includes geological, biological, and cultural resources, such as Great Kobuk and Little Kobuk Sand Dunes; permafrost, glacial, and fluvial features; border zone between boreal forests and tundra; prehistoric migration route for humans and animals; Onion Portage Archeological District; Salmon Wild River.

Kobuk Valley
National Park

Northwestern Alaska

by Sherwood D. Tuttle
University of Iowa

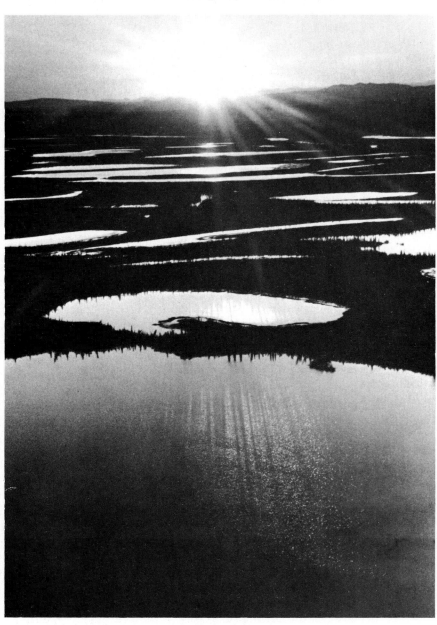

Figure 11.1 A waning Arctic sun sheds its rays over the watery flood plain of the Kobuk River in Kobuk Valley National Park. National Park Service photo by R. Belous.

Geographic Setting and History

Kobuk Valley National Park lies in a mountain-rimmed basin, 270 miles south of the Arctic Ocean. Although this large wilderness park—larger than the state of Delaware—has no roads or facilities, the Kobuk River has been a major transportation route for humans and animals for many thousands of years. For native hunters and fishermen the Kobuk River has been an east-west highway between the ocean and the Alaskan interior ever since people came to this continent. Every autumn caribou herds ford the Kobuk on their way south to the shelter of the subarctic woodlands; every spring they return north to the tundra to graze. The arctic tundra is a treeless, gently undulating, marshy plain that is covered with mosses and low shrubs and underlain by mucky soil and permafrost.

The size of Kobuk National Park, 2672 square miles, reflects the philosophy of the planners to set aside a hydrologic basin and self-sustaining ecosystem within park boundaries. In arctic latitudes, biological support systems are spread thin over large areas. It takes more than one hundred square miles to sustain a single brown or grizzly bear and many times that size habitat for a caribou herd. Most of the national parks established earlier in the century were set aside from leftover lands and had to be fitted in between many kinds of economic and cultural uses and demands. In most of Alaska, broad strokes could be used in delineating land for national parks and preserves to be withdrawn from public sale and mineral leasing. Kobuk National Park, the Salmon Wild River within the park's boundaries, the adjoining Noatak National Preserve on the north, and the Selawik National Wildlife Refuge abutting on the south are all part of the "d-2" or Alaska National Interest Lands that were withdrawn by Act of Congress from "all forms of appropriation" and classified as wilderness. Although the original act that permitted unreserved public lands to be set aside was passed in 1971, the actual units were not created in their present form until 1980 with the passage of the Alaska Lands Bill. In the end a total of 131 million acres of land in Alaska was affected.

The Kobuk River, which rises in the central Brooks Range, flows south from the Arctic Divide and then nearly due west to the Chukchi Sea. (In its upper reaches, within Gates of the Arctic National Park and Preserve, the Kobuk has been designated a part of the Wild and Scenic Rivers System.) The northern and southern boundaries of Kobuk National Park follow drainage divides: along the crest of the Baird Mountains (3500 to 5000 feet high) on the north; and on the south side, along the Waring Mountains that rise to elevations of 1000 to 2000 feet. The larger tributaries, including the Salmon Wild River, enter the Kobuk River from the north side. The tributaries draining slopes on the south side are small creeks. All the streams in the park run clear with low sediment loads because permafrost keeps soil from eroding; and, at the present time, the streams do not carry glacial meltwater. Two small Eskimo villages outside the park are the only settlements—Kiana, 10 miles downriver from the western boundary, and Ambler, a few miles upstream beyond the eastern boundary.

Whalers and traders visited the lower Kobuk River in the eighteenth century, but a party in small boats from the U.S. Revenue Service in 1884 were probably the first Americans to explore the river's upper reaches. In 1903, a U.S. Geological Survey geologist, W.C. Mendenhall, made a reconnaisance of the Kobuk River and its tributaries by canoe. By 1910 geologists were using pack strings in the area, searching for mineral localities. (The Nome gold rush had begun around the turn of the century, and prospectors were looking for "pay dirt" along the Arctic rivers and tributaries.) In the 1950s Arthur Fernald of the U.S. Geological Survey mapped the surficial geology using aerial photographs, float planes, and canoes. Today outboard motors and snowmobiles are the principal means of transportation in the region.

Although the Kobuk Valley has a long and fascinating cultural history, the importance of its prehistoric record has only recently been recognized. When archeologists began to investigate the Kobuk Valley, looking for traces of "Early Man" on this continent, they were astonished to find at Onion Portage, now listed in the National Register of Historic Places, a record of human occupation longer than any other that has been discovered in North America. At this one site (that has become the standard against which all other Alaskan sites are measured), some 30 layers of human habitation document changes of tools and other adaptations indicating eight cultures. Dr. Louis J. Giddings of Brown University began to excavate Onion Portage and other sites in the Kobuk Valley in the 1960s. His investigations and those that followed show that humans have lived in the valley and used its resources for at least 12,500 years (fig 11.7).

The significance of Onion Portage (named for a wild chive that flourishes there) is that it is a caribou crossing, which has made it a magnet for hunters ever since the earliest people of the Paleo-arctic culture discovered it. They were nomadic hunters and gatherers from the Asian mainland who made their way across Beringia, a land mass (now covered by the Arctic Ocean) that connected northeast Siberia and northwest Alaska. These Stone Age people left weapons and tools in the Kobuk Valley. Fashioned of chert and flint, they are very similar to artifacts found in Siberia and Japan. Some groups of immigrants stayed on in the north and adapted to arctic and subarctic conditions.

KOBUK VALLEY NATIONAL PARK

Figure 11.2   Kobuk Valley National Park, Alaska.

In that earlier time, the Kobuk Valley was an ice-free corridor although glaciers still covered large areas of North America. As the Ice Age mammals drifted southward, hunting clans followed. One route may have been an ice-free corridor that presumably opened along the eastern side of the Canadian Rockies. Gradually the hunters and the hunted spread over North America. Several species that they hunted, such as mastodons, mammoths, large-horned bison, and horses, later became extinct on this continent.

A gap in the cultural record (around 6000 B.C.) was followed by a wave of migrants. They were Northern Archaic people, who came about 6000 years ago from forested lands to the south and east. Fish were apparently their main food.

Asian people of the Arctic Small Tool tradition, who were well adapted to arctic conditions, arrived in the Kobuk Valley around 2000 B.C. They were part of the last great migration from Asia that gradually spread eastward as far as Greenland. Their habit was to make seasonal journeys to coastal regions from inland locations, and they never lost their dependence on coastal and marine resources. Around the beginning of this millenium, the Kobuk Valley was left unoccupied for a time. Perhaps the caribou population declined during this period.

By about A.D. 1200 people of the Arctic Woodland culture were living in the Kobuk Valley. They were probably coastal Eskimos who adapted to inland living, and their subsistence activities, such as salmon fishing, caribou hunting, and berry gathering, continued into historic times.

Figure 11.3   Small drainage basin, part of a larger hydrologic system drained by a trunk stream. Interfluves (ridges) separate tributary streams; divides along crests define drainage basins. The drainage pattern is *dendritic*; that is, it resembles veins in a leaf. This kind of drainage pattern develops on a land surface that is fairly uniformly erodible. In Kobuk Valley National Park, where drainage patterns are predominantly dendritic, the bedrock in the highlands on both sides of the valley is largely metamorphic and not severely fractured.

The management plan for Kobuk Valley National Park provides that hunting, fishing, and other traditional activities of the native people will take precedence over tourism. No accommodations are available in the park and none are planned, although campsites exist for primitive camping.

Climatic Conditions

With mean annual precipitation only 16 inches, the Kobuk Valley is semiarid. Most of the rain falls during the short, cool, moist summer seasons. Winters are long, cold, and dry, with an average snowfall of 72 inches. The January average temperature is $-10°$ F, with $-61°$ F being the coldest value on record. With a mean annual temperature of 22° F, the *permafrost* (permanently frozen ground) has little chance of melting, even though the average reading for July is 57° F. Runoff is high in summer and the valley is awash because the permafrost keeps water from soaking into the soil.

The Kobuk Valley's east-west orientation makes it an open funnel for the dry polar easterlies that blow from September to May. Return winds blowing in from the ocean bring the summer rains to the valley. The polar easterlies are strong and blow for long periods without letup because of intensified atmospheric circulation and steep pressure gradients between ice-covered and ice-free regions. Conditions such as these prevailed over a large part of North America during the Ice Age. That is why the climate of the Kobuk Valley is regarded as *relict*, or a remnant, of Pleistocene climates.

The park is in a transition zone between boreal forests that can exist in the northern climate and the treeless arctic tundra. On the south flanks of the Baird Mountains, trees finger up the tributary valleys, thin out, and then stop. Where the tree line ends, the tundra begins. Over the last 10,000 to 15,000 years the tree line has shifted north and south as the climate has changed. Studies of sediment layers and pollens show that when the earliest people came, the climate was colder and drier and most of northwest Alaska was thinly vegetated. The Kobuk Valley was a sheltered, ice-free refuge for both animals and humans.

Geologic Features

The Flood Plain. The broad floor of the Kobuk River Valley is filled with fluvial landforms: sandbars, cutbanks, point bars, oxbows, and sloughs on the flood plain; terraces and fans against the slopes of the encircling hills (box 11.1). Active and stationary sand dunes and dunefields rising south of the flood plain are spread over many square miles and give evidence of the strong winds that sweep down through the valley. Below the surface, at depths of 1 to 5 feet lies the permafrost, the continuously frozen ground of the Arctic. Permafrost features, such as ice-wedge polygons and pingoes, roughen the flood-plain surface. Both are the result of summer freezing and thawing that makes the ground expand and contract. Water fills cracks in the permafrost and refreezes forming *ice wedges*. Intersecting ice wedges form *polygons*. Some are bordered by ridges, and some by troughs. *Pingoes* are conical frost mounds pushed up by ice.

The base of the permafrost has been located at a depth of 390 feet below the flood-plain surface and at more than 600 feet in the Waring Mountains. (The arctic permafrost probably froze before the beginning of the Pleistocene Epoch.) Because water cannot sink into the impermeable permafrost, summer meltwater collects on the surface in pools, marshes, and thaw lakes, making the ground wet and soggy. Conditions are ideal for the annual spring invasion of mosquitoes and black flies that swarm in dense clouds about people and animals.

Box 11.1
Landforms in River Valleys

We can see unusual and unique natural features in the valley of the Kobuk River, but we can also find riverine features that are characteristic of any well developed river system. Onion Portage, for example, is located on a meander loop where canoeists paddling upstream can save five or six miles by portaging across the neck of the meander.

The Kobuk River has built up the relatively flat floor of its *flood plain* season after season by *overbank flow.* Meltwater roars down from the mountains in the spring carrying a fresh load of sediment. The river boils up over its banks and spreads out over the valley floor. Branching channels, called *sloughs,* cut across the flood plain, distributing water and sediment. Then the flood waters slow down, become sluggish, and drop sand and mud. Sloughs become stagnant during the low-flow seasons. *Sandbars* accumulate in the river channel near a riverbank in places where the current slows down enough to drop some of its sediment load. Sandbars build up to water level or to just below the water surface. Oftentimes they are washed away or relocated by the next flood.

Because of their natural hydrologic characteristics and helical (screwlike) flow patterns, rivers tend to create sweeping curves, bends, and loops as they swing from side to side over a flood plain. These sinuous windings are called *meanders.* A stream erodes flood-plain alluvium on the outside of a meander, forming a *cutbank,* and then deposits sand and silt on a *point bar* on the inside of a meander curve. The layers of sediment exposed in an eroded part of a flood plain, such as a cutbank or a bluff, sometimes record a river's history. Geologists have been able to interpret thousands of years of climatic changes in Kobuk Valley by studying deposits exposed in Epiguruk Bluff, upstream from Onion Portage.

The universal habit streams have of eroding on the outside of a meander curve and depositing on the inside causes the channel to migrate laterally on the flood plain. During a flood, lateral movement may amount to many feet. Sometimes meander loops are brought close together in such a way that the narrow neck of one of them may be cut through—or the river may flow over a neck and develop a new channel. When this happens, the meander is cut off and the entrance to the loop fills in with sediment that separates the water in the loop from the river. This is how *oxbow lakes* form. (The name describes the **U** shape.) Oxbow lakes tend to get filled in by sediment from overbank flow and vegetation and eventually become oxbow-shaped swamps, or *meander scars,* on the flood plain.

Natural levees, which are parallel to the riverbanks, are low ridges of sand several feet high but usually many feet wide. They are deposited by flood waters that are slowing down. The size and width of natural levees depend on the river's volume and velocity and the amount of sediment it carries.

A pronounced change in the regimen of a stream, such as an increase in precipitation or a decrease in sediment load, may cause a river to incise its channel into the soft material of its flood plain and then begin to build a new flood plain at the lower level. The remnants of the old flood plain, stretching away toward the valley side slopes, become *alluvial terraces.* Apparently this has happened more than once in the Kobuk Valley. At present, the sediment load of the Kobuk River system is comparatively low, mainly because permafrost stabilizes the soil.

Alluvial fans are low cones or fan-shaped mounds of sediment that accumulate where a tributary coming down from a steeper side valley reaches the flood plain and abruptly slows and spreads out. The sediment load that is dropped at the edge of the flood plain builds up an alluvial fan.

Figure 11.4 Simplified cross section of the Kobuk River Valley. The alluvial terrace on the left, topped by loess deposits and dunes, consists of many layers of older flood-plain deposits. Cross-bedded and horizontal deposits of alluvial material (mainly sand) underlie the modern flood plain on the right. High bluffs beside the present channel expose layers of muck, paleosols, and sand (many beds deformed by solifluction), along with wood fragments, artifacts, bones, tusks, and the like. (Modified after Hamilton, Ashley, Reed, and Van Elten 1982.)

The Sand Dunes

As a geologic agent, wind is generally regarded as having only minor influence in the shaping of landscapes; but here in the Kobuk Valley the wind's erosional and depositional activities are spectacular. Ideal conditions exist: a plentiful supply of sand to be picked up and moved, winds of intensity and high velocity, and a place for the sand to be deposited.

At least 500 square miles in Kobuk Valley National Park are covered with dune sand. The greater amount of this sand is now stable; some is partly stable but occasionally active; and about 30 to 40 square miles of the valley are occupied by active dunes as barren of vegetation as a dunefield in a desert. The two major areas of active dunes, both on the downwind or south side of the river, are Great Kobuk Sand Dunes and Little Kobuk Sand Dunes. The Great Kobuk Sand Dunes have an elevation of about 100 feet near the river and are about 500 feet high where they bump up against the base of the Waring Mountains. NASA scientists have compared these dunes and their characteristics to areas on the surface of Mars that are cold and dry and have strong winds.

The abundance of sand comes from reworked glacial drift and alluvium in sandbars, terraces, and in the dunes themselves. The wind picks up loose grains, a process called *deflation,* and rolls or bounces them along by *traction* and *saltation.* This is similar to the way sediment grains are transported by stream action. *Blowout* and *parabolic dunes* are erosional forms. Their centers are eaten out by the wind so that they become U-shaped with their arms pointing upwind. Dunes of this type are more common on the upwind sides of the dunefields. Toward the center the dune shapes become irregular. In the extreme downwind section, farthest from the source area, the youngest, most active dunes are accumulating. These are the *barchans.* They are half-moon shaped, in map view, with arms pointing downwind. Sand is swept up over the top or car-

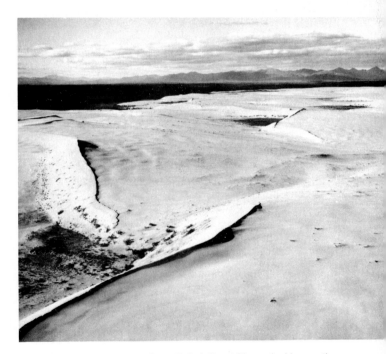

Figure 11.5 View of the Great Kobuk Sand Dunes, looking north toward the Baird Mountains. The dunes migrate over the top of an alluvial terrace above the river. National Park Service photo by R. Belous.

ried along the base, thus lengthening the arms as the middle of the dune moves along. Where there is room to migrate, some dune forms move several miles a year. Internally the dunes show extensive cross-bedding.

The grains of dune sand are well rounded, "frosted" (opaque), and non-uniform in size. These characteristics indicate long-continued abrasion during numerous cycles of erosion and deposition. Much of the work of abrasion was done by glacial scouring of bedrock in the surrounding highlands. The sand fragments consist of resistant quartz, feldspar, and minor amounts of other minerals.

Figure 11.6   Great Kobuk Sand Dunes are encroaching on the boreal forest (foreground). National Park Service photo by R. Belous.

Bedrock

Except for rocky slopes that rim the upwind side and a few knolls of rock that jut up through the dunes on the downwind side, the bedrock of Kobuk Valley is deeply buried. Outcrops are found in the foothills and slopes of the Baird and Waring Mountains within the park's boundaries. Some reconnaissance work on the hard-rock geology has been done, but the area has not been studied in detail. In the Baird Mountains, Paleozoic sedimentary rocks and older metamorphosed rocks (phyllites, schists) crop out, as well as Tertiary conglomerates, cross-bedded sandstones, and shales with thin coal beds. To the south in the Waring Mountains there are Cretaceous conglomerates, sandstones, arkoses, and thick beds of shale. In the Jade Mountains, just outside the eastern park boundary, there are deposits of nephrite jade of gem quality. During the winter the native people move jade boulders by sleds from the talus slopes to the riverbank. In spring, after the ice breaks up, the jade blocks are barged to Kotzebue on the coast, where native artisans make jewelry and jade carvings.

Geologic History

1. Paleozoic and Mesozoic sedimentation.

The marine sediments making up the older rocks in the park region may have accumulated in another part of the globe, along continental margins and continental shelves far to the south. The Kobuk area is a part of Alaska that probably drifted north for millions of years and eventually collided with and became attached to the North American continent. Associated mountain-building, accompanied by volcanic activity and emplacement of igneous rocks in plutons, deformed the rocks, and some were metamorphosed. The jade in the Jade Mountains, for example, is an indication of high-pressure metamorphism of igneous rocks, probably in Jurassic time. In chapters on Alaskan national parks in Part III, this tectonic history is explained more fully.

2. Tertiary sedimentation.

The Tertiary sedimentary rocks in the Baird Mountains accumulated on deltas, tidal flats, and along coastal margins. Interlayered beds of soft coal indicate proximity to sea level and a warm, moist climate. Mountain-building activities that deformed and elevated the rocks probably began early in Tertiary time.

3. Quaternary glaciation.

As global climate cooled late in Tertiary time, glaciers formed on the Brooks Range and grew until a large icecap covered the entire region. The permafrost has been frozen for several million years, probably since late in Tertiary time. Sea level lowered as more and more water became locked up in glacial ice. This episode, named Kobuk glaciation, took place during the Illinoian stage of the Pleistocene Epoch around 200,000 years ago. The deposits correlate with Illinoian glacial material at Cape Krusenstern National Monument about 150 miles to the west.

In Wisconsinan time (the most recent major glacial stage), alpine glaciers advanced and retreated several times. They moved down the valleys of the Baird Mountains and the Brooks Range but did not completely override the Kobuk Valley. Peaty material older than the Wisconsinan glacial outwash in the Kobuk Valley has been dated by radiometric methods as 38,000 to 33,000 years before the present (y.b.p.). More recently, in Holocene time, ice from the east that moved toward the park sent floods of meltwater and glacial drift downstream. Small glaciers still occupy cirques in mountains to the east and northeast of the park.

4. The development of the present landscape.

The extensive areas of stabilized dunes in the Kobuk Valley indicate that eolian activities in the Pleistocene were even more intense than those in recent times, remarkable

Table 11.1 Geologic Column, Kobuk Valley National Park

Time Units			Rock Units	Geologic Events
Era	Period	Epoch		
Cenozoic	Quaternary	Holocene	Alluvial deposits Dunes, loess Glacial drift	Reworking of glacial and fluvial sediments by wind and streams Continued dune building
		(Wisconsinan stage) Pleistocene (Illinoian stage)	Glacial drift, outwash Dune deposits Peat deposits Till and outwash	Several episodes of alpine glaciation in nearby mountains Intense wind erosion and deposition Kobuk glaciation; entire region blanketed by icecap
	Tertiary	(late) (early)	Conglomerates, soft shales and cross-bedded sandstones with interbedded coal	Freezing of permafrost Uplifting of Baird and Brooks Ranges Accumulation of sediments in coastal areas and on deltas
Mesozoic	Cretaceous – – –?– – – Jurassic(?)	(late)	Conglomerates, arkose, shales Jade (nephrite)	Continental shelf, near-shore, and shore deposition, marine and nonmarine High-pressure metamorphism
Paleozoic	(late)		Granitic rocks, volcanic rocks Marine sedimentary rocks	Tectonic activity Extensive deposition

Sources: Smith, 1938; Fernald, 1964; Wahrhaftig, 1965

Figure 11.7   Sketch map showing the location of Epiguruk Bluff and Onion Portage on the flood plain of the meandering Kobuk River. (Adapted from Hamilton, Ashley, Reed, and Van Elten 1982.)

as the presently active dunes are. Dune-building probably began after the retreat of Kobuk glaciation. Along with the dunes are smaller deposits of silt (loess) laid down by wind.

Epiguruk Bluff, an eroded river terrace a few miles upstream from Onion Portage, has revealed in its sediment layers an unusually complete and detailed record of changing environments and life forms in the Kobuk Valley throughout a long segment of Quaternary time. A paleosol (buried soil horizon) at the base of the bluff, which has been dated as 35,000 B.P., contains peat and muck deposited on a flood plain. A younger paleosol (24,000 B.P.), containing loess and alluvium, buried the older layer. Overlying the soils are deposits of cross-bedded channel sands, horizontal sand layers, wind-blown sands, and loess (wind-blown silt). Plant and animal fossils found in some of the deposits help to document climatic changes, as for example, a change from tundra vegetation to spruce vegetation. At times in the past, the river was higher and at times lower than now. Average annual temperatures have been somewhat warmer and at times much colder. Precipitation rates have likewise varied between wetter and drier ranges (fig. 11.7).

As the climate began to warm and the ice melted back around 14,000 years ago, large mammals (horned bison, woolly mammoths, mastodons, and the like) came across Beringia from Asia and gradually spread throughout the Americas. Their tusks and bones wash out of stream banks in the park from time to time. Bands of hunters followed the big game and populated the New World, adapting their cultures to the environmental conditions they encountered. Archeological sites at Onion Portage, along the Salmon Wild River, and elsewhere in Kobuk Valley National Park preserve a unique record of the migrations and cultures.

Bibliography

Belous, R. 1981. Parklands in northwest Alaska, in The Kotzebue Basin, *Alaska Geographic,* 8(3):1–16.

Beikman, H. M. 1980. *Geologic map of Alaska,* 1:2,500,000, U.S. Geological Survey.

Brown, W. E. 1982. *This last treasure, Alaska National Parklands.* Anchorage, Alaska: Alaska Natural History Association, p. 106–113.

Fernald, A. T. 1964. Surficial geology of the central Kobuk River Valley, northwestern Alaska, U.S.Geological Survey Bulletin 1181K, p. 1–31.

Galloway, J. P. and Koster, E. 1981. Comparison of grain size statistics from two northern Alaska dunefields. U.S. Geological Survey in Alaska, Accomplishments 1981. Circular 868, p. 20–21.

Hamilton, T. D., Ashley, G. M., Reed, K. M., and Van Etten, D. P. 1982. Stratigraphy and sedimentology of Epiguruk Bluff, a preliminary account. U.S. Geological Survey in Alaska, Accomplishments 1982. Circular 939, p. 12–14.

Hunter, C. and Wood, G. 1981. Alaska National Interest Lands, *Alaska Geographic* 8(4):8–16, 76–83.

Loney, R. A. and Himmelberg, E. R. 1984. Ophiolitic ultramafic rocks of the Jade Mountain-Cosmos Hills area, southwestern Brooks Range. U.S. Geological Survey in Alaska, Accomplishments 1984. Circular 967, p. 13–15.

McPhee, J. 1977. The encircled river, in *Coming into the country,* New York: Farrar, Straus, Giroux, p. 5–91.

Melham, T. 1984. Kobuk Valley, in *Alaska's Magnificent Parklands,* Washington, D.C.: National Geographic Society, p. 155–167.

Shaver, M. C. and Gilbert, C. 1985. Kobuk Valley National Park, in General management plan/ Environmental assessment, National Park Service, p. 22–54.

Smith, P. S. 1913. The Noatak-Kobuk region, Alaska, U.S. Geological Survey Bulletin 536.

———. 1939. Areal geology of Alaska, U.S. Geological Survey Professional Paper 192.

Wahrhaftig, C. 1965. Physiographic divisions of Alaska, U.S. Geological Survey Professional Paper 482, p. 27–28.

Kobuk Valley National Park

Location: Northwest Alaska
Area: 1,710,000 acres; 2,670 square miles
Proclaimed a National Monument: December 1, 1978
Established as a National Park: December 2, 1980
Address: P.O. Box 1029, Kotzebue, Alaska 99752

Caves and Reefs

© by G. Shimer

Figure II.1 Cave pearls, about the size of cherries, on a cave floor in Carlsbad Caverns National Park, New Mexico. Drawing by G. Shimer from *Field Guide to Landforms in the United States*, J. A. Shimer. The Macmillan Company © 1972 by J. A. Shimer. Used by permission.

"Soft rocks" (an informal term geologists use, meaning sedimentary rocks) are dominant in the seven national parks in Part II. In these parks, as in Part I, most of the stratified bedrock is relatively flat-lying or undeformed. A significant difference, however, is that in the national parks described in Part I, the carbonate rocks (limestones, dolomites, etc.) are less important and abundant than the clastic rocks, such as shales and sandstones. In the caves and reefs of the seven national parks included in Part II, carbonate rocks are overwhelmingly the dominant rock type.

Most natural *caves,* or *caverns,* and the features in them, are the result of erosion, solution, and deposition by ground water (the geologic agent) in carbonate rocks. *Reefs* are rigid, wave-resistant structures built in near-shore environments by marine organisms (corals, calcareous algae, and the like) and infilled by broken corals, shell debris, and sediment. Encrusting calcareous algae, foraminifera, and calcium carbonate (precipitated from seawater in voids and spaces) bind the reef structures together.

Cave and Reefs in the National Parks

Mammoth Cave, Kentucky; Wind Cave, South Dakota; and Carlsbad Cavern, New Mexico (chapters 12–14), are not the only caves protected in the National Park System, but they are without question the largest and most notable. At least two other national parks, Great Basin (chapter 41) and Sequoia (chapter 42) have caves with interesting and unique features, but they are located in mountain areas where rocks have undergone deformation, metamorphism, plutonism, and other activities associated with mountain-building. National monuments set aside to preserve caves are Jewel Cave (near Wind Cave National Park), South Dakota; Oregon Caves, Oregon; Timpanogos Cave in Utah, and Russell Cave in Alabama.

Ancient reefs and modern reefs are dominant features in the four remaining national parks in Part II. Guadalupe Mountains National Park in west Texas, which has ancient reefs, is just south of Carlsbad Caverns National Park, New Mexico, and the geology of these two national parks is closely related. In Virgin Islands National Park in the Caribbean, modern reefs are growing on and around "fossil" reefs. Buck Island Reef National Monument, also in the Virgin Islands, is noted for its underwater trail among "coral gardens." Everglades and Biscayne National Parks, both in Florida, display a great variety of living and fossil reef features, some underwater and some at or above present sea level. Carbonate rocks of various ages in other national parks (for example, Glacier National Park, Montana, chapter 23) preserve the remains of ancient reefs that help geologists reconstruct the geologic history of the past.

Figure II.2   A small tropical island ringed by an outer coral reef, or barrier reef, with an inner fringing reef rimming the shore. A lagoon lies between the outer and inner reefs.

Limestones and Dolomites

By definition, carbonate rocks are composed of more than 50% by weight of carbonate minerals, such as *calcite* ($CaCO_3$) and *dolomite* ($CaMg(CO_3)_2$). Also present in carbonate rocks are varying amounts of clastic material (clay, quartz grains, etc.) plus iron oxide, iron carbonate, and traces of other compounds. Geologists use the word "dolomite" for both the mineral dolomite and the sedimentary rock called *dolomite,* the latter being composed of more than 50% dolomite (the mineral) and most commonly in excess of 90%. Similarly, *limestones* contain more than 50% of the mineral calcite and usually more than 95%.

Limestones and dolomites are often found together in reefs and in caves. Sometimes they are interbedded or laterally gradational. Most dolomites were originally deposited as limestones; but magnesium ions replaced calcium ions during recrystallization and caused the limestone to be partially or completely "dolomitized."

Carbonate rocks occur in many different textures ranging from massive and very fine-grained to coarse and detrital (i.e., made up of rock fragments and organic debris). Limestones are of several types. Biohermal limestones are reefs or reeflike accumulations of organic origin;

Figure II.3   Park Service naturalist exploring passages in Wind Cave National Park, South Dakota. Photo by M. Little.

inorganic (or chemical) limestones are those formed by precipitation in place in seawater; and clastic limestones are accumulations of transported fragments of largely carbonate composition (fossil remains, bits of rock, etc.).

Most of the carbonate rock layers in the national parks in Part II have remained relatively undeformed since they were laid down. Some have been deeply buried, elevated, and subsequently uncovered and reburied as ancient oceans opened and closed and distant mountain ranges were pushed up and then eroded. Many of the fossils and sedimentary structures within these rocks are remarkably well preserved. As was the case in the first eleven national parks of Part I, the geologic structures in the areas described in Part II are, in general, not complicated. With a good eye and only a moderate amount of geologic knowledge, you can "read" or surmise much of the geologic history when you visit caves and reefs. The Virgin Islands (chapter 16) have been most affected by tectonic activities, but even here, the reefs themselves, since they are relatively young, have not been deformed.

The Role of Ground Water

Learning about caves (and visiting a cave, if possible) can help us to better understand and appreciate our *most important geologic resource;* namely, ground water, an indispensable component of Planet Earth's water budget. *Ground water* is the subterranean water that occupies the crevices, pores, and voids in rocks and sediments beneath the surface. More ground water is stored in the rocks of this continent than all the surface water held in rivers, reservoirs, and lakes—including the Great Lakes. Nevertheless, our vital supplies of ground water are presently seriously threatened from overuse (excessive drawdown) and from pollutants that seep underground and spread throughout ground-water systems. Problems are especially acute in regions underlain by carbonate rocks because their high permeability and fairly high porosity make them susceptible to pollution. National park rangers and interpreters try hard to alert park visitors to the need for protection of ground-water supplies and also help the public understand what environmental factors cause problems.

Surface water (from rain, snow, and runoff) is the source of virtually all ground water. Under the influence of gravity, water sinks into soil and percolates downward through cracks, joints, and along bedding planes of bedrock, passing through an unsaturated zone, called the *zone of aeration* or the *vadose zone.* When ground water has filled all the pores and open spaces (the depth is variable), it has reached the *saturated zone,* known also as the *phreatic zone.* An irregular surface separating these two zones is termed the *water table,* which rises in times of plentiful rain and falls during droughts. Sometimes uplift or subsidence may raise or lower the water table (and the base level) of a region over the passage of time.

Figure II.4   How a cave system may develop in carbonate rocks through solution by ground water. The water table (WT) has been higher and, at times, lower than is shown here. From *Landforms and Landscapes*, 3rd edition, S. D. Tuttle. © 1970, 1975, 1980 Wm. C. Brown Publishers. Used by permission.

Carbon dioxide in the atmosphere and from decaying organic matter in soils makes ground water slightly acidic and thus able to slowly dissolve carbonate rocks above and below the water table. Gradually openings form in fractures and along bedding planes that may eventually develop into caves. A general sequence of events is as follows: 1) As ground water moves through carbonate rocks, solution creates openings, tunnels, chambers, pits, or galleries. 2) If the water table drops, due to changes in climate or lowering of base level, air enters the part of the cave system that is in the vadose (unsaturated) zone. Because water that seeps into air-filled caves is very close to saturation with dissolved carbonate, the water loses carbon dioxide to the cave atmosphere and deposits much of its dissolved load as *travertine,* a dense, finely crystalline type of limestone that occurs in an infinite variety of shapes and cave "formations." *Speleothem* is an inclusive term for any secondary mineral deposit formed in caves.

Abrasion by underground stream action (where water acts under conditions of open channel flow), as well as de-position of alluvial fill, may take place in caves intermittently with solutional erosion.

Carbonate rocks in reefs are affected by circulating ground water when they become exposed to erosion by changing sea levels or tectonic activity. Ancient reefs in sedimentary sequences may become buried and subsequently undergo erosion and deposition by solution, as in Guadalupe Mountains and Carlsbad Caverns National Parks. Living and recently constructed reefs, such as those in the Virgin Islands and Florida, are also modified by ground-water processes.

National Parks in Part II
Chapter 12 Mammoth Cave National Park, Kentucky
Chapter 13 Wind Cave National Park, South Dakota
Chapter 14 Carlsbad Caverns National Park, New Mexico
Chapter 15 Guadalupe Mountains National Park, Texas
Chapter 16 Virgin Islands National Park, Virgin Islands
Chapter 17 Everglades National Park, Florida
Chapter 18 Biscayne National Park, Florida

12

Longest known cave in the world, with more than 300 miles of explored and mapped passageways.

Formed in nearly flat-lying Mississippian limestones and dolomites by circulation of ground water, fed by infiltration from the surface.

Great variety of passage shapes, including tubular and canyonlike passages and vertical shafts.

Numerous passage levels show relationship between cave development and the evolution of the landscape on the surface.

The karst topography of the region is the most extensive in the United States.

Cave enlargement continues today; the lowest levels still contain streams.

Mammoth Cave National Park

Western Kentucky

by Arthur N. Palmer
State University of New York, Oneonta

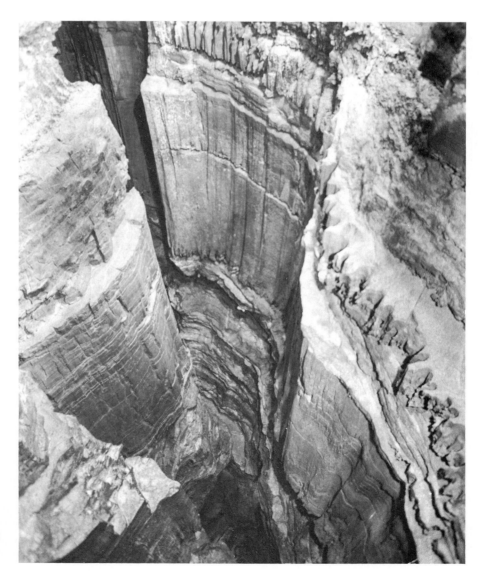

Figure 12.1 Looking down from the top of Mammoth Dome, as seen on the Historic Tour. The eddying and coring of descending water formed this immense vertical shaft. National Park Service photo.

MAMMOTH CAVE NATIONAL PARK

Figure 12.2   Map of surface features of Mammoth Cave National Park, Kentucky.

Regional Setting

Mammoth Cave National Park is located in west-central Kentucky, midway between the cities of Louisville, Kentucky, and Nashville, Tennessee, in the heart of America's most extensive limestone region. Solution of the limestone by underground water has created the longest known series of caves in the world. Mammoth Cave, the largest of these, contains more than 315 miles of surveyed passageways. About 12 miles of the passageways are open to the public.

The park is in a plateau of limestone bedrock capped by insoluble clastic rocks, mainly sandstones, at the southeastern edge of a regional downwarp, the Illinois Basin (a structural basin). The sedimentary rock layers in the plateau dip gently northwestward. Steep-walled valleys have breached the insoluble rocks, exposing 300 feet of cavernous limestone in irregular ridges. This rugged, hilly region is called the Chester Upland. To the southeast, the

insoluble rocks have been removed by erosion, and the limestone is exposed at the surface in an uninterrupted plain, covering thousands of square miles called the Pennyroyal Plateau. Together, the Chester Upland and the Pennyroyal Plateau comprise one of the world's finest examples of *karst topography,* a landscape dominated by solutional features such as caves, closed surface depressions, sinking streams, and large springs. (fig. 12.2).

Ranger-guided cave tours of different lengths and degrees of difficulty are available to visitors in Mammoth Cave National Park. Year-round tours include: (1) the Historic Tour emphasizing pre-Columbian archeology of the cave and saltpeter mining during the early nineteenth century. This tour also gives the clearest view of the limestone stratigraphy and the numerous levels at which the cave has formed. (2) The Half-Day Tour, a five-hour hike through more than four miles of passageways, best illustrates the variety of passages and their interrelationships,

MAMMOTH CAVE INTERIOR
NORTHERN SECTION

Source R. Schlecht and A. Palmer,
National Geographic Society, 1984.

A

MAMMOTH CAVE INTERIOR
SOUTHERN SECTION

B

Figure 12.3 Public areas of Mammoth Cave. A. Underground passages in the Northern section.
B. Underground passages in the Southern section. Adapted from *The Underground World of Mammoth Cave National Park*, R. Schlecht and A. N. Palmer © 1984 National Geographic Society. Used by permission.

as well as the most typical speleothems ("cave formations"). (3) The Frozen Niagara and (4) the Travertine Tours both visit the part of the cave decorated with the best examples of travertine speleothems. Seasonal tours, offered only in summer, include: (1) Echo River Tour, which covers the same route as the Historic Tour, but includes a boat ride on a lake in the lowest level of the cave; (2) Wild Cave Tour, a six-hour, strenuous trip mainly for spelunkers; (3) Great Onyx Cave Tour, a walk through a nearby formerly commercial cave. Hiking shoes and a light jacket or sweater are recommended. The cave temperature is about 55° to 57° F, corresponding to the mean annual temperature of the region.

Local History

Prehistoric Indians of the Woodland Culture, who lived in the area roughly 3000 to 4000 years ago, used the larger cave entrances for shelter and explored several miles of passages in their quest for minerals used as medicines and flint for making tools. For light they made torches of cane, reeds, and twigs, remnants of which have been found in many of the dry upper-level passages of the cave. Other archeological finds include bare foot prints in mud and dust, woven baskets, fiber slippers, and two Indian "mummies" whose desiccated bodies were preserved naturally in dry parts of the cave.

Mammoth Cave became known to white settlers in the years just prior to 1800. Under private ownership, some of its passages were mined for saltpeter (nitrates) to make gunpowder, particularly during the War of 1812. Well preserved remnants of the saltpeter works are shown on the Historic Tour.

The saltpeter operation closed down after the War of 1812, but in its place grew an increasing tourist trade, spurred by many published articles about the vast cave. One of the early guides, a black slave named Stephen Bishop, became America's first truly great cave explorer. He discovered sections of Mammoth Cave far beyond the previous limits of exploration, including Echo River and some of the passages shown today on the Half-Day Tour.

In 1908 the great size of the cave was verified for the first time when a young German visitor, an engineer named Max Kaemper, surveyed 45 miles of passageways. Although he used only a hand-held compass and measured distances by pacing, his map was remarkably accurate.

At that time the known parts of the cave did not extend beyond the flanks of the five-mile-long Mammoth Cave Ridge, one of many flat-topped limestone ridges in the Chester Upland capped by insoluble rocks and separated from neighboring ridges by dry karst valleys. In the late nineteenth and early twentieth century many new caves were discovered and explored in nearby ridges. Some of these were opened for tourists, and commercial rivalry developed among the various cave owners. Although

Mammoth Cave was still by far the most popular attraction of the region, its tourist trade suffered a blow when new entrances into the same cave were excavated beneath neighboring property and its southeastern end was opened to tours in competition with the "historic" northwestern end.

One of the residents of nearby Flint Ridge, to the northeast of Mammoth Cave Ridge, was an enthusiastic cave explorer named Floyd Collins. He discovered a large cave on his family's property, which he called Great Crystal Cave, and opened it for tourists. Of the many tourist caves in the area, his was one of the least accessible by road, so Floyd spent a great deal of time searching for a more conveniently located cave. In 1925, while exploring Sand Cave, a tight, unstable cave, he dislodged a loose rock that fell on his ankle, trapping him. Despite efforts by many people to free him, he died of exposure. Floyd Collins' fate received enormous publicity and brought even more visitors to the Mammoth Cave area.

Mammoth Cave was authorized as a national park in 1926, and by the time it was officially dedicated in 1941, the park boundaries encompassed many of the neighboring caves that had been open to the public earlier. Several of the former operators and explorers of Floyd Collins' Great Crystal Cave formed the Cave Research Foundation in 1957 that was dedicated to the exploration and scientific study of these and other caves within the park. By 1961 the group had discovered connections between all of the major caves in Flint Ridge, revealing a vast cave system that rivaled Mammoth Cave in length. In 1972 a connection was found between the Flint Ridge Cave and Mammoth Cave through a narrow, wet passageway that extended beneath the dry karst valley that separates the two ridges. This joined the two caves into a single enormous system with a combined length of more than twice that of any other known cave in the world. Continued explorations will almost certainly reveal connections with other large caves nearby.

Geologic Features

Stratigraphy and Geologic Structure

The rocks of Mammoth Cave National Park originated as sediments on the floor of a continental sea that covered most of southern North America during the Mississippian Period, approximately 350 to 325 million years ago. Limestone was the major rock type deposited in this vast sea, although sand, silt, and clay were carried in from the land area that lay to the north. The water was very shallow, despite its broad extent, with average depths of only a few tens of feet. The limestones deposited at this time are highly varied in character and possess rather thin and prominent bedding.

Box 12.1
How Caves Are Formed

Virtually all major caves are formed by the dissolving action of underground water within soluble bedrock. Limestone, dolomite, marble, and gypsum are the most common rocks of this type. All but gypsum require that the underground water be at least slightly acidic for the solutional activity to be rapid enough to form sizable caves. Limestone, by far the most common cave-forming rock, is attacked mainly by water containing dissolved carbon dioxide, a natural gas present in the atmosphere and in soil, that combines with water to form weak carbonic acid. The carbonic acid breaks down calcite, the principal mineral in limestone, into soluble components that are carried away by water. The chemical reactions that take place are shown below:

1. H_2O + CO_2 $\rightleftharpoons$ H_2CO_3
   water   carbon dioxide gas   carbonic acid

2. H_2CO_3 $\rightleftharpoons$ H^+ + HCO_3-
   carbonic acid   hydrogen ion in solution   bicarbonate ion in solution

3. H^+ + HCO_3- + $CaCO_3$ $\rightleftharpoons$ Ca^{++} + $2HCO_3-$
   hydrogen ion in solution   bicarbonate ion in solution   calcium carbonate (calcite in limestone)   calcium ion in solution   bicarbonate ions in solution

Water that enters the ground through openings such as fissures in bedrock or cave entrances is able to dissolve limestone throughout its entire underground flow path (though at a decreasing rate with time), provided that the water does not lose its solutional potential by releasing carbon dioxide gas (CO_2). Discrete flows of water, such as those that enter the ground through major depressions, or karst features, at the surface, generally gain carbon dioxide from the subsurface openings rather than losing it, so their ability to dissolve may be sustained for a great length of time.

The water that infiltrates through soil picks up a great deal of carbon dioxide from decaying vegetation and other organic material. Therefore, a given volume of this ground water is able to dissolve much more limestone than water from other sources. However, if the ground water that has seeped down through soil emerges into an air-filled cave, some of the dissolved carbon dioxide is released into the cave atmosphere. When this happens, the solution process is reversed, causing calcium carbonate ($CaCO_3$) to be precipitated in the form of calcite (limestone). This is how many of the cave decorations, or *speleothems,* are formed.

Gravity and capillary forces also affect infiltrating water. Except temporarily and locally, water does not entirely fill the available cracks and pores in the ground. Wherever the water flow is substantial enough to overcome the capillary forces, gravity pulls it downward via the steepest available openings until it reaches a zone where all available openings in the ground are filled with water. In this zone the water flows by a combination of gravitational force and hydrostatic pressure, eventually emerging as springs at lower elevations such as river valleys. The zone of descending gravitational water is called the *vadose zone,* and the zone in which all openings are filled with water is called the *phreatic zone.* The top of the phreatic zone is known as the *water table.*

Cave-forming processes can operate in both the vadose and phreatic zones. Vadose cave features show a gravitational orientation (e.g., stalactites and stalagmites, vertical solution pits, and cave passages with consistent downward trends), whereas phreatic features have no consistent gravitational orientation (e.g., irregular, vaulted rooms).

Mammoth Cave is located within three different limestone formations. From oldest to youngest, these are the St. Louis Limestone, the Ste. Genevieve Limestone, and the Girkin Formation. The St. Louis is about 200 feet thick, although Mammoth Cave extends into only the upper half. The cave occupies the entire thickness of the Ste. Genevieve (100–120 feet) and the Girkin (135–140 feet). The climate was arid while the St. Louis Limestone was forming. As a result, gypsum beds are common at depth below the surface, although the gypsum has been almost entirely removed from the near-surface parts of the formation by ground-water solution. Chert beds and lenses are common, and it is thought that many of these were formed by the replacement of carbonates by silica (SiO_2). The climate grew more humid while the Ste. Genevieve Limestone was accumulating, so this formation contains little or no gypsum. Chert is limited to large rounded nodules and dike-shaped bodies. The upper half of the Ste. Genevieve Limestone contains several thin, impure, silty beds that represent the first stages of encroachment of detrital sediment from the land into the Mammoth Cave area. The Girkin Formation consists of thick limestone members, each about 20 feet thick, interspersed with thin silty limestones and shales.

Above the Girkin Formation lie the younger rock units of the Chester Series, in which the relative proportion of limestone and detrital sediment is reversed (table 12.1). Thick formations of sandstone and shale are interspersed with limestones only 10 to 40 feet thick. The Chester Series forms the resistant cap rock on the limestone ridges of the Chester Upland. This series is truncated by a widespread disconformity with local relief of several hundred feet in places. (The surface of a *disconformity*, representing missing rock layers, is parallel to beds above and below it.) Pennsylvanian conglomerates and other detrital rocks overlie the disconformity in the northern part of Mammoth Cave National Park, but in the southern ridges of the Park, including Mammoth Cave Ridge, the Pennsylvanian rocks have been removed by erosion.

Downwarping of the Illinois Basin occurred while the Paleozoic sedimentary rocks were being deposited, causing the beds to dip gently toward the center of the basin (located in southeastern Illinois). In Mammoth Cave National Park the dip averages one-third to one-half degree toward the northwest, although the direction and amount of dip vary a great deal from place to place because of irregularities in the beds.

Joints are rather small and inconspicuous in the cavernous limestones. Faults with up to 50 feet of displacement occur within the park, but the largest that have been observed in the cave have displacements of less than a foot. Because the fracturing of the limestone is so slight, individual beds and bedding-plane partings exert the main control over the directions of passages in Mammoth Cave.

Surface Features

Unlike Carlsbad Caverns (chapter 14), Mammoth Cave is intimately related to the hydrology and geomorphic evolution of its surrounding landscape. Almost all the runoff in this humid region sinks underground through karst features in the land surface. Such features include *sinkholes*, which are funnel-shaped depressions where soil has subsided into solutionally enlarged cracks in the limestone. Collapse of bedrock into the underlying openings also plays an important role in forming large sinkholes. In the Mammoth Cave area these features number in the hundreds of thousands and range in size up to a thousand feet in diameter and more than a hundred feet deep. Some have open holes in their bottoms or sides that lead directly to caves. Throughout much of the Pennyroyal Plateau the density of sinkholes is so great that there is virtually no surface runoff. Surface streams form where water collects on adjacent insoluble rocks, but where they encounter limestone they disappear underground into sinkholes. These are called *sinking streams*. Despite the large number of karst features, the overall surface of the Pennyroyal Plateau is of remarkably low relief, appearing almost flat when viewed from an overlook. Actually the land slopes gently toward the few entrenched surface rivers that cross the area.

The Chester Upland, in which Mammoth Cave is located, rises abruptly 200 to 300 feet above the Pennyroyal Plateau surface. This steep rise separating the two regions is called the Chester Escarpment. The highly irregular edge of the Chester Upland has many erosional embayments and isolated remnants called *knobs*, which are outliers of limestone capped by insoluble rocks. The Chester Upland is a rugged, hilly region rising to altitudes of nearly 1000 feet. It is extensively dissected by steep-walled valleys, many of which have lost their streams to underground piracy and are now dry, sinkhole-floored karst valleys. The only surface streams in the Chester Upland and Pennyroyal Plateau are major rivers into which ground water emerges through karst springs. The master stream in the Mammoth Cave area is the Green River, which passes westward through both the Pennyroyal Plateau and the Chester Upland.

The Pennyroyal Plateau is lower than the Chester Upland because the erosional removal of the late Mississippian detrital rocks exposed the less resistant limestones beneath. Rocks like sandstone, which make up most of the cap rock of the Chester Upland, are much more resistant in this humid environment than limestone, which is attacked by rapid solution as well as by mechanical weathering. Sandstone is affected very little by solution. The very low relief of the Pennyroyal Plateau resulted from a prolonged period of weathering at or near base level when erosional dissection of the region slowed or stopped. Although the limestones were weathered and eroded almost

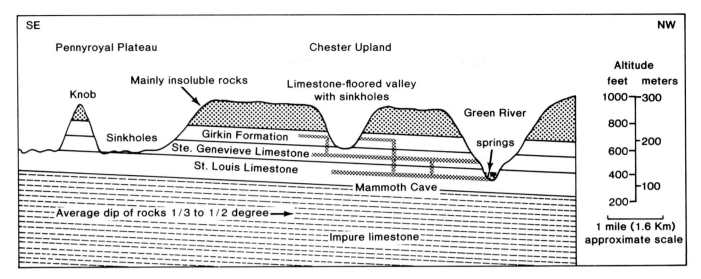

Figure 12.4   Geologic profile through Mammoth Cave National Park, showing the relationship of the cave to local topography and rock units.

to a flat surface at that time, the resistant cap rock prevented all but a modest widening of stream valleys in the Chester Upland. Subsequently renewed dissection of the region has caused major rivers to entrench about 200 feet into the Pennyroyal Plateau and to deepen the valleys in the Chester Upland (fig. 12.4).

Underground water in Mammoth Cave National Park emerges at dozens of springs in the Green River valley. Their flow varies greatly with the seasons and responds quickly to rainfall. Most of the largest springs are fed by cave passages that formed when the Green River valley was deeper than it is now. These passages were flooded when river sediment filled the bottom of the valley to depths of at least 30 feet late in Pleistocene time. At present, in order to reach the river, underground water from these passages is forced to rise upward through funnel-shaped openings in the sediment covering the floor of the Green River valley. In addition, many small springs emerge *above* the Green River level in the walls of valleys, particularly in the thin limestone formations sandwiched between detrital rocks in the cap rock overlying the main cavernous limestones.

Streams in the main part of Mammoth Cave discharge through sediment in the floor of the Green River valley at Echo River Spring. During wet periods the underground water overflows through other passages to River Styx Spring, which is located exactly at the level of the Green River. Both of these springs can be viewed by following short surface trails.

The cavernous limestones have such an enormous permeability that ground water is filtered hardly at all. Contamination of ground-water supplies can, therefore, be a serious problem. Recent dye tests by the National Park Service show that water passes through Mammoth Cave from sources as far as 10 miles away in the Pennyroyal Plateau. With so much water flowing through open conduits in the limestone, great care must be taken to avoid contamination of water supplies in the region by chemical spills and improper waste disposal.

Some of the larger streams in Mammoth Cave are occasionally contaminated when wastes from populated areas to the east and south seep into ground water. Although none of the passages open to the public are affected, National Park Service staff are concerned about the impact on the delicate cave ecology. As a result, they have worked closely with local officials to reduce the threat of ground-water pollution in the park and the surrounding area by controlling waste disposal and by planning a regional waste-water drainage system.

Cave Passages

Mammoth Cave consists mostly of long, winding, nearly horizontal tunnels, many of which can be followed for several miles without interrruption. These passages were dissolved by running water that has infiltrated from the land surface mainly through sinkholes. The various passages join each other as tributaries, forming a crude dendritic pattern. This pattern is not at all apparent from a map of the cave for several reasons:

1. Subsurface water is strongly controlled by the local geologic structure, as will be shown later. As a result, cave passages have much more erratic patterns than those of dendritic surface streams.
2. Many different levels of cave development are superimposed upon each other. As the Green River erodes downward, the water in the cave passages diverts to progressively lower routes and forms new passages.

The abandoned passages are left high and dry. However, at any given time during the evolutionary history of the cave, the passages that are actively forming possess a dendritic pattern.

3. Most passages cannot be followed the entire distance from water input to spring, because they either become too small to explore or are terminated by collapse or sediment fill.

4. Overflow passages can form where periodic flooding of the lowest cave levels causes water to be forced into alternate routes through the limestone, particularly where a stream passage is partly blocked by collapse material or sediment.

5. Many passage junctions have resulted from the random intersection of passages, such as by collapse of bedrock separating two passages where they cross one another on different levels.

As a result of all these factors, Mammoth Cave appears to the visitor as a confusing three-dimensional maze, rather than an orderly dendritic pattern of stream channels.

All of the passages in Mammoth Cave are well adjusted to the patterns of past or present ground-water circulation in the region. Passages emanate from areas of ground-water recharge, such as the Pennyroyal Plateau and karst valleys in the Chester Upland, and terminate at or near the Green River valley, the only significant area of outflow. The chemical character of this water indicates that Mammoth Cave formed by solution of limestone by carbonic acid in water. Measurements in active stream passages show that the limestone is dissolving away at a maximum rate of slightly less than a millimeter each year.

Although this process is responsible for nearly all major caves and karst features, Carlsbad Caverns National Park in New Mexico (chapter 14) is a uniquely interesting exception. Carlsbad Caverns, which formed by the solutional action of sulfuric acid generated by the oxidation of sulfides, does not possess the close association with the land surface and the present patterns of ground-water circulation that Mammoth Cave shows so well; nor does Carlsbad Caverns contain distinctly stream-carved tunnels.

Passages in Mammoth Cave exhibit a variety of shapes determined by the conditions of water flow that formed them. Infiltrating water flows downward through cracks and pores in the limestone through the *vadose zone* and is pulled down mainly by the force of gravity. This water eventually reaches the *phreatic zone,* where all openings in the rock are filled with water. In the phreatic zone, the water moves under the combined influence of gravity and pressure along subhorizontal paths to springs in the nearest available river valley.

A misconception among some geologists is that caves form in the phreatic zone, at or beneath the water table, and are exposed to vadose conditions only when the water table drops with time. This is not true, because vadose water that enters the ground through sinkholes and sinking streams is able to form caves as readily as phreatic water. In fact, the majority of the passages in Mammoth Cave formed in the vadose zone above the water table. Many vadose passages are still being enlarged by underground streams. The sinkholes that feed these passages evolved along with the cave, and, except in rare cases, do not represent points where surface water was diverted underground into preexisting phreatic caves.

Vadose water descends along the steepest available paths, preferably through vertical joints or faults. However, in the Mammoth Cave area, faults are sparse, and most joints are narrow, inconspicuous, and extend through only a single bed. If no vertical opening is present, vadose water follows the next steepest path. Thus the most common passage types in Mammoth Cave are those formed by vadose water that initially flowed not vertically downward, but instead followed the dip of the limestone along bedding-plane partings or favorably jointed beds. When enough solution occurred to allow a discrete stream of water to flow through, such a passage would enlarge by the downward solution and erosion of the floor, thus creating a narrow, sinuous *canyon passage.* A canyon passage looks like the canyon of an arid landscape, except that it is roofed by limestone. Examples include Boone Avenue and Rose's Pass on the Half-Day Tour. Most canyon passages in Mammoth Cave are less than 10 feet wide, but they range up to 100 feet high. However, some of the largest passages in the cave are broad canyons as much as 80 feet wide, such as Kentucky Avenue, Broadway (fig. 12.5), and Audubon Avenue. The pattern of a canyon passage is intimately controlled by the local direction of dip, and minor irregularities in dip cause a marked sinuosity in the passage. Because the direction of dip is so variable, canyon passages are not all oriented in the direction of the northwesterly regional dip.

Where water is able to flow vertically downward along a fracture, the fracture is widened by solution wherever the flow is greatest, and a *vertical shaft* is formed. These features are shaped like the inside of a silo or well with vertical, cylindrical walls. From the bottom, a vertical shaft may look like the inside of a high, narrow dome, but from the top it looks like a pit. For this reason they are sometimes called *domepits.* An example is Mammoth Dome on the Historic Tour (fig. 12.1). Water generally enters laterally through a passage at ceiling level (usually a canyon passage) and exits through a similar passage that leads out through the shaft wall at floor level. In the poorly jointed limestones of the area, most vertical shafts owe part of their vertical extent to the coring action of the descending water upon the limestone floor. Abandoned drain passages, representing former floor levels, are common in shaft walls. Vertical shafts in Mammoth Cave range in

Figure 12.5 Broadway, on the Historic Tour, is a huge canyon partly filled with detrital sediment. This passage is typical of the large upper levels of the cave at an altitude of 600 feet. The wooden pipes are remains of the saltpeter mining operation during the War of 1812. They were used to transport leached nitrates to the Historic Entrance with the aid of large hand-driven pumps. Photo by A. N. Palmer

size up to 30 feet in diameter and 200 feet from top to bottom. They may intersect other passages as they grow, thus providing connections between different levels in the cave. Most shafts are still actively forming and contain water flow ranging from small trickles to major waterfalls.

Tubular passages are tunnels with elliptical or lenticular cross sections formed in the phreatic zone at or below the water table. The long dimension of their cross sections is controlled by bedding planes or, less commonly in Mammoth Cave, by joints or faults. They vary in size up to 30 feet high and 100 feet wide. An example is Cleaveland Avenue on the Half-Day Tour (named for a nineteenth century geologist; the spelling is correct). Tubular passages either form while completely filled by the water flowing through them, or are filled with water only during periods of high flow. In either case the solutional enlargement is not only downward but lateral and upward as well, so that broad, arched ceilings are formed. The overall gradients of these passages are very low, but in places there may be abrupt jogs upward or downward across the bedding for as much as 30 feet. As the Green River erodes its valley downward, the underground water seeks lower paths through the limestone, abandoning the older phreatic passages as relict features within the vadose zone.

Tubular passages show no consistent relationship to the dip of the strata, because phreatic water has no inherent tendency to flow straight down the dip. Instead they develop along the most efficient flow path to the nearest available spring outlet. Because of the prominent bedding and the decreasing openness of partings and fractures with depth, many tubular passages follow a single stratigraphic horizon for thousands of feet, or even miles, commonly with a trend nearly parallel to the local strike of the beds. This trend represents the intersection between the low-relief water table and the dipping beds, where phreatic solution is most intense.

Most large passages in Mammoth Cave have vadose sections in their upstream ends and phreatic sections in their downstream ends. In an actively forming passage, the transition between the two sections is obvious, but the transition is also rather clear in a dry, abandoned passage. The passage may change from a relatively steep canyon to a nearly horizontal tube, and from a predominantly dip orientation (i.e., controlled by dip of beds) to a direction other than down the dip. A fine example of this transition is where the upper level of Boone Avenue, a canyon, changes to the tubular Cleaveland Avenue.

The major cave passages cluster at several significant levels: 600 feet, 550 feet, and 500 feet above sea level. The Green River is presently at 425 feet, and any cave passages below this elevation are completely flooded. The upper level (600 feet) consists of wide canyons and tubes partly filled with sediment. These are the largest and oldest passages in the cave. Audubon Avenue is an example of a wide canyon at this level that is filled with stream-deposited sand and gravel to more than half its height, so its canyonlike shape has been obscured. Apparently the filling took place soon after the passages at this level formed, while they still carried water. These passages correlate in elevation and probably also in age with the surface of the Pennyroyal Plateau. Their large size and sediment fill indicate slow entrenchment by the major rivers in the area, interrupted by at least one period of aggradation (deposition). The two lower levels are formed mainly by large tubular passages that represent periods of rather static regional base level. It is important to note that the cave levels are not the result of favorable zones caused by differences in bedrock solubility. Each level consists of passages occupying different sedimentary layers, depending on the dip of the rocks.

By examining abandoned cave passages, it is possible to reconstruct former patterns of ground-water flow in the region. Although specific flow routes have undergone a great amount of capture (piracy) and diversion throughout the history of Mammoth Cave, the overall pattern of flow has remained rather constant.

Solutional Features in the Cave

In addition to the cave passages themselves, several minor solutional features provide evidence of past flow conditions.

Scallops are asymmetrical solution hollows formed by turbulent water flow. The bedrock surfaces of most canyons and many tubular passages are covered with a continuous dimpling of scallops. The scallops vary in length from roughly one-half inch to several feet, with the length inversely proportional to the velocity of the water that formed them. Their asymmetry also indicates the direction of flow, as with sand ripples or dunes; the steep faces of the scallops are oriented in the downstream direction. Scallops are easily visible in Rose's Pass on the Half-Day Tour and in River Hall on the Historic Tour.

Flutes are parallel grooves formed in cave walls by water dripping vertically or flowing down a steep slope. Aligned in the direction of flow, they resemble the fluting

of columns in classical architecture. Flutes are common in vertical shafts, such as Mammoth Dome on the Historic Tour, and in minor shafts in Boone Avenue on the Half-Day Tour.

Anastomoses are small, winding tubes that interconnect with each other in a braided pattern, normally along bedding-plane partings. The tubes have semicircular to circular cross sections and are dissolved upward into the base of the overlying bed. They range in diameter from about one-fourth inch to a foot. Anastomoses open into the walls of many tubular passages, and in some places they are exposed in the cave ceiling where the underlying bed has dropped or dissolved away. They are usually interpreted as remnants of the initial solution channels from which the cave passages developed. However, recent evidence suggests that they are more readily formed during periodic floods late in the solutional history of a passage, when solutionally active floodwaters fill the passage under pressure, forcing water into partings in the cave walls. Remnants of anastomoses can be seen in Kentucky Avenue on the Half-Day Tour and in Little Bat Avenue on the Historic Tour.

Cave Deposits

Various minerals precipitate in the cave as "cave formations," more properly called *speleothems*. The great majority of speleothems are precipitated by vadose water seeping into the cave. Where water runs over a cave wall, sheets or draperies called *flowstone* are formed (fig. 12.6). Icicle-shaped *stalactites* are deposited by water dripping from the ceiling, and where the drops hit the floor they deposit mounds shaped like totem poles called *stalagmites*. Where a stalactite and stalagmite are produced by the same source of dripping water and grow large enough to meet, a *column* is formed. If the amount of water entering the cave is too small to form distinct drips, it is held to the growing speleothems by surface tension, and the calcium carbonate is deposited in irregular, twisted fingers called *helictites*.

Nearly all of the speleothems are still actively growing. Various forms of travertine ($CaCO_3$) can be seen on the Frozen Niagara Tour and in a few scattered places throughout the other tours. These include flowstone, stalactites, stalagmites, columns, rimstone, and popcorn (explained in chapter 14). The largest of these features is the 75-foot-high flowstone drapery called Frozen Niagara.

Travertine is deposited in caves only when the water infiltrating from the surface is able to lose carbon dioxide to the cave air. This requires that the water first pass through an overlying soil in which the carbon dioxide content is higher than that of the cave. Vadose seepage passes through the soil, picks up a large amount of carbon dioxide, and readily dissolves the underlying limestone. If the seepage takes place in rather small quantities, it quickly approaches saturation with respect to the limestone. When seepage water loses carbon dioxide to the cave air, it becomes supersaturated and precipitates some of its dissolved carbonate as travertine. Large discrete flows of vadose water, particularly those that enter the ground through sinkholes, do not possess such a high carbon dioxide content and may even gain, rather than lose, carbon dioxide in the cave. They remain undersaturated and account for the solutional development of cave passages.

The insoluble cap rock that overlies most of the ridge in which Mammoth Cave is located severely limits the amount of water infiltrating into the cave. Only around the perimeter of the ridge, where the cap rock has been eroded away, is it possible for large amounts of seepage to enter the cave through soil-covered limestone and to deposit large amounts of travertine.

Beneath the insoluble cap rock, passages that have been abandoned by the water that formed them are generally dry. Capillary water is drawn toward them in small quantities from the moist surrounding limestone and evaporates when it emerges into the cave. Many varieties of evaporite minerals are deposited by this water. The most common is gypsum ($CaSO_4 \cdot 2H_2O$), which is derived in part from the oxidation of pyrite (FeS_2) in the limestone and overlying sandstone. The process is similar to the sulfuric-acid reaction and gypsum deposition described in chapter 14, except that in Mammoth Cave the capillary nature of the water produces speleothems quite different from the bedded gypsum of Carlsbad Caverns. In general, the gypsum in Mammoth Cave forms a thin white or golden crust on the cave walls up to several inches thick. Where the deposition of gypsum is particularly intense, delicate *gypsum flowers* grow from the walls as curved fronds and form clusters that resemble lilies or cotton. Many gypsum deposits are seen on the Half-Day Tour, particularly in Cleaveland Avenue.

Mirabilite ($Na_2SO_4 \cdot 10H_2O$) crystallizes as clear needles or stalactitelike forms; *epsomite* ($MgSO_4 \cdot 7H_2O$) forms white, hairlike tufts. Both are deposited in areas of high evaporation where the relative humidity is less than about 80 to 90 percent. They can form remarkably fast, sometimes overnight, but when the humidity rises, they become unstable and may redissolve.

Nitrates are carried into the cave by water that has passed through organic material at the surface. In the saltpeter mining operation of the early nineteenth century, calcium nitrate was leached out of the cave sediment; the resulting solution, when mixed with wood ashes, produced potassium nitrate, one of the essential ingredients of gunpowder.

Deposits in Mammoth Cave resulting from mechanical weathering include sediment carried in by cave

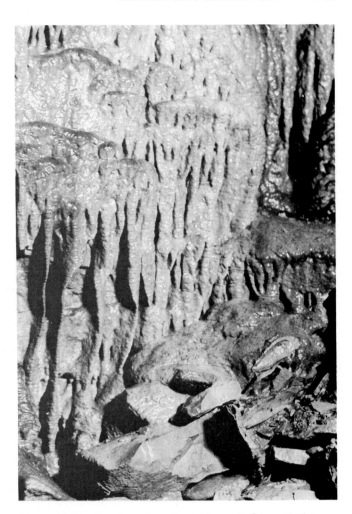

Figure 12.6 Flowstone draperies in Mammoth Cave near the Violet City Entrance. These travertine deposits are formed where water high in dissolved limestone content enters the cave and loses carbon dioxide to the cave air, precipitating calcium carbonate. National Park Service photo.

streams and breakdown blocks that have fallen from the ceiling and walls. The sediment consists of gravel, sand, silt, and clay derived mainly from weathering of the overlying insoluble rocks, plus a small insoluble residue from solution of the limestone. *Breakdown* is collapse material in the form of blocks, slabs, and chips. It is produced in several ways: (1) by solution along the partings and joints that bound a block of limestone; (2) by the loss of buoyant force on the ceiling when a passage is drained of its water by lowering of the water table; (3) by sagging and peeling of limestone beds in the cave ceiling after the underlying beds have been removed by solution; and (4) by the wedging effect of minerals such as gypsum that grow within fractures in the limestone. Breakdown of a passage ceiling usually progresses upward to the top of a weak rock unit such as shaly limestone, and afterward the passage

Table 12.1 Geologic Column, Mammoth Cave National Park

Time Units			Rock Units	Geologic Events
Era	Period	Epoch		
Cenozoic	Quaternary	Holocene	Alluvium	Alluviation; drowning of lowest levels of Mammoth Cave
		Pleistocene		Rapid entrenchment alternating with minor aggradation; lower levels of Mammoth Cave formed
	Tertiary — Neogene	Pliocene		Upper levels of Mammoth Cave formed; nearly flat surface of Pennyroyal Plateau formed during period of relatively static base level
		Miocene		
	Tertiary — Paleogene	Oligocene		
		Eocene		Slow erosion, probably interspersed with periods of aggradation
		Paleocene		
Mesozoic				
Paleozoic	Permian			
	Pennsylvanian	Morrowan	Caseyville Formation (conglomerate and sandstone)	Insoluble rocks deposited; gradual emergence of continent above sea level
			⟨ Disconformity ⟩	⟨ Erosion ⟩
	Mississippian	Chesterian	Leitchfield Formation (shale and sandstone)	Mainly insoluble rocks deposited in continental sea
			Glen Dean Limestone	
			Hardinsburg Formation (sandstone)	
			Haney Limestone	
			Big Clifty Formation (sandstone and shale)	
			Girkin Formation (limestone)	Relatively pure limestones deposited in continental sea
		Meramecian	Ste. Genevieve Limestone	
			St. Louis Limestone	
			Salem Limestone	Impure limestones deposited in continental sea
			Harrodsburg Limestone	

Source: Palmer (1981)

is relatively stable. Where one passage crosses over another, breakdown may occur between them and produce a junction room. Several entrances to the cave have been formed by the intersection of a passage by surface erosion, but in most such places the incipient entrance is totally obscured by breakdown.

Geologic History

1. **Deposition of limestones during the Mississippian Period, 350 to 325 million years ago.**

The shallow continental sea, in which the St. Louis Limestone, the Ste. Genevieve Limestone, and the Girkin Formation were deposited, had covered most of the southern part of the North American craton since early in Paleozoic time. Many beds of older siltstones and impure limestones (not exposed in Mammoth Cave) underlie the 400 to 500 feet of uninterrupted Mississippian limestones. As the limestones accumulated, the climate gradually changed from relatively arid to humid with progressively greater amounts of detrital sediment being carried in from land areas to the north. Downwarping of the Illinois Basin occurred simultaneously with this deposition, imparting a gentle northwesterly dip to the rocks.

2. **Deposition of predominantly detrital rocks late in Mississippian time and in the Pennsylvanian Period.**

Increasing amounts of detrital sediment accumulated in the continental sea and produced the varied rocks of the Chester Series, which today (with the exception of the limestones of the Girkin) form the largely insoluble cap rock overlying the cavernous limestone of Mammoth Cave National Park. A drop in sea level relative to the surface of the continent allowed extensive stream erosion of the Mississippian rocks. With a subsequent rise in sea level early in the Pennsylvanian Period, conglomerate and sandstone were deposited over the erosion surface.

3. **Slow subaerial weathering and erosion, beginning in late Paleozoic time.**

The Pennsylvanian rocks are the last to have been deposited in the Mammoth Cave region. Since then, several hundred feet of Pennsylvanian and late Mississippian rocks have been removed by erosion, exposing Mississippian limestones over broad areas. Karst features began to develop wherever limestone was exposed at the surface. Mammoth Cave belongs to the last stages of this erosional history.

4. **The forming of Mammoth Cave late in Tertiary time, perhaps 30 to 5 million years ago.**

The earliest passages to form in Mammoth Cave were large tubular and canyon passages now located 570 to 690 feet above sea level. The highest of these (in Crystal Cave, not open to the public) lies above the level of the Pennyroyal Plateau and has an irregular ceiling profile, indicating that it originated below the contemporary elevation of the Green River. Therefore, this passage predates the Pennyroyal surface. Later aggradation filled the passage nearly to the ceiling with stratified silt, sand, and gravel.

The major late Tertiary passages originated during the period of rather slow fluvial entrenchment that allowed the nearly flat Pennyroyal Plateau surface to form. Erosion was interrupted by at least one period of aggradation. Cave passages were formed by sinking streams draining from the Pennyroyal Plateau and from karst valleys in the Chester Upland. Audubon Avenue, Broadway, Main Cave, and Kentucky Avenue were formed at this time as a single enormous passage that has since been segmented by breakdown.

5. **Development of lower cave levels during the Pleistocene Epoch.**

Pleistocene glacial advances fell short of the Mammoth Cave region, but their influence on the erosional history of the region was profound. Prior to continental glaciation, the Ohio River was not nearly so large as it is today. Drainage from most of the Appalachian Mountains flowed through what is now northern Indiana into the Mississippi River, bypassing the Ohio River entirely. Early Pleistocene glaciers blocked this drainage and diverted it to the Ohio River, which was located approximately along the farthest advance of the ice. The greatly increased discharge caused rapid entrenchment of the Ohio. The Green River, which controls the levels of passage development in Mammoth Cave, is a tributary of the Ohio River, so it also began to entrench rapidly. As a result, the regional water table dropped and streams in the caves were diverted to lower routes. All but the largest surface streams on the Pennyroyal Plateau responded in a similar way by flowing underground as sinking streams.

Several lengthy pauses in fluvial entrenchment that occurred throughout the Pleistocene were accompanied by minor interludes of aggradation. Each pause allowed the cave streams to remain stable at the same level for a long period of time, thus enlarging the cave passages at that level to comparatively great size. The longest pauses resulted in major tubular passage levels at the present altitudes of 550 and 500 feet. Cleaveland Avenue is the best example at the 550-foot level. It reaches widths of 50 to 60 feet and a height of 15 to 20 feet. Great Relief Hall is part of the 500-foot level. As the water level dropped in the cave, travertine and evaporite speleothems began to accumulate in dry passages.

Terraces formed in the Green River valley during the periods of relatively static base level, and they correlate in elevation with the cave levels of the same age. Only the 500-foot-level terrace is relatively intact today, as the higher terraces have been severely eroded. Tentative dating of the terraces and cave deposits by various methods, including uranium-thorium radiometric dating of travertine and paleomagnetic measurements of sediment, indicate that the 550-foot level is probably of middle Pleistocene age (approximately 700,000 to 800,000 years old) and that the 500-foot level is probably of later Pleistocene age (at least 350,000 years old).

A more recent geomorphic event has been the filling of the Green River valley with 30 to 50 feet of alluvium late in the Pleistocene Epoch, probably during Wisconsinan time when the last major ice advance occurred. As a result, the lowest passages in Mammoth Cave are now completely water filled.

Natural entrances to Mammoth Cave have probably existed at springs and at points where sinking streams entered the ground since the cave first began forming. However, the present entrances are mainly secondary features created by the random intersection of cave passages by surface erosion. The Historic Entrance is an example of an entrance formed where an upper-level passage was intersected by a minor tributary valley of the Green River valley.

The processes that formed Mammoth Cave continue to operate today. For this reason it is an ideal "underground laboratory" for studying the ways in which caves and their related karst features evolve in a humid environment.

Geologic Maps

Haynes, D. D. 1962. *Geology of the Park City quadrangle, Kentucky.* U.S. Geological Survey Map GQ-183.

———. 1964. *Geology of the Mammoth Cave quadrangle, Kentucky.* U.S. Geological Survey Map GQ-351.

———. 1966. *Geology of the Horse Cave quadrangle, Kentucky.* U.S. Geological survey Map GQ-558.

Klemic, H. 1963. *Geology of the Rhoda quadrangle, Kentucky.* U.S. Geological Survey Map GQ-219.

Bibliography

Brucker, R. W., and Watson, R. A. 1976. *The longest cave.* New York: Alfred A. Knopf.

Harmon, R. S., Thompson, P., Schwarcz, H. P., and Ford, D. C. 1975. Uranium-series dating of speleothems. *National Speleological Society bulletin* 37:21–33

Hess, J. W. 1976. A review of the hydrology of the Central Kentucky Karst. *National Speleological Society bulletin* 38:99–102.

Hill, C. A. 1981. Origin of cave saltpeter. *National Speleological Society bulletin* 43:110–126.

Livesay, A. 1961. *Geology of the Mammoth Cave National Park area.* Revised by P. McGrain. Kentucky Geological Survey, Series 10, Special Publication 7.

McGrain, P., and Walker, F. H. 1954. Geology of the Mammoth Cave region, Barren, Edmonson, and Hart Counties, Kentucky. Kentucky Geological Survey Field Trip Log.

Miotke, F. D., and Palmer, A. N. 1972. *Genetic relationship between caves and landforms in the Mammoth Cave National Park area.* Wurtzburg, Germany: Bohler Verlag.

Palmer, A. N. 1981. *A geological guide to Mammoth Cave National Park.* Teaneck, N.J.: Zephyrus Press.

———. 1985. The Mammoth Cave region and Pennyroyal Plateau, in P. H. Dougherty, editor, *Caves and karst of Kentucky,* Kentucky Geological Survey, Special Publication 12, Series XI, p. 97–118.

Pohl, E. R. 1970. Upper Mississippian deposits in south-central Kentucky. Kentucky Academy of Sciences, *Transactions* 31:1–15.

———, and White, W. B. 1965. Sulfate minerals: their origin in the Central Kentucky Karst. *American Mineralogist* 50:1462–1465.

Quinlan, J. F., and Rowe, D. R. 1977. Hydrology and water quality in the Central Kentucky Karst: Phase I. University of Kentucky, Water Resources Research Institute, Research Report No. 101.

Schmidt, V. A. 1982. Magnetostratigraphy of sediments in Mammoth Cave, Kentucky. *Science* 217(4562): 827–829, August 27.

Thrailkill, John. 1968. Chemical and hydrologic factors in the excavation of limestone caves. *Geological Society of America bulletin* 79:19–46.

Watson, P. J., ed. 1974. *The archeology of the Mammoth Cave area.* New York: Academic Press.

White, W. B., Watson, R. A., Pohl, E. R., and Brucker, R. W. 1970. The Central Kentucky Karst. *Geographical review* 60:88–115.

Mammoth Cave National Park

Area: 52,395 acres; 82 square miles
Established: July 1, 1941
Designated: World Heritage Site, October 27, 1981
Address: Mammoth Cave, Kentucky 42259

13

Eighth longest limestone cavern in the world, famous for its "boxwork" and calcite crystals.
The park's location is in the beautiful Black Hills.

Wind Cave National Park

Southwest South Dakota

by Rodney M. Feldmann
Kent State University

Figure 13.1   Boxwork results when a fracture system is filled in by secondary calcite that is left standing in relief, following solution of surrounding original bedrock. Wind Cave is noted for this unusual feature. National Park Service photo.

Figure 13.2 Map of surface features of Wind Cave National Park, South Dakota

Local History

The Sioux Indians (or Dakotas) recounted legends about what they called the "Sacred Cave of the Winds." They believed it was the cave through which their ancestors arrived at the Black Hills. Another myth was that Wakan Tanka ("Great Mystery") sent the buffalo out into the Sioux hunting grounds from the cave.

Many legends persist in the Black Hills region to this day, even as to how Wind Cave was discovered. One story claims a cowpuncher was riding by when the wind blew his hat off and into a small hole. When he went to look for it he discovered the cave. Since the opening was very small, he could not retrieve the hat and was forced to leave without it. The next day he rode that way again and the wind brought his hat back to him, placing it on his head.

The official discoverer of Wind Cave was Tom Bingham, an early settler in the Black Hills. As he was hunting one day, he heard a strange whistling sound and began to search for the source. He found a small hole, through which the wind was rushing and making noise. As the natural opening was too small to permit entrance, another was dug nearby to a depth of six feet that made access to the cave possible.

The region around the cave was open to claims. In 1890 the South Dakota Mining Company filed a claim on the cave and hired Jesse McDonald and his sons, Alvin and Elmer, to take care of the property. Alvin McDonald did most of the exploration. He named many of the features in the cave and kept a diary of all his mapping. Alvin was buried near the cave; a plaque marks his grave site. In 1892, Jesse McDonald and several partners formed the Wonder Wind Cave Improvement Company. They took over the property, developed some passageways, put in stairs, and opened the cave to the public. Magazine stories printed about the cave in 1900 and 1901 brought it to the attention of the government. President Theodore Roosevelt signed a bill to create the national park in 1903.

Wind Cave National Game Preserve, established in 1912, was added to the park in 1935. On their way to the cave entrance, park visitors may see elk, deer, pronghorn, prairie dogs, and bison, as well as a rich variety of birds and plants. In its beautiful setting in the Black Hills, Wind Cave National Park is truly two parks—one above ground and one below.

Geologic Features

Wind Cave is a remarkably ancient cave with many unmapped and still unexplored passageways. In recent years hydrologic investigations and surveying of previously unknown passages have added to our knowledge of the park's subterranean levels. Dr. James Pisarowicz, a Park Service naturalist, and his colleagues have discovered and surveyed many feet of passageways, including an alternate route between the eastern and western parts of Wind Cave. Among the cave features found by the team are five remarkable underground lakes, one of them 150 feet (50 m) long and 60 feet (20 m) wide.

With more than 48 miles of mapped rooms, tunnels, and passages, Wind Cave is now ranked third longest in the United States and eighth longest in the world. Spelunkers (cave explorers) predict that eventually a connection may be found between the subterranean passages of Wind Cave National Park and those of Jewel Cave National Monument that lies (via the land surface) 18 miles west-northwest of Wind Cave. Jewel Cave, which has 73 miles of mapped passageways, is the second largest cavern in the nation and ranks fourth longest worldwide.

Origin of Caves

Caves in limestone are formed when joints and fractures in soluble rock permit passage of slightly acid ground water along selective pathways. These pathways are then enlarged by solution of adjacent bedrock and subsequent precipitation of material as some form of cave deposit. Variations in rock type, jointing pattern, and ground water chemistry result in a wide diversity of forms and a broad spectrum of caves and cave deposits. Each of the caves preserved by the National Park Service illustrates features that are unique. In Wind Cave the dominant precipitate is the unusual boxwork that formed early in the cave's development (fig.13.1). Much less common in Wind Cave are the stalagmites, stalactites, and other dripstone features displayed in most other caves.

Primary and Secondary Features

The development of caves involves an initial period of growth and enlargement, primarily by solution, followed by a period of infilling by precipitation of minerals from ground water. The "formations" observable in caves can be classified on the basis of whether they were formed during the enlargement stage, *primary features,* or during the stage of infilling, *secondary features.*

Primary Features. The most important feature developed in Wind Cave is the *boxwork.* In addition to the regional joint pattern that dictated the orientation of the major passageways in the cave, the Pahasapa Limestone became highly fractured sometime during its history. The orientation of the fractures differs somewhat from the regional joint pattern and is expressed on an entirely different scale. These primary fractures, in which boxwork ultimately formed, are separated from one another by inches or fractions of inches. Calcite was deposited in the fractures as a secondary filling by ground water. Later, the primary limestone was removed by solution, leaving the secondary material behind as thin fins that crisscross one another. Nowhere is boxwork developed in greater volume or in wider variety than in Wind Cave.

WIND CAVE INTERIOR

Figure 13.3   Underground passages open to the public in Wind Cave National Park. Adapted from National Park Service brochure.

Cave earth, a residue of clay particles and other insoluble materials, is found in all caves. In Wind Cave, cave earth ranges from a thin veneer to layers several feet thick. It accumulated either as the insoluble residue from solution of limestone or as material washed into the cave from sources at the surface. The Pahasapa Limestone consists of about 98 percent soluble material, so that, given the large volume of cave earth in Wind Cave, it is likely that most of the material infiltrated the cave during weathering and erosion of overlying rocks.

Collapse blocks are portions of the ceiling and walls that have detached themselves from the surface of the cave and are now lying on the floor. Probably the blocks loosened and fell when the cave was flooded because boxwork on the underside of blocks is undamaged. Had the cave been dry, the impact of the fall would have shattered the delicate structures.

Collapse breccia is found in parts of the Pahasapa Limestone where the cave ceiling collapsed. This caused

portions of the overlying Minnelusa rock to fall, along with the Pahasapa, as a plug into the cavern.

Spongework, a maze of openings, or pores, ranging in size from a few inches to several feet across, is common in Wind Cave. The passages form a randomly oriented, three-dimensional network large enough in some places for a person to crawl through. Spongework appears to form early in the solution process as the most soluble material is selectively removed.

Secondary Features. Wind Cave has a few flowstone and dripstone features (stalactites and stalagmites, for example) but not very many. Such features are more typical of Mammoth Cave National Park (chapter 12) and Carlsbad Caverns National Park (chapter 14).

Cave lining was deposited when the cave was completely under water. The deposit formed during or following boxwork formation and coated all the walls. Composition of the lining can be calcite, hematite, or opaline silica. The cave was drained several times and then resubmerged with the composition of the material coating

A

B

C

Figure 13.4 A,B,C   Speleothems in Wind Cave. *A.* Pieces of cave rafts litter the cave floor around a helictite "bush." The twig-shaped helictites formed above the water level. Cave rafts, which form on a water surface, are left behind when a pool drains away. National Park Service photo. *B.* Cave popcorn, formed by seepage of water, covers the ceiling of a passage. Photo by M. Little. *C.* Delicate frostwork (aragonite crystals) tends to form where air currents move through a passageway. National Park Service photo.

the boxwork varying from time to time. This occurred at some time between 50 million years ago (Eocene) and 10 million years ago (Pliocene). Final draining took place at about the end of the Pliocene (2 or 3 million years ago). Erosional and weathering processes have removed much of the lining in the cave passages near the surface, perhaps as a result of fluctuating relative humidity.

Cave ice features, or calcite rafts, form in bodies of standing water in a cave. They are common in Wind Cave and are also found in Carlsbad Caverns. They are roughly rectangular pieces of calcite so thin that they are supported on top of the water by surface tension. They range in size from 1 to 3 inches across. Disturbance of any type causes them to sink. New ones can form within a year. In an area referred to as the Calcite Jungle, the cave floor is covered with these rafts; they formed at a time when the water level was higher.

Popcorn (globulites) is a series of layered nodules (1 to 10 mm in diameter) supported on a very small stalk (1 mm long). If the nodules are covered with a bristlelike cluster of aragonite needles, they are called *frostwork*. This is the second feature for which Wind Cave is famous. (Jewel Cave is also noted for frostwork.) At first, popcorn was thought to have formed below the water table, but at the present time it appears to be forming adjacent to a wall where dripping water splashes at the boundary between the top of the water and the air. Another possible theory of origin is that the features form where water seeps from the walls at the water line. Frostwork may also form as a coating of needles directly on the wall (fig. 13.4A,B,C).

Table 13.1 Geologic Column, Wind Cave National Park

Time Units		Rock Units		Geologic Events
Era	Period	Formation	Member	
Cenozoic	Quaternary	Alluvium		Development of surface landscape. Solution and precipitation of features in Wind Cave.
	Tertiary			Cave formed.
				Erosion ————————————————
Mesozoic	Upper Lower	Several formations Spearfish		Uplift of Black Hills associated with Laramide orogeny; deposition of marine and nonmarine sediments throughout most of Mesozoic Era.
Paleozoic	Permian	Minnekahta Limestone		Deposition of marine and some nonmarine sediments
	Pennsylvanian	Minnelusa Sandstone		Development of karst terrain on Pahasapa surface
	Mississippian	Pahasapa Limestone	Upper Unit	
			Middle Unit	Gray or red-brown fossiliferous cliff-former, organic fragments and calcareous muds.
			Lower Unit	
		Englewood Limestone		Pink, thin-bedded carbonate rocks.
				Erosion ————————————————
	Cambrian	Deadwood Sandstone		Clastic rocks.

Sources: Feldmann and Heimlich (1980), Palmer (1981).

The needles may be formed of either aragonite or selenite. *Aragonite* is chemically the same as calcite but belongs to a different crystal system; calcite is hexagonal, whereas aragonite is orthorhombic. *Selenite* is a clear, cleavable form of gypsum. Temperature seems to determine which forms develop. For example, aragonite has a tendency to form in areas where the mean annual temperature is between 56° F and 63° F. Presently, the mean annual temperature of the region is about 50° F and the temperature of the cave is only 47° F; so no aragonite is forming. However, Wind Cave is situated in an area of the Black Hills with known geothermal activity. Thermal springs are currently active just south of the park in the city of Hot Springs, and evidence in surrounding areas suggests that before the present day activity was more extensive. Ground water may have ranged through much higher temperatures in the past.

Several other secondary deposits can be identified in the cave, such as *dogtooth spar,* which is well formed calcite crystals, and *hydromagnesite* or "moon milk," a white, earthy magnesium mineral, but none of these forms occurs in Wind Cave in significant quantity.

Geologic History

1. **Deposition of Paleozoic and Mesozoic beds on Precambrian basement.**

During much of the Paleozoic and Mesozoic Eras, the Black Hills region was either slightly below sea level and receiving marine sediments, or was elevated slightly above sea level. Nearly the entire sequence of Paleozoic and Mesozoic rock units found elsewhere in the Black Hills is present in Wind Cave National Park. The Pahasapa Limestone, in which the cave developed, accumulated as a combination of organic fragments and calcareous mud in warm, shallow marine habitats during the Mississippian Period (360 to 320 million years ago). Following deposition of the Pahasapa beds, some erosion and development of karst features resulted in a hummocky surface in the region. The overlying Minnelusa Formation was, therefore, deposited on an irregular surface. Preliminary investigation of the upper levels of the cave suggests that the Pahasapa surface may have had a few hundred feet of relief, but this has not been confirmed.

2. **Uplift of Black Hills from about 70 to 50 million years ago.**

During Cretaceous and Paleogene time, the Black Hills region was subjected to forces that elevated two

Figure 13.5 Physiographic diagram and cross section of the Black Hills showing how the domal structure controls erosion and landscape development. From *The Black Hills*, R. M. Feldmann and R. A. Heimlich. © 1980 Kendall/Hunt Publishing Company. Used by permission.

crustal blocks of the Precambrian basement and the overlying sedimentary rock sequence. The result was an ovoid dome about 150 miles long and 60 miles wide (fig. 13.5). Had the uplift occurred all at once with no accompanying erosion, the Black Hills would have been some 7000 feet higher than they are today. Instead, erosion has stripped sediments from the central part of the Black Hills and redeposited them elsewhere. Some of the sediments were deposited on the flanks of the uplift during Oligocene and Miocene time and are now exposed in the nearby badlands.

3. Formation of Wind Cave in the Mississippian Pahasapa Limestone.

The cave developed in the upper 100 to 150 feet of the Pahasapa Limestone Formation, which has a total unit thickness of 300 to 600 feet in the Black Hills area. Perhaps cave development began as early as middle Eocene (50 million years ago), but as late as Pliocene (10 million years ago) is more likely. In either case, Wind Cave is one of the oldest in the world. Most caves have formed only in the last million years.

As is typical of any cave, Wind Cave developed along the regional joint system and passed through a solutional phase (when it was below the ground-water level) and a depositional phase, generally above the ground-water level. The regional joint system, developed during the Black Hills uplift, trends northwest-southeast and northeast-southwest. Examination of the plan of the cave shows that the northwest-southeast direction has been the primary compass direction of cave development. A third direction, parallel to the bedding in the Pahasapa, is also significant. Because of a combination of structural and stratigraphic conditions, the cave can enlarge only along the strike (trend) of the beds; up or down the dip, pathways intersect either the land surface or the water table. Much of the growth of Wind Cave has been lateral; this is reflected in the observation that the cave is confined to the upper portion of the formation, even though jointing extends throughout this and other sedimentary units in the area. This suggests that the special combination of conditions necessary to cave development (i.e., opening of the joint system, position relative to the ground-water level, and water chemistry) were met only in the upper portion of the Pahasapa.

4. Development of cave zones.

Since the growth of Wind Cave has been dominantly lateral, with the cave floor sloping gently, total relief is only 450 feet. Initial solution of the cave passageways may have occurred simultaneously as rising ground water dissolved limestone along joints and on bedding surfaces. Over time, three "levels," or zones, developed in the cave that are controlled by differences in lithology within the Pahasapa Limestone. The upper zone formed in breccia, the angular debris that accumulated in ancient sinkholes. Crawlways of small and mazelike spongework are characteristic of this zone. Neither boxwork nor cave lining exist. The middle zone contains the biggest caverns, well developed boxwork, and cave lining. In the lowest zones, passages tend to be narrow and strongly controlled by the joint system. The Calcite Jungle with its cave rafts or "ice" is located here, as well as displays of dogtooth spar crystals.

5. Development of boxwork and other features.

The small-scale fracture system that developed parallel to bedding was filled with secondary calcite. Subsequently, the more soluble limestone was dissolved, leaving calcite fins or boxes. The colors are various shades of yellow, brown, pink, and blue. In some chambers (e.g., Fairground) the calcite is fluorescent and glows blue, green, red, and violet under "black light."

The boxwork has preferred directions of N 70° to 80° E and N 10° to 20° W. The boxwork even cuts through geodes formed in the limestone, which shows that the geodes developed first and the fracture systems later.

6. Erosion and formation of topography.

The climate is semiarid today, as is shown by Wind Cave Canyon Creek an *intermittent stream* (i.e., does not flow year round) that drains through Wind Cave Canyon and Red Valley and may flow through the cave at times when it has sufficient water. Evidence of stream flow through the cave can be found near the Garden of Eden. Beaver Creek, the only permanent stream in the park, cuts across the ridges or hogbacks in the region. The landscape is a combination of steep canyons, hills, and flat upland areas. Origin of these flats (e.g., Bison Flats) is uncertain. They could be structural or erosional.

Geologic Map and Cross Section

Robertson, F. *Geologic map portfolio—a laboratory study for historical geology.* Williams & Heintz Pub. Map no. 7.

Bibliography

Bakalowicz, M. J., Ford, D. C., Miller, T. E., Palmer, A. N., and Palmer, M. V. 1987. Thermal genesis of dissolution caves in the Black Hills, South Dakota. *Geological Society of America Bulletin* 99(6):729–738.

Darton, N. H. 1909. Geology and water resources of the northern portion of the Black Hills and adjoining regions in South Dakota and Wyoming. U.S. Geological Survey Professional Paper 65.

Feldmann, R. M., and Heimlich, R. A. 1980. *The Black Hills,* K/H Field Guide Series. Dubuque, Iowa: Kendall/Hunt Publishing Co.

Harpster, J. 1987. Exploring the underworld: seasonal [naturalist] makes discovery at Wind Cave. *Courier,* March, p. 14.

National Park Service. 1979. *Wind Cave National Park.* National Park Service handbook no. 104. 145 p.

Palmer, A. N. 1988. *Wind Cave—an ancient world beneath the hills.* Wind Cave/Jewel Cave Natural History Association.

———, and Palmer, M. V. 1986. Outline of geologic research in Wind and Jewel Caves. National Park Service, unpublished.

Thornbury, W. D. 1954. *Principles of geomorphology.* New York: John Wiley and Sons, Inc.

Wind Cave National Park

Area: 28,292 acres; 44 square miles
Established: January 9, 1903
Address: Hot Springs, South Dakota 57747

14

Largest underground rooms of any cave in the United States.

Formed in Permian limestone by sulfuric acid derived from hydrogen sulfide gas ascending from adjacent oil fields. Gypsum beds in cave are a byproduct of the reaction between sulfuric acid and limestone.

Passages extend through massive reef limestone, bedded back-reef limestone, and fore-reef breccia.

Great diversity in cave "formations," particularly dripstone.

Cave no longer enlarging by solution.

Carlsbad Caverns National Park

Southeast New Mexico

by Arthur N. Palmer
State University of New York, Oneonta

Figure 14.1   The land surface of Carlsbad Caverns National Park is a semiarid plateau, made up of reef limestones deposited in Permian time around the perimeter of the Delaware Basin. Walnut Canyon, shown here, is one of the entrenched valleys of Pleistocene age that dissect the plateau. The prominent bedding of a back-reef limestone is visible in the canyon's far wall. National Park Service photo.

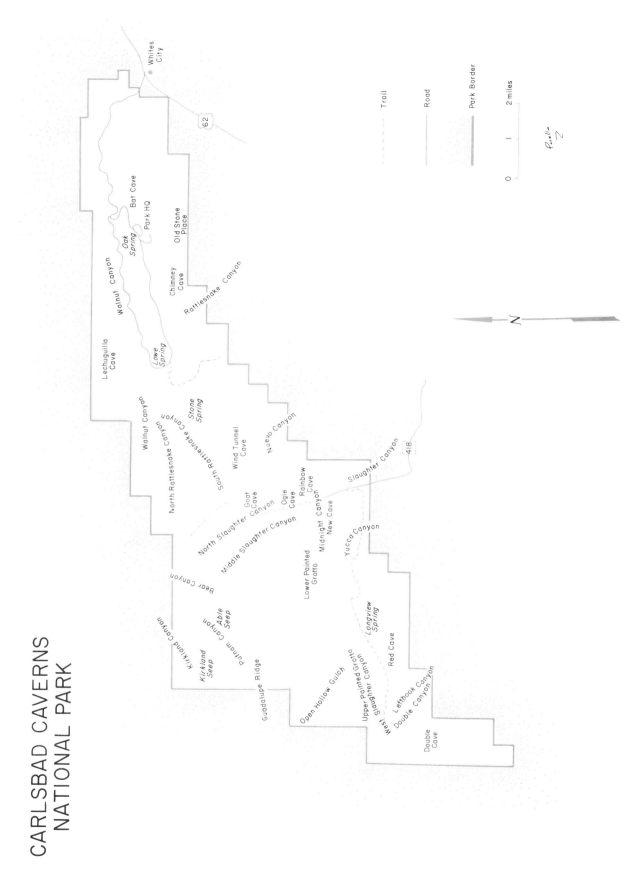

Figure 14.2 Map of surface features of Carlsbad Caverns National Park, New Mexico.

Regional Setting

Carlsbad Caverns National Park is located in the southeastern corner of New Mexico along the prominent southeast-facing Guadalupe Escarpment that extends northeast with diminishing height from Guadalupe Mountains National Park (chapter 15) in Texas. The two parks share nearly the same geologic setting of massive reef and bedded back-reef limestones of Permian age. This is a region of deeply dissected plateaus having an average altitude of 4000 to 6000 feet that overlooks the arid Delaware Basin 1000 feet below.

Carlsbad Cavern,* one of the largest caves in the park, matches the popular image of how a cave ought to look. It has huge, interconnected rooms with shadowy, vaulted openings leading off in every direction; massive and profuse decorations of stalacites and stalagmites; and prodigious length and depth. The cave's largest room, the Big Room, one of the world's largest, covers 14 acres. Its maximum floor-to-ceiling height is 370 feet at the so-called "Bottomless Pit." In the past geologists treated the cave as a grand example of how limestone caves in general must form. It is surprising, therefore, to learn that in its geology and origin Carlsbad Cavern is actually quite different from most other caves.

Public access to the cave is facilitated by well lighted trails. The tour is self-guided with an interpretive program keyed to various locations in the cave through portable radio receivers with headphones. Park Service guides are stationed throughout the cave to answer questions. Exit is provided directly to the surface by a 750-foot elevator that also affords entry for those who wish a shorter and less strenuous tour. The cave temperature is approximately 56° F year-round, equivalent to that of the mean surface temperature of the region.

Local History

The Guadalupe Mountains were inhabited by humans as early as 12,000 years ago, but Indians apparently did not explore Carlsbad Cavern because of a sharp drop-off a short distance inside the cave entrance. By the late nineteenth century, the entrance was well known to local ranchers because of the dense cloud of bats that exits each evening for nocturnal feeding on insects, and then returns to the cave near daybreak. These useful animals are Mexican free-tailed bats and number in the hundreds of thousands. They inhabit a room called Bat Cave (not open to the public), which is just inside the cavern. Each night the bat colony consumes more than a ton of insects.

The first person to enter the cave, lowered on a rope by his father in 1883, was apparently 12-year-old Rolph

*The park's major cave is now referred to as Carlsbad Cavern (without the "s"). The name of the national park remains the same.

Sublett, but he did not explore beyond the limit of daylight. The Bat Cave contains enormous deposits of bat droppings (guano), and the economic potential of the material as a fertilizer did not escape the attention of local speculators. In 1903 Abijah Long filed a claim for the area around the mouth of the cave and established a guano-mining operation. A shaft was constructed from the surface directly into the Bat Cave. Guano was removed in large cable-driven buckets, which also served as the primary means of entry and exit for the miners. The enterprise proved to be an economic failure, although over a 20-year period, six different companies tried unsuccessfully to run the operation. More than 100 million pounds of guano were excavated, lowering the cave floor as much as 50 feet.

A young guano miner named Jim White was fascinated by the huge, steeply descending underground corridor that led from Bat Cave past the natural entrance into the unknown. He made frequent explorations, often alone, deep into the cave and brought back wondrous tales. When the final mining company went bankrupt after World War I, Jim White stayed in the area and led private tours into the cave. At first his enthusiastic reports were scoffed at by most people. However, it was not long before others joined in his praise, including several officials of the federal government. Willis T. Lee of the U. S. Geological Survey led a six-month National Geographic Society expedition to study and map the cave with White serving as a guide. With the size and beauty of the cave substantiated, President Coolidge initiated legislation in 1923 to create Carlsbad Cave National Monument. In 1930 it was established as Carlsbad Caverns National Park.

Today Carlsbad Cavern is the most popular cave in the country and has close to a million visitors each year. More than 20 miles of rooms and passages have been explored and new discoveries are made almost yearly. More than 70 other caves are known within the park boundaries, but only one, Lechuguilla Cave (not open to the public), approaches the size of Carlsbad Cavern. Cave explorers have found spectacular crystalline gypsum speleothems in Lechuguilla Cave, which is longer than Carlsbad Cavern although the latter has larger rooms. Another large cave, New Cave, which contains travertine formations every bit as grand and bizarre as those of its larger neighbors, is open to the public on an unimproved basis. Park Service guides lead groups through with hand-held lanterns.

Geologic Features

Origin of the Cave

Unlike most caves, which are formed by carbonic acid from surface sources, Carlsbad Cavern was dissolved mainly by sulfuric acid created by the oxidation of sulfides. This acid

CARLSBAD CAVERN
INTERIOR CAVE

Figure 14.3   Underground passages open to the public in Carlsbad Cavern. Source: Carlsbad Caverns East quadrangle, U.S. Geological Survey 1957.

is uncommon in ground water except where hydrogen sulfide gas is available from deep sources or where iron sulfides, such as pyrite, are exposed to weathering. Recent evidence suggests that Carlsbad Cavern formed when hydrogen sulfide rising from depth reacted with oxygen at or near the water table and formed sulfuric acid, which readily dissolves limestone. The reactions are as follows:

1. Oxidation of hydrogen sulfide:

$$H_2S + 2O_2 \longrightarrow 2H^+ + SO_4^-$$
$$\text{sulfuric acid}$$

2. Solution of limestone:

$$H^+ + CaCO_3 \rightleftarrows Ca^{++} + HCO_3^-$$

Additional information about the origin of the cave is explained in later paragraphs.

Bedrock and Geologic Structure

Details of the geologic setting are similar to those of nearby Guadalupe Mountains National Park in Texas (chapter 15), so only a brief overview is given here. Three rock sequences, all about the same age but of different lithologies, occur in Carlsbad Caverns National Park. The Guadalupe Escarpment is formed by an exhumed barrier reef that developed around the perimeter of the Delaware Basin during the Permian Period. This reef, called the Capitan Limestone, contains a framework of bryozoan, sponge, and algae fossils in a matrix of unbedded, fine-grained limestone. To the north-northwest of the reef are the back-reef limestones (Artesia Group), which are

prominently bedded. They contain a large percentage of silt eroded from nearby land areas, as well as gypsum that accumulated in areas of restricted water circulation. To the southeast, the Capitan Reef is bordered by an apron of reef talus that merges with dark, fine-grained limestones deposited in the deeper waters of the Delaware Basin, contemporaneously with the other rock units. The basinal limestones are overlain by gypsum and evaporites, mainly of the Castile Formation, which are also Permian in age. Figure 14.4 is a diagrammatic representation of the sedimentary facies that make up the bedrock of the Carlsbad region.

As a result of Mesozoic and Cenozoic uplift, processes of solution and erosion have removed much of the gypsum. The more resistant limestones now stand out in relief, with nearly the same topographic expression they had while they were being deposited in the inland sea 200 to 250 million years ago. Southeasterly tilting and broad folding accompanied uplift. Faults with displacements up to several hundred feet are distributed throughout the park, particularly along the border zone between the reef and basinal facies. Normal faults along the western edge of the Guadalupe Mountains (outside the park) attain a displacement of several thousand feet. Prominent joints in the vicinity of Carlsbad Cavern are oriented in two sets, parallel to and perpendicular to the reef front. Many of the major joints parallel to the reef front have been filled with shallow-water deposits of land-derived sediment. These deposits now show up in the limestone as clastic dikes of sandstone and siltstone.

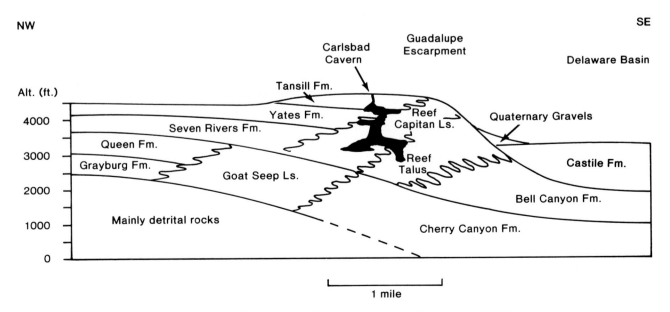

Figure 14.4   Geologic cross section through the Guadalupe Escarpment, showing the location of Carlsbad Cavern.

Cavern Features

A landscape dominated by solutional features is referred to as *karst topography*. A typical karst region contains many solutional fissures, pits, and funnel-shaped depressions called sinkholes where surface water collects and flows underground. Caves show evidence of subsurface solution by underground water. In the semiarid climate of the Guadalupe Mountains, surface karst features are almost entirely absent, so the region is not considered true karst topography, despite its many caves. Karst topography is best developed in a humid climate, such as that of Mammoth Cave National Park in Kentucky (chapter 12).

Most caves consist of long, conduitlike passageways dissolved by underground streams. In such a cave it is easy to distinguish the points through which ground-water recharge took place, as well as the springs through which water emerged at the surface, particularly since many caves are still actively forming today. In contrast, Carlsbad Cavern and most other caves in the Guadalupe Mountains consist of large, interconnected rooms with highly irregular patterns and profiles. Nearly all of these caves are inactive today after being abandoned by the water that formed them. Since clues to the original water sources, patterns of flow, and springs are virtually absent, attempts to describe Carlsbad Cavern in terms of "normal" circulation of ground water and the carbonic-acid reaction do not explain the unusual character of the cave complex.

Carlsbad Cavern is located mainly in the reef limestone, but also extends into the adjacent back-reef beds (in the northwestern part of the cave) and the fore-reef talus (at the south end of the Big Room). Nevertheless, the pattern and character of the cave are remarkably uniform, despite the great contrast in rock type. The main chambers of the cave are broad, arched rooms and linear, fissurelike passages whose major trends are controlled by the prominent jointing of the Capitan Limestone. In places, the rooms are bordered by intricate zones of *spongework,* where the original pores in the limestone have been enlarged by solution to form an interconnected maze of openings similar to that of Swiss cheese. The Boneyard is one of the best examples. The rooms, fissures, and spongework are the result of solution in the phreatic zone (below the water table). Vadose solution features are limited to a few minor rills and drip pockets.

Many of the largest rooms and passages are grouped at three different depths below the entrance: Bat Cave at −160 feet; the Big Room and Lunch Room at −660 feet; and the Lower Cave, King's Palace, and neighboring rooms at −740 feet. These levels are not controlled by differences in rock type and may represent separate stages of cave development. Several small lakes are scattered throughout the lowest rooms, including the lowest point in the cave, Lake of the Clouds, at a depth of 1037 feet

Figure 14.5   Large travertine column in the Temple of the Sun. Cave popcorn encrusts many of the stalactites and stalagmites shown here, giving them a bizarre, knobby appearance. Ceiling height in this part of the cave is about 50 feet. National Park Service photo.

below the entrance. These are merely local zones of ponding in the vadose zone and do not represent the water table. The original shape of the cave was modified in various places by the dropping of large rocks from the ceiling. The enormous Iceberg Rock is the largest of these *collapse blocks.* The presence of vadose travertine on some of the blocks shows that the collapse took place late in the developmental history of the cave; yet the ceilings are quite stable today, for no rocks have dropped in the entire time that people have known about the cave.

The deposits within Carlsbad Cavern are unusually extensive and varied. The most spectacular are composed of travertine, which is calcium carbonate precipitated when the carbon dioxide in ground water is released into the cave atmosphere. The rate at which these features grow can vary a great deal, but maximum rates of about one centimeter per hundred years have been observed in other caves. Stalactites, stalagmites, and helictites are more abundant in Carlsbad Cavern than in Mammoth Cave or Wind Cave (chapters 12, 13). Some of the stalagmites reach heights of 60 feet.

Where water seeps out of pores in the rock, nodular, grapelike protrusions called *cave popcorn* are precipitated. These deposits may form on other speleothems in the same way. Cave popcorn is most abundant in areas where capillary water is drawn from the moist rock by gentle, dry air currents that flow through the cave, as shown by the fact that popcorn growing on stalactites and stalagmites is generally much more profuse on the windward side. However, warm, moist air rising from lakes deeper in the cave has caused water to condense on the colder rock surfaces above. The windward sides of speleothems in such an area are partly dissolved, whereas popcorn is deposited only on the lee sides.

Where water drips into a shallow basin containing small particles such as sand grains, the particles may become coated with calcium carbonate. If the agitation is sufficient to keep the growing particles from being cemented to the floor, rounded *cave pearls* are formed. The largest cave pearls in Carlsbad Cavern are more than an inch in diameter.

Certain speleothems are deposited in ponded water. These include calcite crystals such as dogtooth spar, travertine deposits called *rimstone* around the edges of pools, and bulbous, mammillate wall coatings that look like clouds (hence the name "Lake of the Clouds").

Most of the large speleothems are not actively growing today and are in the process of drying out. Their profusion shows that the climate must have been much more humid in the past, presumably during pluvial episodes of the Pleistocene Epoch. It was discovered recently that a great deal of moisture was being lost as water vapor through the elevator shaft, so air locks have been placed at the elevators to prevent this accelerated drying of the cave.

Sediments in the Cavern

A variety of cave sediments of both mechanical and chemical origin provide evidence for the processes that formed Carlsbad Cavern. Most prominent of these are extensive bodies of gypsum deposited in the middle levels of the caverns. The gypsum deposits, which reach a maximum thickness of 20 feet, have been partly removed by solution, so only isolated blocks remain today. Water dripping from the cave ceiling has dissolved vertical "drill holes" through the gypsum in many places.

Most of the gypsum precipitated on the cave floor from sulfate-rich brine that once filled the cave; but in some areas the gypsum has actually replaced the limestone wall rock to a depth of several feet. Recent analysis of the gypsum shows that it is highly enriched in the light sulphur isotope S^{32}, which indicates a biological influence upon its origin. The gypsum was not derived from the Castile Formation, at least directly, because the Castile is not enriched in S^{32}.

The presence of the clay mineral endellite, overlain by montmorillonite, gives a further clue to the origin of Carlsbad Cavern. *Endellite* is formed by the alteration of other clay minerals in an acidic, sulfate-rich environment, whereas the *montmorillonite* (also a clay mineral) is a typical insoluble residue from the solution of limestone. The combined presence of gypsum and endellite suggests that sulfuric acid played a more significant role than carbonic acid in the origin of the cave. Sulfuric acid can be produced by the oxidation of pyrite, which is abundant in the back-reef limestones; but except for a few rare examples, the cave patterns show no relationship to the distribution of pyrite. Oxidation of sulfides in rising oil-field brines seems to be a more likely source of the sulfuric acid. Hydrogen sulfide gas (H_2S) occurs within rocks of the Delaware Basin that are rich in petroleum. Hydrogen sulfide oxidizes first to sulfur and water, and the sulfur oxidizes further to form sulfuric acid. Remnants of pure sulfur have been found in Carlsbad Cavern and other caves nearby; this supports the view that hydrogen sulfide played a role in the cave's origin. The lack of relationship between the way the cave evolved and the usual patterns of ground-water flow can also be explained. The three levels in the cave may represent average positions of the water table at various times, inasmuch as the greatest rate of oxidation of sulfides occurs at and above the water table but diminishes downward within the phreatic zone.

The bedded gypsum overlies silt that is mainly insoluble material released during the solution of the limestone. Other sediments include sparse, rounded limestone pebbles carried in from the surface during the developmental history of the cave, but they represent only a minor event that has little bearing on the cave's origin.

Geologic History

1. Deposition of limestones in the Delaware Basin during the Permian Period.

The Delaware Basin was a small inland sea 250 to 200 million years ago during the Permian Period. The Capitan Reef was one of many that developed around the perimeter of the basin, with impure, bedded limestones and gypsum accumulating on the landward side of the reef and fine-grained limestones in the center of the basin. Details of the sedimentary history are given in chapter 15, Guadalupe Mountains National Park.

2. First stage of karst development late in Permian time.

Late Permian glaciation in the polar regions caused a drop in sea level and exposed the recently deposited limestones to solution by surface runoff and by circulating ground water. Karst features formed at this time have been largely destroyed by later erosion or filled by detrital sediment.

Table 14.1 Geologic Column, Carlsbad Caverns National Park

Era	Period	Epoch	Rock Units	Geologic Events
Cenozoic	Quaternary	Holocene	surficial gravels	Drying of climate; deposition of cave popcorn
				Deposition of most speleothems in cave
		Pleistocene		Draining of cave
				Erosion of Castile Formation exposing reef front; erosional dissection of limestones
	Tertiary (Neo-gene / Paleogene)	Pliocene	surficial gravels?	Renewed uplift, faulting, tilting; major solutional development of Carlsbad Caverns; deposition of gypsum in cave
		Miocene		Slow erosion
		Oligocene		Early stages of origin of Carlsbad Caverns
		Eocene	surficial gravel and sand?	Laramide orogeny—uplift, faulting, tilting toward northeast
		Paleocene		
Mesozoic	Cretaceous			Minor erosion, low relief
	Jurassic			
	Triassic			
Paleozoic	Permian	Ochoan	NW: Rustler Formation; Salado Formation — SE: Castile Formation	Thin, widespread, terrigenous deposits
			Bell Canyon Formation	Deposition of evaporites in Delaware Basin
			Cherry Canyon Formation	
		Guadalupian	Capitan Limestone (reef); Goat Seep Dolomite; Artesia Group (Tansill Fm., Yates Fm., Seven Rivers Fm., Queen Fm., Grayburg Fm.)	Possible first stage of karst development
				Limestone deposited in subsiding Delaware Basin

Source: modified after Hayes 1957; Hayes and Koogle 1958; Jagnow 1979; and Queen 1981

174

3. Filling of the Delaware Basin with gypsum.

Waning of late Permian glaciers caused sea level to rise. During and after the polar glaciation, the Delaware Basin was nearly isolated from the ocean, and a high rate of evaporation caused gypsum and other evaporites to fill the basin at least to the height of the Capitan Reef.

4. Deposition of detrital sediment in a continental environment near the end of the Permian Period.

Silt and sand, eroded from nearby highlands, covered the area, creating the Salado and Rustler Formations. Some of the fill in the earlier karst features may date from this episode.

5. Tectonic uplift and deformation late in Cretaceous and early in Tertiary time; early stages of cave development.

As part of the Laramide orogeny, the Guadalupe Mountains were formed by a local uplift, which was accompanied by faulting and southeastward tilting. Erosion and karst development began in the higher regions, while sediment derived from these processes may have filled any remaining karst features of Permian age at lower elevations. Mobilization of hydrogen sulfide gas and sulfide-rich brines by the tectonic activity may have begun in mid-Tertiary time, producing sulfuric acid by mixing with oxygenated water that infiltrated from the surface. Some acidity could also have been produced by the mixing of descending water, rich in CO_2, with sulfate brines derived from the Castile Formation. Carlsbad Cavern began to form within the phreatic zone, but the extent of its development during this time is not yet known.

6. Renewed uplift of the Guadalupe Mountains during the late Pliocene Epoch and early Pleistocene Epoch; further enlargement of caves.

The major phase of solutional enlargement of Carlsbad Cavern is thought to date from the time of this uplift. Extensive gypsum deposits in this and many other Guadalupe caves indicate that the solution was caused by sulfuric acid derived from the oxidation of dissolved sulfides released from deeper rocks.

7. Draining of caves during the Pleistocene Epoch; deposition of speleothems.

The major Guadalupe caves gradually emerged into the vadose zone during the Pleistocene Epoch as uplift and erosional dissection continued. Pauses in the emergence, accompanied by continuing solution, may account for the three distinct levels in Carlsbad Cavern. Erosion and solution of the soft Castile gypsum exposed the Capitan Reef, which forms the Guadalupe Escarpment today.

Infiltration of CO_2-rich water from the surface deposited travertine speleothems, particularly during times of relatively humid climate in the Pleistocene. Electron-spin resonance dating of speleothems in the Big Room indicate that Carlsbad Cavern was drained at that level more than 500,000 to 600,000 years ago.

Intersection of the upper cave level by erosion formed the entrance of Carlsbad Cavern. Gradual diminution of infiltrating water and drying of the cave in the semiarid climate since the last glaciation has reduced the rate of speleothem growth. This change has favored the deposition of nodular travertine, such as popcorn, rather than stalactites and stalagmites.

Today Carlsbad Caverns National Park is not only one of the greatest tourist attractions of the Southwest, but it is also the subject of considerable research in the fields of sedimentary processes, climatic history, and the unusual mode of cave origin that has occurred in the region.

Geologic Maps

Hayes, R. T. 1957. *Geology of the Carlsbad Caverns east quadrangle, New Mexico.* U.S. Geological Survey map GQ-98.

Hayes, R. T., and Koogle, R. L. 1958. *Geology of the Carlsbad Caverns West Quadrangle, New Mexico.* U.S. Geological Survey map GQ-112.

Bibliography

Barnett, J. 1980. *Carlsbad Caverns National Park: silent chambers, timeless beauty* Carlsbad, New Mexico: Carlsbad Caverns Natural History Association.

Bretz, J. H. 1949. Carlsbad Caverns and other caves of the Guadalupe Block, New Mexico. *Journal of geology* 57: 447–463.

Davis, D. G. 1980. Cave development in the Guadalupe Mountains, a critical review of recent hypotheses. *National Speleological Society bulletin* 42: 42–48.

Dunham, R. J. 1972. Capitan Reef, New Mexico and Texas: facts and questions to aid interpretation and group discussion. Permian Basin Section, Society of Economic Paleontologists and Mineralogists, Publication 72–14.

Egemeier, S. J. 1981. Cavern development by thermal waters. *National Speleological Society bulletin* 43:31.

Good, J. M. 1957. Noncarbonate deposits of Carlsbad Caverns, New Mexico. *National Speological Society bulletin* 19: 11–23.

Halliday, W. R. 1966. *Depths of the earth.* New York: Harper and Row.

Harmon, R. S., and Curl, R. L. 1978. Preliminary results on growth rate and paleoclimate studies of a stalagmite from Ogle Cave, New Mexico. *National Speleological Society bulletin* 40: 25–26.

Hill, C. A. 1987. *Geology of Carlsbad Cavern and other caves of the Guadalupe Mountains, New Mexico and Texas.* New Mexico Bureau of Mines and Mineral Resources, bulletin 117, 150 p.

Hill, C. A., and Forti, P. 1986. *Cave minerals of the world.* Huntsville, Alabama: National Speleological Society. 238 p.

Jagnow, D. H. 1979. *Cavern development in the Guadalupe Mountains.* Columbus, Ohio: Cave Research Foundation.

McLean, J. S. 1971. *The microclimate in Carlsbad Caverns, New Mexico.* U.S. Geological Survey Open File Report at Albuquerque, New Mexico.

Morehouse, D. F. 1968. Cave development via the sulfuric acid reaction. *National Speleological Society bulletin,* 30:1–10.

Newell, N. D,; Rigby, J. K.; Fischer, A. G.; Whiteman, A. J.; Hickox, J. E.; and Bradley, J. S. 1953. *The Permian reef complex of the Guadalupe Mountains region, Texas and New Mexico—a study in paleoecology.* San Francisco, California: W. H. Freeman and Co.

Queen, J. M. 1981. A discussion and field guide to the geology of Carlsbad Caverns. Preliminary report to the National Park Service for the Eighth International Congress of Speleology.

Thrailkill, J. V. 1971. Carbonate deposition in Carlsbad Caverns. *Journal of geology* 79: 683–695.

Carlsbad Caverns National Park

Area: 46,755 acres; 73 square miles

Proclaimed Carlsbad Cave National Monument: October 25, 1923

Established as Carlsbad Caverns National Park: May 14, 1930

Address: 3225 National Parks Highway, Carlsbad, New Mexico 88220

15

This fault-block range displays portions of the world's most extensive and significant Permian limestones, containing vast numbers of marine fossils.

A major normal fault marks the western side of the mountain mass; the great Capitan Reef escarpment rims its eastern edge.

The karst features of nearby Carlsbad Caverns National Park, New Mexico (chapter 14), formed in a less uplifted portion of the Capitan Reef.

Guadalupe Peak (8749 feet), highest point in Texas, overlooks the Chihuahuan Desert to the south.

McKittrick Canyon, watered by a perennial stream, offers a geologic cross section of the reef, showing types of deposits, bedding, fossils, small caves.

Guadalupe Mountains National Park

West Texas

by Ann F. Budd, University of Iowa
and Katherine A. Goebel
University of Arizona

Figure 15.1   El Capitan, as viewed from Pine Spring Canyon in Guadalupe Mountains National Park. The escarpment of Capitan Limestone, which caps most of the Guadalupe Mountains, exposes part of the ancient reefs that encircled the Delaware Basin in Permian time. National Park Service photo.

GUADALUPE MOUNTAINS
NATIONAL PARK

Figure 15.2   Guadalupe Mountains National Park, Texas.

Local History

Studies of pictographs in caves and other archeological finds suggest that humans inhabited caves in Guadalupe Mountains National Park about 10,000 years ago. They moved with the seasons from the canyon floors to the highlands while collecting plants and hunting animals. Later arrivals in the Guadalupe region were the fierce Mescalero Apache Indians, who discouraged travel or settlement in their domain by any outsiders. From their sanctuary in the Guadalupe Mountains, they sent out raiding parties and kept surveillance over a wide area.

Although the name *Guadalupe* was used on a map as early as 1828, the area was not extensively explored by people of European descent until the mid-1800s when a series of surveys investigated possible wagon routes. The Butterfield Overland Stage set up a station at the Pinery in Guadalupe Pass on the first transcontinental mail route from St. Louis to San Francisco. The station was later abandoned because of Indian troubles, but its ruins are within the park, just off the main highway. Decades of hostilities followed before the Mescalero were subdued and moved to a reservation in 1879. Skirmishing also went on between the United States and Mexico, mainly over the salt flats. An uneasy peace was finally brought about between the two countries in 1878. The population remained sparse although a few ranchers homesteaded in the Guadalupe region early in this century. Bat guano was mined from caves for use as fertilizer for some years. Copper mining on a small scale was carried on in Dog Canyon near the park's northern boundary.

The park was authorized in 1966 after Wallace Pratt, a petroleum geologist and vice-president of Standard Oil Company of New Jersey, donated the North McKittrick Canyon area to the federal government for use as a national park (a total of 5632 acres). An additional 72,000 acres was purchased by the federal government and the park established in 1972.

Weather is sometimes extreme in the Guadalupe Mountains. Frequent high winds, intense summer thunderstorms, occasional snow in winter, and a wide range of temperatures are to be expected. Precipitation is sparse in this desert region. Rigorous conservation practices are preserving the fragile terrain and vegetation.

Geologic Features

The Guadalupe Mountains offer a rare opportunity for geologists to study a completely exposed sequence of evaporite, carbonate, and sandstone transitions from an ancient shelf down into a marine basin. Superbly exposed here is a portion of what is termed the "Permian Reef Complex," which contains one of the largest fossil reefs in the world, the Capitan Reef.

After being buried for millions of years, these reef deposits are now exposed in Guadalupe Mountains National Park (in an uplifted block), in karstic features in Carlsbad Caverns National Park (chapter 14) (in a down-faulted block), and to the south in the Glass Mountains and Apache Mountains. Subsurface drilling shows that the reef extends to the southeast for a distance of approximately 300 miles. It forms the shape of a horseshoe around what was in Permian time a deeper offshore marine basin, called the Delaware Basin (fig. 15.3). The most striking feature of Guadalupe Mountains National Park is the thousand-foot-high El Capitan Cliff, composed entirely of reef limestone (fig. 15.1).

The Capitan Reef

Unlike modern reefs that have been built mainly by a rigid framework of corals, the Capitan deposits consist of calcareous sponges and encrusting algae combined with other lime-secreting marine organisms. A great deal of lime mud is present that precipitated directly from seawater in a vast tropical ocean that covered the region in Permian time. (In general, reefs of Permian age are unusual because they were not built primarily by corals, the principal reef-formers of the middle Paleozoic Era and of the Mesozoic and Cenozoic Eras.)

By definition, a *reef* is a submerged resistant mound or ridge formed by the accumulation of plant and animal skeletons. Although reefs can form in deeper, cooler marine environments, the largest living reefs occur in shallow (not over 425 feet) tropical seas (temperatures over 68° F) where deep sea upwellings provide plentiful nutrients for organic growth. The seawater must also be clear, have normal marine salt content (27–40 parts per thousand), and be free of siliclastic (noncarbonate) sediment or freshwater carried in by streams. Reefs generally form in areas with heavy wave activity so that the upper surface of the reef, the reef crest, is subjected to the constant pounding of breaking waves.

The sediments associated with reefs are quite different from those found in many other depositional settings. Reef sediments, largely composed of calcium carbonate ("lime") and varied amounts of magnesium, consist exclusively of broken fragments of the skeletons of plants and animals that live on or near the reef. These fragments are irregular in size and shape and poorly sorted. They show little evidence of being transported great distances.

Different portions of a reef are subjected to different environmental factors during construction. Relatively high wave action on the seaward-facing section, or *fore reef,* causes both large and small blocks of living and dead skeletal material to roll down the fore reef slope into the deep sea basin and thereby form a slope of debris or talus at

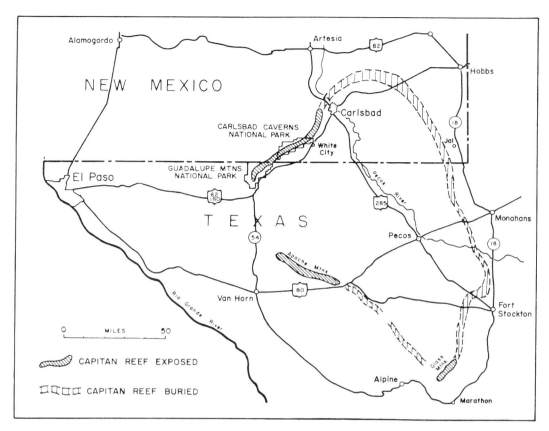

Figure 15.3   Exposed and unexposed parts of the Capitan Reef in Carlsbad Caverns and Guadalupe Mountains National Parks and vicinity. From D. Murphy, *The Guadalupes*. © 1984 Carlsbad Caverns/ Guadalupe Mountains Natural History Association. Used by permission.

the base of the reef. Before they fall, these blocks are severely weakened by bio-eroders (organisms) that rasp and bore through the dead skeletal material. Relatively high light intensity on the top of the reef, the *reef crest,* fosters a rapid upward growth of plants and animals that form the solid framework of the reef. Guadalupe reef crests were sedimentary shoals that received energy from currents. There is relatively low wave or current activity on the shoreward-facing section, or *back reef;* therefore, only fine sediment is carried back into this area, and the water itself is often stagnant and muddy. Salinity is especially high in areas lacking inflowing streams.

With each part of a reef having a characteristic marine environment, it is not surprising that the *biota* (i.e., the living organisms) of the reef area are highly diverse and also different from one zone of the reef to the next. For instance, plants and animals with more robust skeletons tend to inhabit the reef crest and fore reef areas, whereas those that burrow or thrive in muddy places tend to live in the back reef. The fastest growing plants and animals usually occupy the shallow fore reef where competition for space is especially intense. Because of the differences in the physical environment and in the constituent

biota whose skeletons form the reef, distinctive sediment types are found in the fore reef, reef crest, and back reef areas. Each of these sediment types is called a *facies* (different rock types deposited in different environments at the same time). The sediments accumulate gradually as more plants and animals live and die, and their skeletons are washed over the lagoon and fore reef basins. As they are buried, the skeletal remains and the sediments created from them are compacted and recrystallized, becoming lithified as limestones. Sediment deposited in stagnant back reef or lagoon waters often contains high amounts of magnesium. The magnesium combines with limestone to form the rock dolomite. In especially salty water, the salt alone often forms rocks as the water evaporates. Three types of evaporites are associated with the Permian West Texas reefs: anhydrite, gypsum, and halite. As the salt content of the water increases during evaporation, first *anhydrite* (calcium sulfate, commonly as nodules or nodular layers) forms, then *gypsum* (hydrous calcium sulphate that tends to form long slender crystals), and last, *halite* (rock salt, sodium chloride).

During late Permian time (280–230 million years ago) in West Texas and New Mexico, three reef facies formed

Figure 15.4 Location and extent of the Delaware and Midland Basins in Permian time, superimposed on a modern map of New Mexico and Texas. In Permian time, the region was in the equatorial belt, to the southwest of its present position. Modified from *Historical Geology of North America,* 2nd edition, M. S. Petersen, J. K. Rigby, and L. F. Hintze. © 1973, 1980 Wm. C. Brown Publishers. Used by permission.

synchronously along the south-central portion of the North American craton in an inland sea comparable in size to today's Black Sea (fig. 15.4). To the east in the Appalachian region, orogeny took place as the African and North American plates collided. To the west along the edge of the craton through Nevada and Idaho, subduction of an oceanic plate was producing volcanoes. The reef facies formed in an arid tropical region south of the Permian equator. They consisted of: (1) a fore reef talus facies and deeper sea basin facies formed in the so-called Delaware Basin; (2) a reef crest and shallow fore reef facies formed in the surrounding El Capitan Reef Complex; and (3) a back reef and lagoon facies formed in the northeasterly Midland Basin. The rocks of the Delaware Basin consist largely of black limestones with lesser amounts of quartz sandstone. The black color of the limestone is believed to have been caused by high organic content deposited in a reducing environment and thus preserved. The process is similar to that of restricted marine environments of today where the bodies of floating microscopic animals fall to the bottom and are mixed with sediment. Such basin bottoms presumably contain little or no oxygen and thus few bacteria to consume the dead organic material.

The El Capitan Reef Complex consists of light-colored, massive fossiliferous limestones; massive primarily because they were formed by the growth and accumulation of skeletons rather than by deposition of sediment from moving water. The skeletons were stabilized by encrusting organisms that grew over and thereby cemented the solid reef rock. (This contrasts with modern reefs that make a delicate framework of corals.) The Permian organisms accumulated predominantly in shallow water where ample supplies of light existed to provide the energy necessary for algal photosynthesis.

A walk through McKittrick Canyon (fig. 15.5) proceeds through all three major reef facies. The walk begins in the deep basin facies (the Bell Canyon Formation) and moves into the talus slope at the base of the fore reef (the Capitan talus). The Bell Canyon Formation contains numerous beds of *sedimentary breccia* (a rock composed of nonsorted, nonlayered, angular, and often large rock fragments) surrounded by fine-grained basin deposits. The breccias are believed to have been deposited when submarine slides rolled down the steep face of the fore reef and talus slope. The Capitan talus consists almost exclusively of steeply sloping thick beds of this breccia. On top of the Capitan talus lies the shallow fore reef and reef

Figure 15.5   In this view of McKittrick Canyon, the three major reef facies are exposed: deep basin (fore reef), reef crest, and back reef. National Park Service photo.

facies (the Capitan Formation) and then the back reef/lagoon facies (the Carlsbad Group). The reef limestone itself is composed largely of sponges, algae, and bryozoans (tiny colonial animals). Brachiopods (shelled animals), crinoids (sea lilies), and fusulinids (large, single-celled animals) are common in the fore reef and back reef facies. The back reef facies show a progressive change from carbonates (limestones and dolomites) through evaporites to clastic terrigenous (land-derived) rocks. Several specialized rock types can be found: (1) *coquina* ("shell hash"), deposited in the lagoon in water having normal marine salt content; (2) lenses of *pisolites* (spherical particles containing concentric layers, sometimes called "pea-stone"), precipitated inorganically in intertidal areas along the shore; (3) fine-grained dolomite, deposited in intertidal and supratidal areas along the shore in salty water; and (4) red siltstone and sandstone mixed with anhydrite, deposited in supratidal areas along the shore in salty ponds.

The Capitan Reef was originally interpreted by N.D. Newell (1953) to have formed as a barrier reef. *Barrier reefs* are long, narrow, wave-resistant ridges that form parallel to a shoreline and are separated from the shore by a basin of water called a lagoon. Barrier reefs tend to

restrict water circulation between the open ocean and the lagoon.

Recently, controversy has developed over whether the Permian Capitan Reef Complex was actually a wave-resistant structure. Some workers have questioned if the intertwined algal fronds forming the reef could have been strong enough to withstand wave impact. Others have claimed that the large volume of detrital (i.e., composed of broken-up fragments) sediments in all three major facies suggests that the reef and the organisms forming the reef played only a minor role in controlling the sedimentation patterns of the area. Many geologists, therefore, have reinterpreted the Capitan Reef to represent merely a bank of skeletal debris below wave base (Achauer 1969). Despite this re-evaluation, most workers agree that the Delaware Basin sediments were deposited in an offshore enclosed marine basin; that the Capitan Reef Complex was formed on the edge of the basin in shallow marine conditions; and that the Midland Basin sediments were formed shoreward of the skeletal accumulations on an intertidal-supratidal, hypersaline (excessively salty) platform, or shelf.

Table 15.1   Geologic Column, Guadalupe Mountains National Park

Time Units			Rock Units			Geologic Events
Era	Period	Epoch	Back reef, lagoon	Reef crust, fore reef	Offshore basin	
Paleozoic	Permian	Ochoan?	Rustler Formation			Progradation of terrigenous red sandstones and siltstones across evaporites
			Salado Formation		Salado Formation	Widespread deposition of evaporites, filling up the lagoon and back reef areas
					Castile Formation	
	Upper		Carlsbad Group (Seven Rivers, Yates, Tansill Formations)	Capitan "Reef"	Bell Canyon Formation	Reduced basin subsidence and tectonic stabilization Rapid reef growth outward over basin sediments
		Guadalupian	Queen Formation	Goat Seep "reef"	Cherry Canyon Formation	Increased subsidence of deeper offshore basin during which three distinct reef facies (nearshore dolomites, small banks, offshore basin deposits), developed; reef increased in size but most growth was directed upward
			Grayburg Formation	Getaway "Bank"		
			San Andres Formation	Brushy Canyon Formation		Slow subsidence of offshore basin; reef development continued
	Lower	Leonardian	Yeso Formation	Bone Spring Limestone		Infilling of the basin with carbonates and shale lenses to form a gently sloping basin and shelf margin; development of small patch reefs.

Source: modified after Newell et al. 1953

Geologic History

1. Development of the Delaware Basin (West Texas) early in Permian time (Wolfcampian Epoch).

The Delaware Basin formed by earliest Permian time as an inland basin with a narrow outlet to the open ocean (fig. 15.6). More or less oval in shape, the body of water covered an area of around 10,000 square miles. Although the basin was largely enclosed, some ocean water entered periodically to replace water lost by evaporation. The circulation patterns within the basin are believed to have been similar to those in the present-day Mediterranean Sea. During this time, 1600 to 2200 feet of limestones and interbedded dark shale were deposited in the shallow, subsiding basin.

2. Forming of small banks along the basin margin (Leonardian Epoch).

By the beginning of mid-Permian time, the basin temporarily ceased to subside. Alternating beds of gypsum, sandstone, and dolomitic limestone (the Yeso Formation,

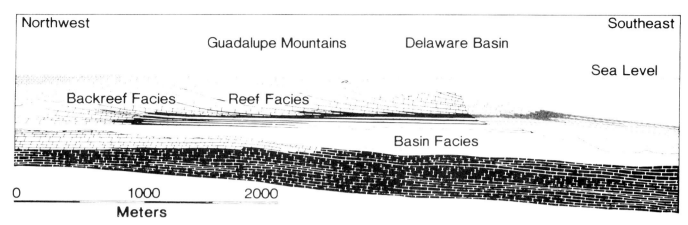

Northwest   Southeast

Guadalupe Mountains   Delaware Basin

Sea Level

Backreef Facies   ~ Reef Facies

Basin Facies

0   1000   2000
Meters
Vertical and Horizontal Scales Are the Same

Figure 15.6   Northwest-southeast cross-sectional diagram of the Delaware Basin showing
Permian structures now exposed in a fault scarp in the western part of the Guadalupe
Mountains. As the basin subsided, the reef grew upward and toward the basin center. Late in
Permian time, as the equatorial sea dried up, the basin filled with evaporites. From *Historical
Geology of North America*, 2nd edition, M. S. Petersen, J. K. Rigby, and L. F. Hintze. © 1973,
1980 Wm. C. Brown Publishers. Used by permission.

the Victoria Peak Member) were deposited in nearshore
areas, while thin-bedded black limestones (the Bone Spring
Limestone) accumulated in the stagnant deeper portion
of the basin. Small, discontinuous patch reefs formed in
shallow water along the rim of the basin.

3. Continued patch reef development early in Upper Permian time (Guadalupian Epoch).

By mid-Permian time, the basin began to subside
again and many larger patch reefs formed along the basin
margin. Cherty dolomites (San Andres Formation) built
up close to shore, while quartz sandstones (Brushy Canyon
Formation), containing scattered patch reefs, formed in
and around the basin.

4. Forming of the first barrierlike reef during the Upper Permian (Middle Guadalupian Epoch).

As the patch reefs developed along the basin margins,
the Delaware Basin itself began to subside rapidly. (This
subsidence may have been related to compressive folding
that resulted from the colliding of North American, South
American, and African plates.) As the basin subsided, reef
growth intensified, predominantly in an upward direction,
forming the Goat Seep reef. Three distinct facies developed: (1) a lagoon and back reef facies composed of sandstones and dolomites (the Queen Formation, the Grayburg
Formation); (2) a reef crest and shallow fore reef facies
containing dolomitized sponge and algae skeletons (the
Goat Seep reef, the Getaway Bank); and (3) an offshore
basin facies composed largely of quartz sandstones (the
Cherry Canyon Formation).

5. Development of an extensive reef complex during the Upper Permian (Late Guadalupian Epoch).

By the end of mid-Permian time, the basin slowly
ceased to subside and achieved a stable position. As this
position was reached, the largest reef, the Capitan Formation, expanded rapidly 350 miles around the rim of the
basin and gradually prograded toward the center of the
basin. The three facies differentiated further. Limestones
and sandstones (the Bell Canyon Formation) were deposited offshore in the basin; massive skeletal limestones (the
Capitan Formation) accumulated in shallow areas encircling the basin; and dolomites and fine-grained sandstones
(the Carlsbad Group) were formed in nearshore areas.

6. Infilling of the basin with evaporites during the Upper Permian (Ochoan Epoch).

Continued sedimentation in the stable basin and possibly a drop in sea level caused gradual shallowing and
ultimate drying up of the basin late in Permian time. As
the basin dried up, extensive accumulations of deep-water
evaporite deposits—the Castile Formation—covered the
lagoon and reef and eventually the basin itself. Anti-
estuarine circulation may have caused the deposits to form.
Laminae (bands of layers $1/16$ inch in thickness) of gray
anhydrite/gypsum, brown calcite, and halite precipitated
first in the basin. Halite and potassium-rich salts (the
Salado Formation) formed synchronously in nearshore
areas and around the basin margin. Eventually the salts
were deposited throughout the entire area.

7. Migration of terrigenous red beds over evaporite deposits during the Upper Permian (Ochoan Epoch).

At the very end of Permian time, terrigenous red siltstones and sandstone (Rustler Formation and Dewey Lake Formation) were deposited on the evaporites as rivers began to migrate across the area.

8. Uplift and tilting of the area during late Mesozoic and early Cenozoic time.

A series of tectonic movements created a major fault on the west side of the Guadalupe Mountains (the upthrown block), separating the range from a downfaulted block, the Salt Flat Bolson. The Guadalupe block was tilted to the northeast exposing the Capitan Limestone in the great rim escarpment with El Capitan at the "prow." The fault line along the Salt Flat Bolson (west side of the park) brings Cretaceous beds in the valley in contact with Permian beds that form the highlands.

9. Erosion during the Cenozoic Era

Stream erosion has removed much of the softer sediment and lowered the region to its present level. Ground water erosion formed numerous caves throughout the limestone units; the largest are those in Carlsbad Caverns National Park (chapter 14). Because of the steep relief and lack of vegetation, landslides have eroded some of the hillsides. This process of mass wasting has continued into the present.

Geologic Map

Roswell Geological Society. 1964. *Geology of the Capitan Reef complex of the Guadalupe Mountains* (field trip guidebook). Roswell, New Mexico: Roswell Geological Society (map and text).

Bibliography

Achauer, C. W. 1969. Origin of Capitan Formation, Guadalupe Mountains, New Mexico and Texas. *American Association of Petroleum Geologists Bulletin* 53:(11) 2314–2323.

Heckel, P. H. 1974. Carbonate buildups in the geologic record: a review. In *Reefs in time and space,* ed. L. F. Laporte, pp. 90–154, Special Publication 18. Tulsa, Oklahoma: Society of Economic Paleontologists and Mineralogists.

Hill, C. A. 1987. Geology of Carlsbad Cavern and other caves in the Guadalupe Mountains, New Mexico and Texas. New Mexico Bureau of Mines and Mineral Resources, Bulletin 117, 150 p.

Kendall, A. C. 1979. Continental and supratidal (sabkha) evaporites (#13) *and* Subaqueous evaporites (#14). In *Facies Models,* ed. R. G. Walker, pp. 145–74. Ontario: Geological Association of Canada Publications.

King, P. B. 1948. Geology of the Southern Guadalupe Mountains, Texas. U.S. Geological Survey Professional Paper 215, 183 p.

Murphy, D. 1984. *The Guadalupes.* Carlsbad Caverns/ Guadalupe Mountains Natural History Association.

Newell, N. D.; Rigby, J. K.; Fischer, A. G.; Whiteman, A. J.; Hickox, J. E.; and Bradley, J. S. 1953. *The Permian reef complex of the Guadalupe Mountains region, Texas and New Mexico.* San Francisco: W. H. Freeman & Co.

Selley, R. C. 1978. *Ancient sedimentary environments.* 2nd ed. Ithaca, New York: Cornell University Press.

West Texas Geological Society. 1960. *Geology of the Delaware Basin and field trip guidebook.*

Wilson, J. L. 1975. *Carbonate facies in geologic history.* New York: Springer-Verlag.

Guadalupe Mountains National Park

Area: 76,293 acres; 119 square miles; wilderness, 46,850 acres
Established: September 30, 1972
Wilderness designated: November 10, 1978
Address: 3225 National Park Highway
Carlsbad, New Mexico 88220

16

Most of the surrounding waters and three-fourths of St. John's land area are protected in Virgin Islands National Park. Of volcanic origin, the Virgin Islands are overlaid by limestone.

Tectonic uplift raised the volcanic islands to their present level. Both onshore and offshore features were modified by changing sea levels during the Pleistocene Epoch.

Coral reefs began to develop early in Pleistocene time.

Virgin Islands National Park

St. John and Hassel Islands
Caribbean Sea, East of Puerto Rico

by Ann F. Budd
University of Iowa

Figure 16.1   Little Cinnamon Bay on the north side of St. John Island. Tropical vegetation (breadfruit, mahogany, palm, mangrove, and kapok) masks bedrock of the Louisenhoj Formation, which weathers to a reddish color, where it is exposed. National Park Service photo.

Local History and Geographic Setting

There are four major groups of island chains or archipelagoes in the Caribbean Sea (fig. 16.2): (1) the Bahama Islands (running northwest to southeast, southeast of Miami); (2) the Greater Antilles including Cuba, Jamaica, Hispaniola, and Puerto Rico (running west to east, south and southwest of the Bahamas); (3) the Lesser Antilles including Antigua, Guadeloupe, Martinique, and Barbados (running north to south, southeast of Puerto Rico); and (4) the Netherland Antilles including Curaçao and Aruba (running west to east along the north coast of Venezuela). The Virgin Islands (about 100 in all) lie at the extreme eastern end of the Greater Antilles between that archipelago and the Lesser Antilles bending southward. The U.S. Virgin Islands include St. Thomas, St. John, St. Croix, and other islands on the west side of the group. The British Virgin Islands, in the northeast part of the archipelago, comprise about 40 islands, including Tortola, Virgin Gorda, and Anegada.

Early inhabitants of the Virgin Islands were peaceful Arawak Indians, who had fled in their canoes from warring tribes along the north coast of South America. They reached the islands around A.D. 300 and settled in small groups along the shores of St. John in Cinnamon Bay, Turtle Bay, Caneel Bay, and Cruz Bay. They were farmers who used a slash-and-burn method for raising *manioc* (a starchy root), which they had brought from South America. They supplemented their diet with birds, fish, shellfish, and sea turtle. During the 1300s, the Carib Indians migrated from the Amazon River Basin and either absorbed or exterminated the Arawaks. The Caribs, who were good hunters, swimmers, canoeists, and fighters, practiced cannibalism.

Christopher Columbus discovered the Virgin Islands in 1493 during his second voyage to America. Columbus called the archipelago *Las Virgenes,* or "The Virgins," in honor of the martyred St. Ursula of Cologne and her 11,000 virgin followers. During the century after Columbus's discovery, explorers came from the Netherlands, England, France, Denmark, and Spain. In the seventeenth century, pirates frequently visited this part of the Spanish Main.

The Danes officially claimed the Virgin Islands in 1687, but St. John Island was not actually settled by Europeans until 1716 when Danish and Dutch planters who lived on St. Thomas moved to St. John, divided the island into estates, and built plantations. The planters brought in slaves (West African blacks) for labor and raised sugar and cotton. By 1726 all potentially usable land was being farmed. A drought triggered a slave revolt in 1733 that resulted in heavy loss of life. After the revolt was finally put down by French soldiers from Martinique, the plantations were re-established and the planters prospered once again. In 1848, however, after a second slave revolt, the

Danish governor emancipated the slaves. Shortly thereafter, many planters and freed slaves left St. John because the soil had become depleted and farming without slave labor was unprofitable. The population of the island fell from 2000 to about 700, and the land went back to bush. Remains of colonial Danish plantations are in the park.

During World War I, the United States bought the Virgin Islands from Denmark for $25 million because of their strategic position alongside the approaches to the Panama Canal. In order to protect the wildlife and natural resources, Laurance Rockefeller bought up most of the land on St. John Island. In 1956 he presented 5000 acres to the United States as a national park. Subsequently, additional land and offshore areas (including Hassel Island in St. Thomas harbor) were incorporated into the park to protect the coral reefs from collection and commercial exploitation. The park was designated a Biosphere Reserve in 1976 by the United Nations.

St. John Island is only 5 miles wide by 9 miles long, and most of the topography is mountainous. Bordeaux Mountain, the highest peak, rises 1277 feet above sea level. The climate is pleasantly mild year round, except for occasional hurricanes. Fahrenheit temperatures seldom reach the high 90s because northeast trade winds and the ocean moderate the tropical heat. Rainfall, supplied by the trade winds and collected in cisterns, is the chief source of fresh water.

Development of Coral Reefs

Much of the modern offshore topography of St. John and other Caribbean Islands developed during the Pleistocene and Holocene Epochs as fringing reefs grew upward and outward along the windward coast on the eastern side of the islands. Such reefs range today from southern Florida (chapters 17, 18) to the southern margin of the Caribbean Sea, throughout an area where temperatures are relatively warm and sedimentation of fine particles is reduced. Although a few atolls have been described in the western Caribbean and some of the reefs in the Bahamas and Florida could be considered fringing barrier reefs, throughout the Pleistocene and Holocene Epochs the vast majority of reefs in the Caribbean were and are either fringing or patch reefs.

Unlike the Permian algal reefs in Guadalupe Mountains National Park (chapter 15), these Pleistocene and Holocene reefs are constructed primarily of coral skeletons. A *coral* is a cylindrical-shaped marine invertebrate that lives either singly or structurally bound together with others in a colony. (Corals belong to class Anthozoa, phylum Coelenterata.) Before a coral matures, it attaches its cup-shaped base to a hard rock substrate where it remains fixed for its entire adult life (sometimes several centuries). Each individual (or *polyp*) then grows upward

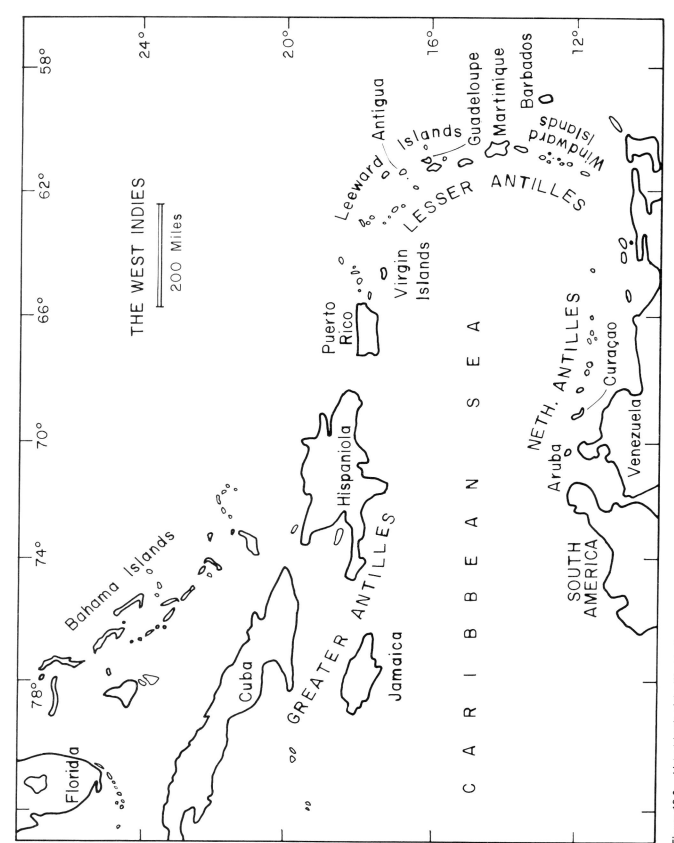

Figure 16.2 Major islands of the West Indies.

VIRGIN ISLANDS NATIONAL PARK

Figure 16.3 St. John Island, Virgin Islands National Park.

Box 16.1
Tectonic Activity in the Caribbean Region

The Caribbean Sea formed during the early part of the Cretaceous Period as Europe and Africa moved away from North and South America (fig. 16.4-I). In the general area now occupied by the Greater Antilles, a system of transform faults (explained below) and spreading centers (A) formed the Caribbean plate and separated North and South America. A subduction zone (B) (in central Venezuela) was active on the boundary between the Caribbean plate and the South American plate. Shortly thereafter, the spreading center system (A) was replaced by a subduction zone (A) (fig. 16.4-II), and a new spreading center (C) became active in the area now between the Greater Antilles and the northern Venezuela coast (fig. 16.4-II). The core of the Greater Antilles consists of rocks extruded from volcanoes in an island arc that developed along the newly formed subduction zone during early to middle Cretaceous time. The Netherland Antilles, which were probably part of the Greater Antilles island arc, lay on the other side of the new spreading center (C) and, therefore, "drifted" away from the Greater Antilles during middle to late Cretaceous time. The older subduction zone (B) through central Venezuela became inactive during the middle of Cretaceous time.

The newer spreading center (C) and subduction zone (A) ceased activity during late Cretaceous or early Tertiary time, and a *transform fault* (A) formed in the location of the subduction zone (A) (fig. 16.4-III). A transform fault is an extensive strike-slip (i.e., having only horizontal movement) fault that occurs at the boundary between two tectonic plates. In this type of plate boundary, no crust is consumed (as in a subduction zone) or created (as in a spreading center); therefore, the North American and Caribbean plates merely slid by one another. A subduction zone (C) and associated volcanic island arc also formed at the same time near the

spreading center (C) along the northern coast of Venezuela. A second subduction zone (D) and island arc became active during the early Tertiary along the line of the modern Lesser Antilles. Subduction along the north coast of Venezuela ceased during late Tertiary time and a transform fault (C) formed along its former position. Subduction of the North American plate beneath the eastern Caribbean plate margin still continues today near the Lesser Antilles (fig. 16.4-IV).

Throughout the last of the Mesozoic Era and in early Tertiary time, no major land barrier separated the Caribbean Sea from the Pacific Ocean. Late in the Pliocene Epoch, the Isthmus of Panama closed the Central American connection between North and South America.

This complicated geologic history is reflected in the structure and rock composition of various islands. Generally, the Bahamas consist of more than 10,000 feet of Cretaceous and Tertiary limestones deposited in Cretaceous time in a shallow-water marine environment between the North American continental margin and the Greater Antilles subduction zone (A) (fig. 16.4-II). The Greater Antilles are composed largely of highly deformed igneous and metamorphic rocks of similar ages that formed in the volcanic portion of the boundary zone between continental and oceanic plates. Volcanic and seismic activity in the Greater Antilles has now largely subsided, and the region is thought to be relatively stable tectonically.

Some scientists believe that the Greater Antilles are related to the North American Cordillera (on the western side of the North American plate), because the Greater Antilles are structurally more similar to northern Central America than is southern Central America (i.e., the Isthmus). Certainly both the North American Cordillera and the Greater Antilles experienced major deformation during Cretaceous time.

The Lesser Antilles, however, consist primarily of Tertiary and younger volcanics. The islands are arranged in two arcs: (1) an inner chain of volcanoes (some dormant and some active) consisting largely of late Tertiary and Quaternary volcanic rocks; and (2) an outer chain of early Tertiary volcanoes capped by late Tertiary limestones. In contrast, the Netherlands Antilles consist of Cretaceous volcanic rock overlaid by widespread Eocene limestones. As in the Greater

Antilles, volcanic and seismic activity in the Netherlands Antilles has largely subsided.

The rock composition and deformation history of the Virgin Islands are more like that of the Greater Antilles than the Lesser Antilles. Rocks forming St. John Island consist largely of volcanic rocks that were intruded and deformed during late Cretaceous and early Tertiary time, when the Laramide orogeny was building the North American Cordillera.

Figure 16.4 I-IV   Schematic representations of major plate movements in the Caribbean Sea during the past 130 million years. Encircled letters indicate zones of significant tectonic activity. (Adapted from Mattson 1977.)

by continuously secreting an external lime skeleton around the lower cup-shaped portion of its body. The polyp(s) thus occupies only the outer surface of a sometimes large (may be more than 5 to 6 feet in diameter) skeleton. Corals feed by capturing microorganisms with a ring of tentacles that encircle the outer edge of the upper surface of each polyp. Each polyp then uses its tentacles to insert the captured prey into a mouth located in the center of its tentacular ring. Some corals contain algae (*zooxanthellae*) within their living tissue that aid in their metabolism. Corals containing numerous algae can secrete skeletons and thereby grow; they grow especially rapidly (as much as 3 inches per year) in well lit habitats. Colonial (not solitary) corals are largely responsible for reef-building. They assume a great variety of shapes, each adapted for a specific wave energy, light intensity, or sedimentation rate.

Like the Permian reefs of Guadalupe Mountains National Park, the corals forming the Quaternary reefs of St. John are distinctively zoned parallel to the reef crest. In offshore areas with high wave energy (for example, across Denis and Coral Bays), the reef crest is dominated by the thick-branched "elkhorn" coral *Acropora palmata* and the mound-shaped brain coral *Diploria strigosa*. On either side of the crest lies a second zone dominated by the thick-branched "staghorn" coral *Acropora cervicornis* (fig. 16.5). Still farther from the crest in both back reef and fore reef areas, one finds a third zone dominated by the knobby "finger" coral *Porites porites* and the mound-shaped "star" coral *Montastraea annularis*. In areas where the wave energy is lower (for example, off Annaberg near Mary Creek), the first zone does not occur. Here the second zone lies on the reef crest and is surrounded only by the third zone. Coral zones like those found today existed throughout the Quaternary Period across the entire Caribbean. Although corals form the reef framework, they do not live alone on these reefs. Coral reefs consist of a diverse, well integrated community of marine animals that may also include sea urchins, fish, shrimp, sea fans, sea whips, snails, worms, and sponges.

Reef Terraces

During the Pleistocene Epoch, several *reef terraces* (horizontal topographic plains bounded on either side by steep slopes) formed along the coasts of all the modern Caribbean Islands. Most of the St. John reef terraces are now submerged; however, on many of the islands (the most spectacular examples can be seen on Barbados), several terraces lie above sea level. These reef terraces have been cut back about 500 feet and relief between terraces is about 50 feet. They are composed of lithified skeletons of corals that lived and formed reefs immediately prior to terrace formation. It is believed that throughout the Caribbean reef development flourished at times when sea level rose in the Pleistocene Epoch. Erosion occurred forming a ter-

race each time sea level was lowered. Terraces formed repeatedly during the Pleistocene Epoch because sea level changed frequently. The ocean rose as much as 60 feet above today's sea level when the glaciers melted during interglacial periods and dropped as much as 300 feet below today's sea level during periods of glacial accumulation. The exact number and timing of Pleistocene terrace formation throughout the Caribbean Islands is not clearly understood because of complications resulting from vertical tectonic movements and isostatic adjustments that affected all the islands during this time. Recently sea level has risen on St. John; therefore, upward reef growth is intensified and many of the ancient reef terraces are drowned. One such terrace occurs off the south shore at depths of 30 to 100 feet.

Wave action has formed the beaches of St. John Island. The reefs and submerged terraces are the main sources of sand. Because no permanent streams exist on the island, runoff does not supply much sediment to the lagoons and beaches except during heavy tropical storms or occasional hurricanes. However, storm waves and "swells" (large waves that travel across many miles of open ocean before breaking on a shore) erode both reefs and shore rocks, thus producing sand and cobbles. Some of the beaches on the island, such as Europa Bay on the southeast shore, are almost entirely made up of coral rubble; but nearby Little Lameshur is mainly fine sand. Cobblestones, well rounded by wave action, cover much of Greater Lameshur, an adjacent beach. Each beach on the island's shore has been shaped by local conditions of the coastal environment, which include bottom topography offshore, character of the reef at that location, tidal currents, and wind and wave dynamics. The National Park Service's intention is to preserve this unique ecosystem of shore, ocean, and reef as much as possible in its natural state.

Geologic History

1. **Submarine extrusion of lava flows and subsequent uplift during early Cretaceous time (Neocomian Epoch).**

The rocks forming the core of the Virgin Islands (the Water Island Formation) consist of more than 15,000 feet of *keratophyre* (a light-colored igneous rock composed predominantly of plagioclase feldspar and quartz) and *spilite* (a mafic green igneous rock composed predominantly of plagioclase feldspar and chlorite). The lavas were extruded over deep sea clay sediments by quiet underwater volcanic flows onto a deep (more than 15,000 feet), relatively flat sea bottom. These rocks are found throughout the southeastern half of the island and can be best seen near Centerline Road and Bordeaux Mountain Road, near Chocolate Hole, and near Reef Bay trail. The lack of pumice and other land-derived sediment in these

Figure 16.5   Staghorn corals (*Acropora cervicornis*), lower left corner of the photo, dominate this portion of the reef in Caneel Bay. Soft corals (upper right) and fish live among the staghorn branches. National Park Service photo.

rocks, as well as their chemical composition, confirms their deep sea origin. The volcanoes that produced the lavas were part of the island arc chain that built up the Greater Antilles Islands. During the Cretaceous Period, these islands lay in the western portion of the Tethyan seaway, an ancient ocean that extended around the central tropics from northern India through the Mediterranean Sea across the then-narrow Atlantic Ocean into the Caribbean Sea.

2. Subaerial volcanic eruption of ash and coarse debris and erosion in Lower Cretaceous time (Albian Epoch).

The Water Island Formation is immediately overlain by a 7000-foot-thick series of blue andesite ash beds, pyroclastic breccias (containing blocks as large as 4 feet in diameter), and brick-red soil layers (the Louisenhoj Formation). These rocks, which cover the western half of St. John, can be seen near Cruz Bay and Rendezvous Bay. The actual volcanic vent through which the lavas and ash were extruded is probably located in Pillsbury Sound between St. John and St. Thomas Islands. The ash beds and soil layers were formed subaerially after the growing volcanoes of the island arc emerged above sea level. The pyroclastics are separated from the Water Island Formation by a pronounced unconformity, which indicates that the Water Island Formation was exposed to air and eroded by running water. Ash and volcanic debris then exploded violently from volcanic craters rising above sea level or from

shallow, underwater vents. Much of the pyroclastic material was later deposited below sea level by submarine slides and slumps on the steep slopes of volcano flanks.

3. Sea floor subsidence, volcanic quiescence, and limestone deposition (Albian Epoch).

Reduced volcanism and gradual sea-floor subsidence followed the eruptive activity. About 600 feet of thin-bedded, dark, silica-rich limestones (the Outer Brass Limestone) were deposited on a moderate slope in water depths of a few hundred feet. The limestones, which can be seen near Mary Creek, are silica-rich for two reasons: 1) the abundance of silica tests (shells) of microscopic floating animals; and 2) the occurrence of thin, interbedded layers of tuff.

4. Faulting and deposition of submarine slides and turbidites (Albian Epoch).

Toward the end of the limestone deposition, tectonic activity intensified and caused vertical fault movement that, in turn, created steep submarine slopes and caused the emergence of portions of the Louisenhoj Formation. Submarine slides and turbidity currents moved material eroded from the Louisenhoj Formation down the steep sea-floor slopes. During this time 6000 feet of flyschlike (widespread sandy and calcareous shales containing sandstone and conglomerate beds) rock suites composed of volcanic wacke (poorly mixed sediment consisting of volcanic

Table 16.1   Geologic Column, Virgin Islands National Park

Time Units			Rock Units	Geologic Events
Era	Period	Epoch		
Cenozoic	Quaternary		Alluvium	Erosion by streams and ground water; formation of submarine limestone terraces as sea level periodically rose and fell during glacial and interglacial stages.
	Tertiary	Eocene	Scattered small dikes and plugs	Emplacement of dikes. Widespread faulting; emplacement of dioritic plutons; localized metamorphism.
			Hans Lollik Formation	Renewed volcanic activity; Folding and strike-slip faulting.
		—?—		
Mesozoic	Cretaceous	Albian	Tutu Formation	Renewed fault movement deposition of terrigenous material primarily as turbidites.
			Outer Brass Limestone	Deposition of deep water limestones during a period of volcanic quiescence and sea bottom subsidence.
			Louisenhoj Formation	Development of red soil horizon and subaerial volcanic activity. Initial formation of deep ocean trench north and east of the Lesser Antilles.
		Neocomian?	--unconformity--	--subaerial erosion--
			Water Island Formation	Submarine extrusion of volcanic material along an east-west high-angle fault onto a relatively flat sea bottom. Gradual drop in sea level and/or tectonic uplift.

particles) and limestone megabreccia (large angular blocks cemented together) (the Tutu Formation) were deposited. The Tutu Formation is exposed from Maho Bay to Leinster Point. The limestone megabreccia occurs at Mary Point, Leinster Point, and in the Annaberg area.

5.   Renewed volcanic activity, faulting, and plutonism from Late Cretaceous time through the Eocene Epoch.

Although they crop out only on islands northwest of St. John Island, at least 10,000 feet of water-laid andesite pyroclastic beds (the Hans Lollik Formation, similar to the Louisenhoj Formation) overlie the Tutu Formation. This suggests that a third period of intense volcanic activity occurred, possibly as late as the Eocene Epoch. This volcanic activity was accompanied by the intrusion of dioritic stocks and dikes and by contact metamorphism. Much of the British Virgin Islands and three islands south of St. John (Buck Island National Monument, Frenchman Cap, and Flanagan Island) were formed entirely by these dioritic rocks. Outcrops on St. John can be found at Leinster Hill and near Mary Point. Extensive vertical faulting occurred concurrently with the dioritic intru-

sions. Later in the Eocene Epoch, numerous dikes were intruded. One can be seen near Caneel Bay on St. John Island. By mid-Eocene time, the Caribbean region had assumed its modern geographic configuration, distinctly separated from the Mediterranean Sea by the widening Atlantic Ocean. Subduction of the North American plate below the Caribbean plate was especially intense during Eocene time, but thereafter movement along the northern margin of the Caribbean was largely horizontal along an east–west transform fault.

6.   Alternations in sea level and reef development during the Quaternary Period (Pleistocene and Holocene Epochs).

As seawater warmed and sea level periodically rose in the Pleistocene Epoch, reefs developed around St. John Island but were later eroded to form terraces when sea level dropped. The cycle was repeated during major advances and retreats of continental glaciers as seawater was withdrawn from global oceans and then released. Over the past 15,000 years sea level has slowly risen and submerged many of the older terraces. Living reefs have become established on the Pleistocene terraces and have grown upward during this last sea level rise.

Geologic Map

Cosner, O. J., and Bogart, D. B. 1972. *Water in St. John, U. S. Virgin Islands.* U.S. Geological Survey, Water Resources Division, Caribbean District, open file, p. 9.

Bibliography

Colin, T. J. 1981. The U.S. Virgin Islands. *National Geographic* 159(2):225–43 (February). Map supplement.

Donnelly, T. W. 1966. *Geology of St. Thomas and St. John, U.S. Virgin Islands.* Geological Society of America Memoir 98, pp. 85–176 (also in Mattson 1977).

———. 1968. Caribbean island arcs in light of the sea-floor spreading hypothesis. New York Academy of Science Transactions, Series 2, volume 30, pp. 745–750 (also in Mattson 1977).

Greenberg, J., and Greenberg, I. 1972. *The living reef.* Miami: Seahawk Press.

Khadoley, K. M., and A. A. Meyerhoff. 1971. *Paleogeography and geological history of Greater Antilles.* Geological Society of America Memoir 129.

Mattson, P. H., ed. 1977. *West Indies island arcs.* Stroudsburg, Pennsylvania: Dowden, Hutchinson, and Ross, Inc.

Robinson, A. H. 1974. *Virgin Islands National Park, the story behind the scenery.* Las Vegas, Nevada: KC Publications.

Virgin Islands National Park

Area: 14,696 acres, 23 square miles; water area, 5650 acres

Established: August 2, 1956

Designated a Biosphere Reserve, 1976

Park headquarters: St. Thomas

Address: Box 806, Charlotte Amalie St. Thomas, Virgin Islands 00801

17

Largest remaining subtropical wilderness in the United States.

Included in the park are freshwater and saltwater areas; offshore islands; mangrove forests; swamps; open, grassy, Everglades prairies; sinkholes; relict coral reefs; and abundant wildlife.

Park wetlands help to protect Florida's threatened water resources.

Everglades National Park

Southwest Florida

Figure 17.1 Sawgrass covers an expanse of freshwater marl prairie in Everglades National Park. A few clumps of trees grow on slightly elevated knobs of limestone, called hammocks. National Park Service photo.

The Wetlands Environment—Past and Present

Although Everglades National Park and Big Cypress National Preserve (a large adjoining tract north of the park) occupy most of southwestern Florida, the Everglades wetlands area is larger still, extending as far north as Lake Okeechobee and originally as far east as the coastal ridge along the eastern shore. Moreover, the Everglades are ecologically linked with a series of lakes, streams, and swamps that reach north from Lake Okeechobee almost to Orlando in the center of the state. The whole lower half of the Florida peninsula is an interdependent hydrologic system with the Everglades wetlands, both inside and outside park boundaries, acting as a delicate and vitally necessary balancing mechanism. Only recently has this function of the wetlands system been recognized, and much is still to be learned about its intricate workings that protect and preserve the state's water resources. The fact that Everglades National Park has been designated a Biosphere Reserve and a World Heritage Site is a significant recognition of the wetlands' importance. The Everglades wetlands are a unique, natural treasure as necessary to the enjoyment of life in southern Florida as the ocean, the beaches, and the benign climate.

The natural rhythms of the Everglades have been evolving for at least 5000 years, during which time climate and sea level have remained relatively unchanged. The water supply of the wetlands is renewed and restored annually by a rising water table and increased circulation of water during the hot, wet summer season. Most of the year's precipitation comes in brief, intense showers and in torrential rains between May and October. The mild winter season is drier; little rain falls between November and April. Most Everglades flora and fauna are not only adapted to, but dependent upon these wet and dry seasons. Their reproductive cycles have become synchronized with times of increasing and decreasing moisture.

In the past, disruptions of the system by periods of drought, lightning-caused fires, and hurricanes were accommodated without much difficulty; in fact, such natural disasters became part of the ecological processes of renewal. Today, controlled burning is a part of the National Park Service management policy in the Everglades because natural communities like the piney woods are restored by periodic burnovers.

The ecological crisis that the Everglades wetlands system has not been able to absorb and recover from is the loss and diversion of a great deal of its fresh water due to increasing residential, commercial, and agricultural development in surrounding areas. Until about a hundred years ago, water from Lake Okeechobee flowed very slowly, but unhindered, through sloughs and swamps toward the south, emptying at last into Florida Bay and the Gulf of Mexico. The interior of the Florida peninsula south of Orlando is a broad shallow basin, tilted just enough toward the south to permit natural drainage to move across the surface as sheetwash and flow sluggishly through irregular channels and stretches of open water (fig. 17.3). Decaying vegetation and fine sediment built up layers of peat and muck on the limestone bedrock. The Everglades vegetation thrives in these muck soils, which were enriched annually by saturation and flooding before the network of canals was excavated.

For several centuries after Spanish conquistadors discovered the Florida peninsula, the Everglades remained a little-known wilderness, uninhabited except for a few Seminole Indians and runaway slaves who took refuge in the swamps and secluded islands in the nineteenth century. Through the years a few hunters and trappers came to know the myriad waterways of the Everglades, but settlers seldom penetrated this formidable and mysterious wilderness. They were discouraged not only by its vastness and the difficulties of traversing its depths, but also by the swarms of ravenous mosquitoes and by several varieties of toxic plants.

The opening up of the Everglades began in this century. Drainage of the wetlands started around Lake Okeechobee with the digging of the first canal in 1906. Muck lands south and east of the lake were cleared for the growing of fruits and vegetables. This set off a land boom that reached its peak in the 1920s. Levees and dikes around the southeast side of Lake Okeechobee confined the water. Drainage ditches kept the fields well watered but not saturated, and bountiful crops were harvested from the rich muck soils. During several drier than normal years in the early 1920s, land promotion schemes brought thousands of people into southern Florida. Several years of abundant rainfall ensued, including disastrous hurricanes in 1926 and 1928 that caused Lake Okeechobee to overflow its dikes. Reclaimed land was flooded, crops were ruined, and several thousand people perished. The land boom of the 1920s collapsed.

Population pressures around the Everglades slowed during the depression years of the 1930s, but picked up again with the influx of military personnel and defense establishments in the 1940s. More canals and drainage ditches were dug without adequate regional controls, and the cycle of disastrous floods, droughts, and fires began again. Finally, after the havoc wrought by the hurricane of 1947, a regional flood control district was set up by local authorities, working with the U.S. Corps of Engineers and the U.S. Geological Survey, to deal with the problems of too much water and too little water. Recognizing that Lake Okeechobee, with an area of 700 square miles (about one-third the size of Lake Michigan), was the key to control of the system, engineers set about making the lake a huge storage reservoir.

Figure 17.2   Everglades National Park, Florida.

Canals had been dug, both to the Atlantic and the Gulf, to improve drainage of excess water to the ocean. These were integrated into an elaborate network with pumps, gates, and levees designed to retain water needed during droughts. During heavy storms, excess water could be back-pumped from the reclaimed farm lands into the lake. Later, when it was needed, the water would be drawn back for irrigation. Another project involved dredging and channelizing the Kissimmee River, which collects water from the north and carries it into Lake Okeechobee. This shortened the river's length and increased the rate of flow into the lake.

The water management plan worked well for a number of years and benefited many people. However, serious problems have recurred because population growth since the 1960s has put enormous stresses on the system. The threat of saltwater encroachment, which began when the first canals were dredged to the Atlantic Ocean—thus disrupting the natural drainage to the south—became more serious. During dry periods, saltwater moved into the

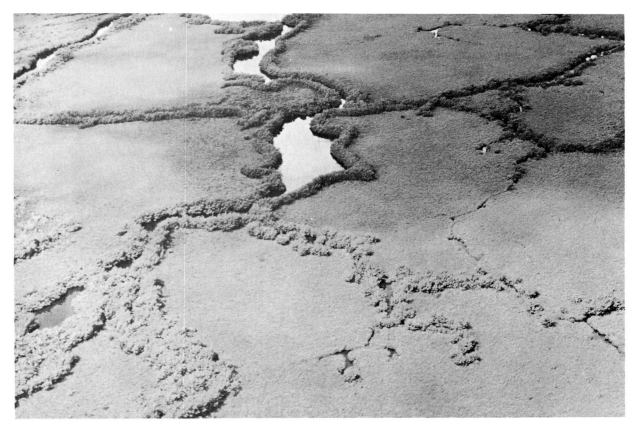

Figure 17.3   Air view of tree-lined waterways winding over a freshwater marl prairie in the Everglades. National Park Service photo.

canals and into the porous bedrock, displacing freshwater. A critical factor here is the level of the water table along the coast, and this level has permanently dropped. Freshwater wells near the coast became brackish, and new wells had to be dug farther inland. Attempts have been made to stabilize the water table by maintaining freshwater pressure high enough to prevent further encroachment of salt water; but whenever rainfall is less than abundant, the water supply problem in coastal areas becomes acute.

In the agricultural lands reclaimed from the Everglades, the muck soils are drying out, oxidizing, and blowing away, the loss being at the rate of about an inch per year. The only way of renewing and regenerating the muck is to allow tracts of land to be flooded for at least half of the year so that more organic material can decay and fine sediment can be deposited. The fact that about 1200 square miles of the Everglades agricultural area has already sunk 6 to 7 feet means that the loss of muck is approaching a critical point.

The channelization of the Kissimmee River lowered lake levels in the northern part of the system and caused the drying out of bottom lands exposed to air. This resulted in more soil loss. When the Kissimmee meandered over its mile-wide flood plain, more water soaked in, the

water table was higher, and during rainy seasons the bedrock aquifer was recharged. (Currently, as a demonstration project, part of the canal system has been blocked so that water flows into old oxbows, refloods the flood plain, and reestablishes part of the wetlands.)

The reduced flow of freshwater into the southern Everglades, including Everglades National Park, has been detrimental to the habitat of a number of plant, animal, and bird species, some of which are now endangered. Sometimes water has been released too late in the breeding season or too early or too much all at once, instead of flowing slowly and continuously through the Everglades during the dry season as in the past. The most serious effect of reduced freshwater on life forms has been the stress on nurseries of shrimp and freshwater and saltwater fish in the southern Everglades and Florida Bay. The brackish estuaries open to Florida Bay have become more salty, greatly reducing the fish and shrimp population and resulting in losses to the fishing industry.

Most of the East Everglades wetlands, adjoining the park, is still an unspoiled area, but it, too, is threatened by residential development. Any further draining of this area will not only be detrimental to the park's ecosystem, but will also deprive the Biscayne Aquifer (box 17.1) of the fresh water it needs for recharge. This is a vital consideration because the aquifer is the chief source of water for more than four million people.

Severe droughts in recent years have alarmed many residents of the Florida peninsula and have given renewed impetus to the conservation movement. The Everglades is now seen as the ecological base for repairs to the hydrologic system. Restoration is impossible; but with prudent and farsighted management, the water resources of south Florida can be maintained, used carefully, and not wasted.

Decades of effort by devoted conservationists, led by crusader Marjory Stoneman Douglas, went into the creation of Everglades National Park. Although the park was authorized by act of Congress in 1934, land acquisition proceeded slowly due to lack of funds. The disastrous hurricane of 1947 provided the spur to action that finally brought enough land together by donation and purchase for the establishment of Everglades National Park. The second big step in the preservation of the wetlands came in 1974 with the setting aside of Big Cypress Swamp as a National Preserve. The federal government continues to acquire wetlands acreage by purchase and donation; the National Audubon Society is creating a wet prairie sanctuary on the former flood plain of the Kissimmee River; and some tracts of land in East Everglades are being withdrawn from development. These measures and more are necessary for the protection of the Everglades wetlands and the preservation of water resources.

A National Park Service brochure that is given to visitors to Everglades National Park explains the wetlands problem plainly and succinctly:

"Water, fresh water, the life-blood of the Everglades, appears to be everywhere, but . . . despite an apparent lush richness, water supplies are critical. Porous limestone underlies the entire park, and rooting plants have only a thin mantle of marl and peat atop this limestone for their support. If not protected, the Everglades' fragile richness would quickly vanish.

". . . Once the benefactor of south Florida's naturally well watered richness, the Everglades now competes at the end of a controlled supply line. To the north, flood prevention, frost protection, pest control, drinking water, and sewage dilution systems siphon off shares. When rainfall is ample, few problems arise. But in drought years arrangements for sharing are required."

The Ecosystems of the Wetlands

Within Everglades National Park are eight zones, or ecosystems, that are dependent upon a balance between wet seasons and dry seasons, between freshwater and saltwater, and, in some cases, flood and fire. Primarily the zones are defined by climatic conditions; elevation in relation to sea level and to the water table; and the geologic aspects of the surface and subsurface materials.

The Marine and Estuarine Zones. Most of Florida Bay is within the boundaries of Everglades National Park. This is a shallow body of water, roughly triangular in shape, lying at the southern end of the peninsula. Of special interest to geologists, Florida Bay is studied as a modern analog for ancient epicontinental seas in which deposition of fine sediment and organic material took place in a brackish environment with little circulation. Sandbars and banks covered by a dense growth of mangroves shut off the eastern end of the bay from tidal interchange with Barnes Sound; and the line of Florida Keys from Key Largo to Key West—along the edge of the continental shelf—keeps out most of the water from the ocean. Between the mainland and the Florida Keys, a lacelike pattern of winding, elongated mudbanks and bars, dotted with hundreds of tiny mangrove islands, spreads over the shallow bottom. Most of the bay is four or five feet deep; maximum depth is only about nine feet. Even on the west side of the bay, which is open to the Gulf of Mexico, daily tides have little effect since they penetrate the bay for only a short distance. It takes a full-fledged hurricane to flush out Florida Bay.

Florida Bay is part of the great "carbonate rock factory" (chapter 18) that produces different kinds of limestone in the various marine environments off the Florida coast. By studying the chemical, biological, and environmental conditions under which carbonate rock is forming in Florida Bay, geologists seek to better understand the processes that formed ancient sediments, which are sources of petroleum. Much of the research being done is supported by petroleum companies.

The fine-grained calcareous mud that is accumulating in Florida Bay (and that may eventually be consolidated into limestone) is less than 5000 years old and lies on top of the oolitic facies of the Miami Limestone, which is Pleistocene in age. The southern Florida Keys (outside the park), which rim the southeast side of the bay, are also made up of oolitic limestone, which covers an older coral reef. Ooids have not formed in this area since Pleistocene time. (For an explanation of ooid formation, see chapter 18.) The age relationship of the bedrock of the bay and the overlying mud suggests that Florida Bay has been under water a relatively short time.

The estuarine environments of Everglades National Park can best be observed along the course of the Wilderness Waterway, a boat and canoe route that goes from the Visitor Center at Flamingo on Cape Sable northwestward to the Gulf Coast Ranger Station (near Everglades City) in the northwest corner of the park. This winding, 100-mile inland waterway goes through long stretches of coastal swamp along an irregular, island-dotted shoreline. The underlying bedrock (rarely exposed because of the thick vegetation) is the bryozoan facies of the Miami Limestone in the southwest and the Tamiami

Figure 17.4   Wave action has piled up coral and shell fragments, pebbles, and coral sand in a berm (beach ridge) on Middle Cape on the Gulf Coast side of Cape Sable. National Park Service photo.

Limestone in the northwest. Saltwater extends several miles up the estuaries.

Both the marine and estuarine zones serve as breeding grounds, spawning grounds, and nurseries for life forms of many kinds, from microscopic organisms to large species, such as crocodiles and manatees, both of which are in the endangered category. Crocodiles, which have a shorter, more tapered snout than alligators and are much less numerous, live mainly in the eastern end of Florida Bay and build their mounded nests on the Upper Key Largo shore and along the Bay shore. Manatees, or "sea cows," are large aquatic mammals that feed on vegetation in mangrove-lined creeks.

The Mangrove Zones. Mangrove trees, which grow in shallow, brackish water, establish themselves in the tidal zone of the marine areas and in the transition zone between the freshwater glades and the salty water in the estuaries. The mangroves have an important geologic role because they extend the land as they spread along the coastline and protect it from storm erosion. In the estuaries mangroves line the banks and hold soil in islands (box 18.1).

In the Ten Thousand Islands region, part of which is within the northwestern section of Everglades National Park, mangroves have created elongate islands from oyster beds in shallow waters offshore. When the upward growth of oyster beds and accumulated sediment reaches the tidal zone, mangrove seedlings are able to gain a foothold on top of the beds; eventually the trees displace the living oysters, and the oysters reestablish beds in new locations farther offshore. Generations of mangroves have built up thick layers of peat over old oyster beds. These layers, representing successive tidal zones, provide a record of gradually rising sea level along the coast.

The Cypress Zones. A cypress stand develops in solution pits and sinkholes in limestone that have been filled with accumulated muds and marls. Dwarf cypress grow in brackish areas inland from the southern coast and near the edge of the sawgrass glades. The larger pond cypress grow where more freshwater is available near the sloughs. Some build up slightly elevated "domes" of mud and vegetation on top of sinkholes. Mixed with other species, pond cypress also thrive on tree islands (bayheads) in the marshes and sloughs.

The Coastal Prairie Zone. Salt-tolerant plants, such as cactus and yucca, grow among grasses in the thin salty soils of the coastal prairie in the southern tip of the park on Cape Sable and at the lower end of Taylor Slough. North of West Lake a stand of hardwoods struggles for survival on an old shell mound left by prehistoric coastal Indians. The shell mound does not get the tree roots high enough out of the salt for the trees to do very well. Patches of coastal prairie are also found between the mangrove-lined shore and the coastal swamps on the west side of the park. Ancient shell mounds mark the sites of camps where Indians feasted on shellfish. One of these at Chokoloskee near Everglades City covers 135 acres. Twenty feet high, this man-made "hill" is the highest point in Everglades National Park. Pots made from marl clay, shark-tooth knives, and other artifacts have been excavated from the archeological sites.

The Freshwater Sloughs and the Freshwater Marl Prairie. Shark River Slough and Taylor Slough are the two main channels that move fresh water slowly southward through the Everglades. Both sloughs occupy broad, shallow depressions in the limestone bedrock and were probably free-flowing rivers at one time. Seen from the air, the sloughs are linear complexes of interconnected waterways and ponds, swamps, and sawgrass glades. The sloughs blend into the freshwater marl prairie on either side, and both prairie and sloughs are dotted with innumerable tree islands, called "bayheads" or "hammocks" (fig. 17.3).

Shark River Slough, the larger, central slough, is more than 40 miles wide during the summer rainy season but dwindles to a few narrow streams during the winter droughts. At its lower end, the slough empties into the myriad waterways and lakes of the Shark River estuary in the Cape Sable area at the southwestern end of the peninsula.

Between Shark River Slough and Taylor Slough, the southern tip of the Atlantic coastal ridge forms an interfluve, not more than three feet high, that divides the drainages of the southern Everglades. Before drainage canals were dug in the East Everglades area, Taylor Slough cleaned and stored water for a much larger part of southeast Florida. At the present time, state and federal authorities are attempting to preserve the wetlands between the park's eastern boundary and the metropolitan area spreading southwest from Miami in order to protect the region's water supply and also keep the feeding and breeding grounds for wildlife in the park from being further imperiled. In recent years heavily populated Miami and its environs have drawn so much water from the Biscayne Aquifer (box 17.1) that the water table has dropped rapidly, permitting saltwater to seep into the aquifer in increasing amounts.

The freshwater marl prairies, which support many of the more than 100 species of grass growing in the park,

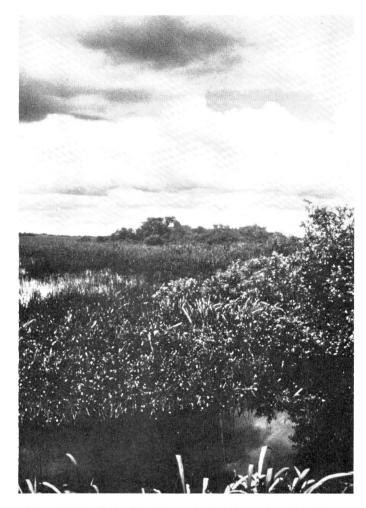

Figure 17.5   Taylor Slough is a natural drainage channel on the southeast side of Everglades National Park. The tree-covered mound in the background is a hardwood hammock near the Anhinga Trail at the southern tip of the Atlantic Coastal Ridge. National Park Service photo by Zimmer.

have by far the greatest extent of the Everglades ecosystems. The seemingly endless expanse of grassy glades around the horizon are what inspired the term "everglades," and this became the name for the whole region. Small ponds and "gator holes" are scattered over the prairies. The latter are pits and solution holes in the limestone bedrock that have been cleaned out by alligators during the dry seasons. They thrash their tails around, clearing out mud and vegetation so that water can collect in the depressions. Such waterholes enable many Everglades species to survive droughts.

The famous sawgrass of the glades and sloughs is really a type of sedge that grows as high as 12 feet and is hardier and tougher than most grasses. (Sedges have solid stems; true grasses have hollow stems.) Sawgrass stems have sharp, toothed edges, hence the name. The Everglades is the largest sawgrass marsh in the world (fig. 17.1).

Table 17.1 Geologic Column, Everglades National Park

Time Units			Rock Units	Geologic Events
Era	Periods	Epochs		
Cenozoic	Quaternary	Holocene	Carbonaceous sands in beach ridges; limy muds, mucks, peat, soil; limy marine muds in bays	Nonmarine accumulations of unconsolidated sediment in shallow stream channels and on bedrock surface
		Pleistocene	Miami Limestone (oolitic and bryozoan facies)	Accumulation of ooids and bryozoan masses in shallow seas
			Fort Thompson Formation (in subsurface of Biscayne Aquifer)	Marine and freshwater marls, limestones, sandstones, shell hash, etc.
	Tertiary — Neogene	Pliocene	/////////	Slight erosion
		Miocene	Tamiami Formation	Marine limestones deposited in shallow water; silts, shelly sands
	Tertiary — Paleogene		/////////	Disconformity
			Underlying marine carbonates of the Coastal Plain (approx. 3 miles in thickness)	Only slight deformation
Mesozoic	Cretaceous			

Source: modified after Hoffmeister 1974

The Pinelands and Hardwood Hammocks. The pinelands in Everglades National Park are on the slightly elevated limestone at the end of the ridge between Taylor Slough and Shark River Slough. Here on mostly bare limestone rock, the pines survive by sending their shallow roots down into pits and crevices in the oolitic rock. The stands of slash pine in the park are about all that is left of the pine forests that once covered the Atlantic Coastal Ridge for some 50 miles from north of Fort Lauderdale southward to the end of the ridge in the Everglades. The plant species in the pinelands have adapted to the periodic natural fires that sweep over the ridge during severe droughts. The slash pine, for example, germinates best after a ground fire has burned off low vegetation, releasing ash that enriches the scanty soil. The thick bark of mature trees protects them from the fire. Controlled burning of the pinelands is now an established part of the park's management.

The hammocks of the ridge area are slightly elevated, isolated clumps of tropical trees growing only in spots not burned over by fires (fig. 17.5). Like the pinelands, the hardwood hammocks were once more extensive on the Atlantic Coastal Ridge. Now they are found only in the park and in a few protected areas. The trees and shrubs of the

tropical hammocks in south Florida are of West Indian origin. The taller, broad-leaved trees form a dense canopy that retains moisture for the forest floor. The hammocks also tend to grow on the edges of sinkholes or other sites where the roots can draw more water. The extra moisture has apparently enabled the groves to withstand the ground fires that have raced through the neighboring pinelands.

The hammocks of dense trees and shrubs that make up the tree islands (bayheads) of the sloughs and freshwater prairie may have less than a foot of elevation above the water surface. Many of them are teardrop-shaped and face into the current, which, although slow and weak, is able to gently mold the island banks. Some hammocks occupy rocky knobs or slight irregularities that jut up above the general flatness of the bedrock surface.

Geologic History

1. Cretaceous and Tertiary carbonate rocks of the Coastal Plain.

The surface bedrock of the Everglades is underlain by a three-mile thickness of marine sedimentary rocks, mainly carbonates, that were deposited in Cretaceous and

Box 17.1
The Biscayne Aquifer

This important water-bearing rock series is a nonartesian aquifer underlying all of southeast Florida. (An *aquifer* is an rock unit—or units—of moderate to high permeability from which ground water can be pumped.) The Biscayne Aquifer is the principal source of freshwater for the densely populated region from Fort Lauderdale to Key Largo.

The aquifer is supplied by rainwater and surface runoff that seep into the ground and fill the pore spaces in the bedrock. The ground water cannot penetrate deeply because the impermeable lower part of the Tamiami Formation holds the water accumulating in the rocks above. The total section of bedrock saturated with water is approximately 10 feet in thickness in western Broward and Dade counties (in the center of the Florida peninsula), and increases to a maximum of about 200 feet along the southeast coast. Throughout most of the area, wells need only to be drilled below the water table in the aquifer in order to tap a supply of potable water.

The best water-bearing unit in the aquifer is the Fort Thompson Formation because of its thickness and high permeability. A good deal of the Fort Thompson is made up of shell hash and contains numerous cavities. The permeable Miami Limestone overlying the Fort Thompson is also part of the aquifer. Because of the solution-ridden characteristics of these limestones, the Biscayne is regarded as one of the most permeable aquifers in the world. What this means is that the aquifer benefits from fairly rapid recharge during abundant rains, but it is also subject to pollution and to encroachment by salt water.

The Biscayne Aquifer has always lost water due to natural causes such as evapotranspiration and ground-water outflow into Biscayne Bay and the ocean. In recent years the amounts of available water have been greatly reduced by (1) the loss of recharge area to industrial, residential, and agricultural development; (2) the extensive system of drainage canals (many of which are polluted); and (3) increasing drawdown by pumping from wells.

The pressure of freshwater in the aquifer is what keeps the saltwater out (fig. 17.6). At the beginning of this century, the peninsular water table was so high and the aquifer was so full that freshwater bubbled up in Biscayne Bay, and freshwater wells could be dug a few feet back from the shore. Now saline water has seeped in

Tertiary time and have undergone little deformation. None of these beds crop out at the surface in south Florida.

2. Miocene deposition of the Tamiami Formation.

In Miocene time, a blanket of cream, white, and greenish gray clayey marl, silt, sand, and sandy marl—up to 100 feet in thickness—accumulated in the south Florida area. These shallow-water sediments were consolidated and cemented into the limestones of the Tamiami Formation, which makes up the bedrock surface of the northwestern part of Everglades National Park. The highly permeable upper part of the Tamiami forms the basal portion of the Biscayne Aquifer. The low permeability of the lower Tamiami members confines the ground water of the aquifer.

3. Pliocene emergence and slight erosion.

Pliocene rocks are not present in Everglades National Park, but sandy and shelly marls in the Okeechobee area to the north suggest that the land there may have been at or slightly above sea level.

4. Pleistocene deposition.

The Lower Pleistocene Fort Thompson Formation does not crop out in the park, but this very permeable unit of marine and freshwater marls, sandstones, and shelly limestones is the main water-bearing component of the Biscayne Aquifer (box 17.1).

When sea level was high during the last Pleistocene interglacial episode (preceding Wisconsinan glaciation), several rock units—deposited under different environalong the base of the aquifer and extends inland underneath most of Dade and Broward Counties, which depend on the aquifer for most of their freshwater. Dams have been constructed in canals to maintain high water levels and prevent loss of freshwater. Other control measures have been put into effect; but the water supply situation remains critical.

Figure 17.6 When stream flow is ample and freshwater levels are high, sea water remains near the coast and cannot move into aquifers (left). During periods of drought or excessive pumping, salty or brackish water moves up tidal inlets and invades aquifers, accumulating beneath fresh ground water (right). From *The Face of Florida*, 2nd edition, C. M. Head and R. B. Marcus. © 1987 Kendall/Hunt Publishing Company. Used by permission.

mental conditions but all about the same age—accumulated in the south Florida area. These rock units make up the surface bedrock of most of south Florida and the Florida Keys, and underlie Biscayne Bay and Florida Bay. The Miami Limestone formed two distinct facies. The oolitic facies makes up the Atlantic Coastal Ridge, the East Everglades region, parts of the southern coast, Florida Bay, and the southern Florida Keys. The oolitic facies interfingers with the bryozoan facies, which makes up the bedrock of southwestern and south central Florida. (The Key Largo Limestone of the same age, which does not crop out in Everglades National Park, is described in chapter 18.)

Examination of a specimen of the Miami Limestone from the oolitic facies shows that its layers consist of bil-

lions of tiny spherical grains, or *ooids,* cemented together to make a rocky mass. Ooids are formed by precipitation of calcium carbonate, and the grains accumulate in a considerable thickness of limey oolitic ooze. Eventually they may be cemented by calcite.

The ooids that make up the Miami Limestone in the Atlantic Coastal Ridge were piled up along a late Pleistocene shoreline by ocean waves. The higher parts of the ridge toward the north, 20 to 40 feet above present coastal elevation, have been exposed to the atmosphere much longer than the southern part of the ridge, which is only about two feet high at its terminus in the southeast part of Everglades National Park.

In the central part of the park, the bedrock surface of the Miami Limestone is made up of bryozoan remains as well as ooids. Bryozoans, which are small marine invertebrates living in colonies, secrete calcium to form the tiny cuplike cells in which the individual organisms live. Byrozoan remains, accumulated over many generations, make up about 70 percent of the rock volume of the Miami Limestone. In Florida Bay modern bryozoans continue to grow in knoblike clusters encrusted with seaweed. These small animals have lived in marine environments and produced carbonate material since late in Cambrian time.

5. Holocene mucks, soils, and sediments.

As the great ice sheets of Wisconsinan glaciation melted, sea level rose again fairly rapidly, the trend continuing into early Holocene time. Then the rate decreased, and sea level has risen much more slowly in the past 4000 to 5000 years. During this time of relative stability, unconsolidated sediments have accumulated on the land surface of southern Florida. Southward-flowing waters spread weathered and eroded material from the Tamiami and Miami Limestones over the saucerlike interior basin now occupied by the Everglades. Mixed with the sediment are abundant organic remains of decayed vegetation. The resulting peat and muck soils are immature but highly fertile, especially under conditions of high moisture and humidity. On land protected by natural vegetation, muck continues to accumulate; but on drained land the surface material becomes powdery during times of drought and is easily eroded.

State and federal authorities, including the National Park Service, continue their efforts to improve the hydrology of South Florida and protect the area's environmental resources. Programs include a combination of canal filling where possible, modification of water control structures, restoration of flow in old sloughs and channels, and land acquisition. The National Park Service has been working on the restoration of the natural systems of Big Cypress National Preserve for a number of years.

Geologic Map

Puri, H. S., and Vernon, R. O. 1964. Geologic map of Florida, plate 2. *Summary of the geology of Florida and a guidebook to the classic exposures.* Florida Geological Survey Special Publication No. 5.

Bibliography

Boyle, R. H., and Mechem, R. M. 1981. There's trouble in paradise. *Sports Illustrated* 54 (6) : 82–96, February 9.

De Golia, Jack. 1978. *Everglades, the story behind the scenery.* Las Vegas, Nevada: KC Publications.

George, J. C. 1972. *Everglades Wildguide.* Natural History Series, National Park Service. 106 p.

Head, C. M. and Marcus, R. B. 1987, 2nd edition. *The face of Florida.* Dubuque, Iowa: Kendall/Hunt. 197 p.

Hoffmeister, J. E. 1974. *Land from the sea, the geologic story of south Florida.* Coral Gables, Florida: University of Miami Press.

Multer, H. G. 1977. *Field guide to some carbonate rock environments, Florida Keys and Western Bahamas.* Dubuque, Iowa: Kendall/Hunt.

National Park Service. 1985. Everglades National Park. Map and brochure.

Everglades National Park

Area: 1,398,939 acres; 2186 square miles
Authorized: May 30, 1934
Established: June 20, 1947
Designated a Biosphere Reserve: 1976
Wilderness designated: November 10, 1978
Designated a World Heritage Site: October 24, 1979
Address: P.O. Box 279, Homestead, Florida 33030

18

A park of reef and water, with 33 keys (islands) forming a north-south chain between Biscayne Bay and the Atlantic Ocean, plus a narrow strip along Biscayne Bay's western shore.
The reef tract is the only living coral reef off continental United States.

Biscayne National Park

Southeastern Florida

Figure 18.1   Islands and channels of Biscayne National Park from the air, looking north. Built up as coral reefs during Pleistocene times of higher sea level, the islands are now covered with vegetation. National Park Service photo.

Figure 18.2   Land and water areas of Biscayne National Park, Florida.

Local History and Geographic Setting

Biscayne National Park is a large marine preserve south of Miami and offshore from the mainland of southeastern Florida. Within park boundaries lies most of shallow Biscayne Bay and a large reef tract, comprising a back reef area and the northern part of an outer reef. Hawk Channel, a seaway between the islands and the outer reef, is also in park waters. Biscayne National Park was established to protect three natural ecosystems: (1) the nursery grounds for many fish and shellfish species in Biscayne Bay; (2) the native vegetation of the upper Florida keys; and (3) the marine plants and animals of the coral reefs.

The term "keys," derived from the Spanish word *cayo* meaning "islet," is given to the coral islands that form a line of fossil patch reefs off the southern Florida coast. A topographic and geologic distinction is made between the lower Florida Keys, which extend southwestward from Big Pine Key to Key West, and the upper keys, which run from Big Pine to Biscayne National Park's northern boundary. The park's southern boundary is between Old Rhodes Key and Key Largo. A causeway connects Key Largo and the rest of the southern keys with the mainland, but access to the keys in Biscayne National Park is by boat only.

Strong currents, high winds, hurricanes, and sudden violent storms, combined with the hidden reefs and shoals, make these waters dangerous for sailing vessels. In the seventeenth and eighteenth centuries pirates lay in wait in the Florida Keys for ships sailing through the Florida Straits between Cuba and the mainland. They attacked ships that were crippled by storms or blown off course onto the shoals and reefs, where they were easy prey. Hawk Channel is marked with the locations of some of the sunken wrecks.

After Florida was purchased from Spain in 1821, the United States Navy undertook a campaign to rid the West Indies and the Florida Straits of pirates. Commodore David Porter organized a "Mosquito Fleet" of light, maneuverable schooners, barges, and other craft well adapted to coastal waters. The small vessels were sent out pirate-hunting among the keys and straits. The sailors and marines spent weeks at a time away from base while they examined the inlets, rocks, reefs, and hidden coves along the many miles of wild and irregular coast. After more than a year of arduous and dangerous work, all of the pirates were captured or driven out. The Florida Keys and Straits were made secure for commercial shipping, but to this day, smugglers operate in the area.

The earliest inhabitants of the Florida Keys were Indian fishermen who left huge shell mounds that actually enlarged some of the islands. In the 1830s bands of Seminoles, driven from their homes in northern and central Florida by white settlers, took refuge in the Everglades and in the Florida Keys, where some of their descendants still live.

In recent years the coastal waters, shorelines, and islands of Florida have suffered severe environmental damage from the pressures of rapidly increasing population and from both commercial and residential overdevelopment. Spurred by conservationists and alarmed by growing problems of water pollution, loss of fisheries, and destruction of wildlife habitat, the Florida legislature passed a Beach and Shore Preservation Act in 1965. This established coastal construction setback lines that have saved some of the state's beaches and inlets from being urbanized. The authorization by the United States Congress of Biscayne National Monument in 1968 stopped the construction of a refinery project on Biscayne Bay, as well as urban development proposed for the northern keys. The monument was expanded and redesignated Biscayne National Park in 1980.

Geologic Features

The narrow strip of shoreline, the keys, the coral reefs, and the ocean floor in Biscayne National Park are all made of carbonate rock, most of it of organic origin. The marine platform on which the keys and the reefs were built is part of the Atlantic coastal shelf, which is at its narrowest here along the southeast coast of Florida. A few miles offshore, just east of the outer reef, the ocean floor drops off steeply into the deep water of the Florida Straits. The Gulf Stream flows through the 90-mile-wide straits, between Florida and Cuba to the south and the Great Bahama Banks to the east.

Landward of the outer reef, the water is shallow and the relief of the keys and the mainland is low. None of the keys rises more than 18 feet above sea level and some are barely above high tide level. The keys and all of southern Florida were under water during the last Pleistocene interglacial episode when sea level was higher than now. The patch reefs that make up the Florida Keys were built by corals that lived and grew on the marine platform during times of high sea level in the Pleistocene Epoch. In the geologic sense, all the rocks, soil, and sediment of the region are very young, and the processes that are significant in the geologic story of the park are still going on.

Box 18.1
The Geologic Role of Mangrove Swamps

In Biscayne National Park, as well as elsewhere on the Florida coast, mangroves have had an important role in the development of the present shoreline. Scientists estimate that despite occasional losses of ground due to hurricanes, the mangrove swamps along the keys, Biscayne Bay, and Florida Bay have in this century added more than 1500 acres of dense forests where formerly open shoals existed. In the park, fringes of mangrove-built shoreline can be seen on both sides of the keys and at Mangrove Point on the mainland.

Among the salt-tolerant trees of the shore zone are the dominant red mangroves (*Rhizaphora*), the only true mangroves; the black mangroves and white mangroves, which belong to two other species although they are called "mangroves"; and the buttonwoods. The buttonwoods grow farther inland than the mangroves and are often found mixed with mahogany and other tropical hardwoods. The mangroves grow at the edge of the ocean in the zone between high tide and low tide. In addition to being salt-tolerant, these remarkable trees have an amazing ability to cling to unstable ground. When storms beat in on the coast and hurricanes bring devastating waves and winds, the mangroves usually survive. Only the most disastrous hurricanes can uproot them. Even then, some of the older, stronger trees that are flattened by the wind manage to send out new anchoring roots; then their trunks bend back upward and start growing toward the sun again. Before the days of advance warning for hurricanes, old-timers often took refuge in firmly anchored small boats in narrow creeks among the mangroves. The protection afforded by the mangroves usually enabled people to ride out the storm.

A full-fledged hurricane always produces coastal changes, but where mangroves grow, such changes are minimized. In fact, if it were not for the mangroves protecting the tidal environment, the Florida coastline would be reshaped by every big storm. Stretches of coastal swamps were drained, filled, and developed as waterfront real

Figure 18.3   Mangrove sapling roots itself in the tidal zone. National Park Service photo.

estate before the protective function of the mangrove swamps was understood. When hurricanes struck, as, for example, Donna in 1960 and Betsy in 1965, the damage was appalling.

Another important adaptation of red mangroves is their ability to colonize. Seeds germinate on the mature trees and then fall as seedlings (fig. 18.3). Some take root in water beneath the parent trees, but other seedlings are spread by tides and currents along the coast and sometimes travel considerable distances before stranding and sending out roots. The young trees anchor themselves by strong roots that sprout both from the trunk and from the lower branches. The roots (1) form a complex network of props and braces that hold up the crown; (2) bring up sap; (3) raise the trunk above the water; (4) furnish oxygen which is unavailable in the watery muck; and (5) catch and hold silt and debris. It is this last function that extends the coastline into the water, continually creating new land.

Various lines of evidence show that for several thousand years, sea level has been rising very

Figure 18.4   A tangle of old mangrove roots holds sediment, helping to protect the shoreline. National Park Service photo.

gradually along Florida shores. Geologists have found mangrove peat accumulations to be a useful indicator of approximate rates of sea level change because mangroves grow only in the tidal zone. In tropical and semitropical latitudes, tidal ranges are low; on Florida coasts the average range between high and low tide levels is only about three feet. Therefore, if sea level were static, mangrove peat accumulations would be about three feet in thickness. However, in protected locations on the shore, peat accumulations have been found to be 12 to 15 feet in depth, which means that sea level must have been rising while successive generations of mangroves grew and peat accumulated (Hoffmeister 1974).

The importance of the mangroves to the nutrient cycles of the coastal environment can hardly be overstated. Besides supporting a large number of rookeries, the mangrove swamps and bordering waters are the habitat of alligators, snakes, turtles, manatees, raccoons, and many other animals. Even the mosquitoes that breed by the millions in the mangrove swamps are a main source of food for many birds, frogs, reptiles, and fish. The tidal inflow and outflow circulate nutrients for the breeding grounds of fish and shellfish in the bays. Nutrients carried out to sea help to supply food for the corals and other organisms living in the reefs.

Figure 18.5   Two views of Elliott Key. *Top:* On the Atlantic Ocean side, the coral rock of the old fore reef drops off steeply. *Bottom:* On the Biscayne Bay side, a wide beach slopes gently toward the mainland across the bay. National Park Service photos.

The "Carbonate Rock Factory"

The marine preserve of Biscayne National Park is part of a large carbonate-producing region that includes the waters around the western Bahamas as well as the Florida Keys. Here in a variety of carbonate environments, a wide range of limestones are accumulating, both organic and inorganic. In the "labor force" are many species of marine plants and animals. The corals and calcareous algae are the main producers. "Manufacturing" processes include photosynthesis (corals and algae cannot work in darkness), precipitation, evaporation, and supersaturation. By these processes the chemical constituents, or "raw materials," are extracted from seawater and transformed into skeletal parts, crusts, and ooids (or ooliths).

Most *ooids* are tiny spheres of inorganically precipitated calcium carbonate. (Noncalcareous ooids, which are less common, are generally formed by replacement processes.) At the present time, ooids are forming only in the western Bahamas part of the carbonate-producing factory. Strong ocean currents there keep the waters agitated and sloshing back and forth over shallow and deep bottom areas. Intense sunlight raises the temperature of the tidal pools and shallows, and evaporation increases salinity, concentrating the calcite and causing it to precipitate around any fragment (a speck of shell, quartz sand grain, etc.) that acts as a nucleus. Precipitation of successive carbonate layers around such nuclei forms ooids.

Although conditions on the Florida side of the Florida Straits do not now permit ooid production, they accumulated in vast quantities in the past in the waters that covered the lower Florida Keys and southern Florida. Ooids were compacted and cemented to form oolitic limestone, some of which has been eroded to produce sand-sized grains and other sediment. Ooids in the form of sediment, either loose or in sedimentary rock, are referred to as *ooliths* or *oolites*. The small beaches on the Florida Keys and the larger beaches along the extreme southern tip of the Florida mainland consist principally of oolitic and coralline sand rather than quartz sand.

Geologists, marine biologists, and other scientists come from all over the world to study the carbonate-producing environments of the coastal waters of Florida and the Bahamas. Much of the research consists of detailed study of the processes of the present in order to better understand how the vast accumulations of carbonate rock on the earth's surface were produced in the geologic past. Because carbonate rocks are an important source of petroleum, a good deal of the research being carried on has significant economic implications.

Figure 18.6   Rapidly growing moosehorn (or elkhorn) corals (*Acropora palmata*) dominate the living outer reef. They are strong enough to withstand storm waves of the open ocean and are major reef builders in Biscayne National Park. Drawn from a National Park Service photo. © 1982 by Kendall/Hunt Publishing Company.

The Living Reefs

The corals of the reef tract in Biscayne National Park are building reefs in the same way their ancestors built the older reefs that make up the keys. The fossil corals of the islands are not different from their living counterparts in the surrounding waters. The same coral species, about 50 in all, live in their chosen habitats throughout the Florida reef tract and the Bahamas. The coral groups are also closely related to those living in the reefs off the Virgin Islands (chapter 16). A few species in the Florida reef tract seem able to thrive in many environments conducive to coral growth; but most of them prefer either (1) the outer reef along the open-ocean side of the marine platform; (2) the numerous patch reefs in the back reef zone (between the keys and the outer reef); or (3) the shoals on either side of the keys.

Each of the three habitats is dominated by a characteristic set of corals (fig. 18.6). The outer reef crest is covered largely by the moosehorn coral, *Acropora palmata,* and below a depth of 10 meters by the staghorn coral, *Acropora cervicornis.* These two corals grow rapidly (10–25 mm/year) in well lit, agitated environments to form dense thickets of branches that shade and thereby inhibit the growth of other corals. The back reef is composed mostly of mound-shaped corals, including the star coral, *Montastraea annularis,* and the brain coral, *Diploria strigosa.* Single mounds of these corals can attain enormous sizes (more than 3 meters in diameter) after

Table 18.1 Geologic Column, Biscayne National Park

Time Units			Rock Units		Geologic Events
Era	Period	Epoch			
Cenozoic	Quaternary	Holocene	Accumulations of thin soils, marls, peat on land. Accumulations of fine bottom sediments, mainly carbonates, in seawater		Development of fringing mangrove swamps Generally rising sea level Growth of coral reefs
		Pleistocene	Miami Limestone (oolitic facies)	} Key Largo Limestone	Fluctuating sea levels, with coral reef growth and oolitic deposits during high sea stands; erosion during low sea stands

Source: modified after Hoffmeister 1974

several centuries of growth. The corals found in the inshore shoals include small (4 cm diameter), almost round star corals, such as *Siderastrea radians,* and knob-shaped finger corals, such as *Porites divaricata.* These corals grow slowly but are able to remove sediment easily and to resist extreme or fluctuating temperatures and salinities.

Geologic History

1. Pleistocene reef-building; Key Largo Limestone

In the upper keys (including those in Biscayne National Park), the Key Largo Limestone is the only bedrock that is exposed. During the last interglacial episode of Pleistocene time, when sea levels worldwide were high, a long, arc-shaped coral reef was built up on the marine platform. Data from drilling show that when it was living and growing the reef had a total length of at least 220 miles and extended from Virginia Bay (part of Greater Miami) to the Dry Tortugas in the Gulf of Mexico, west of Key West. However, the Key Largo Limestone is exposed at the surface only from Soldier Key (northernmost key in the park) to Big Pine Key (about 35 miles east of Key West), a distance of 110 miles (Hoffmeister 1974). The distinction between the upper keys and lower keys is based on the fact that in the lower keys, from Big Pine Key to Key West, the Key Largo Limestone is overlain by the oolitic facies of the Miami Limestone (chapter 17). The differences between these two types of limestone account for most of the erosional and topographic differences between the upper and lower keys.

Apparently the upper keys were constructed as a series of patch reefs in a back reef environment similar to the one that exists today. The principal frame-builders of both the keys and the living patch reefs are the star corals (*Montastrea annularis*), which usually grow in a mod-

erately protected marine environment with limited wave energy. Species such as the moosehorn coral (*Acropora palmata*), which thrive in the surf of the outer reef, have not been found either as fossils in the Key Largo Limestone or as living corals in the back reef area. What seems likely is that an outer fringing reef was growing during Pleistocene time at the same location as that of the present outer reef.

2. Erosion during the last major Pleistocene glaciation.

During Wisconsinan time, as the last advance of the great continental glaciers occurred, sea level dropped more than 300 feet, exposing nearly all of the continental shelf off southern Florida. At the seaward edge of the marine platform, wave erosion lowered the surface of the old outer reef by more than 50 feet. Erosion first by waves and then by streams that began to drain off newly exposed land also stripped off the tops of the patch reefs and the upper parts of the present keys. Channels between the keys, such as Caesar Creek, were widened and deepened. Hawk Channel, which parallels the upper keys, was scoured, and ridges or banks of sedimentary debris were deposited along its sides. (In the back reef area of the present time, tidal currents continue to scour the channels and deposit sediment in shoals and banks.)

Rain and ground water flushed salt from the reef rock of the keys and dissolved some of the calcite. Small pits, sinkholes, and other solution features developed in the rock. Gradually a thin soil developed and the upper keys became thickly wooded.

3. The development of Biscayne Bay.

Biscayne Bay, which lies between the northern part of the upper keys and the mainland, is only 6 feet deep near Miami Beach but deepens toward the south. The shallow bottom next to the mainland slopes off toward the

Figure 18.7   Coral heads and knobs line the mainland shore near University Dock. National Park Service photo.

keys to a 12- or 15-foot depth. When the keys were being built up by coral reefs during the high sea levels in the last Pleistocene interglacial episode, Biscayne Bay was a large coastal lagoon between the Atlantic Coastal Ridge rimming the Everglades (chapter 17) and the reef complex offshore. When sea level was lowered due to Wisconsinan glaciation, the lagoon drained off, exposing the limestone bedrock of the ocean floor. Streams cut valleys in the sides and floor of the troughlike depression. Solution pits developed in the limestone. Freshwater calcitic muds that were deposited below present sea level indicate that for a time a freshwater lake occupied part of the Biscayne basin.

As sea level began to rise again, brackish water entered the lagoon and mangrove swamps developed. Eventually seawater filled the lagoon to its present depth, creating Biscayne Bay. Meanwhile, in the shallows next to the keys, small inshore corals began to grow again. The presence of sediment, constantly stirred up by currents, prevents extensive coral growth in Biscayne Bay, however.

Beneath the bottom sediment, on the bedrock floor of Biscayne Bay, is the contact between the Key Largo Limestone and the oolitic facies of the Miami Limestone, which covers all of southeastern Florida.

Bibliography

Head, C. M. and Marcus, R. B. 1987. *The face of Florida,* 2nd edition. Dubuque, Iowa: Kendall/ Hunt Publishing Company.

Hoffmeister, J. E. 1974. *Land from the sea, the geologic story of south Florida.* Coral Gables, Florida: University of Miami Press.

Multer, H. G. 1977. *Field guide to some carbonate rock environments, Florida Keys and Western Bahamas* (new edition). Dubuque, Iowa: Kendall/ Hunt Publishing Company.

Wheeler, R. 1969. *In pirate waters.* New York: Thomas Y. Crowell Co.

Biscayne National Park

Area: 173,274 acres; land area 4373 acres; 270 square miles
Proclaimed a National Monument: October 18, 1968
Established as a National Park: June 28, 1980
Address: Box 1369, Homestead, Florida 33090

PART III

Landscapes Shaped by Continental or Alpine Glaciation

© by G. Shimer

Figure III.I   Glaciers among the high peaks of the St. Elias Range, Wrangell-St. Elias National Park and Preserve, Alaska. Drawing by G. Shimer from *Field Guide to Landforms in the United States*, J. A. Shimer. The Macmillan Company © 1972 by J. A. Shimer. Used by permission.

Figure III.2   View that John Muir saw when he stood on the rock promontory (left foreground) at Glacier Point and looked north over the rolling Yosemite upland, Yosemite Falls, and the floor of Yosemite Valley, far below. Muir speculated that in the past glaciers had covered the uplands, widened and oversteepened the valley walls, and truncated and scoured the spurs along the lower part of the cliffs. Photo by P. M. Palmer.

Glaciers in national parks are more than fascinating elements of the scenery. Besides storing significant amounts of water in the form of snow and ice, glaciers are also past and present indicators of climatic and environmental change. Knowing something about how glaciers form and how they behave are concerns of vital importance in many parts of the country. In the National Park System we can see fresh examples of landform changes brought about by recent climatic variations and we can also recognize features on the surface of the land caused by glacial action in fairly recent geologic time.

The first three national parks in this section—Voyageurs National Park in Minnesota, Isle Royale National Park in Lake Superior, and Acadia National Park on the Maine coast— are in regions that were covered by great continental ice sheets in Pleistocene time. The remaining 10 national parks in Part III, all in mountain areas, were sculptured by alpine, or valley, glaciers. All the parks in Part III exhibit a variety of erosional and depositional features caused by the mighty glaciers of the Ice Age.

So pervasive were the effects of Pleistocene glaciation on the North American continent that Ice Age indicators of various sorts are discernible in all the national parks of the system. Even the offshore parks, such as Virgin Islands National Park and both national parks in Hawaii, were affected by the cooler temperatures, increased moisture, and rising and falling sea levels characteristic of Pleistocene regimens. At least six of the national parks in volcanic and complex mountain areas (Parts IV and V of this book) still have active valley glaciers, and several more have small remnant glaciers high on northern slopes or in sheltered hollows. One national park in Part I, Theodore Roosevelt in North Dakota, lies on the edge of an area that underwent continental glaciation by Pleistocene ice sheets; and in Kobuk Valley National Park, Alaska, Pleistocene glaciofluvial materials are the main source of the dune sands.

The Theory of Glacial Ages

For many years geologists and other scientists have found studies of glacier characteristics and behavior to be both challenging and rewarding. Oftentimes conclusions drawn from such studies before the theory of glacial ages was universally accepted became the subject of vigorous debates and unyielding opinions. When in the 1870s the renowned John Muir proposed, after months of climbing around Yosemite, that glacial erosion had sculptured that famous landscape, he was challenging the belief of the California state geologist that Yosemite Valley (chapter 25) was the result of violent cataclysm involving faulting. "What does that sheep herder know about it!" exclaimed the eminent authority scornfully. He remained unconvinced even after Muir discovered a small living glacier high above Yosemite Valley in a tributary valley of the Illilouette River.

It was a Swiss geologist, Louis Agassiz, in the middle years of the nineteenth century, who advocated the theory that great ice sheets had moved down from the Arctic and spread over northern and central Europe in prehistoric times. He based the hypothesis on his observations of alpine glaciers and processes of glacial erosion and deposition in his native Switzerland. European naturalists had tried for years to explain such oddities as granite boulders completely unrelated to local bedrock that were scattered on the land surface. (*Erratics* is the geologic term for them.) Also strewn over the surface were deposits of sand, gravel, and cobbles—sorted or unsorted, thick and thin—in landforms of many sizes and shapes. Some thought these materials (called "drift") were left behind as Noachian floods receded. Or, could the erratics have been dropped by melting icebergs tossed up on the lowlands in storms? Another puzzling question was the origin of long scratches and grooves on exposed ledges. The scratches were nearly parallel and, more often than not, trended in a north-south direction. The explanation of the Noachian flood proponents that the scratches were made by flood-swept boulders was not convincing.

Having observed quantities of gravelly material left in front of retreating glaciers in the Alps, Agassiz concluded that the drift deposits and erratics of north-central Europe could have been pushed and carried there by a huge glacier or ice sheet. And since he had seen scratches and grooves (called *striations*) on bedrock in glacial valleys—scratches oriented in the direction from which the ice had come—he theorized that the ice sheet that had advanced over Europe had come down from the polar regions.

Despite some initial opposition, the beautiful logic of Agassiz's ideas gained acceptance. The theory was soon extended to the British Isles and to North America, when Agassiz became a professor of geology at Harvard. Geologists came to recognize glacial ice as a powerful and dynamic agent of erosion and deposition and to understand how glaciers had dramatically created new landscapes over large areas.

Types of Glaciers

During Pleistocene time *continental glaciers* advanced and retreated several times over vast areas of northern Europe and northern North America. These great ice sheets were miles thick and flowed from a central high point downward and outward.

A *valley,* or *alpine, glacier* is confined to a valley (almost always a former stream valley) and flows from higher to lower elevations. Waterton-Glacier International Peace Park (chapter 23), in Alberta, Canada, and Montana, is noted for its many valley glaciers on both sides of the border.

Examples of cirque glaciers can be seen in Rocky Mountain National Park in Colorado (chapter 22). A *cirque glacier* occupies a hollow or basin (called a *cirque*) at the head of a mountain valley and is usually the remnant of a much larger glacier.

When several descending valley glaciers join together at lower elevations and spread out in a wide, bulbous ice mass at the foot of the range, this is called a *piedmont glacier*. The largest piedmont glacier in North America is the Malaspina in Wrangell-St. Elias National Park and Preserve, Alaska (chapter 29).

Figure III.3   How glacial ice moves differently in different parts of a glacier. Note that the firn line marks the transition between the zones of accumulation and wastage. From *Landforms and Landscapes* (3rd ed.) by S. D. Tuttle. © 1980 Wm. C. Brown Publishers. Used by permission.

How Glaciers Form and Move

Glaciers form in regions where more snow falls than melts each year. As more and more snow accumulates, a perennial snowfield builds up, while at the same time compacting and settling under its own weight. Individual snowflakes, formed from water vapor by direct crystallization, lose their characteristic six-pointed shape and become granules due to being compressed closer together. At the stage of transition from snowflakes to glacial ice, the granular snow is called *firn*. Continued compaction expels air and forces the granules more tightly together. Individual grains melt slightly and then recrystallize, filling voids and transforming the whole mass into ice. As soon as the ice mass becomes large enough and heavy enough so that gravity causes it to move or flow downward or outward, it is considered, by definition, a *glacier*.

The high perennial snowfield that nourishes a glacier is called the *zone of accumulation*. Downglacier, where melting and loss of snow and ice exceeds inflow, a zone of wastage begins. Separating the two zones is an irregular, shifting band called the *firn line* that marks the highest level of retreat of winter snow cover during summer.

A glacier's motion is not uniform throughout its mass as it moves downslope. The top ice is brittle and tends to fracture and crack, forming vertical fissures, or crevasses, that may be a hundred or more feet in depth. This outer part of a glacier is the *rigid zone*. Underneath the rigid zone is the *plastic flow zone* where ice deforms without fracturing (plastic behavior), due to pressure from overlying ice. This causes the ice nearer the "sole" of the glacier to move downslope a little faster than the brittle ice at the top (fig. III.3).

Figure III.4   Extent of Pleistocene glaciation. The hachured line that goes from Waterton-Glacier International Peace Park to the Atlantic Ocean off southern New England marks the approximate location of Wisconsinan terminal moraines, the last major advance of continental glaciers. The dotted pattern shows areas covered by earlier Pleistocene ice advances.

The Work of Glaciers

Glaciers are powerful excavators. They have exposed some of the most instructive outcrops that geologists have found in putting together data about mountain-building, geologic structures, and ore minerals. The great weight of a moving ice mass crushes, cracks, and loosens bedrock. In the realm of glaciers, most of this bedrock consists of igneous and metamorphic rocks of varying degrees of resistance because glaciers tend to accumulate at higher elevations where some or all of the sedimentary cover may have already been stripped off. Because the national parks described in this section are in "hard rock country," the chapters include additional explanatory material about igneous and metamorphic rocks.

The work of glaciers is especially effective in mountains, even on tough, resistant bedrock. Chunks of rock are *plucked* or *quarried* by glaciers and moved downslope. Large and small pieces of rock are frozen into the ice and scraped and dragged over underlying rock, scratching, grinding, scouring, pulverizing, and polishing. Glacial striations on bedrock are made in this way. The processes of glacial *plucking* and *abrasion* produce great quantities of sediment of all sizes from *rock flour* (finely pulverized rock debris) to huge boulders. This mostly unsorted rock material, called *till,* is picked up and pushed or carried along on top of, beneath, beside, and in front of a glacier. Ridges of till that accumulate alongside and in front of glaciers are called *moraines.* (Till that is compacted underneath a moving glacier eventually becomes what is called *ground moraine* after the ice melts back.)

In summer seasons, meltwater drains off glaciers, working over and sorting loose till and other debris into coarse and fine fractions, then redepositing them as sediments. *Glacial outwash* is the geologic term for such deposits. A broader term that includes all sorted and unsorted glacial deposits (till, outwash, etc.) is *glacial drift.*

The erosional and depositional processes of active alpine and continental glaciers produce distinctive landforms and other features characteristic of glacial scenery. Rangers in the national parks in glaciated regions usually point out interesting examples to park visitors. However, the key indicators on a landscape of former glacial action are, as in Agassiz's day, the presence of striations, erratics, and glacial drift.

Glacial Advances and Retreats

In Pleistocene time, which lasted about 2 million years, several episodes (or stages) of continental glaciation occurred, separated by ice-free interglacial periods. What brought on a glacial episode was a climatic trend of lower average summer temperatures, possibly for not very many

Figure III.5   Alpine glaciation greatly modifies the topography of mountainous regions. *Top:* The main glacier and its tributary glaciers nearly fill preglacial stream valleys. *Bottom:* In the postglacial landscape, the main valley has become a broad, U-shaped trough, streams tumble from hanging valleys, horns and cols sharpen the skyline, and a cirque is cut in the head of a tributary valley. From *Geology along Going-to-the-Sun Road, Glacier National Park, Montana* by O. B. Raup et al. © 1983 Glacier Natural History Association. Used by permission.

years, accompanied by increased precipitation. The onset of glacial episodes apparently took much less time than the more gradual melting back and decline of the ice sheets.

In the broadest sense, the episodes of global cooling and glacial advance may have been worldwide, but worldwide correlations of glacial features has proved not to be practical because local variations were so great. Intensive study and the use of radioactive dating has shown that advances, standstills, and retreats of glacial ice in different areas, once thought to be contemporaneous, probably occurred at different times in different regions. Glaciers did not reach their maximum extent in Puget Sound (Washington), the Southern Rocky Mountains, or the coast of Maine all at the same time. Accordingly, glacial chronologies and sequences are described, for the most part, in local terms. However, as a general time term, the Wisconsinan stage is still used in North America in referring to the last major glacial advances of the Pleistocene Epoch (fig. III.4).

Carbon-14 Dating

Some of the glacial deposits of the later phases of Wisconsinan glaciation can be assigned absolute ages, mainly within the last 55,000 years, by means of carbon-14 (C_{14}) dating techniques. This method uses short-lived radioactive atoms of carbon-14 produced by cosmic ray activity in the earth's atmosphere. Minute amounts of carbon-14 quickly combine with stable carbon atoms in atmospheric carbon dioxide. As living things incorporate carbon into their tissues, the ratio of carbon-14 to other carbon atoms remains the same as the ratio in the atmosphere. After an organism dies, the carbon-14 contained in its tissues decays radioactively at a fixed rate. When the ratio of carbon-14 to the other carbon atoms in the organic remains is compared, a radiometric date can be calculated that establishes the length of time that has elapsed since the death of the organism. The technique is useful in dating recent glacial and volcanic events, very young sedimentary accumulations, and archeological sites.

Is Another Ice Age Coming?

Scientists have speculated for many years on the ultimate cause (or causes) of global cooling that initiates an ice age and the subsequent warming that brings such an event to a close. No widely accepted explanation has been put forth that would account for all the phenomena observed, but most of those proposed deal with factors affecting precipitation; that is, increases or decreases in mean annual snowfall over time. Among the diverse ideas being considered are the following: (1) "wobbling" of the earth's axis and variations in the earth's orbit, causing shifts in the amount of solar radiation received on the earth's surface; (2) changes in the concentration of atmospheric gases (such as carbon dioxide) that may affect retention of solar radiation; (3) increases in the size and number of volcanic eruptions, creating dust clouds that block the sun's rays; (4) tectonic "drifting" of large land masses toward polar latitudes; (5) changes in the size and extent of the polar icepack, affecting evaporation of moisture by polar winds; (6) and changes in ocean currents, especially those circulating warm sea water into colder regions, resulting in increases or decreases in the amounts of water vapor evaporated from oceans.

The Holocene Epoch, the geologic time unit we live in, is generally regarded as an interglacial stage that has followed the last extensive episode of North American glaciation, which is referred to as the Wisconsinan stage of the Pleistocene Epoch. Since the Wisconsinan stage ended, roughly 12,500 years ago, global temperatures have shifted back and forth and modern glaciers have made relatively minor advances and retreats. It seems likely that at some time in the future, great ice sheets will again form and spread over the northern land masses of the Northern Hemisphere.

National Parks in Part III

19

Interconnected northern lakes, once the route of French-Canadian voyageurs, are surrounded by boreal forests.

A young, glacially scoured landscape exposes ancient Precambrian rocks of the Canadian shield.

Patches of moraine, erratics, lakebed deposits, and peat bogs are scattered over glaciated bedrock.

Pits of small, worked-out, gold mines are scattered along a "greenstone belt" in the park.

Voyageurs National Park

Northeast Minnesota

Figure 19.1 The large, granitic glacial erratic the man is standing on rests on schist bedrock in Cranberry Bay, an arm of Rainy Lake in Voyageurs National Park. National Park Service photo by L. Grim.

VOYAGEURS NATIONAL PARK

Figure 19.3   Air view of waterways and islands in Kabetogama Lake. National Park Service photo.

Geographic Setting

Roads end a short distance beyond the entry points of Voyageurs National Park, which is located in the borderland of islands and waterways between Minnesota and Ontario, Canada. Visitors hike, snowmobile, ski, and fly in Voyageurs National Park, but the main means of traveling among the islands is by canoe or boat. This is an area of water, woods, wildlife, and well exposed, very old igneous and metamorphic rocks, thinly mantled by Pleistocene glacial deposits. The only example of a southern boreal forest in the National Park System is protected in the park; that is, a forest characteristic of the transition zone between the Northern Temperate region and northern boreal forests with subarctic vegetation.

The park is about 40 miles long and extends from International Falls on the west to the Boundary Waters Canoe Area on the east. The width of the park varies from about 3 to 15 miles. During the Pleistocene Epoch, continental glaciers shaped the landscape. As ice sheets advanced and retreated, they scraped, gouged, and pushed away trees, rock, and soil, thus converting the region of huge forests and a few rivers to a barren land of sparse soil, bogs, grassy meadows, and numerous lakes interconnected by a network of streams. The boreal forests that covered the land when the Indians, voyageurs, and fur traders were here grew in the thin soil left by melting glaciers or deposited in temporary lakes after the last ice sheet melted back about 10,000 years ago. Most of the virgin timber was logged off in the years before and after the

turn of the century, but some tracts of the boreal forest remain in Voyageurs National Park, the Boundary Waters Canoe Area, and Quetico Provincial Park in Ontario.

Winters are long and cold in the north woods, and daylight hours are few. Snow may cover the ground from October until May, and ice in the lakes and streams seldom breaks up before the middle of May. Although the summers are relatively brief, temperatures are comfortable and the days are long. Spring and fall are times of intense activity in the wilderness; migratory birds return to the northern breeding grounds as days lengthen, and then head south again as winter approaches.

Local History

"Red Paint People" came into the north woods about 8000 years ago. A few pictographs painted on rock cliffs with red ocher and artifacts, such as slate knives and arrowheads, give clues about their Stone Age culture. Chippewa Indians lived in the park area when the French Canadian trappers and voyageurs traveled back and forth along the waterways. The white men learned Indian tongues and Indian ways of adapting to the wilderness. They made birch-bark canoes for transport and intermarried with Indians. Indian guides taught the voyageurs the safest routes among the myriad streams and lakes.

It was the voyageurs for whom the park was named, and their well established waterway between Lake Superior and Lake of the Woods was made the international boundary by the 1783 treaty ending the American Revolution. Fifty-six miles of that boundary, those passing through the waters north of the Kabetogama Peninsula, also mark the northern limits of Voyageurs National Park.

For 200 years, beginning in the 1600s, the voyageurs plied their trade by paddling and portaging thousands of miles through the wilderness waterways and bringing out many tons of furs. The voyageurs' 3000-mile route between Montreal and Lake Athabasca (northeastern Alberta) was divided into eastern and western stages. Each stage had its own type of birch-bark canoes and group of men. The exchange point was first at Grand Portage, Minnesota, on the western shore of Lake Superior; but after the American Revolution, the traders—wishing to avoid American taxes—moved the rendezvous site to Thunder Bay in Canadian territory.

The larger, 36-foot Montreal canoes were paddled by 10 men and transported trading goods and supplies out and furs back, between Montreal and Grand Portage or Thunder Bay. At the exchange station, cargoes were split up and reloaded into the lighter, 25-foot north canoes, each with a crew of 5 to 6 men, who traveled between Grand Portage and the western end of the line at Lake Athabasca. As soon as the ice began to break up in spring, the voyageurs loaded their canoes with the winter's accumulation of furs that had been traded for goods in remote Indian villages. The whole trip—getting the furs to the rendezvous, picking up the trade goods from Montreal, and returning to Lake Athabasca—had to be completed between spring thaw and autumn freeze-up. In making the 2000-mile journey, the voyageurs paddled and portaged 15 to 18 hours a day. These brave, hardy men, most of them French Canadians, were noted not only for their strength and endurance but also for their songs, their humor, and gusto.

By the middle of the nineteenth century, the fur trade had declined; and by the 1870s, lumbering and mining were assuming more importance. Large-scale logging of the virgin forests went on for about 50 years. Like the voyageurs, the lumbermen had their legendary heroes, exemplified by stories of Paul Bunyan.

The Kabetogama Peninsula was the scene of a gold-mining boom between 1893 and 1898. The boom town of Rainy Lake City (northwest corner of the park) filled up with prospectors and miners but then decreased from a settlement of 500 people to a ghost town in five years. Getting supplies and equipment to the mines and hauling out ore was difficult, with the closest transportation center 100 miles away.

The chief producer was the Little American Mine, located on the island of the same name near the south shore of Rainy Lake. About 500 tons of ore were extracted and crushed, yielding gold that was worth over $4600 in 1894 and 1895. The gold (of hydrothermal origin) was disseminated in a quartz vein 4 to 6 feet wide in a belt of severely metamorphosed rock (sheared chlorite and biotite schist). The open cut of the mine (a shaft 10 feet wide and 44 feet deep) was flooded when the water level of Rainy Lake was raised for the purpose of generating hydroelectric power.

The Big American Mine, the first mine dug in the area, was on Dryweed Island on the homestead of the man who discovered the gold. The pit, 5 feet square and only 5 feet deep, was sunk on a quartz vein. Lyle Mine, on the north side of Dryweed Island, had a shaft 22 feet deep, but Old Soldier Mine, on a very small island off Dryweed, was only a shallow pit. Two other abandoned mine shafts are on the east end of Dryweed.

At Bushyhead Mine on Bushyhead Island, an adit (tunnel) penetrated a shear zone of massive pyrite and quartz just above water level. Gold Harbor and Holman Mines, on a blunt point of the mainland, are shallow pits in a quartz-pyrite vein in a schistose area.

The mining and lumbering industries—although they brought a measure of prosperity to the region—drastically altered much of the natural environment. Lumbering, in particular, left large areas devastated. However, on both sides of the international boundary, people who loved the northern lake country felt that the forests and waterways should be preserved and protected. As early as 1891, the Minnesota legislature sought to have a national park established on the Kabetogama Peninsula. Through the years similar efforts were being made in Ontario, resulting in the establishment of Quetico Provincial Park. The author-naturalist Sigurd Olson, who called the Boundary Waters the "singing wilderness," led the movement for the establishment of Voyageurs National Park, and it finally came about in 1975. Both Canada and the United States have designated the region of boundary lands and waters "to be kept in a state of nature as far as possible."

Geologic Features

The Precambrian Rocks

All of the bedrock exposed in Voyageurs National Park and the surrounding region is Precambrian in age. These ancient rocks are part of a large area of exposed basement rocks called the Canadian Shield, which is, in turn, a part of the North American craton. A *craton* is a part of the earth's crust that has been stable (without orogenic activity) for at least 1500 million years. The oldest cratonic rocks are thought to be the core or nucleus of early continental crust that formed as a cluster of islands and probably later became a single large continent.

Figure 19.4   Migmatitic erratic sitting on chlorite schist of the greenstone belt, Skipperrock Island in Rainy Lake. Pebble shoreline in foreground. National Park Service photo by L. Grim.

Exposures of the oldest rocks, classified as Archean, have been identified in shields in other parts of the world, such as the Fenno-Scandinavian region of northern Europe and in parts of Asia, Australia, South America, Antarctica, and Africa. The fact that the shield rocks are surrounded by platforms composed of somewhat less ancient Precambrian rocks (and early Paleozoic rocks) suggests that the early continents may have grown by accretion. Also the platforms have some characteristics of once-active tectonic belts. Most Precambrian rocks are igneous or metamorphic. Erosion for millions of years brought about the characteristic low relief of the Precambrian shields.

Clusters of radiometric dates show that each Precambrian shield area is divisible into subregions, or structural provinces, based on rock ages. Within each province, the trend of tectonic folds and the style of deformation are distinctive from those of adjacent areas. Thus metamorphic boundaries and fault boundaries, as well as radiometric dates, delineate the various provinces. Both Voyageurs National Park and Isle Royale National Park (chapter 20) are in the Superior Province, which is a large

Precambrian subregion with many bedrock exposures. (The Precambrian rocks in Isle Royale National Park in Michigan are much younger than those in Voyageurs.) Superior Province extends over a large part of south central Canada and dips slightly into the United States in northern Minnesota, Wisconsin, and Michigan. Because of its mineral resources, which include major deposits of iron, gold, copper, and nickel, the Superior Province has been studied by Canadian and American geologists for many years. Much of the pioneer work on radiometric dating was done on Precambrian rock specimens from this area.

The Archean rocks of the Voyageurs region are associated with the Kenoran (Algoman) orogeny, which is regarded as the earliest datable orogeny in North America. Two contrasting assemblages are present in these ancient rocks—the granitoid gneisses, which are highly metamorphosed, and the greenstone belts, which are less highly metamorphosed. The granitoid gneisses were derived from high-silica rocks, possibly of volcanic origin, and sandstones. Because some greenstones are "host rocks" for

Box 19.1
Why Dating and Correlating Precambrian Rocks Is Difficult

Precambrian rocks, wherever they are found, represent geologic activities that took place during more than three-fourths of the entire time of the earth's geologic history. The rocks are "samples of time" that provide tantalizing clues about the origin of continents, early tectonics, and the development of the earth as we know it (fig. 19.5).

Techniques of radiometric dating have yielded a number of absolute age determinations for Precambrian rocks. Such a value may be based on one or more methods of *radiometric dating;* that is, laboratory procedures for measuring how much time has elapsed in the process of radioactive decay since a particular rock formed. *Radioactive*

Figure 19.5   Radiometric ages of Precambrian rocks on the North American craton. Black dots show sites where dates older than 3.2 billion years (Archean) have been obtained. From M. S. Petersen, J. K. Rigby, and L. F. Hintze, *Historical Geology of North America* (2nd edition). © 1980 Wm. C. Brown Publishers. Used by permission.

decay is the spontaneous disintegration of certain unstable atoms, called the parent material, to *daughter products* through gain or loss of isotopes of the radioactive elements present. By carefully determining the ratio of daughter products to parent material in a given sample, the radiometric age of the rock can be calculated because the rates of decay are constant.

Igneous rocks containing mineral grains of radioactive elements tend to yield more precise radiometric values because, having crystallized from a melt, an igneous rock is less likely to be contaminated by previously decayed daughter products. For example, samples taken from a mafic dike on an island in Rainy Lake yielded K-Ar dates of 2130 m.y. and 2040 m.y. (Proterozoic X). K-Ar values are based on the process by which an isotope of potassium, K-40, converts to argon, Ar-40. (*Mafic* refers to igneous rocks with a high proportion of dark minerals that contain iron and magnesium compounds.)

In dating metamorphic rocks, several techniques can sometimes be combined to yield reasonably accurate age determinations. Sometimes radiometric dates for the mineral zircon in metamorphic rocks can be derived through the use of the rubidium-strontium method; that is, ascertaining the proportion of the rubidium radioactive isotope, Rb-87, to its daughter product, the strontium isotope, Sr-87. Age determinations for various rocks sampled in the Rainy Lake area and analyzed by this technique have ranged between 2690 m.y. and 2750 m.y. (Archean).

Deeply weathered rocks are not suitable for radiometric dating because much of the original parent material has been lost or altered.

In studying Precambrian rocks, geologists have not found common factors other than radiometric age that can be used consistently to correlate Precambrian rocks from one region to another. Outcrops are scattered; index fossils are lacking; some rock units are too deeply buried for cores to be obtained; all have been subjected to repeated metamorphism and extensive erosion; and each geographic district has its own geologic history with different events having occurred in various sequences. Logical breaks in Precambrian chronology show up on every continent, but where these interruptions occur varies so much from place to place that using them for purposes of correlation is not feasible. The result has been that scientists on each continent have had to develop their own system for classifying Precambrian rocks and events.

Geologists are in general agreement that Precambrian time can be divided into two eons, Archean and Proterozoic. An *eon* is the longest geologic-time unit, the next in magnitude above an *era*. Because the Archean was a time of high heat flow, the earth's primitive continental crust probably developed fairly rapidly. In Proterozoic time, the amount of heat generated by radioactive decay decreased, continents grew more gradually, tectonic mechanisms similar to those of the present began, and primitive life forms developed.

Obviously, the Archean-Proterozoic boundary is a significant transition zone in the evolution of

continued

Box 19.1 continued

the continents, but where in global geochronology is this boundary to be placed? On the Baltic Shield, the end of the Archean Eon has been designated as 1900 million years ago; but in the Soviet Union, the division between Archean and Proterozoic is drawn at 2600 million years ago.

Lacking a uniform system, geologists in different parts of the world who study Precambrian rocks have experienced communication difficulties when trying to work together on investigations of mutual interest. The U.S. Geological Survey's approach to Precambrian classification is to use the one common factor that does exist; that is, radiometric dating. The Archean-Proterozoic boundary has been set at 2500 million years ago, and the letters X, Y, and Z have been assigned to time units in the Proterozoic Eon. Although these time terms are regarded as being "without specific rank," they are useful for comparing and classifying groups of Precambrian rocks that yield dates falling within these informal divisions.

Table 19.1 compares the U.S. Geological Survey's Precambrian chronology with the

Canadian classification. Although time boundaries in the two systems differ slightly, the two classifications are more similar to each other than either one is to Precambrian chronologies developed for the other continents of the world. Both systems place the end of Precambrian time at about 570 million years ago.

Research now underway on Precambrian life forms may eventually yield information that will be helpful in correlating Precambrian rocks in different parts of the world. In 1954 the first discovery of Precambrian microfossils was made near Thunder Bay on Lake Superior. The Gunflint Chert in which the fossils were found is Proterozoic X in age. Microfossils have not been found in rock units in Voyageurs National Park. However, about 60 miles northeast of the park, near Atikokan, Ontario, algae fossils, 2.7 billion years old (Archean or early Proterozoic), have been found in marine sedimentary rocks of the Steep Rock Formation of the Timiskaming group.

Table 19.1 Comparison of Precambrian Subdivisions Used in the United States and Canada. (Geologic Names Committee, U.S.G.S., 1980)

Canadian Subdivisions		Canadian Orogenies		U.S.G.S. Subdivisions	
Proterozoic	Hadrynian	Grenville-1000 my	800 my	Proterozoic	Z
	Helikian	Hudsonian-1800 my	1600 my	Proterozoic	Y
	Aphebian	Kenoran-2500 my	2500 my	Proterozoic	X
Archean				Archean	

Figure 19.6   Ancient crystalline bedrock, severely scoured by glaciation, dips gently down to the shore of Lake Kabetogama. National Park Service photo by R. Schmid.

valuable metallic ore deposits—in various parts of the world, as well as in the Superior Province—the greenstone belts are of special interest to economic geologists.

The greenstone belts occur in large, elongated down-warps between the more extensive granitoid gneiss accumulations. Dark greenish in color, the greenstones owe their name and appearance to the constituent minerals—chlorite, epidote, actinolite, etc.—that were produced by metamorphism. The presence of pillow lava structures in the greenstone belts indicates that the lavas were extruded in the waters of an ancient sea. Although most of the pillow structures in the Voyageurs area have been destroyed by metamorphism, a few "pillows" up to three feet long have been found on a small islet off Steamboat Island (north of Cranberry Bay). Some poorly sorted mixtures of clay, mica, and quartz grains (graywackes) that were derived from the decomposition of pyroclastic debris (tuff, ash, etc.) suggest that some of the volcanic activity was explosive. Some of the clastic material also has characteristics of rapid deposition by marine currents, possibly in an ocean trench.

The greenstone belt in Voyageurs National Park consists of schistose (i.e., finely foliated) and sheared lavas that have undergone intense metamorphism. In general, the schistosity is so extreme that the composition and characteristics of the original lavas are masked. Major islands in and near the northwest corner of the park, such as Dryweed, Grassy, and Grindstone Islands, are made up of greenstone, as are sections of Neil Point (across the Narrows) and part of Rainy Lake's south shore between Jackfish Bay and Black Bay.

Between the greenstone tracts are large areas of metamorphosed biotite and mica schists (originally sedimentary and granitic rocks). The main portion of the Kabetogama Peninsula and the Black Bay area are in part of the schist belt, and a contact zone between the greenstone belt and the schist belt can be traced in the Rainy Lake area.

The Vermilion batholith, which is associated with the Kenoran (Algoman) orogeny, underlies the Sullivan Bay and Blind Ash Bay area on the south side of Kabetogama Lake. Some of the coarse-grained, pink and gray granites show up in striking exposures along the shore. This large intrusion of granitic magma may have been the source of the gold-bearing quartz solutions that crystallized in veins along an east-west fault zone extending to Neil Point (just outside the western edge of the park).

Glacial Features

Glacial Deposits. In the Voyageurs National Park area, most glacial deposits are associated with Wisconsinan glaciation and consist mainly of four types: lakebed sediments, outwash gravels, peat, and till. The lake deposits accumulated in shallow margins of Glacial Lake Agassiz, a vast, inland body of water that for several thousand years covered hundreds of square miles north and south of the international boundary. The lake formed as water began to collect in the Red River valley in western Minnesota about 12,000 years ago and then spread over eastern North Dakota and adjacent Canadian Provinces. During this time of glacial retreat, meltwater draining the last continental glacier ponded because the north-flowing rivers headed for Hudson Bay were blocked by ice. As climatic conditions changed, lake levels were higher and lower a number of times. Recognizable beach lines recorded stillstands; that is, times when a lake level was constant for a long enough interval for waves to form a notch in the shore or a beach. Deposits of outwash gravels suggest that several outwash channels developed and were subsequently abandoned.

The peat deposits, which have been dated as Holocene in age, formed in hollows or depressions, such as old lake basins, bogs, and swamps.

Till deposits, made up of unsorted glacial debris, occur as ground moraine and end moraine. Patches of *ground moraine,* deposited under the ice sheet, are interspersed with lakebed sediments to form an undulating surface called "swell-and-swale" topography. Till in *end moraines* was left in hummocky ridges dumped along the front of the ice sheet. The sections of end moraine that can be seen in places along the southwest shore of Rainy Lake were left by the last active glacier to occupy the area. This was a lobe of ice that advanced from the northeast.

Erratics. Ranging in size from pebbles to blocks larger than automobiles, erratics abound in the Voyageurs area (figs. 19.1, 19.4). Some were carried by the ice only a few

Figure 19.7   Hardy jack pines cling to crevices and grooves in this glacially polished ledge that slopes down to the shore of Rainy Lake. National Park Service photo by L. Grim.

miles, while others traveled long distances. Erratics plucked by glacial ice from outcrops of Precambrian gneiss and greenstone in the Boundary Waters region are scattered over the landscape of southern Minnesota and are even found in Iowa.

Striations and Glacial Polish. The parallel, linear scratches that show up on outcrops and erratics in the park are glacial *striations,* formed when quartz sand grains and hard, sharp rock fragments, frozen into the bottom of the ice sheet, scratched their way across rock surfaces as the glacier moved along. Striations show local directions of ice movement. In Voyageurs National Park, striations point south, south by southwest, and southwest. On some outcrops, the presence of cross-striations indicates more than one direction of ice movement.

Repeated abrasion by several ice advances has smoothed and polished many rock exposures. *Rock flour,* consisting of fine particles held to the bottom of the ice sheet, put a high gloss on such rock surfaces. Glacial polish on highly resistant outcrops in Voyageurs still looks fresh or only slightly weathered.

Effects of Glacial Rebound

The weight of huge ice sheets moving southward depressed the crust several hundred feet. After the ice retreated, lakes filled the basins and spread over the lowlands. Relieved of the weight of ice, the crust probably regained its normal elevation after each major episode of glaciation. This postglacial upwarp, called *glacial rebound,* is still going on, although probably at a slower rate in the last few thousand years. As the land has risen, some lakes have drained away; in others water levels have dropped. Some lakes have become bogs and swamps, filled in by vegetation and sediment. The lakes of the Boundary Waters region occupy deeper parts of the basins once filled by Glacial Lake Agassiz. Since rebound of the crust has been uneven, old beach lines, representing earlier lake levels in northwestern Minnesota, are now tilted slightly to the south.

Geologic History

1. Early Precambrian (Archean) basement complex.

These intensely metamorphosed rocks of sedimentary, volcanic, and intrusive igneous origin, dated as 2600 million years old, or older, are commonly referred to as "pre-Kenoran" (or "pre-Algoman"). The Coutchiching rocks (which may or may not be the oldest formation) are foliated gneisses and schists rich in mica that may originally have been sandstones. A group of similar rocks of sedimentary origin, designated as Timiskaming (or Windigokan), are thought to be younger, mainly because the conglomerates contain large cobbles of granitic material

that suggest rapid erosion and deposition. The volcanic greenstones, or Keewatin lavas, are largely basaltic but include some clastic material.

2. Kenoran (Algoman) orogeny at the end of Archean time and the beginning of Proterozoic X time.

The emplacement of large granitic batholiths in the Superior Province brought about intense folding and uplift of the earlier rocks and the forming of what were probably mountain ranges of considerable bulk and height. The severe deformation that accompanied the orogeny caused partial remelting of rocks as well as metamorphism. Subsequent erosion of the mountain ranges has exposed a number of ore-bearing structures in the region, such as the gold veins found in the Kabetogama Peninsula area.

3. Erosion of the mountains; minor intrusive activity.

The youngest bedrock in Voyageurs National Park is exposed in mafic dikes, about 2100 million years old (Proterozoic X), that crop out along the lake shore and on some of the islands.

4. Prolonged erosion throughout Paleozoic, Mesozoic, and most of Cenozoic time.

The end of Precambrian time brought to a close the long series of tectonic events that created the basement complex of the Canadian Shield. After many millions of years of erosion, only the roots of the ancient mountains remained. Relief was low. Probably a fairly uniform soil cover, supporting extensive forests, blanketed the ancient bedrock when continental glaciers began to form in Pleistocene time.

5. Pleistocene glaciation.

Ice sheets covered the Boundary Waters region several times, pushing and scraping away virtually all of the old soil cover and gouging out bedrock. The first glacial advance, probably early in Pleistocene time, came from the north-northwest. After a considerable interval, ice from the northeast moved over the area. A brief ice-free period was followed by the advance of a continental glacier from the west. Later, a large lobe of glacial ice came from the northeast, but may not have extended much beyond Rainy Lake. The later ice advances, which are associated with Wisconsinan glaciation, obliterated traces of the earliest glaciations.

6. The postglacial, constructional landscape.

Unconsolidated materials, Pleistocene and Holocene in age, have been superimposed on some of the oldest bedrock in North America. Because relief in the area is low, streams have been slow in reworking and sorting the glacial debris. Lakes occupy the large basins; and small ponds, bogs, and wetlands, interconnected by waterways, are all about. These are indications that drainage in the area is still disorganized.

Table 19.2 Geologic Column, Voyageurs National Park

Time Units			Rock Units	Geologic Events
Era	Period	Epoch		
Cenozoic	Quaternary	Holocene	Alluvium in streams, lakes, bogs, etc.	Continuing glacial rebound; some lakes drained or filled in.
		Pleistocene	Glacial drift	Minor morainal and outwash deposition Severe glacial erosion by four major ice sheets
Paleozoic/ Mesozoic			/////////	Prolonged erosion of the shield
Proterozoic X (Middle Precambrian time)			Mafic intrusive rock	Intrusion of mafic dike
Archean (Early Precambrian time)			Kenoran (Algoman) granitic intrusives	Kenoran (Algoman) orogeny; mountain-building and uplift; metamorphism
			Pre-Kenoran (pre-Algoman) formations (position in column uncertain):	
			Timiskaming (Windigokan)	Rapid deposition of coarse conglomerates, graywackes, silt, etc.
			Laurentian granites	Small granitic intrusions
			Keewatin lavas	Submarine extrusion of mafic lavas; Occasional pyroclastic eruptions
			Coutchiching	Marine deposition of coarse and fine sands, silts, etc.

Source: modified from Pye 1968

Geologic Maps and Cross Sections

Day, W. C. 1985. Bedrock geologic map of the Rainy Lake area, northern Minnesota. U.S. Geological Survey Open File Report 85–0246.

Goebel, J. C. and Walton, M. 1979. Geologic map of Minnesota, Quaternary geology. Minnesota Geological Survey, University of Minnesota.

Sims, P. K. 1970. Geologic map of Minnesota. In *Geology of Minnesota: a centennial volume,* ed. P. K. Sims and G. B. Morey, plate 1. Minnesota Geological Survey.

Bibliography

Day, W. C. and Weiblen, P. W. 1986. Geochemical evidence from the Vermilion Granitic Complex of northern Minnesota. *Contributions to Mineralogy and Petrology* 93(3):283–296.

Harris, F. R. 1974. Geology of the Rainy Lake area, District of Rainy River. Geological Report 115. 94 p., two maps in pocket. Ontario Division of Mines, Ministry of Natural Resources.

Hofmann, H. J. 1971. Precambrian fossils, pseudofossils, and problematica in Canada. Geological Survey of Canada Bulletin 189.

Minnesota Geological Survey. 1969. An evaluation of the mineral potential of the proposed Voyageurs National Park, with map. University of Minnesota.

Ojakangas, R. W. 1972. Rainy Lake area. In *Geology of Minnesota: a centennial volume,* ed. P. K. Sims and G. B. Morey, pp. 163–70. Minnesota Geological Survey.

———— and Matsch, C. L. 1982. *Minnesota's Geology.* Minneapolis: University of Minnesota Press. 255 p.

Petersen, M. S.; Rigby, J. K.; and Hintze, L. F. 1980. *Historical geology of North America.* 2nd ed. Dubuque, Iowa: Wm. C. Brown Company Publishers.

Pye, E. G. 1968. *Geology and scenery, Rainy Lake and east to Lake Superior.* Geological Guide Book no. 1. Ontario Department of Mines.

Southwick, D. L. 1972. Vermilion granite-migmatite massif. In *Geology of Minnesota: a centennial volume,* ed. P. K. Sims and G. B. Morey, pp. 108–118. Minnesota Geological Survey.

Wright, H. E., Jr. 1972. Physiography of Minnesota. In *Geology of Minnesota: a centennial volume,* ed. P. K. Sims and G. B. Morey, pp. 561–78. Minnesota Geological Survey.

Voyageurs National Park

Area: 217,892 acres; 339 square miles land and water

Land area: 138,828 acres; 217 square miles

Established: April 8, 1975

Address: P.O. Box 50, International Falls, Minnesota 56649

20

A forested island, largest in Lake Superior, with an irregular shoreline, long parallel ridges, lakes and bogs in intervening valleys.

Plucking, abrasion, and deep scouring by Pleistocene continental glaciers shaped the landscape.

Left by the last glaciation are drumlins, erratics, ground moraine, and end moraines.

High shorelines, old beaches, and wave-erosion features mark levels of postglacial lakes that preceded Lake Superior.

In late Precambrian time, a thick sequence of basaltic lavas with interbedded sediments and a sedimentary cover subsided to form the Lake Superior syncline.

Copper mines were worked by Indians several thousand years ago and by Americans in the nineteenth century.

Isle Royale National Park

Northern Michigan, in Lake Superior

Figure 20.1 This air view, showing the topographic "grain" of Isle Royale National Park, looks southwest along the strike of major rock units, which dip to the left. Differing resistance of rock units to long subaerial erosion and glaciation produced the narrow bays, peninsulas, and islands of this Lake Superior park. National Park Service photo.

Local History

In northern Lake Superior, the international boundary follows a water passage between Ontario, Canada, and Isle Royale, a long narrow island belonging to the state of Michigan. This wilderness island is 45 miles long and 5 to 8 miles wide. Isle Royale National Park, which comprises the main island, along with its 200 surrounding islets and the waters 4.5 miles out from the shoreline, is accessible only by boat or floatplane. The park is about 50 miles from the shore of Michigan's Upper Peninsula and 15 to 20 miles from the Canadian shore. The island has no roads.

Native copper—the metallic element, uncombined and in a natural state—has long been known on Isle Royale. Indians called the island *Minong,* which means "a good place to get copper." However, Chippewas, who lived in the region when the French explorers came, knew nothing of mining copper or fashioning copper tools. They valued the pieces of "float copper" that they found in loose gravel or on beaches, and they used raw copper fragments for charms or crude cutting and scraping tools. The ancient pit mines that prospectors found on Isle Royale and the Keweenaw Peninsula on the Michigan mainland were a mystery to the Chippewas. Nor did tribal legends preserve the memory of the pre-Columbian people who for many centuries excavated and removed Lake Superior copper.

Archeologists investigating mine pits found charcoal that yielded radiometric dates of 3800 to 1500 years B.P. Ornaments of delicate design made from Isle Royale copper have been discovered in Hopewell mounds in southern Ohio and at other sites, proving that the people of the Copper Culture knew how to soften and shape the nearly pure copper. Artisans developed methods of heating and hammering the copper (pure copper is malleable and ductile) in order to make arrowheads, knives, and other weapons and tools.

Copper Culture mining was time-consuming and arduous. The Indians built a fire on top of a copper vein. When the rock was hot enough, they dashed cold water on it, causing the rock to shatter so that they could pound out chunks of copper with stone mauls. Thousands of the rounded beach cobbles used for this purpose have been found around the mines. Some of the stone mauls are grooved for attaching handles. Although Indian mine pits were left on all parts of the island, the highest concentration is along McCargo Cove on the north side.

When the French occupied the Lake Superior region in the 1600s, they were more interested in fur trading than in mining. After the English gained control of the region a century later, a mining expert was sent to Mackinac but efforts to get mines operating were unsuccessful.

Although Isle Royale was given to the United States by the 1783 treaty with England, the English remained in control until after the War of 1812. The island was important in the fur trade during that time; both English and American companies maintained trading posts. Meanwhile, the Chippewas considered the island to be their territory. Not until after the Chippewas ceded Isle Royale to the United States government in 1843 was the island overrun by prospectors and miners. The famous Ontonagon copper boulder, a glacial erratic long owned by the Chippewas, was the object of a complicated legal wrangle. At last the War Department confiscated the huge specimen, which weighed 3708 pounds, and removed it to Washington where it was placed (and still remains) on display in the Smithsonian Institution.

A report by Douglass Houghton, Michigan's first state geologist, set off the copper boom in the 1840s, both on the mainland and the island. All during the second half of the nineteenth century, mining was carried on at Isle Royale. Because of the severity of the winters, the miners could work on the island only during the short summer season. In order to find ore outcrops more easily, the prospectors and miners burned the forests or cut them down for building settlements. When a promising deposit was found—a contact vein, a dike, or a fissure filled with copper—the miners blasted and excavated. If the site proved out, a shaft was sunk. As a result, in addition to worked-out mines and ancient Indian pits, hundreds of exploration pits are scattered over the island. Many old mines and pits are overgrown with blueberry thickets or have been filled in with dirt.

The nineteenth century miners found three classes of ore on Isle Royale and the Keweenaw Peninsula, which they called "moss," "barrelwork," and "stamp." Moss was large sheets of pure copper that weighed anywhere from a few hundred pounds to several tons. Barrelwork consisted of smaller pieces of copper enclosed in amygdules found at the top of brecciated (rough-textured) flows. An *amygdule* is a filled *vesicle,* or gas cavity, in igneous rock. Bubbles of trapped gas or steam in cooling lava may form vesicles. Later these small cavities may be filled with a secondary mineral (copper in this case) deposited by hot solutions that percolated upward through the rocks. Stamp copper consisted of pebble-sized pieces bound in a rock matrix. The copper was freed by pulverizing the ore in a stamp mill. Occasionally the miners came across masses of native copper weighing several thousand pounds. One mass exhibited at the 1876 Centennial Exposition in Philadelphia weighed 5720 pounds.

The principal mines on Isle Royale (operated at different times between 1843 and 1899) were the Smithwick Mine at Rock Harbor, Siskowit (which produced 190,000 pounds of copper between 1847 and 1855), Minong Mine, and Windigo. Minong Mine, located on Minong Ridge in the midst of ancient mines, was the largest. The Isle Royale

ISLE ROYALE NATIONAL PARK

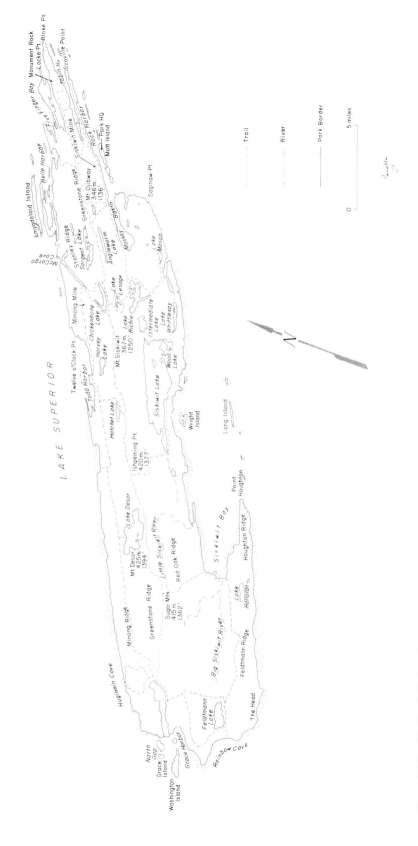

Figure 20.2   Isle Royale National Park, Michigan.

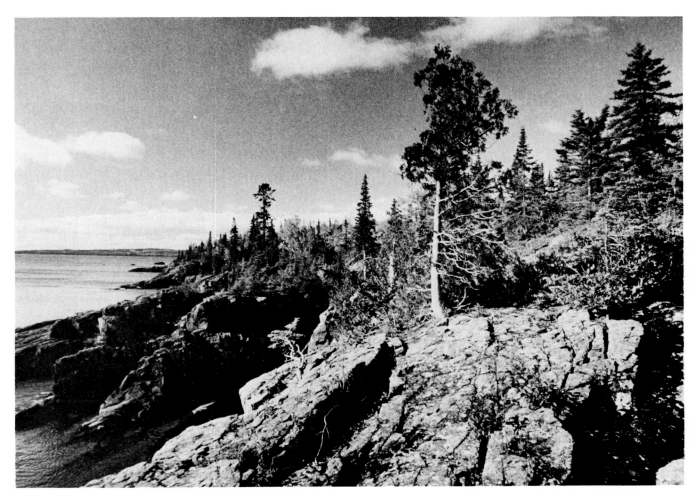

Figure 20.3   Tilted beds of ancient, resistant lava along the shore of Isle Royale. National Park Service photo.

mines were abandoned when the ore was exhausted and excavations became unprofitable. By the end of the century, only a few fishermen and their families were left on the island.

A new era began for Isle Royale when it became a summer resort in the early 1900s and gained a reputation as a haven for hay fever sufferers. The movement to make Isle Royale part of the National Park System was spearheaded by Albert Stoll, Jr., a Detroit newspaperman. Legislation authorizing land acquisition for a national park was passed in 1931, and the park was established in 1940. In 1980 Isle Royale National Park was designated a United Nations International Biosphere Reserve, which makes it part of a worldwide data-gathering system that monitors the impact of global pollution. The recent detection in Lake Siskiwit of airborne PCBs, toxaphene, and dioxins has caused some concern.

Isle Royale is both biologically and geologically interesting due to its isolated location and the geologic events that gave the island its unique character. The long ridges on the island are the result of upfaulting and erosion of late Precambrian volcanics. After prolonged erosion had worn down the surface, Pleistocene ice sheets overrode the whole region and their weight depressed the land. Isle Royale was under water more than once during glacial retreats. Glacial rebound and the lowering of water levels of postglacial lakes exposed the present island surface late in Pleistocene time.

Most of the island was scraped bare of soil and weathered rock, although some glacial till and outwash was left on the western part of Isle Royale. How did plants and animals reestablish themselves on the island? In particular, how did small animals get across the miles of open water between Isle Royale and the Canadian mainland? These are speculative questions to which biologists do not have definitive answers. However, they do know about ecological changes that have come about during historic

Figure 20.4   Schematic cross section showing how Isle Royale, the Keweenawan lavas, Keweenawan Peninsula, and major faults are related to regional structural features. From R. K. and R. A. Paull, *Geology of Wisconsin and Upper Michigan.* © 1977 Kendall/Hunt Publishing Company. Used by permission.

time. Caribou and lynx formerly lived on the island but left after the primeval forests were cut off or burned. The trees and undergrowth that replaced the original stand continue to evolve as a mixture of transitional hardwoods and subarctic conifers.

The Predator-Prey Relationship of Timber Wolves and Moose

Sometime early in this century, moose arrived on Isle Royale. Presumably they swam over from Canada. No predators were on the island, food was plentiful, and the moose population zoomed to several thousand animals, far more than the island could support. By the 1930s, when the water plants, twigs, and leaves on which moose browse were depleted, hundreds of moose starved. A fire in 1936 burned over a quarter of the island, eliminating more vegetation, and more moose died. However, as new growth, stimulated by the fire, renewed the food supply, the moose population increased rapidly. This was followed in about ten years by another starvation cycle.

In the extremely cold winter of 1948–49, Lake Superior froze over, a rare event that occurs only a few times in a century. A pack of eastern timber wolves from Canada crossed over on the ice and established themselves on Isle Royale. Since that time three more packs, either new immigrants or offshoots of the original group, have settled in different parts of the island. The number of wolves on Isle Royale now is between 25 and 40. Each pack kills, on the average, one moose about every three days. Their most likely prey is an animal that is very young, old, sick, or weak. This culling process keeps the moose population healthy and stable at a ratio of about 30 moose per wolf. The fact that the moose herd on Isle Royale has a high proportion of twin calves born each year is a further indication that the stock is productive and vigorous.

A third mammal that helps to maintain the balance of wolves and moose on the island is the beaver. Its role is to build dams that create ponds and beaver meadows to supply browse for the moose and food for other animals in the ecosystem.

Geologic Features

Like Voyageurs National Park (chapter 19), Isle Royale National Park is part of the Canadian Shield and is included in the Precambrian Superior Province. Ancient, mineralized, basaltic lavas are a feature of both parks; but the Precambrian lava flows at Isle Royale erupted millions of years after those in Voyageurs National Park. (Precambrian chronology is explained in box 19.1.)

The Keweenawan lavas of the Lake Superior region poured forth as *flood basalts* from a rift of continental proportions in the craton. Time and time again throughout late Precambrian time, sheets of fluid lava (from the lower crust or upper mantle) spread over thousands of square miles of the ancient landscape. (Flood basalts accumulate in nearly horizontal layers rather than building volcanic cones over vents.) Geologists believe that the fissure from which the lavas emanated originally extended from the Lake Superior region all the way to what is now Kansas. However, south of the Michigan, northern Wisconsin, and Minnesota outcrops, the Keweenawan lavas lie in a subsurface belt under many hundreds of feet of younger rock and glacial drift. The trace of the ancient rift has been mapped by geophysical techniques that detect gravity anomalies beneath the surface. The high densities of the buried Keweenawan basalts register large positive anomalies on gravity meters carried or flown over suspected areas.

The Keweenawan volcanics that crop out on Isle Royale on the north side of Lake Superior and their counterparts on the Keweenaw Peninsula to the south are part of an escarpment of resistant basaltic rock that nearly encloses Lake Superior, and the volcanic sequence is continuous beneath the lake basin (fig. 20.4). The present Lake Superior basin was formed by glacial and preglacial erosion of a large structural trough, called the Lake Superior syncline, that may have existed even before the Keweenawan eruptions began. Subsidence of this trough apparently continued during and after the eruptions. After volcanic activity ceased, sediments were deposited over the lavas. Reverse faulting, probably early in Paleozoic time, along both the north and south limbs of the synclinal

Figure 20.5   Lichen-encrusted trap rock near Rock Harbor, southeast side of Isle Royale. Photo by S. D. Tuttle.

trough, uplifted the Keweenawan rocks. Subsequently, erosion of less resistant sedimentary rocks left the escarpment that is now exposed at Isle Royale and on the Keweenaw Peninsula. The synclinal structure is asymmetrical because although the lava flows in both the north and south arms of the escarpment dip toward the center of the syncline, the flows on the Keweenaw Peninsula dip more steeply than corresponding flows on Isle Royale. The dips of the lavas on Isle Royale are less than 20°.

The Precambrian Rocks

Visitors to Isle Royale National Park can find exposures of volcanic and sedimentary rocks in a rich variety of textures and types, despite a dense ground cover of bushes and trees. The lava flows that were originally stacked more or less horizontally have been tilted by subsidence and faulting so that they are now exposed in strips or layers on ridge tops and along the shore. Thin sedimentary layers between some of the flows represent quiet intervals during which weathering of the flow surfaces and minor erosion and deposition could occur. Individual flows within the volcanic sequence have characteristic textures reflecting

different rates of cooling and varying mineral compositions. Since most flows are basaltic, they are of *mafic* composition; that is, composed chiefly of dark-colored, ferromagnesian minerals. A few flows have an *intermediate* or *felsic* composition; that is, andesites containing an approximately equal amount of light- and dark-colored minerals or dacite, with a higher proportion of light-colored minerals.

The dark, massive, fine-grained flows, called "trap rock," tend to make up the resistant ridges, such as the Minong Ridge. Coarse-grained volcanic rocks, which solidified more slowly, are easily distinguished by their *ophitic texture;* that is, pyroxene crystals surround and enclose laths of plagioclase, giving the rock a mottled appearance. Weathered outcrops often have a knobby surface. The flows with *porphyritic texture* have well defined plagioclase crystals (slow cooling) scattered throughout a fine-grained groundmass (more rapid cooling).

The native copper in the Isle Royale volcanic sequences may have ascended with the basaltic lavas or via channelways between flows. The three classes of deposits in which the native copper was found correspond to the categories that the miners used: 1) copper in a matrix of conglomerate interbedded with lava flows, or what miners called "stamp;" 2) amygdules of copper filling vesicles, the "barrelwork;" 3) and "moss," the copper found in veins, fissure fillings, and fractures.

Michigan's state gem, a mottled-green, semiprecious stone called *chlorastrolite* (meaning "green star") is found in amygdules in amygdaloidal basalt on Isle Royale. This small, unusual gemstone, which fades in sunlight, is seldom found outside the Lake Superior region. Chlorastrolite crystals can be seen in bedrock at Scoville Point, Rock Harbor, and Todd Harbor, but may not be collected within the park.

Similar in composition to chlorastrolite is prehnite, another secondary mineral that occurs in crosscutting veins as well as amygdule fillings. Some of the small pink pebbles on the beaches are prehnite amygdules that have weathered out of lava flows.

Pink Lake Superior *agates*, a banded variety of chalcedony (quartz), formed in amygdules, most notably in the Amygdaloid Island flow. Agates also formed in cavities in volcanic ash (tuff) that occurs on top of the Greenstone flow. Some agates weathered out of the hardened flows and were redeposited with sediments interbedded between flows. Since the agates are resistant, many of them went through a second weathering cycle and now show up on Isle Royale beaches or in glacial till.

Sedimentary units in the volcanic sequences are mainly conglomerates and sandstones that appear to be fluvially deposited. Pebbles in the conglomerates consist of volcanic rock, mostly mafic but some felsic, reflecting the composition of the source rock. Accessible outcrops are in the Chippewa Harbor area, on the south side of Siskewit Lake, and on the north side of Conglomerate Bay. Old mine dumps, such as those at Island Mine, also have accumulations of broken sedimentary rock.

The Copper Harbor Conglomerate. On the south side of Isle Royale National Park, the Portage Lake Volcanics dip conformably and gently beneath the coarse sandstones and conglomerates of the youngest Precambrian formation, the Copper Harbor Conglomerate. About a fifth of the bedrock of the Isle Royale archipelago is made up of this sedimentary unit; the other four-fifths consists of the Portage Lake Volcanics. Exposures of the Copper Harbor Conglomerate crop out along the southwest shore of the main island, on the small islands offshore, and on the higher parts of Feldtmann Ridge and Houghton Ridge, which parallel the southwest shoreline of the main island. Surficial deposits, mainly of glacial origin, cover a good deal of the Copper Harbor Conglomerate. On the Michigan shore, the formation crops out again, lapping up on the Portage Lake Volcanics in the Keweenaw Peninsula. The beds may not be continuous beneath the lake due to erosive activity since their deposition.

Effects of Glaciation on the Preglacial Topography

The topographic "grain" of the Isle Royale archipelago reflects the structures of the bedrock. The main island and nearly all of the small islands are elongate with a northeast–southwest trend. The parallel ridges with valleys or water-filled troughs in between are the result of differential erosion on the uplifted fault-block. In general, massive basalt flows make up the resistant ridges; less resistant amygdaloidal basalts underlie the valleys, producing a "corrugated" preglacial topography that the continental ice sheets delineated even more sharply as they moved over the region (figs. 20.1, 20.2).

Fractures and minor faults in the bedrock also affected preglacial and glacial erosion. Ravines developed along fault zones that cut across the main trend of the ridges. McCargo Cove, a nearly north–south-trending inlet on the north side of the island, is an example of a linear topographic feature that was eroded from shattered rock in a fault zone that cuts across ridges.

The preglacial topography was accentuated by glacial erosion because glacial quarrying and plucking were more effective on the less resistant bedrock underlying the stream valleys. Elongate lakes, such as Angleworm Lake, Sargent Lake, and Hatchet Lake, occupy basins excavated by glacial ice that followed valley trends. Quarrying also deepened fiordlike inlets along the shore and some of the channels between the elongate islands. Some of the crosscutting ravines, especially those on the north side of the island, were gouged out by moving ice. On the ridge tops glacial abrasion smoothed and polished the resistant massive basalts and removed weathered bedrock from all but protected locations. Glacial striations are evident on many outcrops.

The last ice advance (which largely obliterated traces of previous ice sheets) moved southwestward over the eastern half of the island, following the trend of the parallel ridges and valleys. However, the ice veered more to the west over the western end of the island and flowed across the ridges. The change in direction of ice movement is probably the reason for the fact that bedrock on the island's eastern half is more severely eroded, while a number of depositional (constructional) glacial features were left on the western half of the island. Glacial deposits are lacking on the eastern part of Isle Royale except in the bottoms of valleys or small depressions. Glacial erratics are scattered and few in number. However, on the west end of the island, till mantles much of the bedrock, and erratics are abundant.

Constructional features, such as drumlins and ice-margin deposits, are characteristic of continental glaciation. A *drumlin* is an elongate hill or ridge made up of compact glacial till that was carried in the basal part of the glacier and plastered onto the ground surface. The Isle Royale drumlins, on the western end of the main island, are of the "crag-and-tail" variety. The *stoss* end (facing the direction from which the ice advanced) is a steep, ice-smoothed knob of rock (the crag) that obstructed the flow of ice. The *lee* end (the tail), consisting of till, is streamlined and tapering. The alignment of the drumlins shows that on this end of the island the ice moved from east to west.

Morainal ridges (ice-margin deposits) are of irregular sizes and shapes and consist of coarse sand, gravel, and glacial debris. They are strung out across the island, west of Lake Desor, in a more or less northerly direction. They mark the stillstands of an ice front that paused in its retreat.

Abandoned shorelines and old beaches can be discerned in glacial drift on the western and southwestern parts of the main island. Such features record former lake levels when the water was higher (or the land was depressed) as the last ice sheet retreated. On the east end of Isle Royale, where glacial drift is lacking, former shorelines are marked by wave-cut benches in bedrock, and by stacks, arches, and cliffs. Monument Rock (about two miles west of Blake Point) is a stack carved by wave erosion during the existence of Lake Minong, when the water level was about 80 feet higher than it is now. A wave-cut arch on Amygdaloid Island, now high and dry above the shore, was eroded by the waves of Lake Nipissing, a postglacial lake younger than Minong.

Geologic History

1. Middle Keweenawan eruptions during late Precambrian time.

Radiometric and paleomagnetic data collected from the Isle Royale lava flows places them with the Portage Lake Volcanics in the Middle Keweenawan rocks of Proterozoic Y time. All the Middle Keweenawan rocks have *normal polarity;* that is, when the lavas solidified, the earth's magnetic field was the same as it is today. Older lavas (Lower Keweenawan) that crop out in the Lake Superior region (and which may be buried beneath the Isle Royale flows) have *reverse polarity,* as do some younger intrusive rocks (Upper Keweenawan) outside park boundaries. During times of reverse polarity, the direction of the earth's magnetic field was *opposite* to what it is now. (Natural remanent magnetism in rocks and global polarity are explained in chapter 27.) The changes in polarity recorded in the rocks of the region are another indication of the length of time during which the rift in the craton was releasing tremendous volumes of largely mafic lava.

Twelve of the Portage Lake lava flows exposed on Isle Royale have been named and designated as *marker flows* (table 20.1), which means that they have distinctive characteristics and can be traced from place to place. Between the marker flows are unnamed lavas and sedimentary units that are less readily identifiable. Six of the marker flows,

from oldest to youngest, are described briefly below because of their geologic significance and because outcrops are accessible from trails or along the shoreline (Huber 1973a).

Amygdaloid Island Flow. This trap rock, which is more felsic in composition than the other volcanic rocks of the group, forms the resistant ridge that runs the length of Amygdaloid Island.

Hill Point Flow. The thickness and coarse grain of this ophitic rock make it a useful marker. Between Huginnin Cove (west end of the island) and McCargo Cove along the north shore are imposing cliffs eroded from the flow. The type locality outcrop is at Hill Point near the eastern end of Isle Royale. (A *type locality* is a place where a stratigraphic unit is situated and from which it takes its name.)

Huginnin Flow. Porphyritic texture with large, tabular, plagioclase crystals scattered throughout a fine-grained groundmass makes this flow distinctive. The rock is easily recognized in beach pebbles and cobbles as well as in bedrock exposures, such as those near Huginnin Cove.

Minong Flow. This flow, consisting of dark, fine-grained trap rock, is massive, thick, and resistant. It runs the entire length of the main island, forming the Minong Ridge.

Greenstone Flow. Referred to as the "backbone" of Isle Royale, the Greenstone Flow forms the high resistant ridge that runs from east to west through the interior of Isle Royale. Mount Desor, the highest point (1394 feet) on the island, is on Greenstone Ridge, as are Sugar Mountain (1362 feet), Ishpeming Point (1377 feet), and Mount Siskiwit (1205 feet). The thickest part of the Greenstone Flow, about 800 feet, is in the middle of the island. The rock is mainly ophitic in texture.

Scoville Point Flow. Scoville Point, where the flow is about 100 feet thick, is the type locality of this porphyritic rock. Fine plagioclase crystals uniformly distributed in a fine-grained matrix give the rock its characteristic appearance.

2. Mineralization in the Portage Lake Volcanics.

Mineral-bearing solutions may have risen concurrently with the lavas and saturated the rocks of the volcanic sequence during the long cooling intervals. Some metamorphism occurred but not to any great extent.

3. Deposition of the Copper Harbor Conglomerate, completing the Middle Keweenawan sequence.

As volcanism in the Lake Superior region diminished, streams stripped off slopes and deposited sediments over the subsiding basin. The direction of sediment transport is assumed to have been generally eastward because of indications in the sedimentary structures, such as (1) the orientation of current directions in ripple marks and deflections; (2) thickening of the formation and decreasing

Table 20.1 Geologic Column, Isle Royale National Park

Time Units			Rock Units	Geologic Events
Era	Period	Epoch		
Cenozoic	Quaternary	Holocene	Beach deposits Alluvium	Development of Lake Superior Glacial rebound (Lake Nipissing shore features) Stream erosion; wave action
		Pleistocene	Beach deposits Glacial outwash and debris Till	Shorelines formed at various lake levels by wave action (Glacial Lake Minong shore features) Drumlins, ice-margin deposits left by Valdes Stade (last Wisconsinan advance) Severe erosion by repeated glaciations
Mesozoic Paleozoic			//////////	Prolonged erosional interval
Proterozoic Y (Late Precambrian)			Copper Harbor Conglomerate Portage Lake Volcanics *Marker flows:* Scoville Point (porphyrite) Edwards Is. (trap) Middle Point (porphyrite) Long Is. (trap) Tobin Harbor (porphyrite) Washington Is. (ophite) Greenstone (ophite) Grace Is. (porphyrite) Minong (trap) Huginnin (porphyrite) Hill Point (ophite) Amygdaloid Is. (trap)	Isle Royale fault; uplift Deposition of volcanic and clastic sediments Emplacement of native copper and other minerals; mild metamorphism Continuing subsidence of Lake Superior syncline Extrusion of mainly basaltic lavas interbedded with sedimentary deposits

(In the Rock Units column, the Proterozoic Y unit is bracketed and labeled "Middle Keweenawan.")

Source: modified after Huber 1973, 1973a; Wolff and Huber 1973

coarseness eastward; and (3) the composition of boulders and cobbles that appear to have come from older Keweenawan volcanics on the Minnesota shore. Nearly all of the coarser sediments in the Copper Harbor Conglomerate are of volcanic origin.

4. Faulting and uplift, probably in early Paleozoic time.

During the long subsidence of the Lake Superior synclinal trough, more than 40,000 feet of lavas, sandstones, and conglomerates accumulated in its center (fig. 20.4). Then a series of fault movements, possibly beginning before the end of Precambrian time, raised the Keweenawan rocks. The Isle Royale and Keweenaw faults, which are associated with the escarpment around Lake Superior, are two of the four major faults that developed along the north and south edges of the syncline during this time.

5. The forming of the lake basin.

Tectonic activity ceased, followed by many millions of years of erosion. The more resistant volcanic rocks tended to form escarpments, while softer sedimentary rocks were removed from the center of the syncline by streams that probably flowed across the area in a northeasterly direction. Pleistocene glaciers scoured and enlarged the preglacial drainageways, creating the basin now occupied by Lake Superior, the deepest of the Great Lakes. From Isle Royale, on the northwest rim of the lake, the bottom slopes off to a depth of more than 1000 feet.

6. Pleistocene glaciation.

Although Isle Royale was glaciated repeatedly and severely during Pleistocene time, most evidence of earlier ice sheets was destroyed by a late phase of Wisconsinan glaciation about 11,000 years ago. This last advance

Figure 20.6   Pocket beach on the south side of Isle Royale. Storm waves have piled up reworked glacial debris and driftwood. National Park Service photo by R. Frear.

brought ice back over the Lake Superior basin and re-moved surficial material that had previously been deposited on Isle Royale. As the ice began to retreat, the west end of the island was uncovered first. A pause in the retreat allowed accumulation of ice-margin deposits around tongues of stagnant ice, forming irregular morainal ridges. Shoreline features associated with a postglacial lake that existed at that time developed on the western end of the main island. Water level of the lake was about 200 feet higher than the present level of Lake Superior. During the long deglaciation process, a series of lakes occupied the Lake Superior basin. They had different outlets and different elevations. When some outlets were blocked by glacial ice, other outlets opened.

The final retreat of the ice margin was rapid and left only a small amount of glacial debris on the central and eastern parts of Isle Royale. Glacial Lake Minong formed about 10,500 years ago, when the ice margin had melted back to what is now the northern shore of Lake Superior.

Lake Minong, which filled the whole basin, existed long enough for well developed beaches to form on the island, especially in the glacial debris on the southwestern shore.

7. Glacial rebound and the development of the present lake.

Water levels continued to lower for a time but then began to rise again as glacial rebound elevated the outlets so that lake water backed up. Because of the tilting that accompanied rebound, abandoned shorelines are difficult to correlate, especially in the Isle Royale area and the northern shore of the basin where upwarping was irregular and uneven. However, during Lake Nipissing's existence, about 4500 to 3500 years ago, strong wave action produced a number of well defined shoreline features, among them the wave-cut arch that stands in the middle of Amygdaloid Island. Nipissing was the largest of the postglacial lakes.

Continued uplift of the land to the north and the downcutting of the Port Huron outlet to the south brought the Nipissing stage to a close. Then the uplift of Lake Superior's outlet separated that basin from Lake Michigan and Lake Huron, forming the Lake Superior of today. Because glacial rebound is still causing gradual upwarping and tilting of the lake basin slightly toward the south, the shorelines in the vicinity of Isle Royale are rising, while those on the Keweenaw Peninsula on the opposite shore are being very slowly submerged.

Geologic Map and Cross Sections

Huber, N. K. 1973. Geologic map of Isle Royale National Park, Keweenaw County, Michigan. U.S. Geological Survey, Miscellaneous Geologic Investigations, Map I-796.

Bibliography

Books, K. L. 1972. Paleomagnetism of some Lake Superior Keweenawan rocks. U.S. Geological Survey Professional Paper 760.

Dorr, J. A., Jr.; and Eschman, D. F. 1970. *Geology of Michigan.* Ann Arbor, Michigan: University of Michigan Press.

Farrand, W. R. 1969. The Quaternary history of Lake Superior. Proceedings, International Association for Great Lakes Research, 12th conference, 1969. p. 181–197.

Holte, I. 1984. *Ingeborg's Isle Royale.* Grand Marais, Minnesota: Women's Times Publishing. 97 p. (with map)

Huber, N. K. 1973. Glacial and postglacial geologic history of Isle Royale National Park, Michigan. U.S. Geological Survey Professional Paper 754A.

————. 1973a. The Portage Lake Volcanics (Middle Keweenawan) on Isle Royale, Michigan. U.S. Geological Survey Professional Paper 754C.

————. 1983. *The geologic story of Isle Royale National Park* (formerly U.S. Geological Survey Bulletin 1309). Isle Royale Natural History Association. 66 p.

Leskinen, C.; and Leskinen L. 1980. *Copper country history.* Park Falls, Wisconsin: F. A. Weber & Sons, Inc.

Paull, R. K.; and Paull, R. A. 1977. *Geology of Wisconsin and Upper Michigan,* including parts of adjacent states. Dubuque, Iowa: Kendall/Hunt Publishing Company.

————. 1980. *Wisconsin and Upper Michigan,* K/H Geology Field Guide Series. Dubuque, Iowa: Kendall/Hunt Publishing Company.

Rakestraw, L. 1965. *Historic mining on Isle Royale.* Isle Royale Natural History Association.

Wolff, R. G., and Huber, N. K. 1973. The Copper Harbor Conglomerate (Middle Keweenawan) on Isle Royale, Michigan. U.S. Geological Survey Professional Paper 754B.

Isle Royale National Park

Area: 571,790 acres; 893 square miles

Authorized: March 3, 1931

Established: April 3, 1940

Designated a Biosphere Reserve: 1980

Address: 87 North Ripley Street, Houghton, Michigan 49931

21

A mountainous, rockbound coast, scoured by continental glaciers and submerged by an encroaching, postglacial ocean.
Waves beat against an irregular shoreline of cliffs, boulder beaches, tidal pools, bays, a fiord, many islands, and headlands.
Bare, rounded, granite summits, above the forests, overlook the sea.
Paleozoic plutonic rocks emplaced in older metamorphic rocks of suspect terranes during the Acadian orogeny.

Acadia National Park

Eastern Maine Coastal Area

Figure 21.1   The glacially scoured summit of Cadillac Mountain in Acadia National Park is the highest point (1530 feet) on the North American Atlantic coastline. Pitting of the coarse-grained granite is due to weathering. National Park Service photo by R. Frear.

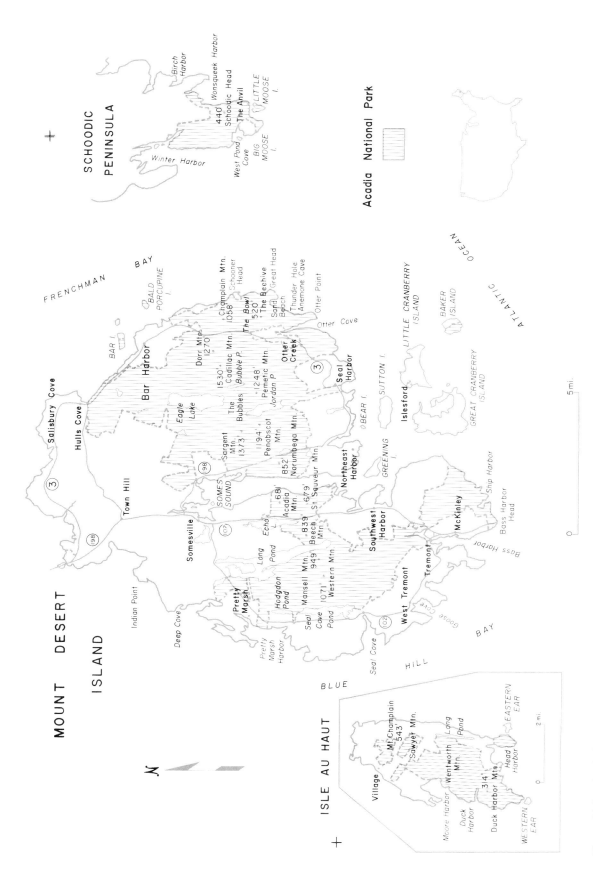

Figure 21.2 Acadia National Park, Maine.

Geographic Setting

The Maine coast extends from New Hampshire on the southwest to New Brunswick in Canada. Ledges, forests, beautiful harbors, and picturesque villages give the Maine coast great beauty and charm. Acadia National Park, located on this coast, is near the international boundary and about 200 miles "down east" from the port of Boston. Of Maine's 2500 miles of coastline, 41 miles are preserved within Acadia's boundaries. Over 4,000,000 people a year visit the park.

Acadia National Park comprises most of Mount Desert Island (highest elevation on the eastern seaboard), part of the Schoodic Peninsula on the mainland, Isle au Haut offshore, and six other smaller islands in the Gulf of Maine Archipelago. This is a deeply indented, rocky coast, with estuaries and scattered islands. The water temperature is cold, and summers are cool, the average air temperature for July and August being 69° F. Precipitation averages 49 inches annually, and rain and fog may be common in the summer months. A tidal range of around 10 feet (between high and low tide) unveils four different life zones, each populated by specifically adapted marine plants and animals, such as starfish, minnows, periwinkles, anemones, and crabs.

Local History

The Red Paint People, so-named because they decorated their faces and bodies with red ocher, were one of the prehistoric groups who inhabited these shores several thousand years before Europeans arrived. The relics of their culture—tools of slate, pottery, and red ocher burials—are similar to those of the Mound Builders of the Middle West.

Explorers and settlers encountered a later group of Indians, the Abnaki, who had developed a birchbark culture. The Abnaki camped in the Somes Sound area each summer to fish, dig for clams, and trap mink until the early years of the nineteenth century. Deep shellheaps indicate that they camped here for hundreds of years. They spent the winters in permanent villages near the headwaters of the Penobscot and other rivers and traveled back and forth to the coast in their birchbark canoes.

Perhaps the Norsemen and certainly the Portuguese explored this coast. A 1529 Portuguese map shows Mount Desert Island and the main estuaries of the region. Since Cadillac Mountain (1530 feet) is the highest point on the Atlantic coast, it was a landmark for sailing ships for the next several centuries.

Samuel de Champlain, the father of New France, was the official discoverer of Mount Desert Island in 1604. Sieur de Monts, Champlain's patron, held a grant from the French king that comprised the entire French claim to North America (all the land between latitudes 40° N and 60° N). The territory was called *La Cadie* (later anglicized to "Acadia"). After establishing a trading post, de Monts sent Champlain to explore and chart the coast and encourage the local Indians to do business with the French. In his journal of the voyage, Champlain described what is now Acadia National Park as: "an island, very high, and cleft into seven or eight mountains, all in a line. The summits of most of them are bare of trees, nothing but rock. I named it *l'Isle des Monts-deserts* ("the Isle of Bare Mountains"). Attracted by the smoke from an Indian camp, Champlain put in to shore near Otter Cliffs and, guided by friendly Indians, made a reconnaissance of the locality. His chart of the coastline shows fairly accurately the islands and estuaries.

Several years later the Marquise de Guercheville acquired the de Monts grant, which had been enlarged by Louis XIII to include the whole of North America to the Gulf of Mexico. The marquise sent a group of Jesuit missionaries to convert the Abnaki to Christianity. The Jesuits set up a mission near the entrance to the Somes Sound. A few weeks later the settlement was spotted by Samuel Argall of the English Virginia colony. He had been ordered by King James I, who disputed the French claim, to clear out any French colonies along the coast. In a surprise raid, Argall wiped out the settlement, sent the French sailors out to sea in a longboat, and carried the priests back to Virginia. Luckily the sailors met a French ship off the Bay of Fundy and got home safely. The Jesuits were taken to England, from whence they made their way back to France. This was the beginning of more than a century of raids, skirmishes, and intermittent warfare between the French and the English over possession of North America.

Even after the Peace of Utrecht in 1713 awarded Maine to the English, the coast was not safe for permanent settlement. Both English and French warships used the safe harbors along the Maine coast for staging and refitting. Finally, when the French and Indian wars ended in 1760, the New England settlements were safe from the danger of raids.

Governor Francis Bernard of the Massachusetts Bay Colony was awarded the grant for the Island of Mount Desert by King George III. In 1761, Bernard helped the Somes and Richardson families to settle on the island. This was the beginning of Somesville, the first permanent settlement, but the island remained an area of remote fishing hamlets for many years.

In the 1840s the beauty of the island scenery was "discovered" by Thomas Cole, Frederick Church, and other American artists. College professors, wealthy summer visitors, and cruising yachtsmen began coming to the island. They were called "rusticators," and some built mansions, which were known as "cottages." Coastal steamers served the area from the 1850s until the advent of the automobile. Although the island was mainly a resort

Box 21.1
More about Metamorphic Rocks

The metamorphic rocks in Acadia National Park have "character." If you observe their distinctive appearance in outcrops along the shore, it's not hard to believe that long ago they were subjected to great heat and pressure far below the earth's surface. Contorted beds of different colors seem to coil and swirl as waves wash over the rock. Dark, banded layers, some thick, some thin, are cut by white quartz veins. Even the more massive rocks that stand in cliffs have subtle gradations in color, inclusions that look like patch pockets, and some have a sparkly texture.

Most of the Acadian metamorphic rocks were originally sedimentary, although a few were first formed as igneous rocks. As you learned earlier, all metamorphic rocks are formed from preexisting rocks, or "parent rocks." The composition of a parent rock has a great deal to do with its response to various metamorphic processes. The other important factors are the degrees of heat and pressure and the amount of water that is present. During metamorphism, parent rocks are transformed in a solid state—without being melted—into new rocks that are texturally and mineralogically distinct from the preexisting rocks.

Nearly all metamorphic rocks form deep within the earth's crust, usually as a result of *regional metamorphism,* which involves a general increase of temperature and pressure over a large area. Tectonic activities, such as subduction and orogeny are associated with regional metamorphism. The most intensely deformed portions of mountain ranges are where metamorphic rocks are found—after the buried mountain cores have been exposed by erosion.

Geologists study metamorphic rocks in order to better understand conditions at depth and what happens during subduction.

Regional metamorphism is usually *progressive;* that is, different conditions of temperature (ranging between 400° C and 800° C) and pressure (encountered at depths from 10 to 35 km) alter rocks in a sequence that may vary from lightly metamorphosed to intensely metamorphosed. Shale, for example, becomes slate when temperature-pressure conditions are low. With an increase in temperature and pressure, implying deeper burial, a phyllite may be formed from the parent shale.

Some rocks, especially when intense, directed pressure is a factor, develop *foliated texture* because the pressure forces the constituents in a rock to become parallel to each other. *Foliation* refers to the parallel alignment of a rock's textural and structural features. Slate, a rock with foliated texture, splits easily along foliation planes because its preexisting, microscopic, platy minerals were pushed into alignment during metamorphism.

The presence of water (from seawater and ground water trapped in crustal rocks) is important in metamorphic processes because it aids chemical reactions that produce new rocks. Water under high pressure forces its way between mineral grains and dissolves and transports ions between minerals so that recrystallization can occur.

In *contact metamorphism,* high temperature is the dominant factor, with less pressure being involved. Magma intruding country rock is usually the source of the intense heat that

for the very rich, an increasing number of tourists who appreciated the uniqueness of the region came to visit. By 1901, the wealthy people who owned most of the land began to fear that lumbering and commercial development would ruin the island's natural beauty. They formed a corporation to protect the area; and by 1913, they had acquired approximately 6000 acres of land on Mount Desert Island, the Schoodic peninsula (across the bay from Bar Harbor), and Isle au Haut. President Charles Eliot of Harvard, John D. Rockefeller, and George Bucknam were among the leaders in the effort to transfer the land to the federal government for a national park. Proclaimed as Sieur de Monts National Monument in 1916, the area was established as Lafayette National Park in February of 1919. Ten years later the name was changed to Acadia National Park.

metamorphoses, or "cooks," country rock. The zone of contact metamorphism (called an *aureole*) is ordinarily narrow. Rocks produced by contact metamorphism tend to be fine-grained but do not have foliated texture because directed pressure is not involved.

Table 21.1 shows how common metamorphic rocks are classified. Note that nonfoliated rocks are classified on the basis of their composition whereas foliated rocks are recognized by the type or degree of foliation.

Table 21.1 Classification of Common Metamorphic Rocks

Nonfoliated Texture (Rock names based on mineral content)

Usual parent rock	Rock name	Predominant minerals
limestone dolomite	MARBLE	calcite dolomite
shale basalt	HORNFELS	plagioclase micas ferromagnesian clays
quartzose sandstone	QUARTZITE	quartz

Foliated Texture (Rock names based on nature of foliation)

Closely foliated	SLATE Splits into thin flat sheets, foliation finer than phyllites or schists
Schistose (platy)	PHYLLITE Crinkly, lustrous surfaces SCHIST Well developed, parallel orientation of minerals
Coarsely banded or layered	GNEISS Banded or layered foliation coarser than phyllites and schists

Geologic Features

Far-Traveled Bedrock: What are Suspect Terranes?

The role of water and ice in the shaping of the dramatic Acadian scenery is significant, but the origin of the rocks that have resisted these forces is equally dramatic. Coarse-grained and medium-grained granites form the moun-tainous core and glacially smoothed summits of Mount Desert Island. Older metamorphic rocks surround the plutons in a roughly circular pattern. The older rocks, originally sedimentary, were long ago deformed and metamorphosed. Later, as part of a minicontinent, they "drifted" northwestward for hundreds of miles, eventually becoming attached to the North American plate within a suspect terrane.

A *suspect terrane* in an orogenic belt can be described as an "internally homogeneous geologic province with features that contrast sharply with those of nearby provinces. They are recognized by contrasts in any or all of the following: stratigraphy, geologic structure, metamorphic and plutonic histories, fossil faunas, mineral deposits, and paleomagnetic signatures. Their boundaries are sharp structural junctions marking discontinuities that cannot be explained by normal gradations . . . conventional facies changes, or standard unconformities." (Williams and Hatcher 1983, p. 34) The same definition can be applied to the term *exotic terrane*, a designation that implies a little more certainty. A more formal term, used in geological literature, is *tectonostratigraphic terrane*.

The Metamorphic Rocks of Acadia

The Ellsworth Schist. This formation is the oldest of the three distinct units that make up the older rocks in Acadia National Park. Several hundred to several thousand feet thick, the Ellsworth Schist consists of mica schist and dense gneiss. Originally the rock accumulated in thinly laminated beds of fine clastic sediments, intermixed with fluvially worked volcanic ash and tuff. The mica-rich schists in the Ellsworth are closely foliated and split readily along surfaces of foliation, but the gneisses, which resist splitting, have poorly developed foliation. Recrystallization by repeated metamorphism has increased the strength and density of the rock. Prominent exposures of the Ellsworth Schist are along the shore at the northern end of Mount Desert Island.

Cranberry Island Series. Many times thicker than the Ellsworth Schist, the Cranberry Island Series was originally composed of interbedded volcanic and sedimentary layers, including pyroclastics and brecciated fragments. Because many of the beds are uniform and well stratified, like the Ellsworth, the assumption is made that the volcanic materials were transported by streams and reworked by waves.

The Cranberry Island units were sharply folded and squeezed not long after deposition. They are strongly unconformable with the Ellsworth Schist, which was crumpled, metamorphosed, and eroded before the Cranberry Island rocks were deposited. Great Cranberry and Little Cranberry Islands, as one might expect, are made up of their "namesake" formation. The rocks also crop out on the southern part of Mount Desert Island in the vicinity of Bass Harbor.

The Bar Harbor Series, which unconformably overlies the Cranberry Island Series, accumulated as beds of gravel, sand, and silt on the floor of a shallow sea. The siltstone, which splits along bedding planes, is the source of much of the shingle on the beaches, especially around Bar Harbor. (*Shingle* is composed of smoothed, spheroidal, or flattened waterworn pebbles and cobbles.) Thin,

white layers in the formation consist of volcanic ash that fell into the ancient sea and was washed by waves and currents.

The Mount Desert Intrusives

The light-colored, barren summits that so impressed Champlain when he sighted them from his ship are made up of massive, coarse-grained granite that is highly resistant to erosion. This is the granite, pinkish in color, that makes up the Mount Desert range and all the high points of the island. Wedged into the northwest section of the pluton is a somewhat younger body of gray, medium-grained granite that is speckled with biotite (black mica). From quarries in this granite (such as those between Somes Sound and Echo Lake), building stone of fine quality was shipped to cities along the eastern seaboard. (The Philadelphia Mint is one of the more famous buildings constructed of this stone).

Along the northwest side of the island, next to the granites, is an *intrusive sheet* of diorite, the third most abundant rock type of the park. Long before the emplacement of the plutons, dioritic magma of intermediate composition came up very hot from great depths and worked its way through layers of crustal rock. Then the magma spread out horizontally between sedimentary bedding planes, forming sills in some places. The intrusive sheet, a large, somewhat irregular body, emanated from an extensive fracture and is sandwiched between the Ellsworth Schist and the Bar Harbor Series. The presence of the metamorphic rock hornfels in the contact zones indicates that the magma was extremely hot when it injected the stratified rocks (fig. 21.3).

A fine-grained granite that crops out between Southwest Harbor and Goose Cove (southwest part of the island) was emplaced after the diorite but before the coarse- and medium-grained granites. This fine-grained granite may be part of a large body of similar rock of the same age that makes up nearly all of the Schoodic Peninsula and also crops out on Isle au Haut.

The Forming of the Ring Dikes

By the time the magma that was the source of the coarse- and medium-grained granites began to rise through the crust, the roof rocks of the Mount Desert area had become seriously weakened, probably due to the length and severity of previous orogenic and intrusive activity. With renewed pressure and heat from rising magma, the roof rocks began to fracture and subside. Eventually a large, plug-like mass of roof rock was incorporated into the melt, with a few remnants being preserved as inclusions in the slowly cooling magma. This magma became the coarse-grained granite of the Mount Desert range. Into circular fractures around the collapsed area, more magma welled up, forming *ring dikes* of medium-grained granite. Curved fractures also developed within the pluton as the coarse-grained granite cooled and subsided. Intensely hot, magmatic liquids under high pressure permeated the curved fractures and recrystallized the rock. These patches of recrystallized granite have a fine-grained matrix with scattered large crystals of quartz and feldspar.

Following further subsidence and fracturing, a second pluglike mass of roof rock collapsed and sank on the western side of the pluton without becoming completely detached. Magma flowed up over this block, filled the curved fissure, formed a large, bulging ring dike in the northwest section of the pluton, and gradually solidified as medium-grained granite.

The Shatter Zone, Pegmatites, Veins, and Dikes

A zone of severely shattered, fractured, and brecciated rock, up to a mile or more in width, lies between the coarse-grained granite and the metamorphosed country rock. Called the *shatter zone,* this band of rock that partially encircles the island (along the eastern shore and across the southern part) marks the edge of the old collapsed area. Shatter zone rock, which is now tough and coherent, is easy to spot because of its brecciated appearance. Angular blocks of diorite, chunks of old sedimentary rock, coarse fragments of various colors, veins, small dikes, pegmatites, and partially blended rubble are all congealed in a matrix of granite.

The pegmatites, quartz veins, and light-colored dikes that fanned out from the granitic magma are most numerous in the shatter zone, but can also be found where cracks, fractures, or joints in the rock gave access to hot hydrothermal solutions and fingers of molten rock forcing their way upward. Small amounts of minerals (epidote, hematite, etc.) are present in the quartz veins and pegmatites, but none are of economic value.

More numerous and more widely distributed than the granitic dikes are the basaltic (diabase) dikes that range in width from less than an inch to more than 50 feet. Some dike outcrops extend straight over mountaintops for thousands of feet. On bare ledges the dark color of the basalt shows up against the lighter granites. Contacts tend to be clear and sharp, which makes the dikes easy to follow. The Schoodic Point area (due east from Mount Desert Island, across Frenchman Bay) is the best place in the park to see basalt dikes. Here large, dark dikes cut through light, fine-grained granite. Where dikes come down to the shore, wave action has eroded straight-walled chasms. The less resistant basalt has been removed, leaving granite on either side (fig. 21.4). The basaltic intrusions apparently came from a very deep source. Radiometric dating techniques have yielded several sets of distinctly different ages for the dikes, indicating that from time to time magma was released and moved up toward the surface.

Figure 21.3   Cliff of massive diorite at Great Head, southeast shore of Mount Desert Island. Glacial plucking by a mile-high ice sheet from the northwest produced the cliffs. The bedrock is part of the shatter zone encircling the granite pluton exposed in Mount Desert. Hylander photo.

The Acadian Orogeny

These "native" and "foreign" rocks, so well exposed in Acadia National Park, became intermingled during a long, intensive series of tectonic episodes associated with the mid-Paleozoic Acadian orogeny. The ancient metamorphic rocks were part of continental crustal material carried by drifting tectonic plates from an area on the globe far to the east and south. When the drifting terranes reached the North American plate, a continent-continent collision ensued, squeezing and crumpling crustal rocks. Eventually, sizable sections of rock were subducted and became molten due to heat and pressure at depth. Accumulated magmas of varying composition began to work their way toward the surface. Some formed plutons and

Figure 21.4   Eroded basalt dike at Schoodic Point. National Park Service photo.

intrusive sheets; some erupted as volcanoes. Again the ancient rocks, subjected to folding and faulting, underwent metamorphism. Some of the intrusives were also altered by the severity of the tectonic events. The scenery that we enjoy today exposes the roots of what must have been grand and imposing Acadian mountains in middle and late Paleozoic time.

Glacial and Wave-Action Features

Before the arrival of Pleistocene glaciers, the coarse-grained granite of Mount Desert Island formed a highly resistant rock ridge, trending east and west and only slightly dissected by the processes of weathering and stream erosion. Possibly the summits of the present range

could be distinguished, although reconstructing the pre-glacial drainage patterns is speculative because most of the evidence has been destroyed. What we see today are six huge troughs, going north and south, cut through this very tough, resistant ridge. The amount of rock that was excavated and the force required to gouge it out and transport it are awesome (figs. 21.5, 21.6). Not much of the rock debris is left on the island. Most of it was pushed or dragged out over what is now ocean bottom.

The Mount Desert ridge must have been a temporary barrier to the great continental glacier moving down from the north, but eventually the mass of ice that was banked up against the north side of the ridge overrode the summits, scouring and rounding them off. At shallow gaps in the ridge or areas of weakness, the erosive effect of the ice was intensified and channeled as flowing ice currents probed for access to the south. As time went on and more and more ice pushed down from the north, troughs developed and became deeper and broader. The deepest trough, Somes Sound, which nearly bisects the island, is a fiord, the only true fiord on the U.S. Atlantic coast. A *fiord* is a long, deep arm of the sea, U-shaped, steep-walled, and always found on a mountainous coast that has been severely glaciated. A fiord represents the seaward end of a glacial trough, partially submerged after melting of the ice. Most of the embayments on the coast of Maine are drowned valleys, and many of them have been modified by glacial action. Somes Sound qualifies as a fiord because notches and grooves were cut into the rock walls by glacial ice, and because the deepest part of the sound is some distance back from the submerged mouth. Characteristically, a fiord becomes shallower near the mouth where glacial scouring was less severe.

Trough lakes, such as Seal Cove Pond, Long Pond, Echo Lake, Eagle Lake, and Jordan Pond, filled the low areas gouged out in the floors of the U-shaped valleys after the ice melted. In the valleys occupied by Long Pond and Jordan Pond, the water backed up to somewhat higher levels because of morainal dams at the south end.

The erosive power of the mile-thick ice sheet altered the shape of the mountains as it carved up the broad, previously unbroken, east-west ridge. Northern (up-ice) slopes became longer, smoother, and rounded; while southern (down-ice) slopes became steep and foreshortened, their surfaces rough and jagged from glacial plucking. The ridges dividing the up-ice and down-ice slopes tend to be perpendicular to the general flow of the ice mass. Two large *roches moutonnées* that resulted from this type of glacial action are The Bubbles that stand between Cadillac Mountain and Penobscot Mountain. Smaller roches moutonées with the characteristic smoothed, up-ice top and plucked, down-ice face jut out from hillsides in numerous places in the park.

Glacial grooves and striations on bedrock have a generally north–south orientation, indicating the direction of

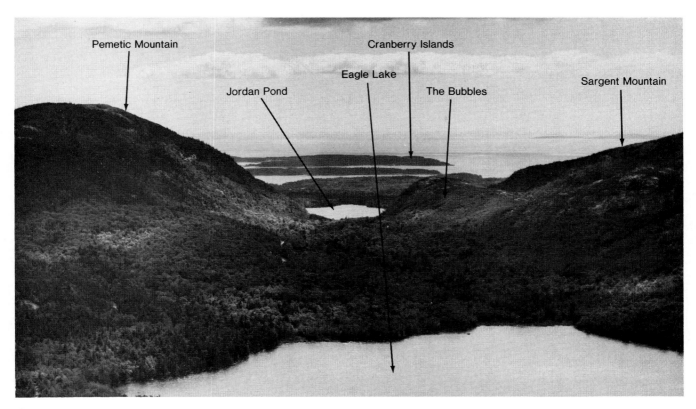

Figure 21.5   Air view looking south (toward the Atlantic) over a broad glacial trough excavated by continental glaciers. The Bubbles are large roches moutonnées. Jordan Pond and Eagle Lake occupy scoured depressions in the floor of the trough. The summits flanking the trough have been rounded and severely abraded. Hylander photo.

Figure 21.6   Looking north over the same glacially eroded trough (fig. 21.5) toward mainland hills in the far distance. Hylander photo.

ice flow. *Chattermarks* (a series of small, close, curved cracks made by vibratory chipping of ice-embedded rock fragments and boulders bearing down against a bedrock surface) show where some of the greatest pressures were exerted. Many glacial erratics, large and small, are scattered over the island.

Jointing in the bedrock influenced both the glacial and wave action features. On the summits and in roadcuts at high locations, particularly around Cadillac Mountain, sheet-jointing, or exfoliation, of the granite is noticeable. The joints that run parallel to the surface are the result of unloading when the rocks covering the pluton had been eroded, long after the granite had cooled. On some of the long-exposed surfaces, *weather pits* have developed in the coarse-textured rock because less-resistant mineral grains weather out first (fig. 21.1).

The geometric pattern of the broad, intersecting vertical joints in the granites is visible from the air and prominent on the ground wherever bedrock is bare of vegetation. On the summits, weathering and frost action have widened the joints and separated the rock into large blocks. On the south-facing slopes of the island, especially where high cliffs stand on the shoreline, the effect of the glacial plucking along joint planes is striking. Frost action and mass wasting of loosened blocks have steepened jagged slopes and in some places undercut cliffs. Cliffs exposed to wave action are being further undercut as the sea continues the attack following the glacier's retreat. On the wave-cut bench formed at the base of a cliff, at and above the water level, rock fragments are rolled around, pounded by waves, and banged up against the cliff face during storms. Intensifying this mechanical action are the salt in the water, the spray, and the mist, which increase the rate of chemical weathering.

One of the many interesting features of the shoreline is the contrast between the effects of wave action on the granites, the shatter zone, and on the older metamorphic (originally sedimentary) rocks. At Bar Harbor, on the eastern side of Mount Desert Island, the rock along the harbor shore is in the Bar Harbor Series, formed from gravel deposits, sand, silt, and some volcanic ash. Beds of purplish gray rock dip toward the water, while the risers lap up in sequence against the shore. Loose pebbles and cobbles, broken off from the beds, tend to be flattened, or shingle-shaped. Most of the angles and corners of the beds have been smoothed and rounded by abrasion. Near the high-tide line is a small *stack* (a free-standing rock column, an erosional remnant), consisting of siltstone layers. Wave erosion (since the retreat of the ice and the rise of sea level) appears to be moderately effective on these rocks.

To the south of Bar Harbor as far as Great Head, the shoreline is mostly in the shatter zone. Along this stretch of jumbled ledges and cliffs, there are few places where one could safely land a boat at either high or low tide.

Most of the rock is fairly resistant, but where weak spots exist, the waves have been able to erode. Anemone Cave is a large sea cave, facing the open ocean, that has been formed by wave action and frost action in a zone of fractured rock.

A short distance farther south, ledges of coarse-grained, resistant granite line the shore. Here the erosive action of the waves has had little effect since the Pleistocene except along joint planes at exposed locations. Thunder Hole, which emits loud booming sounds when waves rush in and out, is a chasm eroded by wave action along large joints in the coarse granite. Frost action, by loosening slabs of rock, helps to enlarge the chasm. At low tide, between waves, one can see large, egg-shaped cobbles swirling and grinding away in the water at the head of the chasm.

One of the few places in the park where post-Pleistocene wave erosion of granite is obvious is on the unprotected ledges and cliffs of the headlands at the extreme southern tip of Mount Desert Island near Bass Harbor Head Lighthouse. Here waves from the open ocean, beating against the rock with full force, have eroded a wave-cut platform of considerable size. A *wave-cut platform,* which includes the lower part of the wave-cut bench, slopes gently seaward from the base of a cliff. At low tide, a wave-cut platform is partially exposed, but is generally covered at high tide (fig. 21.7). On the northern Maine coast, the tidal range is fairly high, normally 8 to 10 feet of rise and fall.

Wave-Depositional Features

Most of the glacial drift and rock material scraped off and excavated from Mount Desert Island is strewn over the ocean bottom out in the Gulf of Maine. Not much sand has accumulated since sea level rose after Pleistocene ice sheets melted back. Wave-depositional features consisting of sand-sized material are minor except for small pocket beaches in coves and a few sandbars in protected areas, such as the one at Bar Harbor that connects the harbor shore with Bar Island at low tide. Most of the pocket beaches have more cobbles and shingles than sand. (Cobbles are generally spherical, while shingles tend to be disk-shaped.) Sand Beach (on the eastern side of the island between Great Head and Otter Cliff) was built across the head of a small bay and receives some sand from a small stream. However, Sand Beach is remarkable for the fact that the "sand" contains a high proportion of finely ground organic material (shells, bones, etc.) that has been washed up from the ocean and reworked by the waves.

Storm waves have built *sea walls* (natural beach ridges) of cobbles and shingle along the high-water line on some of the less rugged parts of the shoreline (where the preglacial slopes were gentle). The sea walls were the source of smooth, even-sized paving stones (called "popplestones" in Maine) that were shipped down to cities on the eastern seaboard for street construction in the nineteenth century.

Geologic History

1. Origin of the Ellsworth Schist, Cranberry Island Series, and Bar Harbor Series.

These metamorphosed (originally sedimentary) rock units are part of the Avalon suspect terrane, or microplate, one of nearly a dozen such blocks that broke off from a Eur-African landmass, drifted north and west, and eventually became welded to eastern North America. Although some of the Avalon rocks appear to have close affinities with certain European and African rock units, the original sites of deposition and lithification of Avalon rocks are unknown. The time during which the sediments accumulated was late Precambrian through Ordovician. The Avalon rocks were deformed, metamorphosed, and eroded before they reached North America.

2. Arrival of the Avalon suspect terrane.

In mid-Devonian time, the Avalon suspect terrane "docked" along the eastern edge of North America. Several other terranes arrived ahead of the Avalon, most of them reaching North America in middle and late Ordovician time during the Taconic orogeny. At least three more terranes followed Avalon late in Paleozoic time. Estimates of the velocities of these movements indicate that suspect terranes drift two or three times faster than the large plates, such as the North American plate, which is moving 2 to 3 centimeters westward per year. The difference in velocities enables the microplates to overtake a large plate drifting in approximately the same direction.

3. Acadian orogeny in Devonian time.

Folding, faulting, and regional metamorphism accompanied plate convergence from mid-Devonian time on through most of the period. Tectonic activity was concentrated in the northern Appalachian Highlands and the Canadian Maritime Provinces.

4. Magmatic and plutonic activity accompanying the Acadian orogeny.

A large igneous complex was emplaced in a series of pulses. The earlier magma was dioritic (intermediate in composition) and the later magmas were more felsic, or granitic. Great quantities of diorite intruded the layers of the Bar Harbor Series as sills, dikes, and an intrusive sheet. Some of the magma probably reached the surface and erupted in lava flows. Contact metamorphism, resulting from the magma's intense heat, baked the shales and the siltstones of the country rock into hornfels. Most of the diorite is fine-grained; but in the thicker intrusions, slower cooling in the center produced a coarse texture. Some of the sills in the Bar Harbor Series are 100 feet thick. The largest body, the intrusive sheet, is about 3000 feet thick.

Figure 21.7 Surf from the open Atlantic beats in on a wave-cut platform on the Schoodic Peninsula. National Park Service photo.

Granitic magma that rose after the diorite had solidified apparently hardened near the surface because the texture is fine-grained. Heat and pressure from the magma shattered the country rock. Many fragments of rock from the Bar Harbor Series and the diorite fell into the liquid magma and became inclusions or blended with the granite. Cracks and fractures in the country rock filled with magma that hardened as dikes, veins, and long stringers (fingers) within the older rock. Uplift of the land and long erosion followed.

5. Climax of Acadian orogeny; intrusions of coarse-grained and medium-grained granites.

In late Devonian time and perhaps extending into the Mississippian Period, crustal disturbances and pulses of intrusive and extrusive activity continued as a great mountain system was built parallel to the eastern coast of North America. Along the Maine coast, the ring dikes, the plutons of coarse- and medium-grained granite, and the patches of recrystallized granite are the roots of what must have been a lofty range. A long period of erosion followed while the mountains were being worn down.

Table 21.2 Geologic Column, Acadia National Park

Time Units		Rock Units	Geologic Events
Eras	Periods		
Cenozoic	Quaternary	Minor wave and stream deposits Minor glacial deposits	Minor wave erosion Crustal rebound Postglacial submergence Pleistocene glaciations: severe scouring, abrasion and excavation of glacial troughs
	Tertiary		Minor uplift, renewed erosion
Mesozoic		~~~~~~~~~~~~~~~~~~~~	Unconformity representing uplift, prolonged weathering, and erosion
Paleozoic	Mississippian ? Devonian	Miscellaneous granites Medium-grained granites Coarse-grained granites Fine-grained granites ~~~~~~~~~~~~~~~~ Diorite (intrusive sheets and sills)	Ring dikes, recrystallization Shattering, contact metamorphism Hydrothermal activity, pegmatites Uplift and erosion ——————— Docking of Avalon suspect terrane Dikes and veins throughout this and succeeding intrusions Metamorphism, volcanism Beginning of Acadian orogeny
	Silurian ? Ordovician	*(Avalon suspect terrane)* ~~~~~~~~~~~~~ Bar Harbor Series ╫╫╫╫╫╫╫╫╫ Cranberry Island Series ╫╫╫╫╫╫╫╫╫ Ellsworth Schist	Uplift and erosion ——————— Deposition of siltstones, shales, conglomerates—mainly marine Deformation, metamorphism, erosion, volcanism Ash, tuff, interbedded sedimentary rocks Deformation, uplift, metamorphism, erosion Marine sedimentation; volcanic ash

After Chapman, 1969, 1970; Bird and Dewey, 1970; Williams and Hatcher, 1983

6. Crustal extension in Mesozoic time.

The repeated plate collisions that brought about the Acadian orogeny finally ceased. (A later Alleghenian orogeny had less effect in this area.) The impinging plates relaxed and slowly began to separate. This caused the development of several Triassic fault basins, one of which lies under the Bay of Fundy (between New Brunswick and Nova Scotia), about 150 miles northeast of Mount Desert Island. By Jurassic time, the plates had separated enough to open an ocean of significant size, which became the North Atlantic. When the eastern landmass drifted away, parts of the microplates were left behind as suspect terranes joined to North America.

7. Mesozoic and Tertiary uplift and renewed erosion.

During plate collision and convergence, the piling up of continental crust, along with shallow subduction, caused the area that is now coastal Maine to became isostatically out of balance, which forced the lighter, less dense crustal material to rise. As uplift continued, erosion removed a mile or more of solid roof rocks that overlay the granite plutons.

In Tertiary time the Mount Desert range was a continuous ridge, trending east and west. The park region, including the offshore islands, was part of the mainland. A minor uplift that tilted the region slightly seaward was due to continuing isostatic adjustment. Renewed erosion, followed by minor subsidence with encroachment of the sea, produced an irregular shoreline, perhaps similar in outline to what we see today. Shoreline fluctuations since that time are associated with glacial episodes that began in the Pleistocene Epoch.

8. Pleistocene glaciation.

The continental glaciers that covered New England and the Maritime Provinces were more than a mile thick. Ice extended southeastward over the Gulf of Maine, as far as the terminal moraine on Georges Banks. Sea level was then much lower because water had been drawn from the oceans to build the enormous ice sheets, and their great weight was depressing the land. When the ice melted, sea level rose rapidly. The ocean came in over the land, drowning the shoreline and filling the estuaries. Georges Banks, which had been a group of islands, was submerged and is now a famous offshore fishing ground. On Mount Desert Island, sea level was probably 300 feet higher than it is today. Old beaches and deltas on the island mark the locations of earlier sea levels. Not long after the unloading of ice, however, the coast of Maine began to rise slowly, a process called *glacial* or *crustal rebound*. Sea level continues to fluctuate slightly and has risen a few feet in the last several thousand years. Recent data also suggest slight tectonic subsidence along the east coast.

Geologic Maps

Osberg, P. H., Hussey, A. M. II, and Boone, G. M. 1985. Bedrock geologic map of Maine. Maine Geological Survey, Department of Conservation, Augusta.

Thompson, W. B. and Borns, H. W. Jr. 1985. Surficial geologic map of Maine. Maine Geological Survey, Department of Conservation, Augusta.

Bibliography

Bird, J. M. and Dewey, J. F. 1970. Lithosphere plate-continental margin tectonics and the evolution of the Appalachian Orogen. *Geological Society of American Bulletin* 81:1031–1060.

Butcher, R. D. 1977. *Field guide to Acadia National Park, Maine.* New York: Reader's Digest Press.

Chapman, C. A. 1968. A comparison of the Maine coastal plutons and the magmatic central complexes of New Hampshire. In *Studies of Appalachian geology, northern and maritime.* Editors: E. Zen, W. S. White, J. B. Hadley, and J. B. Thompson, p. 385–396. New York: Interscience Publishers.

———. 1970. *The geology of Mount Desert Island, Maine.* Chatham Press.

Johnson, Douglas. 1925. *The New England-Acadian shoreline.* New York: John Wiley & Sons, Inc.

Morison, S. E. 1960. *The story of Mount Desert Island.* Boston: Little, Brown and Company.

Shaler, N. S. 1889. The geology of the island of Mount Desert, Maine. U.S. Geological Survey Annual Report 8:987–1061.

Williams, H. and Hatcher, R. D. 1983. Appalachian suspect terranes, In R. D. Hatcher, H. Williams, and I. Zeitz, editors. *Contributions to tectonics and geophysics of mountain chains.* Geological Society of America Memoir 158, p. 33–53.

Acadia National Park

Area: 39,707 acres; 62 square miles
Proclaimed Sieur de Monts National Monument: July 8, 1916
Established as Lafayette National Park: February 26, 1919
Changed to Acadia National Park: January 19, 1929
Address: Route 1, Box 1, Bar Harbor, Maine 04069

22

"Roof of the Rockies," a broad, relatively flat upland, over 11,000 feet high, is crossed by the Continental Divide.
The park encompasses the highest peaks of the Front Range, several more than 13,000 feet high.
Small, living glaciers, rock glaciers, and icefields survive at higher elevations.
Large-scale glacial erosional features, sculptured by mighty alpine glaciers of the Pleistocene past, dominate the mountain scenery.
Moraines document glacial advances and retreats.
Igneous and metamorphic rocks, mostly Precambrian but some much younger, have been uplifted and exposed by faulting and erosion.

Rocky Mountain National Park

Northern Colorado

Figure 22.1   Upper Cascade Creek in Rocky Mountain National Park meanders over an alpine meadow that was buried by hundreds of feet of glacial ice in Pleistocene time. National Park Service photo.

Figure 22.2   Rocky Mountain National Park, Colorado. Map reproduced from R. P. Larkin, P. K. Grogger, and G. L. Peters, *Southern Rocky Mountains*. Used by permission. © 1980 Kendall/Hunt Publishing Company.

Geographic Setting and Local History

The Front Range is a towering mountain mass that rises rather abruptly west of Denver, Boulder, and Colorado Springs. This great topographic barrier extends from the Arkansas River on the south to the north end of the Laramie Range in southern Wyoming. Between the northern and southern boundaries of Rocky Mountain National Park, about 21 miles, air distance, is the highest part of the Front Range, including the highest peaks. The Front Range is the easternmost mountain range of the Southern Rockies and the first major range encountered by travelers coming from the east. Westbound pioneers, with their wagons and animals, avoided the Front Range because it was too much of a barrier. Wagon trains went north across Wyoming or south along the Sante Fe trail through New Mexico.

If you approach Rocky Mountain National Park from Estes Park, a community in the Big Thompson River

Box 22.1
More about Igneous Rocks

In previous chapters, you have become familiar with terms referring to plutonic bodies of various sizes and shapes, such as dikes, sills, batholiths, and stocks. Moreover, you have already learned that igneous rocks are formed from the solidification of magma, or molten rock, that originates under conditions of intense pressure and heat far below the earth's surface. Magma that rises to the surface, erupting as lava, forms *extrusive igneous* rocks . They normally cool so rapidly that individual mineral grains in most specimens cannot be identified without the aid of a microscope. *Intrusive igneous rocks,* on the other hand, tend to be medium- or coarse-grained because they solidify from magma that cools slowly at depth over a long period of time.

In classifying or identifying igneous rocks, determining the texture, or grain size, is an important step (fig. 22.3). Are individual mineral grains in the rock visible to the unaided eye (larger than 0.062 mm)? Or are the grains too small to be seen without the aid of a microscope (generally smaller than 0.062 mm)? If the grains can be seen, is the texture coarse, medium, or fine?

Sometimes changes in a magma's environment during cooling result in the formation of mineral crystals of distinctly different size within a rock, creating *porphyritic texture,* an effect that is sometimes described as similar to rice pudding with raisins. *Pegmatites,* which have very coarse texture, crystallize during the last cooling phase of a magma and are typically found around the margins of a large pluton. Sometimes large sheets of muscovite mica and deposits of rare minerals, such as beryl, can be mined from pegmatite bodies. (Other textures, mostly associated with volcanic activity, are defined in table 22.1.)

Igneous rocks are also classified on the basis of their mineral composition, or mineralogy. The chemical composition of the parent magma determines this aspect of igneous rock classification. From *felsic* magmas, rich in

Figure 22.3   Comparative grain sizes in igneous rocks.

valley, outside the park's eastern boundary, the highway is at an elevation of 7500 feet. Upturned Paleozoic sedimentary rocks are exposed in roadcuts. Then if you drive through the park on Trail Ridge Road, you cross the highway's highest point (between Fall River Pass and Lava Cliffs) at an elevation of 12,183 feet above sea level. As you make the climb, you see displayed on cliffs, peaks, and precipitous valley slopes some of the oldest rocks in North America—contorted and twisted gneisses and schists and speckled granites. Several other highways and many trails lead visitors into the rich scenery and massive grandeur of this mountain park. Some of the trails follow paths that Indians used when crossing the mountains.

Lieutenant Zebulon M. Pike (for whom Pikes Peak, near Colorado Springs, is named) explored the region that is now Rocky Mountain National Park in 1806–07. He aluminum, silicon, sodium, and potassium, come the light-colored rocks that mainly contain the minerals orthoclase, sodium-plagioclase, and quartz. At the other extreme are *mafic,* or iron-rich magmas, that are the source of the dark-colored rocks, made up primarily of the ferromagnesian minerals. Table 22.1 shows the classification of common igneous rocks by color from light to dark, mineral composition, and texture.

Observe that a coarse-textured granite may have essentially the same mineralogical composition as a very fine-textured rhyolite, with both being light in color. Similarly, a dark, coarse gabbro may contain the same minerals as a dark, fine-grained basalt. What is surprising, however, is that the intrusive granite and the extrusive basalt are the most abundant igneous rocks on earth, while rhyolite (an extrusive rock) and gabbro (intrusive), the respective mineralogical equivalents, are relatively rare. For this paradox of nature, involving the most abundant extrusive and intrusive igneous rocks, geologists have as yet no wholly satisfactory explanation. However, in many

Table 22.1 Classification of Common Igneous Rocks by Color, Composition, and Texture

Characteristic rock color	Light gray "salt-and-pepper"	Medium gray medium green	Dark gray to black	Very dark green to black
Mineral content	Quartz; feldspars (white, light gray or pink); a few ferromagnesian minerals	Feldspars (white or gray) and about 50% ferromagnesium minerals	Predominantly ferromagnesian minerals with some plagioclase feldspar	Entirely ferromagnesian minerals/usually olivine and pyroxene
Coarse-grained texture	GRANITE	DIORITE granodiorite monzonite	GABBRO	Ultramafic rocks
Fine-grained texture	RHYOLITE dacite	ANDESITE	BASALT	

Other varieties of igneous rocks classified by texture:
Obsidian: volcanic glass (usually of rhyolitic composition)
Tuff: pyroclastic rock consisting of consolidated volcanic ash (usually rhyolitic or andesitic)
Pumice: light-colored, highly vesicular, pyroclastic material; a form of frothy volcanic glass (commonly rhyolitic)
Breccia: pyroclastic rock consisting of coarse volcanic fragments
Pegmatite: very coarse-grained rock (usually granitic)
Porphyry: large crystals in a ground mass of smaller crystals
Vesicular: rocks that have holes created by gas trapped in cooling lavas

continued

referred to the tallest summit in the range as "Great Peak," but it was renamed Longs Peak in 1820 by Major Stephen H. Long, whose expedition was sent out to map the region of the Great Plains and the Southern Rockies.

Joel Estes and his son Milton found the valley that now bears their name, Estes Park, and built a cabin along Willow Creek in 1860. They left after a few years, and their claim was later transferred to the Earl of Dunraven, an English nobleman who wished to make the valley into a game preserve. Dunraven paid people to file individual claims and then transfer them to him. This method of land acquisition kept out enterprises that might have ruined the region, but the claims were later declared invalid. Dunraven did, however, keep a ranch in Estes Park and built the region's first hotel.

Box 22.1 continued

respects, the distribution patterns of igneous rocks and their great variety can be accounted for by plate tectonic interpretations, based on field data; studies of composition and texture of rocks; and experimental work on artificial magmas.

Rocky Mountain National Park and many other national parks have exposures of various types of igneous rocks. Many of the igneous rock specimens that you see in the parks have mineral grains large enough to be identified by eye using the characteristic physical properties of the constituent minerals. For convenience, diagnostic physical properties—grain shape, color, and cleavage—of common rock-forming minerals are given in table 22.2. *Cleavage* is the breaking of a mineral along specific planes that reflect crystal structure. Not all minerals have cleavage, but minerals that do, break or "cleave" in a characteristic direction or set of directions.

Table 22.2 Characteristic Appearance in Igneous Rocks of Common Rock-Forming Minerals

Mineral	Diagnostic Physical Properties
Orthoclase feldspar	1. Grains generally of uniform size 2. Usually white or pink 3. Two directions of cleavage nearly at right angles
Plagioclase feldspar	1. Lath-shaped grains 2. White, gray, or slightly greenish 3. Two directions of cleavage nearly at right angles; parallel markings on cleavage faces
Quartz	1. Irregularly shaped grains formed between other grains 2. Colorless, glassy appearance 3. No cleavage
Olivine	1. Small, round grains 2. Pale green, glassy appearance 3. No cleavage
Amphibole group: example, hornblende	1. Lath-shaped grains 2. Black or dark green 3. Two directions of cleavage not at right angles
Pyroxene group: example, augite	1. Lath-shaped or blocky grains 2. Black or dark green 3. Two directions of cleavage nearly at right angles
Muscovite mica	1. Flat, flaky grains 2. Silvery white or pale green 3. One direction of cleavage
Biotite mica	1. Flat, flaky grains 2. Black or dark brown 3. One direction of cleavage

The first person to climb Longs Peak was William N. Beyer, who got to the top on August 23, 1868. Among those in his party was Major John Wesley Powell, the geologist who explored the Grand Canyon in 1869. The Front Range has 42 peaks exceeding 12,000 feet in elevation; among them are Longs Peak (14,256), Mount Meeker (13,911), Ypsilon Mountain (13,314), Taylor Peak (13,153), Flattop Mountain (12,342), Snowdrift Peak (12,274), Mount Adams (12,121), and Battle Mountain (12,044).

The person who was most involved in the effort to preserve the region as a national park was Enos Mills. He built a home in Longs Peak Valley in 1866 when he was only 16 years old. After helping a survey party map Yellowstone in 1891, he was convinced that his homesite should be preserved like Yellowstone. He fought long and hard for this until he succeeded. In 1915, Rocky Mountain National Park was established. A skilled mountain climber, Mills was the first to go up the east face of Longs Peak, a very difficult route.

The famous Trail Ridge Road, built in 1932, starts near Estes Park and climbs to the Roof of the Rockies, a broad, relatively flat upland extending over most of the central area of Rocky Mountain National Park. According to the park brochure, in a 40-mile trip over Trail Ridge Road, you traverse life zones equivalent to a round trip to the Arctic Circle. The drive may take you from hot summer weather to wintry conditions with even a snow squall. Above the treeline, the upland is a rolling terrain covered by frost-shattered rock. The fragile vegetation of alpine tundra struggles to survive among the rocks. Nearly continual high winds sweep over the upland, even during the short summer season. At Milner Pass (elevation 10,758 feet) near the park's western boundary, Trail Ridge Road crosses the Continental Divide and then follows the Colorado River south to Grand Lake.

Geologic Features

The Crystalline Bedrock

The ancient igneous and metamorphic rocks exposed in Rocky Mountain National Park are *crystalline;* that is, consisting wholly of crystals (or fragments of crystals), most of them large enough to be seen without a lens. The metamorphic rocks, which originally may have been sedimentary, have been recrystallized by the heat and pressure of intense and prolonged metamorphism. Igneous rocks, emplaced in plutons, crystallized from a melt. Both igneous and metamorphic rocks have been altered in varying degrees by several episodes of *regional metamorphism* (metamorphism caused by high temperature and pressure affecting rocks over a relatively large area). Some of the rocks that were intruded by several generations of granitic magmas show the effects of *contact metamorphism* (metamorphism where high temperature is the dominant factor, usually in a narrow zone between intruding magma and country rock).

The central part of the Front Range has a complex of small plutons fitted between wedges of gneisses, schists, and quartzites. Large batholiths, such as the Pikes Peak Granite in the southern part of the range and the Sherman Granite at the north end, flank the smaller plutons.

Within the complexly faulted, anticlinal arch that makes up the Front Range, there are numerous cross folds and faults. (A *cross fold* is one that intersects a previous fold of different orientation.) The faults and shear zones, which trend generally northwest-southeast, are mainly of Precambrian age. Most of them were reactivated during the Laramide orogeny at the end of Cretaceous time.

The Development of the Rocky Mountains

Although geologists have studied the Front Range and the Rocky Mountains for many years, the relationships of plate movements and episodes of orogeny have not been entirely worked out. One indication of a very long history of tectonic activity in this region is the elevation of the basement rocks—the Precambrian gneisses, schists, and granites—that are now exposed in the Front Range. You can see them thousands of feet above sea level on the Continental Divide and the peaks above the upland.

The present topography of the Front Range is due to a broad upwarping that took place during the Cenozoic Era, when vertical uplift brought Precambrian basement rocks to heights far above their former surroundings. However, the mountains that we see today are but the most recent manifestation of tectonic activities that have gone on in this part of the earth's crust for many millions of years.

In Paleozoic and Mesozoic time, the deeply eroded, crystalline Precambrian rocks were submerged by successive invasions of shallow seas in which many units of marine sedimentary rocks were deposited. Intermittent tectonic activity during the Pennsylvanian Period deformed Paleozoic rocks in central Colorado in an orogenic belt, trending northwest-southeast, that elevated the rocks into a range called the "Ancestral Rockies." These Paleozoic mountains, subsequently eroded away, may have been a westward extension of the Ouchita Mountains. The Ancestral Rockies are in no way related to the present-day Rocky Mountains, but their former presence can be inferred from remnants exposed around the edges of the high mountain ranges.

The Laramide orogeny formed the gross structures of the Rocky Mountains. The mountain-building began as a result of interaction between the North American plate moving westward and the Pacific plate drifting slowly eastward. Compressive forces, generated when the plates converged, intensified from west to east, culminating late in Cretaceous and early Tertiary time. Major orogenic activity, which occurred in the region that is now New Mexico, Colorado, northern Utah, Wyoming, Idaho, and Montana, formed the Northern and Southern Rockies.

Eastward and northeastward thrust faulting, folding, and crumpling of the crust produced a number of elongated mountain ranges where the crystalline basement rock was pushed up to form the cores of very large, faulted, anticlinal structures. The Colorado Front Range illustrates how this happened. Paleozoic and Mesozoic sedimentary rocks lap up against the eastern face of the Front Range (just to the east of the park boundary). High-angle thrust faults pushed the crystalline core thousands of feet upward and outward over the margin of the upfolded Paleozoic rocks.

The uplifting of the Front Range was but one aspect of the Laramide orogeny. Compressive forces deformed and uplifted rocks from Central America to Alaska, creating the North American Cordillera. In order to reach the Colorado area, subducted slabs of oceanic crust must have moved way inland from the western edge of the continent and must have been descending at a relatively low angle. The massive cores of the Rocky Mountain ranges transmitted stresses by large-scale faulting along their eastern margins but without intense deformation of their own structures.

As the mountains were raised higher and higher, erosion intensified. Sediments from summits and mountain slopes filled the valleys and lowlands. A period of crustal adjustment, called the isostatic phase, ensued as the orogenic pressures subsided. The principle of *isostasy* proposes a condition of equilibrium—comparable to floating—involving units of brittle crust on the plastic zone of the *mantle* (the thick rock shell of the earth's interior between the crust and the core). Blocks of crust that are not in equilibrium, due to tectonic forces raising them up or the weight of accumulated sediment pushing them down, tend to compensate by moving up or down in a direction toward equilibrium. Block-faulting and other vertical adjustments of crustal units can sometimes be explained in terms of isostatic compensation. However, this principle cannot account for the much more powerful mechanisms of plate subduction and plate convergence that clearly involve forces of much greater intensity.

Volcanic Activity, Pegmatites, and Hydrothermal Veins

The major faults on both sides of the Front Range and the other faults and folds in the park region were the result of Laramide tectonic activity. In faulted areas, eruptions and intrusions of various kinds occurred repeatedly from the Cretaceous Period to early in the Pleistocene Epoch, with the greatest intensity being in Tertiary time.

Plate movement and increasing volcanic activity during the Tertiary was characteristic not only of the Rocky Mountain areas; it was widespread throughout much of western North America. In the Southern Rockies the more violent eruptions produced great quantities of pyroclastic dust, ash, and other debris that consolidated as tuff. Lava flows were mainly andesitic (of intermediate composition), but some were basaltic or rhyolitic (table 22.1). Some of the lava solidified as obsidian, or natural glass. Where lava flows overran streams, the cold water shattered the volcanic glass so that it consolidated as breccia. Stream gravels mixed with hardening lava and formed andesitic conglomerate. Remnants of these lavas can be seen in the Never Summer Range along the park's western border and in the northwest corner of the park, near where Mt. Richthofen now stands. Lava Cliffs and Specimen Mountain, which were formed by volcanic eruptions of the same age, can be viewed from Trail Ridge Road.

Some of the igneous material solidified below the surface in dikes, sills, and stocks that were later exposed by erosion and uplift. Many of the smaller intrusions are Precambrian in age, but some are much younger. Among these bodies are the pegmatite dikes and pods (small lenses). The coarse-grained pegmatites, made up mainly of large quartz and feldspar crystals, hardened in faults and folds in the country rock around the edges of the main batholiths.

Hydrothermal veins of quartz, some of them associated with pegmatites, are the source of much of the mineral wealth of the Front Range. These veins formed when hydrothermal solutions containing mineral ions escaped from the hot magma and forced their way upward through cracks in the rock. As the liquid cooled, silica precipitated as quartz and other minerals were deposited in the quartz matrix. Most of the mineral-bearing pegmatites and veins are located in the Colorado Mineral Belt, south of Rocky Mountain National Park.

Old Erosional Surfaces

The early Rocky Mountain ranges, uplifted by Laramide mountain-building, were separated by intermontane valleys that rapidly filled with sediment. Interbedded with fluvial deposits washed down from the uplands were deposits of volcanic ash and lava from extensive eruptions. The whole region underwent a lowering of relief during early Tertiary time, and a sloping erosional surface with low hills and wide shallow valleys only a few thousand feet above sea level was produced. The relatively flat summits of Deer Mountain and Flattop Mountain are remnants of this old erosional surface.

The broad upwarping that began in the Cenozoic Era was apparently related to the continuing subduction of the Pacific plate. This affected a wide area that today stretches from Denver to Reno. Episodes of strong vertical uplift were particularly intense in the Front Range. As a result, streams began intensive downcutting, stripping away sedimentary beds and incising canyons in the resistant granitic rocks.

A second erosional surface of moderate relief developed late in Tertiary time as uplift slowed. Remnants of this surface are found on Giant Track Mountain, Old Man Mountain, Castle Mountain, Needles, and Twin Owls. The highest peaks stood as isolated knobs, or *monadnocks,* above the erosional surface. Again, alluvial deposits built fans, filled basins, and spread over the plains to the east. The increasing height of the mountain ranges to the west cut off some of the moisture supply from the Pacific and the climate became more arid.

Inferences that can be drawn from the old erosional surfaces suggest that (1) thousands of feet of rock and sediment were removed from the high areas during and after uplift, and (2) periods of uplift can be related to the erosional history of the region.

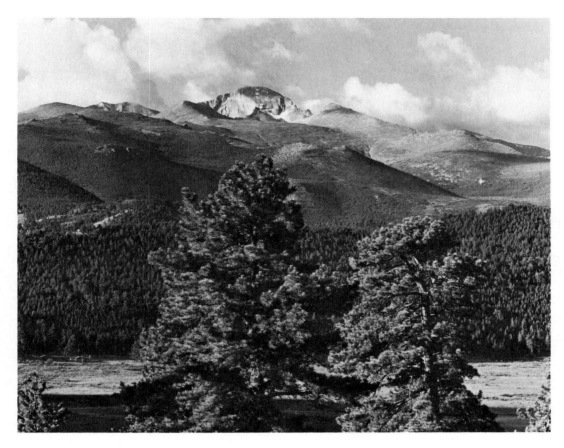

Figure 22.4   Longs Peak (14,256 feet), about 8 miles air distance from Beaver Meadows in the foreground, rises above eroded uplands of the Front Range. National Park Service photo.

A last stage of uplift, near the end of the Pliocene Epoch, brought the mountains of the Front Range to their present height and grandeur. Streams cut deep canyons in the resistant rocks. Along the eastern side of the Front Range, much of the weaker sedimentary rock was eroded away, thus accenting the barrier appearance of the range. This was differential erosion on a grand scale.

For all their toughness, however, the basement rocks that make up the "roof" of the park are not impervious to erosive forces. The extensive joint systems, intensified by uparching, provide conduits for weathering processes. The Needles and Twin Owls are examples of joint weathering. Exfoliation domes, such as McGregor Mountain, show the weathering of massive granite. The results of frost action at high altitudes are clearly evident along mountain trails and highways.

Glacial Features

A worldwide cooling trend that began in Tertiary time culminated in the severe glaciation of the Pleistocene Epoch. In the Rocky Mountains, the increasing elevation of the region brought about conditions favorable for glacier development early in the Pleistocene. Glacial ice displaced running water as the dominant agent of erosion.

Mountain glaciers did the major shaping of the alpine scenery that we see in Rocky Mountain National Park today (fig. 22.5). On the high peaks are prominent cirques, cols, and horns. The crest of the Front Range itself is a serrated ridge. At their maximum extent, the glaciers of the Front Range coalesced to form a small icecap with the high peaks jutting up through the perennial snowfields and glacial ice.

Successive episodes of glaciation left great quantities of *glacial drift* (glacial deposits, debris, outwash, etc.) in the basins, valleys, and foothills. MoraINal segments of various sizes, ages, and types are scattered over the eastern and western sides of the park in mountain meadows and deep glacial troughs. Some of the trails and roads in the park follow moraines. A section of the road west of the park entrance in Fall River valley was constructed on an end moraine. Most of the moraines are forested. Since the lodgepole pine prefers to grow on glacial drift, a good stand of these trees usually indicates the location of a moraine.

Rock glaciers, left by retreating ice glaciers, still contain enough ice so that freezing and thawing enables them to creep downslope. They form continuous, slowly moving strips of talus for thousands of yards along the walls of high valleys. Rock glaciers that formed on some cirque

Figure 22.5   Looking south from Forest Canyon Overlook toward Longs Peak. Intense frost action produced the boulder field in the foreground. National Park Service photo.

floors enlarged as they pushed out over the rim and downward.

Glacial lakes appeared on the landscape as the glaciers departed. Grand Lake, just outside the southwest corner of the park, is dammed by an end moraine. Lakes in morainal kettle holes are Burstadt, Copeland, Bear, Dream, and Sheep Lakes. Cirque Lakes are Blue, Black, Shelf, Lawn, Crystal, Fern, Odessa, Tourmaline, and Chasm Lakes, as well as Lake Mills. Chains of *rock-basin lakes* are found on the "stair treads" of several rocky gorges that rise like staircases. Examples are the Gorge, Loch Vale, Cub Lake, East Inlet (including Lake Verna), and Ouzel Creek Lake. Some lakes, such as Nanita, Nokoni, and Bench Lakes, lie in hanging valleys. Streams that occupy hanging valleys are Roaring River, Chiquita Creek, Sundance Creek, and Fern Creek.

Roches moutonnées, which are characteristic of glacial erosion on massive rocks, are common in the park. A *roche moutonnée* is an elongate knob or hillock of bedrock that has been smoothed and scoured by moving ice on the

upglacier (stoss) side so that the rock is gently inclined and rounded. On the downglacier (lee) side, the rock is steep and hackly from glacial plucking or quarrying. Glacial Knobs is a well known area of roches moutonnées in the park.

Many rock fragments and exposed bedrock surfaces show glacial *striations*. These are furrows or lines that were inscribed on rock surfaces as rock fragments embedded in ice were dragged along with great force by a moving glacier. Striations on outcrops are usually oriented in the direction of glacier movement.

A bedrock surface that has been planed down, striated, and polished by glacial abrasion until it is relatively smooth is called *glacial pavement*. The area around North Inlet has particularly good examples of glacial pavement.

Glacial erratics are abundant at Boulder Field and on the sides of Fern Creek canyon. An *erratic* is a relatively large rock fragment that has been transported by ice from its place of origin that may be a considerable distance away. Boulder Field is also a *felsenmeer* area where frost-

Table 22.3 Geologic Column, Rocky Mountain National Park

Era	Period	Epoch	Rock Units	Geologic Events
Cenozoic	Quaternary	Holocene	Neoglacial deposits, alluvium, talus	Development of modern glaciers; active mass wasting; periglacial processes; stream erosion Melting away of earlier glaciers during post-Pleistocene warming
		? Pleistocene	Glacial tills, moraines, etc., of successive ages; early deposits largely buried or destroyed by later advances	Pinedale glaciation Intermediate glaciation 　Late Bull Lake 　Early Bull Lake Early glaciation (at least two major advances and retreats)
	Tertiary	Paleogene/ Neogene	Volcanic ash Alluvial deposits Rhyolite, welded tuff, obsidian, pyroclastic debris, mudflow deposits	Development of modern drainage patterns Extensive erosion exposing basement rocks on summits, hogbacks on flanks Block-faulting forming basins Uplift along old faults raising Front Range to present height Volcanism (mainly to west and south of park) Mudflows, lava flows, ash falls
	?　　?			Intermittent uplift, faulting, tilting of sedimentary beds; stripping of beds producing erosional surface Laramide orogeny
Mesozoic	Cretaceous Jurassic Triassic		Shales	Last transgression and withdrawal of seas Reduction of relief
Paleozoic	(Cambrian to Permian)		Remnants of sedimentary rock units exposed on eastern and western flanks of Front Range (none in park)	Uplift and erosion of Ancestral Rockies Long deposition, first marine, then terrestrial (due to uplift)
			/////////////	Long erosion exposing granite batholiths
Late Precambrian (Proterozoic Y)			Basaltic dikes, granites, pegmatites	Successive intrusions and emplacement of batholiths; uplift Uplift, orogeny, metamorphisms, intrusions
Middle Precambrian (Proterozoic X)			Granites, schists, gneisses (basement complex)	Long accumulation of sediment interbedded with volcanics

(After Larkin, Grogger, and Peters 1980; Richmond 1974)

shattered, angular blocks of rock cover well jointed bedrock.

Examples of V-shaped stream valleys in the park can be compared with glacially sculptured U-shaped valleys, all of which were originally V-shaped. Fox Creek, Red Creek, and Cow Creek have characteristic V-shapes. Fall River valley, Forest Canyon, Fern Lake valley, and North St. Vrain valley are U-shaped and display other evidence of glacial erosion and deposition.

Mass Wasting Features

Periglacial conditions in the park brought about some modification of glacial topography due to mass wasting on both steep and gentle slopes. On Sundance Mountain and along Trail Ridge Road, *solifluction,* or flowage of soil, has caused slumping and development of hummocky lobes on gentle upland slopes above the treeline and above glacial valleys. Because water cannot percolate down through

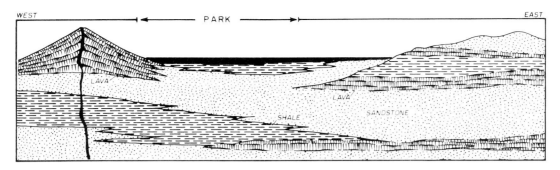

Figure 22.6   Great quantities of volcanic and clastic sediments accumulated throughout early Precambrian time.*

Figure 22.7   Long after the sedimentary layers were buried and lithified, they were folded, metamorphosed, and intruded by granitic magma during Precambrian mountain-building.

the frozen ground that exists at this altitude, the soil becomes saturated during the melt season and begins to slide on even a slight slope. On steeper slopes, especially at lower elevations, landsliding and avalanching may occur seasonally. The abundance of talus debris, talus cones, and rock glaciers also indicate active mass wasting.

Where there are loose rock fragments or soil cover on high, flat surfaces above the treeline, various forms of *patterned ground,* such as stone rings, polygons, stripes, and nets, have developed. Again, this is due to periglacial conditions, intense frost action, and very slow mass movement, which, in combination, tend to assemble stones in symmetrical patterns or designs.

Geologic History

1. Accumulation of sediments during Precambrian time.

The oldest rocks in the park, the schists and gneisses, were deposited over 1800 million years ago. Many thousands of feet of sediment formed shales, sandstones, and carbonate rocks interbedded with volcanic rocks (fig. 22.6).

2. Folding, intrusion of igneous rock, and metamorphism.

Tectonic activity compressed, intensely folded, metamorphosed, and uplifted the sedimentary beds. Intrusions of magma rose from a chamber far below the surface (fig. 22.7). As the orogeny continued, increased heat and pressure intensified the metamorphic processes, changing the rocks further into schists and gneisses. Prominent exposures of these ancient rocks are in outcrops along Trail Ridge Road and in cliffs along valley walls in the western and central parts of the park. Schists and gneisses can also be seen along the Continental Divide in the cirque headwalls.

3. Precambrian uplift and intrusions.

While the earlier magma was cooling, a larger intrusion pushed its way up into the schists and gneisses, eventually creating a granitic mass about 30 miles across. Intensely hot hydrothermal solutions penetrated the country rock surrounding the batholith, and pods and lenses parallel to the banding of the schists and gneisses were formed. In some places the confining pressure was so great that pegmatites formed during the long, slow

*Figures 22.6 through 22.19 are from G. M. Richmond, *Raising the Roof of the Rockies.* © 1974 Rocky Mountain Nature Association. Used by permission.

Figure 22.8   Late in Precambrian time, during a long period of intense mountain-building and regional metamorphism, a large granitic batholith was emplaced.

Figure 22.9   For millions of years, as erosion wore down the uplands, Paleozoic seas spread over the land. Many layers of sedimentary rock covered the crystalline Precambrian basement.

cooling. Radioactive minerals in the intruded rocks have provided an absolute age figure that the event took place about 1450 million years ago (fig. 22.8).

These granites crop out in the western part of Rocky Mountain National Park; in the area east of Shadow Mountain and Grand Lake; and also on the east side of the park, north and south of Estes Park. Additional exposures are on Hagues Peak (near the northern boundary) and on the east face of Longs Peak (near the southeast boundary). The contact between the older schists and gneisses and the granitic intrusions shows up on canyon walls east of the Continental Divide.

4. Erosion of the Precambrian highlands.

The mountains that were formed in the Precambrian orogenies were worn down during a span of about a thousand million years until the tops of the granite batholiths were uncovered. The sands and other clastics that were the products of erosion were lithified as sandstones blanketing the gneisses, schists, and granites. A few remnants of these ancient sandstones have been found overlying granite in the Colorado Springs area southeast of the park, but none within the park itself (fig. 22.9).

A basaltic dike, called "Iron Dike," suggests that another intrusion occurred in late Precambrian time. The dike crops out near Mount Chapin and Storm Pass north of Trail Ridge Road.

5. Uplifting of the Ancestral Rockies.

During early and middle Paleozoic time, shallow inland seas advanced and withdrew, at times covering most of the area that is now Rocky Mountain National Park. For 200 million years Paleozoic gravels, sands, muds, and oozes accumulated and were lithified. None of the sedimentary rocks deposited in this period remain in the park today.

Gradual uplift in the Pennsylvanian Period, about 300 million years ago, created the Ancestral Rockies in the region west of the Front Range. At their maximum the mountains may have been about 2000 feet high. Along with the uplift came increasing erosion and downcutting by streams, processes that stripped away the previously deposited marine sediments, along with some of the Precambrian igneous and metamorphic rocks (fig. 22.10).

6. Mesozoic reduction of relief and transgressions of seas.

By the end of the Paleozoic Era, the Front Range region was a low, swampy area with only low hills remaining where the Ancestral Rockies had stood. Amphibians, reptiles, and dinosaurs roamed over the land. Their footprints and fossilized remains are preserved in sedimentary rocks not far from the park. Jurassic coal beds formed in swamps, and coastal dunes migrated along the

Figure 22.10   The region was uplifted in middle to late Paleozoic time, forming the "Ancestral Rockies." Erosion gradually lowered the highlands.

Figure 22.11   During the Cretaceous period, the sea advanced over the region for the last time. Marine sediments were laid down on the old surface.

shores of an inland sea. Remnants of the dunes can be found in the sedimentary rocks east of the park along the mountain front.

7. Maximum advance of seas in Cretaceous time, followed by uplift.

About 100 million years ago, seawater completely covered the region (fig. 22.11). Many feet of sand and clay accumulated on the shallow sea floor. Remains of some of the shale beds that were deposited at this time are in the northwest part of the park, east of Lead Mountain.

About 70 million years ago, the region that eventually became the Front Range began to rise again, forming a low island in the sea. Uplift continued until by the end of the Cretaceous all of Colorado was above sea level. As the land rose, stream systems developed, eroding the marine rocks and redepositing the sediment in new layers over the surrounding areas.

8. Laramide orogeny; uplift and faulting; stripping of sedimentary beds.

At the end of the Cretaceous Period and the beginning of Tertiary time, a major event in the geologic history of North America took place. During the Laramide orogeny, the North American Cordillera, which extends from Mexico northward into Canada and Alaska, was created. In the Rocky Mountain National Park region, large blocks

were uplifted along deep fractures in the Precambrian basement rocks (figs. 22.12, 22.13).

Periods of volcanism ensued. The lava flows at Mt. Richthofen mark an episode. Near Specimen Mountain, mudflows, lava flows, and obsidian are indications of a later and different type of volcanic activity of Oligocene age. Evidence of volcanic ash can be seen near Iceberg Lake along Trail Ridge Road (fig. 22.14).

During and between the episodes of volcanism and uplift, the processes of erosion went on, lowering relief, stripping away sedimentary rocks, and cutting into the Precambrian schists, gneisses, and granites (fig. 22.15). In all, about 5000 feet of sedimentary rocks were removed.

9. Late Tertiary upwarping and erosion.

In late Tertiary time, successive uplifts that were part of a broad regional uparching brought the mountains of the park to their present dramatic height. The uplifts occurred primarily along the old faults and increased elevations as much as 4000 to 6000 feet (fig. 22.16). The Front Range appeared to rise as a series of steps, particularly on the east side of the Continental Divide. Volcanic activity, mainly the eruption of ash and pyroclastics, was intermittent (fig. 22.17). Most of the intrusive and extrusive activity and accompanying mineralization took place south of Rocky Mountain National Park.

Figure 22.12   At the end of Cretaceous time, differential uplifting, due to Laramide mountain-building, produced fault blocks. Compressive forces caused thrust-faulting to occur.

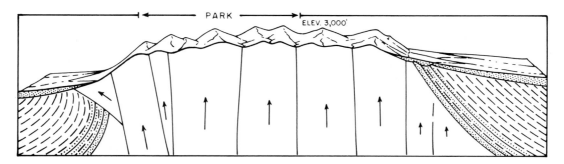

Figure 22.13   As uplift continued into Tertiary time, erosion stripped off sedimentary layers, exposing the rocks of the Precambrian crystalline basement.

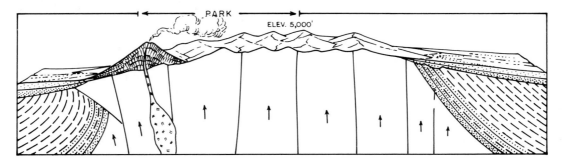

Figure 22.14   Volcanism accompanied uplift and erosion.

Figure 22.15   By middle Tertiary time, erosion had removed so much rock that summits were low and relief was subdued.

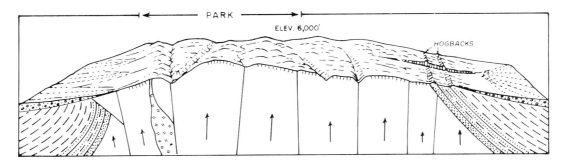

Figure 22.16   Renewed uplift, beginning in middle Tertiary time, raised the mountains again.

Figure 22.17   Volcanoes were built up by eruptions as lavas rose to the surface.

Because of the Tertiary uplifts, the rocks in the park are either very old or very young; that is, they are either Precambrian basement rocks or the remains of young volcanics deposited in valley-fill areas or on protected slopes that "rode up" with the uplands. Virtually all of the rocks of intermediate age, the Paleozoic and Mesozoic sedimentary beds, that once covered the area were stripped away. Erosional remnants of these rocks form the *hog-backs* of the foothill ranges that make up a narrow transitional belt east of Rocky Mountain National Park, between the plains and the Front Range (fig. 22.18).

By the end of the Tertiary, parts of the land surface had been markedly displaced by vertical movement of the large mountain blocks. Basins such as Estes Park were down-dropped as much as a thousand feet while parts of the upland along the Continental Divide were raised to their present height of over 12,000 feet (fig. 22.19).

Drainage patterns were disrupted or altered with each episode of uplift. In the last stage of faulting and uplift, the principal drainage patterns that exist today were established. Waterfalls formed as streams incised their valleys, carved deep canyons, and dissected old surfaces.

10. Pleistocene glaciation.

Glacial ice advanced and retreated a number of times in Rocky Mountain National Park; however, the major episodes can be grouped as follows:

1. Early glaciation (pre-Bull Lake), which involved at least two major advances, began around 1,500,000 years ago.

2. Bull Lake glaciations, between 500,000 and 80,000 years ago, included Early Bull Lake and Late Bull Lake.

3. Pinedale glaciations, during Wisconsinan time, included three distinct episodes. The maximum extent occurred about 15,000 years ago.

   A warm (hypsithermal) interval prevailed from about 9000 to 4000 years ago. The last of the Pleistocene glaciers disappeared in the park region about 7500 years ago.

4. Neoglaciation ("Little Ice Age") began about 3800 years ago and included two advances and retreats.

Traces of the earliest glaciations within the park were obliterated by later ice advances. The oldest glacial deposits in or adjacent to the park are found near the eastern boundary. Some are in roadcuts along Colorado Highway 7 in the southern Tahosa Valley and some are near the Fall River entrance to the park.

Since Bull Lake glaciations were extensive and severe, evidence of their erosion and deposition is distributed throughout the park and adjacent to park boundaries. Both Early and Late Bull Lake advances deepened the same cirque basins, enlarged canyons, joined together tributary glaciers, and left prominent end moraines, usually not far apart. Since the moraines tend to be similar in size, shape, and type of material, only the amount and depth of weathering distinguishes them. Examples of Bull Lake end moraines can be seen in Aspenglen Campground and near the Beaver Meadows entrance. On the west side of the

Figure 22.18   Increased erosion bared the uplifted crystalline summits. Remnants of Paleozoic and Mesozoic sedimentary rocks formed hogback ridges in the foothill ranges to the east.

Figure 22.19   Vertical fault movements raised some blocks and dropped others, causing disruptions of drainage and some displacement of the land surface. The Front Range had attained roughly its present elevation by the beginning of Quaternary time.

park, Bull Lake glaciers that occupied the Colorado River valley left end moraines near Shadow Mountain Lake.

Pinedale glaciation is notable for the sculpturing and quarrying that is conspicuous in the park's alpine scenery. The steep cirque headwalls along the Continental Divide and the nearly vertical east face of Longs Peak are the result of Pinedale glaciation, even though the ice in the valleys did not attain the thickness of the Bull Lake glaciers.

On the east side of the park, Pinedale glaciers began in the perennial snowfields and cirques of the Mummy Range and then flowed down east-trending valleys. One glacier, 13 miles in length, occupying Big Thompson valley, flowed down from the eastern side of the Continental Divide. Another group of glaciers, which originated in the cirque basins of the Never Summer Mountains, flowed down to the Colorado River valley and joined westward-flowing glaciers from the Continental Divide upland to form the park's longest glacier. It extended for 20 miles from an area west of La Poudre Pass southward to the morainal islands in Shadow Mountain Lake.

By scouring and deepening glacially eroded valleys, the Pinedale glaciers gouged out chains of rock-basin lakes, such as the Gorge Lakes and those in Tyndall Gorge. Glacial snouts dug out large basins, such as Horseshoe Park excavated by Falls River Glacier, Moraine Park by Thompson Glacier, and Bartholf Park by Bartholf Glacier.

Figure 22.20   Roadcut near the Estes Park entrance to Rocky Mountain National Park exposes tilted sedimentary beds. Photo by B. Brownlee.

Moraines left by Pinedale glaciers show less evidence of weathering than older glacial drift. Pinedale moraines tend to be prominent ridges with numerous boulders scattered over the surface. The end moraines west of the Fall River valley entrance are examples.

11. Post-Pleistocene glaciers and erosion.

Although all the Pleistocene glaciers melted away during the post-Pleistocene warming period, new glaciers reformed in the old cirques several thousand years ago during Neoglacial advances. The modern glaciers have shrunk to little more than perennial snowfields and *glacierets* (miniature alpine glaciers). In recent years, a few, such as Sprague, Andrews, Tyndall, and Taylor Glaciers, have increased slightly in volume. The moraines built by the post-Pleistocene glaciers are the easiest to identify because they are the least weathered and dissected. The best places to see Neoglacial moraines are near the cirques because the glaciers that built them did not extend far down the valleys.

With glaciers restricted to small areas at high altitudes, the processes of stream erosion and deposition have again become dominant in the shaping and modification of park landscapes. Processes of mass wasting, particularly those associated with periglacial conditions, also continue to play a significant role in landform evolution in Rocky Mountain National Park.

Geologic Map and Cross Section

Braddock, W. A., and Cole, J. C. 1990. Geological map of Rocky Mountain National Park and vicinity, Colorado. U.S. Geological Survey Miscellaneous Investigation series I-1973.

Mallory, W. W. (editor). 1972. *Geologic atlas of the Rocky Mountain region.* Rocky Mountain Association of Geologists.

Tweto, O. 1979. Geologic map of Colorado. U.S. Geological Survey.

Bibliography

Chronic, H. 1984. *Time, rocks, and the Rockies: Geologic guide to rocks and trails of Rocky Mountain National Park.* Missoula, Montana: Mountain Press Publishing Company. 120 p.

———— 1984. *Pages of stone: Geology of western national parks and monuments.* I. *Rocky Mountains and western Great Plains.* Seattle: The Mountaineers. p. 115–127.

Dickinson, W. R., and Snyder, W. S. 1978. Plate tectonics of the Laramide orogeny. In *Laramide folding associated with basement block-faulting in western United States,* ed. V. Matthews, III, Geological Society of America Memoir 151, pp. 255–56.

Larkin, R. P.; Grogger, P. K.; and Peters, G. L. 1980. *Southern Rockies,* K/H geology field guide series. Dubuque, Iowa: Kendall/Hunt Publishing Company.

Richmond, G. M. 1965. Glaciation of the Rocky Mountains, H. E. Wright Jr. and D. G. Frey, editors. *The Quaternary of the United States.* VII Congress of the International Association for Quaternary Research. Princeton University Press. p. 217–230.

———— 1974. *Raising the roof of the Rockies.* Estes Park, Colorado: Rocky Mountain Nature Association, Inc.

Yardell, M. D. et al. (editors) 1975. *National Parkways, photographic and comprehensive guide,* v. III/IV, *Rocky Mountain and Mesa Verde National Parks.* Casper, Wyoming: Worldwide Research and Publishing Company. p. 2–47.

Rocky Mountain National Park

Area: 265,193 acres; 414 square miles
Established: January 26, 1915
Designated a Biosphere Reserve: 1976
Wilderness designated: December 22, 1980
Address: Estes Park, Colorado 94956

23

Precipitous peaks, glacial troughs, lakes, streams, and the Continental Divide, extend over both sides of the international boundary.

Modern alpine glaciers occupy outsize cirques carved by Pleistocene glaciers; waterfalls leap from hanging valleys. Ancient, relatively undeformed, colorful, sedimentary beds are displayed on glaciated, angled summits and steep slopes. Seen in close-up view, along mountain trails and Going-to-the-Sun Highway, are ripple marks, mud cracks, and fossils of primitive Precambrian plants. Horizontal displacement by the great Lewis Overthrust pushed Precambrian strata miles eastward over much younger rocks.

Waterton-Glacier International Peace Park:

Glacier National Park,
Northwest Montana

Waterton Lakes National Park,
Southwest Alberta

Figure 23.1 Looking west across St. Mary Lake toward the glaciated peaks of Waterton-Glacier International Peace Park. St. Mary Lake occupies a glacial trough dammed by a moraine. National Park Service photo by G. A. Grant.

WATERTON/GLACIER INTERNATIONAL PEACE PARK

Figure 23.2   Waterton-Glacier International Peace Park, Alberta (Canada) and Montana.

Geographic Setting and Local History

Although Glacier National Park and Waterton Lakes National Park were set aside separately, Canada and the United States have shared this mountain wilderness for nearly 60 years. The joining together of the two parks to form Waterton-Glacier International Peace Park came about through the grass-roots efforts of Rotary International. The two governments cooperate in keeping the wilderness pristine and largely undisturbed by human activities while each jurisdiction administers the parkland on its side of the border. People move between the two parks almost as freely as the animals. Access roads are limited, but several trails cross the border and an excursion boat on Upper Waterton Lake makes frequent trips in summer between the village of Waterton, Alberta, and a visitor center on the U.S. side at the southern end of the lake.

The map of Waterton-Glacier shows the Continental Divide running from north to south through the U.S. park and along the western boundary of the Canadian park. Streams that drain the western side of the Continental Divide carry water that may eventually reach the Pacific Ocean. Water falling on the eastern side enters streams that end up in either Hudson Bay or the Gulf of Mexico. Triple Divide Peak, south of St. Mary Lake, is unique because water drains in three directions instead of two; namely, west toward the Pacific, northeast toward Hudson Bay, and southeast toward the Gulf of Mexico.

Because of its location straddling the Continental Divide, Waterton-Glacier makes its own weather. The dominant weather pattern is the flow of warm, moist air from the Pacific, which cools and drops rain or snow as air masses rise over the west side of the Continental Divide. The cooled air then plunges down the east slopes, becoming warmer as it descends. Periodically, cold fronts from the Arctic sweep down along the eastern side of the mountains, slide under warmer Pacific air from the west, and create fog, clouds, and gusty winds. Temperature ranges are extreme and may change rapidly. Snow may occur any month of the year. Because of the rain-shadow effect, winter accumulations of snow are much higher on the western slopes than along the eastern edge of the park.

Indians have lived in the region for thousands of years. The Blackfeet Indian Reservation borders the eastern side of Glacier National Park, and for many years the fierce and powerful Blackfeet kept trappers, settlers, and other Indians out of the region. St. Mary Lake and Lake McDonald, the two largest of the more than 200 glacier-fed lakes in the park, were favorite areas of Indian habitation.

Government surveyors marked the 49th parallel as the international boundary in the 1860s. Today only a narrow swath where trees have been cut down marks the border. When the U.S. Geological Survey made a geological re-connaissance along the 49th parallel early in this century, Bailey Willis carried out the first geologic investigation of the Glacier National Park area.

Railroad surveyors came into the Northern Rockies after the Civil War looking for mountain passes suitable for potential rail routes. John Stevens discovered Marias Pass in 1889 and surveyed it for the Great Northern Railroad, which constructed its line through the region in 1892. The railroad now skirts the southern boundary of the park. After Glacier National Park was established, the railroad built hotels and mountain chalets to attract tourists.

In 1885, George Bird Grinnell, a naturalist and conservationist, vacationed in the region. He fell in love with the place and returned many times. As the long-time editor and publisher of *Forest and Stream,* he campaigned for preservation of the area for many years and was influential in the legislation that established Glacier National Park in 1910.

Going-to-the-Sun Highway, named after a legendary Blackfeet chief, was opened in 1933 after a decade of costly and difficult construction. The 36-mile trip between the east and west sides of Glacier National Park is scenically majestic and geologically fascinating.

Geologic Features

The glaciation that produced the spectacular, ice-sculptured landforms in Waterton-Glacier was at its maximum extent during the Pleistocene Epoch. However, the 50 or more active glaciers now existing in the park developed only a few thousand years ago during the "Little Ice Age." All are small cirque glaciers and valley glaciers.

Examples of Glacial Erosion

At the head of a valley glacier, the continual scouring and plucking done by the ice mass produces a cirque basin. Many of the lakes in Glacier National Park, such as Avalanche, Cracker, Ellen Wilson, Helen, Hidden, Iceberg, Kennedy, Pocket, Ptarmigan, and Upper Two Medicine Lakes, are *cirque lakes* that lie in bowl-shaped depressions from which glacial ice retreated.

Where two cirques have eroded headward until they are back to back or adjacent, the thin, blade-shaped wall of rock separating them is an *arête.* Garden Wall and Ptarmigan Wall, which can be viewed from Going-to-the-Sun Highway, are examples. A saddle-shaped depression, termed a *col,* results when glacial ice breaks down an arête between cirques. Jackson and Blackfoot Glaciers are separated by a col, and another can be found behind Haystack Butte. Some cols in the park are mountain passes, such as Logan Pass, Piegan Pass, Marias Pass, and Gunsight Pass. A glacial *horn,* a feature having a pyramidal shape, is produced when cirques cut back into a mountain summit from three or four sides. Some of the horns in

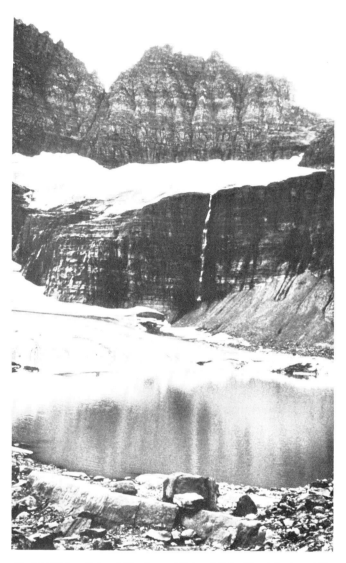

Figure 23.3   Grinnell Glacier and Grinnell Falls. The falls drop 165 feet from the hanging valley. The "comb ridge" (arête) on the skyline, which stood above the levels of former glaciers, is rough and jagged because of intense frost action. Exposed here are rocks of the Grinnell Formation (lower beds) and the Helena (Siyeh) Formation (upper beds). Photo by R. Smith.

Glacier National Park are Clements Mountain, Kinnerly Peak, Little Matterhorn, Reynolds Mountain, Sinopah Mountain, Split Mountain, Mount Wilbur, St. Nicholas, and Haystack Butte.

Valleys that have been heavily glaciated have a characteristic U-shape. This is because valley glaciers tend to straighten bends or curves cut by preglacial streams, smooth and widen the side slopes, and deepen the valley floors. Typical U-shaped glacial valleys in the park are McDonald Valley, St. Mary Valley, Waterton Lake valley, Swiftcurrent and Bowman Lake Valley, Belly River Valley, and Cataract Creek Valley.

Many beautiful hanging valleys, some with waterfalls, are in Waterton-Glacier. The mouth of a *hanging valley* opens out high on the steep side of a larger, glacial valley, many feet above the valley floor. The larger valley was eroded by a trunk glacier, which had much greater erosive power than its smaller tributary glaciers and thus was able to erode its valley more deeply. When the glaciers melted back, the tributary valleys were left hanging far above the main valley. Avalanche Falls is an example of a waterfall joining a stream in a hanging valley to a main stream. Other waterfalls, such as St. Mary, Rockwell, Baring, and Florence Falls, cascade down over glacial steps in bedrock walls. Trick Falls are unique because at the top the falls flow over resistant beds; when stream volume is low, the bottom falls (normally hidden from view beneath the top falls) issue from a cave formed by the enlargement of joints in limestone.

Some glacial valleys have a string of *rock-basin lakes* connected by a single stream. The lakes occupy depressions scoured out of highly fractured or weaker sections of bedrock. An example of this glacial feature can be seen between Many Glacier Hotel and Swiftcurrent Pass. Bullhead Lake and Swiftcurrent Lake are two of the rock-basin lakes in the string.

In late summer natural *ice caves* become accessible at the base of Grinnell and Sperry Glaciers. Glaciologists obtain data on ice pressures and temperatures from instruments placed at the ice-rock interface and left year-round in the cave beneath Grinnell Glacier. Some summers when scientists enter the cave, they find that instruments have been destroyed by the grinding of ice against the rock. Streams drain from the caves beneath the glacial snouts. The characteristic milky white color of the water is due to the heavy concentration of *rock flour* (finely pulverized rock debris) being carried off in suspension.

Glacial Deposition

Glacial erosion produces great quantities of sediment in all sizes—rock flour, sand, gravel, cobbles, huge boulders. Debris is picked up, flattened down, pushed, or carried along on top of, beneath, beside, and in front of a glacier. This largely unsorted material, called *till,* accumulates in *moraines* of various types and sizes. Lakes dammed by moraines are typical of glaciated topography. Examples in the park are Bowman, Josephine, and Kintla Lakes.

Streams work over and sort till and other debris, redepositing the sediment as *glacial outwash.* In the bottom of glacial troughs, glacial outwash usually spreads over till that was brought down by the ice. Eventually the scoured bedrock downvalley from a retreating glacier may be buried by many feet of sediment. Sometimes streams back up, forming lakes held in by natural dams built up by a mixture of outwash and morainal material. Both St. Mary Lake and Lake McDonald, which lie in glacial troughs, were dammed in this way.

Erosion by sediment-laden streams draining off glaciers has developed large, deep *potholes* in some of the less resistant rocks in streambeds. Rock debris that collects in any slight depression or weak spot in bedrock or boulders acts as a drill, swirling around in the water and enlarging the pothole.

Avalanche Chutes

Visible on many of the steepest mountain slopes are straight, groovelike features, called *avalanche chutes,* that run hundreds of feet down the face of a slope. These are the tracks of avalanches that have been worn smooth by repeated slides of snow, ice, trees, and talus that roar down the slopes with incredible speed. In the spring of the year, when conditions cause snow accumulations to become unstable on steep slopes, an avalanche may be triggered by a sudden vibration or sharp noise. Snow, ice, and debris rushing down the mountainside—with a wall of air in front of the slide—can build up such force that trees are torn out by the roots. From Going-to-the-Sun Highway, you can see avalanche chutes on Mt. Cannon on the western side of the Continental Divide.

Glacier Recession

Grinnell and Sperry Glaciers, the two largest glaciers in the park, have shrunk considerably in the last several centuries. Grinnell Glacier, which is now about 275 acres, was twice as large when George Grinnell discovered it a hundred years ago. Precipitation was higher and temperatures were lower when these glaciers were at their maximum extent several thousand years ago, but weather data and measurements of the glaciers have not been collected long enough for current trends to be determined with any certainty. Moreover, glacial studies have shown that normal "lag time" is such that it may take years before changes in average annual precipitation and temperature affect the size and thickness of a glacier.

In Waterton-Glacier International Peace Park, location and elevation have a significant relation to snowfall, temperature, and glacial advance or retreat. For example, at Babb, a village outside the park's eastern boundary and south of the international boundary, average annual precipitation is only about 20 inches, due to the rain-shadow effect. But over Grinnell glacier, 15 miles to the west and about 1700 feet higher in elevation, yearly precipitation is estimated at about 150 inches.

Data collected from a stream gage on the Kootenai River at Libby, Montana, about 75 miles west of the park, shows a downward trend in stream discharge from 1920 to the mid-1940s and an upward trend thereafter. Since the Kootenai drains much of the park area, one might tentatively conclude that for the short term the trend of decreasing rainfall and rising temperatures has slowed. Measurements of glacier shrinkage and climate data collected in other areas have shown a similar pattern. However, long-term predictions of glacial advance or retreat due to climatic variations are extremely tentative.

The Belt Supergroup

Glaciers are such powerful agents of erosion that they modify landscapes and landforms much more rapidly and drastically than streams can. However, glaciers work on the pre-existing topography of a region. In other words, the valleys and divides have already been molded by streams, which, in turn, are mainly controlled by the underlying rock structures. The glaciated alpine topography in Waterton-Glacier International Peace Park has a unique "architectural" character that is not seen in the Rockies farther south. That is because the angled, banded slopes and summits of the mountains in Waterton-Glacier are made up of layered sedimentary and lightly metamorphosed rocks, rather than the granitic and intensely metamorphosed rocks displayed in Rocky Mountain National Park (chapter 22). What is surprising is that the dominant rocks in both parks are Precambrian in age.

The layered rocks in Waterton-Glacier are part of a thick series of sedimentary rocks called "the Belt rocks," or, more formally, the Belt Supergroup. (In Canada, the rocks are known as the Purcell Supergroup.) What makes these rocks so unusual is that they are relatively undeformed and only lightly metamorphosed despite the fact that they were deposited fairly late in Precambrian time, between 1600 million and 800 million years ago (Proterozoic Y).

The Belt rocks are spread over several states and Canadian provinces. In the park area, this great sedimentary wedge is over 18,000 feet thick, and probably totals 45,000 to 50,000 feet some miles to the west. The sands, silts, and carbonates that made up the rocks accumulated in a broad, subsiding basin, which was invaded by shallow seas that spread over a rift between continental landmasses. The rift was probably a spreading center that opened in a Precambrian continent. Eventually, due to deep burial, most of the sandstones, shales, and limestones that formed were lightly metamorphosed to quartzites, argillites, marbles, and dolomitic marbles.

Preserved on bedding surfaces in many of these rocks are clues to the ancient environments in which they formed. *Ripple marks* tell the direction of ancient waves and currents. *Mud cracks,* which formed when wet sediments dried out in the sun, show that that particular bed was above water level for a time. *Rain-drop impressions* indicate that exposed soft mud was pelted by rain from a cloudburst.

The remains of several species of *calcareous algae* record the presence of colonies of single-celled plants in the Precambrian seas. These blue-green algae, collectively known as *stromatolites* in their fossil form, thrived

Figure 23.4   Contorted sandstone beds of the Grinnell Formation near Otokomi Lake (elevation, 6560 feet) in Roes Basin. Photo by R. Smith.

in shallow water where they received abundant sunlight. By removing carbon dioxide from the seawater in which they lived, they released oxygen and secreted or deposited calcium carbonate around their single-celled bodies, forming colonies of various sizes and shapes. Some algae built roughly circular, bulbous masses, called "cabbage heads." Others (*Conophyton*) grew in columns of nested cones. The Proterozoic blue-green algae functioned in the same way as modern, living algae and are essentially the same kind of organisms (fig. 23.5).

Belt Rocks Exposed in Glacier National Park

A geologic guide, prepared by U.S. Geological Survey scientists (Raup et al. 1983), identifies eight formations within the Belt Supergroup that are keyed to stops along the cross-mountain, Going-to-the-Sun road (fig. 23.6). These formations, from oldest to youngest and from east toward the west, are Altyn, Appekunny, Grinnell, Empire, Helena (Siyeh), Snowslip, and Shepard. Prichard, the eighth formation and the oldest one on the west side of the park, lies below the Appekunny, where it is thrust up against younger formations. The Altyn is the oldest formation exposed on the east side of the park. Altyn and Prichard are equivalent in age, but were deposited in dif-

ferent marine environments. Because the ancient ocean bottom sloped to the west, Altyn was deposited in shallow water near shore, while Prichard consists of deep-water deposits. Shepard Formation, youngest of the Proterozoic units described, does not crop out along the highway, but can be seen on the highest peaks in the view from the Logan Pass Visitor Center.

On Going-to-the-Sun Highway, approaching from the east, you go "up-section" to the highest point at Logan Pass (elevation 6680 feet) and then you go "down-section" to the west entrance. This is the geologists' way of saying that the layers of rock get younger toward the top and older toward the bottom. Like a layer cake, the rock units are laterally continuous, so that you see the same formations going down as were exposed on the way up. The outcrops are not at the same elevation, however, because beneath the Continental Divide these Belt rocks are slightly deformed into a broad, open syncline.

Igneous Sill and Lavas

Visible in glacially undercut cliffs and steep slopes is a horizontal dark band that is conspicuous within the light-colored beds of the Helena (Siyeh) Formation. This band is a diorite sill, formed when very hot magma from deep

Figure 23.5   Algal colonies ("cabbage heads") in the Helena (Siyeh) Formation. Photo by R. Smith.

FORMATIONS:
9 Quaternary and Cretaceous
8 Shepard
7 Snowslip
6 Helena
5 Empire
4 Grinnell
3 Appekunny
2 Altyn
1 Prichard (?)

Normal fault—Arrows indicate
direction of movement.

Thrust fault—Arrow indicates
direction of movement of
upper plate.

Figure 23.6   Simplified, west-to-east cross section along the route of Going-to-the-Sun Highway. Here the Belt Supergroup formations have been deformed into an open syncline, and the entire mountain mass has been thrust upward and eastward many miles by the Lewis thrust fault. The compressive stresses also produced the reverse fault on the lower part of the western slope. Adapted from O. B. Raup et al. *Geology Along Going-to-the-Sun Road, Glacier National Park, Montana.* © 1983 Glacier Natural History Association. Used by permission.

in the crust was forcibly injected into Helena bedding planes near the top of the formation. Contact metamorphism "cooked" and bleached the limestone beds above and below the sill. The diorite along the sill margins is fine-grained due to rapid cooling, but the interior of the sill, which cooled slowly, has a fairly coarse texture. Because of their distinctive appearance, the sill and its enclosing beds are easily traced all over the park, and outcrops are marked on both the east slope and west slope of Going-to-the-Sun Highway. The age of the sill has not been determined, but similar sills in the Helena south of the park have yielded radiometric dates of 775 to 725 million years B.P. (Raup et al. 1983).

On several exposures in the park, *pillow structures* in basalt lava flows indicate that magma broke through the sedimentary beds and erupted as "pillow basalt" on the floor of the Precambrian sea. The "pillows," which form when lava flows into water or erupts from a vent beneath the water surface, are close-fitting and tend to form masses or piles. Because the lava cooled suddenly in the water, it is fine-grained and has bubble holes, called *vesicles*. At a later time in geologic history, when the rocks had been elevated above sea level, percolating ground water filled the holes with calcite and formed *amygdules* (similar to those in Isle Royale National Park, chapter 20).

At Granite Park, north of Logan Pass, exposures of pillow basalt are well displayed. Years ago, this dark igneous rock was misidentified as "granite," and the name stuck, although there is no granite in or close to Glacier National Park.

The Lewis Overthrust

The Northern Rockies in the Waterton-Glacier region are described as "mountains without roots" because they consist of ancient Precambrian rocks resting on top of much younger sedimentary beds. The reason for this is that a huge, low-angle thrust fault, known as the Lewis Overthrust, transported the sequence of Precambrian rocks eastward for many miles over the younger rocks. The first geologists to work in the area recognized the structure when they observed and mapped the ancient, tan-and-cream-colored beds of the Altyn Formation lying on top of dark-gray Cretaceous shales of the Montana Group. The trace of the fault surface, which separates the older beds on top from the younger beds beneath, can easily be seen on the face of Wynn Mountain and at other places in the park. The fault surface also crops out along the base of the vertical face of Appekunny Mountain.

Chief Mountain, a peak that stands alone astride the park's eastern boundary, is an erosional outlier, or *klippe* (i.e., a remnant) of the thrust fault, which was of much greater size and extent before being eroded. Making up the top third of Chief Mountain are the horizontal lime-

Figure 23.7   Chief Mountain, located on the eastern boundary of the park, is a klippe, formed by the erosion of a thrust sheet. The trace of the Lewis thrust fault, almost horizontal here, lies about at the top of the line of trees. Rocks of the Precambrian Belt Supergroup are above the fault; those below are Cretaceous in age. Drawing by Bailey Willis in *Geological Society of America Bulletin,* 1902, 13:334.

stone beds of the Upper Member of the Altyn Formation. The lower two-thirds of the mountain consist of the more dolomitic beds of Altyn's Lower Member, which have been slightly tilted and show minor thrusting. Underneath the mountain are the young Cretaceous beds separated from the Precambrian layers by the fault surface of the Lewis Overthrust, which is visible at the mountain's base (fig. 23.7).

How did this come about? The metasedimentary Precambrian beds must have been undisturbed for many millions of years when they began to be uplifted and upwarped by the intense compressive forces of the Laramide orogeny. The huge fault slab, several miles thick, from which the mountain ranges of Waterton-Glacier were carved out, broke loose and very slowly slid eastward some 40 to 50 miles. The relatively strong Precambrian rocks were able to move as a great thrust sheet, with only very slight deformation, over the soft Cretaceous shales.

Along the western boundary of Waterton-Glacier International Peace Park, the North Fork of the Flathead River meanders through a broad, north-south-trending valley, underlain by Cretaceous rocks. However, the Whitefish Range, which rises on the west side of North Fork Valley, is made up of Precambrian strata like those in Waterton-Glacier. The explanation seems to be that the thrust slab cracked or broke over North Fork Valley, like a slab of snow separating into two sections when sliding down a roof.

Erosion, first by streams and then by Pleistocene glaciers removed great quantities of rock from the thrust sheet. The fault trace on the west side of the park is still deeply buried, but in places on the east side (as around Chief Mountain), Mesozoic rocks beneath the fault surface are exposed.

Table 23.1 Geologic Column, Glacier National Park

Time Units			Rock Units		Geologic Events
Era	Period	Epoch	Group	Formation	
Cenozoic	Quaternary	Holocene	Alluvium Glacial deposits		Erosion and deposition by streams. Return of glaciers in Neoglacial time. Present glaciers between 6000- and 9000-foot levels on north and east slopes.
Cenozoic	Quaternary	Pleistocene	Glacial deposits		Valley and piedmont glaciation with several advances and retreats; intense glacial erosion
Cenozoic	Tertiary	/////////////			Major erosional interval
Mesozoic			Montana		Return of seas depositing Cretaceous beds; Laramide orogeny-folding, uplift, thrust faulting, vertical faulting.
Paleozoic	/////////////				Major erosional interval
Late Precambrian (Proterozoic Y)			Belt Supergroup	Kintla (Mt. Shields)	Red argillite, increasing aridity
				Shepard	Red to yellowish brown argillite, stromatolites
				Snowslip	Pale red to green argillite, pillow lava, ripple marks, mudrocks
				Helena (Siyeh)	Tan dolomite, diorite sill, oolite zone, stromatolites
				Empire	Buff to tan argillite and quartzite; copper
				Grinnell	Red to purplish red argillite and quartzite; ripple marks, mud cracks
				Appekunny	Gray to greenish argillite and quartzite; tidal mud flats
				Altyn (east)	Tan- to cream-colored limestone and dolomite, stromatolites, ripple marks (shallow water)
				Prichard (west)	Dark gray to black argillite (deep water)
				Waterton (Alberta)	Red brown to gray dolomite

Sources: Alt and Hyndman (1986), Harrison (1972), Raup et al (1983)

Geologic History

1. Deposition of the Belt Supergroup during late Precambrian time, Proterozoic Y.

The basin in which the Belt Supergroup rock units were deposited held a sea that oscillated over the ancient landmass for millions of years during the span of Precambrian time called Proterozoic Y (explained in chapter 19). The size, shape, and outline of the basin, plus the information derived from the sediment types and facies changes all indicate multiple source areas. The extent of the basin can be determined only approximately because the boundaries have been obscured by younger rocks and by deformation. Brief descriptions of the significant formations follow:

Waterton Formation of Canada. The dolomitic Waterton beds are exposed in Waterton Lakes National Park but not on the U.S. side of the border.

Altyn Formation. Sandy dolomite and limestone beds deposited in shallow water make up this formation, the oldest on the east side of Glacier National Park. Sedimentary features such as ripple marks, mud cracks, and stromatolites (cabbagelike fossil algae) can be easily recognized in many of the outcrops. Thin layers of quartz that filled in minute fractures have weathered out in sharp relief. Altyn rocks form the St. Mary Lake Narrows—the resistant ledge over which Swiftcurrent Falls tumble—and the rim upon which the Many Glacier Hotel is built.

Prichard Formation. This oldest rock unit on the west side of Glacier National Park consists of thin layers of black argillite and siltstone. The beds were deposited in deeper water than any other formations in the park. Cataracts and waterfalls along McDonald Creek above the Lake McDonald inlet flow over resistant ledges of the Prichard Formation. Large potholes have been carved out in the stream bed.

Appekunny Formation. The green mudstones of the Appekunny are darker on the west side than the east side because the beds on the west side were deposited in deeper water. Almost all of the green rocks that crop out in the park are Appekunny. Upstream from the Prichard Formation, McDonald Creek flows over green Appekunny beds. Aster Creek carved its gorge in them. Soil weathered from Appekunny rocks supports a heavy forest cover.

Grinnell Formation. Near the top of the Appekunny Formation, layers of red mudstone are interbedded with beds of green mudstone. The contact between Appekunny and the overlying Grinnell Formation is where red mudstone becomes dominant over the green. The presence of the mineral hematite (iron oxide) gives Grinnell outcrops their distinctive, bright red color. The beds were deposited on tidal mudflats. With repeated wetting and drying, mud cracks and ripple marks formed on the beds. A few striking beds of white quartzite in the Grinnell form prominent ledges between red mudstone layers.

Empire Formation. The argillites, siltstones, and carbonate beds in this rock unit are transitional between the rocks of the Grinnell Formation and those of the overlying Helena (Siyeh) Formation. The depositional environment was changing as the basin subsided and seawater became deeper. Some green carbonate rock layers of the Empire Formation contain "anomalous copper," that is, higher-than-normal concentrations of copper minerals, such as chalcopyrite, chalcocite, and bornite. Anomalous copper (at least 100 parts per million, and much higher in some places) has been found in nearly all parts of the Belt basin. Silver and mercury, also in anomalous amounts, are sometimes associated with the copper. Exploration geologists refer to such copper deposits as "stratabound" because the copper minerals tend to be concentrated along bedding planes and in sedimentary structural features (e.g., mud cracks). Some of the copper in the Belt basin was depos-

ited with the sediments (syngenetic), but probably some copper migrated or was reconcentrated subsequent to deposition (by diagenetic processes). Concentrations of ore grade copper are outside the park.

Helena (Siyeh) Formation. Many algal zones accumulated in the limestones and dolomites of the Helena (Siyeh) Formation. Fossiliferous beds on the Garden Wall and along Going-to-the-Sun Highway have excellent examples of large and small "cabbage heads" and stacked, inverted cones (*Conophyton*). The *Conophyton* colonies formed a zone 100 feet thick in the Helena that is exposed at many places throughout the park (fig. 23.5).

An oolite zone occupies transitional layers between the carbonates of the Helena and the argillites and siltstones of the overlying Snowslip Formation. The *oolites,* which are rocks composed of small, round objects (*ooliths*) resembling fish eggs, are similar to oolites in Everglades and Biscayne National Parks (chapters 17, 18) and probably formed in the same way.

The igneous sill, described earlier, that is sandwiched between limestone beds, near the top of the Helena, ranges in thickness from 10 to 200 feet. The sill and the bleached carbonate beds, altered to marble by contact metamorphism, are exposed in several roadcuts near the top of Logan Pass.

Snowslip Formation. The colorful red and green argillites, purplish siltstones, and pale yellow quartzites of the Snowslip Formation were deposited on tidal mudflats. Land areas along the shore had been elevated enough to increase the supply of sediment being washed by erosion into the sea. Again, one can find ripple marks and mud cracks on outcrops. Some multicolored stromatolite beds are exposed below The Loop beside Going-to-the-Sun road. Iron- and copper-bearing minerals in the argillites have colored the algal heads.

Shepard Formation. The red and tan rocks of the Shepard, which become more clastic toward the east, are a mixture of limestone and siltstone with some dolomite. Well-preserved fossil algae and undisturbed ripple marks and mud cracks suggest that no digging, crawling, or swimming organisms lived in the inland sea. No traces of organisms other than the calcareous algae have been found in these Precambrian rocks.

The climate became more arid, as shown in the increasing redness of the argillites. Oxidation occurring when the beds were exposed to air for longer intervals caused this. The Shepard beds blend into the red argillites of the Kintla (Mount Shields) Formation. Shepard and Kintla rocks form red caps on some of the highest peaks and passes in the northern and middle part of the park.

2. Moderate uplift, erosion of some of the Belt rocks, and probable deposition of Paleozoic rocks.

As the region of Waterton-Glacier International Peace Park was gradually uplifted, the seas withdrew. Stream systems developed and rapidly stripped the land of sediments that were soft, unconsolidated, and easily eroded. Geologists surmise that in time the seas returned to this area because mountains both to the north and south of the park are composed of Paleozoic beds deposited in shallow water. Paleozoic rocks are not exposed at the surface in the park. They probably were eroded during and after the later uplift associated with the Laramide orogeny.

3. Cretaceous deposition.

A gap in the geologic record for most of Mesozoic time suggests that this part of North America remained above sea level during that time. Then a shallow inland sea extended into the Northern Rockies region as the Cretaceous Period was drawing to a close. Because the Cretaceous beds were never deeply buried, they are not so strongly lithified as the Precambrian rocks and thus are more susceptible to weathering. Erosion of Cretaceous rocks east of the park has formed a gently rolling landscape in contrast to the rugged mountains and steep slopes eroded on the rocks of the Precambrian Belt Supergroup. The Cretaceous beds cropping out along the eastern side of the park contain oil, and those in North Fork Valley have natural oil seeps. Where the beds extend under the High Plains of Montana and Alberta, commercial oil fields are being worked. The site of the first oil well in Alberta is now a historical exhibit in Waterton Lakes National Park.

4. The Laramide orogeny and thrust faulting.

As the Cretaceous Period was ending, the upheavals that built the Northern Rockies tilted the flat-lying Belt rocks eastward. As the uplifting increased stream gradients, large stream valleys were eroded. Late in Cretaceous and early in Eocene time, tectonic compression caused massive thrust faulting. Deformation was especially severe in the Northern Rockies where Precambrian rocks in the huge slabs, moved by the Lewis thrust fault, overrode Cretaceous shales. North Fork Valley on the park's western margin formed where a fault slab split apart during its eastward sliding.

5. Cenozoic erosion and deposition; Miocene uplift.

The inland climate was desertlike and vegetation was sparse. Coarse sediments were dumped in valleys by slumps and flash floods. Erosion isolated sections of the overthrust sheet and created klippes, such as Chief Mountain, which is completely surrounded by younger beds.

A renewal of uplift in Miocene time brought the Northern Rockies to their present elevation. The climate became cooler and moister, thus setting the stage for the development of glaciers, which may have begun to form in the high ranges by Pliocene time.

6. Pleistocene glaciation.

In Glacier National Park, evidence indicates extensive glaciation during the middle Pleistocene, two glacial substages during the Wisconsinan glacial stage, and at least two episodes of ice accumulation and advance during post-Pleistocene time. During the most extensive advances, glacial erosion and deposition in the park was done by alpine glaciers that filled all the valleys and piedmont glaciers that coalesced along the eastern mountain front. On the west side, North Fork Valley was covered by a belt of moving ice 10 miles across. Continental ice sheets from Canada, which came up to the boundaries of the park on the northern and eastern sides, joined the piedmont glaciers but never quite overrode the high ranges. The high peaks in the park that stuck up above the ice show a line marking the ice level. Above the line, the cliffs are very rough; below the ice line, the bedrock surfaces are smooth due to the grinding and scouring of glacial ice.

In Glacier National Park (as throughout most of North America) the initial Pleistocene glacial stage was more severe than later glaciations. One indication of this can be seen on the crests of Two Medicine Ridge, Cut Bank Ridge, and Milk River Ridge, all of which have deeply weathered till, obviously older than the blankets of till laid down on the floors of the intervening valleys by later glacial stages. Evidence relating to the earlier glaciation also shows that two large piedmont glaciers—Two Medicine Glacier, that flowed east, and St. Mary Glacier, which flowed northeast—filled valleys with ice 500 to 1000 feet higher than occurred in succeeding glacial stages and also advanced farther eastward onto the plains of Montana. The location of moraines and glacial outwash show that at its maximum Two Medicine Glacier was 20 miles wide and extended some 10 miles out from the mouth of its valley.

7. Final retreat of Pleistocene glaciers about 10,000 years ago; evolving of the present landscape.

Near the end of the Pleistocene Epoch, the climate gradually became warmer. The result was that the park was denuded of glacial ice—except for a few small, stagnant glaciers and snowfields that persisted at high altitudes and on north-facing slopes. No major changes in the landscape have come about since the ice disappeared, but the effects of weathering and erosion continue. Small gorges cut by streams after ice retreat are Sunrise Gorge and the gorge at Hidden Falls. Both were formed as the result of enlargement of the weak zones in the bedrock joint system. The potholes in Avalanche Gorge have also developed over the past 10,000 years. Mass wasting is active under the prevalent freeze-and-thaw conditions, especially on steep slopes. Talus slopes are building up at the foot of cliffs and along the base of the mountain ranges. Avalanches, rockfalls, mudflows, and soil creep all continue to modify the landforms.

Since the melting back of the glaciers took place over a much shorter time than their growth and spread, enormous quantities of debris and sediment were released when the climate warmed. Streams were greatly overloaded and responded, as streams will, by dumping the excess in alluvial fans and outwash features, on top of till beds, moraines, and bedrock. The valleys holding Waterton Lake, St. Mary Lake, Lower Two Medicine Lake, and other lakes in the park remain clogged because the water is dammed by glacial debris.

8. Origin of glaciers that exist today.

The postglacial climatic optimum that brought about the disappearance of glaciers was a warm interval that peaked about 7000 years ago. Average temperatures were warmer and rainfall was ample. The warm interval was followed by a cooling trend, but several climatic fluctuations have occurred. A period cooler than now that prevailed about 5000 years ago was the cause of renewed glacial activity that resulted in the development of the alpine glaciers that presently exist in the park. That the climate will continue to fluctuate seems certain.

The park's active glaciers are found between elevations of 6000 and 9000 feet. The perennial snowfields lie in protected areas, usually on slopes that face north or east. Wind-drifted snow from over the peaks is a main source of their nourishment. Movement of glacial ice may range from 6 to 8 feet per year to as much as 12 to 50 feet per year, depending upon conditions of precipitation and temperature. These velocities are slow in comparison with other valley glaciers that are monitored, especially some of the Alaska glaciers.

Bibliography

Alden, W. C. 1932. Physiography and glacial geology of eastern Montana and adjacent areas. U.S. Geological Survey Professional Paper 174.

Alt, D. D. 1983. Glaciated park: geology, in *Glacier country, Montana's Glacier National Park*, no. 4, Montana Geographic Series, Montana Magazine, Inc. p. 6–25.

Alt, D. D., and Hyndman, D. W. 1973. *Rocks, ice, and water: the geology of Waterton-Glacier Park*. Missoula, Montana: Mountain Press Publishing Co. 104 p.

———. 1986. *Roadside geology of Montana*. Missoula, Montana: Mountain Press Publishing Co.

Beaumont, G. 1978. *Many-storied mountains: The life of Glacier National Park*. Natural History series, National Park Service. 138 p.

Boyer, D. S. 1987. Pride of two nations, Waterton-Glacier International Peace Park. *National Geographic* 171(6):796–823

Chronic, H. 1984. *Pages of Stone: Geology of western national parks and monuments*. I. Rocky Mountains and western Great Plains. Seattle: Mountaineers. p. 75–88.

Harrison, J. E. 1972. Precambrian Belt basin of northwestern United States: its geometry, sedimentation, and copper occurrences. *Geological Society of America Bulletin* 83: 1215–1240.

Johnson, A. 1980. Grinnell and Sperry glaciers, Glacier National Park, Montana—A record of vanishing ice. U.S. Geological Survey Professional Paper 1180, 29 p.

Mudge, M. R. and Earhart, R. L. 1980. The Lewis thrust fault and related structures in the disturbed belt, northwestern Montana. U.S. Geological Survey Professional Paper 1174, 18 p.

Raup, O. B., Earhart, R. L., Whipple, J. W., and Carrara, P. E. 1983. *Geology Along Going-to-the-Sun Road, Glacier National Park, Montana*. Prepared by U.S. Geological Survey and National Park Service; published by Glacier Natural History Association. 62 p.

Roberts, A. E. 1972. Cretaceous and early Tertiary depositional and tectonic history of the Livingston area, southwestern Montana. U.S. Geological Survey Professional Paper 526-C.

Ross, C. P. 1959. Geology of Glacier National Park and the Flathead region, northwestern Montana. U.S. Geological Survey Professional Paper 296.

———. 1962. The Precambrian of the northwestern United States—the Belt Series. In *The Precambrian*, K. Rankama, editor. 4:145–251. New York: Interscience.

Willis, B. 1902. Stratigraphy and structure, Lewis and Livingston Ranges, Montana. *Geological Society of America Bulletin* 13:305–352.

Glacier National Park

Area (U.S.): 1,013,595 acres; 1584 square miles
Established: May 11, 1910
Proclaimed part of Waterton-Glacier International Peace Park: June 30, 1932
Designated a Biosphere Reserve 1976
Address: West Glacier, Montana 59936

Waterton Lakes National Park, established in 1895, has also been designated a Biosphere Reserve by the United Nations.
Park Headquarters and Visitor Center: Waterton, Alberta, Canada, TOK 2MO

24

A wilderness park entirely north of the Arctic Circle, four times the size of Yellowstone National Park.

Straddles the Arctic Divide and the Central Brooks Range.

Jagged peaks, glaciers, broad glaciated valleys.

Forested southern slopes and treeless northern reaches.

Habitat of grizzlies, black bears, wolves, Dall sheep, caribou, moose, wolverines, mosquitoes.

Six streams designated components of the National Wild Rivers System.

A complicated tectonic history partly revealed in complex mountain structures and rocks.

Gates of the Arctic National Park and Preserve

Northern Alaska

by Sherwood D. Tuttle
University of Iowa

Figure 24.1   Walker Lake, which drains the highest part of the Brooks Range in Gates of the Arctic National Park and Preserve, occupies a large, moraine-dammed glacial valley. National Park Service air photo by M. W. Williams.

Figure 24.2 Gates of the Arctic National Park and Preserve, Alaska.

Geographic Setting

Gates of the Arctic National Park and Preserve is a vast wilderness extending east to west for some 200 miles along the crest and slopes of the central and highest part of the Brooks Range in Arctic Alaska. The Brooks Range, named for Alfred H. Brooks, a U.S. Geological Survey geologist who explored Alaska early in this century, stretches like a great barricade from the Canadian border on the east to almost the Chukchi Sea on the west.

From south to north, Gates of the Arctic National Park and Preserve ascends from long, broad, glaciated valleys and rounded foothills to the sharp *arêtes* (narrow, rock ridges) and turrets along the crest of the range, then it goes down the steeper northern slopes to the edge of the Arctic coastal plain. Robert Marshall of the U.S. Forest Service extensively explored the region in the 1920s and 1930s. One day in 1929, as he was following up the North Fork of the Koyukuk Wild River, he saw against the sky two great facing peaks framing the valley he was ascending. He named this magnificent landscape "the Gates of the Arctic." The two peaks he christened Boreal Mountain, the "east portal," and Frigid Crags, the "west portal." Marshall's work and his writings about the region were instrumental in the later establishment of the park/preserve.

When the North Fork of the Koyukuk was placed in the National Wild Rivers System, the Tinayguk, its largest tributary, was also designated. Both rivers drain the south side of the Arctic Divide and the eastern part of the Endicott Mountains. The Endicott Mountains, which are part of the Brooks Range, extend from east to west over most of the park/preserve. The Schwatka Mountains, which have the highest summits in the Brooks Range, occupy the southwest corner of the park/preserve.

The John Wild River, a major stream next west from the Tinayguk and the North Fork of the Koyukuk, rises in the vicinity of Anaktuvuk Pass on the crest of the range and also flows south. The river valley and the pass are a migration route for great herds of caribou that move each spring from the shelter of the forested valleys over the range and down to their summer pastures on the tundra of the North Slope. The Nunamuit Eskimo people, who have hunted caribou in the high valleys for many centuries, live in Anaktuvuk Pass, which is the only year-round settlement in the park/preserve.

Farther to the west, the Alatna Wild River, one of the best float streams in Alaska, bends through rugged country and then flows south along the eastern side of the lofty Arrigetch Peaks. Exposed in this river corridor are some of the complex geologic structures and metamorphic assemblages that challenge field geologists who have worked in the area. South of the park/preserve, the Alatna and the John merge with the Koyukuk, which, in turn, flows into the Yukon. The last two Wild Rivers, the Noatak and the Kobuk, drain the glaciers in the Schwatka Mountains and Arrigetch Peaks. Both rivers flow west and empty into Kotzebue Sound (Chukchi Sea); but the Noatak flows out through Noatak National Preserve on the north side of the Baird Mountains, while the Kobuk passes through Kobuk Valley National Park (chapter 11), south of the Baird Mountains. These three national parklands, Gates of the Arctic National Park and Preserve, Noatak National Preserve, and Kobuk Valley National Park, which encompass most of the Brooks Range, protect this country's last great pristine wilderness.

In addition to the six Wild Rivers, Gates of the Arctic includes two National Natural Landmarks, Arrigetch Peaks and Walker Lake, with additional sites under consideration. The glacially carved Arrigetch, a spectacular mountain area, was given its name by the Eskimos. The English equivalent is "fingers of the hand extended." The sharp, upthrust peaks, knife-edged arêtes, and undercut valleys of the Arrigetch have been eroded from a dark, massive, granitic intrusion—in contrast to the predominantly marine sedimentary rocks exposed in most of the Brooks Range. Mt. Igikpak (8510 feet), highest peak in the Brooks Range, is some 20 miles (air distance) west of the Arrigetch. A granitic pluton, similar to that of Arrigetch, is exposed on Mt. Igikpak. Walker Lake, lying below the Arrigetch and slightly to the southwest, is a long, moraine-dammed lake occupying a huge glacial trough in the headwaters area of the Kobuk Wild River (fig. 24.1).

Also rising in this high headwaters area of marshes and glacial lakes between Igikpak and Arrigetch, but draining to the north, is Noatak Wild River. Anthropologists believe that early hunters used this route to travel back and forth between the Noatak basin and the Kobuk Valley (chapter 11) and that they set up winter camps in the deep, sheltered, glacial valleys.

Present-day Climate. Winters are long and cold in Gates of the Arctic; summers are short and mild. On the southern slopes and in the lowlands, midsummer temperatures occasionally rise to 80° F during the almost continuous sunlight; lows are near 50° F. On the northern slopes, summer temperatures range between 70° and 40° F. August is apt to be rainy. June and July are the worst times for gnats and mosquitoes. Freezing temperatures may occur at any time in summer, but are most likely after mid-August. Snow falls for eight or nine months of the year, with accumulations averaging 60 to 80 inches. January temperatures range from average highs of −3° to lows of −30° F. Ice in the Koyukuk River in the vicinity of Bettles, a town south of the park/preserve, usually breaks up around May 6 and freezes up around October 21.

The ecology and environment of Gates of the Arctic are fragile because of the arctic and subarctic climate of the region. *Permafrost,* the permanently frozen ground of

Figure 24.3   The 83-mile-long Alatna River, a component of the Wild and Scenic Rivers System, lies wholly within Gates of the Arctic National Park and Preserve. The stream winds south from the high mountains of the Central Brooks Range and joins the Koyukuk River. National Park Service air photo by M. W. Williams.

the far north, is at or near the surface throughout the park/preserve. Most of the permafrost is "dry"; that is, the frozen ground holds little moisture below what collects on the surface. The soil is very thin. Under these conditions, plants are slow-growing and easily damaged, especially north of the treeline, despite the apparent lushness of the arctic spring. The northern limit of wooded country (the boundary of the boreal forest zone) passes well below the crest of the Brooks Range along the southern slopes. Above the treeline are barren rocks, varicolored and stark, like those in arid regions of southwestern United States. Because of low annual precipitation, only the higher summits have year-round snow and only the highest areas have glaciers. The northern slopes and the eastern part of the park/preserve are especially deficient in moisture. Average pre-

cipitation in the Mt. Doonerak area, near the park's eastern boundary, is only 8 to 12 inches annually, while Mt. Igikpak and the southwest corner of Gates of the Arctic receive from 12 to 18 inches per year. In the polar desert on the north side of the Brooks Range, only low-growing tundra vegetation survives as stunted shrubs, lichens, mosses, and herbs.

Visiting the Park/Preserve. Access to Gates of the Arctic is by scheduled flights to Fairbanks (200 miles southeast of the park/preserve) and to Bettles, the closest town. Arrangements for charter aircraft and gear can be made at Bettles. No roads, trails, or campgrounds are maintained in the park/preserve, but visitors may backpack or canoe into the wilderness. (The Trans-Alaska pipeline highway that parallels the park's eastern boundary

Figure 24.4   Ice-sculptured Mount Igikpak (8510 feet), highest peak in the Brooks Range. A valley glacier (left foreground) occupies a cirque below sharp arêtes that lead up to Igikpak's turreted horn. Note that the lineated pattern of the metasedimentary rocks strongly influences weathering and mass wasting. National Park Service air photo by R. Belous.

is open during the summer as far north as Dietrich camp.) The long glacial valleys; alpine lakes; float trips on the wild, clear-water rivers; and fishing (permitted in the preserve areas of Gates of the Arctic) are attractions for visitors, plus the experience of being in a wilderness where many of the scenic features are still unnamed.

Local History

The early history of the Gates of the Arctic region is similar to Kobuk Valley National Park (chapter 11). The animals that migrated from Asia to this continent more than 12,000 years ago (when lowered sea levels exposed the Bering land bridge), and the human hunters who followed, spread over what is now northern Alaska and Canada. The numerous archeological sites that have been found within the boundaries of Gates of the Arctic National Park and Preserve mark the routes of the early people and give us clues as to their way of life. Most were nomadic hunters and gatherers. Artifacts of successive

Indian and Eskimo cultures have been found at Anaktuvuk Pass, along the North Fork of the Koyukuk Wild River, among the lakes of the headwaters region between the Noatak and the Kobuk Wild Rivers, and in other areas. The Nunamiut Eskimos and Athabascan Indians who now live in the park/preserve retain ways of life similar to their ancestors. Under the terms of legislation creating the Alaska National Interest Lands (the d-2 lands bill), native people may continue to hunt, fish, and gather berries in the park/preserve for their own subsistence.

Within historic time, the first explorers were American military officers who followed major rivers from the coast into the Arctic interior during the 1880s. Besides tracing the streams back to their sources, they scouted the country for possible mineral locations. One of these intrepid officers, S.B. McLenegan, a navy engineer, set out in a canoe in the summer of 1885, accompanied only by a seaman, to find the headwaters of the Noatak. He recorded in a journal their adventures, hardships, and dangers on the way to and back from what is now the western part of Gates of the Arctic National Park and Preserve. At one point, he wrote (after an "icy bath" when the canoe capsized): ". . . the difficulties encountered only seemed to awaken the stubborn elements of our natures, and with a determination not to be baffled, we prepared ourselves to meet anything short of utter annihilation." (*Alaska National Parklands,* p. 104.)

In the 1890s the search for gold lured prospectors and adventurers to the rivers of the far north, and in 1893 gold was discovered in the Middle Fork of the Koyukuk, east of the present park/preserve boundary. However, Alaska's main gold rush began in 1898 after gold in significant amounts was found in the Klondike. Mining camps sprang up at Bettles and along the Middle Fork of the Koyukuk. Some prospectors tried to find gold along the Kobuk and the Noatak Rivers and in the headwaters area between the two drainage systems. Their high, but unfulfilled, hopes are shown by names such as Midas Creek and Lucky Six Creek and by the ruins of a few cabins in the park/preserve.

During these years, the U.S. Geological Survey began sending expeditions into the region to do reconnaissance field work and mapping. Geologic investigations continue to this day, but a number of scientific questions about the area are unresolved.

Geologic Features

Effects of Glaciation

Erosion and deposition by a large system of Pleistocene valley glaciers produced most of the landscape features of the Brooks Range in Gates of the Arctic National Park. Horns, cols, cirques, cirque lakes, and other alpine landforms are characteristic of the higher parts of the range. Below the summits, the glacial troughs deepen and widen as they descend; among the foothills, tongues of glacial outwash extend in valley trains or spread over lowlands. Till and stratified drift make up moraine ridges and kame terraces. Some of the sediment is reworked and deposited as alluvial fans and lake fill. Many of these glacial deposits and valley fills have been prospected as possible sources of placer gold, but "shows" of significant size have not been found.

Most of the glaciers have disappeared because the lighter snowfalls of the present drier climate cannot support large active glaciers, like those in the national parks of southern Alaska. The few active glaciers that remain in Gates of the Arctic National Park and Preserve do not produce enough pulverized rock, or *rock flour,* to make the streams cloudy or "milky" even during the melt season. Because most of the water comes from rain and clean snow and runs down over permanently frozen ground, the arctic rivers are remarkably clear.

The Rocks

Many different kinds of rocks are exposed in outcrops in Gates of the Arctic. Some are carbonates and fine-grained clastics that accumulated in ancient marine shelf or deltaic environments. Some are volcanic, mostly basaltic, and others are plutonic and granitic. Most of the rocks have undergone moderate regional metamorphism, but some have been altered by local or contact metamorphism. Sedimentary rocks that indicate their tectonic origins include conglomerates, flysch sequences, and graywackes. (A *flysch* is a thinly bedded, rapidly deposited marine sedimentary facies consisting of different sizes and types of sediments.) Original thicknesses of the sedimentary sequences exceeded thousands of feet but are difficult to determine because great thrust sheets and nappes have shortened, overridden, and compressed the beds.

Fossil correlations have yielded ages of sedimentary rocks ranging from Cambrian through Cretaceous. Radiogenic dates derived chiefly from plutonic rocks indicate Precambrian through early Tertiary igneous activity.

Geologic Structures

Structures of all sizes, from minute foliation to great regional thrust sheets, suggest that for a considerable span of geologic time a converging zone was active in this northern region between the North American continental plate and one or more oceanic plates. Intense folding occurred as plates collided. East–west trending structures were overturned to the north, and reverse faults were upthrust from south toward the north. Thrust faults moved rocks many miles. So intense was this thrusting that nappes (thrust sheets) lie one over another, stacked like pancakes.

Oceanic crust was overridden and mangled as some of the marine sedimentary rocks were scraped off. Then the plate dipped down, carrying the remaining sediments as it sank. Under heat and pressure at depth, the mixture began to melt, allowing globs of molten rock to work up toward the surface. This process produced the plutons, stocks, dikes, and sills in the Brooks Range, along with some volcanic rocks. Meanwhile, large-scale regional metamorphism was altering most of the rocks undergoing compression at various depths.

As tectonism continued, the mountains were uplifted, causing erosional processes to intensify. Nonmarine sedimentary rocks accumulated in mountain basins. Complicating the picture are the blocks of *exotic terranes* (*tectonostratigraphic terranes*) that were "rafted" to the northwestern continental margin, presumably from south of the equator. These pieces of continental crust ("minicontinents") that rode on an oceanic plate like floating islands were involved with the younger nonmarine rocks during later compression, faulting, igneous activity, and uplift that produced the Brooks Range (fig. 24.5).

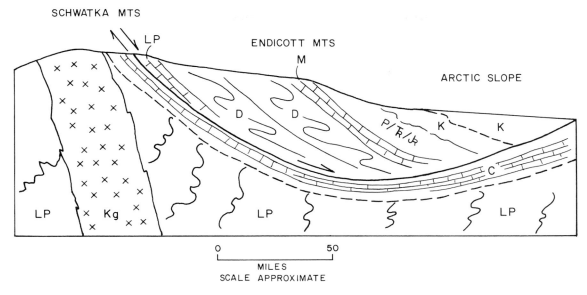

Figure 24.5   Generalized cross section through the Brooks Range, from southwest to northeast, showing geologic relationships. Key: LP=Lower Paleozoic; D=Devonian; M=Mississipian; C=Mississippian/Pennsylvanian; P/TR/JR=Permian/Lower Mesozoic; K=Upper Cretaceous; Kg=Cretaceous granite. The rocks above the thrust fault are mostly exotic (Endicott Terrane); northward-moving subduction not shown. Upper Cretaceous sediments (right) deposited on nappe during and after thrusting. The intrusive mass and folding are probably younger than the thrust faulting. Based on Mull et al., 1975; Connor and O'Haire, 1988.

Geologic History

Because of the remoteness of the area and the complexity of the rocks and structures, the geologic history of the Brooks Range and Gates of the Arctic National Park and Preserve is still being pieced together. In general, however, the patterns and sequence are similar to those of other complex mountains discussed in this section and in Part V.

1. **Deposition of island arc and coastal plain sediments during Precambrian (?) to late Paleozoic time.**

    Conformable, partially metamorphosed, sedimentary rocks at the bottom of exposed Paleozoic sequences in the "Doonerak anticlinorium" (a very large composite anticlinal structure, regional in extent) in the eastern part of the Brooks Range may be Precambrian or early Paleozoic in age. Their age is inferred from the fact that rock units above these beds have yielded fossils (trilobite fragments and a few brachiopods) correlated as middle Cambrian in age. Other Paleozoic rocks include the Kanayuk Conglomerate (about 7500 to 8000 feet thick and part of the Endicott Group), which is one of the most extensive fluvial units in North America. The Kanayuk Conglomerate, made up of river flood-plain deposits, Late Devonian and Early Mississippian in age, crops out over an area about 30 miles north/south and 450 miles east/west (Moore and Nilsen 1984). Geologists interpret the formation as being

part of a large, compound delta, consisting mainly of coarse sediments washed down from the North American landmass. This delta prograded (built out its front) southwestward late in Devonian time but subsequently retreated, probably because the sea was advancing, early in Mississippian time.

Deformed coastal-shelf and deltaic rocks related to the North American plate have been cut by intrusive bodies dated as 470 and 350 million years old. Therefore, we can assume that compression and some subduction occurred early in Paleozoic time.

In the heart of the Endicott Mountains, rocks of the Lisburne Group include limestones and marbles of Mississippian and Pennsylvanian age. The Kanayuk Conglomerate and the Lisburne Limestone have been used in ongoing paleomagnetic studies.

2. **Sedimentation, intrusions, and thrusting in late Paleozoic and early Mesozoic time.**

    In this interval, more rocks were deposited locally, while larger and more complex geologic structures developed in the Brooks Range. Granitic plutons were emplaced under Mt. Igikpak and the Arrigetch Peaks in middle Devonian time. The late Paleozoic Nuka Formation was thrust north toward what is now the Arctic Coastal Plain.

Table 24.1 Geologic Column, Gates of the Arctic National Park and Preserve

Time Units Era/Period/Epoch			Rock Units	Geologic Events
Cenozoic	Quaternary	Holocene	Glacial drift Alluvium	Neoglacial episode Periglacial activity
		Pleistocene	Talus Glacial drift	Four glacial stages
	Tertiary			Continuing uplift Intense erosion
Mesozoic			Nonmarine rocks	Uplift, accretion of terranes, compression, thrusting, subduction, plutonism
Paleozoic				Compression, thrusting, subduction, plutonism, metamorphism
			Nuka Formation Lisburne Group Kanayuk Conglomerate Marine sedimentary rocks of various ages Cambrian sedimentary rocks	Deltaic sedimentation, metamorphism, folding, plutonism
Precambrian		?	Plutonic rocks	Igneous activity

Sources: Dutro et al., 1981; Hamilton et al., 1986; Mull et al., 1982

3. Major orogenic activity from late Mesozoic to early Tertiary time.

Geologic activity speeded up, becoming even more spectacular toward the end of the Mesozoic Era. Events that took place in the Brooks Range are related to tectonic activity occurring more or less contemporaneously to the south in the Canadian and U.S. Rocky Mountains. Here in the north, several continental, oceanic, and island-arc crustal blocks reached the margin of the Arctic-Alaska region and were accreted. These tectonostratigraphic units probably rotated counterclockwise in varying amounts as the Arctic-Alaska block moved southward.

Early in Cretaceous time, the Lisburne Group was pushed some 50 miles northward. Granitic intrusions, dated as Late Cretaceous, uplifted the Schwatka Mountains. Widespread thrusting resulted in crustal shortening on a grand scale as rock units were carried from the southern foothills to the northern foothills. One field geologist has described five major thrust sheets, or nappes, made up of many thousands of feet of metasedimentary rocks, with lower Paleozoic units now on top, upper Paleozoic beds in the middle, and upper Paleozoic and Cretaceous strata at the bottom! Each sequence has been telescoped northward in relation to the underlying sequence (Mull 1979). Intermixed with the thrust sheets are rocks from exotic terranes and from the North American plate.

Subduction with accompanying plutonism (dated at 92 to 86 m.y. by the potassium-argon method) and volcanism was ongoing. At first, subduction was apparently due to north-moving rocks dipping downward, but this activity was succeeded by the plunging of south-moving rocks under the already complexly distorted beds. During this time, injection of mineral-bearing solutions was widespread. Although traces of gold, lead, zinc, copper, antimony, and tungsten have been found, their exploitation has not been economically profitable. Of the several tectonostratigraphic terranes that were "sutured" to the Brooks Range during these orogenic spasms, some apparently came from the North American continent (perhaps from western Canada or from the "Lower Forty-eight") while others have been traced to possible source areas in the Southern Hemisphere, perhaps even the southwestern Pacific. The marbles, schists, and quartzites in the Brooks Range were regionally metamorphosed by compression, folding, and thrusting during the mountain-building activities of this time.

4. Tertiary uplift and erosion.

Isostatic uplift that followed the orogenic activities elevated the Brooks Range still more and accelerated the processes of weathering and erosion. Sheets and wedges of sediment were shed from the mountains, accumulating on the Arctic Coastal Plain and spreading over the central Alaska lowlands. Many of these deposits have been buried by younger glacial and fluvial sediments.

5. Glaciation in late Tertiary and Quaternary time.

The processes of glacial and fluvial erosion and deposition dominate the more recent geologic history of the Gates of the Arctic region. In these northern latitudes frost action is intense, especially at the moderately high elevations of the Brooks Range. Frost-shattered rocks move downslope, breaking up more and more as they fall, roll, and slide. Eventually, streams haul the rock debris away.

The Brooks Range was thoroughly glaciated in the Pleistocene Epoch by at least four episodes of glaciation—two in late Pleistocene, one in middle Pleistocene, and one in early Pleistocene (Hamilton 1977). These glaciations produced the alpine topography of the summits, the large troughs at mid-elevations, and the innumerable glaciofluvial deposits of the lowlands and foothills. The small glaciers on the higher mountains are remnants of Neoglacial ice that has retreated during the last 200 years.

Bibliography

Brown, W. E. et al. 1982. Far North Parklands, in *Alaska National Parklands, this last treasure.* Anchorage: Alaska Natural History Association. p. 89–120.

Churkin, M. Jr., Nokleberg, W. J., and Huie, C. 1978. Tectonic model for the western Brooks Range, Alaska. *U.S. Geological Survey in Alaska: Accomplishments during 1978.* Circular 804B, p. B22–B25.

Detterman, R. L. and Miller, J. W. 1983. Kaguyak Formation, an Upper Cretaceous flysch deposit. U.S. Geological Survey Circular 945, p. 49–51.

Dutro, J. T. Jr., Palmer, A. R., Repetski, J. E. Jr., and Brosge, W. P. 1981. The Doonerak anticlinorium revisited. *U.S. Geological Survey in Alaska: Accomplishments during 1981.* Circular 868, p. 17–19.

Grybeck, D., and Nokleberg, W. J. 1978. Metallogeny of the Brooks Range, Alaska. *U.S. Geological Survey in Alaska: Accomplishments during 1978.* Circular 804B, p. B19–B21.

————, Cathrall, J. B., LeCompte, J. R., and Cody, J. W. 1983. Buried felsic plutons in Upper Devonian redbeds, central Brooks Range. *U.S. Geological Survey in Alaska: Accomplishments during 1983.* Circular 945, p. 8–10.

Hamilton, T. D. 1977. Late Cenozoic stratigraphy of the south central Brooks Range. *U.S. Geological Survey in Alaska: Accomplishments during 1977.* Circular 772B, p. B36–B38.

————, Reed, K. M., and Thorson, R. M. 1986. *Glaciation in Alaska, the geologic record.* Anchorage: Alaska Geological Society. 265 p.

Hillhouse, J. W., and Gromme, S. 1980. Cretaceous overprint revealed by paleomagnetic study in northern Brooks Range. *U.S. Geological Survey in Alaska: Accomplishments during 1980.* Circular 844, p. 43–46.

Jones, D. L., Silberling, N. J., Berg, H. C., and Plafker, G. 1980. Tectonostratigraphic terrane map of Alaska. *U.S. Geological Survey in Alaska: Accomplishments during 1980.* Circular 844, p. 1–5.

Moore, T. E., and Nilsen, T. H. 1984. Sedimentology of meandering-stream cycles, upper part of the Ear Peak Member of the Kanayuk Conglomerate, central Brooks Range. *U.S. Geological Survey in Alaska: Accomplishments during 1984.* Circular 967, p. 15–21.

Mull, C. G. 1976. Apparent south vergent folding and possible nappes in Schwatka Mountains. *U.S. Geological Survey in Alaska: Accomplishments during 1976.* Circular 751B. p. B29–B30.

———— 1979. Nanushuk Group deposition and late Mesozoic structural evolution of the central and western Brooks Range and Arctic Slope. *U.S. Geological Survey in Alaska: Accomplishments during 1979.* Circular 794, p. 5–14.

————, Tailleur, I. L., Mayfield, C. F., and Pessel, E. H. 1975. New structural and stratigraphic interpretations, central and western Brooks Range and Arctic Slope. *U.S. Geological Survey in Alaska: Accomplishments during 1975.* Circular 733, p. 24–26.

————, ————, ————, Ellersieck, I, and Curtis, S. M. 1982. New late Paleozoic and early Mesozoic stratigraphic units, central and western Brooks Range, Alaska. *American Association of Petroleum Geologists Bulletin* 66:348–362.

National Park Service. 1983. Environmental overview and analysis of mining effects, Gates of the Arctic National Park and Preserve.

———— 1985. Draft: General management plan, environmental assessment, land protection plan, wilderness suitability review.

Gates of the Arctic National Park and Preserve
Area: 9,164,000 acres; 14,819 square miles
Proclaimed a National Monument: December 1, 1978
Established as a National Park and Preserve:
    December 2, 1980
Wilderness designated: December 2, 1980
Address: P.O. Box 74680, Fairbanks, Alaska 99707

25

Yosemite National Park includes 1189 square miles of the central Sierra Nevada from deep glacial valleys on the west to the crest of the Sierra on the east. The park has magnificent scenery and the highest and most spectacular waterfalls in the United States.

Batholithic intrusions that make up the greatest part of the bedrock in the park represent the root zone of a volcanic-plutonic complex that formed in Mesozoic time as a function of subduction on North America's western margin.

The history of uplift and erosion of the range has been deciphered from studies of the batholith unroofing, volcanic activity, and the area's geomorphology.

A series of glacial advances are recorded by sculpted topography and thick sections of glacial tills and moraines.

Yosemite National Park

East Central California

by Donald F. Palmer
Kent State University

Figure 25.1   Yosemite Falls on the north wall of Yosemite Valley descends 2425 feet in three steps to the flat-floored valley below. The upper fall pictured here is 1430 feet high. The lower fall is 320 feet and the intermediate cascades between the main falls are 675 feet. The cliffs to the right of the upper falls show a series of spires and pinnacles characteristic of the closely jointed sections of the El Capitan Granite and the Sentinal Granodiorite. U.S. Geological Survey photo by R. E. Wallace.

N

Excelsior Mt.
12,240'

Moraine Ridge

Jack Main Canyon

HETCH
HETCHY
RESERVOIR

120
Tioga
Pass
Entrance

GRAND CANYON OF THE
TUOLUMNE RIVER

TUOLUMNE   RIVER

Tuolumne
Falls

Lembert
Dome

Tioga   Road

Mt. Hoffman
10,836'

Mammoth Peak
12,117'

Big Oak
Flat
Entrance

Lake
Tenaya

120

Lost
Lakes

Tuolumne
Grove

Big

Oak

Merced Grove

Flat

Road

Moraine Dome
8,012'

Mt. Florence
12,561'

See Inset

Merced Lake

Big
Meadow

Turtleback
Dome

Glacier Point Rd.

140

Red Peak
11,699'

Arch
Rock
Entrance

Mariposa Grove
of Giant Sequoias

41

South
Entrance

Park Boundary

0   5 mi.

YOSEMITE VALLEY

Watkins
Pinnacles

North Dome

Washington Column

Upper
Yosemite Falls

Royal Arches

Eagle Peak

Three Brothers

Mirror
Lake

Split Pinnacle

Half Dome

Ribbon Falls

El
Capitan

MERCED

Lower Yosemite
Falls

Mt. Broderick
Liberty Cap

"Map
of N. America"

Sentinel
Rock

Vernal
Falls

RIVER

Sentinel Dome

Illilouette Falls

Nevada Falls

Bridalveil Falls

Cathedral Spires

0   5 mi.

Tynal

Figure 25.2   Map of Yosemite National Park, California

Local History

Yosemite National Park is located approximately in the center of the Sierra Nevada, the largest continuous mountain range in the United States. Early settlers and travelers found the range to be an impediment to communication and progress, so the region was of little interest until the discovery of gold in the foothills of the Sierra Nevada in 1848. The resulting gold rush brought miners and traders into direct competition with the Miwok and Paiute Indian tribes that had lived for many decades in the region between the foothills and the valleys east of the Sierra Nevada.

It was this confrontation that led to the opening of Yosemite Valley and the high Sierra to tourists and nature lovers. In 1851, the Mariposa Battalion, under the direction of Major James Savage, entered the valley from the western end in pursuit of a group of perhaps 200 Indians of the Ahwaneechee tribe led by their chief, Tenaya. Major Savage's plan was to force the Indians, who had been responsible for a number of raids against trading posts in the area, out of their stronghold and to convince them to give themselves up to the Indian commissioners. While the majority of the battalion thought primarily of the immediate military objectives, the company physician, Dr. Lafayette Bunnell, was nearly immobilized by the grandeur of the valley before him. It was Bunnell who suggested that the valley be named Yosemite, an anglicization of *Uzumati,* the name of the Indian tribe, meaning "grizzly bear." Bunnell, whose impressions were later recorded in his book *The Discovery of the Yosemite,* also named many of the major topographic features and lakes of the area. His enthusiasm for the scenery was infectious among the company and their reports on their return to the foothill towns interested many in the region. By 1855, the area had become tamed enough so that James Hutchings, of San Francisco, could visit the valley with three others, including the artist Thomas Ayres. Ayres had been sketching landscapes since he had arrived in northern California in 1849, and Hutchings felt that the artist's talents would be well employed in bringing to the public the first accurate pictures of the wonders of Yosemite Valley. Ayres' drawing of the Yosemite Falls was published in the fall of 1855, and in July 1856, the first issue of *Hutchings' California Magazine* presented four drawings of Yosemite by Ayres.

Ayres returned to the valley in 1856 and visited the high country near Tuolumne Meadows. His sketches and personal accounts were published throughout the country and a show of his drawings was given in New York City. Other landscape artists, notably Albert Bierstadt, Thomas Hill, and William Keith, were to follow in the next decade. In 1859, Hutchings brought to Yosemite its first photographer, Charles Leander Weed, whose photographs were shown in San Francisco in September of that year.

Weed was followed soon by the great photographer Carleton Watkins who had the technical mastery and the artistic sensibility to make Yosemite come to life for those who had not seen it. It is impossible to overestimate the importance of the early artists in establishing Yosemite as the first wilderness park. Their work illustrated a world that could only be imagined by the general populace through reading the accounts of writers and visitors.

In the first decade after discovery, a vigorous tourist industry was built in Yosemite Valley. A camp on the south fork of the Merced was built in 1856, and horseback tours were conducted through the Valley. Two hotels were opened on the valley floor by 1859.

The education of the public, through the combined efforts of artists, explorers, scientists, and entrepreneurs, aided the political efforts of those who sought to preserve the area. In June 1864, Congress granted Yosemite Valley and the Mariposa Grove of Big Trees to the state of California in a proclamation signed by President Lincoln. Thus, Yosemite became the first public park under the administration of a state.

Galen Clark, who had campaigned for the preservation of the area, was appointed the first guardian of the park. He served his first term from 1866 to 1879 and a second term from 1889 to 1896. Clark's management recognized the two-fold mission that still confronts the National Park Service today: to protect and preserve the wilderness and at the same time to allow reasonable access for public recreation and enjoyment. Clark worked to make the valley more pleasant and accessible to tourists. In 1878, in order to reduce the swampy conditions in parts of the valley, he dynamited part of a recessional moraine that dammed the Merced River. This resulted in fewer mosquitoes and a dryer, more hospitable valley floor of pleasant meadows. He cleared forest litter from some areas in order to allow clear pathways through the groves of Big Trees. Hiking and horse trails were constructed and maintained, and a variety of access roads were built into the valley and high country.

John Muir, the Scottish-born naturalist and conservationist, came to California as a young man in 1868 and made the state his home. He wrote many articles about Yosemite and campaigned tirelessly for the expansion of the park beyond the valley itself. He persuaded numerous influential people to visit the park. Ralph Waldo Emerson, then in his late sixties, spent ten days with Muir, camping and traveling on horseback throughout the Yosemite region in 1871. Emerson's admiration for Yosemite was conveyed to others in the eastern United States. Years later in 1903, President Theodore Roosevelt spent three days with Muir in Yosemite.

Largely through Muir's efforts, Yosemite National Park was established around the original Yosemite Grant in 1890. Then in 1905 the state of California returned its original grant to the federal government, in order to unify the administration of the park. Several additions to the park were made in subsequent years, bringing it to its present size.

Wagon roads were laid out in the park in 1878; before that, travel by horseback and on foot were the only methods of getting around in the park. In 1914, the first year in which automobiles were admitted, 127 cars passed through the area. In recent years, with over 2 million people visiting the park annually, automobile travel has had to be restricted because of traffic problems and pollution.

While the aim of the National Park Service is to preserve the wilderness of Yosemite, the effects of man's development of the region were far-reaching. The introduction of domesticated animals and plants in the 1860's led to changes that were largely reversible, but the actions of the first guardians of the park brought about changes that were sometimes unexpected and often irreversible. When the early explorers and tourists first entered the Valley, they beheld grand vistas across large open meadows. Over 750 acres of meadows were measured on the Valley floor in the 1860s. Since then, thick forest growth has invaded the original meadows, decreasing the number of unobstructed views of the cliffs and waterfalls. This stark change in the vegetative cover of the valley floor was caused in part by the aim of preservation which kept people from clearing areas and cutting smaller trees for firewood, as the Indians had done, and in part by the actions of Galen Clark who sought to make the Valley more hospitable. In blasting out the recessional moraine to reduce the swampiness of parts of Yosemite Valley, Clark accelerated the spread of the forests into the newly dried soils of the meadows. Today, the approximately 340 acres of meadow can be maintained only by judicious clearing of trees and shrubs.

A more permanent change was effected in part of the park by the construction of a dam across the Tuolumne River. The dam resulted from the needs of the rapidly growing city of San Francisco for water and electric power as it was recovering from the great earthquake of 1906. Interests focused on the Hetch Hetchy Valley as an "ideal" reservoir site. This valley was similar to Yosemite Valley with cliffs over 2000 feet high and with great waterfalls (fig. 25.3). In spite of opposition by conservation groups, Congress gave permission for the flooding of the Hetch Hetchy Valley in 1913. The dam, completed in 1923 and enlarged in 1938, flooded scenery that once rivaled Yosemite Valley in grandeur and beauty. The loss of this area is acutely felt today, for the population pressures exerted by tourists on Yosemite Valley could have been alleviated to some extent by the existence of an alternate valley within the park.

Geologic Features

Within Yosemite National Park, a complex of igneous rocks of varying composition was intruded over a period of more than 100 million years. These batholithic rocks were intruded into older metasedimentary and metavolcanic rocks now preserved as isolated masses and at the eastern and western margins of the batholith. Because the granites, granodiorites, quartz diorites, and monzonites that make up the batholith in this area have different properties, the rocks have responded differently to geologic processes since their emplacement. Thus the character of the igneous structures has influenced, and to a large extent controlled, the development of the magnificent and varied scenery that we see in the park today.

Finally, the morphology of the surface has been sculpted by glaciers that steepened the river valleys and modified the upper surfaces of the Sierra Nevada by erosion and deposition. Perhaps nowhere else in North America is there a better place to see the effects of glaciation on an igneous rock terrane.

Bedrock Geology

The oldest rocks in the park and surrounding area are Paleozoic units occurring on the western and eastern margins of the park and as isolated masses suspended within the batholithic intrusions. These rocks fall into two main groups. The first—and older—group consists of middle Paleozoic metasedimentary rocks deposited along the old continental margin located in what is now eastern California and northwestern Nevada. These rocks were derived largely from continental sources and included limestones, sandstones, and shales that have been metamorphosed to marbles, quartzites, and schists. Above this continental sequence the appearance of marine volcanic and silica rich ash flows indicate the beginning of plate collision in Upper Devonian time. This first group crops out most extensively far north of Yosemite but probably overlay most of the region at one time.

The second group consists largely of a somewhat younger oceanic and island-arc assemblage of metavolcanic rocks with associated serpentinite and chert; fine-grained, clastic metasedimentary rocks; and some blocks of limestones. This latter group of Permian to Pennsylvanian rocks has been called the Calaveras Complex. In the northwestern foothills of the Sierra, the two groups were juxtaposed along a major thrust fault in late Permian to early Triassic time.

Along the crest of the Sierra near Mt. Dana, on the eastern boundary of the park, an apparently different group of Permian metavolcanic rocks overlap an earlier Paleozoic section of metasedimentary units. In almost all cases, the age determinations have been questioned because of later metamorphism which has destroyed fossils and reset the radiometric ages to some extent.

Figure 25.3   Wapama Falls descended 1500 feet in a series of cascades to the floor of Hetch Hetchy Valley in Yosemite National Park. This photograph was taken prior to the 1923 flooding of the valley and shows the 2000-foot high cliffs of Rancheria Mountain Granodiorite, which were steepened by the action of the Tuolumne glacier system. Retreating glaciers left the irregular topography of the valley floor seen in the foreground. In Yosemite Valley, thick sequences of lake sediments created a smooth valley floor. Photo by W. A. Fiske, from the archives of the Department of Geology, Occidental College.

The Mesozoic rocks that crop out within the park occur mostly on the eastern margin along the crest of the Sierra. They occur as elongate roof pendants parallel to the mountain range. This sequence of Mesozoic rocks is exposed at the top of Mt. Dana as a series of volcanic tuffs and flows interbedded with sandstones of volcanic origin. These rocks are Jurassic in age and may be the volcanic expressions of the Jurassic intrusions that are the early stages of the Sierra Nevada Batholith. North of Tioga Pass, the eastern entrance to the park, a third section of metavolcanic rocks suggests nearly continuous igneous activity from Triassic to early Jurassic time (225 to 185 million years before present). In the southeastern part of the park an early Jurassic metasedimentary and metavolcanic sequence has been found that correlates with that of the Boyden Cave sequence near Kings Canyon National Park (see chapter 42).

Beyond the western boundary of the park and west of the Calaveras Complex lies a series of rocks that throw considerable light on the evolution of the Yosemite region. Here thick sections of island-arc volcanic and deformed ultramafic rocks attest to the existence of a subduction zone immediately west of the Yosemite region from late Paleozoic through early Mesozoic time.

Mesozoic Batholiths

The tectonic events occurring west of the Yosemite region led to the fusion of deep crustal material and the intrusion of the Sierra Nevada Batholith. In the park, dozens of plutons of varying sizes, ages, and compositions were emplaced. Ages of the plutons tend to fall into two groups with a large gap of time between them. The first group of dates, ranging from 210 to 190 million years before present (mybp), indicates that a period of plutonism occurred

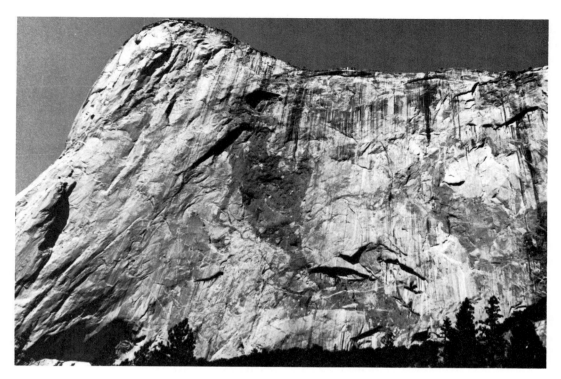

Figure 25.4   The face of El Capitan is made up of massive El Capitan Granite with few through-going joints. The dark irregular mass in the center of the cliff is called "the Map of North America" because of its shape. This mass is part of a complex of diorite dike intrusions that invaded the plutons of the earlier Western Intrusive Suite. The very thin white lines crossing both the diorite and the El Capitan Granite are later pegmatite dikes. The outcrop gives an excellent example of the crosscutting relationships geologists use to determine the relative ages of intrusive rocks. From *Field Guide to Northern California* by John Harbaugh. © 1975 Kendall/Hunt Publishing Company, Dubuque, Iowa. Used by permission.

during late Triassic and early Jurassic time. This period of activity was intimately associated with the volcanic activity discussed previously. The second group marks intrusions over a period of 50 million years and represents the main batholithic event in the Yosemite region. In Yosemite Valley, the early Cretaceous intrusive phase lasted from roughly 136 to 126 mybp. The main late Cretaceous phase occurred from about 90 to 80 mybp. In Yosemite Valley, minor episodes occurred between these two. The southeastern part of the park, however, is underlain by major plutons and associated metavolcanic rocks that date from 100–95 mybp. The distribution of ages shows that major episodes of plutonism were often separated by large intervals of time. During early Cretaceous time, the Jurassic and older sedimentary rocks were intruded by a sequence of magmas. In Yosemite Valley this sequence is called the Western Intrusive Suite, the members of which have been generally named for prominent exposures (table 25.1). The sequence has been worked out based on the rules of crosscutting relationships and the inclusion of fragments of the earlier intrusions in later plutons. The most plentiful, or widespread, of this group is the El Capitan Granite, which crops out over a large area.

The next sequence within Yosemite Valley is known as the Minor Intrusive Suite. These intrusions are known to come after the early intrusive group because the Minor Intrusives cut the older intrusions. One of the most notable of the Minor Intrusives is the diorite of the "Map of North America," which is crudely shaped like a map of North America and appears as a dark mass on the light-gray cliff face of El Capitan (fig. 25.4). Some thin, light-colored pegmatite dikes, younger than the diorite, cut across the maplike intrusion and further demonstrate the order of intrusion in the area.

The youngest of the Cretaceous intrusions are the rocks of the Tuolumne Intrusive Suite (table 25.1). The earliest members of the suite crop out in Yosemite Valley; but the youngest members (Cathedral Peak Granite and Johnson Granite Porphyry) crop out north and east of the valley toward Tuolumne Meadows. This suite of intrusions is a nested group of plutons that were intruded as parts of a single magmatic system over a relatively short period of time. The term "nested" refers to the concentric zonation that developed as later intrusions were emplaced into the cores of the earlier plutons. The Tuolumne Suite shows a gradual change in the composition and texture of the rocks throughout time. The earliest and outermost rocks are richer in the mafic minerals hornblende and biotite and have a higher percentage of the sodium-calcium feldspar, plagioclase. The younger, central units are poorer in the mafic minerals and contain more of the potassium feldspar, orthoclase.

Texturally, the earlier rocks are *equigranular,* meaning that the different minerals are approximately the same size throughout the rock. On the other hand, the central intrusives in the suite are *porphyritic,* a texture in which large crystals occur in a matrix of smaller grains. Porphyritic textures develop in two stages. First the magma cools and crystallizes slowly at depth, leading to large crystals suspended in the magma. Later, the partially crystallized magma is remobilized and emplaced at a higher, cooler level in the crust. At this level, the magma solidifies more rapidly into the finer grained matrix around the early formed crystals.

The combination of chemical and textural changes in the Tuolumne Suite is consistent with the evolution of a single magmatic system brought about by *fractional crystallization.* Fractional crystallization occurs as minerals that crystallize at higher temperature, such as hornblende, biotite, and plagioclase, are separated from the magma. The residual liquid, therefore, is changed in its composition, and rocks that crystallize later from the residual magma will be lighter colored and may be richer in potassium feldspar and quartz.

An alternative hypothesis for the origin of the intrusive suites is that they represent sequential formation of magmas by melting of rock at deep crustal levels. By this hypothesis, as the region cooled, only lower temperature magmas could be mobilized and intruded. Whatever the explanation, the existence of the intrusive suites does demonstrate the episodic nature of igneous activity in a region that was hot and active over many millions of years.

Late Cenozoic Volcanic Activity.

Most of the late Cenozoic volcanics are outside park boundaries, mainly to the east in the vicinity of Mono Craters. However, some volcanic and intrusive activity occurred in the northern and eastern parts of the park. North of the Grand Canyon of the Tuolumne River, lava flows of intermediate (andesite) composition as well as mudflows of volcanic ash (lahars) have been found. On the flanks of Mt. Dana, dikes and plugs occur, some of which appear to have been intruded along old fault zones. These isolated bodies are remnants of extensive activity that occurred throughout the Sierra. The evidence shows that rhyolite ash flows covered much of the northern part of the range about 30 mybp. Ten million years ago andesitic and basaltic lavas were extruded during part of the uplift of the range. Finally, some remnants of flows and intrusions which date at 3.5 mybp are found in the eastern part of the park.

Joint Systems and Landform Development

The surficial processes that have shaped the features of this geologically young mountain range are weathering, stream erosion, mass wasting, and glaciation. The unique and distinctive character of the topography, however, was determined by preexisting structures within the rocks. In Yosemite National Park, the types and systems of jointing of the various granitic rocks have been a key factor in the development of the landforms.

Joints are planar surfaces in rocks along which separation will occur. The motion of the two blocks of rock on separate sides of the joint are away from one another. Joints differ from faults in that faults have motion of the rock parallel to the surface of the fracture. The joints that occur in the batholithic rocks in Yosemite are formed by pressure release associated with the uplift of the range and the unloading of the area by erosion of the rocks that once overlay the intrusions.

Joints usually occur in sets of variable spacing. Large continuous joints separated by considerable distances are termed "master joints." They act to control drainage patterns and occasionally lead to parallel cliff faces. Lesser joint sets involve smaller joints, which are not as continuous, but usually occur more closely together.

In some areas, massive, unjointed rocks form large cliffs such as El Capitan, which rises 3604 feet above the Valley and is made up of massive granitic rock that is highly resistant to weathering and erosion because it has almost no joints or fractures. Close by, however, are intricately jointed granitic rocks that form castlelike landforms of striking appearance.

Intersecting joint sets occur in many of the Yosemite granites with joints oriented in two or more different directions. Where the joints are fairly closely spaced, pillars, columns, and pinnacles have been formed by erosional processes. Examples in Yosemite Valley are Watkins Pinnacle, Split Pinnacle, Washington Column, and Cathedral Spires. A larger feature that is bounded by more widely separated joints is the Three Brothers, with its asymmetrical, "gabled-roof" summits. At a number of locations, rock with widely spaced vertical joints occurs beside rock with close, intersecting joints. An example is Cathedral Rocks (massive features) next to Cathedral Spires (columns and pinnacles).

Sheet jointing, or *exfoliation,* is a type of physical (mechanical) weathering that is related to vertical pressure release from unloading. As the great weight of rock overlying a batholith is removed by erosion, the upper part of the batholith is able to expand slightly because of the reduced confining pressure. Cracks, or sheet joints, may develop parallel to the outer surface as a result of expansion. As the rock is exposed, mechanical weathering causes the thin layers of rock to spall off or break loose in concentric slabs from the granite mass, a process usually aided

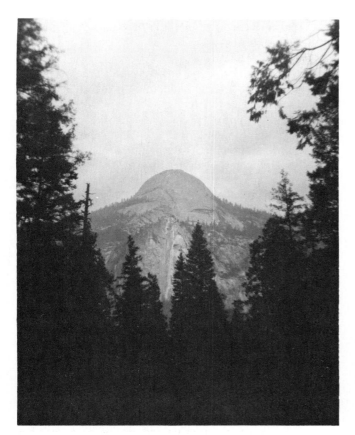

Figure 25.5   View of North Dome showing the effects of exfoliation in the Half Dome Granodiorite. The curved lines on the lower part of the dome mark the edges of exfoliation slabs. Below North Dome is the steep 1500-foot cliff of Washington Column. Photo by P. M. Palmer.

Figure 25.6   The planar north face of Half Dome is defined by a nearly vertical set of joints along which rock was easily removed. The stepped character of the jointed face can be seen in the photograph as straight or broken white lines such as those on the upper left or lower center of the cliff. The dark vertical lines are water drip lines. The photograph was taken from Mirror Lake at an elevation of 4082 feet. The crest of Half Dome is 8852 feet and the steepest planar part of the upper face is nearly 2000 feet high. Photo by M. B. Palmer.

by gravity (mass wasting). In massive granites devoid of much vertical jointing *exfoliation domes* have developed. Exfoliation of concentric shells of rock can be readily observed at Sentinel Dome, Turtleback Dome, or North Dome (fig. 25.5)

Half Dome developed due to both vertical jointing and exfoliation, with some help from glacial activity. This famous landform was never a whole dome that was cut in two by a tremendous glacier, as it may appear; in fact, the "other half" was never there. The steep face of Half Dome (fig. 25.6) began to take shape as thin slabs of rock between parallel vertical joints were loosened by weathering and mass wasting and fell away, layer by layer—a process that is continuing at the present time. The orientation of the face of Half Dome is parallel to a line of cliffs south of the Merced River, thus it may represent one of a set of "master joints" in this region of the Sierra. Half Dome was modified in the Pleistocene when the Tenaya Glacier

moved down the valley, reaching within 500 feet of the top of Half Dome. The glacier cleared away the exfoliation debris and talus; and because it was flowing in a direction parallel to the trend of the vertical joints, it removed more sheets of rock by plucking, thus leaving the precipitous face that we see today. The rounded back of Half Dome developed more gradually as curved shells of rock were exfoliated (fig. 25.7).

In some of the massive granites, exfoliation produced arches on the steep side of the domes. Royal Arches near the head of Yosemite Valley is the classic example. These are not "see-through" arches, but are, rather, a series of recessed arches set back into the main arch that curves out over the dome's base, a thousand feet below the top of the overhang. Here the Tenaya Glacier probably removed the lower part of the concentric shells of rock, leaving the upper parts unsupported.

Figure 25.7 The view of Half Dome from Glacier Point shows the exfoliated top and south side of Yosemite's most famous landmark. The vertical north face was controlled by joints that intersected the horizontal sheeting. National Park Service photo.

Effects of Glacial Erosion

Some domes such as Liberty Cap, Lembert Dome, and Mount Broderick were actually overridden by glaciers. These domes took on the characteristic profile of giant *roches moutonnées*. They have a smooth, rounded, and polished side where the glacier approached and overrode the resistant outcrop, and a steep often precipitous face on the side where the glacier flowed away from the dome (fig. 25.8). On the down-flow side the glacier was able to pluck slabs of rock from the face by refrozen meltwater and by ice that worked its way into joints in the outcrop.

Abrasion by rocks and gravel carried at the back of the glaciers was severe and glacial polish occurs on many of the surfaces in Yosemite. The comparative freshness and sheen of the polish shows the degree to which weathering has not affected it. Polished surfaces tend to roughen due to long-term weathering so that no polish remains on glacial pavement scoured by the earliest glaciers. Thus the polish can be helpful in determining the extent of the various stages of glaciation.

Glacial erratics, some quite sizable, are a common sight in the park. Glacier Monument, which is 1500 feet high, has erratics on top that were brought by the ice from a mountain 12 miles to the east. Erratics, along with patches of till, are found mostly on the uplands above the valleys. Morainal remnants are numerous in the park. The ones that are deeply weathered and dissected are associated with earlier glacial stages. Nunataks, such as Half Dome and El Capitan, provide clues as to the size and extent of glaciers. (*Nunataks* are mountain peaks that projected above the glaciers and were not overridden by ice.)

In the high basins where the glaciers originated, cirque lakes formed following the retreat of the ice. Some still exist but others have silted in, becoming subalpine marshes and meadows.

Glacial Lake Yosemite, which was impounded by a moraine near El Capitan, was about 5.5 miles long when the ice was melting back. The lake rapidly filled with sediment and disappeared, leaving the swampy meadows found by the first explorers. Deep drill holes and seismic surveys show that the bedrock is 2000 feet below the floor of the valley near Glacier Point. The glacial sediments filling the valley include thick sections of glacial outwash and till and at least six separate layers of lake sediments. This sedimentary succession shows the complexity of the glacial history of Yosemite Valley. It suggests that early glaciations gouged out the valley floor and then a long sequence of glacial advances and retreats yielded glacial and interglacial sediments that refilled and smoothed the valley floor.

The floors of Little Yosemite and Tenaya Canyons (fig. 25.9) are covered by similar deposits of drift and outwash. Most of the moraines that were left in Yosemite Valley by retreating glaciers were washed out by swiftly flowing meltwater or were covered over by lake deposits. The destruction and burial of moraines makes them difficult to correlate; therefore, age dating of glacial advances is obscure. Furthermore, the scattered occurrence of moraines in isolated valleys and in highland basins contributes to the uncertainty in the details of glacial history.

Waterfalls

The jointing of the bedrock, the stream erosion associated with the uplift, and Pleistocene valley glaciers all influenced the development and location of Yosemite's spectacular waterfalls. Most of the highest waterfalls in the country are in Yosemite National Park. Among the highest are:

Yosemite Falls—2425 feet (740 m) (fig. 25.1).
Ribbon Fall—1612 feet (491 m).

Figure 25.8 Lembert Dome stands as a resistant mass of Cathedral Peak Granodiorite above the Tuolumne Meadows. The gently sloping right side of the dome was overridden and polished by glaciers moving to the west (from right to left). The steep western face was formed by glacial plucking of slabs of rock, which were defined by steeply inclined joints. Photo by D. F. Palmer.

Figure 25.9 View from the east end of Yosemite Valley looking eastward up the glacially scoured Tenaya Canyon (named for the Yosemite Indian Chief). The valley shows the typical steep-sided U-shape caused by glaciation. In the foreground, the flat floor of Yosemite Valley is built on thick glacial outwash and lake sediments deposited over the scoured bedrock during several periods of glacial retreat. The granite bedrock is almost 2000 feet below the sediment surface at this point. Photo by M. B. Palmer.

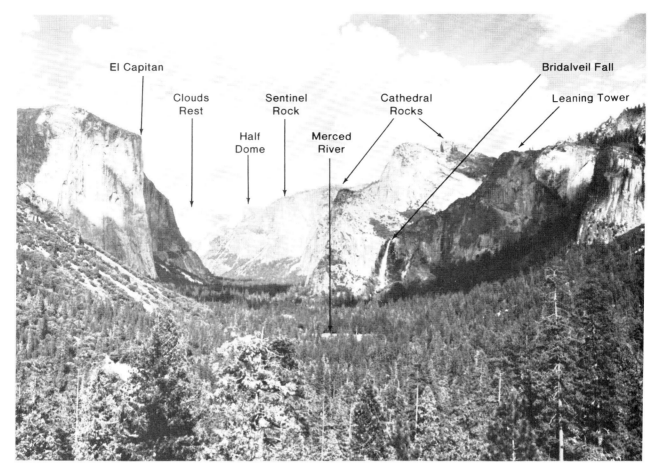

Figure 25.10   View of Yosemite Valley looking east from the Wawona Tunnel parking lot. The Mariposa Battalion first saw the valley from near this vantage point. The present morphology of the valley has evolved through uplift and stream erosion followed by steepening of the valley walls by glaciation. This left a series of hanging valleys such as that of Bridalveil Creek which joins the Merced River after the 620-foot drop of Bridalveil Fall. National Park Service photo.

Wapama Falls (Hetch Hetchy Valley)—1500 feet (457 m) (fig. 25.3).

Staircase Falls—1300 feet (396 m).

Tueeulala Falls (Hetch Hetchy Valley)—1000 feet (304 m).

Bridalveil Fall—620 feet (189 m) (fig. 25.10).

Nevada Fall—594 feet (181 m) (fig. 25.11).

Illilouette Falls—370 feet (113 m).

Vernal Fall—317 feet (97 m) (fig 25.11).

Yosemite, Ribbon, Wapama, Tueeulala, Bridalveil, and Illilouette Falls all issue from hanging valleys that existed before glaciation and were heightened by glacial action. The regional uplift and tilting that raised the Sierra Nevada Range to its present height disrupted the earlier drainage systems of the whole area and caused the main streams, such as the Merced River, to flow generally parallel to each other in a southwesterly direction. In the later stages, the uplift was so rapid that the increased stream gradients allowed these main streams to erode their valleys at an accelerated rate. This erosion was much faster than that of the tributary streams, which flowed parallel to the range with gradients that were little changed by the uplift. By the time the Ice Age began, the tributary streams of Yosemite were already entering the main valley in steep cascades, tens to hundreds of feet above the floor of the main valley. When the glaciers moved into the valleys, the erosive action of the ice straightened and deepened the main valley, making its walls even steeper and accentuating the "hanging" relationship of the tributary valleys. Many of the cascades then became free-leaping falls.

Nevada Fall and Vernal Fall have a somewhat different geomorphic history. They descend over giant glacial steps created as ice moved downvalley, plucking out sections of closely jointed bedrock between strips of massive granite and forming a giant glacial staircase (fig. 25.11). The action in this case was very similar to that which steepened the down-flow faces of overridden domes such as Lembert Dome.

Figure 25.11   Vernal Fall in the foreground and Nevada Fall in the center right are caused by the Merced River flowing over a stepped surface formed by glacial plucking. Above Nevada Fall, the bedrock of Little Yosemite Valley shows large scale exfoliation parallel to the broad U-shaped cross section of the valley. In the distance are the high peaks of the Cathedral Range and of Mt. Florence and Mt. Lyell on the eastern boundary of the park. Photo by W. A. Fiske, from the archives of the Department of Geology, Occidental College.

Geologic History

The Yosemite region developed as a result of a long and complex series of events that affected the entire Sierra Nevada range and, indeed, the whole western United States. The events can be looked at as those that led to the formation of the rocks of the area and those that affected the evolution of landforms in the Sierra.

1. Plate collision and the volcanic-plutonic complex

The early geological evolution of the region has been largely obscured by the great batholiths of Jurassic and Cretaceous time. To understand the early history of the region, geologists have had to look to the areas west and east of the batholith and at the isolated masses of metamorphosed rock in the roof pendants of the plutons.

The rocks around the Yosemite region show that the area was close to the continental margin during much of the Paleozoic. During the late Paleozoic, a subduction system developed that accreted slabs of material onto the western edge of the continent. East of the Sierra, thick sections of Precambrian and lower Paleozoic sedimentary rock occur that are representative of sedimentation on the continental margin.

As subduction continued into the Mesozoic, volcanic and plutonic activity began to be important within the Sierra Nevada province. Thick sections of oceanic crust and upper mantle were accreted onto the western coast of North America in what are now the foothills of the Sierra Nevada. In late Jurassic time, regional metamorphism and structural deformation marked the Nevadan orogeny. The rocks were later remetamorphosed and (in many places) obliterated by the intrusions of the Cretaceous batholith.

Within the Sierra, a great volcanic-plutonic complex developed as a result of the subduction. Geochemical studies suggest that the magmas were derived in part from the down-going slab and in part from continental crustal

Table 25.1 Generalized Geologic Column, Yosemite National Park

Era	Period	Epoch	Hetch Hetchy Valley	Yosemite Valley	Geologic Events
	Time Units		Rock Units		Geologic Events
Cenozoic	Quaternary	Holocene	Glacial deposits, alluvium, landslides, rockslides, talus		Erosion and deposition by glacial, fluvial, and mass wasting processes
		Pleistocene	Glacial tills, moraines, etc.; some early deposits buried or destroyed by later ice advances		Severe erosion during glacial advances (Sherwin, Tahoe, Graveyard, Tioga stages). Sculpturing of giant staircases, hanging valleys, basins, etc.
	Tertiary — Neogene	Pliocene Miocene	Mudflows (lahars) Andesite lava		Volcanism (northern part of park) Series of uplifts with faulting and tilting of Sierra Nevada block.
	Tertiary — Paleogene				
Mesozoic	Cretaceous		*Tuolumne Intrusive Series* Aplite Yosemite Creek Granite Mt. Gibson Diorite	Johnson Granite Porphyry Cathedral Peak Granite	Major intrusions into country rock and igneous rock of the composite batholith Dikes, veins
			Half Dome Quartz Monzonite Sentinel Granodiorite		
			Minor Intrusive Series	Diorite of "Map of North America" Quartz-mica diorite Bridalveil Granite Leaning Tower Quartz Monzonite	Minor intrusions of composite batholith
			Western Intrusive Series Ten Lakes Alaskite Rancheria Mt. Granodiorite Double Rock and Mt. Hoffman Granodiorites	Taft Granite	Major intrusions into country rock and emplacement of composite batholith
			El Capitan Granite		
			Tamarack Creek Quartz Diorite	Granodiorite of the Gateway Arch Rock Granite Diorite of the rockslides	
	— ? — Jurassic		South Fork of Tuolumne River Quartz Diorite	Mariposa Slates	Deformation, metamorphism, beginning of intrusions Interbedded sediment and volcanics
	Triassic		Diorite		
Paleozoic			Metasediments and metavolcanics (mostly outside park; some preserved in roof pendants)		Prolonged sedimentation followed by deformation, metamorphism, erosion

Source: modified after Smith, 1962; Peck, Wahrhaftig, and Clark, 1966; Harbaugh, 1975 material, which was melted and assimilated by the magmas coming from greater depth. The depth at which the batholiths were emplaced has been estimated to be about six miles, although those closely related to volcanic rocks may have been shallower. As the activity continued, the subduction center moved to the west and the center of batholithic intrusion shifted slightly to the east. This latter shift is seen in the "younging" of the batholiths towards the crest of the Sierra in Yosemite, Sequoia, and Kings Canyon National Parks (see also chapter 42).

2. Uplift and erosion

The main batholithic phase ended 80 million years ago and was followed by a period of uplift and extensive erosion during the late Cretaceous and early Cenozoic. Cenozoic volcanic activity covered much of the northern Sierra. The fact that the remnants of these once extensive lavas and ash flows are directly over the Cretaceous plutons shows that by the middle of Cenozoic time, erosion had stripped away most of the metamorphic rocks into which the batholith was emplaced. Further uplift and erosion continued episodically throughout the late Cenozoic. The region was alternately one of steep and of subdued topography. The latest period of uplift involved the development of the asymmetric mountain range we see today. Tilting increased with successive episodes of uplift so that by the Pleistocene, the eastern side of the range was a very steep, fault-bounded face and the western slope had become steeper than it was originally. The range is geologically young, and historic earthquake activity and faulting show that the mountains are still rising.

3. Development of the Merced River valley.

As the regional slope to the west developed due to uplift and tilting, the Merced River (and the other main streams) began to flow toward the west and southwest. During this time, the Merced valley lay between rolling hills several hundred feet high. With the gradient fairly low, the Merced meandered back and forth, forming a flood plain. As uplift continued, the gradient increased, causing the river to incise its channel in the broad valley floor.

Late in the Cenozoic, the continued tilting of the Sierra block began to have a major effect on the landscape. Valley walls became steeper as the incised channel of the Merced became deeper. On the other hand, the tributary streams, which were oriented more nearly parallel to the mountain range, did not experience the increased gradients that affected the westward-flowing streams. Because of their lower gradients, the tributaries on the upland were not able to cut their valleys as rapidly as were the main streams. Thus hanging valleys began to develop, and the water in the tributaries flowed down steep cascades into the narrow V-shaped gorge of the main valley.

In Yosemite Valley where tributaries such as Bridalveil Creek flowed on resistant, unjointed granite, erosion was slow and the streams cut down slowly. As a result, renewed or continued uplift accelerated downcutting of the Merced River valley and waterfalls began to develop on the tributaries flowing over resistant rock.

Major faults forming on the east side of the range permitted the block to rise even higher, and Owens Valley was formed as the downdropped block. By the onset of the Ice Age, the highest part of the Sierra had reached an elevation of at least 14,000 feet.

4. Pleistocene glaciation.

Details of the glacial history of Yosemite National Park have been difficult to work out because of the problems of dating and correlating successive moraines from one isolated valley or basin to another. Chief among these problems has been the correlation of glacial sediments from the western to the eastern side of the Sierra Nevada Mountains. There is still considerable disagreement, but four major periods of glaciation with numerous advances and retreats have been defined in the area. From oldest to youngest these stages are called: pre-Tahoe, Tahoe, Tenaya, and Tioga. The last three stages have been correlated with the Wisconsinan glaciation, which was the last major advance of continental glaciers in North America. The dates of the advances are only approximately known. However, the Tioga glacial stage is thought to have begun about 30,000 years ago, to have reached its maximum perhaps 20,000 years ago, and to have ended 10,000 years ago. The timing of the maximum is hard to determine in part because there appear to have been many fluctuations in the position of the ice margins during the glacial maximum. The age and definition of Tenaya glaciation is very much in question, but the age of the Tahoe glacial stage is generally agreed to have reached a maximum about 70,000 years ago.

The last advance of the Tioga glaciation was the least extensive, and during this period ice did not fill Yosemite Valley. The pre-Tahoe advance, on the other hand, inundated virtually all of the landmarks around the valley. The advances of the Tenaya and much of the early Tiogan stages did fill the valley as far as Bridalveil Fall. It was following these advances that the most recent of the several Lake Yosemites was formed. The glaciers that advanced, retreated, and readvanced during the earlier stages lasted the longest and were the most erosive. The great Tuolumne Glacier, for example, was fed by the snowfields of such mountains as Dana, Gibbs, Lyell, and Maclure. Miles wide, it spread over the uplands and then flowed around Mount Hoffman and divided into two trunk glaciers. Snows were so abundant that glacial ice accumulated in the high areas faster than the trunk glaciers could carry it away. One filled Big Tuolumne Canyon and Hetch Hetchy Valley. The other crossed a 500-foot divide that separated the Tuolumne and Merced basins and then flowed through Tenaya Canyon into the Yosemite Valley. Branch glaciers filled and coalesced until finally only the highest peaks and domes, such as El Capitan, Eagle Peak, Sentinel Dome, and Half Dome, stuck up through the ice. In the narrow canyons and gorges, spurs and projections were truncated by the moving ice, which also gouged out the bedrock of the valley floors.

The 2000-foot-thick glaciers of Tenaya Canyon and Little Yosemite joined together and filled the main Yosemite Valley. Tenaya Glacier moved parallel to one set of bedrock joints, carving and deepening a canyon with a gently sloping floor. However, the ice flowing down Little Yosemite Valley, which is underlain by massive granite with widely spaced joints, created the giant staircase over which Vernal and Nevada Falls now descend.

5. Deglaciation and the present landscape

Glaciers are great excavators; but when the active digging stops and the melting processes take over, the basins and troughs that have been scooped out make convenient receptacles for water, glacial debris, and sediment. Yosemite Valley was filled in stages as glaciation waned and glacial advances became less extensive. Streams are still filling in the basins and lakes, as well as eroding their valleys. Exfoliation of domes and arches and active mass wasting continue. Landslides dammed Tenaya Creek near the mouth of the canyon, forming Mirror Lake. Near the inlet a delta is building up in the shallow water. Rockslides (possibly triggered by earthquakes) have formed long talus slopes in the valleys.

Many small glaciers exist in the park; the two largest are Lyell and Maclure Glaciers. These produce glacial features today on a small scale. However, the present active glaciers are not remnants of the earlier glaciers. They are new glaciers formed during the latest of a number of Holocene glacial advances. These "Little Ice Age" glaciers have advanced and retreated several times in the last 3000 years.

Probably between 3000 and 4000 years ago, the giant sequoia trees reestablished themselves in scattered groves on the western slopes of the Sierra. Once widespread in temperate North America, the sequoias and redwoods were nearly exterminated by the multiple glaciations of the Pleistocene. Since in their native stands, giant sequoias grow only on sandy loams (or loamy sands), the soil derived from granitic rocks is favorable to their establishment. The sequoia groves in the park are Mariposa (200 trees), Tuolumne (25 trees), and Merced (20 trees).

Geologic Maps and Cross Sections

Bateman, P. C., Kistler, R. W., Peck, D. L., and Busacca, A. 1983. Geologic Map of the Tuolumne Meadows Quadrangle, Yosemite National Park, California. U.S. Geological Survey Map GQ 1570.

California Division of Mines and Geology. 1967. *Geologic atlas of California,* Mariposa sheet.

Calkins, Frank C., and others. 1985. Bedrock Geologic Map of Yosemite Valley, Yosemite National Park, California. U.S. Geological Survey Map I-1639.

Huber, N. King, and Rinehart, Dean C. 1965. Geologic Map of the Devils Postpile Quadrangle Sierra Nevada, California. U.S. Geological Survey Map GQ-437.

Kistler, Ronald W. 1966. Geologic Map of the Mono Craters Quadrangle Mono and Tuolumne Counties California. U.S. Geological Survey Map GQ-462.

Kistler, Ronald W. 1973. Geologic Map of the Hetch Hetchy Reservoir Quadrangle, Yosemite National Park, California. U.S. Geological Survey Map GQ-1112.

Krauskopf, Konrad B. 1985. Geologic Map of the Mariposa Quadrangle, Mariposa and Madera Counties, California. U.S. Geological Survey Map GQ-1586.

Matthes, F. E. 1930. Geologic history of the Yosemite Valley. U.S. Geological Survey Professional Paper no. 160. Plate 29.

Peck, Dallas L. 1980. Geologic Map of the Merced Peak Quadrangle, Central Sierra Nevada, California. U.S. Geological Survey Map GQ-1531.

Bibliography

Bowen, Ezra. 1976. *The High Sierra,* American Wilderness Series. Rev. ed. New York: Time-Life Books.

California Division of Mines and Geology. 1962. *Geologic guide to the Merced Canyon and Yosemite Valley, California.* California Division of Mines and Geology Bulletin 182.

Ditton, R. P.; and McHenry, D. E. 1973. *Yosemite road guide, Yosemite National Park, California.* Yosemite Natural History Association.

Harbaugh, J. W. 1975. *Field guide to Northern California,* K/H geology field guide series. Dubuque, Iowa: Kendall/Hunt Publishing Company.

Heady, Harold F., and Zinke, Paul J. 1978. Vegetational Changes in Yosemite Valley. National Park Service Occasional Paper Number Five.

Hill, Mary. 1975. *Geology of the Sierra Nevada.* Berkeley: University of California Press.

Huber, N. King. 1987. *The Geologic Story of Yosemite National Park.* U.S. Geological Survey Bulletin 1595. 64 p.

Jones, W. R. 1971. *Yosemite, the story behind the scenery.* Las Vegas, Nevada: K. C. Publications.

Keeney, N. T. 1974. The other Yosemite. *National Geographic* 145: 762–780.

Matthes, F. E. 1930. Geologic history of Yosemite Valley. U.S. Geological Survey Professional Paper 160.

———. 1950. *The incomparable valley,* ed. F. Fryxell. Berkeley: University of California Press.

———. 1958. *The story of the Yosemite Valley.* U.S. Geological Survey. Text on back of topographic map.

Muir, John. 1961 reprint (originally published in 1894). *The mountains of California.* Natural History Books.

Peck, D. L.; Wahrhaftig, C.; and Clark, L. D. 1966. Field trip guide to Yosemite Valley and Sierra Nevada batholith. In *Geology of Northern California,* ed. E. H. Bailey, pp. 487–502. California Division of Mines and Geology Bulletin 190.

Robertson, David. 1984. *West of Eden: A History of Art and Literature of Yosemite.* Yosemite Natural History Association and Wilderness Press, 174 p.

Russell, C. P. 1968. *100 years in Yosemite.* Yosemite Natural History Association.

Schaffer, J. P. 1977. Pleistocene Lake Yosemite and the Wisconsin Glaciation of Yosemite Valley. *California Geology,* v. 30 p. 243–248.

Smith, A. R. 1962. Petrography of six intrusive units in the Yosemite Valley area, California. California Division of Mines and Geology Special Report 91.

Yosemite National Park

Area: 761,170 acres; 1189 square miles
Established: October 1, 1890
Designated a World Heritage Site: October 31, 1984
Address: Box 577, Yosemite National Park, California 95389

26

The North Cascades Range supports 318
valley and cirque glaciers within North
Cascades National Park and about that
many more in the surrounding wilderness.
High, jagged peaks intercept moisture-
laden winds that nourish glaciers,
waterfalls, lakes, rushing streams, and lush
forests.

Deep glacial troughs and glacially
dissected uplands were left by the greater
glaciers of the Ice Age; frost-riven rock
towers above active alpine glaciers of the
present.

Exceedingly complex igneous and
metamorphic rocks and geologic structures
suggest a tempestuous tectonic history
involving plate collisions, subduction,
suspect terranes, uplift, and continuing
volcanism.

North Cascades National Park

North Central Washington

Figure 26.1   "Sawtooth" skyline of the Picket Range, North Cascades National Park. Frost action and
mass wasting have emphasized foliation patterns of the intensely metamorphosed rock exposed in the cliffs.
National Park Service photo.

Local History and Geographic Setting

A tribal legend explaining the Cascades Range illustrates the Indians' intimate awareness of their natural world. When the earth was young, the story goes, the Cascades region was flat. Rain never fell, but forests and plants drew up the moisture they needed from underground water that kept coming to the surface. After a while, water stopped coming up in eastern Washington. In desperation, the tribe sent a delegation west to Ocean to plead with him to send rain. Ocean agreed to help the Indians and sent his children, Clouds and Rain, eastward with a good supply of rain. All went well for a while, but the thirsty Indians would not let Clouds and Rain go back home. The Indians kept digging large pits for Clouds and Rain to fill with more and more water.

Ocean promised the Indians that if they would let Clouds and Rain come home, he would send water regularly, but still the Indians refused to let Ocean's children go. Ocean then appealed to the Great Spirit to punish the Indians. So the Great Spirit scooped up a huge mass of dirt and rock and molded it into the Cascades. Ocean quickly flowed into the ditch where the earth had been, and this made Puget Sound. East of the Cascades the land dried up and has remained fairly dry to this day. The only water left to the Indians was what was in the pits, the largest pit being Lake Chelan. (Lake Chelan, in the National Recreation Area south of the park, is 55 miles long, steep-sided, narrow, and deep.)

Presumably the Indians were struck by the contrast between the mist, fog, rain, and snow typical of the western slopes of the range and the dryness of the land on the east side. They appear to have understood the principle of the *rain-shadow* effect of a high mountain range that lies in the path of prevailing winds that are moisture-laden from having passed over water.

Hunters and trappers were the first non-Indians to explore the North Cascades, and they were followed by miners. Before the 1860s, gold, copper, molybdenum, and zinc were mined here on a small scale. Today mining is not an important enterprise, although some commercial mining has been carried on in the North Cascades region in recent years.

The prospectors who looked for "strikes" in the North Cascades, especially near the Canadian border, must have endured great hardships. The same could be said for the geologic field parties who did the original reconnaissance work in these mountains. Travel was by foot, with pack mules carrying equipment and supplies. Even today, the only road that crosses the park is Washington Highway 20, which goes through Ross Lake National Recreation Area between North Unit and South Unit. Most geologists backpack their own supplies when working in remote areas, although mining companies and the U.S. Geological Survey sometimes use helicopters for supplying field parties. Under the best of conditions, field work in high areas is difficult. Some snow normally falls during summer months. Rain and mist reduce visibility much of the time. Below the treeline, the forest cover is heavy. But for those hardy people who appreciate wilderness beauty, hiking or backpacking in the North Cascades is truly rewarding.

North Cascades National Park was created to preserve the remaining wilderness areas of the range and protect the watershed. Following the recommendations of a study team, the park was established in 1968 with four units placed under National Park Service jurisdiction: (1) North Unit, including the Picket Range Wilderness; (2) Ross Lake National Recreation Area, comprising a hydroelectric dam and reservoir holding the Skagit River drainage and separating North and South Units; (3) South Unit, including Eldorado Peaks Wilderness; and (4) Lake Chelan National Recreation area.

Geologic Features

Water, ice, and jagged peaks are the scenic elements for which the North Cascades are called "the American Alps." Adding interest to the skyline—although they are not inside national park boundaries—are two volcanoes of the Cascade volcanic chain; namely, Mount Baker and Glacier Peak. Both are several thousand feet higher than the ranges of the North Cascades, which average around 5000 feet in elevation south of the park and rise to more than 8000 feet within the park.

The active Cascades glaciers within and around the park, covering in all more than 100 square miles, make up this country's largest concentration of glaciers outside Alaska. Also in this wild and rugged region are many spectacular arêtes, horns, cirques, and deep glacial troughs that were sculptured by Pleistocene glaciers. Why do these mountains of only moderate height bear so much ice? The answer is water—water as rain, fog, mist, snow, and more snow, all adding up to the average annual precipitation on the western side of the range of 110 inches. A winter's accumulation of 46 feet of snow is not unusual, and 4 to 5 feet of snow may pile up in a single storm. In the North Cascades, where the snowline on the western slopes is between 6500 and 7500 feet in elevation, glaciers are more numerous and extend lower than in Waterton-Glacier International Peace Park (chapter 23), which is at the same latitude and has higher peaks. The difference is because Waterton-Glacier is several hundred miles farther inland from the moisture source, the Pacific Ocean.

In the *rain-shadow zone*, east of the North Cascades, annual precipitation rates drop sharply to only 35 inches, for example, around Lake Chelan, and a mere 12 inches in the Pasayten Wilderness. The prevailing winds, carrying moisture from the Pacific Ocean, are forced upward over the North Cascades. Expanding and cooling, the

Figure 26.2   North Cascades National Park, Washington.

Mount Shuksan   Mount Baker

Figure 26.3   Mount Baker, a volcano, rises above the Chilliwack Range in the far distance. Mount Shuksan is in front of Baker and slightly to the left. The western boundary of North Cascades National Park passes between the two mountains. Silver Lake lies in the cirque in the foreground. National Park Service photo by M. W. Williams.

rising air loses moisture. Descending air on the eastern side of the Cascades compresses and warms, thus holding moisture and causing high evaporation. Therefore, the downwind side of the range receives scant rainfall.

Development of the North Cascades

The North Cascades Range is a region of great scenic beauty that is also geologically and geographically complicated and challenging. The spectacular Cascade mountains are carved into uparched bedrock from a part of the earth's crust that was uplifted along a north-south axis late in Cenozoic time. Intrusive and volcanic activity accompanied uplift. Yet the mountains we see represent only the latest and relatively mild event in a region that has been in the midst of intense geologic activity since Precambrian time. The record of orogenies involving severe compression and crustal shortening is exposed in rocks throughout the North Cascades. The latest uplift may be largely a reflection of isostatic adjustments.

In the North Cascades we have in close proximity very old rocks and some of the youngest intrusive and extrusive rocks in North America. Mount Baker and Glacier Peak, the two volcanoes just outside park boundaries and only about 60 miles apart, were formed from Quaternary lavas that rose through many layers of much older rocks.

Glacier Peak, a relatively inaccessible cone in the middle of the range (in Glacier Peak Wilderness), is built on top of a high ridge of pre-Cretaceous metamorphic rocks (slate, phyllite, and greenschist) between two plutons of quite different age. Andesite lavas and, later, dacite lavas that make up the volcano have no apparent relation to the older rocks that compose the mountains of the range. Late in the Pleistocene Epoch and in Holocene time, the dacitic lavas produced a great deal of ash and pumice because this type of lava (more silica-rich than andesite) tends to erupt explosively. Glacier Peak's most recent ash eruption was only about 200 years ago (Beget 1982).

Mount Baker, in the foothills near the western edge of the range, has a much larger cone (elevation, 10,775 feet) that was built on top of older sedimentary and volcanic rocks (fig. 26.3). In shape and construction, Mount Baker is a typical *composite volcano*. Volcanoes of this type are long-lasting and tend to construct large, lofty cones, built up over time by alternating layers of andesite lava and pyroclastics. Because of emissions of steam and sulfurous ash for several months in 1975, Mount Baker (along with other Cascade volcanoes, chapters 32, 33, 34) is being continuously monitored.

Box 26.1
Why Are the Cascade Volcanoes Hazardous?

The glacier-covered summits of Mount Baker and other Cascade volcanoes greatly increase the hazards of any eruption because of the possibility of a *lahar,* or volcanic mudflow, being triggered when snow and ice, suddenly transformed to water and steam, mix with hot ash and other debris making a slurry. A lahar roars down a mountain valley destroying everything in its path. When thermal activity on Mount Baker increased, the level of Baker Lake, a reservoir at the foot of the mountain, was lowered because U.S. Geological Survey geologists feared that a mudflow might cause the dam to break, which would devastate communities farther down the river valley. In 1980, a lahar set off by an explosive eruption at Mount St. Helens (170 miles to the south, air distance) was not so destructive as it might have been because reservoir levels had been lowered when earthquake swarms and steam eruptions indicated that the volcano might erupt.

Composite volcanoes, like those in the Cascades, sometimes have explosive eruptions, depending upon the amount of gas in the lava and its *viscosity,* or resistance to flow. The viscosity of a lava controls the ease or difficulty with which the gas escapes into the atmosphere. Composite cones are built up by lava that is predominantly intermediate, or andesitic, in composition. Andesitic lava is ordinarily more viscous than basaltic lava; but if the temperature of an andesite lava is extremely hot, then it becomes more fluid, or nonviscous, and is able to flow easily from a crater or vent. If a vent becomes clogged and gas pressure builds up, then an explosion occurs that ejects pyroclastic debris, clouds of ash, and lava. Eruptions of composite volcanoes tend to be intermittent, with periods of violent activity followed by years of apparent dormancy.

Sudden release of intense heat is the immediate danger when a volcanic explosion

occurs. All living things on the volcano's flanks may be incinerated by the blast. Glaciers and snowfields turn to hot water and steam that mix with ash, forming dangerous lahars and mudflows. Water storage reservoirs that have been constructed on the lower slopes of most of the Cascade volcanoes may serve as catch basins for mudflows—providing enough warning has been given so that the water can be drawn down. Whether or not the dams can hold is uncertain. A large enough mudflow could overtop a dam, thus endangering communities downstream by creating a catastrophic flood. Hot falling ash carried rapidly downwind may start fires over a wide area, contaminate water supplies, collapse roofs, render highways impassable, clog storm drains, endanger aircraft, and damage crops.

Can eruptions be predicted? The monitoring program of the U.S. Geological Survey, in cooperation with state surveys, seeks to interpret data and issue warnings based on several types of indicators: (1) increases in thermal activity; (2) increases or changes in seismic activity; (3) gravity and magnetic data; (4) remote sensing and infrared studies; (5) gas and temperature monitoring; and (6) precise measurements of changes, such as bulging or tilting, in a volcano's outward conformation.

In the Cascades, volcanic eruptions have been infrequent in terms of human life spans. Before Mount St. Helens erupted, many residents did not believe the geologists and other scientists who warned of potential hazards. However, the 1980 explosion and subsequent less intense eruptions made believers out of skeptics. What is now recognized is that major evacuations are really the only adjustment that human beings can make when a volcano becomes threatening.

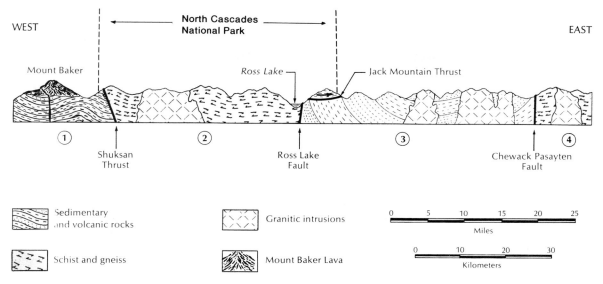

Figure 26.4 Schematic west-east cross section through the North Cascades. Major blocks are 1) the western foothills— folded sedimentary and volcanic rocks; bounded on the east by the Shuksan fault; 2) Chilliwack Batholith intruding metamorphosed schists and gneisses; 3) the Methow graben, a down-faulted block between Ross Lake and Chewack Pasayten faults; 4) the eastern highlands. Adapted from Bates McKee, *Cascadia* (fig. 7.3). © 1972 McGraw-Hill Book Company. Used by permission.

North Cascade Rocks and Structures

Distinct structural blocks, or "suspect terranes," can be identified from west to east across the North Cascades Range. Steeply dipping, north or northwest fault systems, approximately parallel with the trend of the range, separate the terranes (fig. 26.4). Within each block, the rocks are fairly consistent in age and type; but the rocks in each block differ markedly from the rocks in adjacent blocks. Field interpretation of the fault systems shows the presence of thrust faults and strike-slip faults that resulted from compressive and lateral movements. Younger vertical faults have been mapped within blocks.

The suspect terranes, or *tectonostratigraphic terranes* (the more formal term), arrived as microplates in late Mesozoic and early Tertiary time. First as easterly moving and subsequently as northerly moving, the plates were driven against older, complex rocks (already deformed by past orogenies) that lay along the continental margin. The incoming rocks as well as the rocks already there were profoundly affected by the intensity of the collisions. The fault systems between the terranes and the old continental margin represent boundaries or junctions where vertical and horizontal movements of tremendous force occurred, along with oblique subduction, folding, metamorphism, volcanic activity, and the emplacement of plutons.

For convenience, the rocks across the range from west to east are described in four major zones of rock distribution (fig. 26.4). Most of North Cascades National Park is within the second zone, where the Chilliwack Batholith intruded gneisses and schists. Small areas of the park extend into the first zone, on the west, and the third zone, on the east. The four rock distribution zones are described briefly below, but the emphasis throughout the rest of the chapter is on the rock units exposed within park boundaries.

Zone 1 is the western foothills region, bounded on the east by the Shuksan thrust fault. Mount Baker, a young volcano, rises from a platform of folded Late Paleozoic and Mesozoic sedimentary and volcanic rocks in this zone.

In Zone 2, the composite Chilliwack Batholith lies between the Shuksan fault and the Ross Lake fault. The granitic magmas of the batholith intruded older, severely deformed and metamorphosed schists and gneisses.

In Zone 3, the Methow graben, a downdropped fault block, lies between the Ross Lake fault and the Chewack Pasayten fault. The Methow graben consists of Upper Mesozoic sedimentary beds that have been intruded by Cenozoic granites. Near the western edge of the block, the Jack Mountain thrust, a wedge of rock forming the tops of several peaks in the park, rests on top of younger Upper Mesozoic beds. The park's eastern boundary is in this zone.

Zone 4, the eastern highlands, is outside North Cascades National Park on the east side of the Chewack Pasayten fault. This block consists mainly of schists, gneisses, and granites.

The processes of regional and contact metamorphism in the North Cascades produced a wide range of *metamorphic facies* that provide clues about physical and

Figure 26.5 A roadcut at Ross Lake Overlook displays a spectacular exposure of Skagit Gneiss. Photo by S. D. Tuttle.

chemical conditions in the various subsurface environments in which metamorphism occurred. Each metamorphic facies represents a particular combination, or assemblage, of minerals that are associated with certain pressure and temperature relationships. By identifying the metamorphic facies of different outcrops of schists, for example, geologists can infer the approximate depth at which metamorphism took place. In a metamorphic facies series, a greenschist suggests an environment of relatively low temperatures and pressures, while an amphibolite schist indicates that the parent rock was subjected to higher temperatures and pressures at a greater depth.

Under some conditions of metamorphism, *migmatites,* which are interlayered or intermixed rocks, partly igneous and partly metamorphic, are produced. The North Cascades migmatites were probably formed at great depth during regional metamorphism as very hot water, bearing mineral ions, was forced at high pressure through planes of weakness (foliation) in the metamorphosed rocks, changing them to granite along the paths where the water traveled. The well known "Skagit Gneiss," a striking migmatized schist, crops out in roadcuts along the highway (Washington 20) that crosses the range between North and South Units. Veins and dikes of pegmatites and other igneous rocks swirl, crosscut, and interfinger the banded gneiss in colorful, intricate networks. At the Diablo Overlook, where large specimens of North Cascades rocks are displayed in a "rock garden," the following "recipe" appears on a plaque identifying Skagit Gneiss:

> Place schist and assorted granitic rocks deep underground. Allow time to metamorphose at somewhat below melting point. Be patient!
> Add more granitic rocks, hot liquids. Increase pressure.
> Stew and shake for a few million years.
> Allow to cool, and let rise for ten million years.
> Result: The very complex, layered, folded, and intruded rock formation visible in the highway cuts nearby.

The North Cascades are not rich in mineral deposits, but some gold, copper, and silver grains were carried by mineral-bearing hydrothermal solutions into quartz veins during cooling of the intrusions. Some of the gold was subsequently freed by weathering and erosion from its quartz matrix and washed into stream gravels. Since gold has a high specific gravity, it tends to sink to the bottom of a stream and the grains ("gold dust") and nuggets become concentrated in sand bars and outwash. Such an accumulation is called a *placer deposit.* Before the filling of the Ross Lake reservoir, placer gold was taken from glacial outwash and stream gravels in Ruby Arm, but this site is now under water.

Figure 26.6   Arêtes and horns on the serrate crest of the ice-sculptured Picket Range (North Unit). National Park Service photo by M. W. Williams.

Glacial Features

From the air, looking down over North Cascades National Park, one sees an awesome terrain—snowfields from peak to peak; ice spilling out of cirques and tumbling down valleys; sharp, jagged ridges that extend for miles; steep-sided horns; precipitous mountain slopes; deep U-shaped troughs; and bubbling cascades of meltwater falling into wooded valleys. High country and low country are inextricably mixed. Some of the names that prospectors gave to the rugged terrain near the 49th parallel (the U.S.-Canadian boundary) evoke images of difficulties they must have had in trying to get around in this region on foot—names like Devils Dome, Terror Ridge, Fury Ridge, and Ragged Ridge (fig. 26.6).

Excavation by Pleistocene glaciers provided some access into these formidable mountains. Washington Highway 20 crosses the North Cascades by following large glacial troughs through the Ross Lake National Recreation Area and the national forest on the east side of the park. The Skagit River and its tributary, Granite Creek, now drain the deep, steep-walled valleys that were excavated by an enormous, branching trunk glacier that flowed from the uplands far down into the Puget Sound lowland. The long straight stretches of the river's course attest to the weight and power of the moving ice. Another means of access to the interior of the range is a well known trail. Once used by Indians, prospectors, and pioneers, and today by backpackers, the trail goes from fiordlike Lake Chelan on the southeast side of the range, up the glacial valley of the Stehekin River, across Cascade Pass (a glacial col), down the Cascade River, and then down the Skagit River.

Many examples of alpine erosional features are in the park. Above Cascade Pass (elevation 5392 feet) towers Eldorado Peak, 8868 feet high, a sharply eroded glacial horn rising above active glaciers. Part of the Boston Glacier, longest glacier in the park, is on Eldorado (fig. 26.7). Numerous active glaciers are in the vicinities of Mounts Spickard, Challenger, and Shuksan, Bacon Peak, and Colonial Peak. The two volcanoes outside the park, Mount Baker and Glacier Peak, have extensive glaciers. Glaciologists have pointed out that the Cascade volcanoes—even though they are higher and tend to support more glaciers—show the effects of glacial sculpturing less than the

Figure 26.7   Forbidden Peak rises above the large compound cirque of the Boston Glacier near Cascade Pass (South Unit). Masses of glacial ice and thin streams of meltwater cascade down over nearly vertical slopes in the foreground. National Park Service photo by M.W. Williams.

nonvolcanic peaks of the North Cascades. The inference is that since volcanic eruptions continued during and after the Pleistocene, the volcanoes were able to keep "repairing" their cones.

Mount Shuksan (fig. 26.8), near Mount Baker, is a glacial horn with glacial cirques indenting its flanks. The Picket Range, with its thin, serrated ridge, was named for its resemblance to a picket fence extending along the skyline. Mount Johannesberg, at the headwaters of the Skagit River, is another knife-edged glacial arête. To the south, paralleling Lake Chelan, is the aptly named Sawtooth Ridge. Three Fools are glacial horns on the crest of a nearly vertical headwall that drops thousands of feet to the cirque floor.

Of the glacial troughs, Lake Chelan (in Lake Chelan National Recreational Area and Wenatchee National Forest) is the most remarkable. The bottom of Lake Chelan lies 400 feet below sea level, and its 1500-foot depth makes it one of the nation's deepest lakes. The Chelan trough was excavated by an 80-mile-long glacier that came down the Stehekin River valley from the uplands above Cascade Pass. Many of the flat-floored glacial troughs, such as Ruth Creek valley and Nooksack valley, contain braided streams because of the low gradient and high sediment load. Hanging valleys, most of them containing waterfalls, are numerous throughout the park. The most spectacular is a 1500-foot waterfall from the cirque on Bacon Peak that holds Green Lake. Ann Lake and Rainy Lake are also examples of cirque lakes.

Except for moraines on valley floors and sidewalls, some outwash deposits, and similar accumulations, glacial depositional features in the park are less significant than erosional landforms. Most of the glacial debris was moved southward by the Cordilleran ice sheet that pushed down from Canada. Erratic boulders, however, from British Columbia were left sitting on some of the lower summits that were scoured by the ice sheet.

3. Middle and late Paleozoic deposition and volcanism.

West of the park, the metamorphosed rocks of the Chilliwack Group (not to be confused with the much younger Chilliwack Batholith) were deposited as alternating volcanics and sedimentary beds, mostly in shallow water. In some places the Chilliwack beds are associated with Mesozoic clastics (conglomerates, slates, etc.), suggesting that deposition was more or less continuous in ocean trenches between plates from the end of the Paleozoic Era into Mesozoic time. An episode of metamorphism is dated, on the basis of zircons, at 415 million years in age (Silurian). Another date, derived from zircons in metamorphosed sedimentary rocks places the Marblemount belt in early Triassic time, 220 million years ago.

4. Mesozoic deposition and volcanism.

Part of the Ross Lake glacial trough, Mount Hozomeen, and the eastern boundary of North Unit are within a belt of rocks of the Hozomeen Group that are believed to be Triassic (possibly Permian) in age. However, most of the thick sequences of Mesozoic volcanic and sedimentary rocks are outside park boundaries in the western foothills and the eastern highlands. The rock units are typical ocean trench deposits that accumulated as microplates drifted toward the continental margin.

5. Late Mesozoic to early Tertiary docking of tectonostratigraphic terranes, with accompanying subduction, faulting, metamorphism, and mountain-building.

Geological activity in the North Cascades region increased in complexity and intensity as microplates were squeezed between the converging Pacific and North American plates. As they were "stitched" onto ancient complex rocks of the continent, the microplates became suspect terranes. During the 100 million years since the onset of these geologic events, segments of continental and oceanic rocks of various types and origins were buried and uplifted; heated and cooled; folded, crushed, contorted, and shattered; intruded by granitic magmas; covered by lavas and ashfalls; and intricately combined to produce a "collage" of interesting and puzzling "metamorphic suites" and associated rock units.

The rocks of the Skagit Metamorphic Suite and the Shuksan Metamorphic Suite (zone 2, fig. 26.4) are prominently exposed in North Cascades National Park. The term *suite,* as used here, means an assemblage of associated plutonic or metamorphic rock units. (In a stratigraphic sense, a suite is comparable to a group.)

During the movement of the Shuksan thrust fault on the west side of the range, rocks were sheared, shattered, and metamorphosed, producing the complex structures of the Shuksan Metamorphic Suite, which trends north–northwest in the western foothills and along the western edge of the park. The unit consists of the Darrington Phyllite (parent rock, shales) overlain by the Shuksan

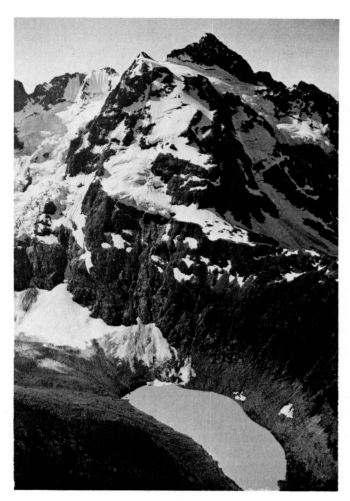

Figure 26.8   Mount Shuksan (9131 feet) is on the crest of an uplifted fault block of highly metamorphosed schist and gneiss. Price Lake, in the rock basin below the peak, is fed by meltwater from Price Glacier. National Park Service photo.

Geologic History

1. The oldest rocks, the Yellow Aster Complex.

The term "complex" is appropriate because the early geologic history of the Yellow Aster rocks, which consist of crystalline basement material, has been obscured by repeated igneous and tectonic activity, as well as long erosion. The Yellow Aster Complex is thought to be old continental crust, clearly pre-Devonian in age and probably made up of orogenically deformed early Paleozoic and Precambrian sedimentary and igneous rocks. Zircons from the Yellow Aster Complex have been dated as older than 1650 million years. Yellow Aster rocks crop out in the western foothills (Zone 1 in fig. 26.4).

2. Early and middle Paleozoic erosion.

As earlier tectonic activity decreased in intensity, long periods of erosion produced unconformities.

Table 26.1 Geologic Column, North Cascades National Park

| Time Units | | Rock Units (W = west side, E = east side) | Geologic Events |
Era	Period		
Cenozoic	Quaternary	Glacial and alluvial deposits Cascade extrusives (W) (Mount Baker, Glacier Peak)	Fraser glaciation (latest of successive major episodes) Severe glacial erosion Construction of composite volcanoes
	Tertiary	Tertiary volcanics (W and E) Chuckanut Formation (W)	Uparching along north-south axis, block faulting, erosion, sedimentation, volcanism
Mesozoic	Cretaceous	Late Cretaceous and Tertiary plutons and volcanics, including Chilliwack Composite Batholith and the Eastern Intrusive Belt	Oblique subduction, thrust faulting, compression, metamorphism, migmatization, vertical and lateral fault movements
		Shuksan Metamorphic Suite (W) (including Shuksan Greenschist and Darrington Phyllite) Jack Mt. Phyllite (E)	Docking of suspect terranes Multiple intrusions (batholiths, stocks, dikes, veins, mineralization)
		Eldorado Gneiss Skagit Metamorphic Schist (including Skagit Gneiss/Cascade River Schist)	
	Triassic	Elijah Ridge Schists (E) (derived from North Creek Volcanics) Hozomeen Group (E)	
Paleozoic	Upper	Chilliwack Group (W) not exposed in park	Accumulation of sediments and volcanics; several orogenies
		Major unconformity	Uplift and erosion
Precambrian		Yellow Aster Complex (W) (Pre-Devonian crystalline basement)	Highly deformed cratonic rocks

Sources: Misch 1966, 1977, McKee 1972, Brown 1987

Greenschist (parent rock, submarine basalt flows). Unusual blueschists occur with the greenschists. The schists were altered in a low-temperature, high-pressure environment (i.e., a greenschist metamorphic facies). Tectonic slices of ancient Yellow Aster rock were brought up from deep within the crust, along with slices of ultramafic rock (largely olivine) that may have been derived from upper mantle rock, just below the earth's crust).

The Skagit Metamorphic Suite, which is older than the Shuksan Metamorphic Suite, dominates the main part of the range. The Skagit rock units show varying degrees of metamorphism from greenschist metamorphic facies to amphibolite facies; that is, from low temperatures with high pressures to deep burial with conditions of high temperatures and high pressures. The Cascade River Schist (part of the Skagit Metamorphic Suite) crops out mainly in the western part of the park's South Unit. The principal component of the Skagit Metamorphic Suite, the Skagit Gneiss, crops out in both North Unit and South Unit and is the migmatized equivalent of the Cascade River Schist. (The spectacular exposures of Skagit Gneiss at Diablo Overlook, with the recipe for gneiss, are discussed earlier in the chapter.)

Most of the rock units in the Skagit Metamorphic Suite are part of very old continental crust that drifted across the Pacific for millions of years before colliding with North America. Zircon dates of 2000 to 1452 million years ago have been derived from the Skagit Gneiss. Several granitic plutons were emplaced before, during, and after the arrival of the microplate and were themselves metamorphosed or remetamorphosed. The Eldorado Gneiss, for example, consisting of metamorphosed granitic rock, has yielded a zircon date of 92 million years.

On the east the Ross Lake fault separates the Skagit Gneiss from a long, thin belt of metamorphosed rocks. In this belt is the Elijah Ridge Schist (probably derived from the North Creek Volcanics), overlain by the Jack Mountain Phyllite. The Jack Mountain thrust fault moved older rock units eastward over younger beds (fig. 26.4). Parts of these fault zones have been destroyed by younger intrusions.

6. Late Mesozoic and early Cenozoic deposition; minor deformation and continuing uplift.

In late Cretaceous and early Tertiary time, highlands (precursors of the present mountains) that were elevated by the Cretaceous orogeny became a source of sediment for deposits in marine embayments on the west and east

Figure 26.9   A meltwater stream has cut a narrow channel through a gravelly end moraine. National Park Service photo.

sides of the range. Then, continental beds (mixed sedimentary and volcanic) were laid down over the marine units. Patches of these beds, such as the Chuckanut Formation and Tertiary volcanics, exist in the park, but mainly the sequences are found outside park boundaries.

7. Multiple Cenozoic intrusions, including the Chilliwack Composite Batholith and the Eastern Intrusive Belt.

Emplacement of a large, composite batholith that along with associated older metamorphics forms the core of the Cascade Range began with the intrusion of stocks having a diorite or gabbroic composition and was followed by a series of major intrusions, most of which were quartz diorite. Later in the sequence they tended to be granodiorite. Next, stocks of very light-colored granitic magma pushed up into the plutons placed earlier. Small intrusions of granitic composition continued through the rest of the Tertiary.

An Eastern Intrusive Belt, emplaced east of the Ross Lake fault, includes the Ruby Creek Heterogeneous Plutonic Belt, the Black Peak Batholith, the Goldenhorn Batholith (comprising the only true granite in the North Cascades), and the Perry Creek intrusives of granodiorite and quartz diorite.

8. Late Tertiary block-faulting and uplift.

Uplift along a north-south axis was greater toward the north, with maximum uplift occurring in the area of the 49th parallel. Concurrent and subsequent erosion exposed the older metamorphic rocks and deep-seated intrusions in the North Cascades. Rocks that old are not exposed in the Southern Cascades (southern Washington, Oregon, and Northern California).

9. Pleistocene and Holocene composite volcanoes.

The continuing subduction of oceanic crust and sediments far below the North Cascade range generated lavas that rose through thousands of feet of metamorphic and igneous rocks to form the composite volcanoes, Mount Baker and Glacier Peak (outside the park's western boundary). Eruptions in the Cascade volcanic chain have continued into the present (chapters 32, 33, and 34). Frequent earthquakes in central and western Washington are another indication of ongoing tectonic activity. Movement of the oceanic crust (the small Juan de Fuca plate) has been determined as 35 mm per year of northeasterly motion, relative to North America. Calculations based on seismic data suggest that the subducting slab under Puget Sound has a dip of 9° (Tabor 1986).

10. Pleistocene and Holocene glaciations.

Glaciers have been present in the North Cascades since before the beginning of the Pleistocene, and have been extensive during most of this time. Again and again, glaciers in the mountains of western Canada grew until they formed a Cordilleran ice sheet that moved down over northern Washington, expanding and at times overrunning the alpine glaciers of the North Cascades. The latter enlarged into piedmont glaciers, accumulated great ice fields, and eventually merged with the ice sheet from the north. During warmer interglacial periods, the process was

Figure 26.10   Trapper Lake and glacial valley (South Unit). Pleistocene glaciers filled this valley and joined a trunk glacier occupying the Stehekin valley and the Lake Chelan basin. Now during the melt season, remnant glaciers and the snowfields among the peaks furnish much-needed water to ranches and communities in central Washington. National Park Service photo by M. W. Williams.

reversed as the ice sheet melted back, leaving glaciers occupying the same valleys where they are found today.

In the high mountains, evidence of early and middle Pleistocene glaciations was destroyed by later glacial advances. Fraser glaciation, the last major advance of the Cordilleran ice sheet, began about 25,000 years ago and attained its maximum stand about 50 miles south of Seattle between 15,000 and 13,500 years ago. The ice sheet's recession was followed by a readvance into northern Washington about 11,000 years ago, after which it disappeared. Since that time, the North Cascades glaciers have advanced and retreated many times in response to climatic changes. Significant advances occurred in Neoglacial time, between 3500 and 2000 years ago and about 1000 years ago. In the last 800 years, the glaciers have made a number of minor advances and retreats.

The Cascade glaciers are an important water resource in the Pacific Northwest because they store ice and snow in the winter season and release meltwater in summer when it is needed for irrigation and other purposes.

Geologic Maps and Cross Sections

Brown, E. H., Blackwell, D. L., Christenson, B. W., Frasse, F. I., Haugerud, R. A., Jones, J. T., Leiggi, P. A., Reller, G. J., Rady, P. M., Sevigny, J. H., Silverberg, D. S., Smith, M. T., Sondergaard, J. N. and Ziegler, C. B. 1986. Geologic map of the Northwest Cascades. Geological Society of America Map and Chart series.

Misch, P. 1979. Geologic map of the Marblemount quadrangle, Washington. GM-23. Washington Division of Geology and Earth Resources.

Staatz, M. H., et al. 1972. Geology and mineral resources of the northern part of the North Cascades National Park, Washington. U.S. Geological Survey Bulletin 1359. Plate I.

Bibliography

Adams, J. 1984. Active deformation of the Pacific Northwest continental margin. *Tectonics* 3:449–472.

Alt, D. D. and Hyndman, D. W. 1984. *Roadside geology of Washington.* Missoula, Montana: Mountain Press. 282 p.

Beget, J. E. 1982. Recent volcanic activity at Glacier Peak. *Science* 215:1389–90, March 12.

Brown, E. H. 1987. Structural geology and accretionary history of the Northwest Cascades system, Washington and British Columbia. *Geological Society of America Bulletin* 99(2):201–214.

Crandell, D. R. 1965. The glacial history of western Washington and Oregon. In H. E. Wright, Jr., and D. G. Frey, editors. *The Quaternary of the United States* (VII INQUA Congress). Princeton, New Jersey: Princeton University Press. p. 341–354.

Malone, S. D., and Frank, D. 1975. Increased heat emission from Mount Baker, Washington, *EOS Transactions.* American Geophysical Union. p. 679–85.

Mattinson, J. M. 1972. Ages of zircons from the North Cascade mountains, Washington. *Geological Society of America Bulletin* 83:3769–3784.

McKee, B. 1972. *Cascadia.* New York: McGraw-Hill Book Company.

Misch, P. 1966. Tectonic evolution of Northern Cascades of Washington State. H. Gunning, editor. *A symposium on the tectonic history and mineral deposits of the Western Cordillera,* special volume 8. Canadian Institute of Mining and Metallurgy.

————. 1977. Bedrock geology of the North Cascades. E. H. Brown and R. C. Ellis, editors. *Geological excursions in the Pacific Northwest.* Department of Geology, Western Washington University.

National Park Service. 1985. *North Cascades, a guide to the National Park Complex, Washington,* Handbook 131, 112 p.

Staatz, M. H., et al. 1972. Geology and mineral resources of the northern part of North Cascades National Park, Washington. U.S. Geological Survey Bulletin 1359.

Tabor, J. J. 1986. Crustal structure of the Washington continental margin from refraction data. *Bulletin of the Seismological Society of America* 76:1011–1024.

———— and Smith, S. W. 1985. Seismicity and focal mechanisms associated with the subduction of the Juan de Fuca plate beneath the Olympic Peninsula, Washington. *Bulletin of the Seismological Society of America* 75:237–249.

Tabor, R. W., and Crowder, D. F. 1969. On batholiths and volcanoes—intrusion and eruption of late Cenozoic magmas in the Glacier Peak area, North Cascades. U.S. Geological Survey Professional Paper 604.

North Cascades National Park

Size: 504,781 acres; 789 square miles
Established: October 2, 1968
Address: 2105 Highway 20, Sedro Woolley, Washington 98284

27

Olympic National Park

Northwest Washington

The highest annual precipitation rates in conterminous United States nourish some 60 active glaciers and support a temperate rain forest on the lower western slopes.

By contrast, a rain shadow keeps northeastern slopes nearly as dry as the Baja California coast.

Very young sedimentary rocks, intermixed with submarine lavas, were scraped off the ocean floor by plate convergence and pushed up into a chaotic, mangled heap on the Olympic Peninsula.

Hot springs bubble up from deep faults.

The western shores are exposed to surf as violent and stormy as anywhere in the world.

The largest herd of Roosevelt elk left on earth is protected in the park.

Figure 27.1   Air view of the Hoh Glacier on the summit of Mount Olympus, Olympic National Park. Glaciers actively erode resistant sandstones and conglomerates of the Olympic massif. National Park Service photo.

OLYMPIC NATIONAL PARK

Figure 27.2   Olympic National Park, Washington.

329

OLYMPIC
NATIONAL PARK
(COASTAL SECTION)

WASHINGTON

Ozette Indian
Reservation
Cape Alava

Sand
Point

Swan
Bay

Ozette
Lake

Allens
Bay

OCEAN

Quillayute

River

La Push
Quateata

Quillayute Indian Reservation

Second
Beach
Teahwhit Head

Third
Beach

Taylor Point

Toleak Point

PACIFIC

Hoh Head

Hoh Indian Reservation

Hoh   River

Abbey
Island

Destruction
Island

Kalaloch   101

N

Park boundary

0   4 mi.

TYNAC

Figure 27.2   Continued

Local History and Geographic Setting

The natural features of Olympic National Park are remarkable in so many ways it is difficult not to use too many superlatives when discussing the scenery, the wildlife, and the geology. Originally, the land was set aside for the elk and the forests were thus preserved. A naturalist, C. Hart Merriam, discovered the Olympic elk and named the new species for Theodore Roosevelt in 1897. In 1909, just before leaving office, President Roosevelt signed the proclamation that created Olympic National Monument. President Franklin D. Roosevelt, 29 years later, approved legislation that established Olympic National Park and protected a much larger area, nearly a million acres. Olympic National Park was designated a Biosphere Reserve in 1976 and a World Heritage Site in 1981.

Preservation of the wilderness habitat of the elk meant that some of the Olympic rain forest was also saved. Among the unique and valuable trees are towering Douglas fir of record size, Sitka spruce, giant western hemlock, and red cedar. Some trees have diameters of 12 feet or more; some attain heights of 300 feet; many are between 100 and 200 feet tall. Beneath the umbrella of the giant trees, ferns, mosses, and smaller plants grow in profusion on the forest floor (fig. 27.3).

Higher up are forests of thin-spired, subalpine firs that fringe beautiful alpine meadows near the treeline. On the crest of the range are bare ledges, glaciers, snowfields, and the peaks of the massif.

The Olympic Peninsula, with its snow-capped mountains looming up, was a landfall for ships exploring the Pacific in the seventeenth and eighteenth centuries. The English navigator Captain James Cook searched these waters for a "northwest passage" in 1778. Although he failed to find the fabled route, he took on board a load of sea otter pelts that he sold for great riches in China, thus establishing the maritime fur trade. Captain Robert Gray (for whom Grays Harbor is named) was the first American trader to explore the northwest coast. During his

Figure 27.3   Mosses drape giant trees in the Hoh Rain Forest. Precipitation in the area averages around 150 inches a year. The couple standing on the trail (center) look tiny in comparison to the trees on either side. National Park Service photo.

second voyage, from 1790–93, he met Captain George Vancouver, who had been sent from England to chart the coast. Near Destruction Island, the two captains exchanged information about the coastal waters and harbors. In subsequent years, American and English fur traders established trading posts along the Washington and Oregon coast.

The Indians in the region were brave hunters and fierce fighters. The Makah Indians were greatly admired for their courage because they hunted whales in dugout canoes using spears and harpoons with the points fashioned from stone. Today much is being learned about the Makah culture because a few years ago at Cape Avala in the Ozette Indian Reservation, a hillside of clay slumped down, exposing a series of Makah villages that had been built on the site at different levels. Apparently when one village was covered by a slump or slide, the Makahs built a new village on the ruins of the old. Many artifacts, such as paddles, baskets, weapons, carvings, and other items of the Indians' daily life, were preserved by the clay. Anthropologists from Washington State University and Makah descendants living in the area have been excavating the site and cleaning and cataloguing the artifacts; many are now on display at a tribal museum in Neah Bay, a village in the Makah Indian Reservation. The museum and the site being studied are adjacent to the northern end of the coastal strip belonging to Olympic National Park.

The interior of the Olympic Peninsula remained an unexplored wilderness for many years after the northern coast and the Puget Sound area were settled. On published maps, the central mountainous area was left blank as late as 1890. Highway maps today show roads skirting the perimeter of the Peninsula but none crossing Olympic National Park from east to west or from north to south. A few paved deadend roads lead to points of interest, such as the visitor centers at Hoh Rain Forest and Hurricane Ridge, but only logging roads and trails provide access to the park's interior.

A satellite photo of the Olympic Peninsula shows ridges in disarray, appearing to trend in every direction, with few passes through the mountains. Drainage is radial from the high area near Mount Olympus, the highest peak (elevation 7965 feet). Streams rush down narrow valleys between steep interfluves. Getting from one drainage to another is difficult, particularly where glaciers block the way. Below the treeline, dense forests impede one's passage.

The first official exploration of the mountains was led by Lieutenant Joseph O'Neill, U.S. Army, in 1885. He and his party managed to reach the top of Hurricane Ridge above the Elwha River, which flows north to the coast near Port Angeles (now the site of Park Headquarters).

An expedition sponsored by the *Seattle Press* set out in December 1889 to cross the Olympic Range. The five men in the party had planned to go up the Elwha River by boat, but that proved to be impossible. Next, the sledges they built to carry their supplies bogged down in heavy snow. Then they tried mules, but one fell off a cliff and the other dropped from exhaustion. So they backpacked. With their supplies nearly gone after 14 weeks of climbing and searching for a divide that would get them to a southward-flowing stream, they found a river going to the southwest. Following this stream down, they met a hunting party, who took them by canoe to Lake Quinault. From there, they hiked to the Pacific shore and walked south along the beach to Aberdeen. The trip from Port Angeles to Aberdeen took six months! What they had proved was that the mountains could be crossed. They had marked a trail of sorts and had named many previously unknown peaks and rivers.

In the summer of 1890, Lieutenant O'Neill led a scientific expedition into the Olympic Mountains. In the party were a naturalist-cartographer and a botanist, as well as soldiers to assist them. With great effort, the party constructed a mule trail up the North Fork of the Skokomish River and down the Quinault. They mapped the South Fork of the Skokomish; the Wynoochee River, which joins the Chehalis River; the Humptulips, which empties into Grays Harbor; and several minor streams. Finally, they traveled down the Queets River and reached the Pacific at the southern tip of the park's coastal section.

The Making of Geologic Maps

The geology of the Olympics remained obscure for a long time, even after much of the topographic mapping had been done. Undertaking field work in the rugged and densely forested terrain was difficult; moreover, the geologic structures and stratigraphic relationships are extremely complex. A noted geologist, who in his youth was with a U.S. Geological Survey party in the rain forests, described field work as "going sideways through the trees day after day, soaking wet." Another geologist in an early field party, who collected hand specimens, measured bed thickness, and took strike and dip of beds, recalls "crawling up creek beds through the icy-cold water, next to the bank and beneath overhanging bushes, because that was the only place where bedrock was exposed!" Field garb in those days consisted of heavy, long-sleeved shirts; "tin pants" (duckcloth impregnated with paraffin); hobnail boots; and leather gloves—the latter worn as protection against "devils club," a bushy plant with poisonous barbs—very painful if touched by bare skin. By this kind of arduous, close-to-the-ground work, and hiking and climbing for months at a time, geologists finally accomplished the task of collecting field data necessary for the preparation of geologic maps of the Olympic Peninsula.

In making a geologic map, all outcrops visited are marked on a base map, and the type of rock, the strike

PACIFIC OCEAN

JUAN DE FUCA RIDGE

OLYMPIC PENINSULA

PUGET BASIN

Sea Level

JUAN DE FUCA PLATE   Sediments

NORTH AMERICAN PLATE

OCEANIC CRUST (Volcanics)

Figure 27.4   How plate convergence may have produced the complex structures of the Olympic Peninsula. Oceanic crust, bearing marine sediments, moved eastward for millions of years from the Juan de Fuca ridge, a Pacific spreading center. When heavier oceanic crust met less dense continental crust of the North American plate, the oceanic plate was subducted. Some sedimentary material was dragged down on the oceanic plate; some was scraped off and piled up against the North American plate, enlarging the continent by accretion. From Rau 1980, *Washington coastal geology between the Hoh and Quillayute Rivers*, Bulletin 72, State of Washington Department of Natural Resources.

and dip, and other pertinent observations are entered in a field notebook. Later, field notes are collated; and hand specimens are examined for fossils, bedding characteristics, size of grains, color, and the like. Eventually patterns of rock distribution are developed and their relative positions are studied. Using geologic maps based on field data and laboratory studies, together with scientific knowledge about nearby areas, geologists develop a model for the geologic history of a region. Such a model is by no means finite, for it is revised and refined from time to time as new data and new knowledge become available.

Geologic Features

From early on, geologists have realized that the rocks of the Olympic Peninsula are not old, but only in recent years has it been shown how very young some of them are. For example, we might look at sea stacks (erosional remnants) of Miocene rock—only about 15 million years old—on the western Olympic shore and be awestruck by the fact that some beds are vertical and some are upside down. What force since Miocene time could have stood on end or overturned beds that had been deposited and lithified on the sea floor in a horizontal position?

Not far away, at Point of Arches in the Makah Indian Reservation (outside the park), the oldest rocks of the Peninsula jut out into the ocean. These are pre-Cenozoic outcrops of gabbro, at least 80 million years older than the oldest rocks in the Olympic Mountains.

A Plate Tectonics Explanation

What seems evident from the data available, is that most of the sedimentary rocks of the Olympics were deposited on basaltic oceanic crust, although some were laid down on partly submerged continental rock that may have been on the leading edge of the North American plate. Another possibility is that the continental rock formed in another location (perhaps to the south) and rode along on the oceanic crust as a microcontinent or block of "suspect" terrane.

When an eastward-moving segment of oceanic crust (called the Juan de Fuca plate) encountered the westward-moving North American plate, the heavier basaltic oceanic crust was subducted or overridden by the lighter continental crust (fig. 27.4). In the process, during a period of several million years, accumulations of marine sedimentary rock were scraped off the plate, folded, compressed, thrust downward or squeezed upward, fractured, partially metamorphosed, sliced, and uplifted. The creation of the Olympic Mountains was one result of this powerful tectonic activity. Farther east the effects can be inferred from the upwarping of the North Cascade Range (chapter 26) and the construction of the northern Cascade volcanoes (chapters 32, 33).

While the sediments were still accumulating on the ocean floor, great quantities of basaltic submarine lava began welling up and spreading over or interfingering with the sedimentary layers. In some places lava flows piled up so high that offshore islands may have been formed. Most of the flows cooled quickly under water as pillow lavas or brecciated rocks, but some solidified more slowly in dikes and sills between sedimentary beds or in previously erupted lavas (fig. 27.5).

The chemical composition of the basalt indicates that the lava rose from very deep in the earth and that it had not been previously mixed with surface or marine rocks. However, alteration of the basalt soon occurred as the lavas were subjected to the same forces of deformation as the sedimentary rocks.

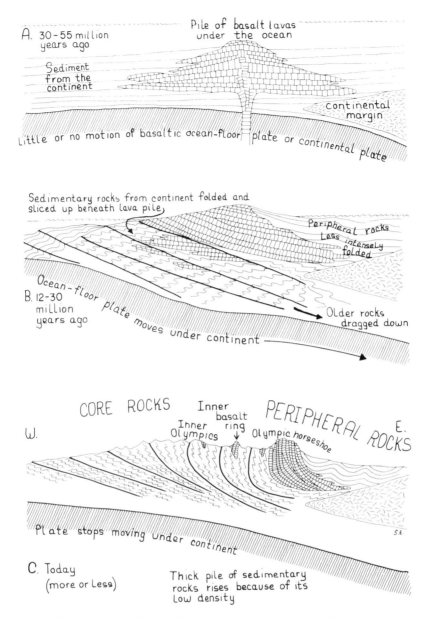

Figure 27.5   How the Olympic Mountains developed. *A.* Accumulation of basaltic lavas, interbedded with sediments, on the ocean floor. *B.* Subduction of the oceanic plate, causing deformation of sedimentary and volcanic rock layers. *C.* Uplifting of sedimentary rocks due to their lower density. From R. W. Tabor, *Geology of Olympic National Park: Hiking and Motorist Guide.* © 1987 Pacific Northwest National Parks and Forests Association. 3rd printing. Used by permission.

Between eruptions of pillow lava, during quiet intervals, a few beds of fossiliferous red limestone were deposited on the ocean floor. The fossils in the rock are Foraminifera (one-celled protozoa that secrete calcite), whose tiny shells can barely be seen with a magnifying glass. The red color of the rock is from iron leached out from cooling lava by hot sea water. Associated with the iron are manganese and copper. Some copper and manganese ores were mined early in this century, but the deposits were too small to be of much commercial value. The upper North Fork of the Skokomish River, one of the areas where the red limestones are associated with pillow lavas, was well explored by prospectors and miners. A claim staked near Black and White Lakes (named for a brand of whiskey) yielded over a hundred tons of manganese ore that was carried out of the mountains by mule train during World War I.

Box 27.1
Paleomagnetism in Iron-Bearing Rocks

Basalts and many other igneous rocks contain iron-bearing minerals, such as magnetite, that recorded the strength and direction of the earth's magnetic field *at the time* they crystallized in a cooling lava flow. The atoms in the crystals responded to the earth's magnetic field by "pointing" toward the location of the north magnetic pole. Unless the rocks are remelted, they permanently retain the magnetic alignment of the time of their crystallization, and this time can be ascertained by techniques of radiometric dating (chapter 19).

The polarity of the present day places the magnetic north pole close to the geographic North Pole. This is what we call *normal polarity.* However, paleomagnetic studies of ocean floor basalts and basalt flows stacked up on land have revealed that some lavas cooled with a magnetic orientation that is directly opposite to the earth's present magnetic alignment. This means that time after time during the geologic past *magnetic reversals* occurred when the magnetic north pole and the south pole exchanged positions so that the magnetic lines of force ran in the opposite direction (i.e., from north to south) in *reversed*

polarity, instead of from south to north, the normal polarity of the earth's magnetic field today.

Scientists have determined that the youngest basalts of the ocean floor are those most recently erupted from a spreading center between two oceanic plates. Outward from a spreading center, the basalts get progressively older on either side as the plates move like two conveyor belts taking the rocks farther and farther away from the point of eruption. Times of magnetic reversal show up in a banded pattern in the ocean-floor basalts, each side (outward from the spreading center) is a mirror image of the other.

When it was discovered that the patterns of progressively older magnetic reversals recorded in the ocean basalts could be matched up with reversals of polarity shown in lavas on land, a powerful tool became available to geologists for determining ages and structural relationships of Olympic rocks. Much of the scientific basis for the concept that oceanic crust plunged beneath the North American plate near the present Olympic coast is derived from study of magnetic patterns in basalts on land and on the ocean bottom.

Because people were interested in mining prospects, the basalts of the Olympic Peninsula were mapped first. Geologists found that most of the basalts cropped out in an arcuate or horseshoe pattern, partly ringing the Peninsula, with the arc portion of the horseshoe on the east side and the arms extending westward. This rock unit, named the Crescent Formation (for exposures at Crescent Bay), is also referred to as the "basaltic horseshoe." The pulloffs and switchbacks on Hurricane Ridge Road, which goes from Park Headquarters to an overlook at about the 6000-foot level, display excellent outcrops of the horseshoe basalts and sedimentary beds, with the oldest rocks at the summit of Hurricane Ridge and the youngest at the base.

Since the fossil Foraminifera in the red interbedded limestone were dated as Eocene in age (Rau 1964), the associated lavas were also presumed to be Eocene. No older bedrock has been found on the peninsula except for the

small outcropping of pre-Cenozoic intrusive rock on Point of Arches. In fact, the sedimentary layers between the Crescent Formation and the western coast become progressively younger from east to west.

The plate tectonic explanation is that some basalts peeled off and were wedged up against the older rocks of the continent as the ocean floor was underthrust and bent downward (fig. 27.5). Rocks that had been deposited seaward of the basalt pile were crammed accordionlike against the more rigid arch of basalt, and some beds were dragged down by the subducting plate. All the layers were severely deformed, fractured, and sliced by the process, as well as partially metamorphosed by heat and pressure. The younger rocks farther west were also deformed, but they were not so contorted and mashed as the beds jammed inside the basaltic horseshoe. Microfossils (mainly Foraminifera) are more common in the younger beds. The lack of fossils in the older rocks is probably due to destruction of the shells by severe deformation and recrystallization.

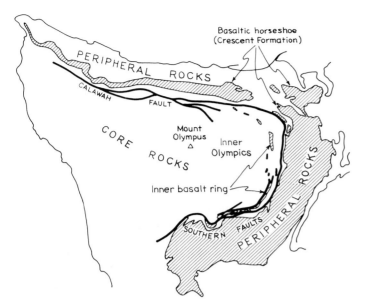

Figure 27.6   Locations of dominant rock groups and major faults on the Olympic Peninsula. From R. W. Tabor, *Geology of Olympic National Park: Hiking and Motorist Guide.* © 1987 Pacific Northwest National Parks and Forests Association. 3rd printing. Used by permission.

Study of the fossil Foraminifera helped geologists to better understand the structural and stratigraphic relationships of the Olympic sedimentary rocks, but no fossils existed in the lavas. Fortunately, the study of *paleomagnetism,* the record of ancient magnetic fields, which was developed in conjunction with the combined theories of sea-floor spreading and plate tectonics, proved to be helpful in discerning age relationships and interpreting structures.

Faults and Hot Springs

During the convergence and subduction of plates, the rock layers in the Olympics that had been spread over a large expanse of sea floor were severely sliced up, stacked in a thick pile, turned on edge, and curved into arcuate belts that are roughly concentric with the basaltic horseshoe. Faults separate the rock layers from each other and from the basaltic horseshoe. The most intense disruption and metamorphic alterations occurred in the rocks in the center of the bend of the horseshoe (Tabor 1975, p. 30).

As time went on, isostatic adjustments, or "rebounding," allowed the compressed and at least partially subducted rock layers to move upward. The sedimentary rocks, being less dense, tended to rise higher than the heavier basalt. Eventually, as fault movement continued, the Olympic core rocks rose to their present height.

The Calawah fault zone, a major fault on the Olympic Peninsula, follows a wide curving pattern within the basaltic horseshoe and shows evidence of strike-slip motion. The fault zone of sheared and shattered rock extends westward from Crystal Ridge (west of the Elwha River) to the Soleduck River and continues out to the ocean. To the east, the Calawah fault zone bends southward; it parallels the inner curve of the Crescent Formation, and then interlinks with a group of smaller faults and slate belts, called the "southern fault zone" (fig. 27.6). In combination, the two fault zones clearly mark the separation of the simpler structure of the peripheral rocks from the highly deformed, complex rocks of the inner core.

The hot springs in the park, which are located on or close to the fault zones, are not considered a sign of recent or current fault activity. Active faulting probably took place millions of years ago when the rocks now at the surface were deeply buried. The water that issues from the hot springs is not noticeably different in chemical composition from surface water in the vicinity. The most likely explanation for the existence of the hot springs is that ground water percolates deep into the earth, is heated by the higher temperatures at depth, and then rises to the surface via the easiest, most direct route, which is usually through the shattered rock of the fault zones.

The hot water in Olympic Hot Springs issues from rock in the Calawah fault zone in the northern part of the park. The Sol Duc Hot Springs (south of the Soleduck River) do not flow directly from the Calawah fault zone, but reach the surface by way of fractures in sandstone beds adjacent to the fault. Elk Lick, which used to be a hot spring but is now a seep or boggy area used by elk as a wallow, is in the southeastern part of the park near a branch of the Calawah fault zone.

Turbidites, Tectonic Mélanges, and Piercement Structures

So striking is the appearance of some of the mélange rocks, a hiker is apt to stop and ask, "What's *that?*" To geologists, however, the outcrops may provide significant clues when they are studying a locality and attempting to reconstruct past geologic events.

Turbidites are rocks (or sediments) that were deposited in a marine environment by settling out from muddy water that flowed along a sloping ocean bottom. Such a sediment-laden flow is called a *turbidity current.* Storm waves may start them, or an earthquake that causes undersea sliding, or perhaps a flooding river pouring quantities of sediment into the ocean. As a turbidity current slows down, the sediments tend to sort or grade themselves by size so that the coarsest particles drop first and the finest are deposited last. Turbidite sequences are usually deposited one on top of the other, each sequence grading from coarse to fine, then to coarse again, and so on. When consolidated, the beds become siltstones, sandstones, and conglomerates.

In Olympic National Park, turbidite beds are well exposed in vertical and overturned positions on the western

Figure 27.7   Compressive forces overturned turbidite beds of the Hoh Formation in Miocene time. Intense wave action on the Olympic coast has eroded the rocks. The dashed line marks an earlier wave-cut surface eroded when sea levels were higher. Above the line are reworked glacial, stream, and beach materials deposited on top of the old surface. Compare with figure 27.10. Photo courtesy W. W. Rau and State of Washington Department of Natural Resources.

shoreline (fig. 27.7) and in the high mountains. Some have been fractured, as well as folded and overturned, and the fractures have been subsequently filled with white veins of calcite.

Since tectonic mélanges are associated with intense deformation and shearing, their presence in the Olympic rock sequences is not surprising. By definition, a *tectonic mélange* is a body of rock (large enough to be mapped) that includes fragments and blocks of numerous sizes and kinds of rock embedded in a generally sheared matrix of more malleable rock, such as a claystone or shale. Some of the Olympic mélange blocks are as large as houses, which gives an indication of the tremendous force involved in their formation. The park's western beaches are the best areas to see the mélange rocks. Since moisture causes the clay minerals of the matrix to swell, this part of the rock becomes crumbly, making it susceptible to wave erosion and separable from the resistant inclusions. Locally these rocks (which are part of the Hoh rock assemblage) are called "smell rocks" or "smell muds" because of their strong petroleum odor. Microfossils in the mélange silt-

stones indicate that the depositional environment was a deep-water ocean basin beyond the continental shelf.

Piercement structures are associated with tectonic mélange rocks. Under conditions of high pressures, certain rocks become "plastic," or able to flow; they are squeezed upward into overlying rocks, doming them or even breaking through them. In this way, older rocks may be forced up through younger beds; the result is called a piercement structure.

The Shaping of the Olympic Landscape

Running Water. All the while that great masses of rock from the ocean floor were being pushed higher and higher, the agents and processes of erosion were actively shaping the surface of the newly risen land. Moisture-bearing clouds from the Pacific discharge rain and snow copiously on the western slopes. Precipitation averages 140 to 150 inches now and was probably at least that much in the past.

Water running off in every direction from the central massif initiated the radial drainage, and swiftly de-

Box 27.2
Oil on the Olympic Coast?

Oil seeps, traces of oil in coastal streams, and a natural gas seep on the western coast of Washington prompted efforts aimed at recovering petroleum in the early years of this century. Wells were drilled in the marine shales and sandstones of the Hoh rock assemblage, but none of these attempts was a commercial success.

The early drillers had trouble with the high clay content of subsurface beds. They called such rock "heaving shale" because of its aggravating tendency to swell up when wet, closing off drill holes and buckling pipes. Then there were horrendous problems in getting equipment to drill sites on a coast where an overland approach was impossible, safe natural harbors were few, and ships foundered in heavy surf on dangerous offshore rocks.

The first well was drilled in 1900 at the top of a mélange rock bluff on the shore. A barge carrying equipment and supplies was destroyed during the landing attempt, but most of the machinery was salvaged. A donkey engine was dismantled and carried piece by piece up to the drilling site where it was reassembled and used to haul the rest of the equipment from the beach to the top of the cliff. Things didn't get much better when the crew finally began to sink a hole. Drilling in the mélange rocks was fraught with difficulties. A slight "show" of petroleum must have buoyed their hopes for a while, but drilling conditions became worse and worse. Finally the site was abandoned, along with the machinery and equipment that had been so arduously brought in. An old steam boiler and other equipment are still rusting away in the underbrush, close to one of the park trails.

The Jefferson Oil Seep, also within the present boundaries of Olympic National Park, was a more promising site. It was explored initially in 1913–14 and again in the 1930s. Being near the mouth of the Hoh River, the locality was more accessible. Eleven wells were drilled in the 1930s, but production was never high enough to be profitable. Today the wooden derricks have tumbled down, but a bit of oil still seeps out at the surface and natural gas bubbles up.

What are the conditions that make recovery of petroleum possible? First, there has to be a *source rock* of marine sediment containing organic matter buried during sedimentation. From the decayed organisms, droplets of oil (hydrocarbons) form and migrate through pores in the rock, generally upward because oil is lighter than water. Unless a geologic structure is present that can trap and retain migrating droplets, they may rise all the way to the surface, seep out, and be lost forever. A favorable structure might be an anticline with a layer of impermeable rock that could prevent the escape of droplets. A third requirement is a permeable *reservoir rock,* below the impermeable rock, with enough porosity to hold accumulating petroleum. The Olympic coast sedimentary beds have a source rock, but the right combination of reservoir rock and appropriate geologic structure may not be present.

---

scending streams cut steep, narrow valleys. The differing hardness of rock types has made modifications in the original radial drainage. The basalt ridges tend to be higher and sharper in relief than the less resistant sedimentary beds surrounding the basalt. Since the sandstone beds are harder than the shale beds, the tributaries in the latter tend to cut more deeply than the tributaries that flow over sandstone. As a result, some of the streams in the deeper valleys have captured the drainage of the streams on more resistant rocks.

Also affecting how the radial drainage evolves is the contrast between the copious moisture on the western slopes and the amount that falls in the rain-shadow zone on the eastern slopes. In the city of Sequim, for example, outside the northeast corner of the park, precipitation is only 16 to 18 inches per year, or a little more than one-tenth of what falls on the western slopes. Tributaries on the east side of the Olympic Mountains have steeper gradients, are less numerous, and carry smaller volumes of water than those on the west side.

Figure 27.8   This view looks south toward Mount Olympus, about ten miles away (air distance). The steep slopes of the ridge, underlain by tilted beds of late Tertiary sandstone, are mantled by thick sheets of talus resulting from frost action and mass wasting. The forest cover, supported by ample precipitation, has begun to stabilize the slopes. National Park Service photo by G. A. Grant.

Glacial Ice. The Olympic landscape shows evidence of severe glaciation in the past; but at the present time, glacial ice, while significant, is secondary to running water as an agent of erosion. What is most remarkable about existing Olympic glaciers is their low elevation on the western slopes due to the abundance of snow that keeps them well nourished. Above the snowline, at around 6000 feet, winter snows last through the summer season. Some of the glaciers extend down their valleys to as low as 4500 feet. The Hoh glacier, descending three miles, is the longest.

Largest in size is the Blue Glacier, which has been closely monitored by glaciologists since 1938. A University of Washington glacier research station, built close to the Blue Glacier in 1957–58, collects data continuously as part of a study to discern relationships between regional climate, accumulation and evaporation of snow and glacial ice, melting rates, and velocity of ice flow. Measurements show that in recent years the Blue Glacier has grown slightly.

Moraines, U-shaped valleys, cirques, rock-basin lakes, erratics, and other glacial features beyond the limits of the existing valley glaciers attest to the more extensive alpine glaciation of Pleistocene time. Because of rapid weathering and erosion in the Olympics, some of these landforms are not so long-lasting as similar glacial features in more resistant rock or drier climates.

Rock-basin lakes tend to have short lives in the high Olympics. As a lake silts in, vegetation spreads over the shallows. Soon the lake is a bog; and then it is a meadow, with perhaps a stream winding through it. If the basin is below the treeline, the forest takes over the open space.

Till and glacial outwash overlie the floors of the glacially excavated valleys. One of the most beautiful glacial troughs in the park is the flat-floored Enchanted Valley, lying between steep mountain walls that lead up to hanging glaciers on either side of Mount Anderson. This is the headwaters region of the Quinault River. Farther down, just south of the park boundary, is Lake Quinault (in the Quinault Indian Reservation), dammed by a glacial moraine. These are but a few examples of interesting alpine glacial features that can be visited in the Olympic wilderness.

Part of the Olympic Peninsula was covered by advances of continental ice. Granite and gneiss pebbles in the stream gravels on the north and east slopes are among the indications of continental glaciation. Presumably the pebbles were dumped by invasions of Cordilleran ice from British Columbia, because no granite or gneiss bedrock is exposed in the Olympic Peninsula. Also foreign to the Peninsula are some granitic and metamorphic boulders and cobbles dropped by icebergs in fiordlike lakes that at one time filled valleys on the north and east sides of the mountains. The Cordilleran ice sheet dammed these valleys after the alpine glaciers that carved them melted back. Icebergs broke off ("calved") from the waning ice sheet, floated up the lakes, and melted, dropping the rock debris.

Processes of downslope movement. Mass wasting was most effective during times of glacial retreat when freshly excavated valley slopes, unprotected by vegetation, were left in an unstable condition. Sometimes a ridge collapsed when it was no longer supported by glacial ice. Rock units with tilted beds are particularly susceptible to avalanching (fig. 27.8).

A recent rockslide occurred near the terminus of the Hoh Glacier, which has retreated more than a mile since the early 1800s. Some 20 or more years ago, cracks formed on the ridge crest, and the unsupported valley wall gave way; this left a fresh scarp on the valley side slope and a pile of rock debris on the moraine below the glacier. Throughout the park, blocks of rock on oversteepened cliffs or cirque walls fall or slide down.

Some lakes in the park backed up behind natural dams that resulted from landsliding. Three examples are Jefferson Lake, Elk Lake, and Lena Lake in the southeastern part of the park. On the Elwha River, in the fairly recent past, a large rockslide dammed a narrow canyon and formed a lake. In 1967 the dam gave way, releasing the impounded lake water, which washed out a trail bridge and spread gravel over the flood plain downriver. The river had caused the rockslide in the first place by undercutting the base of an unstable slope.

A great landslide mass (that probably came down soon after the last Cordilleran ice sheet retreated) separates Lake Crescent, the largest lake in the park, from Lake Sutherland. The two lakes were originally one body of water that occupied a long glacial trough drained by Indian Creek, which flows eastward into the Elwha River. The landslide, which was caused by the collapse of the valley side slopes near Mount Storm King, dammed the valley and cut off Lake Sutherland. Lake Crescent cut a new outlet, forming the Lyre River which flows north into the Juan de Fuca Strait. Indians may have witnessed the catastrophic landslide because an Indian legend describes the spirit of Mount Storm King becoming so angry over tribal warfare going on in his domain that he threw down great rocks that blocked the valley and put an end to the fighting.

Figure 27.9   At Third Beach, near Taylor Point, blocks of conglomerate (Hoh Formation) fall to the beach from sea cliffs undercut by wave erosion. Rounded pebbles and cobbles and fragments of a variety of rock types make up this poorly consolidated conglomerate. Photo courtesy W. W. Rau and State of Washington Department of Natural Resources.

The Olympic National Park Coastal Strip and Its Shore Features

Between the west coast of Washington and the islands of Japan are 6000 miles of open ocean. Impelled by westerly and northwesterly winds, waves travel unimpeded across this vast stretch of water. Throughout the year, a great deal of wave energy is available to erode the beaches and cliffs and transport beach material. During winter storms, waves 15 to 20 feet high beat on the unprotected shore for days at a time. Comparison of coastal surveys reveals that the Washington Pacific coast near Hogsback retreated 225 feet in 60 years. At this rate, 375 feet of shore are being eroded back every 100 years!

The contrast between the high rate of coastal retreat on the Olympic coast and the relatively low rate of erosion of the coast of Acadia National Park (chapter 21) is remarkable. The North Atlantic waves that beat on unprotected parts of the northern Maine shore are not significantly less powerful than the North Pacific waves. What makes the difference is the character of the bedrock. The contrast we see is between the effects of wave erosion on very old, highly resistant, igneous and metamorphic rocks in Acadia National Park and wave action on very young, relatively nonresistant sedimentary rocks on the Olympic coast (fig. 27.9).

Figure 27.10   Simplified cross section showing uplift, sea level changes, deposition and erosion during Quaternary time on this tectonically active coast. The position of the shoreline has changed several times, to elevations both higher and lower than at present, due to changes in sea level associated with Pleistocene glacial episodes as well as tectonic uplift of the Olympic Peninsula. The older land surface (top), built by glacial and fluvial sediments, extended seaward for many miles beyond the present shoreline. As the coast rose, sediments were eroded and a younger wave-cut surface developed. This surface was, in turn, covered by a thinner series of sediments that extended westward of the present shoreline. Wave action, following uplift, has cut back the shoreline. Alexander Island is a remnant of the former mainland. From Rau 1980, *Washington coastal geology between the Hoh and Quillayute Rivers,* Bulletin 72, State of Washington Department of Natural Resources.

The coastal section of Olympic National Park is a narrow strip that extends about 40 miles from the mouth of the Queets River on the south to Cape Alava and the Ozette Indian Reservation on the north. Two small Indian reservations are enclaves in the coastal strip: the Hoh Indian Reservation beside the Hoh River and the Quillayute Indian Reservation at the mouth of the Quillayute River. A coastal highway (Route 101) follows the shore from Queets to the Hoh Indian Reservation. The rest of the park's coastal section is a wilderness area.

In the last 8 million years or so (since early in Pliocene time), the Olympic coast has been at times higher and at times lower than present sea level. Some wave-cut platforms are several hundred feet higher than sea level today, and some old wave erosional surfaces are covered by alluvial deposits that once extended to a shoreline hundreds of feet to the west. Because sea level fluctuated considerably, the shoreline has also shifted eastward and westward a number of times. The raising and lowering of sea level was partly due to Pleistocene glaciation that drew moisture from the oceans when the glaciers were enlarging and then released water as they melted back. But tectonic warping and minor uplift of this coast was also occurring during the Pleistocene, making interpretation of features of emergence and submergence difficult (fig. 27.10).

Rapid erosion also alters the position of the shoreline. Cliff retreat produces a widening, submerged, wave-cut platform that extends several hundred feet out from the water's edge. Stacks and small islands of more resistant material that were recently part of the mainland now sit on the wave-cut platform surrounded by water. Sea arches collapse and become stacks. Caves in cliffs are eroded into sea arches (figs. 27.11, 27.12).

As cliffs are undercut and landsliding occurs, the forests that grow right to the edge of the bluff begin to sag and slump. On the north side of the Hoh River mouth is a "drunken forest" where live trees are attempting to bend and grow vertically again after being tilted by a slide. Other drunken forests can be seen along the coast.

Beaches are littered with dead trees that have fallen and great logs that have floated down the rivers to the ocean and then washed ashore. When storms come, these logs act as battering rams hurled against the rocks by the waves.

The beaches between headlands tend to be narrow and have fairly steep slopes. In some areas the steep cliffs come right down to the water. Nevertheless, the Olympic beaches have been the most convenient "pathway" from one place to another along the coast since prehistoric times because of the difficulty of traveling inland through the thick vegetation and the rain forests. Today beach hikers can follow well marked trails over the cliffs between stretches of beach. However, the warning still holds that hikers must guard against becoming trapped on a narrow beach at high tide, because most of the cliffs are too steep to climb. The beaches must also be avoided when heavy surf or storm waves are running.

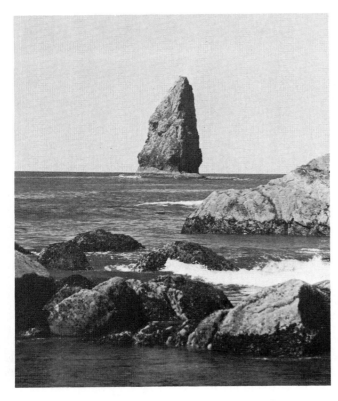

Figure 27.11   Sea stack of massive sandstone (Hoh Formation) off Taylor Point. Weathering and wave action in the surf zone produced the wave-cut notch around the base of the stack. Photo courtesy W. W. Rau and State of Washington Department of Natural Resources.

Figure 27.12   Waves attacking a fracture zone in massive sandstone of the Hoh Formation eroded the sea arch at Second Beach. Photo courtesy W. W. Rau and State of Washington Department of Natural Resources.

Most of the rocks and cliffs in the park's coastal section are in the Hoh rock assemblage of Miocene age. The Kalaloch Rocks, offshore from the Kalaloch Ranger Station, are the southernmost exposure of the Hoh turbidites along the coast. They are mainly sandstones and conglomerates, but they are called "graywackes" because they contain fragments and grains of other rocks. Most of the beach sand is derived from these rocks or from sediment washed down from the mountains. Blocks of tectonic mélange Hoh rocks form many of the stacks and other strange-shaped erosional remnants sticking up in the water close to shore. The blocks are jumbled masses of resistant basaltic breccias, conglomerate, sandstone, and siltstone. Much of the original matrix of softer rock has been removed by weathering and erosion. Some of the mélange rocks, such as those in the bluffs at Jefferson Cove, are thought to be part of piercement structures.

Overlying the Hoh rocks along the shore are Pleistocene deposits of sand, gravel, and silt. Most of the deposits are alluvial and derived from glacial outwash that poured out of the mountains as the valley glaciers retreated. The youngest Pleistocene deposits consist of fine, wind-deposited silt, called *loess,* which tends to form bluffs because the silt grains, being coherent, cling together. The loess at Quateata headland (First Beach) stands directly on an old wave-cut platform of bedrock and gravel; and at Taylor Point (Third Beach), the wind-blown deposits rest on older alluvial sediment. The loess accumulated near the end of the Pleistocene Epoch when the ice was melting back rapidly but sea level was still quite low. Large areas of land, unprotected by vegetation, were uncovered and left exposed to unusually high winds. Fine surface sediment, picked up by the winds, was redeposited over the coastal area.

The beaches between the headlands are made up of sand, gravel, and cobbles of various sizes and compositions. The sediments are derived from the erosion of local bedrock, from glacial drift, and from material brought down by rivers from the Olympic Mountains. Here and there patches of red sand have accumulated. The red sand grains are tiny garnet crystals that have weathered out of erratics or been transported in glacial debris and then eroded by streams. Being heavier than quartz grains, the garnet crystals tend to be concentrated in stream beds and beach terraces. Presumably the red sands have been transported, concentrated, re-eroded, and reconcentrated several times on their journey to the shore from their places of origin.

Box 27.3
How Waves Do Geologic Work

Wind-driven water waves erode, transport, and deposit material in a narrow coastal zone where land, water, and air meet. The source of wave energy is wind. Energy is transferred from the moving air into wave shapes that move through the water. Note that the water itself does not move very far; rather, it is the waves, impelled by the wind, that transmit the wind's energy, releasing it when they strike land. The size of waves is a function of (1) wind velocity, (2) the length of time that the wind blows in the same direction, and (3) the *fetch,* or distance, of open ocean over which the wind blows.

Great waves may be several tens of feet high and several hundred feet long. Once wave sets develop in an open ocean, they may move across hundreds of miles of sea before they break against a coast, their power and size unaffected by lighter, or changeable local winds.

When a deep-water wave begins to "feel" a shallowing bottom near shore, its lower part begins to flatten due to drag, while the upper part becomes shorter and steeper. Within a few seconds, the top of the wave falls forward, becoming a curling, breaking wave, or *breaker,* that slams against a cliff or rushes up a beach. When forward motion stops, the water runs back down the beach and slides under the curl of the next breaker. This upwash and backwash (called "swish and swash") breaks up and abrades rock and sediment, sorting coarser fragments from finer particles and rolling them back and forth. Coarser sediment tends to accumulate along the top of the beach while finer materials tend to stay in the *surf zone* (the strip of shore where the waves break and roll up the beach or beat against a cliff).

The surf zone shifts up and down the beach twice a day as tides rise and fall. Thus normal wave erosion is most intense on the *foreshore,* the area between mean low-water level and mean high-water level. Storm waves attack any part of the shore they can reach with devastating effect. Sizable rocks may be picked up and tossed high into the air, often shattering as they hit. Tons of sand may be removed from a beach in a single storm.

Most beach sediment is only temporarily at rest. Sand is moved along beaches by *beach drift.* When waves come in obliquely, as they often do, instead of head on, sand is carried up the foreshore at an angle. But the grains of sand roll back, under the influence of gravity, perpendicularly, progressing along the beach in a series of looping movements.

Sometimes water "piles up" just outside the surf zone and starts moving in a *longshore current* carrying sand parallel to the shore. When the current slows down or is deflected by a headland, the sand drops and may form a bar or *spit* (a fingerlike extension of a beach, with the far end terminating in open water). This process is called *longshore drift.* Sediment entering the ocean at the mouth of a river may encounter a longshore current and be carried up (or down) the shore, along with beach sediment. Meanwhile, the river has to keep changing its outlet to get around the sandbar or spit built by the longshore current. Records show that the Quillayute River entered the Pacific at several different locations from the late 1800s until 1931, when the U.S. Army Corps of Engineers constructed a harbor at La Push and stabilized the mouth of the river with dikes.

Table 27.1 Generalized Geologic Column, Olympic National Park

Time Units			Rock Units	Geologic Events
Era	Period	Epochs		
Cenozoic	Quaternary	Holocene	Glacial drift, fluvial and landslide deposits, talus Beach materials	Glaciation Erosion and deposition by streams, mass wasting, and wave action
			////////// Unconformity	Unconformity
		Pleistocene	Younger glaciofluvial deposits and loess Older glaciofluvial deposits	Wind deposition on and near coast Fluctuation of sea level; uplift and warping Repeated episodes of alpine and continental glaciation
	Tertiary	Neogene — Pliocene	////////// Unconformity	Unconformity
			Quinault and Quillayute Formations (western Olympic coast, not exposed in park) Largely marine sedimentary rocks; slightly tilted.	Marine deposition in coastal areas Folding and upwarp
		—?—	////////// Unconformity	Unconformity
		Miocene	Hoh rock assemblage (western Olympic coast), late Eocene to Miocene. Steeply tilted, overturned sedimentary rocks; turbidites; tectonic mélange rocks	Faulting and uplift Plate convergence and subduction accompanied by intense folding, shearing, slicing, partial metamorphism
	Paleogene	Oligocene	Core rocks (central and western) Eocene to Miocene highly folded sandstones, shales; some highly disrupted. Basaltic horseshoe (Crescent Formation); Eocene to Miocene peripheral sedimentary rocks	More or less continuous marine sedimentation
		Eocene	Inner Olympic rocks Eocene to Oligocene severely deformed sedimentary rocks and basalts	Submarine eruption of lavas
		Paleocene	////////// Unconformity	Unconformity
			Pre-Cenozoic rocks of the old continent (not exposed in park)	

Source: modified after Tabor 1975; Rau 1980

Geologic History

1. The Crescent Formation and the peripheral rocks.

In early Eocene time the lavas of the Crescent Formation began to pile up on the ocean floor, probably not far from the mainland. Eruptions continued into middle Eocene time, even as subduction of the oceanic plate proceeded. As the plate moved down, the thick mass of basalt acted as a buttress, protecting the sandstones, shales, and conglomerates of the peripheral rocks deposited between the accumulated lavas and the continental margin, so that they were only moderately deformed. The sedimentary peripheral rocks are Eocene to Upper Miocene in age (figs. 27.13, 27.14).

2. Eocene to Oligocene rocks of the inner Olympics.

While protecting the peripheral rocks, the Crescent Formation resisted the eastward movement of the sedimentary rocks that were deposited on the seaward side of the lavas. Therefore, the marine basalts, sandstones, phyllites, slates, and conglomerates within the arc of the horseshoe are more thoroughly metamorphosed, foliated, and recrystallized than the younger beds to the west. All the innermost rocks have been highly disrupted by slicing, shearing, and fracturing.

3. The core rocks from the inner Olympics westward.

The inner Olympic rocks and the core rocks are grouped in "lithic assemblages," rather than formations,

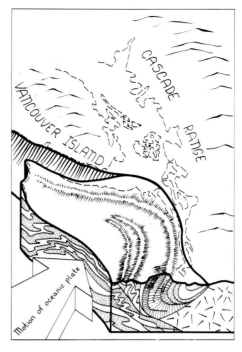

Figure 27.13   Opposing movement of oceanic and continental plates may have jammed the rocks that make up the Olympic Mountains into a tight arc between Vancouver Island and the Cascade Range. From R. W. Tabor, *Geology of Olympic National Park: Hiking and Motorist Guide.* © 1987 Pacific Northwest National Parks and Forests Association. 3rd printing. Used by permission.

Figure 27.14   Glacially eroded bedrock (Crescent Formation) and a small remnant glacier on Mount Anderson (7321 feet) near the southeastern end of the Basaltic Horseshoe. National Park Service photo.

because boundaries and specific rock types are not well defined. As is evident on a geologic map, the lithic assemblages tend to have an arcuate pattern, similar to that of the Crescent Formation. The core rocks range in age from Eocene to Oligocene; the beds closer to the inner Olympics rocks are more intensely folded and disrupted. Many of the core rock units show a well developed slaty cleavage, which indicates some degree of metamorphism. They consist of sandstones and shales with minor occurrences of conglomerate and basalt.

Mount Olympus and the other high peaks in the center of the peninsula are in what is called the "Western Olympic lithic assemblage," a generally younger, less deformed group primarily made up of thick-bedded sandstone, with elongated patches of slate, phyllite, foliated sandstone, and semischist.

4. The rocks of the Olympic Peninsula western coast.

The greater part of the rocks on the Pacific coast are in the Hoh rock assemblage, which is mostly Miocene in age. (Some Eocene volcanics interbedded with siltstones are exposed at Point Grenville, south of the park's coastal section.)

5. The Quinault and Quillayute Formations (not exposed in the park).

The Quinault rocks, Pliocene in age and largely of marine origin, are separated from the Hoh rock assemblage by an unconformity. Quinault outcrops can be seen on the shore in the Quinault Indian Reservation, just south of the park's coastal section. For about 40 miles northward along the coast from the outcrops, the Quinault Formation extends seaward, underlying Quaternary sediments on the continental shelf. The formation is of interest because it is regarded as a potential petroleum reservoir. The Quinault beds, which are only mildly tilted, were deposited after the Hoh rocks had been raised above sea level, eroded, and again partly covered by the ocean.

The Quillayute Formation, which is very late Miocene in age and similar to the Quinault Formation, crops out in the Quillayute River valley (upstream from the coast) in a few places where thick Pleistocene deposits have been eroded away.

6. Pleistocene glaciation and deposition.

Alpine glaciers have probably existed on the highest mountains of the Olympic Peninsula since the onset of the Pleistocene and have advanced and retreated many times. At their maximum extent, the largest valley glaciers on the west side reached as far as the present coast and perhaps beyond, although parts of this coastal region were left unglaciated.

The far northwest coast and all of the north and east sides of the peninsula were heavily glaciated, because the great Cordilleran ice sheet from British Columbia encountered the Olympic glaciers as it moved southward and then divided into two lobes. One lobe filled the Puget Sound

lowland to south of Tacoma, while the other spread over the Juan de Fuca Strait and the northern coast. At least four times, possibly six, the ice sheet advanced and retreated. Cordilleran ice on the northern side of the peninsula, at least 3500 feet thick, scraped the weak peripheral rocks on the outer side of the basaltic horseshoe, destroyed the preglacial topography, and lowered relief along the northern coast.

During warmer interglacial periods, when the glaciers melted back, great quantities of glaciofluvial deposits were laid down over the lowlands and coastal areas. The Pleistocene sands and gravels on the west side and part of the south side are better preserved, because they were not reworked (except in the glacial valleys) by later glaciations. Exposed on the Pacific coastal region are several sequences of Pleistocene and Holocene deposits. The oldest ones, which are semiconsolidated, originally extended for a considerable distance farther to the west than now, but they have been cut back by wave erosion.

Sea level has fluctuated on the Olympic Peninsula, during and since Pleistocene time. Some fluctuations have been due to depression by glacial ice and subsequent release by melting; some changes have been the result of tectonic uplift and warping.

7. Modern glaciers and erosional processes.

In recent years Olympic glaciers have shown slight increases in volume. Stream erosion at present is dominant over glacial erosion; mass wasting is also significant. In the coastal section, wave erosion is unusually rapid.

Geologic Maps and Cross Sections

Rau, W. W. 1975. Geologic map of the Destruction Island and Taholah quadrangles, Washington. Washington Division of Geology and Earth Resources Geologic Map GM-13.

————. 1979. Geologic map of the vicinity of the lower Bogachiel and Hoh River valleys and the Washington coast. Washington Division of Geology and Earth Resources Geologic Map GM-24.

Tabor, R. W., and Cady, W. M. 1978. Geologic map of the Olympic peninsula, Washington. U.S. Geological Survey Miscellaneous Investigations Series Map I-994.

Bibliography

Alt, D. D., and Hyndman, D. W. 1984. *Roadside geology of Washington*. Missoula, Montana: Mountain Press. p. 249–276.

Cady, W. M. 1975. Tectonic setting of the Tertiary volcanic rocks of the Olympic peninsula, Washington. U.S. Geological Survey *Journal of research* 3(5):573–582.

Crandell, D. R. 1965. The glacial history of western Washington and Oregon. In *The Quaternary of the United States,* VII INQUA Congress, H. E. Wright, Jr., and D. G. Frey, editors, pp. 341–8. Princeton, New Jersey: Princeton University Press.

Kirk, R. 1980 (3rd ed.). *Exploring the Olympic peninsula.* Seattle: University of Washington Press.

Rau, W. W. 1964. *Foraminifera from the northern Olympic peninsula, Washington.* U.S. Geological Survey Professional Paper 374-G.

———. 1973. *Geology of the Washington coast between Point Grenville and the Hoh River.* Washington Division of Geology and Earth Resources Bulletin 66.

———. 1977. General geology of the southern Olympic coast. In *Geological excursions in the Pacific Northwest,* E. H Brown and R. C. Ellis, editors, p. 63–83. Bellingham, Washington: Department of Geology, Western Washington University.

———. 1980. *Washington coastal geology between the Hoh and Quillayute Rivers.* Washington Division of Geology and Earth Resources Bulletin 72.

Richards, B. 1984. The Olympic Peninsula, *National Geographic* 165(5):644–673.

Scharf, J. G., and Wilkerson, J. n.d. *Geologic guide to the Hurricane Ridge area, Olympic National Park.* Port Angeles, Washington: Pacific Northwest National Parks and Forests Association. 21 p.

Snavely, P. D., and Wagner, H. C. 1963. Tertiary geologic history of western Oregon and Washington. Washington Division of Geology and Earth Resources, Report of investigation 22.

Stewart, R. J. 1971. Structural framework of the western Olympic peninsula, Washington. Geological Society of America Abstracts with Programs 3(2): 201.

Tabor, R. W. 1987, 3rd ed. *Guide to the geology of Olympic National Park. Hiking and motorist guide.* Seattle: Pacific Northwest National Parks and Forest Association.

———. 1987. A Tertiary accreted terrane: Oceanic basalt and sedimentary rocks in the Olympic Mountains, Washington. *Geological Society of America Centennial field guide—Cordilleran section.* p. 377–382.

———, and Cady, W. M. 1978. The structure of the Olympic mountains, Washington—Analysis of a subduction zone. U.S. Geological Survey Professional Paper 1033.

Warren, H. 1982. *Olympic, the story behind the scenery.* Las Vegas, Nevada: K.C. Publications. 64 p.

Note: For visitors who want to know more about the geology of Olympic National Park, Tabor (1975) and Rau (1987) are helpful guides to take along on excursions.

Olympic National Park

Area: 914,576 acres; 1408 square miles
Proclaimed Mount Olympus National Monument:
    March 2, 1909
Established as a National Park: June 29, 1938
Designated a Biosphere Reserve: 1976
Designated a World Heritage Site: October 27, 1981
Address: 600 East Park Avenue, Port Angeles,
    Washington 98362

28

Icefields, valley glaciers, tidewater glaciers.
Deep fiords, high coastal mountains.
Stormy outer coast, protected inner shorelines.
Glaciers advancing; glaciers retreating.
Glaciated landscapes—outwash plains, drumlins, eskers.
Reforestation after retreat of ice.
Active tectonism, uplift, earthquakes.
Exotic terranes.

Glacier Bay National Park and Preserve

Southeastern Alaska

by Sherwood D. Tuttle
University of Iowa

Figure 28.1   The crevassed terminus of Margerie Glacier towers above a passing cruise ship at the northern end of Glacier Bay. Chunks of ice, large and small, are "calved" from the ice front. A medial moraine rides on the glacier in the middle distance. National Park Service photo.

GLACIER BAY NATIONAL PARK AND PRESERVE

Glacier
Park Border
Preserve Border

0   10 mi.

Figure 28.2   Glacier Bay National Park and Preserve, Alaska.

Figure 28.3   Looking east over Lituya Bay, a large fiord, toward the Fairweather Range. In the foreground, where the fiord opens into the Gulf of Alaska, morainal material in the spit is reworked by waves and currents. Part of an end moraine is plastered over Cenotaph Island (middle distance). Slopes on spurs projecting into Lituya Bay are bare because huge waves, generated by a 1958 earthquake, washed away trees and soil. National Park Service photo by R. Belous.

Geographic Setting

In southeastern Alaska, at the upper end of the Alaska panhandle, Glacier Bay National Park and Preserve occupies a wild and rugged region of ocean, bay, and inlets; lofty, snow-capped peaks; fiords, glaciers, and rain forests. The park/preserve is on the northern side of Cross Sound and Icy Strait, a waterway between the Inland Passage and the Gulf of Alaska. The Alexander Archipelago extends to the south. Heading north, the park/preserve's western boundary is drawn a few miles offshore, parallel to the ocean coast up to Cape Fairweather where it turns into the shoreline and continues on to Dry Bay and the Tongass National Forest. British Columbia lies on the northern border of the park/preserve. Juneau, Alaska's capital city, is about 50 miles east of Glacier Bay on the other side of the Inland Passage. Neither Juneau nor Glacier Bay National Park and Preserve is accessible by road. Transportation is by airplane, ship, or small boat.

Glacier Bay proper is a large, branching fiord, surrounded by mountain ranges and occupying a central lowland in the park/preserve. The huge glaciers that completely covered Glacier Bay 200 years ago have re-

treated. The snout of Grand Pacific Glacier, which was the largest and longest, is now approximately on the international boundary, some 65 miles inland from the mouth of Glacier Bay.

The settlement of Gustavus on Icy Strait, at the mouth of Glacier Bay, is both airport and seaport for the park/ preserve with park headquarters nearby on Bartlett Cove. During the tourist season National Park Service concessionaires operate a lodge at Bartlett Cove and run small-boat trips daily up through Glacier Bay to the snouts of tidewater glaciers, such as Johns Hopkins Glacier, which descends from the Fairweather Range and "calves" (breaks off and releases icebergs) into an arm of the bay. Cruise ships and planes bring thousands of visitors to the park/preserve every summer.

The mountain ranges in Glacier Bay National Park and Preserve extend north and northwest (across a small wedge of British Columbia) to connect with the ranges in Kluane National Park, Yukon Territory, and those in Wrangell-St. Elias National Park and Preserve (chapter 29). The coastlines of Wrangell-St. Elias National Park and Preserve and Glacier Bay National Park and Preserve are separated by a narrow stretch of shore in the vicinity of Yakutat Bay, where the Alaska panhandle joins the main part of Alaska.

Mount Fairweather (elevation 15,300 feet), the highest peak in southeastern Alaska, is at the northern end of the Fairweather Range, a south-trending extension of the St. Elias Range. Mount Fairweather and its neighbor, Mount Quincy Adams (elevation 13,560 feet), mark adjacent corners on the international boundary between Alaska and British Columbia. East of the park/ preserve, also near the international boundary, is Klondike Gold Rush National Historical Park at Skagway, Alaska.

Glaciers flourish in the region because conditions are ideal. Storms sweeping in from the Gulf of Alaska dump rain and snow on the high coastal mountains the year round. Precipitation is estimated to be as high as 180 inches per year midway up the slopes along the western side of the Fairweather Range. The ocean coast is cold, foggy, and virtually uninhabited, with the temperature seldom rising above 75° F. Strong winds from the gulf blow in during all seasons. Heavy ocean waves beat against the shore almost constantly. The combination of high mountains, a prolific source of moisture, cool temperatures, and many stormy or overcast days is what brings about active glaciers and an extensive rain forest along the outer coast of the park/preserve.

At the present time only one tidewater glacier calves directly from the park/preserve into the Gulf of Alaska. This is LaPerouse Glacier, between Lituya Bay and Icy Point. Grand Plateau Glacier (south of Dry Bay) descends to sea level but terminates in a bay.

High tides are characteristic of the inland waterways, with a tidal range of up to 20 feet in many bays and inlets. The high tides are due to a combination of high latitudes and the constricting effect of bays and narrow channels. Tidal currents run at velocities of 5 to 7 miles per hour in the passages, even higher in Lituya Bay. Navigation is made more difficult in the foggy channels of the inner bay by the presence of rocks, floating ice, and icebergs.

Local History

When Russian explorers accompanying Vitus Bering reached the Glacier Bay area in the summer of 1741, the bay was blocked by a solid front of glacial ice. Probably they did not suspect the bay's existence. However, the Tlingit Indians, the native inhabitants, knew about the bay. Recorded in their legends are stories about prehistoric retreats and advances of glaciers. Before the glaciers advanced and they had to move southward, the Tlingits camped on the islands and shores of Glacier Bay to fish and gather plants. Many years later, as glaciers retreated, the Tlingits resumed their traditional activities throughout the Glacier Bay area.

The Tlingits were a fierce, hardy people who had developed a complex culture with an elaborate social system. They fashioned cedar canoes that they navigated with great skill through the treacherous tides and rough waters of the region.

The Russian sailors did not have that skill and knowledge. A landing party of 11 men in a small boat rowed up an inlet from Icy Strait and were either lost, wrecked, or possibly killed by the natives. The Russian captain sailed off, after several days of waiting, to rejoin Bering.

Captain James Cook, a British naval officer, named Cross Sound, which he discovered on Holy Cross Day in 1778. He described it as a bay appearing to branch into several arms, the largest of which turned north; but glacial ice blocked further entry. Cook and other explorers of that time were searching for the fabled Northwest Passage across North America. He sailed on up the coast and named Mount Fairweather. Perhaps he saw it on a rare sunny day.

A French explorer, Comte de Laperouse, put into Lituya Bay in 1786 and spent nearly a month there after his longboat was swamped. He became acquainted with the Tlingits who were living there and "purchased" Cenotaph Island from one of the chiefs. Mt. LaPerouse and LaPerouse glacier are named for him.

In 1794 Captain George Vancouver, who had served under Captain Cook on the earlier voyage, returned to Glacier Bay with his own expedition. He again explored Cross Sound, sailed into Taylor Bay, discovered the Brady Glacier, and found glacial ice at tidewater a few miles inside the opening of Glacier Bay. Yet only 85 years later,

when the intrepid John Muir came to study the glaciers, the ice had retreated 48 miles up the bay!

Aside from Russian fur traders and a few prospectors, few outsiders had visited the Glacier Bay region before Alaska was purchased from Imperial Russia in 1867. But John Muir, who made several visits to the area, aroused the interest of both scientists and venturesome tourists. He moved about in a small boat, made temporary camps along the shores, journeyed to the glacier snouts, and climbed up on top of the glaciers and on some of the ridges and peaks above the ice. He reported his observations of glaciers calving and retreating in his dispatches to California newspapers.

Commercial steamships began bringing tourists to Glacier Bay to view glaciers in action, and parades of scientists came along to study the glaciers' behavior and the reclamation by vegetation of recently uncovered land. Many of the place names on topographic features honor these early scientists, especially the geologists. Some names were assigned by Professor H. F. Reid, who mapped glaciers and made careful measurements in the 1890s. Some names are those of eminent glaciologists who visited Glacier Bay during a field trip sponsored by the 12th International Geological Congress in 1912. All agreed that glacial rates of retreat in Glacier Bay are the fastest recorded anywhere on earth.

The glacial phenomena, rugged mountains, the great natural beauty of the region, the marine mammals, birds and other wildlife, and the plant successions all inspired a strong movement to protect Glacier Bay. A national monument was proclaimed in 1925, and after several boundary changes, Glacier Bay National Park and Preserve was established in 1980. Because of intense lobbying by individuals and commercial interests, shortly after the national monument was established, Congress passed legislation permitting mining in the area. Over the years extensive prospecting has been done, and several small mining operations have been carried on without much profit.

Recently a test drilling program verified the presence of a sizable nickel deposit beneath the Brady Icefield and claims have been staked out by a mining company. The mining interests believe that ore can be extracted without serious degradation of the environment, but some scientists claim that the environment is too fragile to permit this kind of exploitation.

Geologic Features

Structurally Controlled Topography

The topography of Glacier Bay Park and Preserve has a pronounced "grain" inasmuch as most of the mountain ranges, glacial valleys, and waterways trend northwest to

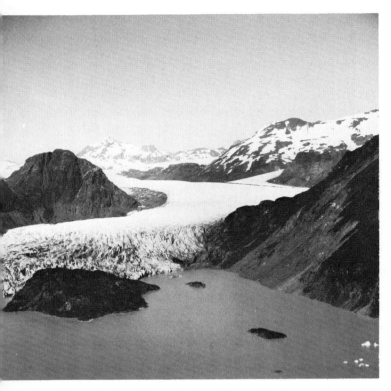

Figure 28.4   Rodman Glacier, at the northwestern end of Glacier Bay National Park and Preserve, calves into Ustay Lake. An ice-smoothed bedrock island in the lake was recently uncovered by the retreating glacier. National Park Service photo by R. Belous.

southeast. Beginning with the Gulf of Alaska coastline on the west and going inland, the major features that follow this trend are the Alsek Glacier; St. Elias and Fairweather Ranges; Brady Icefield and Brady Glacier; Glacier Bay and its extending arms, Tarr and Muir Inlets; the Beartrack and Excursion Rivers; the Takinsha Mountains; and—just outside the park/preserve's eastern boundary—the Chilkat Range. Between these major topographic features are a number of smaller landforms that parallel the main trend, such as the Carroll Glacier.

Two major features with an east-west orientation that cut across the regional trend are the Alsek River at the northern edge of the park/preserve and the Cross Sound/ Icy Strait waterway marking the southern boundary. The Alsek River flows south from Yukon Territory, then turns west and flows from British Columbia across the park/ preserve to empty into Dry Bay. The Alsek is of interest to geologists because it is a transverse stream, and it is the only stream that crosses the St. Elias Range. A few smaller landforms, such as Lituya Bay (opening on the Gulf of Alaska) and Geikie Inlet (on the west side of Glacier Bay) also have an approximate east-west trend. One has only to glance at these features on a map to conclude that the topography of Glacier Bay Park and Preserve is structurally controlled.

The long, northwest-southeast-trending valleys follow fault lines that are zones of less resistance to weathering and erosion. Branching faults and faults that cut across the fault blocks probably underlie the prominent transverse features. Weathering and preglacial stream erosion deepened the fault zones; valley glaciers scoured and enlarged them. These valleys are magnificent, large-scale examples of how structural features control the locations and orientations of drainage lines and ice flow. Only when the ice became deep enough to bury the divides between stream valleys, as with the Fairweather Glacier, did ice tend to flow across the dominant valley trends. Later as glaciers retreated, flow directions again were confined by the controlling structural patterns.

Three major faults trending northwest-southeast in the Glacier Bay area show up on the geologic map of Alaska as part of major fault systems that extend southward along the Canadian and United States west coast and northward into the Alaska Range, joining the Denali fault system (chapter 31). The Border Ranges fault, for example, which includes a segment in the park/preserve called the Fairweather fault, runs from the St. Elias Range southward to the Alexander Archipelago. It is described as a late Mesozoic plate boundary where an entire rock complex became welded to the continental margin by widespread deformation, metamorphism, and plutonism during late Cretaceous and Tertiary time.

Recent studies show the faults as boundaries between tectonostratigraphic terranes in southeastern Alaska (Brew and Ford, 1982). (A *tectonostratigraphic terrane* is a fault-bounded region characterized by rocks and geologic history differing markedly from those of neighboring regions.) Since the southeast Alaska terranes are believed to have been "rafted" into their present position rather than formed on the site, they are referred to as *exotic,* or *accreted terranes.* Complicating our understanding of these features is evidence of overthrusting brought about by compression plus considerable lateral displacement caused by impinging and pivoting tectonic plates in the area (fig. 28.8).

Glacial Features

Since many of the valley glaciers in the Glacier Bay area come down to sea level with their snouts extending out into salt water, they are called *tidewater* glaciers. They may float free if the water is deep enough, or they may ground themselves on inlet floors. The irregular forward movement of brittle ice at the snouts and the action of waves and tides cause the ice to fracture and break. Large blocks and crumbling masses of ice break or fall continually from glacier snouts, a process called *calving* that produces icebergs of all sizes and floating, shattered pieces of ice called "growlers." The ongoing crumbling and falling

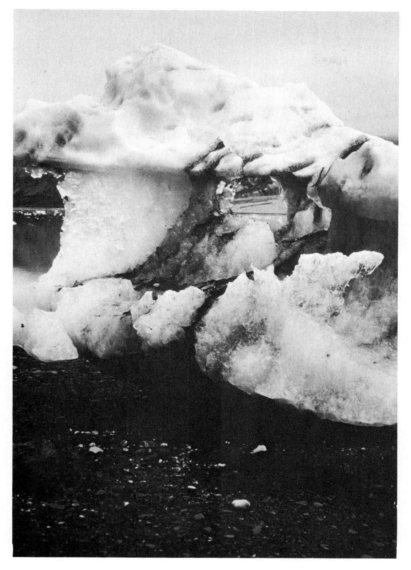

Figure 28.5 "Growler," (small iceberg) stranded on a pebbly beach in Tarr Inlet. Photo by G. Haselton.

of ice creates enormous waves and splashes, accompanied by loud booming and cracking. When ice hits the water, it goes under the surface, then bouyantly bobs up again, bouncing and flopping around. The calving process stirs up the water, bringing feed to the surface and attracting birds that shriek as they swirl and dip over the water. It is this whole dramatic and noisy display that continues to enthrall visitors to Glacier Bay Park and Preserve.

Glacial erosion in the park/preserve has tended to produce two quite different mountain shapes. The higher peaks, such as Mt. Fairweather and Mt. Salisbury, are horn-shaped due to dissection by cirques and frost action. At lower elevations, especially alongside inlets and narrow bays, ridges and peaks have smooth, rounded, domelike outlines. Abrasion and scouring by overriding glacial ice

have left these bedrock masses bare of soil and vegetation. Smaller knolls that are oriented parallel to the direction of ice flow are termed *roches moutonnées*. *Nunataks* are isolated, rounded bedrock knobs or peaks protruding through an icefield or a glacier. A splendid example of such a feature is The Nunatak, an isolated knob left standing in Muir Inlet after the ice surrounding it had retreated.

Glacial deposits, whether laid down by advancing or retreating glaciers, take many forms in the park/preserve. Among subglacial deposits are *drumlins*—oval, elongated hills or mounds of till; crevasse fillings; cross-valley ridges of till; *crag-and-tail* features—like roches moutonnées at the stoss end but overlain by a tapering, streamlined "tail" of till; and *eskers*—narrow, sinuous, steep-sided ridges made up of layers of sand and gravel

Figure 28.6   Grand Plateau Glacier (northwestern section of the park/preserve) calves into a lagoon that is separated from the Gulf of Alaska (foreground) by an end moraine, the knobby, forest-covered ridge with an ocean shoreline. Not too long ago, this large glacier, fed by snow and ice in the Fairweather Mountains (background), emptied into the Gulf of Alaska. National Park Service photo by R. Belous.

deposited by meltwater in ice tunnels in a stagnant glacier. Moraines and outwash plains along with temporary lakes, ice-block holes and the like are found on lowlands near the locations of present and former glacier snouts.

Studies of Glacial Advances and Retreats

When John Muir first came into Glacier Bay in a canoe with Tlingit guides, he studied the area around their first encampment (near Geikie Inlet) with growing excitement. Around the small rocky beach, he saw a landscape being born. Everything was new and raw, and all the signs indicated that ice had only recently left the campsite. A few sparse mosses were the only living plants, but stumps

of sizable trees stood among the wastes of sand and gravel. Muir thought these were petrified and was surprised to see the Indians using them for firewood. Actually the stumps were the remains of forests that had been overwhelmed by an earlier ice advance.

On this first trip Muir and his party reached the tip of Russell Island, which was where the glacier front was in 1879. Muir predicted that an island and a large fiord would soon be exposed for he noted that the sea was undermining the glacial ice. Muir was a keen observer. On his next expedition in 1880, he deduced that in Taylor Bay (which opens into Cross Sound) a large glacier was advancing, in contrast to the other glaciers that they had visited, all of which were retreating! By the summer of 1890, steamships were bringing supplies and tourists to Glacier Bay, and Muir established a camp at what is now called Muir Point. There a geologist, Professor H. F. Reid, leading a scientific party, joined Muir. The rest of the summer was spent studying, mapping, and exploring the area around Muir Glacier. Reid was the first to plot the positions of glacial ice using a theodolite and plane table. W.O. Field, Jr., over a span of nearly forty years, mapped the termini of glaciers in the Glacier Bay area and recorded their rates of retreat (or advance). More recently, under the direction of R.P. Goldthwait, geologists from the Institute of Polar Studies at the Ohio State University continued Field's work by carrying on detailed studies of glacial deposits and glacial retreats and advances in the fiords and on the slopes of Glacier Bay National Park and Preserve.

Basically the aim of scientists is 1) to measure and survey the melting back of glacial ice and from this determine rates of retreat; 2) to search for evidence that might explain the rapid changes observed; 3) to find out how subglacial landforms are created and how lakes appear and disappear; 4) to study the return of plants and animals to recently uncovered, barren landscapes; and 5) to measure and study Glacier Bay's exceedingly rapid rate of *glacial rebound*. As melting releases the weight of ice, the shorelands and islands rise and expand in size.

In the last two centuries, the glaciers in Glacier Bay have retreated about 15 times faster than most glaciers elsewhere in the world. Rates of retreat per year in Glacier Bay have been from 600 to 1100 feet horizontally, and ice surfaces have been lowered at rates of around 12 vertical feet per year. The rapid unloading of glacial ice has caused the land surface to rise at rates greater than anywhere else in southeast Alaska. At Bartlett Cove the land rises at a rate of about an inch and a half per year (4 cm), and farther up in the bay, where ice left more recently, the rebound rate is even higher. The creation of new beaches, tidal zones, and shoals makes marine charts short-lived! Tectonic activity in the Glacier Bay area may be a contributing factor in these rebound rates.

A few glaciers in the park/preserve have been advancing. Sometimes this has been due to "surging" caused by excess water within the glacier that acts as a lubricant between ice and bedrock permitting faster sliding as well as rapid movement along ice fractures. In other cases glacial advances may have occurred because of high snowfall, snow avalanching, increased debris cover, or perhaps changes in weather patterns that result in less runoff and melting.

No clear-cut, simple explanations have been developed to explain the extreme fluctuations and rapid movement of the glaciers in the Glacier Bay area. Several factors are probably involved, with individual glaciers re-

sponding differently to each factor and often at different rates. It appears, for example, that the Brady Icefield has been much more stable than glacial ice on the east side of the bay. Because of its elevation and nearness to the coast, Brady Icefield has a steady supply of snow, and it is also fed by large glaciers descending from the Fairweather Range.

The large-magnitude changes that occur in the eastern part of the park/preserve and the Takhinsha Mountains may be due to generally rising mean annual temperatures since the 1700s and may also be the result of a partial rain shadow effect produced by the Fairweather Range. Less snow accumulates east of the bay. Another factor is elevation of the source area. Advancing glaciers generally come down from snowfields at higher elevations that receive and retain more snow.

The tidewater glaciers are the most variable in their behavior with some advancing, some retreating, and some changing very little. In general, snouts of glaciers that discharge in deeper water or tidal estuaries fluctuate more than those on dry land or those that rest on shoals.

Since southern Alaska is a seismically active area, earthquakes "shake down" snow accumulations in the high areas, resulting in increased pressure from the added load all down the line. An earthquake may also devastate a tidewater glacier. This happened to Muir Glacier in 1899 when an earthquake broke up the terminus and caused the glacier to begin a rapid retreat. Glacier Bay became choked with a mass of floating ice within a few hours, and for several years ships could not get into Muir Inlet.

Reforestation

Muir and the scientists who followed him were fascinated by the way in which plant zones followed glacial retreat. Within a few years after melting ice uncovers bare rock, hardy species such as mosses, dryas, dwarf fireweed, horsetails, and lichens take root. Alder and willow follow within 60 years. Sitka spruce begins to grow in the young forest and eventually dominates the stand, but as time passes, is displaced by western hemlock—except on relatively sterile gravels and exposed beaches.

Between 2500 and 5000 feet in elevation, the species of the arctic alpine zone flourish, and rich alpine meadows appear. At higher elevations, vegetation shrinks to heath saxifrage and crustal lichens (fig. 28.7).

Bedrock

Exposures of bedrock ledge are prevalent where retreating ice has left ground open to air and the advancing cover of vegetation is still a hundred or more years behind. Geological mapping shows these rocks to be a remarkably diverse lot, a partial confirmation of the theory that Glacier Bay Park and Preserve is made up of exotic terranes.

Figure 28.7   Vegetation returns to shores uncovered by glaciers retreating up Tarr Inlet. Photo by G. Haselton.

One finds carbonate, volcanic, metamorphic, and gabbroic rocks in no particularly logical arrangement. Recognizable sedimentary rocks range in age from early Silurian through late Tertiary, about 20,000 to 30,000 feet of Paleozoic rock and over 12,000 feet of Tertiary beds. Detailed mapping in the area east of Mt. Fairweather shows a typical assemblage containing, in ascending order: Willoughby Limestone, 5000 feet; Tidal Formation, 10,400 feet; Upper Tidal Formation, 2100 feet; Pyramid Peak Limestone, 2300 feet; Pendleton Formation, 2000 feet; Black Cap Limestone, 4500 feet. Plutons associated with the Coast Range Batholith of Mesozoic age intrude these stratified rocks. (The Coast Range Batholith is a large-scale pluton of multiple intrusions that crops out from Washington state to Alaska.) Metamorphism of low to intermediate intensity has altered some of the rocks. Numerous small and marginal mineral deposits are scattered throughout the park/preserve.

As described earlier, several northwest/southeast-trending faults cut through the rocks of the park/preserve. The faults divide and rejoin; some lie offshore; others pass beneath the Brady Icefield and Glacier Bay. Similar fault structures are beneath the Lynn Canal, a natural waterway close to the park's eastern boundary.

Field geology in the Glacier Bay area has always been challenging. Geologists must make rain-soaked climbs in fog and drizzle and cross uneven, blocky stretches of snow and ice to reach critical outcrops. Transportation is usually by small boat, but in recent years the U.S. Geological Survey has used helicopters for transporting scientists and supplies from a floating barge to mapping sites.

Geologic History

1. Possible origin of terranes in Precambrian and Paleozoic time

Fossil and paleomagnetic evidence, although far from complete, seems to indicate that some of the terrane rocks were formed in the Southern Hemisphere. The terranes may have been islands, seamounts, microcontinents, or pieces broken off from larger land masses. At this point we can only speculate about the route of their "journey," but it must have taken many millions of years. What is now the eastern side of the plate boundary was probably located somewhere to the south and east of the present position. Indications are that the region had long been a continental margin.

Figure 28.8   Simplified representation of major tectonostratigraphic terranes in Glacier Bay National Park and Preserve and vicinity. (After Brew and Ford 1982.)

2. Mesozoic and Cenozoic accretion

The terranes probably "docked" from middle Mesozoic through middle Tertiary time and were pushed with great force against the North American plate. The three terranes that are the largest and most recognizable are (from west to east) Chugach, Wrangellia, and Alexander (fig. 28.8). The long, prominent, roughly parallel faults are the *sutures,* or junctions, between these microplates. As the Pacific plate pushed against the North American plate, thrusting occurred from the southwest toward the northeast. Strike-slip faulting moved the southwest block toward the northwest (in relation to the northeast block) as the Pacific plate rotated counterclockwise while pressed against the western edge of North America. Because oceanic crust was subducted as the plates converged,

folding and reverse faulting occurred and plutons of granitic, dioritic, and gabbroic composition were emplaced, some as late as Miocene time. Regional and contact metamorphism, along with deposition of minerals, also took place in the converging zone.

3. Late Tertiary and Quaternary tectonic activity

The forces that drove exotic terranes against southeastern Alaska are still active and are causing the mountains to rise. Major earthquakes continue to occur. In 1958 an earthquake registering 8 on the Richter scale caused millions of tons of rock to break loose from the rock walls enclosing Lituya Glacier. A great rockslide tore off the glacier snout and the chaotic mass of rock and ice crashed into Lituya Bay (fig. 28.3). A giant wave, 1740 feet high,

Table 28.1 Simplified Geologic Column, Glacier Bay National Park and Preserve

Time Units			Rock Units	Geologic Events
Era	Period	Epoch		
Cenozoic	Quaternary	Holocene	Neoglacial drift accumulations	Rapid glacial rebound Rapid rates of glacial retreat
Cenozoic	Quaternary	Pleistocene	Wisconsinan drift Thick older drift in offshore marine sections	Numerous episodes of glacial advance and retreat (since late Tertiary time) Ongoing tectonic activities (since Mesozoic time): 　Strong earthquakes 　Horizontal displacement with Pacific side sliding northwest 　Vertical uplift 　Thrust faulting 　Deformation by compression 　Subduction with plutonism 　Regional metamorphism
Mesozoic		?	Wide variety of rock types (Silurian to late Tertiary in age): 　Carbonates 　Granite, diorite, granodiorite, gabbro, lavas 　Marble, schist, gneiss, quartzite, hornfels, granulite	Accretion (Jurassic through Eocene time) of at least three exotic terranes: Alexander, Wrangellia, Chugach
Paleozoic	Silurian		Willoughby Limestone	

Sources: Beikman 1980; Brew and Ford 1982, 1986; Mackevett 1971; Nokleberg, James, and Silberling 1985; Rossman 1959, 1960; Seitz 1959

surged over the ridge, fell back, and then swept down to the ocean. Waves sloshed back and forth in Lituya Bay for 20 minutes. When it was over, all the trees in a wide swath along the sides of the bay were gone. Tlingit legends warn of similar earthquakes and destructive waves at Lituya Bay in the prehistoric past.

4. Pleistocene glaciation

In this part of Alaska only sketchy evidence for pre-Wisconsinan glaciation exists. However, thick tillites have been found in offshore drilling in the Gulf of Alaska. Much of this glacial detritus is believed to be late Tertiary and early Pleistocene in age. Apparently the Fairweather Range was being uplifted throughout Pleistocene time with uplift continuing into Holocene time. (Estimated rate of uplift on the outer coast is now 0.7 cm per year.) As the mountains rose, precipitation on the slopes increased and glaciers enlarged. Possibly the Fairweather fault, which skirts the range front, blocked or split most of the glaciers, keeping all but the largest away from the coast. Evidence

of middle and late Pleistocene glaciation is missing in many of the outer coast areas.

Wisconsinan glaciation dominated the region for thousands of years, attaining maximum extent between 21,000 and 15,000 years before the present. Ice sheets accumulated in three areas: 1) the Brady Icefield, west of Glacier Bay; 2) the Takinsha Icefield along the northeast edge of the park/preserve; and 3) on the northern highlands along the international boundary. Ice in the northern highlands may have coalesced into an intermontane ice sheet or a piedmont glacier.

5. Neoglacial advances during Holocene time

In late Pleistocene and Holocene time, a combination of cooler climate and the long-continued uplift of the Fairweather Range increased snowfall markedly and brought on glacial advances. Apparently the main flow within the piedmont ice sheet moved southward over Glacier Bay, turned seaward, and passed through Cross Sound. Holocene warming came slowly to Glacier Bay, but about 10,000 years ago, ice retreated and the land rose. The trend

Figure 28.9   Medial moraines ride on the surface of an upper tributary of Johns Hopkins Glacier. Crevasse patterns show how brittle ice is strained as the glacier negotiates the curve (background) on its way down to Johns Hopkins Inlet and Glacier Bay. National Park Service photo.

reversed some 4000 to 5000 years ago, and icefields again expanded. Rejuvenated glaciers overwhelmed forests, buried the stumps in till and drift, and deposited till and outwash in lowlands and valleys. Glaciers continued to grow until about 500 years ago, at which time retreat began. The retreat picked up speed between A.D. 1660 and 1760 and has maintained a fast rate of meltback since then, surpassing rates measured elsewhere.

Bibliography

Anderson, P. S., Goldthwait, R. P., and McKenzie, G. D. (editors). 1986. *Observed processes of glacial deposition in Glacier Bay, Alaska.* Columbus, Ohio: Institute of Polar Studies, Ohio State University. Miscellaneous publication 263. 167 p.

Beikman, H. M. (editor). 1980. Geologic map of Alaska (scale 1:2,500,000) U.S. Geological Survey.

Boehm, W. D. 1975. *Glacier Bay.* Anchorage: Alaska Northwest Publishing. 134 p.

Bohn, Dave. 1967. *Glacier Bay, the land and the silence.* Anchorage: Alaska Natural History Association. 160 p.

Brew, D. A., and Ford, A. B. 1982. Tectonostratigraphic terranes in the coast plutonic-metamorphic complex, southeastern Alaska. In *The United States Geological Survey in Alaska: Accomplishments during 1982.* Circular 939, p. 90–93.

———. 1986. Southeastern tectonostratigraphic terranes revisited. *Bulletin American Association of Petroleum Geologists,* 69:657.

Clarke, J. M. 1979. *The life and adventures of John Muir.* San Francisco: Sierra Club. 326 p.

Field, W. O. Jr. 1942. Glacial studies in Alaska. *Geographical Review,* 32:154–155.

———. 1947. Glacier recession in Muir Inlet, Glacier Bay, Alaska. *Geographical Review,* 37:369–399.

Haselton, G. M. 1966. Glacial geology of Muir Inlet, southeast Alaska. Institute of Polar Studies, Ohio State University, Report 18, 34 p.

———. 1979. Some glaciogenic landforms in Glacier Bay National Monument, southeastern Alaska. *Moraines and Varves* (Ch. Schluchter, editor), Proceedings of an INQUA symposium on genesis and lithology of Quaternary deposits, Zurich, September, 1978, p. 197–205.

MacKevett, E. M., Brew, D. A., Hawley, C. C., Huff, L. C. and Smith, J. G. 1971. Mineral resources in Glacier Bay National Monument, Alaska (geologic map included). U.S. Geological Survey Professional Paper 632, p. 1–89.

Mercer, J. H. 1961. The response of fiord glaciers to changes in the firn limit. *Journal of Glaciology,* 3:850–858.

National Park Service. 1983. *Glacier Bay: a guide to Glacier Bay National Park and Preserve, Alaska.* National Park handbook 123. 128 p.

Nokleberg, W. J., James, D. L., and Silberling, N. J. 1985. Origin and tectonic evolution of the Maclaren and Wrangellia terranes, eastern Alaska Range, Alaska. *Geological Society of America Bulletin,* 96:1251–1270.

Plafker, G., Jones, D. L., Hudson, T., and Berg, H. C. 1975. The Border Ranges fault system in the St. Elias Mountains and Alexander Archipelago. In *The United States Geological Survey in Alaska: Accomplishments during 1975,* circular 733, p. 14–16.

Plafker, G., Hudson, T., and Rubin, M. 1975. Late Holocene offset features along the Fairweather fault. In *The United States Geological Survey in Alaska: Accomplishments during 1975,* circular 733, p. 57–58.

Reid, H. F. 1894. *Glacier Bay and its glaciers.* U.S. Geological Survey 16th annual report, p. 421–461.

Rossman, Darwin, 1959. *Geology and ore deposits of the Reid Inlet area, Glacier Bay, Alaska.* U.S. Geological Survey Bulletin 1058B, p. 33–59.

———. 1963. *Geology and petrology of two stocks of layered gabbro in the Fairweather Range, Alaska.* U.S. Geological Survey Bulletin, 1121F, p. 1–99.

———. 1963. *Geology of the eastern part of the Mt. Fairweather quadrangle, Glacier Bay, Alaska.* U.S. Geological Survey Bulletin 1121K, p. 1–55.

Seitz, J. F. 1959. *Geology of Geikie Inlet area, Glacier Bay, Alaska.* U.S. Geological Survey Bulletin 1058C, p. 61–119.

Glacier Bay National Park and Preserve

Area: 3,935,269 acres; 6149 square miles
Proclaimed a National Monument: February 26, 1925
Established a National Park and Preserve: December 2, 1980
Address: Gustavus, Alaska 99826

29

In this largest unit of the National Park System are North America's largest assemblage of glaciers and the greatest collection of peaks above 16,000 feet. Mount St. Elias, elevation 18,008 feet, is the second highest peak in the United States.

Spanning three of Alaska's four climatic zones, the park/preserve encompasses diverse and spectacular landscapes: rugged coasts; surging tidewater glaciers; the continent's largest piedmont glacier; high icefields on massive ranges, supporting valley glaciers in grand array; wild rivers in deep canyons; expanses of tundra; a rich concentration of wildlife.

Major faults bound tectonostratigraphic terranes containing a remarkable variety of rocks.

Earthquakes and volcanoes are signs of ongoing tectonic activity.

Old mining works recall days when riches in copper, silver, and gold were taken out of the mountains.

Wrangell-St. Elias National Park and Preserve

South Central and Southeastern Alaska

by Monte D. Wilson
Boise State University

Figure 29.1 Mt. Blackburn (16,523 ft.) and the upper Nabesna Glacier in Wrangell-St. Elias National Park and Preserve. One of the longest in Alaska, the Nabesna Glacier flows northeastward for roughly 80 miles and drains into the Nabesna River. Mt. Blackburn, an eroded volcano, has been inactive since Pleistocene time. National Park Service photo by M. W. Williams.

WRANGELL - SAINT ELIAS NATIONAL PARK AND PRESERVE

Figure 29.2   Wrangell-St. Elias National Park and Preserve, South central and southeastern Alaska.

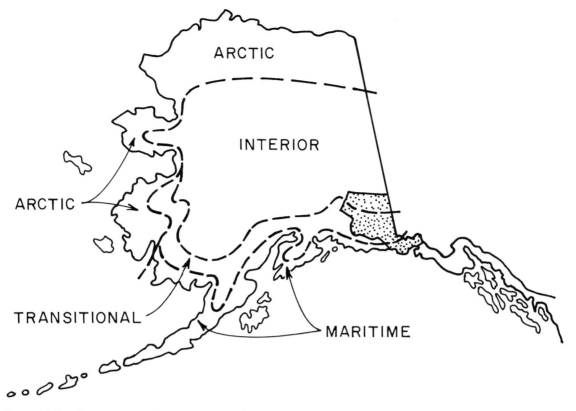

Figure 29.3   Climatic zones of Alaska. Wrangell-St. Elias National Park and Preserve, the shaded area, spans three of the state's four climatic zones.

Geographic Setting

Wrangell-St. Elias National Park and Preserve is by far the largest unit in the National Park System; it is larger than several of the smaller states. In conjunction with the adjacent Kluane (kloo-ah-nee) National Park in the Yukon Territory of Canada, this is the world's largest parkland and is, in fact, much larger than the combined size of all national parks in the other 49 states and Puerto Rico.

Creation of a national monument in the Wrangell-St. Elias Mountains and adjacent areas was recommended in 1938 by Ernest Gruening, who was then director of U.S. Territories and later served as Alaska's governor and U.S. Senator. Forty years after Gruening's recommendation, national monument status was proclaimed by President Jimmy Carter. In 1979, the United Nations Educational, Scientific, and Cultural Organization (UNESCO) designated Wrangell-St.Elias National Monument and Kluane National Park as a World Heritage Site because the area's natural and cultural resources are of such value that they constitute a part of all human heritage. Wrangell-St. Elias was established as a National Park and Preserve in 1980, also under President Carter's administration, but due in large part to the support and effort of Interior Secretary Cecil Andrus. The area extends from Yakutat Bay and Malaspina Glacier on Alaska's south coast north to the upper Tanana River Valley in the interior and from the Canadian border west to the Copper River.

Wrangell-St. Elias National Park and Preserve displays a great diversity of spectacular landscapes ranging from rugged coastlines to high mountain ranges and includes deep canyons, large glaciers, tundra areas and whitewater rivers. The National Weather Service recognizes four climatic zones in Alaska: Maritime, Transitional, Continental, and Arctic (fig. 29.3). Wrangell-St. Elias lands span three of these zones, missing only the Arctic zone. The Gulf of Alaska coast and the adjacent Chugach and St. Elias Mountains are in the moist Maritime climate. Along the coast, mean annual precipitation exceeds 130 inches and moderate temperatures prevail, ranging from winter lows around 0° F to summer highs in the 70s. The St. Elias Mountains are the world's highest coastal range; the land rises in only 15 miles from sea level at Icy Bay to an elevation of 18,008 feet at the peak of Mt. St. Elias. This uniquely steep topographic gradient causes heavy precipitation to fall from moist maritime air masses and produces average annual snowfalls of more than 600 inches (50 feet) at higher elevations. These conditions have resulted in the largest assemblage of glaciers in North America.

Most of the Copper River basin, inland from the coast, is in the Transitional climate zone. Precipitation at various elevations in this area ranges from about 10 to 25 inches per year, mostly delivered as snow. The Transitional climatic zone has colder winters and warmer summers than the Maritime zone. The Continental climatic zone, north of the Wrangell and St. Elias Ranges, lies in the rain shadow of these high mountains. Villages just north of the park/preserve are Northway, Alaska, and Snag, Yukon Territory; both have mean annual precipitation of 8 to 10 inches and temperatures that range from −70° F to the high 80s.

Local History

Four distinct groups of Indians currently occupy the Wrangell-St. Elias region in areas established by their ancestors during prehistoric times. Most anthropologists agree that all of these people descended from Asian hunters who migrated across the Bering Strait between 8000 and 12,000 years ago near the close of the last great ice age (Wisconsinan glaciation).

In the coastal zone the Eyak tribe settled the Copper River Delta area and people from the Tlingit (pronounce the first syllable halfway between "cling" and "thling") tribe came to occupy Yakutat Bay. The extensive interior areas of the Wrangell-St. Elias region were settled by two groups of Athapascan peoples; valleys of the Copper and Chitna Rivers are the homeland of the Ahtna, or "ice people," while valleys of the Nabesna and Chisana Rivers are homeland for the Nabesna or Northway Indians. Linguistic similarities indicate that the Ahtna and Nabesna Indians are related to several Indian tribes of California and other southwestern states. Anthropological theory holds that all American Indians are descendants of the Asian peoples who migrated across the Bering Land Bridge and, over many generations, down an "ice-free corridor" that existed just east of the Canadian Rockies. Debate over the possible existence and timing of such an ice-free corridor has continued for decades due in part to incomplete and in some cases contradictory archeological and geological evidence.

The abundance of both marine and land resources, as well as the mild coastal climate, enabled the Eyak and Tlingit Indians to meet life's basic needs relatively easily. This left time for them to develop elaborate social customs and beautiful art forms, including the totem poles and cedar houses of the Tlingit. Subsistence in the interior was much more difficult due to the rigorous climate and limited diversity of food resources; famine, particularly in late winter and early spring, was an occasional occurrence.

High-grade copper deposits in the southeastern part of the Wrangell Mountains were exploited by Ahtna Indians living along the Copper and Chitna Rivers. Some of these people became skilled at using heat to work the copper into sharp projectile points and cutting implements. Copper items were traded among native groups for at least 1400 years prior to contact with whites. In the spring of 1885, Lieutenant Henry T. Allen of the U.S. Army found Chief Nicolai's Copper River Indians, part of the Ahtna group, using bullets of a copper-silver alloy produced from their own mines. Chief Nicolai showed Lt. Allen the location of a rich vein of copper that was being worked by his people.

The Wrangell-St. Elias Park and Preserve contains many significant prehistoric and historical archeological sites. Those of Athapascan origin are most numerous. Others are from the Eyak and Tlingit Indians as well as from the Chugach Eskimos, who were occasional users of the area.

The earliest recorded entry of Europeans into Alaska was by Vitus Bering, a Dane, who in 1741 explored this area for Czar Peter the Great of Russia. Bering named Cape St. Elias (on Kayak Island) in honor of the Saint upon whose feast day the cape was first sighted. He also recorded sightings of the rugged, glaciated mountains including the prominent peak that came to be called St. Elias.

Over the next several decades, this part of the Gulf of Alaska coastline was visited and claimed, in part or total, by explorers from several nations. In May 1778, Captain James Cook visited Yakutat Bay, named several topographic features, and claimed these regions for Great Britain. Captain Salvador Fidalgo named Valdez Bay in 1790 and claimed that area for Spain. Another Spanish land claim was made by the Italian Captain Alejandro Malaspina who in 1791 sailed into Icy Bay; although disappointed in finding that the bay did not lead to the elusive Northwest Passage, he nevertheless had a rock monument built and laid claim to the port for his employer, the King of Spain.

It was the Russians, however, who, by establishing permanent settlements and a local government in the area, were able to hold their land claims. In 1795, Aleksandr Baranov, the Russian governor of Alaska, had a convict labor colony established on Yakutat Bay to engage in agriculture and shipbuilding. After ten years of conflict with local Indians, the town, called "Gloria of Russia", was razed and 40 of its inhabitants were killed by a Tlingit war party. Russian fur traders also traveled into the interior along the Copper River, but hostilities between Ahtna Indians and Russians led to bloodshed in that area as well.

In 1867, William H. Seward, President Andrew Johnson's Secretary of State, arranged the United States purchase of Alaska (then called Russian America) for $7,000,000, or about 2 cents per acre. The unpopular purchase, called "Seward's folly," ended formal Russian influence in Alaska.

Lt. Allen's 1885 expedition was the first American reconnaissance in the area. He explored the Upper Copper River Basin, the Chitna River Basin, and the Wrangell Mountains before going north over the divide to the Tanana River Valley. In addition to establishing friendly relations with Chief Nicolai and his Copper River group of Ahtna, Allen named the village of Chitna and the Chitistone River (both using the Indian word "chiti," meaning copper) plus Mt. Blackburn, Mt. Drum, and Mt. Sanford.

In 1898, following the discovery of rich placer gold deposits along Canada's Klondike River, the Copper River Exploring Expedition was organized under command of the U.S. Army. Included in the expedition were two U.S. Geological Survey geologists, Oscar Rohn and T. E. Schrader. Rohn named the Chitistone Limestone and Nicolai Greenstone in the Chitna River drainage. He pointed out similarities to the Lake Superior Greenstones that suggested rich mineral potential. In 1899 the Nicolai mine was located (with assistance from Chief Nicolai) and in 1900 the incredibly rich Bonanza Mine was located eight miles west on a steep ridge, just east of the lower Kennicott Glacier. The mines did not go into full production until 1911 when the completion of a railroad allowed ore to be hauled from the mining town of Kennicott, via a route along the Chitna and Copper Rivers, to the port of Cordova. When the high-grade ore was worked out in 1938, mines in the Kennicott-McCarthy area shut down. By that time they had produced over a billion pounds of copper plus more than nine million ounces of silver, which was recovered as a byproduct. For many years there were plans to use the old railroad bed as the base for a road to link Cordova with Alaska's highway system at Chitna. These never realized plans were kept active until the 1964 earthquake destroyed one span of the bridge across the Copper River.

On the north side of the Wrangell Mountains, rich lode deposits (veins in bedrock) of gold were worked in the 1920s and 1930s at both Nabesna and Chisana (pronounced "shoe-shana"). Prior to 1932, most supplies were hauled to these mines by horses in the summer, but dog teams delivered mail in the winter. A particularly accomplished individual of this era was Harry Boyden, a man of modest stature and impeccable manners. He worked as a packer and meat hunter for the mining companies and for several years carried the mail by dog team from McCarthy over the Wrangell Mountains and across glaciers to Nabesna. In 1931 the contract for mail and small freight pieces was taken over by Bob Reeve who came to be known as the "Glacier Pilot."

Within Wrangell-St. Elias National Park and Preserve, three of the numerous sites associated with the mining industry—one each at Kennicott, McCarthy, and Nabesna—are in the National Register of Historic Sites. In 1985, a total of 441 patented mining claims were under private ownership within park/preserve boundaries. In order for a mining claim to become patented, government regulations concerning discovery and development of mineral resources must be met; this usually requires the investment of much time, effort, and money with no guarantee of return. On patented claims, the land and mineral rights have been transferred from government to private ownership. Active placer mining areas in 1985 included Dan Creek and Rex Creek, while bedrock mining was being conducted at Upper Kotsina and Nabesna. Mineral exploration continues within several areas of the preserve.

The Wrangell and St. Elias Ranges include more high mountain peaks than any other part of North America. Mt. Logan, located only 15 miles outside the United States border, is Canada's highest point at 19,850 feet, and second only to Mt. McKinley at 20,306, as the highest peak in North America. Mt. St. Elias with its summit of 18,008 feet situated on the United States-Canada border, is the second highest peak in both countries. Mountain climbers have challenged these peaks for over a century. Climbing the numerous high peaks in these ranges is extremely difficult, partly because of the area's inaccessibility and very rugged topography, but also because the combination of high latitude and high altitudes produces a rigorous climate that supports extensive snowfields and many glaciers.

Beginning in 1886, several attempts were made to climb Mt. St. Elias before it was conquered in 1897 by an Italian expedition under the leadership of His Royal Highness, the Duke of Abruzzi. Mt. Blackburn (16,390 feet) was first climbed in 1912 by Dora Keen and her party of Americans. In 1925 the summit of Mt. Logan was reached by the MacCarthy-Lambart party of Canadians. Aircraft support of mountain-climbing parties in this region began in 1937 when Bob Reeve, the Glacier Pilot, landed scientists Bradford Washburn and Robert Bates at the 8500-foot level to begin the first ascent of Mt. Luciana (17,147 feet). The following year, Washburn and Terris Moore were the first to climb the volcanic cone of Mt. Sanford (16,237 feet). The second successful climb of Mt. St. Elias, and the first by an American party, was accomplished in 1946 by a Harvard Mountaineering Club expedition led by glaciologist Maynard Miller. Beginning in 1949 many climbing expeditions to the area made first ascents of major peaks; McArthur Peak (14,400 feet), the last unclimbed peak higher than 14,000 feet, was finally climbed in 1961.

Box 29.1
Plate Tectonics and Accreted Terranes

Different parts of the Wrangell-St. Elias area have drastically different types of bedrock and geological structures. A thousand geologists could work for a thousand years in this area, but unless they understood plate tectonics (particularly the accreted terranes aspect of plate tectonics) they would probably not reach any generally acceptable interpretation of regional geologic history. It seems that the drastically different portions of the Wrangell-St. Elias area can best be explained as being accreted terranes.

Geologists use the word *terrain* for its common meaning of general landscape characteristics or "lay of the land." Pronounced the same, but spelled differently, the term *terrane* refers to a block of crustal rock (having surface dimensions of tens to a few thousand kilometers) with a unique sequence of rock units that have resulted from its particular geologic history. Because of its history, one terrane might have massive granites at its base overlain by a thick sequence of Paleozoic carbonate sedimentary rocks with warm-climate fossils and an uppermost zone of basalt flows interbedded with terrestrial sediments. Another terrane might consist of oceanic crust at the base overlain by Mesozoic island-arc basalts and associated sediments, which are then capped by a sequence of clastic sediments containing cold-water fossils.

Finding such dissimilar rock sequences on adjacent ridges in southern Alaska leaves the field geologist with a very difficult problem in geologic history so long as it is assumed that this area has always shared the general geologic history of the North American continent. However, such major differences in bedrock can easily be accounted for if it is assumed that the various blocks, or terranes, originated in different parts of the globe and had completely different geologic histories until they "drifted" via the mechanism of plate tectonics to collide with, and accrete to, North America. Such terranes apparently existed either as "microplates" or as islands of continental materials within the upper part of an oceanic plate moving toward, and subducting beneath, the North American continental plate. When they collided with North America, the terranes were "scraped off" and attached to the main continental mass. It now appears that a wide strip of western North America from Alaska to California is made up of accreted terranes, including all, or portions of, several national parks*. The concept of accreted terranes is still new, and it is generally agreed that there is a lack of detailed understanding of the mechanism. Lacking also are adequate factual data derived from geological mapping, fossil collections, and absolute ages.

Even in its immature state, however, the concept of plate tectonics and accreted terranes helps us to understand the rock sequences, volcanoes, and earthquakes of Wrangell-St. Elias. Among the terranes in Wrangell-St. Elias National Park and Preserve are: Yukon-Tanana, Wrangellia,

*Further discussions of accreted (tectonostratigraphic) terranes are in chapter 21, Acadia National Park; chapter 26, North Cascades National Park; chapter 28, Glacier Bay National Park and Preserve; chapter 30, Kenai Fjords National Park; chapter 31, Denali National Park and Preserve; chapter 35, Katmai National Park and Preserve; chapter 36, Lake Clark National Park and Preserve; chapter 47, Shenandoah National Park; and chapter 48, Great Smoky Mountains National Park.

Chugach, Alexander, and Yakutat (fig. 29.4). Paleomagnetism and fossil evidence indicate that some of these terranes originated in equatorial regions. Boundaries between terranes are partially delineated by major faults trending northwest-southeast, such as the Denali fault, Totschunda fault, Border Ranges fault, and St. Elias fault.

Figure 29.4   Major faults and accreted terranes in Wrangell-St. Elias National Park and Preserve.

Geologic Features

Bedrock Sequences

The northernmost part of Wrangell-St. Elias National Park includes a small portion of the Yukon-Tanana terrane. That terrane consists of varied metamorphic and igneous rocks with the oldest being Precambrian in age. Future geologic mapping may show the Yukon-Tanana terrane to be a composite of several smaller microplates.

The Wrangellia terrane has been studied in more detail than most accreted terranes, so its major rock units are well known. A thick assemblage of island-arc volcanic rocks of late Paleozoic age forms the base of the Wrangellia terrane. This volcanic unit is overlain by fossil-bearing limestones, shales, and cherts of Permian age. The lowest Triassic rocks are cherty shales; resting upon them is a very thick sequence of greenstones that were metamorphosed from basalts of both submarine and subaerial origin. The Nicolai Greenstone is of particular importance within this sequence because it contains deposits of copper similar to those of Precambrian age in Isle Royale National Park (chapter 20). Uppermost Triassic rocks are limestones and argillites of marine origin. Jurassic and Cretaceous rocks of the Wrangellia terrane are mostly sandstones with some shale and conglomerate. They are overlain by Lower Tertiary volcanics, which, in turn, are partially covered by Upper Tertiary and Quaternary lavas that issued from volcanoes of the Wrangell Mountains.

A small part of the Alexander terrane, consisting of Paleozoic marbles, lies within Wrangell-St. Elias National Park and Preserve, just on the Alaska side of the international boundary. Other rock types are present in the more extensive Canadian portion of the terrane.

The Chugach terrane, which forms a belt across Southern Alaska, makes up most of the southern part of Wrangell-St. Elias National Park and Preserve. Much of the Chugach terrane consists of Upper Cretaceous gray-wackes and slate that have been intensely folded and weakly metamorphosed. Interspersed within these rocks are discontinuous assemblages that include chert, gabbro, pillow basalts and tuff of Late Jurassic and Early Cretaceous marine origin.

The Yakutat terrane, in the southeastern part of the park/preserve, is made up largely of Upper Mesozoic graywacke and shale with interleaved, complexly faulted units of chert, argillite and volcanic rocks. Eocene age basalts and organic-rich shales are the youngest rocks of the Yakutat block.

Surficial deposits of Late Tertiary and Quaternary age are found on all the terranes. The most voluminous of these surficial deposits are glacial and fluvial materials, both of which occur in great variety and are widely distributed throughout the area. Other surficial deposits include beach and shallow marine sediments along the coast, widespread mass gravity materials (especially from rock glaciers and solifluction), and two fields of sand dunes.

Earthquakes and Volcanoes

Wrangell-St. Elias National Park and Preserve is located in one of the world's most active zones with regard to plate tectonics. In this region the Pacific Plate is being subducted beneath North America and the Yakutat terrane is accreting to North America. The associated volcanic belts, active faults, frequent earthquakes, and high mountains are expected results of the high heat flows and stresses in crustal rocks of this plate tectonic setting. The major zone of volcanoes is in the Wrangell Mountains with Mt. Wrangell (14,163 feet), one of the world's largest andesitic volcanoes, being the park's only active cone. Its last reported eruption was in 1930. Carl Benson of the University of Alaska reported an increase in heat flow in the early and middle 1980s near the summit of Mt. Wrangell.

Quaternary lavas make up several other spectacular, but presently inactive, volcanoes in the park, such as Mt. Bona (16,421 feet) in the St. Elias Range, Mt. Blackburn (16,390 feet), and Mt. Sanford (16,237 feet), both in the Wrangell Mountains. Another inactive volcano in the Wrangells is Mt. Drum (12,010 feet), which is of particular interest because of three large thermal springs, known as the Klawsi Mud Volcanoes, on its western flank.

Some of the world's strongest earthquakes have occurred in this area. In September 1899, two earthquakes of Richter magnitudes 8.3 and 8.6 were recorded with epicenters near Yakutat. The several earthquakes with magnitudes in the 7 and 8 range that have occurred in the area during this century give ample evidence that the Yakutat microplate is actively colliding with, and accreting to, North America. *Tsunami* (earthquake-generated sea waves) are a major hazard of strong earthquakes that have epicenters offshore or nearshore.

Mountains and Canyons

We have already referred to mountain climbing and volcanoes of the area, but a general topographic description is also needed. From north to south, the major ranges in the park/preserve are the Mentasta, Nutzotin, Wrangell, St. Elias, and Chugach Mountains. Although these ranges have separate names, not all of the boundaries between them are obvious, and all of these mountains are parts of larger regional features. The Mentasta and Nutzotin Mountains, along the northern margin of the park/preserve, are actually part of the more extensive Alaska

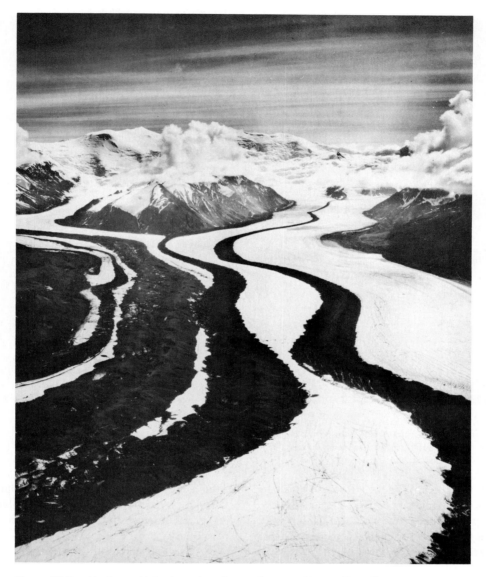

Figure 29.5   Median and lateral moraines form a sinuous pattern on Russell Glacier, which drains northward off the western part of the St. Elias Range. The Russell's meltwater forms the White River, which flows northeasterly to join the Yukon River in Canada. Mt. Churchill (15,638 ft.) and the inactive volcano, Mt. Bona (16,421 ft.), are on the skyline. National Park Service photo by M. W. Williams.

Range. Skolai Pass is the boundary between the Wrangell and St. Elias Mountains, which together make a southeast-trending, continuous range. The Chugach Mountains run approximately east-west and merge with the larger Wrangell-St. Elias Range near the park's southeastern corner. Nine of the 16 highest peaks in the United States, including Mt. St. Elias, the second highest, are in Wrangell-St. Elias National Park. Mt. Logan and many of Canada's highest peaks are in the adjacent Kluane National Park.

Deep valleys and canyons cut the high mountains in several places. Particularly spectacular in this regard are the sheer walls and plunging waterfalls of Chitistone and Nizina Canyons, both of which are much larger than Yosemite Valley.

Glaciers and Rivers

Relatively warm waters of the Japanese current wash the shores of Southern Alaska. Air masses moving over these warm currents take up water vapor until they become completely saturated. As these air masses move inland, the high mountains force them to rise. While the air is rising, it cools and releases precipitation, mostly as snow. Much of the snow is converted to ice with the result that

Figure 29.6   The clean Regal Glacier joins the dirty Rohn Glacier to form the Nizina Glacier. Chimney Mountain stands between the tributary glaciers. The Nizina flows south off the Wrangell Mountains. National Park Service photo by M. W. Williams.

the Wrangell, St. Elias, and Chugach Ranges have the largest concentration of glaciers in North America. Bagley Icefield, which runs along the crest of the Chugach Range for about 150 miles, is the largest subpolar icefield in North America. It covers most of the core of the Chugach Mountains and nourishes dozens of valley glaciers that drain down both sides of the range.

Malaspina Glacier, in the southernmost part of the park/preserve, is the largest piedmont glacier in North America. It is nearly 40 miles across and is larger in area than the state of Rhode Island. The word *piedmont* means "foot of the mountain." The term is applied to more or less flat-lying glaciers that form by the coalescence of two or more alpine glaciers where they flow beyond a steep slope. Several large alpine glaciers drain southward off the St. Elias Range to nourish the Malaspina Glacier, which covers most of the coastal plain between Yakutat Bay and Icy Bay.

A particularly interesting aspect of the glaciers in this area is their high level of activity. Many of them sustain rather high rates of flow; Hubbard Glacier has a continuous flow rate of about 10 meters per day. Additionally, many of the glaciers in this area tend to undergo periodic surges. These surges are times of rapid advance that last

from a few weeks to a year or more and are usually followed by rapid meltback. Variegated Glacier surges every 20 years, but most surging glaciers do not have an obvious periodicity.

At several places in the park/preserve, glaciers tend to create glacier-dammed lakes on an almost regular basis. Kennicott Glacier blocks a stream drainage to form Hidden Lake; each summer, raised water levels due to snowmelt cause the lake to overtop the glacier and break out, the result being intense flooding on the Kennicott River. Oily Lake and Malaspina Lake are also glacially dammed, but their outburst floods are not always annual.

In 1986, a particularly dramatic glacial damming event that occurred on the margin of the park/preserve got a great deal of public attention. Hubbard Glacier, which flows into Yakutat Bay at the southeastern corner of the park, has an unusually high continuous velocity of about 10 meters per day in some parts of the glacier. This large alpine glacier, after advancing for months, blocked off the 30-mile-long Russell Fjord on May 29, 1986. The ice dam transformed Russell Fjord from an arm of the sea into a salty lake that began to rise as it received runoff from many streams. Public attention was focused on the possible starvation of many marine mammals, mostly seals

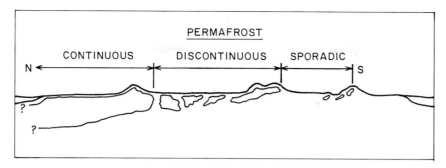

Figure 29.7 Diagrammatic north-south cross section through Alaska showing permafrost zones. In the continuous zone, permafrost extends to depths exceeding a thousand meters; the active layer (which thaws in summer) ranges from 10 to 50 centimeters. In the discontinuous zone, the active layer ranges from 25 centimeters to 3 meters in thickness while maximum permafrost depth is usually less than 100 meters and taliks (nonfrozen areas) exist. Within the sporadic zone, permafrost exists only in particularly favorable spots. The southern part of Wrangell-St. Elias National Park and Preserve is in the zone of no permafrost; much of the park preserve is in the sporadic zone, while the northern part lies in the zone of discontinuous permafrost.

and porpoises, trapped in the lake by the ice dam. Over the next few months, Lake Russell accumulated 5.3 cubic kilometers of fresh water and rose to a height of about 25 meters above sea level. Local residents were concerned about two possible hazards: (1) the potential for sudden failure of the ice dam with ensuing flooding and iceberg damage at Yakutat; (2) the possibility that the ice dam might remain stable and cause Russell Lake to rise and overflow its southern end, which would damage the salmon spawning beds of the Situk River, a very important economic resource in the area. Early in the morning of October 8, the dam failed and the lake drained out in only 36 hours. The tremendous flooding and enormous icebergs brought some changes to the shoreline and bottom of the bay, but no major damage to the town of Yakutat. During the summer of 1989, the glacier again advanced and was threatening to block off Russell Fjord.

The Copper River and its tributaries, especially the Chitna River, form the largest stream drainage system in the park/preserve. Waters in this system run generally southward into the Gulf of Alaska. The Nabesna, Chisana and White Rivers, which drain the north slopes of the ranges, flow north and westward through the Yukon River system into the Bering Sea. Rivers in this area tend to have large annual fluctuations. They have very high discharge rates during summer runoff from melting snow and glaciers, but very low discharge rates during the frozen times of mid- and late winter.

Melting glaciers form the headwaters of all major streams in the park/preserve. The great volume of sediment in glacial meltwaters causes the streams to carry large amounts of silt during the summer. Because of their muddy waters, the streams do not support large populations of resident fish, but do serve as migration routes be-

tween the ocean and spawning beds, and also between various clearwater lakes and tributaries. Three streams, the Tebay River, Hanagita River, and Beaver Creek, that do not start from melting glaciers have clear water and larger resident fish populations. The park/preserve offers excellent whitewater boating and fishing opportunities. Information on river hazards and fragile resources, such as salmon-spawning beds and swan-nesting areas, can be obtained from the National Park Service.

Periglacial and Permafrost Features

Wrangell-St. Elias National Park and Preserve lies mostly within the discontinuous permafrost zone. Areas at high elevations or in the northern part of the park/preserve are underlain by permafrost. However, permafrost is absent or spotty at lower elevations, particularly in the southern part of the park/preserve along the Copper and Chitna River Valleys and near the coast. Areas of intermediate elevations tend to have permafrost where there is a layer of sphagnum moss or other protection from summer heat. Distribution of permafrost and thickness of the active zone (the surface layer that is frozen in winter and thawed in summer) is a primary control of vegetation in such areas. For example, in the Copper and Chitna River Valleys stands of "scrubby" black spruce usually indicate permafrost at shallow depths.

Periglacial features in Wrangell-St. Elias are similar to those of Denali National Park and Preserve. Soils in periglacial areas tend to be thin, discontinuous, and coarse-grained. In such cold temperatures, chemical weathering of minerals is extremely slow so the breakdown of rocks is accomplished by mechanical weathering—primarily by frost action. Silt- and sand-size particles are produced in abundance by frost wedging, but this process cannot pro-

Table 29.1   Simplified Geologic Column, Wrangell–St. Elias National Park and Preserve

Time Units			Rock Units	Geologic Events
Era	Period	Epoch		
Cenozoic	Quaternary	Holocene	Holocene drift, alluvium, colluvium and marine sediments	Holocene glaciation, fluvial action and (W) volcanism
		Pleistocene	Wisconsinan drift	Accretion of Yakutat Terrane, ongoing faulting, earthquakes, (W) Quaternary volcanism
	Tertiary	Pliocene	Older drift	Numerous glacial advances and retreats since Late Tertiary
		Miocene		Ongoing tectonic activities, especially faulting and uplift
		Oligocene		
		Eocene	(W) Andesite	(W,Y) Volcanism
		Paleocene	(Y) Basalt and shale	
Mesozoic	Cretaceous		(Y) Graywacke and shale	Accretion of four terranes during Jurassic through Eocene: (Y-T), (A), (W) and (C)
			(C) Graywacke and slate	
			(C,W) Clastic sediments	
			(C) Pillow basalt	
	Jurassic / Triassic		(W) Limestones and shale, Nicolai Greenstone, cherty argillite	
Paleozoic	Permian		(W) Limestones, shale, argillite, and chert	(W) Island-arc volcanism (Pennsylvanian-Triassic) with ongoing marine sedimentation
	Pennsylvanian		(W) Volcanic breccia, water-laid tuffs, minor limestone	
	Mississippian Devonian Silurian Ordovician		(A) Marble	
			(Y-T) Wide variety of metamorphic rocks Precambrian to Mississippian: Schist, gneiss, quartzite, marble, and plutons	
	Cambrian			
	Precambrian			

Explanation: (A) Alexander Terrane; (C) Chugach Terrane; (W) Wrangell Terrane
(Y) Yakutat Terrane; (Y-T) Yukon-Tanana Terrane
Sources: Burns, 1983; Csejtey, et al., 1982; Jones, et al., 1982; Silberling and Jones, 1984

duce finer clay-sized particles. This is why cold-climate soils are usually sandy and silty, but have little clay. Bacterial decay of organic material is also slow in cold climates. Because of this, humus and peat have accumulated to considerable thicknesses at some favorable localities in Wrangell-St. Elias Park and Preserve.

Geologic History

1. **Complex metamorphic rocks of Paleozoic and possibly Precambrian age in the Yukon-Tanana terrane.**

The oldest rocks in Wrangell-St. Elias National Park and Preserve are exposed north of the Denali fault in the Yukon-Tanana terrane. Foliated metamorphic rocks such as quartz-mica schist and gneiss are common; within the nonfoliated metamorphic rocks, quartzite is common and marble occurs in a few places. Metamorphosed rhyolite and plutonic rocks are also abundant in this terrane. Based on limited field studies and radiometric dating, it appears that the metamorphic rocks developed either from sediments deposited during Precambrian time, or from sediments that were derived by erosion of Precambrian rocks during the Paleozoic Era. These marine deposits accumulated either at the outer edge of the North American continental shelf or in a marine basin far removed from the continent. In any case the sediments were lithified, and by mid-Paleozoic they had been deformed, metamorphosed and intruded by granite plutons.

2. Mid-Paleozoic accumulation of carbonate sediments on the Alexander Terrane.

During the Devonian Period, carbonate-rich muds were deposited on the flanks of a volcanic island arc located far from the North American continent. Lithification changed these muds into limestone; subsequent metamorphism transformed the limestone into marble. The island arc was transported thousands of kilometers by crustal plate movements until, much later, it accreted to North America as the Alexander terrane. Most of the Alexander terrane occurs in Canada as an elongate area approximately parallel to the west coast. Although the portion of the Alexander terrane within Wrangell-St. Elias National Park contains primarily Devonian marble, other parts include diverse marine volcanics and sediments that have undergone varying degrees of metamorphism.

3. Late Paleozoic through early Mesozoic volcanism and sedimentation in the Wrangell terrane.

An arc of volcanic islands developed in Wrangellia early in the Pennsylvanian Period. After producing lavas for many millions of years, the volcanic arc became inactive. Cooling, subsidence, and accumulation of sediments followed, continuing into the Triassic Period. An episode of rift-zone volcanism occurred during the middle part of the Triassic.

Rocks of the Wrangell terrane reflect this complex geologic history. The basal rock unit consists of Upper Paleozoic volcanic breccias, lava flows, and associated clastic sedimentary rocks. As plate movements transported the arc toward North America and volcanic activity ceased, the rocks slowly cooled, became more dense, and subsided. This subsidence caused the older rocks to be covered first by shallow and then progressively deeper marine waters in which were deposited a sequence of limestones, shales, argillites, and chert beds. By mid-Triassic time, the rifting of the Wrangell terrane produced thousands of meters of pillow basalts and lava flows, which built up into an oceanic platform and later an island. In time, as volcanic activity ceased, a second sequence of cooling, subsidence and sediment accumulation occurred on Wrangellia.

4. Mid-Mesozoic to early Cenozoic accretion of exotic terranes to western North America.

Absolute dates for accretion of the several exotic terranes in the Wrangell-St. Elias area have not yet been determined; however, the chronological sequence can be interpreted. The first microplate to accrete must have been the Yukon-Tanana terrane because it adjoins the "original" North American continent. This "docking" of the Yukon-Tanana terrane was followed by accretion of the Wrangell, Alexander, and Chugach terranes. Actually there is good evidence that these three microplates collided and joined together as a composite terrane long before

they accreted to North America as a single large unit early in the Cenozoic Era.

As the microplates moved together and then collided with North America, there was a considerable amount of volcanic and tectonic activity. Evidence of submarine volcanism is preserved in Jurassic pillow basalts of the Chugach terrane, whereas andesite flows indicate Eocene terrestrial volcanism on the Wrangell terrane.

The accretion of terranes onto the North American continent, which occurred over tens of millions of years, brought about a great deal of uplift, folding, faulting, and metamorphism of rock units in the region. This tectonic activity produced the basic structures of the several mountain ranges in the region.

5. Late Cenozoic and ongoing accretion of the Yakutat terrane.

Details of the origin and early history of the Yakutat terrane are not known. The terrane is composed of both continental and oceanic rocks, some as old as late Mesozoic, and it originated thousands of kilometers from its current position (fig. 29.8).

Figure 29.8   Travel path of the Yakutat block with respect to North America. The double lines indicate the approximate position of a spreading center; the single line extending westward from the Yakutat block is a magnetic anomaly interpreted as a plate boundary. The viewer should keep in mind that both North America and the Yakutat block were and are moving. (Modified from Burns, 1983.)

Collision of the Yakutat block with North America began in Pliocene time and is still going on. The Yakutat terrane occurs in the southern portion of Wrangell-St. Elias National Park and Preserve. Forces generated by its accretion are responsible for uplift of the Chugach and St. Elias Ranges to their present elevations.

The Yakutat terrane is still actively accreting. This is one of the reasons why there are so many earthquakes in the region, including several very strong earthquakes during this century.

6. Quaternary volcanism.

Some of the most impressive peaks in the park/preserve owe their present forms to volcanic activity in the Quaternary Period. Within the St. Elias Range, Mt. Bona is the highest peak built up by Quaternary volcanism. Great outpourings of andesite lavas in the Wrangell Mountains produced Mt. Sanford, Mt. Blackburn, and Mt. Drum. Mt. Wrangell, which also formed from Quaternary volcanism, has the distinction of being the only active volcano in the park/preserve.

7. Late Cenozoic glaciation.

Since late in the Tertiary Period there have been several major episodes of glacial advance and retreat in the Wrangell-St. Elias region. Most evidence of earlier glaciations was destroyed by subsequent glacial advances, so a detailed history has not been worked out.

Rates of glacial advance and retreat in this region are rapid. Historic records document glacial recession of tens of kilometers within several decades. The residents of Yakutat understand the possibilities for rapid glacial advance, and this causes them to be quite concerned about how Hubbard Glacier might affect their future.

References

Burns, Terry R. 1983. Model for the origin of the Yakutat block, an accreting terrane in the Northern Gulf of Alaska, *Geology* 11: 718–721.

Csejety, Bela, Jr., and others, 1982, The Cenozoic Denali fault system and the Cretaceous accretionary development of Southern Alaska. *Journal of Geophysical Research* 87:(B5) 3741–3754.

Eliot, John L. 1987. Glaciers on the move. *National Geographic* 171:104–119.

Janson, Lone E. 1975. *The Copper Spike.* Edmonds, WA: Alaska Northwest Publishing Co. 175 p.

Jones, David L., and others. 1982. The growth of Western North America, *Scientific American* 247:(5) 70–84.

————. 1982. Character, distribution, and tectonic significance of accretionary terranes in the Central Alaska Range, *Journal of Geophysical Research* 87:(B5) 3709–3717.

Motyka, R. J., and Benson, C. S. 1983. Fluctuations of heat flow from the North Crater, Mt. Wrangell, Alaska, *EOS,* 64:(9) 90.

National Park Service, 1985, Wrangell-St. Elias National Park/Preserve, Alaska. Draft general management plan/environmental assessment/land protection plan/wilderness suitability review, 203 p.

Sharp, R. P. 1942. Soil structures in the Saint Elias Range, Yukon Territory. *Journal of Geomorphology* V:(4) 274–301.

Wrangell-St. Elias, International mountain wilderness, 1981. *Alaska Geographic* 8:(1) 144p.

Geologic Maps

Beikman, Helen M. 1980. Geologic map of Alaska, U.S. Geological Survey.

Howell, D. G., and others. 1985. Preliminary tectonostratigraphic terrane map of the Circumpacific region. American Association of Petroleum Geologists, Tulsa.

Silberling, N. J. and Jones, D. L. 1984. Lithotectonic terrane map of Alaska (west of the 141st meridian), U.S. Geological Survey Open-File Report 84–523 Part A, map and 12-page explanation.

Wrangell-St. Elias National Park and Preserve

Area: 13,200,000 acres; 20,625 square miles (park— 8,945,000 acres; preserve—4,255,000 acres)
Wilderness area: 8,700,000 acres
Proclaimed a National Monument: December 1, 1978
Established as National Park and Preserve: December 2, 1980
Wilderness designated: December 2, 1980
Designated a World Heritage Site: October 24, 1979
Address: P.O. Box 29, Glennallen, Alaska 99588

30

The geologic story of Kenai Fjords National Park is centered in the ice-capped southern Kenai Mountains where dynamic interplay between a mile-high icefield with outward radiating glacial tongues, and the fjords below creates spectacular scenery sculptured by earthquake, ocean, and ice.

Kenai Fjords National Park

South Central Alaska

by Donald S. Follows
Interpretive Geologist
Anchorage, Alaska

Figure 30.1   Skiers on the Harding Icefield, Kenai Fjords National Park. Granite nunataks jut up through the icefield surface. National Park Service photo by W. Resor.

KENAI FJORDS NATIONAL PARK

Seward
Phoenix Pk.
1,571 M.
5155'

Resurrection River

Exit Gl.

Resurrection Bay

Bear Gl.

Cheval Island

Aialik Pen.

Aialik Gl.

Pederson Gl.

Aialik Bay

Holgate Arm

Harris Peninsula

Harris Bay

Chat Is. Aialik
Harbor Cape
Natoa Is
Twin Is
Matushka Is
Granite Is

Harding Icefield

Skilak Gl.

HARDING MOUNTAINS

KENAI MOUNTAINS

Northwestern Gl.

Two Arm Bay

Cloudy Mt.
552 M.
1810

Lone Rock
Island
Seal Rock
Island
Chiswells

GULF OF ALASKA

Tustumena Gl.

Dinglestadt Gl.

McCarty Fjord

Pye
Islands

Truuli Peak
2,015 M
6,612

Tustumena Lake

West Arm

Nuka
Bay

Nuka
Island

Storm Mt.
1156 M.
3793

Iceworm Pk.
1767 M.
5800

Yalik Gl.

Petrof Gl.

KACHEMAK BAY

Homer

COOK INLET

Town

Glacier

Park Border

Road

10 mi.

0

Figure 30.2   Kenai Fjords National Park, Alaska.

Geographic Setting

Kenai Fjords National Park was established by Congress in 1980 to recognize the 936-square-mile Harding Icefield, with its associated glacial system of at least 38 glaciers, and to protect representative wildlife populations, onshore and offshore, that are attracted to emerging and deglaciated land. The Kenai National Wildlife Refuge complements the national park ecosystem along its west and northwest boundaries.

The major portion of the national park is located in the mountains above the wild south coast of the Kenai Peninsula. The area known as "Kenai Fjords" includes an indented, drowned coastline from approximately Port Dick on the west to Resurrection Bay near the port of Seward. Part of this coast is in the national park, part is in the Alaska National Maritime Refuge, and a small part is privately owned. The state of Alaska also owns some of the coastal lands. Although only 80 airline miles from Anchorage, the park is isolated and undeveloped.

Kenai Fjords National Park celebrates a true "geoecosystem" where living and nonliving components of the environment respond to the gradual retreat of the glaciers. Each year tens of thousands of breeding seabirds and marine mammals colonize embayments and offshore islands to feed from nutrient-rich ocean waters upwelling from the continental shelf along the Kenai Fjords coast. Fractured granite islands, steep cliffs, and sea stacks create ideal habitats for colonial-nesting seabirds. Steller sea lions haul out on wide rock benches at the tideline, while harbor seals doze on drifting ice floes. Dall porpoise, sea otters, humpback, minke, and killer whales ply the productive waters of the fjords.* Ling cod, black bass, halibut, salmon, red snapper, shrimp, and crab are harvested by both commercial and recreational fishermen.

Farther back in the fjords, bald eagles soar over black bear, moose, coyote, and mountain goat foraging thinly vegetated slopes and coves between the icefield and the ocean shore. River otter play along the short, swift streams that connect ocean to mountainside. Dominating the large embayments are seven tidewater glaciers that calve into the ocean fjords.

Although glaciers are retreating, the climatic conditions that create a dumping ground high in the mountains for massive snowfalls have changed little since the Ice Age. As relatively warm air masses rise from the ocean and cool over the Kenai Mountains, the moisture falls as rain and snow. Average precipitation is estimated at 200 inches per year, including 500 to 1000 inches of snow before compaction. Two to four times as much precipitation falls as snow on the Kenai Mountains than as rain along the fjord coast.

*The spelling of "fjord" as a common noun as well as a proper noun in this chapter conforms to preferred usage in Kenai Fjords National Park.

On the Kenai coast the ocean is seldom calm. Storm waves and heavy swells beat against the sea stacks, arches, and cliffs. Safe landings for boats are few on this outer fringe. Summer temperatures range from 40° F to 60° F, but freezing weather occurs on the icefield in even the warmest months.

Topography

The Harding Icefield rests atop the southern half of the Kenai Mountains, the range that forms the backbone of the Kenai Peninsula. Rising 3000 to 6000 feet above sea level, the Kenai Mountains are lithologically and structurally part of the Kenai-Chugach Mountains, an arcuate coastal range that bends eastward at the top of Kenai Peninsula and stretches alongside the Gulf of Alaska. To the southwest, many miles from the Kenai Peninsula, the same mountains rise from the ocean to form the spine of Kodiak Island.

Changes in relief are abrupt in the southern Kenai coastal area where elevations of a few feet are close (horizontally) to features 5000 feet above sea level. In McCarty Fjord a 6300-foot peak is less than two miles from tidewater, while the ridge along the fjord is nearly 5000 feet in elevation. Farther north and west, the 6612-foot summit of Truuli Peak ranks as the highest point on the entire Kenai Peninsula. ("Truuli" was the Kenatzie Indian name for the Kenai Mountains.)

As glaciers continue to retreat, uncovering the higher ridges and summits, the rectangular, blocklike appearance of the Kenai Mountains becomes more evident; but over half the park's area is still covered by glaciers. The thickness of the icefield over ancient preglacial stream valleys may be more than a thousand feet. When Pleistocene ice began to accumulate, the glaciers followed the previously eroded stream courses, scouring the mountain valleys into U-shaped troughs and lowering passes across the trend of the range.

The geoecosystem of Kenai Fjords National Park extends several miles beyond its present shoreline out onto the continental shelf. The 300-foot depth contour on the ocean bottom seems to correlate with glacial features and deposits left during the advance of valley glaciers before the Holocene rise in sea level that flooded the troughs.

Local History

Geological investigations of the southern Kenai Mountains began early in this century, but the lack of major economic discoveries, coupled with the ruggedness and extreme remoteness of the outer coast, soon terminated general interest. In 1909, when geologists first explored the south coast, it was virtually uninhabited from English Bay to Seward and has remained that way until the present day, except for a few hardy gold miners in Nuka Bay. During a National Park Service field reconnaissance of

Figure 30.3   Edge of Harding Icefield, rolling down to Kenai Fjords a mile below. When the picture was taken, fresh ash from the 1976 eruption of Augustine Volcano partially covered the icefield surface. Augustine, over 50 miles to the west on the far side of Cook Inlet, is south of Lake Clark National Park and Preserve (chapter 36). National Park Service photo by D. Follows.

some 600 nautical miles of south coast shoreline in 1976, the investigators did not encounter a single inhabitant.

Although Captain Cook in 1778 and Captain Vancouver in 1794 sailed past these shores, the rugged, uncharted coast, rough weather, and treacherous ice floes kept the explorers from putting in. However, Russian and American fur traders may have hunted sea otter in the coastal waters in the early nineteenth century. Native and Russian fur traders regularly used Portage Pass at the northern terminus of the Kenai Mountains for access to and from Prince William Sound. The portage avoided the long, difficult passage around the wild south coast of the Kenai Peninsula.

The park's interior was even less well explored than the coast. It is believed that the first party to cross the entire Harding Icefield did so in 1968 after climbing Truuli Peak at the southwest edge and dropping off the icefield at Exit Glacier in the northeast corner.

In 1970 a local entrepreneur set up a flightseeing and snow machine tour on Harding Icefield, complete with an overnight cabin in which fly-in visitors could take shelter. While the interest level was high and tourists were attracted from all over the world, the continual burial of the cabin by snow made trips increasingly difficult. Because the Harding Icefield receives much more snow than is melted during the summer months, it became impossible to maintain the cabin on each season's new snow layer. After the second winter, the cabin was buried too deeply to be dug out again and had to be abandoned.

Geologic History and Features

1. Oldest rocks, Mesozoic in age.

Great gaps exist in the geologic record, making the regional sequence of events difficult to discern. No Paleozoic rocks have been identified in the park. The oldest rocks in the region crop out northwest of the park in a highly

Table 30.1 Geologic Column, Kenai Fjords National Park

Time Units			Rock Units	Geologic Events
Era	Period	Epoch		
Cenozoic	Quaternary	Holocene	Alluvium, glacial drift	Coastal subsidence; subduction Earthquakes, sporadic faulting At least two glacial advances
		Pleistocene	Wisconsinan drift	Wisconsinan and earlier glaciations; Cook Inlet and Prince William Sound covered by ice sheets; deep erosion and scouring
	Tertiary	Pliocene		Uplift of Kenai Mountains; increased erosion
		Miocene		Earthquakes Accretion of Yakutat block (eastern Gulf of Alaska) Uplift, erosion
		Oligocene	Granitic intrusives	Emplacement of Harding Icefield batholith Subduction; regional metamorphism
		Eocene	Orca Group (dark gray flysch rocks, turbidites)	Accretion of composite Chugach-Prince William terrane (central Alaska)
		Paleocene	?	Thrust faulting Strike-slip faulting
Mesozoic	Upper Cretaceous		Valdez Group (metasedimentary flysch—sandstone, siltstone, conglomerate; major unit in park) ?	
	Jurassic		McHugh Complex (mélange of siltstone, sandstone, conglomerate, tuff, pillow basalt, chert, limestone, argillite)	Deformation along subduction zone
	Triassic			

Sources: Plafker et al., 1977; Nilsen, 1981; Péwé, 1975; Dumoulin, 1987.

deformed mélange belt that parallels the Border Ranges fault system along the western Kenai-Chugach Mountains. (A *mélange* is a mappable body of rock characterized by a lack of internal continuity of contacts or beds.) Included in the mixture are blocks and fragments of all sizes, both exotic and native, embedded in a fragmented matrix of finer-grained material. Known as the McHugh Complex, these rocks are believed to be Triassic to Cretaceous in age. Exposures have not been found within the park.

2. The flysch rocks of the Valdez Group.

A *flysch* is a thinly bedded sequence of marine sedimentary rocks consisting of sandy shales, marls, muds, conglomerates, graywackes, etc. Upper Cretaceous flysch rocks of the Valdez Group are the dominant rock series in the park. Their lack of fossils has for years challenged geologists trying to date the sequence. The bulk of the Kenai Mountains is composed of this great series of dark-colored, metamorphosed, marine sediments that consist generally of a monotonous series of rhythmically in-terbedded sandstones, siltstones, and minor mudstones and pebble conglomerates that have been intensely folded and regionally metamorphosed. The sequence forms a flysch (turbidite) unit several thousand feet thick.

The distribution of rock types in the region is broadly parallel to the major faults that border the southern Kenai Mountains to the east and west. The Border Ranges fault, lying west of Kenai Fjords National Park, marks the topographic break between mountains and lowlands and is believed to represent a "seam" joining together distinctly different bodies of rock of regional size, called *terranes* (*tectonostratigraphic terranes*). In south central and southeastern Alaska the terranes that lie in a great bend along the Gulf of Alaska are believed to be *exotic terranes;* that is, the rocks formed in another location far to the southwest and were subsequently "rafted" to their present site, possibly late in Cretaceous and early Tertiary time. The McHugh Complex and the Valdez Group are part of a belt of rocks that extends from southeast Alaska for hundreds of miles along the Gulf of Alaska coast to the Shumagin Islands in the Aleutians.

Figure 30.4 On the rugged south coast of the Kenai Peninsula, a snow-melt stream cascades a thousand feet from the Harding Icefield to the ocean. Slates and graywackes of the Valdez Group stand on end in the cliff. High precipitation supports the lush vegetation clinging to the bedrock. National Park Service photo by D. Follows.

3. Accretion and tectonic activity early in Tertiary time.

To the east of the park, in Prince William Sound, rocks of the Orca Group were deformed and accreted in Paleogene time along the south side of the Valdez Group. Both the Valdez and Orca Groups were partially subducted and underwent severe regional deformation. In some areas intense folding and faulting shortened and compressed the rock section by as much as 50 percent or more!

Recent petrographic studies of the Orca and Valdez rock sequences across Prince William Sound suggest that there is no terrane boundary or break between these two groups that separates discrete packages of rocks with distinct geologic histories. Instead, the Orca and Valdez probably form a single package derived from a common source during Late Cretaceous into Eocene time (Dumoulin, 1987).

4. Tertiary uplift, erosion, and intrusions.

By the beginning of Tertiary time, practically all of what is now Alaska had emerged from the sea. During Eocene time, vast quantities of sand and mud were removed by streams from highlands bordering Cook Inlet and the Susitna lowlands west of the park and from the rising Kenai Mountains as well. Small plutons, dikes, and the Harding Icefield granitic batholith, all Eocene in age, were emplaced in the deformed rocks of the southern Kenai Mountains. Contacts are sharp and steeply dipping where the batholith intruded slate and sandstone beds of the Valdez Group. Most of the batholith is buried beneath the Harding Icefield, but the granite crops out in cirques, glacial troughs, and in the isolated peaks jutting up through the ice, called *nunataks* (fig. 30.1). The Aialik and Harris Peninsulas in the park are part of the batholith, as are the offshore islands, except for Nuka Island just outside the park's southwest corner. The Pye Islands, across Nuka Bay from Nuka Island, are a southern extension of the batholith. (Nuka Island, the Pye Islands, and several other islands offshore from Kenai Fjords are in the Alaska National Maritime Refuge, south of the national park.)

A zone of thermally altered granitic rocks that weather to a reddish brown color is exposed in the inner part of Northwestern Fjord, near the head of Harris Bay. Also associated with the batholith are felsic dikes composed of plagioclase feldspar with minor amounts of muscovite, quartz, and orthoclase. Where these light-colored rocks intrude the dark metamorphic rocks of the Valdez Group, a pattern of black-and-white stripes is formed on cliff walls.

The granite islands (Cheval, Chat, Chiswell, Lone Rock, Seal Rocks, Matushka, Granite, etc.) offshore from the park and in the Alaska Maritime National Wildlife Refuge are hauling and nesting places for marine mammals and seabirds. During extensive wildlife surveys, National Park Service teams have found that coastal seabirds have an affinity for granitic rocks, perhaps because the fractured nature of these rocks allows underground nesting sites. However, exposures of pillow basalt on Cape Resurrection provide rookery sites for a large population of kittiwakes. By contrast, Nuka Island, composed of a large outcrop of graywacke, is relatively lacking in seabird colonies.

From Eocene through Pliocene time, no sedimentary record was left in the park region although erosive processes continued to carry detritus from the uplands to the sea. At that time the general distribution of mountain mass and lowlands was about the same as what can be seen today. The Kenai-Chugach Mountains had been raised to approximately their present elevations. However, the landscape did not look like the glaciated uplands we see in the Kenai Mountains today. Because of the relatively rapid uplift that occurred in Tertiary time, the preglacial streams cut steep, narrow slot valleys that followed weak planes and fractures in the rocks. The mountain range had a somewhat more massive, blocky appearance.

Figure 30.5 Granodiorite sea stack on the sinking Kenai coast. National Park Service photo by D. Follows.

5. The onset of glaciation.

In most of Alaska the Ice Age began earlier, was more intense, and lasted longer than in the lands to the south. As world temperatures cooled, the snowline on mountains crept down to lower elevations. Small glaciers expanded, grinding and gouging their way down valleys. Eventually they filled stream courses brim to brim and overtopped ridges.

The Kenai-Chugach Mountains became the nourishing grounds and source area for a multitude of alpine glaciers that gradually coalesced into one massive icecap. At the time of the glaciers' maximum extent, ice was over 3000 feet thick in many valleys. Only the highest peaks escaped the effects of glacial abrasion and scouring. Prince William Sound and Cook Inlet were entirely covered by expanding glacial systems.

Throughout the several glacial stages in Pleistocene time (identified in areas outside the park), the high Kenai Mountains were source and supply of glacial ice as storms continued to bring snow to the uplands. The great icefield, which then encompassed the entire backbone of the southern Kenai Mountains, remained relatively intact while its long glacial arms advanced and retreated through the valleys below, swelling or shrinking like a giant glacial octopus. These climatically controlled advances and retreats of the ice edge in the valleys and over the lowlands created the landforms and magnificent scenery that we see today.

6. Naptowne Glaciation and Neoglacial time.

The Naptowne episode, the last major glaciation of southern Alaska, took place 14,000 to 10,000 years ago. The Harding Icefield expanded to the north and west, completely filling the Resurrection and Kenai River valleys. Glacial ice then flowed into the Kenai lowlands to merge with a large piedmont glacier occupying the western flank of the Kenai Mountains. The glaciofluvial deposits left as the ice melted back about 10,000 years ago provided a habitat of great diversity that today burgeons with thriving life forms of many kinds. This geoecosystem is now protected in the Kenai National Wildlife Refuge west of the national park.

Figure 30.6   Three-Hole Arch in Aialik Bay shows effects of intense wave action. Note that beaches are lacking on this drowned coast. National Park Service photo by D. Follows.

The advance of Neoglacial ice began about 4000 years ago and ended about the middle of the nineteenth century. Since that time glaciers in southern Alaska have retreated fairly rapidly. In Kenai Fjords National Park rocks very recently uncovered by the melting back of ice are barren of vegetation; but in other places land exposed to natural processes a little longer is slowly becoming habitable for new generations of plants and animals.

7. Ongoing tectonism and a look ahead.

The boundaries of Kenai Fjords National Park were drawn to include the icefield and its glaciated forelands, the deep-water fjords, drowned cirques, igneous intrusions, and an *active tectonic zone along an axis of maximum subsidence.* Geologic evidence suggests that for at least the past 900 years and perhaps for as long as 1360 years, the coastal region within this tectonic zone has been sinking into the ocean. The axis of subsidence has been located roughly along the crest of the coastal mountains. During the 1964 Good Friday earthquake (which affected the entire south central coast of Alaska), tectonic move-

ment dropped the Kenai Fjords coastline 7 to 9 vertical feet!

Tectonic movements associated with the 1964 quake were but one pulse in an ongoing trend of regional deformation that may have begun in late Pliocene time. More than half of Alaska is seismically active at present. Quaternary tectonism can be detected by the presence of faulted deposits, uplifted glacial terraces, offsets of stream courses, and altered marine terraces. As the Kenai coast is submerged by ongoing geologic processes, glacial cirques become half-moon bays; peaks that once topped coastal mountains are now islands, battered by storm waves.

Kenai Fjords National Park with its glaciers, deep fjords, rain forests, wildlife, and sinking coast is a place of subtle beauty as wild and mysterious as can be found in Alaska. For life forms, clinging to the raw and broken edges of an ice age, this park preserves an ever-challenging environment—a place where one step back in time seems like a sweep through eternity!

Bibliography

Capps, S. R., 1940. The geology of the Alaska Railroad region. U.S. Geological Survey Bulletin 907. p. 51–64.

Dumoulin, J. A. 1987. Sandstone composition of the Valdez and Orca Groups, Prince William Sound, Alaska. U.S. Geological Survey Bulletin 1774.

Field, W. O., Editor. 1975. *Mountain glaciers of the Northern Hemisphere,* v.2. New York: American Geographical Society.

Follows, Donald S. 1978. Glacial landscapes and geology of the Harding Icefield complex, southern Kenai Mountains, Alaska, in Landscapes of Alaska Seminar. Alaska Resource Library, Anchorage.

Grant, U. S. and Higgins, D. F. 1913. Coastal glaciers of Prince William Sound and Kenai Peninsula, Alaska. U.S. Geological Survey Bulletin 526. p.52–56.

Karlstrom, T. N. V., 1964. *Quaternary geology of the Kenai lowland and glacial history of the Cook Inlet region, Alaska.* U.S. Geological Survey Professional Paper 443. p. 56–57.

Lethcoe, J. R. 1989. *An observer's guide to the geology of Prince William Sound.* Valdez, Alaska: Prince William Sound Books.

Miller, D. W. 1984. *A guide to Alaska's Kenai Fjords.* Seward, Alaska: Wilderness Images.

———. 1986. The Kenai Fjords. Where mountains meet the sea, Alaska's Gulf Coast. *Alaska Geographic,* v. 13, no. 1. p. 63–81

Nilsen, T. H. 1981. Accretion model for the Cretaceous Chugach terrane, southern Alaska. U.S. Geological Survey in Alaska: 1981 programs. p. 93–97.

Péwé, T. L. 1975. *Quaternary geology of Alaska.* U.S. Geological Survey Professional Paper 835. p. 31–34.

Plafker, G. 1969. *Tectonics of the March 27, 1964, Alaska earthquake.* U.S. Geological Survey Professional Paper 543-I. p. I-1/I-74.

———, Jones, D. L., and Pessagnon, E. A. Jr. 1977. A Cretaceous accretionary flysch and mélange along the Gulf of Alaska margin. U.S. Geological Survey in Alaska: Accomplishments during 1976. p. B1–B112.

Rice, B. 1987. Changes in the Harding Icefield, Kenai Peninsula, Alaska. Unpublished M.S. thesis, University of Alaska, Fairbanks.

Kenai Fjords National Park

Area: 670,000 acres; 1047 square miles
Proclaimed a National Monument: December 1, 1978
Established as a National Park: December 2, 1980
Address: Box 1727, Seward, Alaska 99664

31

Large glaciers, 20 to 30 miles long, descend more than 15,000 vertical feet from the high peaks of the Alaska Range to the flats below.

Mount McKinley, North America's highest mountain, which rises to 20,306 feet from its base at the 2000-foot level, has the greatest vertical relief of any mountain in the world.

Plate collisions, accretion of suspect terranes, and subduction, pile up crustal material and push the McKinley massif higher.

As the mountains rise, vast amounts of sediment, eroded by grinding ice and swift streams, choke valleys and spread over lowlands.

Caribou, Dall sheep, moose, grizzly bears, and timber wolves roam the Denali wilderness.

Denali National Park and Preserve

South Central Alaska

Figure 31.1 North side of Mount McKinley as seen from the Park Road, which skirts the base of the massif for some 90 miles between Denali National Park and Preserve Headquarters and Kantishna. Cotton grass (beside thaw lake) and other tundra plants thrive on glacial outwash and alluvium in areas of discontinuous permafrost. National Park Service photo by R. McIntyre.

DENALI NATIONAL PARK AND PRESERVE

Figure 31.2 Denali National Park and Preserve, Alaska.

Geographic Setting and Local History

The Athapascan name for Mount McKinley is Denali, "the High One." (This is the name most Alaskans prefer, although McKinley remains the official name.) Although Mount McKinley's two peaks are often shrouded in clouds in summer, its great bulk dominates the landscape. When the clouds part, revealing the magnificent massif, the summit, South Peak (elevation 20,306 feet), towers over the lower North Peak (elevation 19,470) and over all the mountains of the Alaska Range.* On clear days, McKinley's majestic, snow-covered summit can be seen from ships in the Gulf of Alaska. From the north, its height is even more impressive. The steep north face of Mount McKinley rises 15,000 feet from its base. Everything in Denali seems to be on a grand scale.

The Alaska Range is a curved mountain chain about 600 miles long that forms a topographic barrier and drainage divide between the coastal lowlands around Cook Inlet and the Yukon lowlands of the Alaskan interior. Denali National Park and Preserve lies like a saddle across the central and highest part of the range. In the massif section, where the Alaska Range is about 60 miles wide, the mountains trend in a generally northeast–southwest alignment. Nearly 20 peaks in this part of the range are over 10,000 feet in elevation; elsewhere in the Alaska Range, the mountains are mostly between 7000 and 9000 feet high. Mount McKinley and its lofty, white-crowned companions in the park/preserve—Mounts Mather (12,050 feet), Silverthrone (13,169 feet), Crosson (12,772 feet), Hunter (14,580 feet), Foraker (17,395 feet), and Russell (11,676 feet)—are challenged only by the highest peaks in Wrangell-St. Elias National Park and Preserve (chapter 29), some 200 miles (air distance) to the east.

The lands added in 1980 to create Denali National Park and Preserve surround the original McKinley National Park on three sides and ensure the protection of the entire massif, its glaciers, and its ecosystem. The Kichatna Mountains on the southern flanks of the massif, including the geologically unique Cathedral Spires, are now in the park/preserve. Eight major glacial troughs radiate from Cathedral Spires headwalls that rise nearly a mile from their bases. The historic Kantishna mining district on the northwest side of the original park has also been placed under National Park Service protection along with migration routes and wintering and calving areas of caribou.

Mount McKinley was named in 1896 by a gold prospector in honor of the 25th President of the United States,

who was elected in that year. In 1898, George H. Eldridge, a geologist, and Robert Muldrow, a topographer, were sent by the U.S. Geological Survey to explore the southern approaches to the range. Eldridge and Muldrow determined the map location and calculated the elevation of Mount McKinley with an accuracy close to currently accepted figures, which was an amazing achievement considering the difficulties of mapping a region without (at that time) topographic control points. The two men also plotted the route for the Alaska Railroad, built 25 years later from Seward to Fairbanks. The railroad, which skirts the eastern boundary of the park/preserve, is a principal means of access to the Denali area.

Lt. Joseph S. Herron led a U.S. Army expedition in 1899 that was the first to cross the Alaska Range from south to north. On their way from Cook Inlet to the Yukon River valley, north of the range, they discovered and named Mount Foraker, the second highest peak of the Alaska Range, the mountain that Indians call "Denali's Wife."

The first expedition to reach the base of Mount McKinley was led by Alfred H. Brooks, a pioneer geologist who took a U.S. Geological Survey party to the northern side of the Alaska Range in 1902. Brooks climbed a shoulder of Mount McKinley but was forced to turn back at about the 7500-foot level because of icy slopes and steep cliffs. He placed a note in a cartridge shell in a rock cairn at the spot. The note and shell were recovered in 1954 by geologists of the U.S. Geological Survey and are now on display in the Park Museum.

The scaling of Mount McKinley, to reach either of the two peaks (two miles apart with a difference in elevation of more than 800 feet), is a hazardous undertaking that requires many hours of continuous ice climbing. Because of its high latitude location, as well as its high elevation, the air at the summit is thinner and usually colder than the air on high summits in low latitudes. Moreover, less rock is exposed and the snow and ice are thicker and more treacherous than on most other high mountains. Only experienced climbers who have National Park Service permission may attempt an ascent. Climbing back down the mountain is equally dangerous.

Weather on the mountain is unpredictable. On a clear, bright day a blizzard may come up within 15 minutes, or dense fog and freezing mists may suddenly overtake climbers. Temperatures as low as −90° F and winds as high as 100 miles per hour have been recorded. "Normal" temperatures are around 0° F, with winds 25 to 30 miles per hour. Even in July, the best month to attempt a summit climb, conditions are often perilous. More than 42 climbers have died on Mount McKinley, and over 150 have been in serious accidents (Heacox 1986).

In 1903 Judge James Wickersham and a party of four from Fairbanks made the first attempt to reach the top of Mount McKinley. Unfortunately, the route they chose was

---

*University of Alaska-Anchorage scientists, using gravity meters and satellites, have determined Mount McKinley's elevation as 20,306. This figure is 14 feet lower than the elevation of 20,320 feet calculated by Bradford Washburn in 1954. McKinley's great mass exerts enough gravitational pull to distort standard surveying techniques like those Washburn used. (Reported by Associated Press, July 1989.)

by way of Peters Glacier to the north face and a sheer escarpment, called Wickersham Wall, which is almost impossible to climb. The highest elevation they reached was 10,000 feet. A map of the Kantishna Hills that they filed with a mining claim after their return was the impetus for a gold rush. Rich placer diggings were discovered in the Kantishna mining district. Later in life, when he was the territorial delegate to Congress, Judge Wickersham introduced the bill that provided for the establishment of Mount McKinley National Park.

In 1910, four miners in the Kantishna district, calling themselves the Sourdough Party, boasted that they could get the job done. Having prospected in the foothills, the miners knew the approaches to McKinley well and had already decided that the Muldrow Glacier on the north side offered the best access to the summit. They discovered McGonagall Pass, leading to the upper Muldrow Glacier, and established a base camp on a ridge at 11,000 feet. Two miners reached the top of North Peak, believing it to be the higher of the two summits. Once there, they realized their mistake, but planted a spruce pole to mark their achievement. They descended North Peak to the saddle, and started toward South Peak but turned back because of bad weather. Because the pole was not visible by telescope from Fairbanks, the miners' accomplishment was discredited.

The Muldrow Glacier route the Sourdoughs used is still a principal route by which climbers approach Mount McKinley's summit. The West Buttress route, which is shorter and safer and ascends by way of Kahiltna Glacier, was pioneered in 1951 by the distinguished explorer-scientist Bradford Washburn, who led a number of expeditions into the high areas of the McKinley massif. In 1947, his wife, Barbara Washburn, became the first woman to climb to Mount McKinley's summit.

The Stuck-Karstens Expedition, a party of four men, achieved the honor in 1913 of making the first successful ascent of Mount McKinley's South Peak. The leader was Archdeacon Hudson Stuck, an Episcopal missionary. Co-leader Harry Karstens had been a packer for naturalist Charles Sheldon and had spent many months with him, roaming the region that is now Denali National Park and Preserve. After reaching the top of South Peak, the climbers saw the pole on North Peak that the Sourdoughs had left three years earlier, and this vindicated the miners' claim.

Harry Karstens later became the first superintendent of Mount McKinley National Park in 1917. The park's establishment was largely due to Charles Sheldon's efforts. This noted naturalist lived in a remote cabin on the Toklat River through summer and winter seasons, between 1906 and 1908, in order to make the first scientific study of the region's wildlife.

Visitors can reach Denali National Park and Preserve via the Alaska Railroad, by Denali Highway (#8), George

Figure 31.3 Sketch map showing access to the park/preserve and principal drainageways. The Alaska Railroad and the paved Anchorage-Fairbanks highway follow essentially the same route and are open year-round. The Denali highway, approaching from Tok Junction (southeast), is closed in winter. The Alaska Range forms a drainage divide between north-flowing streams to the Nenana and Yukon Rivers and south-flowing streams that join the Susitna River.

Parks Highway (#3), and by private plane and charter aircraft (fig. 31.3). Private vehicles are not allowed on the park roads (except to get to the campgrounds), but shuttle buses operate daily during the summer season. During the winter, visitor facilities are closed, but the park/preserve is open to skiers and dog-sledders. Winter seasons, drier and less cloudy than summer months, last for about half the year, but snow may fall in any month. Rivers are frozen from late October to late April. Summer temperatures may range from 40° F to 80° F. January has less than four hours of light each day, but in June, skies may be brilliantly colored at midnight.

Geologic Features

Mount McKinley and the high neighboring peaks—Foraker, Hunter, Mooses Tooth—are carved from the tough granitic core of the Alaska Range. But the height of the McKinley massif, standing as it does at the apex of Alaska's geology, is due more to its unique position than

Figure 31.4   The Denali fault system arches across southern Alaska. Principal movement along the fault system has been horizontal, with the south side moving west and northwest relative to the north and northeast side. Vertical movement has also occurred on several fault segments. Both types of movement still go on.

to its resistant bedrock. In order to understand the geology and the geologic history of the Alaska Range, we need to consider (1) the great faults, (2) the tectonostratigraphic terranes, (3) the ancient rocks that the terranes brought to Alaska, (4) movements and convergence of tectonic plates (Cretaceous into the present), and (5) the forming in place of Mesozoic and Tertiary sedimentary and igneous rocks.

The Great Faults

The Denali fault system, with its segments and branches, which cuts through Denali National Park and Preserve, is the dominant geologic structure of the southern Alaska region (fig. 31.4). This complex system forms an arcuate belt from south of the Alaska Panhandle northwestward, then west, then southwest out into Bristol Bay and the Bering Sea. The Alaska Range lies approximately along the middle of the arc.

In Denali National Park and Preserve, two east–west-trending segments are the major elements of the Denali fault system. These are the Hines Creek fault on the north and the McKinley strand, lying some 20 miles to the south.

Both are north of Mount McKinley. Movement along the faults has been mainly by complex strike-slip displacements (horizontal), with the south side moving westward in relation to the north side. Two parts of a small pluton, split by fault movement, were offset 19 miles. In some places lateral displacement may have been much greater.

In the Denali region, strong vertical uplifts have allowed the Alaska Range to rise dramatically, especially along the currently active McKinley strand where it skirts the north side of the McKinley massif. Here the rocks on the south side of the fault have been raised many thousands of feet and uplift is continuing. The Wickersham Wall on the northwest side of Mount McKinley is the result of relatively recent movement. The trace of the McKinley strand is marked by a linear, topographic depression where the rocks in the fault zone have been crushed, pulverized, and eroded. Between Gunsight Pass (below Wickersham Wall) and Anderson Pass, a distance of about 15 miles, the Muldrow Glacier follows the fault zone northeastward before it makes a sharp left turn and flows northwest, down through the foothills to where its terminus drains into the McKinley River.

Tectonostratigraphic Terranes

The east-west faults represent boundaries between several suspect terranes that docked over a span of time from late in the Cretaceous Period into early Tertiary time. Horizontal and vertical movements along the faults have occurred sporadically ever since.

In the central Alaska Range, the terranes that have been identified are (1) Yukon-Tanana, (2) Wrangellia, and (3) McKinley. The Yukon-Tanana terrane, which lies north of the Hines Creek fault, consists largely of ancient, repeatedly metamorphosed, crystalline rocks. Wrangellia is a large terrane on the south side of the Hines Creek fault, mostly outside the eastern boundary of the park/preserve. This terrane, which covers much of eastern and southeastern Alaska (chapters 28, 29), was originally a volcanic island arc near the equator in the Pacific Ocean.

The McKinley terrane occupies the area between the Hines Creek fault on the north and the McKinley strand on the south. Less metamorphosed than the Yukon-Tanana and Wrangellia terranes, the McKinley rocks were originally marine and volcanic sediments. Old seamounts rode along on the microplate with the sediments.

Rocks That the Terranes Brought to Alaska

For more than 300 miles along its strike, the Denali fault system separates rock units that are grossly dissimilar in their lithologies, ages, and degrees of metamorphism. In Denali National Park and Preserve, the contrast is clearly evident along the Hines Creek fault, which crosses the Alaska Railroad near Park Headquarters and runs westward through a series of fault valleys for about 50 miles. The oldest rocks of the region are those of the Yukon-Tanana terrane, north of the Hines Creek fault. They are highly deformed and metamorphosed micaceous schists, quartzites, and marbles that were originally sedimentary and volcanic rocks deposited in Paleozoic and probably Precambrian time. They crop out in an east–west band across the northern part of the park/preserve. Mount Healy, Mount Wright, the Wyoming Hills, and the Kantishna Hills are made up mostly of these old crystalline rocks. Good exposures can be seen near Park Headquarters and the Savage River Campground.

South of the Hines Creek fault are groups of somewhat younger rocks, also metamorphosed and deformed, that are part of the McKinley terrane and, to the east, part of Wrangellia. The rocks are obviously of different origin than those north of the fault. One group of gray slates and marbles, dark gabbros, and altered basalts is exposed along the crest of the Alaska Range between Windy Pass and Mount McKinley. They can be seen more easily from the park road on the slopes above the Toklat River. A second group consists of interbedded marbles and slates extending from Savage River westward to Stony Creek. A third group, made up of weakly metamorphosed sandstones and shales, is found in a band between the other two groups.

Movement and Convergence of Tectonic Plates

The North American plate and oceanic plates or microplates from the Pacific region have been in deformational contact since mid-Paleozoic time and are still actively colliding, sliding by one another, and subducting. Because of North America's shape, what began as a more or less direct collision became more and more of an oblique contact. An apparent counterclockwise rotation affected much of the Pacific basin and is now evident in the ongoing strike-slip displacement of the Denali fault system (similar to that occurring to the south along the San Andreas fault of California's coastal region).

Plate collisions have been accompanied by orogenic activities, including folding, faulting, metamorphism, subduction, and uplift. Plutons were emplaced in the core of the Alaska Range and volcanoes and fissures spewed out ash and lavas on the flanks. Portions of the sedimentary and volcanic material riding on the subducting plate were scraped off and piled up, producing mélange rocks. Numerous large-magnitude earthquakes and volcanic eruptions (at the present time, mostly east and southwest of the Denali area) are indications that tectonic and deformational events that began in Cretaceous time in southern Alaska still occur in this dynamic region.

Younger Rocks Formed in Place

Unconformably overlying the terrane rocks, on both sides of the Hines Creek fault, are young formations that were deposited along the margins of mountainous coasts and in downfaulted basins. As mountain blocks were uplifted and shifted around, erosional processes produced great volumes of sediment. The oldest of the rock units, the Cantwell Formation, consists of sandstones, shales, and conglomerates that have been tilted and folded by compressive forces. Deformed beds show up clearly on slopes above the East Fork River, and outcrops can also be seen from the park road. Overlying the Cantwell rocks are rhyolitic and basaltic lava flows of the Teklanika Formation and the andesitic Mount Galen Volcanics. The bright hues of ash layers interbedded with clastic sediments and lava flows are strikingly displayed in Polychrome Pass and on Polychrome Mountain on the south side of the park road.

The granitic rocks that make up the highest, most rugged peaks of the range were emplaced in plutons of varying ages. On the southeast side of Mount McKinley in the area around Ruth Amphitheater and along the Great Gorge enclosing the upper part of Ruth Glacier, great spires and walls of granite soar thousands of feet above the ice. At the southwest end of the park/preserve are the

Figure 31.5   Eight major glaciers and several smaller ones radiate from Cathedral Spires (southwest section of the park/preserve). Kichatna Spire (8985 feet), North America's greatest single stand of vertical rock, was first climbed in 1966. National Park Service photo by M. W. Williams.

granite Cathedral Spires, probably the highest stand of vertical rock in North America and famed as a challenge to the most intrepid and skillful mountain climbers (fig. 31.5). Accompanying the plutonic activity was the injection of a few ore deposits, including gold that subsequently was eroded and accumulated in placers.

Glacial Features

The alpine scenery of Denali National Park and Preserve is large-scale and spectacular. When weather is favorable, charter aircraft fly visitors over the massif so that they may see the glacier systems and the sculpturing of rock by ice. Granite spires, arêtes, huge cirques, and gorges are impressive examples of the work of glaciers.

The more extensive glaciers and snowfields are on the southeastern side of the range because moisture-bearing winds from the Gulf of Alaska drop more snow on this side. The five largest south-facing glaciers are the Yentna (20 miles long), Kahiltna (30 miles), Tokositna (23 miles), Ruth (31 miles), and Eldridge (30 miles).

On the drier north side, except for the Muldrow Glacier, which is 32 miles long, glaciers are smaller and shorter. The Muldrow Glacier's length is due to the fact

that it follows the northeast-trending Hines Creek fault zone to Anderson Pass, then bends sharply and flows northwestward down to the lowlands. Twice in the last hundred years the Muldrow glacier has "surged" (moved forward for a short time at a greatly increased rate of speed), flowing over lower stagnant ice and making a jumble of broken ice-blocks. Higher up, ice ledges on the granite walls were left hanging when sections of the glacier collapsed and sank. A build-up of water between the bottom of the glacier and the bedrock channel may lubricate the rock-ice interface and cause surging. In the case of the Muldrow, movement along the fault at or near the bend may trigger a surge.

The glaciers of the Alaska Range carry enormous amounts of rock debris on, in, and beneath the ice as they move downslope. Loose fragments and blocks that fall from cliffs or are scraped from rock walls accumulate as low ridges of till that ride along on the edges of the glaciers. These are *lateral moraines*. Where tributary glaciers merge with each other or the trunk glacier, adjacent lateral moraines join to form *medial moraines* that are also carried down on the surface of the moving ice. Some of the large trunk glaciers, like the Yentna (fig. 31.6), have accumulated lateral and medial moraines that, seen from above, look like lane markers on a superhighway.

From the snouts of the glaciers, braided meltwater streams, heavily loaded with rock debris, continually shift and intertwine their channels over the valley floors. New channels open up as old channels are filled. Valley trains build up as streams drop quantities of poorly sorted sediment. *Valley trains* are long, narrow accumulations of glacial outwash, confined by valley walls. In some places coalescing outwash fans at the mouths of valleys have built up *outwash plains* over lowlands on both sides of the range.

Sometimes when glaciers are retreating and melting is rapid, blocks or masses of ice buried by till and outwash may persist for many years far beyond a glacier's terminus. When the ice finally does melt, the sediment slumps in, forming depressions called *kettles*. When filled with water, the depressions become *kettle lakes*. Wonder Lake, near Kantishna and the end of the park road, is the result both of glacial erosion and ice-block melting. About 4 miles long and 280 feet deep, the lake occupies a depression cut in weak rock by a glacier flowing off a resistant ridge. After the glacier retreated and the ice left in the cut melted, Wonder Lake was created.

Old moraines and kettles on the outwash plain are indications that the glaciers extended to lower altitudes in the past. Also, the fact that the present glaciers are "underfit" in their valleys shows that many of the glacial features that we see in the park/preserve today are the result of more extensive glacial action in the past.

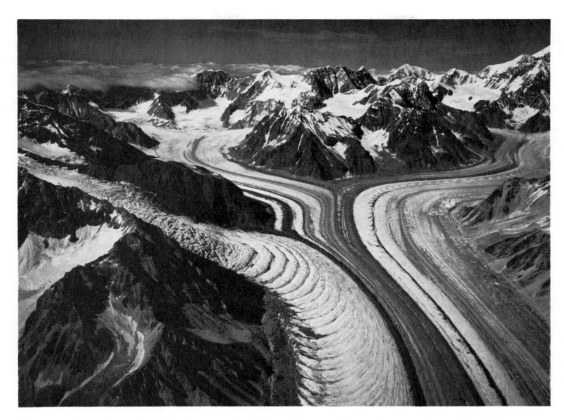

Figure 31.6   Yentna Glacier, flowing south from the McKinley massif. Lateral and medial moraines from tributary glaciers form a wide band of medial moraines being carried downslope. Joining the Yentna from the left is a small glacier cascading over a rock rim. National Park Service photo.

Permafrost and Periglacial Activity

During colder Pleistocene climates, all the ground in the park/preserve area became solidly frozen. On the north side of the range, much of the ground is still frozen because average annual temperatures have not risen enough since the Ice Age for melting to get below the thin topsoil. Permanently frozen ground, or *permafrost,* is *discontinuous* in the Denali area. The mean surface ground temperature, $-1°$ C to $-5°$ C (a few degrees below freezing), has allowed some thinning of the permafrost and a great deal of local variability. (North of the Arctic Circle permafrost is *continuous* and generally more than 500 feet in thickness.)

Permafrost is covered by an active layer that freezes and thaws seasonally. The depth of the active layer is controlled by variable factors, such as vegetation, surface material, topography, amount of moisture, and so on. In the Denali area, the active layer ranges from less than an inch to 10 feet or more, depending upon local conditions. Thicknesses of permafrost below the active layer have been measured between 30 and 100 feet at various locations around the Alaska Range.

The *periglacial milieu* is the active layer above the permafrost where freeze-thaw oscillations are more or less continual. Water changing back and forth between its liquid and solid states produces periglacial phenomena. In the active layer, water is available (and usually plentiful) because it cannot seep down into the impermeable permafrost below. (Voids and pores in permafrost are filled with ice.)

Simple frost action—a process of mechanical weathering that breaks up rock—is not usually dominant in temperate regions. In subpolar climates, however, the freeze-thaw cycles are predominant over other processes most of the time. Periglacial effects are noticeable around the margins of melting glaciers, on slopes, in river valleys, and on the tundra plains. Under periglacial conditions, *frost splitting* of wet rocks produces pinnacles like Cathedral Spires in massive, resistant rocks and thick accumulations of talus in fissured rocks. Sometimes talus masses in draws or gullies retain moisture, freeze, and become *rock glaciers* that move slowly downslope.

Frost heaving disrupts unconsolidated material, both sorted and unsorted. Soil volume increases irregularly when it freezes and then decreases irregularly when thawing occurs. Thus the surface of the soil keeps heaving up and falling back a little differently each time. As this

Figure 31.7   Horseshoe Lake, a mile north of Park Headquarters, is in a lowland area close to the Nenana River. Details of the land surface are constantly altered by periglacial activities and by streams working over glacial till and outwash. National Park Service photo by V. Watson.

process goes on, grains, pebbles, and cobbles on a flat surface tend to rearrange themselves in polygonal patterns and strips. Underneath the surface, lenses of coarse and fine material slump and sag chaotically.

On slopes, the active layer may become saturated during summer melt seasons due to the interaction of rain, meltwater, and low evaporation. The saturated soil (usually carrying a cover of vegetation) slides and slumps down a slope, exposing fresh permafrost underneath. This process is called *solifluction,* or soil flow. A stand of white spruce growing on a lower slope of Mount McKinley is called the Drunken Forest because the oddly leaning trees appear to be staggering as lobes of sliding soil slowly carry them down the mountainside. Marshy flats south of the park road also have areas of drunken trees leaning in all directions as the thawed soil layer slides over the permafrost.

Many of the shallow ponds in the park/preserve, like those in the tundra plains northwest of the range, are *thaw lakes* and *cave-in lakes* (fig. 31.1). They form where sun-

warmed water has melted basins in the underlying permafrost. The lakes gradually deepen from summer to summer, as warmth penetrates the frozen ground, and they enlarge as the rims collapse.

In winter, the freeze-thaw cycles slow and then become suspended, but ice, having 10 percent greater volume than water, continues to exert pressure. Thermal expansion and contraction at depth can cause cracks to develop in the ice-cemented permafrost. During the melt season, water gets into the cracks, forming veins of ice called *ice wedges,* that enlarge with successive seasons of freezing and thawing. In periglacial regions, excavations or landslides sometimes reveal very large ice wedges that have been buried in sand and gravel for centuries.

Periglacial activity and the presence of permafrost cause engineering problems for highways, railroads, landing fields, and buildings because structural failures tend to occur whenever the natural thermal regime is disturbed. Uneven thawing may cause roadbeds and foundations to tilt or collapse.

Table 31.1 Geologic Column, Denali National Park and Preserve

Time Units			Rock Units		Geologic Events	
Era	Period	Epoch				
Cenozoic	Quater-nary	Holocene	Glacial, alluvial and landslide deposits		Glacial and periglacial activity Stream erosion and deposition Maximum glacial stages	Vertical uplift Strike-slip faulting Earthquakes
		Pleisto-cene	Extensive glacial deposits			
			~~		Erosion ———	
	Tertiary		Nenana Gravel		Sedimentation in valleys	Uplift Block faulting
			~~		Erosion ———	
			Coal-bearing Group			
			~~		Erosion ———	
			Mt. Galen Volcanics Teklanika Formation Cantwell Formation	Granitic rocks	Ashfall Lava flows Nonmarine deposition in lowlands	Deformation Episodes of plutonic intrusion
			~~		Erosion ———	
Mesozoic and Paleozoic					Arrival of suspect terranes Intense orogenic activity, plate convergence Thrust faulting, uplift, folding, metamorphism	
			~~		Erosion ———	
			Sandstones, shales (S) Pillow basalts (S) Slates, marble (S)		Materials of the McKinley and Wrangellia suspect terranes	
			~~		Erosion ———	
?			Metavolcanic quartz schists, marble (N) Gneisses, quartzites, mica-quartz schists (N)		Materials of the Yukon-Tanana suspect terrane	
Precambrian						

S = South side of Hines Creek fault
N = North side of Hines Creek fault
Sources: Connor and O'Haire 1988, Gilbert 1979, Wahrhaftig 1958

Geologic History

1. Late Mesozoic and early Tertiary arrival of tectonostratigraphic terranes.

Microplates composed of oceanic and continental rocks of various types and ages drifted on oceanic plates toward western and northern North America from low latitudes far to the south and west. They traveled at velocities of 20 to 40 centimeters per year, or more than twice as fast as the usual velocity of larger tectonic plates. As the microplates crashed into the far northern margin of early Alaska, they became firmly attached to the rocks that were already there.

Many of the ancient crystalline rocks north of the Hines Creek fault that make up the Yukon-Tanana terrane are probably Precambrian in age. South of the Hines Creek fault and mainly north of the McKinley strand are the rocks of the Wrangellia and McKinley terranes, which are a little younger (probably early Paleozoic) and less highly metamorphosed.

Figure 31.8   Block diagram showing Pacific plate subduction beneath the North American plate. Along the Denali fault system, a change in plate motion, through a series of horizontal displacements, causes the plates to slide by each other. Where the plate boundary bends south, Mount McKinley is being pushed upward. Modified from W. G. Gilbert, *A Geologic Guide to Mount McKinley National Park.* © 1979 Alaska Natural History Association. Used by permission.

2. Mesozoic and Cenozoic plate convergence.

The rocks of the suspect terranes—the Yukon-Tanana and the later arriving Wrangellia and McKinley terranes—along with rocks that had earlier become part of the Alaskan margin, were all engulfed in intense, long-lasting tectonic events. In late Mesozoic time, an oceanic plate converging against the North American plate was subducted. Rock units were further deformed by compression and uplift. Great volumes of granitic rock were emplaced in the Alaska Range by a series of plutonic episodes that lasted through most of Cenozoic time. Lava flows poured out at the surface and volcanoes erupted. Sedimentation was controlled by tectonic events.

3. Accumulation in place of late Mesozoic and Tertiary sedimentary and igneous rocks.

As the land rose, sediments washed down into basins between the highlands and were interbedded with ash and lava. In this way the Cantwell and Teklanika Formations and the Mount Galen Volcanics accumulated. The Tertiary rocks are thinner and less deformed than the Mesozoic strata and have widely varying thicknesses and lithologies. Continued plate convergence caused the sedimentary rocks to be tilted and folded.

4. Deposition of coal and gravels in middle Cenozoic time.

As erosion wore down the mountains, plant material accumulated in swamps and muddy lakes. Sand and gravel buried the plant remains, and eventually they became compacted to a low grade of coal. The youngest rocks in the park/preserve are the units of the Coal-bearing Group and the poorly consolidated Nenana Gravel. At one time coal was mined near Sable Pass in the northeastern part of the park/preserve.

5. Uplift and erosion in Pliocene time.

Uplift caused renewed erosion. Nenana Gravel, 4000 feet thick in some places, filled downfaulted basins and spread over the lowlands north of the Alaska Range.

6. The pushing up of the McKinley massif.

As plate motion changed at the end of Pliocene and the beginning of Pleistocene time, strike-slip displacement occurred all along the Denali fault system. In the park/preserve, rocks from the southeast moved westward along the McKinley strand. In the Mount McKinley massif vicinity, the fault system bends sharply southwest toward the Aleutian Range. Subduction is now occurring under southwestern Alaska and the Aleutian Islands (fig. 31.8).

Some geologists have suggested that continental crust riding on the leading edge of the Pacific plate, on the south side of the McKinley strand, is too lightweight and too bulky to subduct and that it tends to pile up rather than slide under the overriding plate. If that is so, plate movement in this area tends to keep pushing Mount McKinley upward.

7. Pleistocene glaciation.

Deciphering the record of past glaciations in the Denali area has been complicated by the relatively rapid uplift that has been occurring. At the time of the first glaciation in late Pliocene time, the mountains in the McKinley part of the range were of moderate height. By the time the second glaciation began, deep stream valleys had been cut in the rising massif. Glaciers flowed down these valleys, widening and deepening them. On the snowier south side, ice covered nearly all the land between the Alaska Range and the Gulf of Alaska. North and west of the range, glaciers filled valleys and flowed outward but did not cover the entire area. (The interior of Alaska between the Alaska Range and the Brooks Range was never glaciated.) Third and fourth glaciations followed the same pattern of ice advance, but were not so extensive.

An *icecap* (small ice sheet) that developed on the Alaska Range during each of the last three glaciations did not connect with the great Canadian ice sheets to the east. The Alaskan icecap originated locally and had its own supply area, direction of movement, and persistence. Several minor advances and retreats of valley glaciers have occurred since Pleistocene time, but even the warmest interglacial period did not greatly reduce the size of the glaciers.

8. Recent tectonic activity.

Along the Denali fault system, the Pliocene Nenana Gravel has been folded and faulted. Moraines, river terraces, and outwash features have been offset. Some of these displacements can be seen from the Alaska Railroad and the north-south highway skirting the park/preserve. These are all indications of recent fault movement. Frequent earthquakes in southern Alaska are also a sign that tectonic activity in the region is ongoing.

Bibliography

Alaska Travel Publications Editors. 1976. *Exploring Alaska's Mount McKinley National Park,* 2nd ed. Anchorage: Alaska Travel Publications, Inc.

Beikman, H. M. 1980. Geologic map of Alaska. U.S. Geological Survey and Alaska Department of Natural Resources.

Buskirk, Steve. 1978. *Mount McKinley, the story behind the scenery.* Las Vegas, Nevada: KC Publications

Churkin, M., Jr., Carter, C., Trexler, J. H., Jr. 1980. Collision—deformed Paleozoic continental margin of Alaska—Foundation for microplate accretion. *Geological Society of America bulletin* 91:648–654, November.

Connor, C., and O'Haire, D. 1988. *Roadside geology of Alaska.* Missoula, Montana: Mountain Press Publishing. p. 192–198.

Gilbert, W. G. 1979. *A geologic guide to Mount McKinley National Park* (geologic map in pocket). Alaska Natural History Association.

Heacox, K. 1986. *The Denali road guide, a roadside natural history of Denali National Park.* Alaska Natural History Association.

McIntyre, R. 1986. *Denali National Park, an island in time.* ARA Outdoor World Ltd. Santa Barbara, California: Sequoia Communications. 80 p.

Péwé, T. L. 1975. *Quaternary geology of Alaska.* U.S. Geological Survey Professional Paper 835.

Reed, B. L., and Lanphere, M. A. 1974. Offset plutons and history of movement along the McKinley segment of the Denali fault system, Alaska. *Geological Society of America bulletin* 85:1883–1892, December.

Reed, J. C., Jr. 1955. Record of the first approach to Mt. McKinley. *American Alpine Journal* 9:78–83.

Tricart, J. 1968. Periglacial landscapes. R. W. Fairbridge, editor. *Encyclopedia of geomorphology,* p. 829–32. New York: Reinhold Book Corporation.

Wahrhaftig, C. 1958. *Quaternary geology of the Nenana River and adjacent parts of the Alaska Range.* U.S. Geological Survey Professional Paper 293A

———, and Black, R. F. 1958. *Engineering geology along part of the Alaska Railroad.* U.S. Geological Survey Professional Paper 293B.

Washburn, B. 1963. Map of Mount McKinley (Special topographic map). Boston, Massachusetts: Museum of Science.

Denali National Park and Preserve

Area: 6,030,000 acres; 9422 square miles

Established as Mount McKinley National Park: February 26, 1917

Separate Denali National Monument proclaimed: December 1, 1978

Both incorporated into and established as Denali National Park and Preserve: December 2, 1980

Wilderness area (1,900,000 acres) designated: December 2, 1980

Designated a Biosphere Reserve: 1976

Address: Box 9, Denali National Park, Alaska 99755

PART IV

Volcanic Features
and Volcanic Activity

© by G. Shimer

Figure IV.1 A flow of blocks, cinders, and ashes near Cinder Cone in Lassen
Volcanic National Park. This type of lava flow is called "aa," a Hawaiian term.
Drawing by G. Shimer from *Field Guide to Landforms in the United States,* J. A.
Shimer. The Macmillan Company © 1972 by J. A. Shimer. Used by permission.

Volcanoes, wherever they are, dominate the surrounding landscape. Volcanic eruptions, like earthquakes, are compelling events to human beings, many of whom may be unaware of quieter, more ordinary geologic processes. To geologists, a volcano erupting is "doing its thing." To monitor an eruption, a geologist may drop everything and hurry to observe the activity. As much as any tourist, a geologist regards a volcanic eruption as awe-inspiring, magnificent, and perhaps even frightening; but primarily, geologists see an eruption as a fascinating scientific opportunity. Volcanoes and volcanic eruptions are significant geologically far beyond their contribution to landscapes.

Volcanic Features in the National Park System

The varied examples of volcanism in the national park system are remarkable combinations of uniqueness, scientific merit, and landscape beauty. Of the eight national parks in Part IV, Hawaii Volcanoes National Park has had the most eruptions during recent years, but all eight have had volcanic activity of one sort or another during Holocene time. Moreover, eruptions have occurred within the last few centuries in all but one of these national parks. The exception, Yellowstone National Park, has notable geothermal features; more active volcanism may resume in the not-too-distant future.

Mount Wrangell in Wrangell-St. Elias National Park (chapter 29, part III) is an active volcano. Several other national parks, not included in part IV, are located in regions of volcanic activity. North Cascades National Park (chapter 26) is one example.

At least seven national monuments that highlight volcanic features are protected within the national park system: Alaska's Aniakchak National Monument and Preserve, with its immense caldera and an active volcano (fig. IV.2); Chiricahua National Monument, Arizona, a maze of columns, spires, and pedestals eroded from volcanic rocks; Craters of the Moon National Monument, Idaho, a desolate landscape of volcanic cones, craters, and caves; Devils Postpile National Monument, California, where tall, pipelike columns of basalt form imposing cliffs; Lava Beds National Monument, California, with its lava flows, lava tubes, and cinder cones; Pinnacles National Monument, California, a Miocene volcanic field, split in two and its parts separated by movement of the San Andreas fault; Sunset Crater National Monument, Arizona, a cinder cone. The nation's first national monument, Devils Tower in Wyoming, was long regarded as the neck or core of an eroded volcano; but most geologists now believe that the 865-foot pillar of columnar rock is a remnant of an intrusive igneous body, possibly a sill.

Mount St. Helens in Washington, the youngest Cascade volcano, erupted violently in May 1980, destroying a large part of its beautiful cone. Since that time, quieter eruptive activities have built a dome in its crater. In 1982,

Figure IV.2   A new volcano rises in the huge Aniakchak caldera that exploded and collapsed less than a thousand years ago. Aniakchak National Monument and Preserve is on the Alaska Peninsula, about 130 miles southwest of Katmai National Park and Preserve (chapter 35). Both Aniakchak and Katmai are on a volcanic arc that marks the plate boundary where the Pacific plate is being subducted beneath the North American plate. National Park Service photo by M. W. Williams.

Congressional legislation established the 110,000-acre Mount St. Helens National Volcanic Monument in order to protect and preserve this evolving volcanic feature and its surroundings. The monument is not a unit of the National Park System, but it is administered by the U.S. National Forest System.

Volcanism and Extrusive Rocks

The variety of volcanic landscapes and landforms in nature suggests that the igneous rocks that formed them must have originated from different sources in many different ways. *Extrusive igneous rock* (which includes all types of volcanic rock) is the surface manifestation of hot magma that has risen through the crust. (*Magma* is mobile, liquid rock material generated within the earth. Magma that hardens in the crust below the surface becomes *intrusive igneous rock.*) What produces the many different kinds of igneous rocks, whether intrusive or extrusive, is mainly a matter of a magma's chemistry and temperature and the crustal conditions it encounters as it works its way upward. If it reaches the surface before it solidifies, magma then becomes lava, which may erupt quietly as hot liquid or be ejected violently as *pyroclastic debris,* or *tephra.* Pyroclastic material (tephra) includes fragments of all sizes and shapes—pumice, ash, cinders, bombs and blocks—that are thrown out from a volcanic vent during an explosive eruption.

Characteristics of Lava

Why do some lavas erupt explosively while others are relatively "quiet"? The amount of gas in lava and whether or not gas can easily escape into the atmosphere are the key factors; and a lava's *viscosity,* or resistance to flow, determines how gases are released. A highly viscous lava with a large volume of gas trying to escape creates a violently explosive eruption. A less viscous lava that releases gas as it rises and flows out may not explode at all. Most of the gas released during eruptions is water vapor in the form of steam, with small amounts of hydrogen sulfide, carbon dioxide, and hydrochloride acid also given off.

The degree of viscosity depends on a lava's chemical composition and its temperature. Silica (SiO_2) is the major component in all volcanic rocks, with the amount ranging from about 45% to 75% of the total weight. Lavas with high percentages of silica (65% or more) have high viscosity and tend to solidify at much lower temperatures (roughly 700° C). Silica-poor lavas, which solidify at around 1500° C, have low viscosity and usually erupt as very hot fluid, sometimes with the consistency of molasses. The wide range of possibilities in composition and temperature combinations accounts for differences in the behavior of lavas, as well as for the many varieties of extrusive rock that are produced by volcanic activity.

How Lavas and Volcanic Rocks Are Classified

Silica-poor volcanic rocks and their parent lavas (around 50% silica by weight plus oxides of aluminum, calcium, magnesium, and iron) are considered *mafic.* Because they are composed of mostly dark green or greenish black minerals, mafic extrusive rocks tend to be dark in color. Basalt is the most common rock in this group. Rocks and lavas that are silica-rich (65% silica or more, with aluminum, sodium, and potassium oxides plus very small amounts of the oxides of calcium, magnesium, and iron) are classified as *felsic.* The most abundant volcanic rock with a felsic composition is rhyolite, which is generally light-colored because it contains mixtures of white, pink, light gray, and pale green minerals. *Intermediate extrusive rocks* are

Figure IV.3 Relationship between lava composition and common volcanic rocks.

those that are transitional in chemical composition between mafic and felsic. *Andesite,* which typically has a dark to medium grayish color, is the most common intermediate volcanic rock.

Types of Volcanoes and Volcanic Belts

Not surprisingly, the kinds of lavas that reach the surface and the ways in which they erupt have a great deal to do with the shape and size of the volcanic features and landforms that are produced. Lavas that are predominantly mafic tend to construct *shield volcanoes,* which have broad, convex-upward cones, built up by layers of basaltic lava flows. Kilauea, in Hawaii Volcanoes National Park (chapter 37), is a shield volcano built up from the ocean floor.

A *composite cone,* such as Mt. Rainier (chapter 32), tends to have a slightly concave profile and is constructed of alternating layers of mainly andesitic lavas and pyroclastics.

Small, steep, conical hills, formed by the accumulation of ash, cinders, and other pyroclasts, are called *cinder cones.* Usually of basaltic or andesitic composition, cinder cones may be the result of local concentrations of gas that cause pyroclasts to erupt through a volcanic vent. A typical example is Cinder Cone in Lassen Volcanic National Park (chapter 34).

Very viscous, felsic lavas (usually rhyolite) may solidify in or just above a volcanic vent, forming a steep-sided *volcanic dome.* This is a less common type of volcano but is apt to be hazardous due to its explosive potential. Some of the most destructive volcanic explosions in recorded history have been the result of the eruption of volcanic domes. Lassen Peak (chapter 34) is an unusually large volcanic dome.

Volcanoes have a tendency to change their eruptive patterns over time. Some start out with explosive eruptions of silica-rich lavas and then change to a long period of fluid, basaltic eruptions. Others build andesitic cones of lava and ash and then blow their tops off when rhyolitic lavas rise to the surface. Since volcanologists don't have much specific information about the chemical mixtures in the magmas that are "cooking" below the crust, making predictions is apt to be more of an art than a science. Nevertheless, considerable progress has been made in recent years as more data are collected and observations made of worldwide volcanic activity.

In general, volcanoes are located in belts that parallel the boundaries of tectonic plates. When plates are separating and pulling apart at a *spreading center,* the rifting allows hot lavas (mainly basaltic) to reach the surface. Sometimes volcanoes form, such as the Icelandic volcanoes in the Atlantic Ocean, which emanate from the rift along the mid-oceanic ridge. The volcanoes along the North American west coast (northern California to Alaska) are part of the circum-Pacific volcanic belt and are related to the subduction of oceanic crust beneath the continent (fig. IV.4). The heat and pressure resulting from plate convergence causes rocks to melt, producing lava that works its way to the surface (box 32.1, chapter 32).

Figure IV.4   Plate convergence boundary where an oceanic plate and a continental plate move toward each other. The continental crust, being more buoyant, overrides the denser oceanic crust, which sinks along a subduction zone. Rocks subjected to the compression of the converging plates are folded. Intense pressure at depth and heat generated by friction where the plates grind against each other causes rocks to melt and creates magma. Rock near the subduction zone that does not melt is metamorphosed. Some of the magma works its way up through the deformed material of the continental plate and is erupted, building volcanoes at the surface, above the subduction zone.

Volcanic activity *not* related to plate boundaries may be due to the presence of a *hot spot* at depth that causes molten rock material to rise to the surface. The Hawaiian volcanoes (chapters 37,38) and Yellowstone volcanic features (chapter 39) are believed to be located over hot spots in the upper mantle, far below the surface.

The national parks included in Part IV display volcanic features that are easy to recognize. Each has a story to tell of how it was formed—in hot clouds of incandescent dust and steam, or bubbling out from a lake of boiling lava, or squeezed up as gobs and spatters, or in any number of ways. An active volcanic area is like a laboratory showing how rocks are formed and is also a constant reminder that we live on a dynamic and changing earth.

National Parks in Part IV

32

Mount Rainier National Park

Cascade Range, Washington

Mount Rainier, a large composite volcano, is the highest mountain in the Cascade Range.

Explosive eruptions in Holocene time destroyed and reshaped the summit crater and upper part of the cone.

Steaming hot mudflows (lahars) have roared down the volcano flanks in the recent geologic past and may reoccur in the future.

Radiating from the summit slopes of Rainier's massive cone is the greatest single-peak glacial system in the United States.

Glacial erosion has modified the volcano's surface, creating beautiful alpine scenery. Downslope from ice and snow, subalpine meadows give way to dense forests.

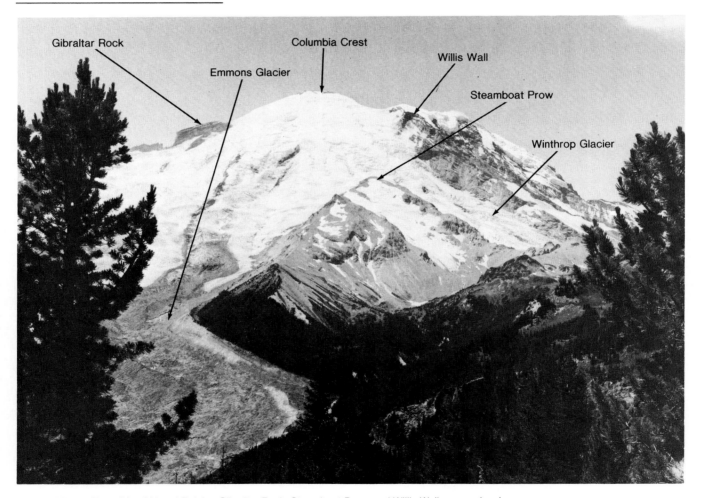

Figure 32.1 East side of Mount Rainier. Gibraltar Rock, Steamboat Prow, and Willis Wall are erosional remnants of the volcano's cone when the summit was at its maximum elevation, some 75,000 years ago. Glaciers on the east side descend to lower elevations because they are shaded from the afternoon sun. Emmons, the longest glacier in the park, extends into the White River valley. U.S. Geological Survey photo by R. E. Wallace.

Geographic Setting and Local History

At the beginning of this century, when Mount Rainier (elevation, 14,410 feet) was believed to be the highest mountain in the United States, the name "Columbia Crest" was given to what was thought to be the highest point on the crater's rim. More recent surveys have shown Register Rock to be the highest point at the summit. The grand old title of Columbia Crest remains, although the distinction of being "the highest" has long since passed to loftier peaks.

Few mountains anywhere dominate their surroundings as does Mount Rainier. Viewed from Puget Sound on the west, the volcano appears to rise from the sea. And from any side, Rainier's cone towers over all other peaks on the skyline. Yet the hills and ridges around the base of Mount Rainier are not insignificant. Several of these lesser peaks are over 6000 feet high.

To the Indians who inhabited the region for thousands of years before nonnative people arrived, the mountain was Tahoma, or "Snow Mountain," a holy and fearsome place. Out of respect and awe, Indians hunted on the lower slopes only, not daring to approach the summit. A myth of the Nisqually Indians personifies Mount Rainier as a Monster Spirit, who sometimes spewed out her venom in great torrents. This may have been the Indians' explanation for occasional volcanic mudflows (lahars) that roared down the mountain, filling valleys for many miles and overspreading lowlands. Indian encampments were destroyed by prehistoric mudflows, and some of the inhabitants perished.

Certainly Mount Rainier must have seemed impressive to Captain George Vancouver of the British Navy when he sailed his ship into Puget Sound in 1792 on his mission to explore and chart the North Pacific coast. He named the mountain for his friend and superior officer, Rear Admiral Peter Rainier, who never saw the great volcano that bears his name.

The demand for furs for the China trade brought sea captains, traders, and mountain men to "Oregon country," as it was known then. By 1818, a treaty between the United States and Great Britain provided for joint occupancy and the development of trading posts by fur companies. When the Hudson Bay Company established Fort Nisqually (near Tacoma's present waterfront) in 1833, William Fraser Tolmie, a young Scottish physician and botanist employed by the company, set about exploring the region and was apparently the first white man to enter the park area. In that year, with five Indian companions, he went into the northwest section of what is now the national park and climbed the mountain named for him, Tolmie Peak (5030 feet).

Mount Rainier, a difficult climb under the best of circumstances, challenged early explorers. The first officially recorded attempt to reach the summit of Rainier was made by Lieutenant A. V. Kautz and his party in 1857. Kautz, Dr. R. O. Craig, four soldiers, and a Nisqually Indian guide almost made it to the top but were turned back by high winds and violent storms. Their approach was from the south and up the Kautz Glacier.

Thirteen years later, in August 1870, General Hazard Stevens and P. B. Van Trump carefully planned an approach to the summit up a narrow arête, the Cowlitz Cleaver, that separates the snowfields of Cowlitz and Nisqually Glaciers and leads to Gibraltar Rock, which is close to the top (fig. 32.1). When Sluiskin, the Indian guide, realized that Stevens and Van Trump really meant to get to the top, he refused to accompany them because he believed that an evil spirit lived in a lake of fire at the summit. He promised to wait three days and asked for a letter absolving him of blame for their deaths if they did not come back. The two climbers did reach the top but had to spend the night there because of darkness and bad weather. They thought they would freeze to death until they discovered ice caves and steaming, foul-smelling vents on the summit. They spent an uncomfortable night roasting by the vents and shivering in the caves, but survived and descended the next day. When the anxious Sluiskin saw the two ice-encrusted figures emerging from the fog that had enveloped the mountaintop, he fled in terror, believing them to be ghosts. Finally they convinced him that they were alive, and the party returned to the settlements safely. The Stevens route to the summit via Gibraltar is still considered the safest and most convenient approach.

James Longmire, an early settler in the area, assisted Samuel F. Emmons of the U.S. Geological Survey in the early geologic reconnaissance of Mount Rainier and the surrounding region. Longmire himself finally climbed the mountain in 1883. On that trip he discovered mineral springs on the southwest side, and he later built a resort hotel at the site. The Park Headquarters and Visitor Center are now close to Longmire's springs, where the first settlement in the park area was located.

Considerable disagreement arose in the late 1880s over what to call this mountain area that was fast developing into a park—Tahoma, Tacoma, or Mount Rainier. The last name won out, and in 1899 the Mount Rainier area became the fifth national park to be established. At the present time, 80 miles of paved roads within the park provide access to notable viewing points and campgrounds, as well as to many miles of trails into wilderness areas.

Figure 32.2 Mount Rainier National Park, Washington.

Figure 32.3   Cross section of Mount Rainier, with an approximate profile of the volcano's maximum size superimposed over its present outline. From S. L. Harris, *Fire Mountains of the West: The Cascade and Mono Lake Volcanoes.* © 1988 Mountain Press Publishing Company. Used by permission.

Geologic Features

Mount Rainier, the highest volcano in the Cascade Range, has the greatest single-peak glacial system in the country. From its broad summit, glaciers radiate in all directions. During the volcano's existence, less than a million years, fire and ice (volcanism and glaciation) have continually interacted in molding, altering, and reshaping its form and appearance (fig. 32.3).

Matthes (1928) described Rainier's cone as "resembling an enormous tree stump with spreading base and irregularly broken top." Yet, lofty and massive as the summit is now, it was, as recently as 6000 years ago, more than a thousand feet higher than today. A series of eruptions and explosions began about 6600 years ago, culminating in an extremely violent explosion 5700 years ago that destroyed the summit. Enormous blocks of rock and glacial ice were tossed up in the air and landed miles away. Great avalanches of ice and rock roared down the mountain. Burning, steaming mudflows plunged with great speed down the valleys. Dark clouds of ash and steam must have blotted out the sun, and torrential rains probably began to fall soon after the explosion. The Indians' fear of the mountain god Tahoma is not surprising since stories of this disaster must have been passed down through many generations.

Snow and glacial ice again accumulated and the glaciers rebuilt themselves, beginning again their work of erosion. Then, a mere 2500 years ago, lava flows and ash falls began to repair the summit of Mount Rainier. Within the old *caldera* (a large basin produced by explosion and collapse) a new cone rose. Later, lava opened another smaller crater just to the east of the earlier vent. The two craters intersect at Register Rock, the highest point of the volcano. Within historic time (which in the Pacific Northwest does not go back very far), minor events in the late nineteenth century (smoke, steam, and ash emissions) were reported by Longmire, F. G. Plummer, and other observers; but since then the mountain has been relatively quiet. However, the volcano is continuously monitored by seismographs placed at various locations and elevations; infrared images that record variations in summit temperatures; and other observations, direct and indirect. Data from these sources have shown recent increases in thermal activity and some movement of magma at depth. Volcano watchers emphasize that such indications may be harbingers of renewed volcanic activity on a mountain that has had a remarkably varied history since its beginning in the Quaternary Period.

Box 32.1
The Cascade Volcanoes, a Part of the Pacific "Fire Zone"

Mount Rainier is one of more than a dozen geologically young volcanoes with composite cones (stratovolcanoes) that are perched on older rocks of the Cascade Range. In chapter 26, North Cascades National Park, you learned something of the geologic history of the Cascade Range and the development of two of these volcanoes, Mount Baker and Glacier Peak. In this chapter and the two that follow—33, Crater Lake National Park

and 34, Lassen Volcanic National Park—we take the view that in the Cascade region, times of general quietness are merely lulls during a half million years of fairly intense volcanic activity.

The Cascade volcanoes are remarkably aligned in a north-south direction, roughly paralleling the Pacific coastline for about 500 miles, from northern California to southern British Columbia (fig. 32.4). They are a small segment of

Figure 32.4   The small, eastward-moving Juan de Fuca plate is subducted beneath the westward-moving North American plate. The Cascade volcanoes are aligned at the surface above a melt zone at depth. From C. C. Plummer and D. McGeary, *Physical Geology*, fourth edition. © 1979, 1982, 1985, 1988 Wm. C. Brown Publishers. Used by permission.

the circum-Pacific volcanic belt, or "Ring of Fire," that encircles the Pacific Ocean. This major volcanic belt marks converging plate boundaries where oceanic crust is subducted beneath overriding continental crust or an island arc. As friction and pressure heat rock to the melting point, magma accumulates at depth, works its way toward the surface, and may erupt as lava to build a volcano. Since plates converge at different rates in different regions, not all volcanoes in a major volcanic belt erupt at the same time. However, any area of active subduction is considered a "fire zone" and should be regarded as potentially dangerous.

The volcanoes of major volcanic belts are almost entirely composite cones and andesitic in rock composition. Composite cones are so named because they are constructed of alternating layers of lava flows and pyroclastic eruptions (ash falls, etc.) The Cascade volcanoes are typical composite cones in that the prevailing rock type is andesite and each volcano has grown, over thousands of years of time, while brief eruptive episodes (lasting years or decades) were interspersed with long periods of quiesence.

Andesite is classified as an intermediate igneous rock because it is compositionally between the light-colored, silica-rich rocks (e.g., rhyolite) and the dark, basaltic rocks that are relatively deficient in silica. Andesite, commonly dark gray or greenish black in color, is made up of approximately equal amounts of light-colored minerals (plagioclase feldspars) and dark minerals (hornblende, olivine, pyroxenes). To geologists studying the Cascade volcanoes, variations in chemical composition, texture, and structure of the andesites are important inasmuch as very minor differences often enable investigators to identify and date lava flows and ash falls. Volcanic emissions are like fingerprints; that is, no two are exactly alike when samples are examined under a microscope and analyzed chemically.

Although andesite is clearly the predominant rock of the Cascade volcanoes, it is not the exclusive rock type; for at some time in its history, each volcano has had basalt and rhyolite eruptions. Several of the more explosive volcanoes have had mainly dacitic eruptions, dacite being a light andesite with a higher percentage of quartz (i.e., closer to rhyolite). Lavas containing higher percentages of silica are more viscous and tend to erupt more explosively.

Why andesitic rock is so prevalent in the Cascade volcanoes is a topic of debate among geologists. The oceanic crust that is being subducted is largely basalt. Where does the andesite come from? According to plate tectonic theory, the basalt originated from submarine eruptions along a spreading zone in the Pacific and has moved like a conveyor belt away from the oceanic ridge to the far boundary of the plate. Along the Pacific Northwest coastline, where the westward-moving North American plate is moving against a segment of eastward-traveling oceanic crust, the less dense continental crust overrides the denser oceanic crust. The oceanic crust, which is relatively young (10 million years old) is subducted beneath North America at a rate of approximately 4 centimeters per year. The subducted slab of basalt, along with some of the overlying sediment, descends at an angle to progressively greater depths, becoming hotter and hotter as it goes down—partly due to the geothermal gradient and partly because of the friction of the slab grinding past the overlying crust. (The *geothermal gradient* is the rate of temperature increase associated with increasing depth beneath the surface of the earth.) At some critical depth, the temperature is high enough so that some of the rock melts and forms magma. As pockets of magma coalesce, they tend to rise— perhaps as large blobs, shouldering aside rocks of the continental crust, or perhaps ascending fairly rapidly through fissures and erupting on the surface.

continued

Box 32.1 continued

Why is this rising magma andesitic? Several lines of speculation have been suggested. Perhaps a silica-deficient basaltic melt is formed along the subduction zone and is contaminated by silica-rich rocks of the continental crust to produce an intermediate magma; or perhaps only part of the basaltic crust beneath the subduction zone melts, giving a melt higher in silica than the solid residue of oceanic crust remaining in the earth's interior. Or, possibly silica-rich sediment and sedimentary rock carried downward by the descending slab are incorporated into the molten basalt to form andesitic magma.

The remarkable alignment of the Cascade volcanoes can be explained because the plate convergence zone is parallel to the coastline in this area, and all along this zone the subducting oceanic plate bends downward at a fairly shallow angle. Thus material on the descending slab reaches melting depth below the surface at about the same time and at the same distance inward from the coast. The line of volcanoes is believed to be located approximately above the zone of melting deep within the crust. They are supplied by magma rising from this zone and erupting at the surface as lava. Does it follow, then, that all the Cascade volcanoes have had a monotonously similar history? The fact is that they are more like a large family of brothers and sisters. They share a common background and have similar characteristics, but each grew and developed in a highly individual way.

Is there a potential for hazards other than volcanic eruptions in the Cascade region? Specifically, what are the estimations of seismic risk? Historically the Pacific Northwest coastal area has had no large-scale earthquakes for about 200 years; but severe earthquakes accompanied by tsunamis have occurred in modern times in subductions zones such as Columbia and southern Chile that are similar to the Cascade subduction zone. Seismologists suspect that the Cascade subduction zone may be storing strain energy that may be released by catastrophic earthquakes in the future. Any severe shaking of the Puget Sound or Willamette Valley area would cause destructive local tsunamis along the Washington and Oregon coasts. Indian legends suggest that such catastrophic events have occurred in the past (Heaton and Hartzell, 1987).

Mount Rainier differs from other Cascade volcanoes in that about 90 percent of its eruptions have been in the form of lava flows. Having only a tenth of its bulk made up of pyroclastics (*tephra*) is abnormally low for a composite cone. Moreover, some of those pyroclastic layers were originally lava that was blasted apart by exploding gases and later reconsolidated as volcanic breccia.

By contrast, Mount St. Helens has had a more violent history with few lava flows and a high volume of pyroclastics. In fact, much of the ash and pumice on Rainier's slopes came from St. Helens during explosive episodes not unlike the 1980 explosions that littered the region with pyroclastics. (*Pumice* is a frothy rock resulting from the sudden explosion of lava containing a large amount of dissolved gas and steam. Due to the explosive release of pressure, tiny gas bubbles cause the solidified lava to have a spongy or frothy texture.)

About 6800 years ago, the explosion of Mount Mazama in Oregon (chapter 33) spread several inches of ash over Mount Rainier and left a blanket of ash on the entire Pacific Northwest, parts of Canada, and even western Montana. This was the most extensive ash fall in postglacial time. Ash that erupted before and during glacial stages is difficult to trace or date with accuracy because processes of glacial erosion and deposition have left very few undisturbed patches.

The postglacial pumice and ash beds on Mount Rainier are identified by letter; i.e., the fall from Mount Mazama is called Pumice Layer O. The fall of maximum thickness, 20 inches, is Pumice Y from Mount St. Helens. Pumice Layer X, which erupted from Mount Rainier itself between 100 and 150 years ago, has a maximum thickness of only one inch. More recently, the 1980 eruption of Mount St. Helens deposited fresh ash on Rainier's slopes.

Glacial Features

The cone of Mount Rainier, from summit to base, has been deeply furrowed and scarred by severe and long-continued glacial erosion. Remnants of the volcano's outer layers show on the mountain flanks as sharp crags and ridges (the "cleavers") that jut through the ice and separate the radiating glaciers descending from the snowy summit. One of these residual rock masses, Mount Tahoma, is described by Matthes (1928) as "a sharp, triangular tooth on the east flank" of Rainier that rises to an elevation of over 11,000 feet. "In its steep, ice-carved walls one may trace ascending volcanic strata aggregating 2000 feet in thickness that point upward to the place of their origin, the former summit of the mountain" (more than a thousand feet higher than the present top).

The caldera rim left by the explosion that destroyed the summit has been breached and the rock crumbled by overriding ice cascades. Even the new cone is beginning to show minor indications of erosion by glacial ice. Several hundred feet of snow fill the two young craters. Beneath the snow are the famous steam caves. An extensive network of tunnels is in the east crater. The west crater has a smaller cave system but contains a pool of meltwater at the present time under a canopy of ice. If only a small increase in vent temperature occurred for any length of time, lakes might form on the snowpack surface in the craters. (This has happened from time to time in the past.)

Storm clouds from the Pacific Ocean, borne by prevailing westerlies, drop a large part of their load at the lower and middle elevations on Mount Rainier, giving these zones more precipitation than the summit. Estimated annual snowfall at middle levels on the western slopes of the mountain exceeds 100 feet. Added to high snowfall are great volumes of snow that are blown off the summit or come down in avalanches. And since summer temperatures remain relatively low between the 8000- and 10,000-foot elevations, the middle zone is the most favorable for glacier development.

Rainier's "atmospheric layering" is familiar to those who live near the mountain or come to climb it. On many days a climber on the higher ridges can be in bright sunshine and look back and see thick white clouds covering everything below; valleys and all the lower mountains have disappeared from view. The converse is true (as climbers should be so cautioned); sudden storms may envelop the summit while the valleys below are in sunshine.

The glaciers that originate at the summit—Kautz, Tahoma, Winthrop, Emmons, Ingraham, and Nisqually—are "refueled" after they get 4000 feet or so down from the top and reach the high snowfall zone. Another group of glaciers—among them Carbon and Russell, North and South Mowich with Edmunds in between, Puyallup, South Tahoma, Paradise, Cowlitz, Ohanapecosh, Fryingpan, and several others—start from cirques in the mountain flanks at the middle elevations where more snow accumulates. A few smaller glaciers in the Mount Rainier system still exist at around 6000 feet on north-facing slopes. Here the winter snowfalls are heaviest, and snowfields are protected from summer sunshine. In all, 26 named glaciers, occupying around 40 square miles of land surface, are in Mount Rainier National Park.

Periodic variations in climate affect the volume and length of the glaciers. Recently the glaciers have been advancing slightly after about a century of shrinking. The larger glaciers are surveyed and measured carefully at regular intervals because they are the source of major streams (the Nisqually, Puyallup, White, Carbon, and Cowlitz Rivers) that supply hydroelectric power to the region. Emmons Glacier, largest in the park, supplies the White River. Flow in these rivers tends to be low in winter and high in the summer melt season. The effect of meltwater on the streams is perhaps most striking near the headwaters. In the early morning a small stream issuing from a glacier can be crossed easily on foot; but by midafternoon on a sunny summer day this same stream is a torrent of milky-white water, banging cobbles along its bed and far too swift and dangerous to walk or wade across. The white color is due to the pulverized fragments or grains of *rock flour* scoured and abraded by moving glaciers.

Glacial erosion on a large scale is most evident in the precipitous cirque headwalls, such as Willis Wall, the largest cirque on the mountain, with a headwall of 3600 feet and a diameter of a mile and a half. In this great basin, opening to the north, is stored the snow and ice that enable the Carbon Glacier to descend to an altitude lower than any other glacier on the mountain (fig. 32.5).

Conspicuous lateral moraines descending with the glaciers give us some idea of the quantity of rock being removed from the valley sidewalls, both by abrasion and by mass wasting. On the valley sidewalls, above the rock debris moving along with the ice, are the remains of earlier lateral moraines built when the climate was colder and the glaciers larger. Most of the present glaciers of any size have medial moraines as well as lateral moraines. The medial moraines trail down from a junction point where a branch joins the main glacier.

Drift and till deposits of various ages are found mostly at lower elevations in moraines built during more severe glaciations. A large moraine from the last major glacial stage (about 11,000 years ago) dams Mystic Lake at the foot of the mountain on the north side. *Rock glaciers,* now inactive, were also left by the last glaciation. The fronts of these large talus accumulations are abrupt and steep and from 25 to 100 feet high. Also built of talus are *protalus ramparts,* which are ridges 5 to 15 feet high of coarse, angular rock pieces that fell or rolled down from heights and accumulated at the bottom of large perennial snowbanks piled against the base of a cliff. When the climate warmed and snowbanks finally melted, a protalus rampart remained, standing apart from the toe of the slope.

Valley trains, made up of outwash deposited by meltwater streams, cover the floors of the narrow valleys that descend from the ends of the glaciers. Valley trains that come down from Mount Rainier have been modified by ash falls and mudflows. Some valley trains open out onto outwash plains at lower elevations.

Lovely cirque lakes, like Tipsoo Lake, Crescent Lake, Mowich Lake, and others, can be seen in Mount Rainier National Park. Some tributary streams that flow in hanging valleys have waterfalls. Christine Falls and Comet Falls are examples. Ice caves, formed by meltwater, are usually visible at the bottom of Paradise, Carbon, and several other glaciers. Some ice caves have been enlarged by steam vents.

Figure 32.5   Morainal debris surrounds the melting snout of the Carbon Glacier, while glacial outwash and debris extend for miles down the Carbon River valley. The Carbon Glacier descends to the lowest elevation (approximately 3000 feet) of any United States glacier, outside of Alaska. Eocene coal deposits, far downstream from the glacier, gave the Carbon River its name. National Park Service photo.

The Role of Mass Wasting Processes on Mount Rainier

Rockfalls, rockslides, avalanches, debris flows, mudflows, and other processes of downslope movement, singly or in combination, operate more or less continually in the Mount Rainier environment. Moreover, the interaction of processes of mass wasting, weathering, volcanism, and glaciation has created a most powerful group of mechanisms in the eroding, transporting, and depositing of debris far and wide. Their potential for seriously disrupting the environment is a matter of grave concern to volcano-watchers. Consider the following items.

On almost any clear day, those who visit the Paradise or Sunset overlooks (or other vantage points) can see dust clouds almost hourly rising from small rockfalls and avalanches near the summit. And scanning the snowfields of major glaciers with telescope or field glasses, one sees fresh, dark debris that has fallen since the last snow came, perhaps the previous night or within the past few days.

Geologists have mapped more than 55 *lahars* (flows of pyroclasts, volcanic debris, mud, etc., usually mixed with hot water) that have flowed down Mount Rainier in Holocene (postglacial) time. The longest and largest of these, the Osceola Mudflow (± 5700 years ago) ran about 70 miles to the northwest, altered the course of the White River, and filled in an arm of Puget Sound.

In October 1947, within a span of about 8 hours following heavy rains, successive mudflows roared down Kautz Creek to the Nisqually River, leaving behind several cubic miles of muddy debris in the Kautz valley. Witnesses reported that large trees and boulders 13 feet in diameter were borne along by the mudflows, which had the consistency of wet cement.

In December 1963, a series of rockfalls and avalanches, apparently triggered by a steam explosion, fell in quick succession from the north face of Little Tahoma Peak. Plummeting downward, this great volume of broken rock struck the Emmons Glacier with great force and then bounced and flew down the White River valley. While dropping some 6200 feet in altitude, the avalanche traveled nearly 4 miles from its source. Some of the boulders that were transported are as large as buildings.

These examples are by no means unusual events in the geologic story of Mount Rainier. Mixtures of ash, pumice, fragmented bedrock, and solid blocks blanket the mountain flanks and are found as thick fill in the valleys

Figure 32.6   Tephra deposits, lava flows, and mudflows from Mount Rainier in Holocene time. River valleys radiating from the north side of the volcano appear to be at greatest risk. From S. L. Harris, *Fire Mountains of the West: The Cascade and Mono Lake Volcanoes.* © 1988 Mountain Press Publishing Company. Used by permission.

that lead away from the cone. The action of hot water and steam on the lavas and ashfalls at the summit and on the slopes causes any consolidated material to break up both mechanically and chemically, producing quantities of debris for mass wasting processes, glaciers, and streams to carry away.

Glaciers are very efficient at breaking and grinding up rock. On Mount Rainier active glaciers oversteepen slopes, expose fresh surfaces, and pulverize loose fragments. Continual freeze-and-thaw action at high altitudes on rock unprotected by soil or vegetation tends to break up and shatter the rock. Loosened fragments fall off cliffs and roll down slopes. Tremors or even slight explosions shake loose more rocks.

The fact that andesite, the dominant rock, is more susceptible to chemical weathering than, say, granite is significant. The feldspars and ferromagnesian minerals that make up andesite tend to weather out as clay minerals and soluble salts, especially in the presence of hot water, which hastens the chemical action.

With all this fragmentation of rock going on, plus the intermittent addition of ash, lava, and pyroclastics, it's no wonder that it doesn't take much to trigger slides and flows. Given the availability of great quantities of loose (or easily loosened) material, the dynamics of slides and flows can be explained as follows: (1) Precipitation and runoff on Mount Rainier are high. Rocks and debris are water-soaked much of the time. Material that is wet tends to move, slip, and slide faster than dry material. (2) Slopes at high altitudes are generally steep and unprotected by vegetation. Heavy snow accumulations tend to set off avalanching and slides. (3) Eruptions drop ash and pumice on snow accumulations. Mix in steam and hot water and all at once you have a mudflow. Add melting glacial ice and rock debris, and you have a disaster roaring down valleys (fig. 32.6).

Note that the distinction between debris flows and mudflows is gradational; that is, a *mudflow* is a variety of *debris flow* in which the solid component is composed of

Table 32.1 Generalized Geologic Column, Mount Rainier National Park

Time Units			Rock Units	Geologic Events
Era	Period	Epoch		
Cenozoic	Quaternary	Holocene	Surficial deposits: avalanches, lahars, ash, pumice, glacial and alluvial material Andesite lava: pumice, tuff	Mudflows, debris flows, ash falls Growth and retreat of modern glaciers New cone constructed in old caldera Osceola mudflow and lahars Explosive destruction of old summit
		Pleistocene	Ash, pumice, lahar deposits Interbedded andesite lavas and breccias Glacial deposits throughout epoch Andesite lavas Intracanyon lavas	Building of composite cone to maximum height Glacial and interglacial stages Beginning of Mount Rainier Filling of valleys by lavas
	Tertiary — Neogene	Pliocene		Beginning of glaciation Uplift and erosion of Cascade Range
		Miocene	Tatoosh Granodiorite — — — — — — — — Fifes Peak Formation — ? — ? — ? — ? —	Intrusion of Tatoosh pluton with associated volcanics, dikes, sills Andesite eruptions; basalt lava flows; pyroclastics
	Tertiary — Paleogene	Oligocene	Stevens Ridge Formation ///////////////// Ohanapecosh Formation — ? — ? — ? — ? —	Extensive ash flows, ash falls; nuées ardentes; welded tuff Unconformity Folding, erosion; weathering of red soil Andesite lava eruptions; volcanic breccias, mudflows, ash falls deposited in shallow water and on land
		Eocene	Puget Group (not exposed in park) — ? — ? — ? — ? —	Sand, clay, swamp deposits; coal beds

Source: modified after Fiske et al. 1963; Crandell 1969

at least 50 percent sand, silt, ash, and clay. The water content critically affects velocity and mobility. By gaining water along the way, a mudflow becomes more of a slurry; and if it runs into a good-sized river, then that river becomes an overloaded stream that floods lowlands and deposits the mountain's debris on flood plains and, in the Pacific Northwest, in harbors.

According to estimates, lahars and mudflows travel at speeds of about 50 miles per hour. This doesn't allow much time for people and vehicles to get out of the way. However, authorities at Mount Rainier National Park carefully monitor daily conditions and have set up early warning systems and evacuation plans. Campgrounds have been relocated on higher ground and other sensible precautions taken. The park would be closed immediately if thermal or seismic data indicated any possibility of an eruption. Meanwhile, visitors are able to enjoy the park and its many natural wonders without being overly apprehensive.

Geologic History

1. Eocene lowland.

In the early part of the Cenozoic Era, the area now occupied by the Cascades was a coastal plain. Over several million years, sand, clay, and organic swamp deposits accumulated in subsiding tidal marshes. These sediments became a 10,000-foot thick sequence of sandstone, shale, and coal (some of which has been mined) known as the Puget Group. Arkose of the Puget Group interfingers with the younger Ohanapecosh Formation near the park's western boundary, but is not exposed in the park.

2. Eocene volcanism, continuing into Oligocene time.

About 40 million years ago, volcanic activity began as a local manifestation of large-scale tectonic activities that affected western North America from late Mesozoic time on into the present. The eruptions marked the beginning of subduction of oceanic crust by overriding continental crust. Evidence for this volcanism, which occurred millions of years before Mount Rainier's cone was built, is in the Ohanapecosh Formation, overlying the Puget Group. The Ohanapecosh Formation contains breccia from volcanic explosions and material from lava flows, mudflows, and ash falls, all of which accumulated in shallow water. Sinking continued to keep pace with the build up of deposits on the ocean floor.

3. Oligocene uplift.

As convergence of plates continued, uplift accompanied by compression raised the Puget and Ohanapecosh rocks above sea level and gently folded them during early Oligocene time. During this period, a warm, wet climate prevailed, resulting in deep weathering and extensive erosion. A thick red soil developed, and streams carved gullies between low hills.

4. Oligocene volcanism; Stevens Ridge Formation.

Later in the Oligocene, perhaps 25 or 30 million years ago, volcanism began again. Ash falls and flows covered the area. Violent, exploding clouds of fine pumice (*nuées ardentes*) erupted. As these glowing avalanches settled, ash particles with still molten edges bonded together to form a *welded tuff*. Ash flows filled the valleys, and eventually a plateau was built. This rock unit became the Stevens Ridge Formation. Outcrops of the light-gray welded tuff, containing darker inclusions of angular pumice fragments, can be seen in the southern part of the park along the highway through Stevens Canyon.

5. Continued volcanism in Miocene time; Fifes Peak Formation.

The Fifes Peak Formation shows that the character of the volcanism was changing. Many flows of both andesite and basalt lava, ranging from 150 to 500 feet thick, piled up to a total thickness of 2500 feet. Rocks of the Fifes Peak Formation make up many of the cliffs and peaks in the northwestern part of the Mount Rainier National Park.

6. Miocene deformation and uplift; intrusion of granodiorite.

As Fifes Peak volcanism ceased, the region was again uplifted and gently folded. (The already folded, underlying Ohanapecosh Formation was compressed into tighter folds.) Some faults, with displacements of thousands of feet, resulted from the compression.

A body of magma welled upward through the crust. A small amount reached the surface and erupted as lava, but most of the magma solidified slowly underground, forming the Tatoosh pluton, which is composed largely of coarse-grained granodiorite having a "salt-and-pepper" appearance (i.e., light-colored and speckled with large dark mineral grains). Numerous dikes and sills are associated with the pluton, and some contact metamorphism occurred in the country rock. The Tatoosh pluton has been dated radiometrically as being about 12 million years old.

7. Pliocene uplift and erosion of the Cascade Range.

The Cascade Range continued to rise, and deep erosion carved the region into peaks and ridges separated by stream-carved valleys. The rocks overlying the Tatoosh pluton were eroded, and the pluton was "unroofed." The granodiorite of the pluton is exposed in the Carbon and White River valleys and part of the upper Nisqually River valley. Plutonic rocks also form the central part of the craggy Tatoosh Range in the southern section of the park, near the base of Mount Rainier.

8. The building of Mount Rainier, a composite cone; repeated episodes of Pleistocene glaciation.

The climate was already very cold and the Tatoosh Mountains were undergoing glaciation when construction of the cone began with a series of relatively fluid lava flows early in the Pleistocene Epoch. The lavas flowed into stream-carved canyons, eventually filling them. Continued eruptions built a low, broad volcano on top of the mountainous terrain. Mudflow and glacial deposits were interlayered between lava flows, indicating that glaciation was ongoing throughout this time. Then a large number of short lava streams from a central crater, interspersed with ejections of breccia and ash, began constructing the cone, which rose to a height of perhaps 15,500 to 16,000 feet. A plug of solidified magma filled the central vent and dikes radiated from the center, like spokes in a wheel. The peak probably attained its greatest height about 75,000 years ago.

9. Erosion of the cone and destruction of the summit.

Intense glacial erosion steepened the mountain flanks, and by the end of the Pleistocene, perhaps a third of the mountain's bulk had been removed. However, steam and

Figure 32.7   Southwest side of Mount Rainier. Glaciers have deeply excavated the cone and the old crater rim. National Park Service photo.

occasional eruptions at the summit prevented glaciers from eroding at the top of the mountain, so the summit remained high.

Beginning about 6000 years ago, a series of violent explosions blasted away an enormous portion of the eastern side of the mountain and, later, the summit itself. It was probably this last explosion that caused the catastrophic Osceola Mudflow that traveled as far as Puget Sound. Rainier's lowered summit area had become a caldera about two miles across, the highest points on the rim barely 14,000 feet above sea level. Glacial ice filled the caldera and eroded the rim and caldera basin (fig. 32.7).

10. Reconstruction of the summit by lava flows and pumice.

Reconstruction began about 2500 years ago when pyroclastic eruptions, followed by lava flows, built a new cone within the caldera. In time, eruptions shifted somewhat, creating a double crater for this "volcano upon a volcano." The mountain probably erupted at least once between 1820 and 1854 while Mount St. Helens was having a series of pyroclastic eruptions. Minor eruptions may have occurred again on Mount Rainier in the 1870s and 1880s. Fumaroles (small vents emitting vapor and gas) and occasional steam explosions on the flanks of Mount Rainier indicate continued high heat flow. *Microseisms* (very small earth tremors) also tell us that eruptions may begin again.

Geologic Map and Cross Section

Fiske, R. S., Hopson, C. A., and Waters, A. C. 1964. Geologic map and section, Mount Rainier National Park, Washington. U.S. Geological Survey Misc. Geological Investigations. Map I-432.

Bibliography

Crandell, D. R. 1969. Surficial geology of Mount Rainier National Park, Washington. U.S. Geological Survey Bulletin 1288 (text plus map in folder).

———. 1969. The geologic story of Mount Rainier, Washington. U.S. Geological Survey Bulletin 1292.

———. 1971. Postglacial lahars from Mount Rainier volcano, Washington. U.S. Geological Survey Professional Paper 677.

Crandell, D. R. and Mullineaux, D. R. 1967. Volcanic hazards at Mount Rainier, Washington. U.S. Geological Survey Bulletin 1238.

Fiske, R. S.; Hopson, C. A.; and Waters, A. C. 1963. Geology of Mount Rainier National Park, Washington. U.S. Geological Survey Professional Paper 444.

Harris, S. L. 1988. *Fire mountains of the west: the Cascade and Mono Lake volcanoes.* Missoula, Montana: Mountain Press Publishing Co.

Heaton, T. H. and Hartzell, S. H. 1987. Earthquake hazards on the Cascadia subduction zone. *Science* 236: 162–168, April 10.

Hopson, C. A., Waters, A. C., Bender, V. R., and Rubin, M. 1962. The latest eruptions from Mount Rainier volcano. *Journal of geology* 70:635–647.

Matthes, F. E. 1955 (revised). Mount Rainier National Park, Washington. Text on back of quadrangle map. U.S. Geological Survey.

———. 1928. Mount Rainier and its glaciers. National Park Service.

Plummer, F. G. 1899. Mount Rainier Forest Reserve. U.S. Geological Survey 21st Annual Report, part V.

Sigafoos, R. S., and Hendricks, E. L. 1972. Recent activity of glaciers, Mount Rainier, Washington. U.S. Geological Survey Professional Paper 387-B.

Tuttle, S. D. 1980. *Landforms and landscapes.* 3rd ed. Dubuque, Iowa: Wm. C. Brown Company Publishers.

Mount Rainier National Park

Size: 235,404 acres; 368 square miles
Established: March 2, 1899
Address: Tahoma Woods, Star Route, Ashford, Washington 98304

33

Crater Lake, 27 square miles in size and over 1900 feet deep, lies in the caldera of a collapsed volcano.

Mazama, one of the aligned composite cones that mark an active subduction zone beneath the High Cascades, exploded in climactic eruptions about 6845 years ago. In multicolored walls, encircling the lake and rising 500 to 2000 feet from the water, are the records of intense volcanic and glacial activity during Mazama's construction and destruction.

Pumice pinnacles, a pumice "desert," plug domes, satellite craters, cinder cones, U-shaped valleys, and other volcanic and glacial features add to the geologic story of this remarkable national park.

Crater Lake National Park

Southwest Oregon

. . . The color of the water is ultramarine, bordered by turquoise along the shores. Set in majestic cliffs, it is a most attractive natural jewel . . .
J. S. Diller, U.S. Geological Survey, 1911.

Figure 33.1   Looking north over Crater Lake and Wizard Island. Llao Rock, jutting out from the caldera rim, aligns with Wizard Island in this view. Mount Thielsen, on the far skyline, is an extinct volcano. Oregon Department of Transportation photo.

Local History

If you have seen Crater Lake, you can imagine how astonishing and mysterious it must have seemed to early visitors to Oregon Territory. John Wesley Hillman, a young prospector, came upon the place quite by accident in 1853 as he was following his mule up a ridge in search of a lost mine. When the mule suddenly stopped on the brink of a cliff, Hillman looked down over the edge and saw below the indigo water of the lake in its great caldera, a good five miles from rim to rim. He called it "Deep Blue Lake." The spot where he stood, on the southwest side of the rim, is now Discovery Point. Later, settlers in the region decided "Crater Lake" was the ideal name. This name has been retained, despite assertions by teachers, geologists, and park rangers that it should be called "Caldera Lake"!

The lake's remarkably intense blue color, which continues to amaze visitors, is due to the purity of the water and the great depth of the caldera. Crater Lake's officially measured depth is 1932 feet, which makes it the deepest lake in the United States. The water, which has accumulated from rain and snow, contains very little dissolved or suspended material. Measurements of visibility in this transparent water have reached depths of over 325 feet. The lake was devoid of vegetation and aquatic life until some 60 years ago when fish were introduced. Freshwater shrimp had to be brought in first so that when the lake was stocked with trout, the fish would have something to eat.

Except for minor seasonal changes, the level of the water remains fairly constant from year to year; the intake from precipitation balances loss from evaporation. Springs on the outer slopes of the caldera are fed by local ground water, not by subterranean discharge from Crater Lake. Maximum summer temperatures of the lake water average 65° F; winter temperatures approach freezing, but the lake seldom freezes over, perhaps because of a slight heat source at depth.

Crater Lake was a place of mystery to the Indians as well as to the first settlers who came into the region. A Klamath Indian legend is suggestive of the geological story, as we know it, and may be a tribal memory of catastrophe many generations ago. We know that Klamath Indians were living in the region, perhaps as long as 7000 years ago (before the climactic eruptions), because their moccasins and other artifacts have been found beneath Mazama ash layers.

The Klamath Indians told of two great volcanoes in mortal combat: Llao, the chief of the Below-World; and Skell, chief of the Above-World. Llao lived in Mount Mazama, and Skell's home was Mount Shasta, about a hundred miles to the south. They thundered at each other, threw barrages of fiery rocks, and sent clouds of flaming ash toward each other. As the two volcanoes fought on, a terrible darkness covered the land for days. (Were Mazama and Shasta erupting at approximately the same time? The legend suggests this, but geologic evidence cannot confirm the timing with any exactness.)

In an attempt to pacify the volcano gods and expiate the sins of the tribe, the Klamath medicine men climbed to the top of Mount Mazama and threw themselves into the crater. Impressed by their sacrifice, Skell fought even harder. Finally he drove Llao far down into the Below-World and pushed the top of the volcano down, imprisoning Llao forever.

An Oregonian, Judge William Gladstone Steel, was instrumental in the movement to preserve the Crater Lake region as a national park. As a boy in Kansas, he had read about a mysterious, deep blue lake in Oregon. When he was 18 years old, his family moved to Oregon, and soon young Steel began searching for the lake, asking all the oldtimers where it was. Finally, in 1885, after 13 years of searching, he stood on the caldera's rim and looked down at Crater Lake. Then and there, he decided that it must be protected and preserved. After 17 years of effort, his wish became reality in 1902 when Crater Lake National Park was established. It was Judge Steel who first put fish into the lake. He hiked nearly 50 miles, carrying minnows from the Rogue River to Crater Lake.

Highways into the park lead to a scenic drive around the caldera rim. Probably the best vantage point to see many features of geologic interest is the top of Mount Scott, reached by a trail from Rim Drive. Looking down one can see both volcanic and glacial landforms around the caldera, as well as an incomparable view of the lake. To the north and south rise the High Cascades, topped by snow-capped volcanoes. To the west is "a sea of peaks and ridges"—the Western Cascades. Beyond them, the more distant Klamath-Siskiyou Mountains are seen. To the east stretches an immense platform of basalt (the Columbia River basalts) that goes farther than one can see into Idaho. The lava plateau is cut by canyons, and a few steep-scarped, tilted, fault-block mountain ranges break the flatness of the surface in the distance.

Geologic Features

Mazama's Volcanic Rocks, a Brief Description

Basalt, andesite, trachyte, and *rhyolite* are compositional varieties of dense-textured, extrusive igneous rocks, ranging from the darker basalts (higher in iron and magnesium, lower in silica) to light-colored rhyolite (lower in iron and magnesium, higher in silica).

Dacite, a compositional variety containing principally plagioclase and quartz, is difficult to distinguish from rhyolite without the use of petrographic techniques. Much of the rock at Crater Lake is described as *rhyodacite.*

Pumice is a textural term for a frothlike volcanic glass that cooled from liquid lava during eruption. A typical

CRATER LAKE NATIONAL PARK

Figure 33.2   Crater Lake National Park, Oregon.

pumice contains so much trapped air that it can float. The composition is usually that of rhyolite or trachyte.

Tuff is a general term for consolidated rocks made up of pyroclastic fragments of various sizes. A *tuff breccia,* for example, consists of about half ash and half larger fragments. In a *welded tuff,* glass shards, hot ash, pumice, and cinders have been fused by hot gases and the weight and heat of overlying materials. Uneven compaction causes laminations and banding. The composition is highly siliceous.

Scoria refers to cinders or a cindery crust that has a vesicular texture (full of holes). Escaping volcanic gases during the eruption and solidification of lava produce vesicular texture. Scoria, which is usually darker and heavier than pumice, is andesitic or basaltic in composition.

Volcanic Features

The large caldera occupied by Crater Lake plus the encircling flanks and ridges are about all that is left of once-mighty Mazama. Its early history was similar to that of the other Cascade volcanoes. Like its contemporaries—Mount Rainier (chapter 32) and Lassen (chapter 34)—Mazama grew during the Pleistocene, while alpine glaciers advanced and retreated on its slopes. From its foundation rocks at the base (between 5000 and 6000 feet in elevation), Mazama rose another 6000 feet to an elevation of approximately 12,000 feet above sea level. The cone was broader and less steep than Shasta's or Rainier's be-cause the subsidiary vents and overlapping cones gave the volcano a bulky, irregular shape. The present caldera rim is 7000 to 8000 feet above sea level, and geologists have estimated that the original cone probably rose 2500 to 3500 feet above that.

Associated with the construction and destruction of Mazama are a variety of extrusive and intrusive features. Thick pumice and ash deposits, relics of the violent explosive eruptions, lie on the flanks of the mountain, relatively fresh and only mildly weathered by the elements during the short (in geologic terms) time since they fell. Some of the deposits are welded tuffs produced by fiery clouds and ash flows (*nuées ardentes*). The Pumice Desert, which stretches away to the north from the rim, is a very thick blanket of ash that is mostly devoid of vegetation. The pyroclastic material is too loose to hold water and too siliceous to develop soil or support much plant growth.

The Pinnacles, an area of tall spires about 5 miles southeast of the rim, are erosional remnants of tuff deposits, 200 to 300 feet thick, ejected in flaming ash flows during the climactic eruptions. Steam clouds rose for years after the hot pyroclastics blanketed the slopes and filled the canyons. Where steam was localized, as in fumaroles, particles became cemented together in the shape of pinnacles, or spires, and were more resistant to erosion. Thin, hollow tubes, where gas escaped, still go down through the centers of some of the pinnacles (fig. 33.3).

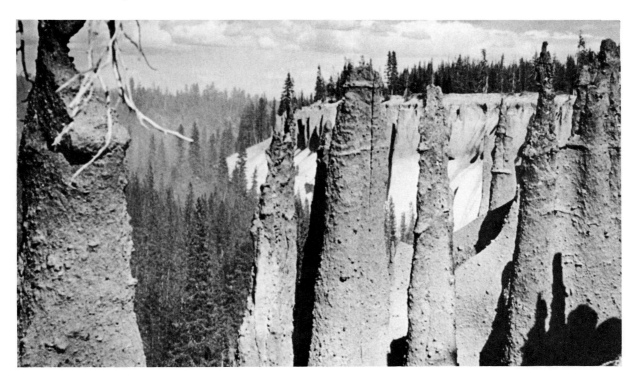

Figure 33.3   The Pinnacles, some five miles southeast of the caldera rim, were fused by hot gases escaping through many feet of ash after the climactic eruption over 6000 years ago. Erosion has removed the thick cover of loose ash, leaving the resistant pillars standing in the deep canyon of Sand Creek. National Park Service photo.

Associated Composite Cones

Mount Bailey and Mount Thielsen, both a short distance north of the park, probably became inactive early in the Pleistocene, perhaps around the time that Mazama started up. Mount Scott, highest point in the park (8926 feet), developed as a parasitic cone from Mazama's eastern base. Mount Scott's very steep cone is made up of layers of andesite lava, closely related chemically to some of the earlier flows of Mazama. Evidently Mount Scott became extinct before the end of the Pleistocene. A glacial amphitheater (large cirque) on the northwest side indicates severe glacial erosion that was not "repaired" by postglacial volcanic activity. Not much is left of Union Peak in the southwest corner of the park except its *volcanic plug,* an erosional remnant of lava that solidified to resistant rock in the neck of the volcano.

Shield Volcanoes and Cinder Cones

Basalt lavas and andesitic basalts built a number of satellite cones and craters around Mazama. Crater Peak formed as a shield volcano of andesite and basalt flows and later erupted pyroclastics of andesite and dacite. Another shield volcano, Timber Crater, in the northeast corner of the park, is also made up of basalt and basaltic andesite flows, but its shield is capped by two cinder cones. None of the cinder cones are more than a thousand feet high from base to summit. Examples are Desert Cone, Bald Crater, Red Cone, and Diller Cone. Volcanic *bombs* (fragments of lava, or ejecta, usually rounded, ropey, or spiral, that hardened while in the air) lie on the cinder cones in myriad sizes and shapes. Some of the cinder cones are aligned on what seems to have been a fissure system tapping a magma chamber.

Wizard Island, the large cinder cone in Crater Lake that resembles a wizard's hat, for which it was named, formed in the caldera from eruptions after Mazama's collapse. The cone rises nearly 800 feet above lake level. Its well developed crater is 400 feet across and about 90 feet deep (fig. 33.1).

Dikes and Other Features of the Caldera Walls

Most of the large andesite dikes exposed in the park are old conduits and radial fissure fillings that probably served as feeders for Mazama's flank eruptions. Devils Backbone, which runs up the cliffs of Crater Lake from the water's edge to the west rim, is the most conspicuous dike (fig. 33.4). A similar dike forms a high vertical wall immediately beneath the Watchman and merges upward into the Watchman lava flow (western rim). Phantom Ship, a much older eroded dike or conduit remnant, sticks out of the water near the lake's south shore (fig. 33.5).

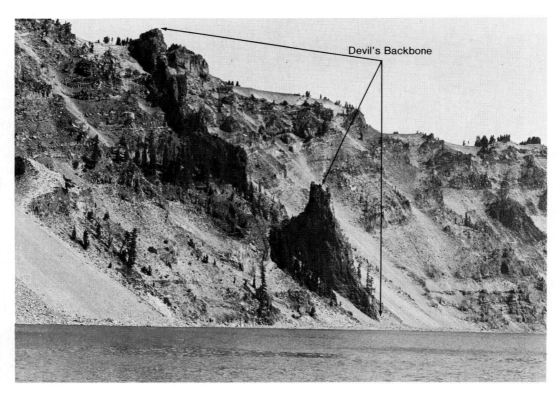

Devil's Backbone

Figure 33.4   Devils Backbone, a large andesite dike exposed by erosion of softer material in the caldera's west wall. National Park Service photo.

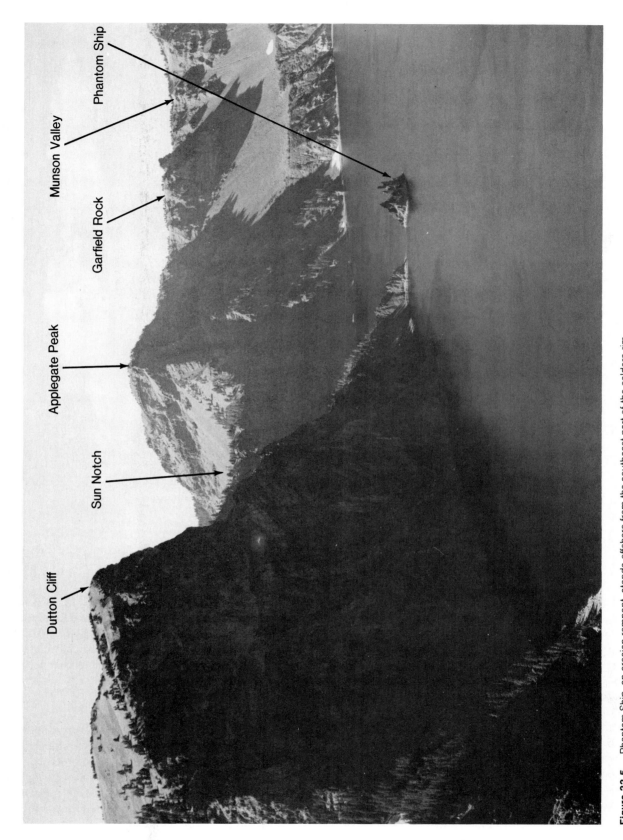

Figure 33.5 Phantom Ship, an erosion remnant, stands offshore from the southeast part of the caldera rim. Sun Notch is a U-shaped glacial trough beheaded by the explosion of Mazama. National Park Service photo.

421

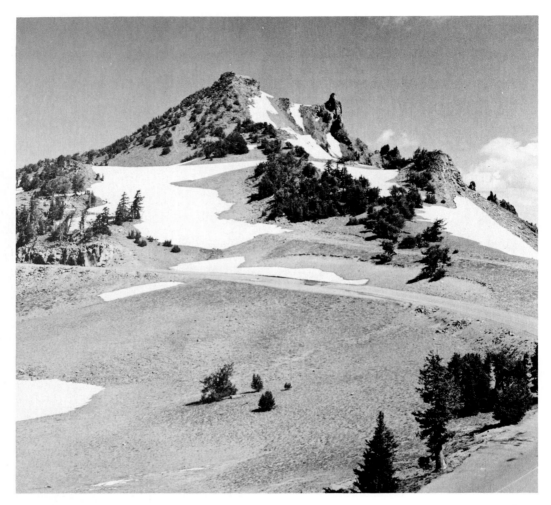

Figure 33.6   Hillman Peak (elevation 8100 feet), on the caldera's west rim. Pumice deposits and blocks of lava blanket the slope between the peak and Rim Drive. Oregon Department of Transportation photo.

Dacite dikes, several of which are clustered near Llao Rock, are smaller and less prominent. Llao Rock (fig. 33.1) was formed by a massive rhyodacite flow that came from near the bottom of a satellite cone that exploded, not long before Mazama's climactic eruptions. Because of its shape, Llao Rock was earlier thought to have been formed by a lava flow that filled a glacial trough. But more recent evidence indicates that the flow was emplaced in an explosion crater (Bacon, 1983).

Feeder dikes of Hillman Peak are exposed in the remaining half of the old conduit that was laid bare by explosions that removed the eastern half of its cone (fig. 33.6). Hillman Peak, highest spot on the rim, was one of the parasite cones of Mazama and now reveals a nearly perfect cross section of its interior due to the blasting out of the caldera.

The unusual colors displayed in the caldera walls—hues of yellow, buff, brown, orange, and faint purple—are the result of hydrothermal weathering of the volcanic rocks by ascending ground water and chemically charged gases. Park visitors like to watch the colors change during the day as the light changes.

Dacite Domes

Dacite flows that began erupting toward the end of Mazama's development tended to become thicker and more viscous. Some earlier dacite flows are overlain by andesite or by glacial drift. The later, more viscous dacites formed domes. Some, such as Lookout, Pothole, and Dry Buttes, are isolated domes. Others, like the group south and east of Mount Scott, are in clusters.

Figure 33.7   The grooves on these rocks perched on the caldera rim are glacial striations incised by glaciers on Mount Mazama, long before the climactic eruptions. National Park Service photo.

Glacial Features

Between eruptions, when the mountain was quiet, glaciers rebuilt themselves and carved out deep valleys on the flanks of the volcano. When eruptive activities resumed, the glacial troughs provided convenient channels for hot lavas, ash flows, lahars, and glowing avalanches. The culminating eruptions and explosions truncated, or "beheaded," glacial valleys that extended from the high slopes to elevations lower than the present caldera rim. Kerr Valley and Sun Notch on the southeast side are the two most spectacular U-shaped troughs beheaded by the explosions and now exposed in the precipitous caldera wall. Also exposed in the cliffs rising from the lake are horizons of glacial debris interspersed with layers of lava and pumice. Grooves of glacial striations come to the very brink of the cliffs; and where layers of andesite jut out slightly from the cliff walls, more striations can be discerned on the angular tops. This indicates that after a lava flow hardened, glaciers had time to abrade the surface (fig. 33.7).

The largest glacier on Mazama flowed down Munson Valley on the southwestern flank by the present location of Park Headquarters. Munson Glacier went on into Annie Creek Valley and down beyond the park's southern boundary, perhaps as far as Fort Klamath. Meltwater streams from the extensive glaciers of the southern and eastern slopes carried large quantities of outwash into the Klamath Marsh. On the west side, several glaciers joined and went down the valley of the Rogue River; while on the north slopes, the Mazama glaciers joined those of Mounts Bailey and Thielsen.

Moraines are of many ages and types. Some are almost completely buried by pumice and by the ash of the culminating explosions. The youngest moraines were deposited after the explosions had ceased, indicating an interval of colder climate sometime within the past 6000 years. No glaciers exist in Crater Lake National Park at the present time, but heavy snows fall each winter, about 50 feet between September and May.

Table 33.1 Geologic Column, Crater Lake National Park

Time Units			Rock Units	Geologic Events
Era	Period	Epoch		
Cenozoic	Quaternary	Holocene	Andesitic ash, lava Rhyodacite lava Ash-flow deposits Intracaldera debris Wineglass Welded Tuff Air-fall tephra, pumice Rhyodacite lava and pumice	Cinder cones within caldera: Wizard Island, Merriam Cone Lava platform in caldera Minor glacial advance Climactic eruptions 6845 y.b.p. and collapse of caldera Cleetwood, Grouse Hill, Llao Rock, and Redcloud flows
		Pleistocene	Dacitic and rhyodacitic volcanics interbedded with glacial deposits, debris flows, reworked ash, etc. Older volcanic rocks: andesite, rhyolite, dacite	Early, middle, and late Wisconsinan lava flows and ash falls building up Mazama cone; parasitic cones, flank eruptions Intense alpine glaciation concurrent with volcanism throughout epoch Construction of composite cones near Mazama: Union Peak, Mounts Bailey, Thielsen
	Tertiary	Pliocene	Western Cascade volcanics	Build-up of lava plateau; uplift Western Cascade volcanoes Regional uplift

Sources: Bacon, 1983; Cranson, 1982.

Geologic History

1. Development of the Western Cascades, middle and late Tertiary time.

Along the western part of the Cascade Range lies a belt of older lavas and pyroclastics that make up the Western Cascades. These mountains were folded and deeply eroded long before the Cascade volcanoes began to erupt. Both the Western Cascades and the younger High Cascades are the result of the continuing subduction of oceanic crust beneath the westward-moving North American plate (chapter 32). None of the Tertiary volcanics are exposed in Crater Lake National Park; however, some of these rocks may underlie the base of Mazama.

2. Miocene uplift and Pliocene volcanism; building of the High Cascades.

As the volcanic mountain system was uplifted, it became a barrier to the moisture-carrying winds from the Pacific that dropped rain and snow on the slopes. Glaciers had probably begun to form at the higher elevations by the end of Miocene time.

The High Cascades evolved as a north-south belt east of the Western Cascades, with younger volcanics overlapping older Tertiary rocks. Pliocene lava flows, mostly basalts and andesites, built up lava plateaus and the older volcanoes near Crater Lake, such as Mounts Bailey and Thielsen and Union Peak.

3. Construction of Mazama in middle and late Pleistocene time.

The Mazama cone began to grow on top of Pliocene and Pleistocene lavas approximately 75,000 years ago. Mazama's summit may have attained about the same elevation as Lassen, Hood, and Baker—around 11,000 to 12,000 feet—but never grew as high as Rainier and Shasta. Like the other Cascade composite volcanoes, Mazama was built up by alternating lava flows and pyroclastic eruptions, mostly andesite in composition. The volcano had a number of parasitic cones and numerous flank eruptions from its gently sloping sides. Mount Scott is the largest surviving satellite cone.

4. Pleistocene glaciations before, during, and after volcanic eruptions.

Throughout much of its history, Mazama was capped by snow and ice. In the caldera wall, glacial horizons (layers of glacial deposits) are interbedded with lavas down to and below the lake surface. From this record, we can see that glaciers advanced and retreated a number of times, sometimes because of climatic variations and sometimes because of destruction by volcanic activity. Since all evidence higher up was destroyed, cross sections in the caldera wall cannot be used for correlation of regional advances and retreats.

When Mazama was at its maximum height, high peaks and ridges stuck up through the glacial ice. The absence of till at the top of Hillman Peak and Watchman suggests that these peaks were never entirely covered by ice. Hillman Peak probably acted as a "cleaver," or arête, similar to Mount Rainier's cleavers that divide glaciers coming down from the summit (chapter 32).

Long periods of quiescence followed times of active eruptions. Bands of soil containing charred vegetation at the top of till deposits, later covered by lava flows and pumice, record such quiet times when reforestation of the slopes could take place.

It seems likely that the time of maximum glaciation was near the end of the andesitic eruptions and after the early dacitic flows that erupted mainly from fissures along the north side of the mountain. This activity may be the reason why glacial erosion was more severe on the south side. The large U-shaped valleys on the south side, Munson, Kerr, and Sun, were deeply excavated during this time. Glacial ice was probably a thousand feet thick in the troughs.

5. Quiet ending of Pleistocene time; differentiation of magma.

After Mazama had reached its maximum height, large-scale volcanic activity on the main cone ceased for about 40,000 years. In the interim, the Wisconsinan stage of glaciation came to a close. The discovery of carbonized stumps of trees that had been growing in till in a glacial col near Discovery Point implies that a long period of quiescence preceded the last catastrophic explosions. The pumice that buried the trees can be correlated with those explosions. The presence of trees on lower slopes also suggests that at the time of the explosions, the glaciers had receded to higher elevations on the mountain.

Several miles below the surface of the High Cascades, magma collected and became concentrated in the magma reservoir while the volcanoes were relatively quiet. Mazama's magmatic system was reorganized and strengthened. Gradually a kind of chemical zoning took place in the magma chamber, with layers of lighter, more siliceous magma rising nearer the top and the denser, more mafic magma sinking toward the bottom.

6. The beginning of the end of Mazama.

During the last few hundred years before the climactic eruptions, several vents along the northern slopes of Mazama began to "leak out" volcanic material. These dacitic eruptions of lava and pumice included the Llao Rock, Redcloud, and Cleetwood flows plus several others. During the brief, but very intense, Cleetwood flow, which directly preceded the culminating explosions, quantities of hot tephra were ejected, and blocks of old andesite from inside the summit were thrown out with the pumice.

7. Mazama's climactic eruptions 6845 ± 50 years ago.

The single-vent phase. The initial tremendous explosion was sudden, thunderous, and violent. A huge column of fiery, gaseous pumice shot 5 to 10 miles high in the air at nearly twice the velocity of sound (fig. 33.8A). The column may have been a mile in diameter and probably kept on "spouting" for several days. As it began to collapse, glowing avalanches sped down from the vent. Some became the intensely welded tuff of the Wineglass (northeast part of the caldera rim). The tephra ejected in the initial explosion was almost entirely rhyodacitic in composition.

On the slopes and around the base of Mazama, many feet of ash and pumice fragments of various sizes accumulated. Finer ash particles were carried many miles by the prevailing winds, principally toward the east and north. Deposits of various thicknesses fell over a considerable range, landing as far away as Alberta and Saskatchewan to the north; Nevada and northern California to the south; Idaho and Montana to the east, and even to the northwest corner of Yellowstone National Park in Wyoming (fig. 33.9).

Radiocarbon dating of the Mazama ash layer places these eruptions at about 6845 years ago, with a plus-or-minus factor of 50 years. However, some pollen studies have yielded dates that nearly pinpoint the eruption. Dates from a bog in western Montana, in which pollen is associated with Mazama ash, place the ash fall within a maximum time "window" of only three years around the postulated 6845 years ago. Wherever a Mazama ash deposit is found, it provides a significant date of reference

Figure 33.8A   A great eruption column burst from a vent on the north side of Mazama. Falling ash spread to the northwest for hundreds of miles.

*Figures 33.8, A through E, are reproduced from S. L. Harris, *Fire Mountains of the West: The Cascade and Mono Lake Volcanoes.* © 1988 Mountain Press Publishing Company. Used by permission.

Figure 33.8B   As the column collapsed, fiery pyroclastic flows sped downslope.

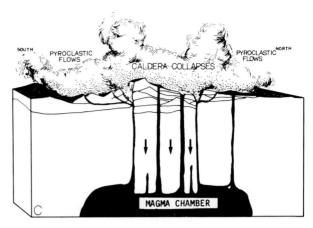

Figure 33.8C   The whole summit area caved in on itself, new vents opened, and more pyroclastic flows roared down the slopes on all sides of Mazama.

Figure 33.8D   Dust-filled steam clouds boiled up from the caldera; choking jets of gas rose from canyons clogged with ash; a toxic volcanic haze hung over the incinerated surface of the land.

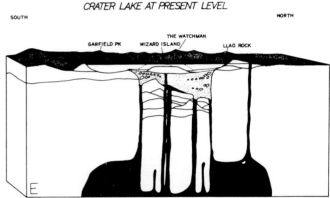

Figure 33.8E   Sporadic lava flows and eruptions emanating from the caldera floor built Wizard Island and several small cones, now submerged. Over the centuries, as the volcanic rocks cooled, rain and melting snow collected to form a deep lake in the caldera basin.

for geologists and archaeologists. Indian artifacts (mats, points, scrapers, etc.), preserved by Mazama ash, have been found at sites many miles to the east of Crater Lake. Geologists have been able to determine average rates of sedimentation in a number of lakes in the Pacific Northwest since the time of the Mazama ash falls. They can also tell which glacial moraines in the Cascades are younger than Mazama ash and which ones are older.

The ring-vent phase: multiple vents. Due to the release of pressure and the rapid extrusion of large volumes of tephra, the cone's upper section could no longer support its own weight. The whole summit area fractured and collapsed back into the volcano's center. A series of large glowing avalanches boiled out from vents all around the collapsing crater and roared down the mountain flanks, branching into the glacial valleys and spreading destruction everywhere. The whole mountain was set ablaze as trees and anything else combustible were incinerated. One fiery ash flow traveled 40 miles down the Rogue River valley. Another went north in the valley between Mounts Bailey and Theilsen, crossing Diamond Lake and then dumping pumice bombs into the valley of the North Umpqua River. Flows that went east carried large boulders of pumice many miles across the flats east of the volcano. To the south, some flows ended in the Klamath Marsh. Floating chunks of pumice, resembling styrofoam, were later washed down into the Klamath Lakes (fig. 33.8C).

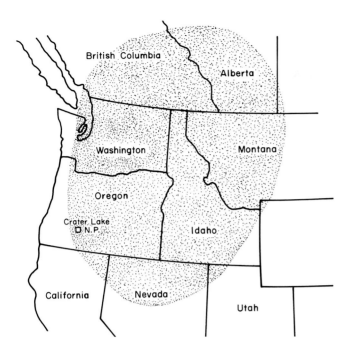

Figure 33.9   The speckled area shown on the map was blanketed by varying thicknesses of Mazama ash from the climactic eruption. Dust and ash, borne on prevailing winds, also fell farther to the east and northeast, far beyond the Pacific Northwest. (Modified after Williams 1942)

During the ring-vent phase, as collapse continued and the caldera began to form, some of the hot ash and lava flowed back into the volcano's center. The Cleetwood flow, which preceded the culminating explosion, had not completely cooled when it was flattened and rifted by backflow oozing down the caldera wall.

The last eruptions apparently came from very deep in the magma chamber. Dark-colored scoria covered some of the light-colored pumice of the earlier avalanches. On the north side, dark scoria created Pumice Desert. Light and dark pumice layers are especially noticeable in the Pinnacles.

After the eruptions had ended, the air was full of choking, acrid volcanic smog and dust. Heavy rains fell and clouds of steam and fine dust continued to billow upward. Pumice-filled canyons remained hot and smoking for years. When the dust finally settled and the air cleared, only a great pit was left where the cone had been. The caldera was 5 miles wide and 4000 feet deep (fig. 33.8D).

How much of Mazama's previous bulk disappeared due to explosion and collapse? According to estimates based on a number of measurements, approximately 15 cubic miles of brecciated rock and pyroclastic material were engulfed by the collapse of the cone into the caldera. The volume of ejected and erupted material was even greater. In all, about 25 cubic miles of tephra were thrown out by the ash falls and ash flows.

8. Postcollapse activity.

Volcanic features younger than the climactic eruptions developed only within the caldera and were by no means extensive. Lava flows oozed out over the caldera floor. Two cinder cones, Wizard Island and Merriam Cone, grew upward as more lava erupted. Merriam Cone was later submerged by water rising in the lake. Wizard Island is about 1000 years old. The youngest volcanic materials in the park are lava flows extending out into the water from the base of the Wizard Island cone. These postcollapse features were probably fed by magma that rose up along the ring fractures that developed during the second phase of the climactic eruptions.

Has all volcanic activity in Mazama ceased? Possibly not. Areas of high heat flow have been measured on the caldera floor, near where ring fractures are believed to be located. The heat may be from thermal springs that carry small amounts of chlorine and sodium dissolved in hot water from a magmatic source deep below the lake.

After the caldera finally cooled down, rain and snow gradually filled the pit, perhaps over a period of 700 years, or as long as 1500 years, depending on hydrologic conditions. Eventually, the deep blue lake came to exist in the quiet caldera (fig. 33.8E).

Around the caldera, the walls were attacked by the forces of erosion—landsliding, mud flows, freezing and thawing. A brief recurrence of small glaciers at higher elevations left drift patches on top of pumice. Running water reestablished a radial drainage pattern as it tried to clear out the debris-choked valleys. After a time, dense forests grew around the base and up the slopes, hiding some of the scars.

Bibliography

Alt, D. D. and Hyndman, D. W. 1978. *Roadside geology of Oregon.* Missoula, Montana: Mountain Press Publishing Co.

Bacon, C. R. 1983. Eruptive history of Mount Mazama and Crater Lake caldera, Cascade Range, U.S.A. *Journal of volcanology and geothermal research* 18:57–115.

———. 1988. Crater Lake National Park, Oregon; map and text on back of map. U.S. Geological Survey.

Baldwin, E. M. 1981. *Geology of Oregon.* 3rd ed. Dubuque, Iowa: Kendall/Hunt Publishing Company.

Cranson, K. R. 1982. *Crater Lake, gem of the Cascades, the geological story of Crater Lake National Park,* 2nd edition. Lansing, Michigan: KRC Press.

Harris, S. L. 1988. *Fire mountains of the west: the Cascade and Mono Lake volcanoes.* Missoula, Montana: Mountain Press Publishing Co.

Williams, H. 1942. *The geology of Crater Lake National Park, Oregon* (with geologic map). Washington, D.C.: Carnegie Institution of Washington Publication 540.

———. 1976. *The ancient volcanoes of Oregon,* 7th edition. Condon Lectures. Eugene, Oregon: Oregon State System of Higher Education.

Crater Lake National Park

Size: 183,227 acres; 287 square miles
Established: May 22, 1902
Address: P.O. Box 7, Crater Lake, Oregon 97604

34

Lassen Peak, the world's largest volcanic dome, was constructed in the ruins of an earlier composite cone that collapsed toward the end of Pleistocene time.

Lassen has not erupted for about 70 years; but hot springs, steaming fumaroles, mud pots, and sulfurous vents indicate volcanic activity at depth.

Park visitors can see a wide range of volcanic phenomena and extrusive rocks, including very young lava beds and an active cinder cone.

Glacial erosion and deposition modified some of the volcanic features.

Lassen Volcanic National Park

Northern California

Figure 34.1   Bumpass Hell, an active thermal area in Lassen Volcanic National Park, has bubbling mud pots, boiling springs, steam vents, and, overall, the pervasive odor of sulfur vapors. From J. W. Harbaugh, *Northern California.* © 1975 Kendall/Hunt Publishing Company. Used by permission.

LASSEN VOLCANIC
NATIONAL PARK

Figure 34.2 Lassen Volcanic National Park, California.

Local History

When President Theodore Roosevelt proclaimed Lassen Peak and Cinder Cone National Monuments in May 1907, the generally held assumption was that these little known areas in the northern California wilderness contained interesting volcanic phenomena, obviously extinct, but worthy of preservation for future generations to study and enjoy. The Indians who lived in the area had told settlers that the mountain was not "dead." The Indians said that Lassen Peak was full of fire and water, and that one day the mountain would blow itself to pieces.

Seven years after the monument was proclaimed, in May 1914, Lassen Peak awoke with a roar, opening up a new crater and emitting lava and ash. After several hundred intermittent eruptions of varying length and severity between 1914 and 1921, the volcano quieted down again. It was during the period of activity, after some impressive outbursts, that Lassen Volcanic National Park was established on August 9, 1916.

Lassen Peak, the southernmost volcano in the Cascade Range, is near the northern end of the Sacramento Valley. The 10,457-foot high mountain was a landmark for immigrants coming to the Sacramento Valley in the mid-1800s. The peak was named for Peter Lassen, a Danish blacksmith who settled in Northern California in the 1830s. As a sideline, Lassen guided parties of immigrants to the Sacramento Valley. In later years, California-bound wagon trains followed Nobles Emigrant Trail, parts of which can be seen in the park today as deep ruts and wagon tracks (fig. 34.2).

Although Nobles Emigrant Trail winds along the base of Cinder Cone in the northeast corner of the park, apparently only a few miners, homesteaders, and travelers witnessed the most recent eruptions of ash, steam, and fragmental material from Cinder Cone in 1850–51. Observers reported that fire on the top of the cone could be seen for many nights from points as far distant as 40 miles and from even a hundred miles away. Population in the region was still sparse in 1914 and 1915, when the more violent Lassen eruptions occurred. Some property damage resulted from the explosions and mudflows, but people were able to get out of the way in time. It was, perhaps, sheer luck that enabled some of the photographers and observers who recorded the events at close range to survive with only minor injuries.

At the recommendation of the U.S. Geological Survey, the Visitor Center and accommodations near Manzanita Lake were closed in 1974. These facilities would be in the path of an avalanche from Chaos Crags should movement of the unstable slopes be triggered by seismic activity or renewed volcanism.

Geologic Features

Unlike Shasta (80 miles to the north) and the other major volcanoes of the Cascade Range, Lassen Peak is not an andesitic composite cone. Lassen, which is composed of dacite, is a volcanic dome, the largest in the world. A *volcanic dome* is characterized by an upheaved, pluglike conduit-filling that forms when viscous lava is forced through a vent onto the earth's surface. Like toothpaste squeezed out of a tube, the viscous lava is unable to flow freely, so it builds a bulbous dome over its vent.

The Lassen dome is in a sense an offspring of a once-mighty composite volcano, called Tehama, more than a thousand feet higher than Lassen, with a diameter of 11 to 15 miles at its base. Late in Pleistocene time, Tehama (also known as Brokeoff Volcano) collapsed, forming a caldera more than two miles in diameter. Preceding its collapse, vast quantities of fluid dacite streamed from vents on Tehama's northeast flank. Subsequently, several small dacite domes rose (Eagle Peak, Vulcan's Castle, Mount Helen, and Bumpass Mountain) in the area just south of Lassen Peak, perhaps at the same time that the Lassen dome was beginning to form. The outpouring of so much lava probably weakened Tehama's structure and may have caused its collapse.

About 11,000 years ago the volcanic dome that became Lassen Peak began pushing up through a vent in the pre-collapse dacite flows on Tehama's northern flank. The huge dome's rise appears to have taken place with great force and in a remarkably short time. The quantities of talus that surround and partially bury Lassen's dome were produced by the fracturing of older rock as the plug of dacite forced its way to the surface.

How Eruptive Activity Is Related to Type of Lava

Dacite, the rock type of the Lassen dome, is intermediate in composition between rhyolite and andesite. Because dacite has a higher percentage of silica than andesite, it tends to flow less easily, hence its propensity for constructing domes. Dacite is associated with more explosive eruptions because gases are not able to bubble out freely through the viscous magma. Gas is trapped until enough pressure builds up to blast a way out.

Volcano-watchers at Mount St. Helens, which has also had recent dacite eruptions, have noted many similarities between its renewal of eruptive activities (spring 1980) and the first year (1914) of Lassen's eruptions. Activity began at both volcanoes with a series of *phreatic eruptions* (explosive, steam-propelled ejections of ash and mud caused by heating and expansion of ground water) that shot ash and steam columns thousands of feet into the air.

How is Lassen related to the other Cascade volcanoes? According to plate tectonic theory, a spreading center, lying some 300 miles offshore from northernmost California, Oregon, and Washington, adds about an inch of new crust each year to a small plate of oceanic crust, thus pushing it toward the Pacific Northwest coast. As this small plate collides with the westward-moving North American plate, the denser oceanic crust bends down and slips beneath the leading edge of the continental plate. (To the south, where the large Pacific plate and the North American plate slide past one another along the San Andreas fault, subduction is not occurring.)

The Cascade volcanoes are products of the subduction of oceanic crust and marine sedimentary rock. The marine sediments, some of which were washed down from uplifted continental crust, accumulated in an offshore basin. As the plates came together, sedimentary beds were squeezed, deformed, and pulled down into the subduction zone. Intense heat and pressure at depth melted the rock, creating a mixed magma of basaltic and felsic constituents.

Why is dacite the rock type rather than andesite? One possible explanation is that being farther inland from the continent-ocean boundary, the magma comes from deeper in the subduction zone. Moreover, a high proportion of sedimentary material in the mix may produce dacitic magma. Magmas that are more silica-rich seem to be associated with such sources. Alternatively, the relative age of volcanic activity may explain the dacitic composition. Late-stage eruptions of composite volcanoes seem to be more silica-rich. The fact that Lassen Peak formed on Tehama's flanks suggests that the construction of Lassen represents a late episode in continuing volcanic activity.

Magma mixing has presumably been characteristic of volcanism in the Lassen-Tehama area throughout most of its history. The wide range in chemical composition of the volcanic rocks in the park suggests disequilibrium of magmas due to mixing of materials of various crustal and mantle origins. At least two episodes of magma mixing in the 1914–21 eruptions can be deduced from accumulations of banded pumice in which flows change color from light to dark from top to bottom. Some geologists think that quartz crystals found in recent lava flows at Cinder Cone may be the result of magma mixing.

Volcanic Features Related Mainly to Tehama

South of Lassen Peak, on the opposite rim of the ancient caldera, Brokeoff Mountain (elevation, 9235 feet) stands as an erosional remnant. Brokeoff Mountain is the second highest peak in the park. Other prominent remnant peaks, outlining the rim of the old caldera, are Mount Diller, Pilot Pinnacle, and Mount Conrad. Lava pinnacles, once part of the caldera, show up on the skyline. Much of this severe erosion was due to a glacial advance that followed Tehama's collapse.

Thermal features in the southwest corner of the park, in or close to the old caldera, are associated with Tehama. Sulphur Works, with its hot springs, fumaroles, and strong, rotten egg odor (indicating the presence of hydrogen sulfide), is thought to be at the center of the former cone. Between Sulphur Works and Little Hot Springs Valley (an area of hydrothermal activity) is Diamond Point, an old conduit through which lava moved upward during the construction of Tehama.

As a result of hydrothermal alteration, many of the rocks at the Sulphur Works and Little Hot Springs

Figure 34.3   Simplified cross section showing the relationship of Lassen Peak to its predecessor Tehama (Brokeoff Volcano). The dashed line shows Tehama's projected former profile. From S. L. Harris, *Fire Mountains of the West: The Cascade and Mono Lake Volcanoes.* © 1988 Mountain Press Publishing Company. Used by permission.

Figure 34.4   Northeast slope of Lassen Peak (elevation 10,457 feet), showing Devastated Area and Survivors Hill. National Park Service photo by C. Peters.

Valley—which were formerly hard, gray-green andesite lavas—have been transformed into bright-colored clays. Other rocks are rapidly decaying and have become yellow, buff, and red in color from iron oxides. The hot water coming from the fissures contains both sulphurous acid and sulphuric acid, which react with the minerals of the original rock to produce clays and iron oxides. Nearly pure opal (not gem quality) forms where the acidity of the hot water is very high.

Just beyond Little Hot Springs Valley is Bumpass Hell, a 16-acre tract of boiling springs, steaming sulphur vapors, and bubbling mud pots that are associated with old caldera fissures. Bumpass Hell is named for a cowboy who worked in the area in the 1860s. When he broke through the crust over a mud pot and scalded his feet, he told friends that he had been "in hell." Later, when he was escorting a newspaper editor through the area, Bumpass broke through the crust again. This time he was so badly burned that he lost a leg. The superheated steam in the springs at Bumpass Hell and the fumaroles are an indication of the late stages of volcanism in the Tehama area (fig. 34.1). Present-day visitors get a close look at Bumpass Hell from the safety of stout boardwalks and are strongly cautioned not to step off!

Devils Kitchen is another area of hot sulfurous springs, located in Warner Valley about seven miles southeast of Lassen. The springs are so strongly acidic that holes and pits have been eaten into the bedrock. Note on the map (fig. 34.2) that the hot springs and thermal features are clustered along a northwest–southeast trend and appear to be associated with an old fault area.

The Lassen Eruptions

The volcanic dome of Lassen Peak, with its surrounding talus, represents about a cubic mile of material (fig. 34.4). This makes a very large summit area for a dome, but its bulk is small compared to the composite cone of Mount Rainier (chapter 32). An unusual feature of Lassen Peak was its development of summit craters from which pyroclastics and lava were ejected, as happened in the phreatic eruptions in 1914. Most domes do not have summit craters because the explosions are emitted from the base of the plug. Chaos Crags, for example, is a complex dome on Lassen's northwest flank that is made up of several dacite plugs, but it does not have a crater.

In the crater bowl on Lassen's summit, a pre-1914 vent and a 1914-15 vent were filled and covered by a large dacite lava flow in May 1915, and a third summit crater,

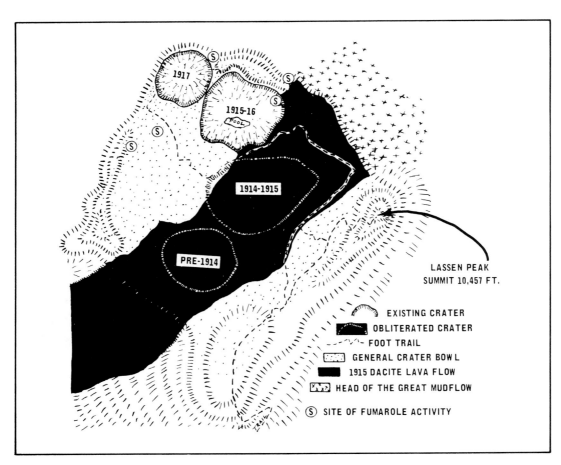

Figure 34.5   Summit area of Lassen Peak, showing sites of eruptions and fumaroles. From National Park Service.

or vent, was opened. A fourth crater was blasted out of the northwest corner of the summit by a series of steam and ash explosions in April, May, and June of 1917 (fig. 34.5). These were the last of the major eruptions of the 1914–1921 period of activity. Slight fumarolic activity at the summit (steam escaping from small fissures) has persisted to the present.

The dacite flow on May 19, 1915, poured down the western side of Lassen about 1000 feet before it hardened. Observers some 20 miles away to the west and south reported that beginning about 9 o'clock that night the mountain was "boiling over" for two hours. The photographer B. F. Loomis and the postmistress of nearby Manton had described a fire at the summit a few days earlier and then two days before the eruption, a large black mass had begun to rise at the top. This "black mass" was the plug, or column of lava, that was pushing up through the conduit, breaking the rock, and opening a new vent.

Hot lava and ash flowed down the steep northeast side onto packed snow left from the previous winter, and quickly combined with the snow to form a thick, pasty lahar that slid rapidly down the slope, burying a mile-wide swath of

forest in its path, filling Lost Creek, then gliding up over a divide and down into Hat Creek. In the valleys north of Lassen, fences, bridges, farm buildings, and fields were totally devastated. Fortunately, homesteaders were awakened by the roar of the oncoming mudflow, just in time to flee to high ground and safety. The worst was yet to come.

A few days later, on May 22, 1915, at 4:30 P.M., a violent explosion shot a great cloud of ash thousands of feet into the air. Photographs of the event, referred to as "the Great Eruption," show an enormous cloud in the shape of a mushroom over Lassen Peak. Prevailing westerly winds carried fine ash particles for many miles to the east. Much of the force of the blast was deflected downward as a glowing avalanche swept down the northeast slope, snapped off many more trees (some with trunks 6 feet in diameter), and threw them many feet from their stumps. Hot ash—mixed with melting snow—reactivated the mudflow and the whole scalding mass sped down the same valleys, split in front of Mount Raker, and rushed by, the two branches extending beyond the park's northern boundary. Smaller mudflows streamed down the north and northwest slopes.

Figure 34.6   Chaos Jumbles (foreground) and Chaos Crags (background). Several hundred years ago, a series of explosions in Chaos Crags triggered high-velocity avalanches that left angular dacite blocks strewn over the Chaos Jumbles area. From J. W. Harbaugh, *Northern California.* © 1975 Kendall/Hunt Publishing Company. Used by permission.

The tract of greatest destruction is a wide swath about 4 miles long, called the Devastated Area, which marks the path of the mud and ash flows that roared down the mountain during the May 19 and 22 eruptions. Visitors who drive through the park can see that scars on the landscape are being covered by natural reforestation, but the trees in the Devastated Area are sparse and stunted. The siliceous ash soil is lacking in nutrients and too porous to hold sufficient water for normal tree growth.

Chaos Crags (north of Lassen Peak) is younger than the other volcanic domes. It was formed by four or more dacite domes less than 1200 years ago. Chaos Jumbles, a rock avalanche that extends toward the northwest corner of the park from the base of Chaos Crags, was set off by the explosion and partial collapse of one of the domes about 300 years ago. Avalanching transported large angular blocks a mile or more from their source. At speeds of around 100 miles per hour, the rock debris apparently traveled some of the distance on a cushion of compressed air that reduced friction (a phenomenon called "air-layer lubrication"). The damming of Manzanita Creek by the avalanche formed Lake Manzanita (fig. 34.6). Soil development and regrowth of trees has been slower on the blocky landslide debris of Chaos Jumbles than in the Devastated Area. The unstable Chaos Crags slopes are monitored very carefully because it wouldn't take much to start the loose rock moving again.

Inactive Shield Volcanoes

Four shield-type volcanoes, built up by successive flows of fluid andesite, are located in the park. These volcanoes are Raker Peak, north of Lassen; Prospect Peak in the extreme northeast corner; Red Mountain, next to the south-central boundary; and Mount Harkness in the southeast corner. All are between 7000 and 8400 feet in elevation. They have been modified by glacial erosion and by ash falls from Lassen Peak and Cinder Cone. Although apparently constructed entirely of lava without interbedded pyroclastic deposits, each of the four shield volcanoes is topped by a cinder cone rising from a central summit crater. The cinder cones and perhaps the top layers of lava appear to be postglacial in age. On Raker Peak, a large dacite dome has pushed up on the southwest side, forming a line of remarkable dacite cliffs with columnar structure.

Cinder Cone

Located about 10 miles northeast of Lassen Peak, Cinder Cone has been built up during the past few centuries by a succession of ash eruptions. It rises as a symmetrical cone from the flatlands, 750 feet high and about a half-mile in diameter. A cinder sheet of varying thickness surrounds the cone. Two circular craters are at the top. After cinder eruptions had built the cone, several lava flows oozed out from the base of the volcano and spread north and south, damming streams that formed Butte Lake at the

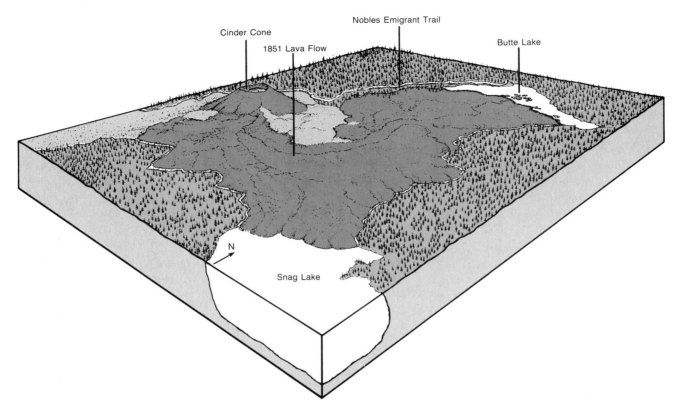

Figure 34.7   In 1851 (and prior to that time), basaltic lava erupted from vents near the base of Cinder Cone. Flows dammed drainageways, creating Snag Lake and then Butte Lake. Water from Snag Lake seeps through porous lava beds and keeps Butte Lake from drying up. Nobles Emigrant Trail skirts Butte Lake and follows the edge of the lava flow around Cinder Cone. Drawing made from a National Park Service photo by C. Peters.

north and Snag Lake at the south. The area of the young, unweathered flows is called Fantastic Lava Beds because of their blocky, jagged, rough surface (figs. 34.7, 34.8).

An unusual feature of the basaltic rock that cooled from the young lava flows at the base of Cinder Cone is the presence of tiny, sparkling quartz crystals that can be seen with the unaided eye. Quartz basalt, a rare combination, is considered a "hybrid rock." The quartz crystals are partially assimilated *xenocrysts,* defined as "foreign" crystals in a body of igneous rock. They may have been picked up by the basalt as it rose toward the surface, or they may have been produced by magma mixing.

Glacial Features

In Lassen Volcanic National Park, glacial debris and volcanic debris are found mixed together and reworked in mudflows, lahars, landslides, and piled up in a few moraines. Some moraines and other glacial features were buried by tephra or covered by lava flows.

Some of the volcanoes, such as Red Mountain, Raker Peak, and Lassen Peak, which was a center of ice radiation, were more severely modified by glacial erosion than others. Small valley glaciers persisted at the higher alti-

Figure 34.8   Blocky front of a fresh, unweathered lava flow beside the route of Nobles Emigrant Trail, near Butte Lake. Photo by S. D. Tuttle.

Table 34.1 Geologic Column, Lassen Volcanic National Park

Time Units			Rock Units	Geologic Events
Era	Period	Epoch		
Cenozoic	Quaternary	Holocene	Dacite; pyroclastics, lavas, mudflows	Lassen Peak eruptions, 1914–1921
			Quartz basalt flows Pyroclastics, cinders	Eruptions from base of Cinder Cone, 1570–1871 Construction of Cinder Cone
				Chaos Jumbles produced by avalanching from Chaos Crags 300 years ago
			Dacite breccia Dacite domes	Eruption of Chaos Crags, 1000–1200 years ago Construction of Lassen Peak on flank of Tehama 11,000 years ago Eruption of smaller domes, south side of Lassen Peak
		—?—	Pre-Lassen fluid dacite flows	Columnar dacite at base of Lassen Peak Erosion of Tehama caldera Decline and collapse of Tehama
		Pleistocene	Fluid andesite lavas; some basalt flows	Construction of shield volcanoes
			Glacial deposits interbedded with volcanics	Episodes of glaciation before, during, and after volcanic eruptions
			Andesite flows and pyroclastics	Construction of composite volcano, Mount Tehama
			Eastern basalts and pyroclastics	Lava ridge; later eroded to rugged hills on east side of park
	Tertiary	—?—	Flatiron Andesite Juniper Lake Andesite; Twin Lakes Andesite	Construction of Central Plateau and lava plain
		Pliocene	Willow Lake Basalt	Flowed across southern portion of park
			Tuscan Formation (not exposed in park)	Pyroclastics, lahars

Source: modified after Macdonald 1963, 1969; Kane, 1980; Harris, 1988

tudes and in protected locations well into post-Pleistocene time. Lakes formed in some of the cirque basins. Two examples are Emerald Lake and Lake Helen. The latter was named in honor of Helen Tanner Brodt, whose ascent of Lassen Peak in 1864 was the first by a woman. Summit Lake, where campgrounds are located, is dammed by a moraine. Other lakes of glacial origin include Feather, Silver, Big Bear, Little Bear, Swan, Echo, and Twin Lakes.

Geologic History

1. The pre-Cenozoic rocks.

Inasmuch as Paleozoic and Mesozoic rock units, intruded by granite, crop out only 15 miles southeast of Lassen Volcanic National Park, it seems reasonable to assume that Sierra Nevada basement rocks extend north-

ward under the park. Moreover, similar rocks of the same age make up the Klamath Mountains that rise west of Lassen and extend northward up the coast into Oregon. Gravity studies have shown that the lighter (less dense) volcanic rocks occupy a deep basin and that very far below them are the basement rocks. Presumably, rocks linking the Sierra Nevada and the Klamath Mountains are buried beneath the Southern Cascade volcanoes and the great lava plateau on which they stand.

2. Cenozoic uplift and the beginning of volcanism.

The region began to be uplifted, and the Sierra Nevada became a westward-tilted fault block (chapter 25). As uplift proceeded, extensive volcanism was generated by a subduction zone at depth. Huge lahars poured down stream channels on the gentle western slopes of the Sierra.

These became the Tuscan Formation of Pliocene age. Although the Tuscan rocks are not exposed in Lassen Volcanic National Park, they lie just below the surface in a number of places.

3. Willow Lake basalt flows in Pliocene time.

Beginning in the southern portion of the park, basaltic flows poured out of fissures and vents and covered an increasingly wide area. These flows, and many subsequent outpourings, built up the lava plateau.

4. Extensive andesite flows in Pliocene and Pleistocene time.

The basalt flows were covered over by a thick series of very fluid andesite flows called the Jupiter lavas. The Twin Lakes lavas of black porphyritic andesite flooded a large area at about the same time. The Twin Lake andesites, although much older than the quartz basalt lavas at Cinder Cone, are also rich in xenocrysts of quartz. The Flatiron andesites erupted and spread over the southwestern part of the park. By this time the landscape had become a fairly flat lava plain stretching over a wide area. No pyroclastic debris had accumulated.

A renewal of basalt flows (the Eastern basalts) built low hills along the park's eastern boundary, which were later eroded to a rugged topography. With the advent of pyroclastic eruptions, some cones began to appear in the northern part of the park.

5. Development of Tehama (Brokeoff Volcano) during the Pleistocene Epoch.

In the southwestern corner of the park, Tehama became larger and larger from successive andesite flows, interbedded toward the top of the cone with increasing amounts of tuff and breccia. The great composite volcano became an impressive feature above the lava plain, rising to over 11,000 feet.

6. Eruption of the four shield-type volcanoes.

Eruptions of fluid andesite formed Raker and Prospect Peaks, Red Mountain, and Mount Harkness during the late stages of Tehama's development. These shield-type volcanoes rose at the four corners of the Central Plateau area.

7. Late Pleistocene destruction of Tehama; eruption of pre-Lassen dacites.

Near the present location of Lassen Peak, on the subsiding northeast flank of Tehama, a new vent opened and extruded flows of fluid, black, glassy dacite, 1500 feet in thickness. These flows hardened as columnar lavas that now encircle the base of Lassen Peak. At about the same time, Tehama's cone collapsed in on itself, forming the huge caldera.

The pre-Lassen dacites had erupted during an interglacial period. When the glaciers returned, they enlarged the caldera by severely eroding the soft, decaying rocks of the old volcano.

8. Construction of Lassen Peak about 11,000 years ago.

A pause in volcanic activity followed Tehama's demise. Then, close to the end of the Pleistocene Epoch and after an ice sheet had melted away, a large volcanic plug of partly solid, partly viscous, dacite lava began to rise through the northeastern vent on the old cone's flank. The rock enclosing the vent shattered and cracked as the lava kept on rising. This produced an extensive cover of talus on the steep slopes of the emerging volcanic dome. Within a comparatively brief time, Lassen Peak attained its present height. Because the ice sheet had gone, the talus accumulations are relatively undisturbed. Later, during a return of colder climate, valley glaciers reformed and eroded shallow cirques on the north and northeast sides of the Peak. Small moraines mark the extent of these glaciers (fig. 34.9).

Smaller, satellite domes of dacite pushed up, forming Chaos Crags north of Lassen Peak. Tuff and pumice were thrown out by pyroclastic explosions. The most recent of

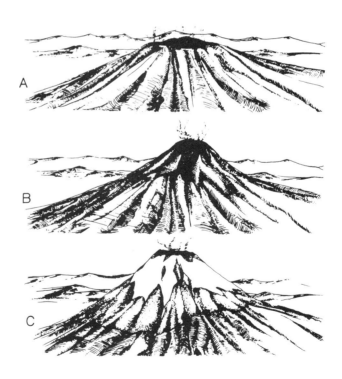

Figure 34.9 Development of Lassen Peak. *A.* Dome of viscous dacite lava rises in the eroded crater of Tehama, toward the end of Pleistocene time. *B.* As the dome kept pushing up, some of the lava overflowed. *C.* Talus blocks from the plug dome and the fractured vent accumulated on the slopes. During a return of colder climate, small glaciers reformed on the north side of Lassen and eroded shallow cirques. From S. L. Harris, *Fire Mountains of the West: The Cascade and Mono Lake Volcanoes.* © 1988 Mountain Press Publishing Company. Used by permission.

Figure 34.10   Summit area of Lassen Peak showing 1915 dacite flow (foreground) and a small crater, formed by subsequent eruptions between the dacite flow and Lassen Peak. From J. W. Harbaugh, *Northern California.* © 1975 Kendall/Hunt Publishing Company. Used by permission.

these explosive eruptions caused the rock avalanche that created Chaos Jumbles about 300 years ago.

9. Construction of Cinder Cone by pyroclastic eruptions.

Cinder Cone began several centuries ago with violent pyroclastic eruptions that spread ash over 30 square miles. After the cone was built, quartz basalt flows erupted from vents at the base. The most recent pyroclastic activity occurred in 1851.

10. Eruptive activity of Lassen Peak, 1914–1921.

Before the eruptions began, the pre-1914 crater contained a fairly deep lake. This crater was totally destroyed and the depression filled by the explosions and subsequent lava flows of the active period. Two craters at the summit can be seen today.

Nearly 400 outbursts, a few major but most of them minor, were recorded by observers during the eruptive period. Scientific data collected at that time and subsequent studies of Lassen's activity have been helpful to geologists who study the Cascade volcanoes and try to anticipate what they will do next. Monitoring of volcanic features in the park by seismometers, tiltmeters, and inclinometers is carried on by the U.S. Geological Survey in cooperation with the National Park Service.

Geologic Maps and Cross Sections

Macdonald, G. A. 1963. Geology of the Manzanita Lake quadrangle, California. U.S. Geological Survey map GQ 248.
———. 1964. Geology of the Prospect Peak quadrangle, California. U.S. Geological Survey map GQ 345.

Bibliography

Amesbury, R. 1967. *Nobles' emigrant trail.* Privately published by R. Amesbury. (Distributed by Loomis Museum Association) 42 p.

Bullard, F. M. 1984 (2nd revised edition). *Volcanoes of the earth.* Austin, Texas: University of Texas Press. p. 565–573.

Bumpass Hell nature trail, Lassen Volcanic National Park. 1985 revision. Loomis Museum Association leaflet.

Chronic, H. 1986. *Pages of stone, geology of western national parks and monuments,* v. 2, Sierra Nevada, Cascades, and Pacific Coast. Seattle, Washington: The Mountaineers. p. 59–70.

Cinder Cone nature trail, Lassen Volcanic National Park. 1984. Loomis Museum Association leaflet.

Crandell, D. R. 1972. Glaciation near Lassen Peak, northern California. U.S. Geological Survey Professional Paper 800-C pp. C179-C188.

Harbaugh, J. W. 1975. *Northern California,* K/H Geology Field Guide Series. Dubuque, Iowa: Kendall/Hunt Publishing Company. p. 86–109.

Harris, S. L. 1988. *Fire mountains of the west: the Cascade and Mono Craters volcanoes.* Missoula, Montana: Mountain Press Publishing. 320 p.

Heiken, G. and Eichelberger, J. C. 1980. Eruption of Chaos Crags, Lassen Volcanic National Park. *Journal of volcanology and geothermal research,* 7:443–448.

Kane, P. S. 1980. *Through Vulcan's eye, the geology and geomorphology of Lassen Volcanic National Park.* Loomis Museum Association. 118 p.

———. 1982. Pleistocene glaciation, Lassen Volcanic National Park. *California geology* 35 (5):95–105, May.

Loomis, B. F. 1971 (3rd revised edition). *Eruptions of Lassen Peak: pictorial history of Lassen.* Loomis Museum Association. 96 p.

Macdonald, G. A. and Katsura, T. 1965. Eruption of Lassen Peak, Cascade Range, California, in 1915: example of mixed magmas. *Geological Society of America bulletin,* 76:475–482.

Pakiser, L. C. 1964. Gravity, volcanism, and crustal structure in the southern Cascade Range, California. *Geological Society of America bulletin,* 75:611–620.

Schultz, P. E. 1983 revision. *Road guide to Lassen Volcanic National Park.* Loomis Museum Association. 40 p.

Williams, H. 1929. The volcanic domes of Lassen Peak and vicinity, California. *American journal of science,* 18:313–330.

———. 1931. The dacites of Lassen Peak and vicinity. *American journal of science* 22:385–403.

Lassen Volcanic National Park

Size: 106,372 acres; 166 square miles

Wilderness area, 78,982 acres

Proclaimed Lassen Peak and Cinder Cone National Monuments: May 6, 1907

Established as Lassen Volcanic National Park: August 9, 1916

Address: Mineral, California 96063

35

Site of the most cataclysmic volcanic event of the century, the violent eruption of Novarupta in 1912, which created the ash-filled Valley of Ten Thousand Smokes. Katmai's active volcanoes are in the Alaskan-Aleutian volcanic belt. A plate convergence zone associated with the volcanic belt has been active since late in Cretaceous time.

Powerful waves attack an emergent coast of fiords, cliffs, and waterfalls.

Wildlife abounds in an ecosystem of mountains, glaciers, lakes, streams, forests, tundra, and marshes.

The Alagnak Wild River, rising in a northern lake in the preserve, teems with salmon.

Katmai National Park and Preserve

Southwest Alaska

Figure 35.1   A heavy snow cover surrounds the crater of Mount Martin, while steam continues to rise from heat below. This volcano in Katmai National Park and Preserve erupted small amounts of ash in 1951. National Park Service photo.

Figure 35.2 Katmai National Park and Preserve, Alaska.

KATMAI NATIONAL PARK
AND PRESERVE

Glacier

Park Border

Preserve Border

0   10 mi.

Figure 35.3   Active volcanoes of the upper Alaska Peninsula. The line of volcanoes marks the inner edge of an ongoing subduction zone. In this part of the Alaska-Aleutian volcanic arc, the magma that builds the volcanoes rises through continental crust, then erupts as lavas of variable composition. (Modified after Fierstein 1984).

Geographic Setting

Located in the northern part of the Alaska Peninsula, overlooking the Shelikof Strait, Katmai National Park and Preserve is 290 air miles southwest of Anchorage. The park/preserve extends along the coast from Kamishak Bay south to the Becharof National Wildlife Refuge and westward nearly to the community of King Salmon. To the north, at the head of the peninsula, is Lake Clark National Park and Preserve (chapter 36). The active volcanoes in the two parks are part of the Alaskan-Aleutian volcanic belt (fig 35.3).

The Katmai volcanoes, like those in the Lake Clark region, trend from northeast to southwest, parallel to the eastern coast of the Alaska Peninsula. In this part of the peninsula, the volcanoes rise over 7000 feet from the shores of Shelikof Strait. Inland from the coast, the mountains of the Aleutian Range have elevations of 3000 to 5000 feet; and the whole range, including the volcanoes, is about 100 miles wide from east to west. Several hundred miles

down the peninsula, the range breaks up into the Aleutian Island chain, which stretches in an arcuate pattern westward for another thousand miles out into the North Pacific. Throughout its length, the Aleutian Range has 60 volcanoes, of which 47 are known to have been active since 1760. In the spring and summer of 1986, Augustine Volcano, which rises from the floor of Kamishak Bay (between Katmai National Park and Preserve and Lake Clark National Park and Preserve), erupted large quantities of ash (chapter 36).

The Katmai climate is wet and stormy. From whatever direction the wind blows, moisture-laden winds are forced to rise over the mountainous Alaska Peninsula. Light rains can last for days, and strong winds and gusts ("williwaws") frequently sweep the area. In summer, skies are clear perhaps 20 percent of the time. However, summer temperatures (averaging 50° to 60° F) are mild by Alaska standards.

The world's largest predatory land mammals, the Alaska brown bears, thrive in the wild, remote Katmai area. The huge size of these Ice Age survivors, weighing from 600 to 1500 pounds, is attributed to fish oils in their rich diet. The bears feed on the red salmon that spawn in the lakes and streams of the Alaska Peninsula.

Facilities in the park/preserve are open only from June to September, although charter flights over Katmai can be made during the winter. In summer, access to Katmai is by air from Anchorage to King Salmon, where the park headquarters is located, outside the western edge of the park/preserve. Float planes ferry visitors from there to the Brooks River Camp on the shore of Lake Naknek. A four-wheel-drive bus makes daily trips to an overlook on the rim of the Valley of Ten Thousand Smokes. This is the only road in the park/preserve.

Local History

The Aleutian Range was such a difficult barrier to cross that separate native cultures developed on the Gulf of Alaska and Bristol Bay sides of the Alaska Peninsula. Russian fur traders, based on Kodiak Island across Shelikof Strait from Katmai Bay, were the first Europeans to live in the area. They were succeeded by American traders after the Alaska purchase in 1867. The establishment of salmon canneries on the Bristol Bay side of the Alaska Peninsula and the Nome gold rush brought more people into the area in the late nineteenth century. In order to avoid the long, stormy passage around the Aleutian Islands, traders and prospectors followed the old Katmai Trail, used since prehistoric time as a shortcut across the mountains of the Alaska Peninsula. This arduous route went from Katmai Bay through Katmai Pass and down to Bristol Bay, where prospectors could board sailing ships to Nome and the goldfields. The 1912 eruption destroyed the trail. Parts of the route have been restored on the west side of Katmai Pass and can be used by backpackers, but the section from Katmai Pass to Katmai Bay has not been opened because of heavy brush, braided streams, quicksand, quickmud, and frequent slides.

The 1912 Eruption

Apparently no lives were lost during this catastrophic event, probably due to the sparse population and the fact that a week-long series of earthquakes, prior to the explosion, prompted natives in mountain villages to flee to coastal areas on either side of the Alaska Peninsula.

Around noon of June 6, 1912, people on the northern and eastern sides of the Gulf of Alaska knew that an extraordinary volcanic event had occurred because they heard loud booms and saw in the distance great clouds of ash. Around midnight of that day, a severe earthquake occurred, estimated to be at least magnitude 6.4 on the Richter scale.

The ashfall lasted for three days in the immediate vicinity of the volcano, disrupting wireless communications. No word reached the outside world until June 9. More than a week passed before a rescue party, paddling through chunks of floating pumice, could reach Katmai Bay from Kodiak Island. They found the native village deserted and ash from 3 to 15 feet deep over the whole area. Kodiak Island was covered by a foot of ash and many homes were destroyed.

A U.S. revenue cutter happened to be in Kodiak Harbor (100 miles east of the volcano) on June 6. The crew took local residents on board in complete darkness during the ashfall. The ship lay at anchor until the skies cleared and it was safe to go ashore on the morning of the third day after the first explosions.

Although much of the ash and pumice fell in the Gulf of Alaska, over 3000 square miles of land were covered by about a foot of pyroclastic debris. Ash fell in Juneau on the Alaska Panhandle, in British Columbia, and as far north as Fairbanks. Ash reaching the upper atmosphere was carried around the globe, causing the average annual temperature of the Northern Hemisphere to be reduced by 1.8° F for more than two years.

Later that summer, the U.S. Geological Survey and the National Geographic Society sent a geologist, George C. Martin, from Washington, D.C. to Katmai to make the first scientific assessment of the disaster. During the next few years, the National Geographic Society sponsored four more scientific expeditions to the eruption area. All but one were led by Robert F. Griggs, a botanist, who originally went to devastated Kodiak Island to study the effects of the eruption on plant regeneration. In the course of his first three expeditions to the mainland (1915, 1916, and 1917), Griggs explored, under difficult and often dangerous field conditions, the calderas of Katmai and Novarupta and the Valley of Ten Thousand Smokes.

As a result of these investigations, Katmai was proclaimed a national monument in 1918 by President Woodrow Wilson. The Society sponsored two subsequent expeditions in 1919 and 1930, both led by Dr. Griggs. His illustrated account of his scientific adventures (Griggs 1922) vividly portrays the Katmai scene. Griggs and his fellow scientists accumulated a remarkable body of data and observations. Their work, and studies done since, have finally enabled scientists to reconstruct a fairly complete account of what happened during the Katmai eruptions and how they came about.

Geologic Features

The Katmai-Novarupta Connection

Katmai Volcano, which was probably about 7500 feet in elevation before its summit collapsed, and its lower neighbor, Novarupta (elevation, 2757 feet), were involved in the eruptive sequence in a rather unusual way. The first visitors to the area thought that the Katmai vent was the main source of the approximately 4 to 6 cubic miles of volcanic debris ejected during the eruptions. The entire top of Katmai had collapsed, leaving in place a great caldera. Not until 1953 was it determined that all the ash flows and pumice had been ejected from Novarupta's vent rather than from the Katmai outlet. Why, then, did Katmai collapse in on itself?

The earthquakes and fault movements preceding the eruptions may have allowed magma from Katmai to flow laterally through fissures and connect with conduits containing magmas of differing chemical composition rising beneath Novarupta. As the magma masses mingled, they frothed and exploded, ejecting from Novarupta great clouds of fine, white rhyolitic pumice, followed by dacitic ash, and finally andesitic scoria. Fissures opened in the smaller volcano, releasing colossal pyroclastic flows that roared northwestward down the valley of the Ukak River. The incandescent mass devastated everything in its path, buried the valley floor, and covered 40 square miles with ash as much as 700 feet deep. When Griggs named this area the Valley of Ten Thousand Smokes in 1916, countless fumaroles of steam, some a thousand feet high, still issued from the plain of ash. At the present time, only a few active vents remain.

As the magma chamber emptied, the unsupported summit of Katmai collapsed, forming a caldera three miles long, two miles wide, and 3700 feet deep from the rim down. Only minor explosive eruptions have occurred in these volcanoes since 1912, the most recent activity on Katmai being in 1931. The posteruption elevation of the rim of Katmai is 6715 feet (fig. 35.4).

Crater Lake formed in the Katmai caldera not long after the initial cooling. In recent years the lake level has risen, but the lake probably won't run over the caldera rim because water seeps through the porous pyroclastics making up the caldera walls.

The Valley of Ten Thousand Smokes

On his second expedition in 1916, Robert F. Griggs went up the Katmai River valley. On their way to Katmai Pass, he and his companions made the first ascent of Katmai to the caldera rim and looked down at the lake, more than 1000 feet below. They saw hundreds of steam jets shooting into the air, roaring "like a great locomotive when the safety valve let go," as Griggs described it.

Wanting to press on to the Ukak River valley, they continued toward Katmai Pass. But when Griggs crossed the divide, he saw a most awesome sight. Thousands of columns of steam rose 500 feet and more from the ash-filled valley below. Escaping steam hissed and roared, and the sulfurous odor was overwhelming. Brightly colored encrustations sparkled around the spouting vents. The river was no more, but hot water bubbled up and ran off in innumerable rivulets. He knew that in this desolate place, which he named the Valley of Ten Thousand Smokes, no living thing could have survived the eruptions.

Scientists who later studied the Valley of Ten Thousand Smokes have found that the heat is "rootless"; that is, the fumaroles, escaping gases, steam jets, and hot water were generated by the gradually cooling ash-flow deposits rather than by magmatic heat at some depth below the surface. Fumarolic activity in the valley had died out by the 1930s. Near the Novarupta caldera, however, a few fumaroles still give off steam as ground water percolates down to the hot rocks below the volcanic vent.

The flat surface of the Valley of Ten Thousand Smokes is still barren of vegetation. However, new growth now covers much of the ash-fall layers outside the valley, and clumps of grass and low plants struggle to survive in sheltered crannies in the ash-flows.

Tephra in the valley consists of fine ash, *lapilli* (small fragments), and *volcanic bombs* (larger fragments). Massive pyroclastic flow deposits fill the valley. Air-fall deposits cover the ash flow deposits and are spread over the

Figure 35.4   Cross section of Katmai after its summit collapsed in 1912. The dashed line shows its projected earlier profile and volume. Accumulating meltwater formed Crater Lake in the caldera basin. (Modified after Griggs 1922).

Figure 35.5 At the end of the road from Brooks Camp to Three Forks Overlook, visitors stand on an ash bluff above the Ukak River, the principal stream draining the Valley of Ten Thousand Smokes. The lowering clouds over the mountains in the background are characteristic of Katmai weather. National Park Service photo by R. Nichols.

region. The chemistry of the ejecta shows that the magma was of mixed composition—40 percent rhyolite, 50 percent dacite, and 10 percent andesite.

Novarupta

In 1917, Griggs discovered what he recognized as a new caldera west of Katmai. He named the feature Novarupta, meaning "newly erupted." Novarupta's pre-1912 configuration is unknown, but Griggs hypothesized that a crater did not exist at that location before 1912. The explosion of Novarupta produced a shallower, smaller caldera, less impressive than Katmai's broad, steep-walled caldera. It is not surprising that the early investigators were convinced that most of the ejected pumice came from Katmai.

In later years, detailed mapping and study of ash deposits by Garniss Curtis, Howel Williams, and other geologists show that the ash layers thicken away from Katmai and toward Novarupta, despite changes in wind direction that occurred during the eruptions. Clearly, Novarupta was the true outlet for the tephra thrown out in the 1912 explosions and eruptions.

The eruption probably began with deafening booms and roars as a great high plume of gas and ash shot up into the atmosphere, like a jet stream from the nozzle of a giant fire hose. It "played" for perhaps for 60 hours. Curtis (1968) has estimated the rate of expulsion of ejecta during this phase as 35 billion cubic feet per hour. As the plume began to subside, pyroclastic flows, or glowing avalanches, swept down the slopes, and falling ash blanketed all the surrounding terrain. Some of the ash and pumice welded into dense rock due to great heat and pressure from overburden. The mixing of the magmas also produced blocks of "banded pumice" with striking black and white swirls.

In a late phase of the 1912 eruption, a rhyolitic dome was pushed up from the floor of Novarupta's collapsed caldera, and possibly a dacite dome emerged on the floor of Katmai's caldera.

Related Volcanoes and Eruptions

Kaguyak, about 40 miles northeast of Katmai, has a caldera and a lake, but it has been inactive in historic time. Fourpeaked Mountain, farther to the northeast, has had no volcanic activity in historic time and is almost completely covered by an icefield and glaciers. Neighboring Mt. Douglas, also ice-covered except for its 7063-foot peak, steams intermittently. Between Kaguyak and Katmai are four inactive volcanoes—Devils Desk, Stellar, Denison, and Snowy—and one, Kukak, that is fumarolic. Mount Griggs, which rises above the Valley of Ten Thousand Smokes, across from Novarupta, has been inactive in historic time but has two sulfurous fumarole fields (*solfataras*). Trident, Mageik, and Mount Martin, which trend southwest from Katmai, have been active in historic time. Trident has had a number of small eruptions. The most recent was in 1974 when a dome formed. Between 1953 and 1968, several flows of blocky andesitic lava moved down Trident's flanks, covering 1912 ash. Trident's neighbor, Mageik, had its most recent ash eruption in 1953, simultaneously with Trident. Another feature of Mageik is the "boiling lake" in its crater. The warm, acidic water in the crater bubbles vigorously due to escaping gases. Mount Martin steams intermittently and had a small ash eruption in 1951.

Glacial Features

The active glaciers in the park/preserve, all of which are on the flanks of the coastal volcanoes, are nourished by moisture-bearing winds and protected from melting by the nearly continual cloud cover. Serpent Tongue glacier, between Mt. Denison and Snowy Mountain, is the largest glacier in the park/preserve. The Knife Creek glaciers, which flow down Katmai's western slopes, are so well insulated by their cover of ash that they have changed little

Table 35.1 Geologic Column, Katmai National Park and Preserve

Time Units			Rock Units	Geologic Events
Era	Period	Epoch		
Cenozoic	Quaternary	Holocene	Volcanic rocks: basalt, andesite, rhyolite, dacite, pumice, welded tuff, etc.	1912 eruption, with ash falls, ash flows, explosions
		Pleisto-cene		Volcanic activity preceding and following climactic eruption, ranging from steam eruptions, minor ash falls, lava flows to violent eruptions
			Glacial drift	Glacial advances and retreats from late Tertiary to Holocene
	Tertiary		Volcanic rocks Small plutons	Build-up of cones
				Subduction, deformation, uplift
				Lava flows; intrusive activity
				Docking of suspect terranes
		Eocene	Nonmarine sediments; coal beds; volcanics	Erosion of highlands
				Compaction of coal in swamps and coastal marshes
				Intermittent volcanism
Mesozoic	Creta-ceous		Kaguyak Formation	Fossiliferous marine sediments Ongoing subduction
	Jurassic		Naknek Formation Shelikof Formation Granitic plutons	Accumulation of marine sediments in subsiding basin Emplacement of Aleutian batholith
Paleozoic	Permian		Lava flows, breccias, fossili-ferous limestones (adjacent to south edge of park/preserve)	Erosion

Sources: Keller and Reiser, 1959; Buck, 1965

since the 1912 ejecta blanketed them. They look more like mud slides than glaciers.

In the foothills on the western side of the Aleutian Range, a half-dozen large, elongated lakes and many small lakes occupy glacial troughs, basins and cirques excavated by the mightier glaciers of the Pleistocene Epoch. Moraines dam most of the lakes. Naknek Lake, which is the largest, was for a time connected to Bristol Bay. Kukaklek Lake, in the preserve (northwest side of the park), is the source of the Alagnak Wild River, noted for whitewater rafting and sport fishing.

Geologic History

1. Paleozoic rocks.

North of Puale Bay, about five miles south of the park/preserve's southern boundary, limestones bearing mid-Permian fossils have been found, interbedded with volcanic flows and breccias. The limestones are part of a 4000-foot sequence of sedimentary and volcanic rocks.

2. Mesozoic rocks.

Thick sequences of Jurassic and Lower Cretaceous rocks, consisting of clastic and volcanic sediments, make up the main mass of this part of the Aleutian Range. In

early to middle Jurassic time, the accumulation of more than 15,000 feet of such material suggests the long-continued presence of an area of subsidence, which was probably an oceanic trench. Uplifting and rapid erosion of coastal areas to the northwest brought arkosic sediments, which are interbedded with a considerable volume of volcanic debris, presumably from an island arc. These rocks make up the Shelikof Formation and the overlying Naknek Formation that are widespread throughout the region. The Cretaceous Kaguyak Formation is frequently found overlying the Shelikof and the Naknek, sometimes conformably and sometimes unconformably. Fossils are abundant in these Jurassic and Cretaceous formations.

Granitic plutons, part of the Aleutian batholith, began to intrude the sedimentary rocks early in Jurassic time. Plutonic activity continued in a series of pulses through most of the period. Late in Cretaceous and early in Tertiary time, tectonostratigraphic terranes, rafted by a subducting oceanic plate, docked in the area, causing intermittent folding, uplift, volcanism, and renewed intrusive activity.

3. Cenozoic rocks.

In the Katmai area, mainly in the southern coastal section of the park/preserve, nonmarine Eocene beds, including coal deposits, are exposed. The coal was probably laid down in coastal swamps. Lava flows are interbedded with the sedimentary rocks. Small intrusions of quartz diorite and quartz monzonite, with radiometric ages ranging from early to middle Tertiary time, have cut the Eocene rocks. Younger Tertiary rocks are absent in the park/preserve and were probably stripped off by erosion during and following uplift of the area.

Convergence of plates and subduction of oceanic crust has continued to the present time. A major fault system, the Bruin Bay fault, which trends northeast–southwest down the Aleutian Range, probably represents a suture, or "seam," between the Peninsular terrane on the southeast side and older rocks on the northwest side. Strike-slip displacement has moved rocks on the southeastern side of the fault to the southwest. Vertical movement has also occurred.

4. Quaternary volcanic activity.

Fifteen young volcanoes rise from a volcanic pile on top of Tertiary rocks in Katmai National Park and Preserve. Some cones are presently inactive, but six have erupted since 1912. Studies of Holocene ash layers have revealed at least 12 major volcanic events that preceded the 1912 eruption. The oldest ash layer that has been identified and dated fell in 5400 B.C. Other ash layers in the marine sediments of the Gulf of Alaska carry the record back 30,000 years.

With more than 6 cubic miles of ash and pumice discharged, the 1912 Novarupta eruption was the most voluminous of this century. The volume of ejecta thrown out by Katmai was about 12 times that of Mount St. Helens in 1980. Steam plumes, heat, and occasional eruptions from the more active volcanoes in the park/preserve, plus frequent earthquakes, are indications that we can expect more volcanic events to occur in Katmai National Park and Preserve in the future.

5. Quaternary glaciation.

Wisconsinan glaciation in Katmai National Park and Preserve (and in its neighbor to the north, Lake Clark National Park and Preserve, chapter 36) was extensive and severe. Most evidences of earlier glaciations were destroyed. In some places glacial deposits and volcanic deposits are interbedded, indicating that glaciers were melted by eruptions and then reformed. Holocene glaciers advanced and retreated as patterns of climate changed. The most recent major advance ended about 4000 B.C. However, the fact that two small glaciers reformed on the inner walls of Katmai's caldera since 1912 provides evidence that at high elevations, given sufficient precipitation, glaciers continue to thrive in Katmai National Park and Preserve.

Bibliography

Buck, C. A. 1965. *Geology of the Alaska Peninsula: Island arc and continental margin* (with geologic map). Geological Society of America Memoir 99.

Bullard, F. M. 1984, 2nd edition. *Volcanoes of the earth.* Austin: University of Texas Press. p. 159–176.

Curtis, G. H. 1968. The stratigraphy of the ejecta from the 1912 eruption of Mt. Katmai and Novarupta, Alaska. In *Studies in volcanology: a memoir in honor of Howel Williams.* Geological Society of America Memoir 116, p. 153–210.

Dennen, W. H., and Moore, B. R. 1986. *Geology and engineering.* Dubuque, Iowa: Wm. C. Brown Publishers. p. 256.

Fierstein, J. 1984. *The Valley of Ten Thousand Smokes, Katmai National Park and Preserve.* Anchorage, Alaska: Alaska Natural History Association. 16 p.

Griggs, R. F. 1922. *The Valley of Ten Thousand Smokes.* National Geographic Society. 340 p.

Hildreth, W. 1983. The compositionally zoned eruption of 1912 in the Valley of Ten Thousand Smokes, Katmai National Park, Alaska. In S. Aramaki and S. Kuchivo (editors). Arc Volcanism, *Journal of Volcanology and Geothermal Research* 18:1–56.

Hildreth, W., Fierstein, J. E., Grunder, A., and Jager, L. 1981. The 1912 eruption in the Valley of Ten Thousands Smokes, Katmai National Park: A summary of the stratigraphy and petrology of the ejecta. U.S. Geological Survey Circular 868, p. 37–39.

Keller, A. S. and Reiser, H. N. 1959. *Geology of the Mt. Katmai area, Alaska* (with geologic map). U.S. Geological Survey Bulletin 1058G, p. 261–298.

Montague, R. W. et al., editors. 1975. *Exploring Katmai National Monument and the Valley of Ten Thousand Smokes.* Anchorage, Alaska: Alaska Travel Publications. 276 p.

Snyder, G. L. 1954. Eruption of Trident Volcano, Katmai National Monument, Alaska, February-June 1953. U.S. Geological Survey Circular 318. p. 1–7.

Katmai National Park and Preserve

Area: 4,090,000 acres; 8809 square miles Wilderness, 3,473,000 acres

Proclaimed a National Monument: September 24, 1918

Established a National Park and Preserve: December 2, 1980

Address: P.O. Box 7, King Salmon, Alaska 99613

36

Lake Clark National Park and Preserve

South Central Alaska

Two active composite volcanoes, over 10,000 feet high, overtop the rugged Chigmit Mountains in the southern Alaska Range; two more cones rise to the north and south of the park/preserve.

Aligned over a subduction zone, the volcanoes are part of a belt that trends southwestward through the Aleutians.

Earthquakes and eruptions signify ongoing tectonic activity.

Lake Clark, 42 miles long, occupies a deep glacial trough, carved out along a fault zone.

Dozens of active glaciers and hundreds of waterfalls line the walls of major glacial valleys.

Jewel-like rock-basin lakes are in the foothills that rim the western side of the Chigmit Mountains.

Fiords open into Cook Inlet in the coastal section.

Three streams in the park/preserve are protected in the National Wild and Scenic Rivers System.

Figure 36.1   Air view, looking northeast, near the north end of Lake Clark in the remote wilderness of Lake Clark National Park and Preserve. Lake Clark, occupying over 40 miles of a long glacial trough, drains into Iliamna Lake, south of the park/preserve, and eventually Bristol Bay on the west side of the Alaska Peninsula. National Park Service photo by M. W. Williams.

LAKE CLARK NATIONAL PARK AND PRESERVE

Figure 36.2 Lake Clark National Park and Preserve, Alaska.

Geographic Setting and Local History

Lake Clark National Park and Preserve, which extends inland from the western shore of Cook Inlet, occupies a rugged, mountainous area southwest of Anchorage. The park/preserve is predominantly wilderness in character. Here at the head of the Alaska Peninsula, the granitic rocks of the Chigmit Mountains link the Alaska Range, curving to the north and east, with the Aleutian Range, which stretches to the south and west. Four distinct physiographic regions, from east to west, give the park/preserve its diversity in ecosystems, scenery, and climate: 1) the coastal strip along Cook Inlet; 2) the abruptly ascending Chigmit Mountains, topped by Iliamna and Redoubt Volcanoes; 3) the foothills of the Chigmit Mountains' western flank, descending among lakes and wild rivers to 4) a boreal forest and a tundra plain.

Maritime elements influence the coastal climate, but a continental climate prevails throughout most of the interior. June through August temperatures on the coast average between 50° and 65° F. Around 60 inches of precipitation falls annually on the eastern slopes. Summers in the interior of the park/preserve are warmer and drier, with temperatures occasionally as high as 80° F; but interior winters are long and severe. Temperatures may plummet to −40° F. Along the coast, winters are more moderate. Strong winds (especially in mountain passes), sudden storms, and subfreezing temperatures may occur at any time of year throughout the park/preserve.

The diversity and richness of plant and animal life reflect the widely differing—but generally favorable—natural conditions within the region's ecosystems. Large mammals like brown and black bears, caribou, moose, and Dall sheep forage in uplands and lowlands along with wolves, lynx, foxes, and smaller creatures. The lakes and streams that drain west and then south into Bristol Bay are important spawning grounds for several species of salmon and trout. On the eastern flanks of the range, rivers cascade to the sea, and coastal cliffs provide rookeries for thousands of puffins, cormorants, and other seabirds. Fish abound in the coastal waters; seals and whales are often seen offshore.

Despite the severe climate and the difficulties in getting around, humans have lived in the Lake Clark region for more than 10,000 years. Migrating clans found sheltered camp sites and an abundance of fish and game. Some sites and artifacts of the earliest cultures have been located. Rock paintings, done with red hematite mixed with animal fat, were discovered in a rock shelter at the head of Tuxedni Bay. The bird, animal, and human figures are arranged in eight main groups, determined by the fracture planes of the rock. The paintings, which are of great antiquity, are similar to some found in Siberia. Much archeological work remains to be done on ancient and more recent sites.

Relics of European origin found in old villages and campsites reveal that Russian fur traders and missionary priests had made contact with Native Alaskans by the 1700s. The Russians controlled the fur trade until the Americans took over in the 1870s following the Alaska Purchase. In 1897, John W. Clark, an agent of the Alaska Commercial Company on Bristol Bay, led an expedition, sponsored by *Leslie's Illustrated Weekly,* into the interior. He followed up river systems to a large inland lake, called Qizjeh Vena, or "many-peoples-gathered-lake" by the Indians, but later named Lake Clark in his honor. Native Alaskans of Eskimo and Indian descent still live in small settlements in the Lake Clark area.

The salmon industry attracted settlers to the region early in the century. Prospectors and miners looked for gold and other minerals in the Chigmit Mountains without much success, although a few patented claims are still held. The first organized geological investigations were not undertaken until the 1920s, when Stephen R. Capps led U.S. Geological Survey expeditions into this area. He described Lake Clark, fed by hundreds of waterfalls from its rimming mountains, as "one of the most beautiful bodies of water in the world."

No highway access to the park/preserve exists. When weather conditions permit, visitors to the Lake Clark region come by small aircraft, mainly from Anchorage. Sometimes gusting winds in Lake Clark Pass and Merrill Pass are too hazardous for small planes to fly into the interior of the park/preserve. Aside from field headquarters at Port Alsworth on Lake Clark, no National Park Service facilities are available in the park/preserve. A few licensed operators offer cabins, guiding and outfitting services; primitive camping is allowed in certain areas. Sport fishing and float trips are permitted in the Chilikadrotna Wild River, which drains Twin Lakes; the Mulchatna Wild River, which flows out of Turquoise Lake; and the Tlikakila Wild River, which flows from Summit Lake (near Lake Clark Pass) southwestward until it empties into the northern part of Lake Clark.

Geologic Features

Even though glaciers, both past and present, have shaped the park/preserve landscape, it is the active volcanoes that dominate the scene. Except for Augustine (4025 feet), which rises from the floor of Cook Inlet, the volcanoes are several thousand feet higher than the Chigmit Mountains. The other volcanoes are Mt. Spurr (11,070 feet), northeast of the park/preserve boundary; Redoubt Volcano (10,197 feet), north of Tuxedni Bay; and Iliamna Volcano (10,016 feet), north of Chinitna Bay. Augustine Volcano, occupying Augustine Island, is south of the park/preserve. All four volcanoes have had violent eruptions and all have erupted recently: Augustine in 1986, Iliamna in

1947, Redoubt in 1966, and Mount Spurr in 1953. Since 1700 (roughly the beginning of recorded history in Alaska), Augustine and Redoubt have each erupted seven times, Iliamna six times, and Mt. Spurr once.

Shallow earthquakes occur frequently in the park/preserve area; since 1972, there have been 13 with a Richter magnitude of 5 to 6. Epicenters of the quakes were determined to be between 55 and 110 miles below the surface in the area between Chinitna Bay and Tuxedni Bay, more or less aligned with the four volcanoes of the Lake Clark region.

Why Are the Volcanoes Here?

The Lake Clark region volcanoes are a major link in the Alaska volcanic belt that extends from Mount Spurr eastward to Wrangell-St. Elias National Park and Preserve (chapter 29) and, in the other direction, southwestward to Buldir Island in the western Aleutians. Part of the chain, and closely related, are the volcanoes in Katmai National Park and Preserve (chapter 35), some 50 to 75 miles (air distance) southwest of Lake Clark. The volcanic belt marks a subduction zone at depth that has been active for millions of years. The recent eruptions represent only the latest phase in a history of tectonic collision and volcanism since late in Mesozoic time and throughout the Cenozoic Era. In the Aleutian trench, parallel to and in front of the 1600-mile arc of volcanoes, oceanic crust is being pulled down (subducted) beneath the North American plate at the rate of 2.6 inches per year. The subduction zone of the Aleutian trench lies about 30 miles beneath the surface of the Kenai Peninsula (chapter 30), but steepens abruptly at the western edge of Cook Inlet to a depth of more than 65 miles. Beneath Redoubt and Iliamna Volcanoes, the depth of the subducting plate is greater than 100 miles. This steep, rapid descent causes frequent, strong earthquakes. Intense heat, generated by the subduction process, melts rock at depth, forming magma, which subsequently rises, erupts, and builds volcanoes.

Volcano Hazards in the Lake Clark-Cook Inlet Area

Most of the eruptions of the Lake Clark volcanoes are explosive and pyroclastic, involving ash flows, ash falls, lava flows, and lahars (mud flows). The volcano-watchers of the U.S. Geological Survey, basing their conclusions on studies of ash deposits over the past 10,000 years, warn that inhabitants of the Cook Inlet region can expect a measurable ash flow on an average of every 30 years. In recent years, as noted above, eruptions of ash and lava have occurred at a more frequent rate.

The hazard potential of ash falls depends on local winds and upper-air conditions. But the presence of the mountains aligned near the coast tends to channel wind-blown ash up or down Cook Inlet or eastward over the Kenai Peninsula. During some of the 1986 Augustine

eruptions, aviation flights to and from Anchorage were canceled or diverted for several days at a time because of wind-carried ash. Ash in the atmosphere also causes temporary radar blackouts and reduces the electrical generating capacity of turbines.

Tsunamis, caused by sea-floor movement or avalanching (above or below the waterline), are a major potential hazard in the coastal regions. In 1883, Augustine's most violent eruption in historic time produced a tsunami, due to avalanching, that crossed Cook Inlet and inundated Port Graham on the tip of the Kenai Peninsula. No lives were lost, but fishing boats and homes were wrecked.

The presence of glaciers on the slopes of active volcanoes is cause for concern, as we have seen in our discussions of the Cascade volcanoes (e.g., Mount Rainier National Park, chapter 32). Because of their height and topographic position (in the path of moisture-carrying winds from the Gulf of Alaska), Mount Spurr, Iliamna, and Redoubt all have sizable glaciers.

In prehistoric time, *lahars* (volcanic mudflows that are pasty masses of snow, ice, volcanic debris, rocks, vegetation, etc.) have roared down valleys on both Redoubt and Iliamna. Some reached Cook Inlet. Lahars are associated with eruptions. On the other hand, *glacier outburst floods* (also called "jokulhlaups," an Icelandic term) may or may not be triggered by volcanic activity. An outburst flood is a sudden, catastrophic release of glacial meltwater, which may be due to the melting of a drainage channel, the buoyant lifting of ice by meltwater, or an increase in subglacial heat emanating from a volcano's interior. A sudden release of steam from a vent may also trigger a glacier outburst flood.

The Redoubt eruptions that took place between 1965 and 1968 were the cause of several outburst floods. Heat from explosive ash flows melted and broke up the glaciers. Huge blocks of ice and uprooted trees were carried down the Drift River valley. Up to six feet of water backed up behind the ice at the river's mouth, where it meets Cook Inlet. A Canadian seismic crew (prospecting for oil) had to be hastily evacuated. (Inexplicably, an oil storage facility with a capacity of 800,000 barrels has since been built on this very site, just outside the park/preserve. If an eruption-caused flood or an earthquake should bring about the rupture or collapse of these tanks on the shore of Cook Inlet, the effect on marine life could be devastating.)

Glacial Features

Redoubt's glaciers have recovered somewhat since the volcano's last eruptions, but Iliamna's are larger and more numerous. Tuxedni Glacier, which descends 16 miles from Iliamna's cone to within a few miles of Tuxedni Bay, is the longest glacier in Lake Clark National Park and Preserve.

Innumerable glaciers in the Chigmit and Neacola Ranges continue to sculpt the mountain landscape of the park/preserve. Most of these glaciers have been in retreat for several hundred years. The large glaciers on the volcanoes, which form at higher elevations and receive more precipitation, are not retreating at the present time. A nearly continuous cloud cover during summer months keeps the glaciers near the coast from melting back.

The legacy of the more extensive glaciers of the past is the spectacular scenery of today. Lake Clark lies in a glacially eroded trough of massive proportions. The lake, 42 miles in length and 1 to 4 miles wide, covers an area of 110 square miles. Its maximum depth is 860 feet. Its location on a fault zone indicates structural control. Glacial ice, the excavating agent, moved along the "path of least resistance," where rock was already broken up by faulting (fig. 36.1).

Several other lakes in or just outside the park/preserve lie in glacially excavated fault troughs that in the past conveyed glacial ice as far as Bristol Bay. Morainal dams, dumped by retreating glaciers, now separate these lakes from each other and from the ocean. Some small lakes occupy kettles in glacial drift. *Kettles* are slumped-in depressions located where a buried ice-block has melted.

In the foothills region west of the Chigmit Mountains, many jewel-like lakes fill scoured basins left after glaciers had melted back. The turquoise-green color of the lakes comes from the fine, glacial powder (rock flour) that washes down from glaciers in the mountains. Some examples (from north to south) are Two Lakes, Telaquana, Turquoise, Twin Lakes, Portage, Lachbund, Kontrashibuna, and Tazimina.

The fiords, cliffs, marine terraces, lowlands, and other features in the coastal part of the park/preserve show the effects of changing sea levels (both emergence and submergence) in Pleistocene and Holocene time. In this region of recent glaciation and active tectonism, both glacial rebound and tectonic movements of the sea floor are important factors in sea level rises and drops that have occurred.

Landsliding and Rockfalls

Mass wasting processes are actively modifying the landscape in the park/preserve. Much of the volcanic rock is weak and forms unstable slopes. Valley walls and cliffs, oversteepened by glacial excavation, tend to fall or slide downslope, especially when loose rock and soil are saturated by water. Sometimes an earthquake triggers a series of landslides. Landslide deposits, some fresh, some stabilized and tree-covered, are numerous.

On slopes covered with tundra, solifluction lobes are conspicuous features. (*Solifluction* is the slow creep of moist, fine-grained sediment caused by more or less continual freezing and thawing.)

Figure 36.3   Twin Lakes, on the western side of the Aleutian Range, are the headwaters of the Chilikadrotna River, which is in the National Wild and Scenic Rivers System. The lakes occupy glacially scoured basins. National Park Service photo by M. W. Williams.

Rocks and Structures

A park naturalist described the Chigmit Mountains as a "frenzy of peaks thrown up by colliding mountain waves." Simply put, granitic rocks, emplaced in older sedimentary rocks, were uplifted and faulted, covered by lavas and pyroclastic debris, and deroofed by erosion. The combination of intrusive, metamorphic, volcanic, and sedimentary rocks, all of which have been disarranged by tectonic activity, presents a complex puzzle.

Here, as in the other national parks located along the Gulf of Alaska, the concept of tectonostratigraphic terranes helps to explain the juxtaposition of discrete, seemingly unrelated crustal blocks, bounded by major faults (sutures). In the Lake Clark region, the Peninsula terrane is the predominant terrane, as it is also in Katmai National Park and Preserve to the south (chapter 35).

Major faults are the Castle Mountain fault and the Bruin Bay fault (fig. 36.4). Both faults trend northeast–southwest, but the Bruin Bay fault branches south from the Castle Mountain fault and skirts the coastline, east of Redoubt and Iliamna Volcanoes. The Castle Mountain fault passes on the west side of the volcanoes through the

Figure 36.4   Sketch map showing the association of major faults and volcanoes in Lake Clark National Park and Preserve and immediate vicinity.

trough occupied, in part, by Lake Clark. High-angle reverse movement has occurred on the Bruin Bay fault, which continues down the coast into Katmai National Park and Preserve. Movement on the Castle Mountain fault has been chiefly horizontal. The segment of the Castle Mountain fault that runs through the park/preserve (commonly referred to as the Lake Clark fault) follows a topographically prominent glacial trench that extends for many miles northeastward beyond Lake Clark and leaves the park/preserve after going through Lake Clark Pass. Shear zones and foliation in the granitic and metamorphic bedrock along the fault occupy a belt from 60 feet to as much as 725 feet wide. Tlikakila Wild River, which flows along the fault zone before emptying into the northern end of Lake Clark, is closely flanked by perpendicular cliffs and alpine glaciers descending from 10,000-foot-high, snow-capped mountains. Fault structures control the flow of a number of other rivers and glaciers in the park/preserve.

Geologic History

1. Jurassic intrusions.

Granitic magma, in a series of pulses (from 180 to 160 million years ago), was emplaced as a huge batholithic mass beneath a chain of volcanoes. The batholith was 800 miles long and 10 to 20 miles wide. The country rock, beds of Triassic age and older, were metamorphosed into marbles, quartzites, schists, and argillites. Andesitic lavas erupted, forming composite volcanoes. The Jurassic granitoids now form a major part of the Chigmit Mountains.

2. Uplift and erosion.

As the solidified batholith was uplifted, thousands of feet of volcanic debris and underlying granitic rocks were stripped off, transported by streams, and redeposited, mostly in an ocean. At Fossil Point, north of Tuxedni Bay, Jurassic marine fossils have been found, mainly ammonites and pelecypods. The area between Tuxedni Bay and

Table 36.1 Generalized Geologic Column, Lake Clark National Park and Preserve

Era	Period	Epoch	Rock Units	Geologic Events
Cenozoic	Quaternary	Holocene	Alluvial deposits, ash, glacial drift, landslide debris, talus, etc.	Glacial advances and retreats, stream erosion, mass wasting
				Ongoing volcanism, earthquakes, subduction
		Pleistocene	Glacial deposits	Several episodes of severe glaciation
			Andesites, breccias, pyroclastics	Construction of composite cones
	Tertiary	Late	Volcanic rocks, mostly andesitic flows	Beginning of glaciation
				Accumulation of volcanic rocks, pyroclastic debris
				Subduction, uplift, tilting
		Early	Conglomerates, sandstones, siltstones	Accumulation of marine and nonmarine sediments
			Basaltic and andesitic flows	Uplift, widespread volcanism
				Subduction
Mesozoic	Cretaceous		Regional unconformity 〰〰〰	Arrival of tectonostratigraphic terranes
	Jurassic	Late	Naknek Formation	Accumulation of thousands of feet of fossiliferous marine sediments, coarse- and fine-grained, mostly clastics
			Chitnitna Formation	
			Tuxedni Group	Uplift
		Middle	Quartz diorites, quartz monzonites, granodiorites, and other granitic rocks	Emplacement of Aleutian Range batholith
		Early	Talkeetna Formation	Andesitic tuffs, flows, breccias, mixed with clastic sediments
	Triassic		Undivided metamorphic rocks	Complexly folded marbles, quartzites, etc.

Sources: Detterman and Hartsock, 1966; Detterman and Reed, 1973, 1980; Péwé, 1975

Iniskin Bay (just south of the park) is one of the most complete sequences of Jurassic rocks in the United States. These rocks, which lie east of the Bruin Bay fault, are a mixture of volcanic breccia, tuff, siltstone, and gray-wacke.

3. Late Cretaceous and early Tertiary intrusions and volcanism.

Between 60 and 80 million years ago, coarse-grained igneous rocks were emplaced north of Iliamna Lake and west of the Jurassic batholith. A cluster of volcanoes were built up just outside the park/preserve in the Iliamna Lake area. The volcanic rocks in the western part of the park/preserve are early Tertiary (50 to 60 million years) in age. Thousands of feet of ash and welded tuff were explosively erupted from volcanic vents.

4. Accretion of terranes.

At the end of Mesozoic time and early in the Tertiary Period, terrane blocks, or microplates, that had "drifted" in from the south began to collide with the stable Alaska edge on the North American craton. Some of the material was subducted but most of these plates were "sutured" to Alaska, with major faults marking the plate boundaries. The Peninsular terrane is the most prominent block in the Katmai-Lake Clark area.

5. Middle Tertiary volcanism; construction of the present volcanoes, mainly in Quaternary time.

A period of quiescence followed the early Tertiary volcanism. Volcanic activity resumed in middle Tertiary time (about 35 million years ago) and has continued into the present. The volcano belt, and presumably the plate boundary, shifted to approximately the present position.

The volcanoes in the Lake Clark region, Mt.Spurr, Redoubt, Iliamna, and Augustine Volcanoes, all developed comparatively recently on top of older granitic rocks. (Only Redoubt and Iliamna are within park/preserve boundaries.) The explosive eruptions of highly siliceous ash and lava that has characterized these volcanoes in historic time is expected to continue. Dome building, especially in Redoubt and Augustine, is a typical eruptive pattern.

6. Pleistocene glaciation.

Wisconsinan glacial advances toward the end of the Pleistocene Epoch, which were extensive in this region, destroyed most of the evidence of earlier glacial stages. Many of the moraines that have confined or separated lakes were left by ice of Wisconsinan age. At their greatest extent, valley glaciers coalesced to form large lobes and piedmont glaciers. Glacial ice dammed Cook Inlet and extended to the Kenai Peninsula.

7. Neoglacial advance in Holocene time.

Glaciers in the park/preserve began readvancing around 4500 to 4800 years ago, and were still advancing up until about 200 years ago. Some glaciers have retreated about a quarter of a mile in the past 25 years. In the late 1930s, Lake Clark Pass was occupied by a glacier about five miles wide, but only parts of this former ice field remain.

Bibliography

de Laguna, F. 1974. Magical cave paintings from prehistoric Alaska. In M. Sherwood, editor. *The Cook Inlet collection.* Anchorage, Alaska: Alaska Northwest Publishing Company. p. 1–7.

Detterman, R. F., and Hartsock, J. K. 1966. *Geology of the Iniskin-Tuxedni region, Alaska.* U.S. Geological Survey Professional Paper 512. p. 1–78.

Detterman, R. F., and Reed, B. L. 1973. Surficial deposits of the Iliamna quadrangle, Alaska. U.S. Geological Survey Bulletin 1368A, p. A1–A64.

———. 1980. Stratigraphy, structure, and economic geology of the Iliamna quadrangle, Alaska. U.S. Geological Survey Bulletin 1368B, p. B1–B86.

Harris, D. and Kiver, E. P. 1985. *The geologic story of the national parks and monuments,* 4th edition. New York: John Wiley and sons. p. 65–69.

Hirshman, F., editor. 1986. Lake Clark and Iliamna country. *Alaska Geographic* 13 (4).

National Park Service. n.d. Geology of Lake Clark (unpublished).

———. 1984. Lake Clark National Park and Preserve, General management plan, environmental assessment.

Péwé, T. L. 1975. *Quaternary geology of Alaska.* U.S. Geological Survey Professional Paper 835.

Reed, B. L., Miesch, A. T., and Lanphere, M. A. 1983. Plutonic rocks of Jurassic age in the Alaska-Aleutian Range batholith: chemical variations and polarities. *Geological Society of America Bulletin* 94:1232–1240.

Yount, E. M., T. P. Miller, B. M. Gamble. 1986. The 1986 eruptions of Augustine Volcano, Alaska: Hazards and effects. In Hamilton, T. D. and J. P. Galloway, editors. *Geologic studies in Alaska by the U.S. Geological Survey during 1986.* Circular 998. p. 3–13.

Lake Clark National Park and Preserve

Area: 3,653,000 acres; 5708 square miles
Proclaimed a National Monument: December 1, 1978
Established as a National Park and Preserve: December 7, 1980
Address: 701 C Street, Box 61, Anchorage, Alaska 99513

37

Magma, melted by a "hot spot," rises from reservoirs beneath the ocean floor and builds Hawaiian volcanoes. Cones attain elevations above sea level of more than 10,000 feet. As oceanic crust passes slowly over the hot spot, old volcanoes become dormant; new ones form and grow from the ocean depths.

Frequent, "gentle" eruptions are characteristic of volcanic activity on Hawaii Island.

Lava may squirt in 100-foot-high fountains from volcanic vents, or cascade down slopes in rivers of incandescent rock.

Mauna Loa and Kilauea erupted together in 1984; Kilauea's eruptions have been numerous and at times continuous since a quiet period ended in the late 1950s. A new volcano, Loihi, is being born in the ocean south of Hawaii.

Earthquakes and precise measurements on upper slopes of the volcanoes yield data that allow forecasts of eruptive behavior.

Hawaii Volcanoes National Park

Island of Hawaii, Hawaii

Figure 37.1   Waves erode sea arches in lava flows on Hawaii's southeast coast, Hawaii Volcanoes National Park. National Park Service photo.

HAWAII

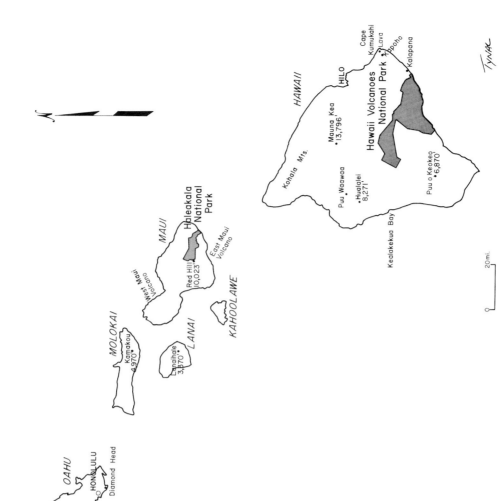

Figure 37.2 Islands of the state of Hawaii.

Box 37.1
The Geography and Geology of the Hawaiian Archipelago

The Hawaiian Archipelago in the Pacific Ocean is a group of shoals, atolls, and islands, stretching 1700 miles on a northwest-southeast linear trend, and a little over 2400 miles west of Los Angeles. The southeast part of the chain, made up of eight main islands and numerous islets, includes: Hawaii, at the southeast end, largest island and location of Hawaii Volcanoes National Park; Maui, the second largest island and site of Haleakala National Park (chapter 38); Kaho'olawe; Lana'i; Moloka'i; Oahu, where Honolulu, the state capital, is located; Kaua'i; and Ni'ihau (fig. 37.2).

Extending toward the northwest from Ni'ihau are the small, mostly uninhabited atolls and shoals of the Hawaiian Islands National Wildlife Refuge. The Midway Islands and Kure, at the far end of the chain, are some 1100 miles northwest of Honolulu. The state of Hawaii includes all the islands and reefs in the chain from Hawaii to Kure, except the Midway Islands, which remain under the administration of the U.S. Navy. Ka Lae (South Cape) at the southern tip of Hawaii is the southernmost point in the United States.

Climate. The subtropical climate is pleasant and mild. Air temperatures range from below freezing at high elevations to the high 90s on the lee coasts. Because the Hawaiian Islands lie in the path of the northeast trade winds, rainfall is heavy on the windward sides of the islands. Throughout most of the year, the trade winds blow across the islands; but occasionally in winter, winds from the south or southwest ("kona winds") bring heavy storms and may dump as much as 24 inches of rain in four hours. Maximum precipitation on the islands occurs between 2000 and 6000 feet above sea level on windward slopes (i.e., facing northeast), but rain shadow zones reduce precipitation on some islands. Annual rainfall ranges from about 10 inches on lee slopes to about 450 inches in the wettest belts. Stream densities are high and stream erosion is intense on the windward sides of the islands.

Geologic Development. The islands in this long chain consist mostly of inactive volcanoes built by basaltic lavas that punched through the crust beneath the ocean floor. Repeated eruptions of fluid, relatively thin lava flows (2 to 3 feet thick) brought the volcanoes up to the surface from depths of some 2.5 to 4 miles below sea level in less than a million years. Above water, their evolution as volcanic islands has followed a consistent pattern. Some reached greater heights than others; but regardless of how high above sea level the volcanoes grew, each one evolved into an "alkalic cap stage," which means that the basaltic lavas became poorer in silica and richer in sodium and potassium. Eruptions were less frequent and more explosive during this stage and lava flows became more viscous.

As eruptive activity waned, erosion cut deep valleys into the soft rock of the volcano slopes. On the older islands, Ni'ihau, Kaua'i, Oahu, Moloka'i, Maui, and Kaho'olawe, erosion was interrupted by renewed volcanism of even more alkalic and explosive lavas. Diamond Head on Oahu, for example, is the result of a period of renewed volcanism. As each volcanic island aged and was eroded, it sank as the oceanic crust beneath adjusted to the rapid growth of newer volcanoes in the chain. Coral reefs, like those on Midway, encircled the older volcanoes. With continued subsidence, waves eroded the tops of extinct volcanoes, and fringing reefs became atolls surrounding lagoons.

Radiogenic dates derived from rock samples taken at various places along the Hawaiian chain confirm that the volcanoes get progressively younger toward the southeast. Ages of principal lava flows, in millions of years, beginning with Kaua'i (the oldest) are 5.6 to 3.8; Oahu, 3.4 to 2.2; Moloka'i, 1.8 to 1.3; Maui, 1.3 to 0.8; and Hawaii, less than 0.7 and still growing (Tilling et al. 1987).

A seamount named Loihi, about 20 miles south of Hawaii Island, is expected to be the next island

continued

Box 37.1 continued

in the chain. Loihi has risen about 10,000 feet from the ocean floor and has grown to within 3100 feet of water level. Recent surveys show its shape as similar to Mauna Loa and Kilauea, with a flat summit area, a caldera, and two probable rift zones.

What caused the Hawaiian chain to form this linear pattern of older to younger volcanoes across the ocean floor? Since the volcanoes are erupting near the middle of the Pacific plate, more than 2000 miles from the nearest plate boundary, the volcanism cannot be associated with a subduction zone or tectonic activity occurring at plate margins. Moreover, the oceanic crust in that part of the Pacific is much older (Cretaceous) than the lava flows that constructed the Hawaiian volcanoes. Most of them are younger in age than Miocene.

The most likely hypothesis suggests the existence of an immobile *hot spot,* a thermal high at depth, perhaps as much as 200 miles across, that melts rock material just below the overriding Pacific plate. The magma produced accumulates in pockets that bulge up and eventually break through a stretched or weakened place in the oceanic crust. Hot lava then rises rapidly through conduits and fissures to the ocean floor. The release of great quantities of basaltic lava builds up shield volcanoes that rise above the water level in a surprisingly brief span of time. As the Pacific plate drifts slowly northwestward, successive volcanoes form over the hot spot, are rafted away on the moving plate, and become dormant as they are cut off from the source of fresh magma. The hot spot theory thus explains the basic pattern of a volcanic chain with active volcanoes at the southeast end and volcanoes becoming inactive or extinct toward the northwest.

Local History

Polynesian voyagers from the South Pacific settled the Hawaiian Islands. These courageous, seafaring people began arriving at the islands about 2000 years ago, their final migration being around A.D. 750. Volcanoes, believed to be the abodes of deities, were central to Polynesian culture and religion. Especially sacred were the islands of Maui and Hawaii, on which many temples were built. Pele, the goddess of all volcanoes, lived in Halema'uma'u, the fire pit in Kilauea caldera on Hawaii. Hawaiian legends about the origin of the islands are suggestive of modern scientific explanations. According to myth, Pele carried a magic stick that she used for digging when she wanted to start volcanic eruptions that would build up an island. After making Kaua'i, she went on to dig fire pits on Oahu, then Moloka'i, next Haleakala. Halema'uma'u on Kilauea was her final and eternal fire pit.

Captain James Cook discovered the islands in 1778 on a voyage of exploration for the British Admiralty. Because he arrived at the time of a religious ceremony, the Hawaiians thought he was a god; but when he visited the islands again on his return voyage, they realized that he was not a god and relations were less pleasant. At Kealakekua Bay on Hawaii, Cook got into an argument with some Hawaiians and was killed.

These were troubled times in the islands, then ruled by warring kings. In 1810, Kamehameha I became ruler of all the islands by conquering the other kings. Encouraged by Kamehameha I, American traders, whalers, and missionaries came to the islands in increasing numbers. The new arrivals brought with them infectious diseases, which reduced the Hawaiian population and brought about a decline of native cultural traditions.

The Americans encouraged Kamehameha III, who became king in 1825, to establish a constitutional monarchy. When G. P. Judd, an American missionary, became prime minister in 1842, the United States recognized Hawaii's independence, and American influence increased. The last Hawaiian monarch, Queen Liliuokalani, was overthrown in 1893, and seven years later Hawaii became an American territory. Statehood was attained in 1959.

From Captain Cook on, visitors to the islands have been fascinated by the Hawaiian volcanoes. Travelers from all over the world came to see the amazing spectacle of Halemaumau, the boiling lake of lava; among them was Mark Twain, who said that, by comparison, Vesuvius was a "soup kettle"!

James Dwight Dana, a young geologist attached to a U.S. Navy expedition commanded by Lieutenant Wilkes, conducted the first scientific investigation of the Hawaii volcanoes in 1840–41. Dana recognized the trend of decreasing age of the volcanoes from northwest to southeast because he observed differences in the degree of erosion that had taken place on the islands since eruptions had

Figure 37.3   Hawaii Volcanoes National Park, Island of Hawaii.

occurred. He theorized that a great fissure zone, from which lava erupted, might have opened on the ocean floor.

The movement to protect and preserve the land around the volcanoes on Maui and Hawaii Islands was led by L. A. Thurston, editor of the *Honolulu Advertiser.* Hawaii National Park, including land on both islands, was established by Congress in 1916. Haleakala National Park (chapter 38) on the island of Maui was set aside as a separate park in 1960. Hawaii Volcanoes National Park was designated in 1961.

As founder of the Hawaiian Volcano Research Association, Thurston was also active in helping to establish the Hawaii Volcano Observatory. It was formally organized in 1911 under the auspices of the Massachusetts Institute of Technology and the Carnegie Institute of Washington. Professor T. A. Jagger, a geologist and volcanologist at Massachusetts Institute of Technology, chose Kilauea as the site for the proposed observatory because of the constancy of the volcano's activity and because investigators could get close to active lava without exposing themselves to too much danger. Continuous recording of seismic and volcanic activity has been carried on at the Hawaii Volcano Observatory since 1912 under the successive sponsorship of the Massachusetts Institute of Technology, the U.S. Weather Bureau, the National Park Service, and (since 1956) the U.S. Geological Survey. The Observatory buildings are located in the park on the rim of Kilauea Crater, overlooking Halema'uma'u.

Included in the park is the summit area of Mauna Loa, a strip of land between Mauna Loa and Kilauea, and most of the southern flank of Kilauea, from the caldera to the coast (fig. 37.3). Mauna Loa and Kilauea make ideal natural laboratories for studying volcanic activity. They erupt frequently but seldom explosively; they emit enormous amounts of lava; and they have large, relatively simple structures.

Geologic Features

Five shield volcanoes make up the island of Hawaii, called "Big Island." Its area is 4030 square miles. Mauna Loa and Kilauea are both active volcanoes; Kohala, at the northern end of the island, and Mauna Kea, on Kohala's south side, have been inactive in historic time. Haulalai, on the west coast of Big Island, erupted last in 1801. Mauna Kea, the highest cone (elevation 13,784 feet), is also the world's highest peak on an island. At 13,680 feet, Mauna Loa is a close second in elevation.

Approaching the island of Hawaii, whether on board ship or in a plane, travelers see the summits of Mauna Loa and Mauna Kea from great distances if the weather is clear. The closer one gets, the more impressive their bulks appear as they rise from the ocean. Yet much more of their great size is hidden beneath the water; the ocean floor on

which the volcanoes stand is 18,000 feet below. Thus Mauna Loa's true height is 32,000 feet, or more than 6 miles! Kilauea, whose highest point is 4090 feet above sea level, is nearly 10,000 feet lower than Mauna Loa and Mauna Kea. All of the rock above sea level on Big Island has been erupted during the past million years, which gives us an idea of the rapidity with which this tremendous volume of lava has been extruded.

Because Mauna Loa and Kilauea have been fairly continuously active during historic time, many changes in their features have been documented. During the past century and a half, Mauna Loa has produced a lava flow on the average of every seven years and has exhibited minor eruptive activity in the years between. Its shield-shaped dome is about 70 miles long and 30 miles wide. Even more impressive is its volume of about 10,000 cubic miles, which makes it the world's largest active volcano. At the summit the Moku'aweoweo caldera coalesces with adjacent pit craters. Formed by collapse rather than explosion, the caldera has changed its size, shape, and depth frequently during historic time as lava flows have spread out over the caldera floor or breached the rim.

Rising from the southeastern flank of Mauna Loa is Kilauea, which has a shield dome about 50 miles long and 14 miles wide. Its caldera is 3 miles long and 2 miles wide (fig. 37.4). Within the caldera is the fire pit (active crater) of Halema'uma'u that from time to time contains a lake of liquid lava. A *lava lake* forms when magma rises within the volcano and floods the pit in the floor of the caldera. Sometimes a lava lake lasts for years, bubbling and glowing.*

If gas pressure raises the level of the molten rock high enough, lava spills out over the rim of the fire pit and spreads over the caldera floor. More often, lava breaks out through the flanks of the volcano above or below sea level. As lava pours out of the flanks, the level of the lake falls and the lake may disappear, perhaps for years at a time. If the whole plumbing system of the volcano is drained, then the caldera floor collapses. After a time, more magma rises into the reservoir, pushes up through the conduit, and the whole process—with modifications—may be repeated.

Volcanic Rocks that Form from the Lavas

The Hawaiian basalts are of two general types, or suites—*tholeiitic* and *alkalic,* both believed to be derived from the upper mantle, but perhaps from different levels. The tholeiitic basalts make up the bulk of the shields (99%). The

---

*In *Roughing It,* Mark Twain describes in awed terms seeing the lava lake after dark: "The greater part of the vast floor of the . . . [caldera] under us was as black as ink and apparently smooth and level; but over a mile square of it was ringed and streaked and striped with a thousand branching streams of liquid fire! . . . Imagine a coal-black sky shivered into a tangled network of angry fire." Mark Twain visited Halema'uma'u in 1866. The fiery "branching streams" he described were lava fountains.

Figure 37.4   Kilauea (foreground) and Mauna Loa (background) are active shield volcanoes. U.S. Geological Survey photo.

alkalic basalts (1%) are found mainly on the caps of the shields. The predominant phenocrysts in both types of Hawaiian basalts are olivine. Tholeiitic basalts commonly contain low- and high-calcium pyroxenes, both poor in aluminum. Pyroxene phenocrysts are rare in these rocks. Alkalic basalts contain only one pyroxene, a high-calcium and moderately high-aluminum variety that occurs commonly as phenocrysts as well as groundmass grains.

Types of Lava and Lava Features

Since most of the Hawaiian lavas are fluid, they flow from vents relatively quietly and congeal with a smooth, billowy or ropy surface called *pahoehoe* (pah-hoay-hoay). Typically, lava erupts in a series of pulses that build up layers of pahoehoe, one on top of the other. Sometimes a flow hardening as pahoehoe changes to another common type of congealing termed *aa* (ah-ah), which has a rough, jagged, or clinkery surface. The basalt in the interior of aa flows is dense and thick with many stretched vesicles left by escaping gas. The front of the flow moves along as a red-hot, clinkery wall. Massive aa flows may show columnar jointing. The change from a pahoehoe flow to

aa involves the initial gas content of the lava and changes in lava viscosity. An aa flow never reverts to pahoehoe, but a fresh pahoehoe flow may be erupted on top of an earlier aa flow (fig. 37.5).

When a lava lake is active, or when a flank vent is first erupting, gases bubbling up from depth produce *lava fountains* that may spout from a few inches to 30 feet in height, rising and falling on the hot lava surface. Occasionally lava fountains shoot up much higher, as happened at Kilauea Iki in 1959 when a lava fountain reached a height of 1900 feet, the highest ever observed anywhere in the world. Pu'u 'O'o, Kilauea's newest cone, which began erupting in 1983, delighted volcano-watchers with its exceptionally high and frequent fountaining.

During such high fountaining, *thread-lace scoria,* or *reticulite,* which is a light, feathery type of pumice, may form and be carried many miles downwind. Tiny strands and droplets of fountaining lava, also caught by the wind, harden in the air and fall. These glassy pellets and strings, called *Pele's tears* and *Pele's hair,* are fragile and short-lived.

Figure 37.5   Lava blocking the coast highway during 1986 eruptions from Pu'u 'O'o. *Top:* Pahoehoe lava; pens and white centerline give scale. *Bottom:* Aa lava crossing the highway a short distance away. Note that lava in the left foreground appears to be a mixture of pahoehoe and aa. Photos by M. S. Tuttle.

Lava tubes and *lava tunnels* beneath the congealing surface of a pahoehoe flow convey fresh, hot lava out to the moving front. Later, when the whole mass has cooled, the tunnels and tubes remain. A lava tube serves as a large conduit for the movement of lava down a slope far from the vent. The tube's outer crust acts as an insulator, keeping lava from cooling. The Thurston lava tube on the east side of Kilauea, which is 20 feet high and 22 feet wide in places, was formed by the crusting over of a pahoehoe flow from one of the Twin Craters. Some lava tubes contain *lava stalactites,* produced as liquid lava hardened while dripping from the tube roof.

Tree molds and *lava trees* may form in pahoehoe lava when a flow inundates a forest. Hot lava, chilled by its initial contact with a tree, either sticks up against the bark, hardening quickly, or flows around the trunk. After the charred tree rots away, the lava "trees" are left standing. Tree molds, resembling empty pails in the flow, show where tree trunks stood. Tree molds are numerous in Hawaii Volcanoes National Park and in Lava Tree State Park.

Mounds or hillocks on the surface of pahoehoe lava are called *tumuli* (singular, *tumulus*). They tend to form parallel to the flow front in places where lava is confined, especially in craters. Solid crust at the top of the flow is dragged along and domed up by the faster-moving fluid lava underneath. Sometimes fresh lava dribbles or spatters out, hardening into *driblet spires* of fantastic shapes.

Lava balls are accretionary features characteristic of aa flows and range in size from a few inches to 10 feet or more. They form when a fragment of solidified lava rolls along, enlarging as it picks up sticky lava. (Children in snowy regions make snowballs large in much the same way by rolling them in sticky snow until they're too big to push.)

An Hawaiian term, *kipuka,* meaning "opening," refers to islandlike areas of older land surrounded by more recently erupted lava. Kipukas, ranging in size from a few square feet to several square miles, are usually the result of irregularities in the land surface. A kipuka surface may be higher or lower than the level of the surrounding lava, the latter case being more common. New lava is sterile and barren of vegetation, but some kipukas have weathered long enough to have soil that can support plant life, making them islands of greenery in a landscape of bare volcanic rock. Several kipukas are in the park. The best known is Kipuka Puaulu, or Bird Park, northwest of Hawaii Volcano Observatory.

Ash Explosions and Pyroclastics

Kilauea has been an ash producer, probably because more moisture is trapped on its windward slopes. Ash explosions are commonly associated with the sudden mixing of ground (phreatic) water and hot lava or magma. Violent explosions have occurred when water drained down into newly emptied vents or conduits and reached magma. The result has been a *phreatomagmatic explosion.* In 1790 such an explosion killed a band of Hawaiian warriors who were marching across the Ka'u desert (a rain shadow zone on the southwest slope of Kilauea). Despite years of weathering, the warriors' footprints are still faintly visible on an ash surface close to the park's western boundary.

Phreatic explosions occur any time that hot lava encounters ground water and turns it to steam. The most explosive steam explosions in historic time occurred during the 1924 eruptions from Kilauea when large blocks of old lava (that can be seen scattered around the summit area) were ejected from the volcano's main vent. More typically, phreatic explosions produce ash, cinder cones, and pyroclastics of various sizes, especially at or near the shore where moisture is plentiful. Most cones, such as those along Kilauea's east rift zone, are a combination of cinders and spatters.

Rift Zones on the Volcanoes

Hawaiian shield volcanoes are built up not only by summit eruptions but also by eruptions from *rift zones* that radiate from the summit calderas and extend down into the ocean. Their orientation is influenced by the buttressing effects of neighboring volcanoes and by gravitational stresses. The rift zones are scored at the surface by many cracks, some of which have served as vents for erupting lava. Below the surface, dikes of congealed lava fill fissures. Each upwelling of magma from depth causes swelling of the summit and flanks, which creates new stresses on the rift zones and opens new cracks and fissures or reopens old ones. An eruption relieves pressure in the magma reservoir, and the volcano then deflates.

Mauna Loa and Kilauea each have two principal rift zones (fig. 37.6). Mauna Loa's main growth has been along the two principal rift zones that extend from the caldera to the southwest and to the northeast. Kilauea's south flank is pushing into the sea, growing a few inches per year due to repeated intrusions of magma into the southwest and east rift zones. Since it began erupting in 1983 from the east rift zone, Pu'u 'O'o has built its cone to a height of more than 830 feet, making it the highest volcanic landform that has grown and been monitored by scientists in historic time. In 1986, after the Pu'u 'O'o conduit ruptured, lava erupted to the east of the cone from a new vent and began to build up a shield. Flows from the vent, named Kupaianaha, formed a system of lava tubes that carried fresh lava over a fault scarp and down to the ocean.

Figure 37.6   Major rift zones associated with Mauna Loa and Kilauea and the Hilina fault system. Note location of Loihi, offshore and still beneath the surface. (From Tilling, Heliker, and Wright 1987).

Fault Systems

Rift zone eruptions tend to make the volcano flanks out of balance or unstable. Earthquakes accompanied by sudden movements along fault lines release the accumulated strain. A fault system between Mauna Loa and Kilauea appears to act as an "expansion joint" between the two volcanoes. A jolting 6.6 earthquake on this fault system in November 1983 damaged nearby roads, trails, and some buildings. Along Mauna Loa's southeastern base, step faulting, by which the mountain side has moved downward in response to gravity, has produced a series of fault scarps (called "palis") facing southeast.

Similar step faulting takes place from time to time in the Hilina fault system along Kilauea's southern slope. Here the fault blocks form a series of steps down to the coast, with the trend being parallel to the shore. Measurements after a magnitude 7.2 earthquake in November 1975 revealed that some rocks along the Hilina fault had dropped 11 feet and shifted southward as much as 24 feet (Tilling 1987). The fault scarps of the Hilina fault system show up on Kilauea's south flank as steep cliffs overlooking the sea. Lava flows have cascaded down over the fault scarps into the ocean.

Shore Features and Submarine Features

Lava flows that reach the ocean surface produce cinders and ash when the hot lava mixes with the sea water. Cinder cones and benches are formed when lava enters the ocean. Waves erode and transport volcanic sediment, depositing "black sands" on beaches. Since the source of supply is limited and some sand is washed out to sea each year, the black sands may last only two or three centuries.

Sea cliffs and sea arches (fig. 37.1) are undergoing wave erosion and are retreating. In some places along the shore, recent lava flows protect the cliffs and have halted their retreat temporarily.

Most of the submarine part of Hawaii volcanoes consists of pillow lavas. When fluid lava erupts under water or flows from a tube that opens in the ocean, *pillow structures* may form. With water removing heat rapidly, the

Box 37.2
Forecasting Eruptions

Crisscrossing the island of Hawaii is a network of scientific instruments for monitoring the volcanoes. Collecting and analyzing data and making direct observations, are the responsibility of volcanologists at the Hawaii Volcano Observatory and the University of Hawaii. Tied in with this network are seismic stations and investigators in other parts of Hawaii, Japan, New Zealand, and many locations throughout the world.

What are changes or signs that scientists look at? What combinations of factors indicate an impending eruption? In suggesting lines of investigation, Macdonald (1972) emphasizes that "no single tool is adequate to predict an eruption except in the most general way" (p. 418).

1. *Periodicity.* What is the eruptive history of the volcano? How long are its quiet periods? Such information is useful in a general way, although eruptive patterns of volcanoes may change unexpectedly.
2. *Changes in the behavior of fumaroles and in the output, composition, and temperature of gas.* Steam and gas emissions are affected by atmospheric temperatures and humidity, but observations in this category are useful. On Kilauea, for example, an increase in the hydrogen content of gas emissions usually signals the onset of an eruption.
3. *Shifts in the orientation of the volcano's magnetic field and reduction of magnetic attraction of rocks due to increasing heat.* Changes in magnetism are difficult to interpret, but they appear to indicate movement of magma within the interior of a volcano. Often an eruption does follow such movement, but

sometimes a volcano quiets down without an extrusive event occurring.

4. *Infrared photographs and remote sensing.* Increases in heat and changes in the locations of hot areas show up when these techniques are used. Any eruptive activity is invariably preceded by an increase in heat on the volcano's surface.
5. *Geoelectrical changes.* Variations in electrical currents in a volcano have been observed just before explosions or eruptions, suggesting that geoelectricity may be useful as an indicator.
6. *Tumescence and tilting.* Kilauea and Mauna Loa have histories of swelling and tilting before eruptions, but occasionally swelling goes down without an eruption occurring. Land deformation on any active volcano is considered an ominous sign. The cause of bulging, rising of the summit, and tilting is assumed to be magma filling the volcano reservoir and rising in the conduits. Tiltmeters on Kilauea and Mauna Loa continuously monitor land surface deformation as well as migration of the center of a tumescent area.
7. *Changes in seismicity.* From earliest times, people who lived near volcanoes associated earthquakes with eruptions. On Hawaii Island, certain kinds of earth tremors and rumblings were called "Pele's thunder." These were regarded as warning signals of her anger.

Earthquakes are frequent on Big Island; many of them are so slight that they are imperceptible to residents. Most of the earthquakes are shallow, local, and indicate a shifting around of rocks in

continued

lava quickly crusts over and solidifies as a pillowlike or tonguelike protrusion. More lava squeezes out (like toothpaste from a tube) and stacks another "pillow" on top of the first. The process is repeated many times as the lava flow advances under water. Because the lava is under

pressure from the weight of the overlying water, explosive interactions seldom occur between the cold ocean water and the hot lava. In recent years scuba-diving scientists have observed and filmed the forming of pillow structures during underwater eruptions of lava near Big Island.

Box 37.2 continued

response to stresses in the volcanic structures. On Maui and the other Hawaiian Islands, earthquakes are less common and are probably caused by isostatic adjustments of the oceanic crust to the load of overlying volcanoes.

Seismic data collected on Hawaii have been studied for more than 70 years by scientists seeking clues as to the imminence of eruptions. Several significant patterns have shown up, but they must be combined with other observations to be meaningful. Earthquake "swarms," consisting of thousands of little earthquakes, are recorded on seismographs before eruptions occur on Mauna Loa and Kilauea. These quakes, called *harmonic tremors,* are probably the result of minor faulting that accompanies tumescence or the inflation and pushing up of the volcano top. Fissures and troughs may develop along the rift zones. After the magma has drained and the summit starts to shrink, quakes begin again, sometimes stronger than those that accompany the swelling. Evidence from tiltmeters may confirm whether the volcano is shrinking or possibly swelling again. This is important in Hawaii where eruptive periods last for weeks, months, or years.

Forecasts regarding eruptions are necessarily based on several lines of evidence and are expressed in probabilistic terms. Nevertheless, even though anticipating volcano behavior remains an inexact science, forecasts can be made that can help to prevent loss of lives when dangerous areas need to be evacuated.

The active Hawaiian volcanoes, which are regarded as less hazardous than most other active volcanoes, are seldom explosive. The Cascade and Alaskan volcanoes, as we have seen in previous chapters, tend to be erratic, more explosive, and much less predictable. However, perhaps at some future time, scientists may be able to understand and forecast their behavior if the body of data about the West Coast volcanoes becomes comparable in scope to observations of Hawaiian volcanoes.

Geologic History

1. Initial eruptions on the sea floor, probably early in Pleistocene time.

The age of the rock at the base of the volcanic pile has not been determined. Putting down drill holes in order to get core samples for age determination is a difficult and costly operation. Moreover, in some locations, drill bits might hit magma that would melt them.

2. Construction above sea level of Hawaii Island.

Five shield volcanoes, each with its own magma reservoir but all having the same magma source in the upper mantle, built up the island during Pleistocene time. Kilauea, the youngest volcano, is still enlarging at the southeastern end of the island. South of Hawaii, Loihi seamount is rising. It is regarded as the newest surface expression of the hot spot beneath the Pacific plate.

3. The development of Mauna Loa.

Rock units. Mauna Loa's rocks are mainly olivine basalts. The oldest that are exposed are the lavas of the Ninole Formation that make up the core of the mountain. Ninole rocks crop out in low hills on Mauna Loa's southeast slope. Deep canyons were cut in these lavas during an erosional period, producing an unconformity between Ninole beds and the overlying lavas of the Kahuku Formation. The Kahuku lavas partially filled the canyons and built up the slopes of Mauna Loa. Pahala ash, 5 to 50 feet thick, mantles the Kahuku rocks. Most of the Pahala ash in this area came from Kilauea and was probably the result of violent phreatomagmatic explosions that accompanied the collapse of Kilauea's caldera. Much of the Pahala ash has been covered by younger lavas of the Ka'u Formation. The fairly fresh and relatively unweathered rocks of the Ka'u include lavas from eruptions during historic time. Mauna Loa had probably attained its present height by the end of the Pleistocene Epoch.

Eruptions in historic time. Mauna Loa's eruptions since early in the nineteenth century have been mainly of two types: (1) summit eruptions without flank flows and (2) the more typical flows from the rift zones that may be accompanied by brief summit eruptions. Eruptions generally begin with harmonic tremors and the release of clouds of steam and gas. Lava fountains may shoot up many feet in Mokuaweoweo, the summit caldera. Strands of Pele's hair are blown out and away by uprushing winds. During a summit eruption, lava pours from cracks in the caldera wall and spreads over the floor. Pit craters may develop or enlarge in and around the caldera.

When eruptions begin in the rift zones, lava and hot rising gases shoot up, making a "curtain of fire" along the

Table 37.1 Geologic Column, Hawaii Volcanoes National Park

Time Units			Rock Units and Geologic Events	
Era	Period	Epoch	Mauna Loa	Kilauea
Cenozoic	Quaternary	Holocene — 'Historic" / "Prehistoric" / ? — Pleistocene	*Younger Ka'u Formation* Summit eruptions, 1975, 1984 Numerous summit and rift zone eruptions, 1832–1950 Continuing shield construction by flows of both aa and pahoehoe Enlargement of caldera by pit craters *Older Ka'u Formation* Shield construction by aa and pahoehoe; cinder cones, spatter cones, etc. Development and collapse of Mokuaweoweo caldera; pit craters *Kahuku Formation,* capped by Pahala ash Building up of shield by thick accumulations of aa and pahoehoe	*Younger Puna Formation* Construction of satellite cones Mauna Ulu, Kilauea Iki, Pu'u 'O'o since 1968 Numerous summit and east and west rift zone eruptions, 1790 to present Continuing shield construction by flows of aa and pahoehoe Explosive eruptions; 1790, 1924 Appearance, disappearance, reappearance of lava lakes; enlargement of caldera *Older Puna Formation* Shield construction by aa and pahoehoe flows; ash falls, cinder cones, spatter cones Development and collapse of Halemaumau in summit caldera; pit craters *Hilina Formation,* capped by Pahala ash Beginning of shield construction by many flows, both aa and pahoehoe, building up thick accumulations; some ash layers, indicating past explosive eruptions
			Great erosional unconformity Fluctuations of sea level and submergence; intense erosion of shield; canyon cutting	
			Ninole Formation Construction of Ninole volcano Volcanic "basement" rocks (not exposed at surface); age unknown	Volcanic "basement" rocks (not exposed at surface); age unknown

Source: modified after Macdonald and Abbott 1970; Tilling et al. 1987

length of a fissure opening. Clots, blobs of pumice, and threads of volcanic glass harden in the air and then fall glistening to the ground. During the first few days, very fluid pahoehoe pours down over the slopes like a slow-moving river (10 to 25 miles per hour). Later flows may change to aa.

The 1859 eruption, which lasted for 10 months, produced a 33-mile flow, the longest recorded in historic time. Several eruptions from the northeast rift zone have endangered the city of Hilo and its harbor on the eastern coast of the island. The closest call was a pahoehoe flow 29 miles long in 1881 that stopped on Hilo's outskirts. In 1935 and again in 1942, planes dropped bombs on flows that appeared headed for Hilo. The flows stopped well before reaching Hilo, but geologists believe that this was because the supplies of fresh lava were exhausted.

A major eruption of Mauna Loa took place in June 1950. Lava first began erupting near the caldera at the top of the southwest rift zone. In minutes the erupting fissures had opened up the southwest rift zone for several miles down the slope. Within 24 hours, lava fountains blazed in a curtain of fire down the mountainside for 8 miles. On both sides of the rift ridge, rivers of lava poured down the slopes. Three flows reached the ocean. The volume of lava (600,000,000 cubic yards) that was extruded during the three weeks of the 1950 eruption equaled the amount of lava that erupted in the 10-month eruption of 1859. In terms of volume, these two eruptions are the largest that have been recorded on the island of Hawaii. Although Mauna Loa has been quiet most of the time since 1950, major eruptions might still occur.

4. The development of Kilauea.

Rock units. Eruptions of Hilina Formation lavas from the southeast flank of Mauna Loa began building Kilauea in about the middle of Pleistocene time. The Hilina lavas are presumed to be roughly equivalent in age to the

Kahuku Formation on Mauna Loa. Hilina basalts are exposed in the fault scarps along the southern coast. Flows of both pahoehoe and aa can be distinguished.

Pahala ash overlies the Hilina Formation. Above the Pahala ash are rocks of the younger Puna Formation. Puna lavas were erupting at about the same time as the Ka'u lavas on Mauna Loa, and the flows interfinger in the area between the two volcanoes. Puna flows were erupted largely from vents in Kilauea's caldera and from the rift zones. Spatter and cinder cones formed on both the southwestern and eastern rift zones. Those on the eastern rift zone were more numerous.

Eruptions in Historic Time. The types of activity have been (1) flank flows, more often than not preceded by summit activity or eruption; (2) shield construction; (3) rising and falling of lava lakes; (4) brief summit eruptions; and (5) violent explosive eruptions.

The lava lake in the fire pit on Kilauea's summit apparently filled not long after the violent explosion of 1790. Throughout the 1800s and the first quarter of this century, the lava lake was almost continuously active; although some sinking, followed by refilling, occurred from time to time after lava had been drained by flank eruptions below sea level. The lava lake was at a high level at the beginning of 1924 when a series of earthquakes began. They first occurred at the summit and then shifted outward along the eastern rift zone, suggesting that magma was moving in that direction. Meanwhile, the level of the lake dropped several hundred feet. The explosions began in May and lasted for about three weeks. Ash-filled clouds of steam and gas shot up. Sulfur fumes killed vegetation over a wide area. Blocks of caldera rock were thrown a mile into the air. Local earthquakes, lightning, and heavy downpours of rain accompanied the explosions. Afterwards Halema'uma'u had increased in diameter from 500 feet to over 3000 feet, and the floor of the caldera had sunk 11 feet. Even the caldera rim was lowered a few feet. What caused the explosions and collapse was ground water pouring into the hot vents as lava drained away. Instantaneous boiling resulted, and then the violent escape of the steam. The explosions were not magmatic since no fresh lava was ejected with volcanic debris.

Since 1924, the lava lake that had bubbled and flamed for so many years has refilled partially, but only for very brief intervals. The fire pit is quiet except for wisps of steam. Summit eruptions have been short and infrequent. Flank eruptions, on the other hand, have been numerous and sustained. One summit eruption, however, was spectacular. In 1959, Kilauea Iki, a large pit crater close to the caldera, had a major eruption. The lava fountains that shot up were the highest on record, and cinders and pumice devastated a nearby forest.

Before the lava lake disappeared, the southwest rift zone was more active than the eastern rift zone. Mauna Iki is a secondary lava cone southwest of the caldera that was built by an eruption in 1920. Fresh lava flowed out near Mauna Iki in 1971.

The east rift zone has been exceedingly active for many years. From the caldera, the east rift zone goes southeastward and then takes a sharp bend almost due east to Cape Kumukahi and on out into the ocean. The Chain of Craters, on the upper part of the eastern rift zone, used to be a row of 13 pit craters coming down from the caldera. A highway went through the area from the Crater Rim road southeastward to the coast. The road was buried by lava in 1969 when an entirely new vent opened up and Mauna Ulu, a satellite cone, began to grow. Mauna Ulu continued to erupt until 1974, and many of its flows reached the ocean. Next to Mauna Ulu is the Alae shield, which was built up over an old pit crater that was 440 feet deep. Other nearby craters were partly filled or buried, and some reopened and began pouring out lava or filling up and forming lava lakes.

The following year, in 1960, lava erupted 28 miles away from the summit near Cape Kumukahi at the extreme eastern tip of the island. Lava flowing from the eastern rift zone destroyed the village of Kapoho and created some 500 acres of new land where the flow went into the ocean. (The question of who owns new land has been in litigation.)

The Pu'u 'O'o eruptive series, which has been continuously active since 1983, has become the longest and largest eruption in Kilauea's history. Flows from Pu'u 'O'o and its associated vent Kupaianaha have constructed a new shield and added new land to the island where lava has entered the ocean. The coastal community of Kalapana, which lay in the path of the flows, has been wiped out by repeated lava flows. More than 160 homes and businesses have been destroyed, as well as a National Park Visitor Center, archeological sites, roads, trails, and other structures. Eruptions continue as of this writing (1991).

5. Glaciation.

A small icecap existed on Mauna Kea, Hawaii's highest volcano, during the Pleistocene Epoch. Glaciers may have been present on Mauna Loa at the same time, but all evidence has been destroyed or buried by recent lava flows. In today's climate, snow falls every winter on Mauna Kea and Mauna Loa, but does not last long.

6. Changes in sea level.

Old shorelines, some higher than the ocean's present stand and some lower, testify to the fluctuating ocean level around the Hawaiian Islands during their comparatively brief geologic history. The most significant long-term trend

Figure 37.7   Pahoehoe lava from a 1986 eruption "paves" a street and isolates a mailbox near the park.
Photo by M. S. Tuttle.

Figure 37.8   Children play on a tractor that was caught in lava from one of the 1986 eruptions. Photo by M. S. Tuttle.

is land submergence. As the volcanic pile has gotten higher and heavier, the sea floor has tended to sink. However, older sea levels are not consistent from place to place because local tilting and warping have occurred. Thus tectonic changes have affected the relative position of land and ocean. The islands' shorelines were also affected by worldwide changes in sea level that occurred during Pleistocene time.

Geologic Maps and Cross Sections

Macdonald, G. A. 1971. Geologic map of the Mauna Loa quadrangle, Hawaii. U.S. Geological Survey Map GQ-897.

Peterson, D. W. 1967. Geologic map of the Kilauea Crater quadrangle, Hawaii. U.S. Geological Survey Map GQ 667.

Bibliography

Dalrymple, G. B., Silver, E. I., and Jackson, E. D. 1973. Origin of the Hawaiian Islands. *American Scientist* 61 (May–June):294–308.

Decker, R. W., Wright, T. L., and Stauffer, P. H. (editors). 1987. *Volcanism in Hawaii*. U.S. Geological Survey Professional Paper 1350, 2 vol. 1667 p.

Decker, B., and Decker, R. W. 1987, 2nd edition. *Road guide to Hawaii Volcanoes National Park*. Mariposa, California: Double Decker Press.

Hazlett, R. W. 1990. *Geological field guide, Kilauea Volcano*. Hawaii Natural History Association.

Heliker, C. 1990. *Volcanic and seismic hazards on the Island of Hawaii*. U.S. Geological Survey general interest publication.

————, Griggs, J. D., Takahashi, J. T., and Wright, T. L. 1986. Volcano monitoring at the U.S. Geological Survey's Hawaiian Volcano Observatory, *Earthquakes and volcanoes* 18 (1):3–71. U.S. Geological Survey.

Herbert, D.; and Bardossi, F. 1968. *Kilauea: case history of a volcano*. New York: Harper & Row Publishers.

Kaye, G. 1976. *Hawaii Volcanoes, the story behind the scenery*. Las Vegas, Nevada: KC Publications.

Macdonald, G. A. 1972. *Volcanoes*. Englewood Cliffs, New Jersey: Prentice-Hall.

————, and Hubbard, D. H. 1989, 9th edition (revised by Jon W. Erickson. *Volcanoes of the national parks in Hawaii*. Hawaii Natural History Association and National Park Sevice.

————, Abbott, A. T., and Peterson, F. L. 1983, 2nd edition. *Volcanoes in the sea: The geology of Hawaii*. Honolulu: University of Hawaii Press.

Stearns, H. T. 1966. *Geology of the state of Hawaii*. Palo Alto, California: Pacific Books.

————. 1978, 2nd edition. *Road guide to points of geologic interest in the Hawaiian Islands*. Palo Alto, California: Pacific Press.

————, and Macdonald, G. A. 1946. Geology and ground water resources of the Island of Hawaii. Hawaii Division of Hydrography Bulletin 9.

Swanson, D. A., and Peterson, D. W. 1972. Partial draining and crustal subsidence of Alae lava lake, Kilauea Volcano, Hawaii. U.S. Geological Survey Professional Paper 800-C.

Tilling, R. I., Heliker, C., and Wright, T. L. 1987. *Eruptions of Hawaiian volcanoes: Past, present, and future*. U.S. Geological Survey, general interest publication. 55 p.

Hawaii Volcanoes National Park

Size: 229,177 acres; 358 square miles
Wilderness area, 123,100
Established as part of Hawaii National Park:
    August 1, 1916
Redesignated Hawaii Volcanoes National Park:
    September 22, 1961
Wilderness designated: October 20, 1976
Designated a Biosphere Reserve: 1980
Listed as a World Heritage Geologic Site: 1990
Address: Hawaii National Park, Hawaii 96718

38

Spectacularly eroded summit "crater" of 10,023-foot-high Haleakala Volcano.
Youngest lava flow from Haleakala erupted c. 1790.
Kipahulu Valley, with its unique and fragile ecosystem.
Scenic pools in 'Ohe'o Gulch.

Haleakala National Park

Island of Maui, Hawaii

The island of Maui is made up of two old volcanoes. The taller of the two, with its mid-slopes wrapped in clouds and its lower levels lost in the haze above the sea, often seems to be floating.

"The Maui Scene" (Kyselka and Lanterman 1980)

Figure 38.1   Haleakala Crater in Haleakala National Park is not a true crater or caldera because it was formed by erosion rather than by volcanic explosions. The view is from an overlook on Haleakala's west rim. National Park Service photo.

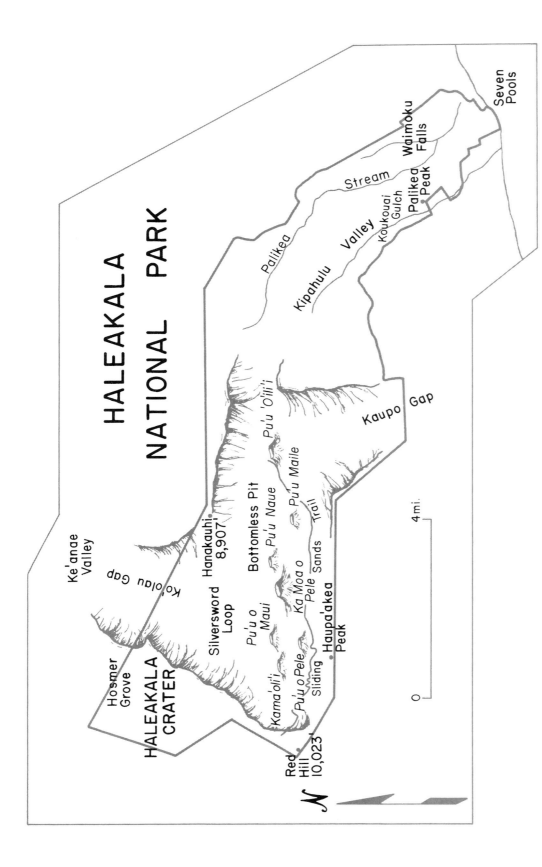

Figure 38.2 Haleakala National Park, Island of Maui, Hawaii.

Local History

Haleakala National Park is on Maui, the next island northwest of Hawaii Island, where Hawaii Volcanoes National Park is located. (See box 37.1 in chapter 37.) Maui was a religious and cultural center for the Polynesian people who settled the Hawaiian Islands. In the Hawaiian language, *Haleakala* (ha-lay-ah-ka-lah) means "House of the Sun."

Lahaina, a port on the west side of Maui and the site of the first nonnative settlement in the Hawaiian Islands, was an important whaling center in the nineteenth century. Humpback whales spend the winter months in the channel waters between Maui and the neighboring islands of Lana'i, Kaho'olawe, and Moloka'i.

Originally part of Hawaii National Park, Haleakala National Park was redesignated as a separate unit in 1960. Most of the park is contained within Haleakala Crater; but in 1969 the Kipahulu Valley was added, which connects the crater section with the ocean. This narrow corridor was set aside so that the fragile ecosystem of Kipahulu Valley, with its rare and endangered species of plants and birds, could be protected. At least 25 bird species that lived on Maui before 1900 became extinct after mosquitoes carrying bird malaria arrived on a ship from Central America. The forested slopes of the Kipahulu Valley are not open to the public but are kept in a natural state for scientific study only. The park was designated a Biosphere Reserve in 1980.

"Climatic layering" on Maui produces unique weather patterns and tends to control the prevalence of or lack of vegetation. Maximum precipitation is greatest on northeastern slopes between the 2000- and 6000-foot levels, where the trade winds drop more than 300 inches of rain per year. Precipitation, clouds, and fog often envelop the middle-level slopes of Haleakala and are responsible for the luxuriant growth in the rain forests. The Kipahulu Valley is a notable example. Above 6000 feet and on the leeward slopes, the environment tends to be semiarid. From the air, one sees heavy forests on the lower and middle flanks of the volcano and barrenness at the higher elevations around the summit and crater.

An additional factor causing layering of vegetation is the extreme porosity of the volcanic rocks in the crater. Rain water drains away so fast that not much soil develops and vegetation is sparse. Runoff moves loose volcanic debris down the inner slopes of the crater and into the stream valleys.

Among the rocks in the desertlike crater and on the rim grows the tall silversword, a rare, yuccalike plant, native to Hawaii, that has a mass of silvery, saberlike leaves. At maturity, the plant produces hundreds of purple flowers and then dies after the seeds ripen. A hundred years ago, the silversword covered acres of the crater floor, but nearly became extinct before coming under National Park Service protection. Also protected is a flock of nene (nay-nay) (Hawaiian geese), which were close to extinction early in the century. The nene is Hawaii's state bird.

Geologic Features

Maui's two eroded volcanoes, Haleakala and West Maui, were once part of a large volcanic massif made up of six major volcanoes. When sea level was lower, during the coldest times of the Pleistocene Epoch, the six volcanoes formed one island. Now there are four islands—Maui, Kaho'olawe, Lana'i, and Moloka'i—separated by channels.

Maui's volcanoes are older and more deeply eroded than the great volcanic shield complex making up the island of Hawaii (chapter 37). In map view, Maui's appearance is like an asymmetrical dumbbell, with the smaller West Maui volcano on one side and the larger East Maui volcano, Haleakala, on the other side. Joining the two volcanoes is a low isthmus constructed of lavas from Haleakala and sandy sediments that accumulated in dunes during the Ice Age and have since been lithified. A pumping station located on the isthmus supplies Maui's water, which is drawn from a fresh-water lens beneath Haleakala and the isthmus. The fresh water floats on top of the denser salt water that saturates the entire base of the island.

West Maui is 5788 feet high and 19 miles across. By contrast, on East Maui, the rim of Haleakala Crater rises to over 10,000 feet and the volcano is about 33 miles across. Maui's total area is 728 square miles. The highest point on the rim, Red Hill (elevation 10,023 feet), is one of three prominent cinder cones at the summit of Haleakala. The National Park Service maintains geological exhibits at overlooks on the rim. Nearby, but outside the park, is Mees Observatory with telescopes operated by the University of Hawaii Institute for Astronomy.

The so-called crater, the dominant topographic feature of the island, covers 19 square miles and is several thousand feet deep. Its inward-facing slopes are much steeper than its outer flanks. Because it looks like a crater or caldera, this impressive landform will probably always be called "Haleakala Crater," even though by geologic definition it is neither a crater nor a caldera (fig. 38.1).

Haleakala had a real crater originally and was once much higher. But after the volcano became inactive, very rapid stream erosion removed the youngest lava flows, ash, and cindery material. Because rainfall is heaviest on north- and east-facing slopes, due to the trade winds, streams draining those areas were able to erode faster and capture the headwaters of lesser streams. Several master valleys developed; two of them—by cutting headward into the mountain's core—finally merged, creating a huge depression on the summit. A low dividing ridge across the floor

Figure 38.3   Small cinder cones erupted on the floor of Haleakala Crater when old rift zones reopened, long after the volcano's summit had been deeply eroded. National Park Service photo.

of this calderalike feature separated the two drainage basins, the Kaupo Valley on the southeast and Ke'anae Valley toward the northeast.

Volcanic activity resumed after the erosional period. The old rift zones of the volcano reopened. Lava pouring from fissures on the sides of the surrounding walls and from the crater floor flowed down the eroded stream valleys. An enormous mudflow, along with cinder blocks and lava, went through the Kaupo Gap and down to the sea. Cinder cones, constructed of lava, pumice, cinders, blocks, ash, and obsidian, built up on the floor of the depression, with some cones rising as much as 600 feet in height (fig. 38.3). Sizable volcanic bombs, formed from molten fragments that cooled before striking the ground are strewn about. Bizarre spatter cones formed, such as Pele's Pigpen and Bottomless Pit, the latter named for its deep vent. Lava tubes like Holua tube were left when pahoehoe lava cooled on the outside while fresh, hot lava on the inside drained away. Bubble-shaped caves in the lava are the result of gases expanding and lava hardening over the bubble instead of collapsing.

One of the striking features of Haleakala Crater is that the rocks in the steep walls and in many of the features of the floor are multicolored. One reason for this is that when the sun's rays strike the rocks at different angles during the day, the colors vary as the sun's position changes. Also, volcanic gases have left varicolored crusts on the rocks due to precipitation and sublimation. Chemical weathering of the basaltic rocks adds still more colors to the landscape as iron-bearing minerals are oxidized and hydrated. The prevailing warm temperatures and frequent mists and rains intensify the effects of weathering processes on these easily weathered rocks.

Most of the streams on Maui flow only during and right after rains. These are *ephemeral streams*. On the windward slopes, however, streams cascading down over volcanic cliffs and ledges have created beautiful pools and waterfalls that are cut in the valley-filling lavas. Hikers and swimmers who visit the Seven Pools of 'Ohe'o Gulch in the lower Kipahulu Valley on the eastern shore of Maui, can see interesting lava outcrops, birds, and flowers. (Actually, the number of pools—usually *more* than seven—depends on the amount of water in 'Ohe'o Stream. After heavy rains, the stream is a torrent, and no pools can be seen!) In the rugged sea cliffs along the shore, older and younger lavas are exposed, as well as partially buried cinder cones and other volcanic features.

Table 38.1 Geologic Column, Haleakala National Park

Time Units			Rock Units	Geologic Events
Era	Period	Epoch		
Cenozoic	Quaternary	"Historic time" Holocene —?—	Hana Formation	Small lava flows c. 1790
				Valley-filling lavas, crater lavas, ash, cinder cones, spatter cones
			Mudflow debris	Kaupo mudflow
		Pleistocene	Accumulations of sediment	Landsliding, canyon-cutting; development of amphitheaters and Haleakala Crater by stream action, frost action, mass wasting
			Great erosional unconformity	Fluctuation of sea level and submergence
	—?— —?— Tertiary / Neogene	—?— Pliocene	Kula Formation	Cinder cones, ash; alkalic cap construction Extensive aa and pahoehoe flows
			Kumu'iliahi Formation	
			Honomanu Formation	Build-up of shield by accumulations of pahoehoe and aa flows
			Volcanic "basement" rocks (not exposed at surface)	Lava flows building up from the ocean floor

Source: modified after Stearns 1966; Macdonald 1978

Geologic History

1. Construction of the volcanic massif.

In late Pliocene time, the shields of six volcanoes were built up above sea level from a basaltic lava foundation on the ocean floor. As the volcanoes continued to erupt, they coalesced as a great volcanic complex.

2. Eruption of Honomanu Formation lavas on East Maui.

In earliest Pleistocene time, flows of the Honomanu Formation, many thousands of feet thick, constructed the bulk of the East Maui volcano. West Maui, which was older, was built up by similar outpourings. Rocks of both volcanoes consisted of tholeiitic basalts and olivine basalts that hardened from pahoehoe and aa flows. Honomanu rocks above sea level were buried by the flows of later lavas and are now exposed only in the cliffs along the northeast shore of the island and in the walls of the deeper stream valleys.

3. Eruption of Kumu'iliahi Formation lavas in Pleistocene time.

At the base of the south wall of Haleakala Crater (north and northeast of Kumu'iliahi Peak) and in a small knob at the east edge of the Ko'olau Gap, are thin-bedded pahoehoe and aa lavas formerly assigned to the Honomanu Formation but now regarded as a discrete rock unit on the basis of chemical composition. Kumu'iliahi rocks are too alkalic to be considered tholeiitic, but are less alkalic than the overlying rocks of the Kula Formation.

4. The Pleistocene Kula Formation.

Kula lavas erupted along two rift zones that trend to the southwest and to the east from the summit of Haleakala. Minor eruptions occurred along the third rift zone that extends northward from the summit. These outpourings of lava buried the shield formed by the Honomanu lavas and formed a thick cap (2500 feet) over the summit, raising the top of the volcano some 2000 to 3000 feet above its present elevation. The rocks of the Kula Formation include alkalic basalts, but hawaiite,* an andesitic rock, is the predominant type.

The Kula flows, mainly aa, contain interbedded layers of ash and pyroclastics. Since many of the eruptions were explosive, large cinder cones formed with most of them astride the volcano's rift zones. Lenses of stream gravel between some Kula lavas indicates that quiet intervals separated eruptive periods. Exposed in the walls of Haleakala Crater are a few dikes of Kula age that parallel the rift zones. The dike rocks, which have a higher proportion of feldspar minerals, tend to be more resistant than the surrounding lavas. The ancient Hawaiians made adzes from an even more resistant, fine-grained, igneous rock exposed in a few intrusive masses emplaced in the west wall of the crater.

*Not to be confused with the mineral hawaiite, a gem variety of olivine.

5. The great submergence; weathering and erosion.

While volcanism subsided, sea level rose until at times the ocean met the shore at levels more than 100 feet higher than at present. The isthmus between East and West Maui was under water. Throughout the Hawaiian Islands, extensive beach deposits and deltas are graded to high sea stands, which were probably the result of warmer climatic intervals with melting of Pleistocene glaciers. Stream erosion and younger lava flows have obliterated most of the erosional and depositional features produced by waves on Maui.

When cooler climates prevailed during the Pleistocene, more rain and snow than usual fell on Maui. Glaciers probably did not develop; but frost action may have been severe, and snow accumulated in winter. Stream erosion and mass wasting accelerated, and erosion was intense. The height of Haleakala was reduced, perhaps by more than 2000 feet. The south slope of Haleakala became steeper and shorter because seaward slip faults triggered huge landslides on that side of the volcano. On the northeast flanks, where precipitation is heavier due to the trade winds, the larger stream valleys cut downward several thousand feet. Because of fluctuations of sea level, the mouths of some stream valleys were scoured out below present sea level and later partially filled with gravel and mudflow debris. In the summit area, intensive headward erosion, downcutting, and the rapid removal of rock and debris produced large, amphitheater-shaped basins at the heads of the major stream valleys.

6. Development of Haleakala Crater.

Harold T. Stearns, who for many years was District Geologist for the U.S. Geological Survey in the Hawaiian Islands, developed the theory that the calderalike depression at the summit of Haleakala was mainly the result of coalescence of amphitheaters of the Ke'anae and Kaupo Valleys. The Ke'anae Valley trends northeastward from the summit area, and the Kaupo Valley runs southeastward and southerly. The large size of Haleakala Crater is due to the valley heads being offset, instead of joining in a straight line across the top of the mountain. Throughout the crater area, evidence is clear that a great deal of erosion took place, although some of the erosional features have been buried under younger volcanic lavas and mudflows.

7. Late Pleistocene to Holocene volcanic activity.

The lava flows, ash falls, and cinder cones of the Hana Formation that erupted after the erosional period are conspicuous on the floor of the crater, on its western rim, and on parts of the volcano's flanks. The surface of the most recent flows in the crater area is fresh and only slightly weathered, indicating that at least some eruptions occurred a few hundred years ago. Hana rocks are alkalic basalt and, like the Kula rocks, are distinctly lighter in color than the dark tholeiitic basalts that built the primitive shield.

Many of the Hana flows traveled much farther from their vents than Kula lavas did; this suggests that, in general, the Hana lavas were more fluid. Some Hana lavas went down the Ke'anae and Kaupo valleys to the ocean, and some erosional valleys were nearly filled. The Kipahulu Valley, for example, was virtually filled with Hana lava; the stream cut a new gorge along the north side of the filled valley; and then fresh Hana lava partly refilled the canyon again. During one eruptive episode, an enormous mudflow came down the Kaupo Valley and reached the ocean.

8. Volcanic activity during historic time.

Carbon-14 ages for charcoal associated with unweathered Hana lavas from the southwest rift zone of Haleakala have been determined as A.D. 1070 ($\pm$170 years), A.D. 1310 ($\pm$140 years), and A.D. 1370 ($\pm$120 years). The most recent eruption from the southwest rift zone took place C. 1790. At the present time, Haleakala is considered inactive. Earthquakes that occur from time to time on Maui are probably due to subsidence of the crust from the weight of the volcanoes.

Bibliography

Culliney, J. L. 1988. *Islands in a far sea.* San Francisco: Sierra Club Books. 410 p.

Daws, G. 1989, 3rd edition. *Hawaii, the islands of life.* The Nature Conservancy of Hawaii. Honolulu: Signature Publishing. 156 p.

Decker, R. W., Wright, T. L., and Stauffer, P. H., editors. 1987. *Volcanism in Hawaii.* U.S. Geological Survey Professional Paper 1350, 2 vol., 1667 p.

Kyselka, W. and Lanterman, R. 1980. *Maui, how it came to be.* Honolulu: University of Hawaii Press. 159 p.

Macdonald, G. A. 1978. Geologic map of the crater section of Halaeakala National Park, Maui, Hawaii (with accompanying summary). U.S. Geological Survey Map I-1088.

———, and Hubbard, D. H. 1989, 9th edition, revised by Jon W. Erickson. *Volcanoes of the National Parks in Hawaii.* Hawaii Natural History Association and National Park Service. 65 p.

———, and Abbott, A. T. 1983, 2nd edition. *Volcanoes in the sea.* Honolulu: University of Hawaii Press.

Mack, J. 1979. *Haleakala, the story behind the scenery.* Las Vegas, Nevada: KC Publications.

Reber, G. 1959. Age of lava flows on Haleakala, Hawaii. *Geological Society of America bulletin* 70:1245–6, September.

Stearns, H. T. 1942. Origin of Haleakala Crater, Island of Maui, Hawaii. *Geological Society of America bulletin* 53:1–14, January.

———. 1966. *Geology of the state of Hawaii.* Palo Alto, California: Pacific Books.

Haleakala National Park

Size: 28,655 acres; 45 square miles

Wilderness area, 19,270 acres

Authorized as part of Hawaii National Park: August 1, 1916

Redesignated Haleakala National Park: September 13, 1960

Wilderness designated: October 20, 1976

Designated a Biosphere Reserve: 1980

Address: Box 369, Makawao, Hawaii 96768

39

Over 10,000 geysers, hot springs, mud pots, and colorful travertine terraces make Yellowstone the largest, most spectacular, and most complex thermal area on earth. Mineralized hot water, spouting and evaporating, constantly changes the thermal features.

A "hot spot" below the continental crust is believed to be the source of heat. Volcanic breccia, ash, and lava have built up the Yellowstone Plateau, over a mile above sea level. At least three explosive eruptions in the last 2 million years have developed giant calderas. Subtle geophysical indicators suggest that eruptive activity may resume in the not-too-distant future.

Lakes, waterfalls, cirques, high mountain meadows, petrified trees, Obsidian Cliff, the Grand Canyon of the Yellowstone River, and large mammals are also wonders of this first national park.

Yellowstone National Park

Northwest Wyoming, Eastern Idaho, Southern Montana

by Sherwood D. Tuttle
University of Iowa

Figure 39.1 Castle Geyser, in Upper Geyser Basin, erupts water and steam. The cone building up around the vent is made of geyserite, an opaline silica. Photo by W. H. Parsons.

Yellowstone and the National Park Image

As the oldest and best known national park, Yellowstone symbolizes to many Americans all that is awe-inspiring, beautiful, and wonderful about our national parks. Old Faithful has fascinated millions of visitors. Every day during the summer, thousands of people gather to watch its eruptions. Parking lots fill, empty, and then refill as this world-famous geyser puts on performance after performance. Mammoth Hot Springs, its great colorful terraces ever growing and changing, impresses visitors with its size. In the Grand Canyon of the Yellowstone—a deep gash in the Plateau—are the strange "yellow rocks" for which the park was named. Black bear, grizzlies, bighorn sheep, moose, elk, mule deer, and bison roam in the forests and glades.

The administrative structure developed by the National Park Service at Yellowstone National Park has served as a model for national parks throughout this country and in other parts of the world. Many individuals contributed to the development of the national park idea, and no more worthy place could have been chosen to work out the concept and put it into effect than Yellowstone National Park. International recognition of Yellowstone's value and uniqueness was formalized by its designation as a Biosphere Reserve in 1976 and a World Heritage Site on September 6, 1978.

Discovery and Exploration

Bands of Indians hunted and fished in the Yellowstone region for thousands of years. In summer they camped near the hot springs and geysers, but most groups did not stay through the winters because of heavy snows and severe cold. The Indians made tools and weapons for their own use and for bartering from the plentiful obsidian. Large piles of obsidian chips near Obsidian Cliff and the finding of arrowpoints made from Yellowstone obsidian in the Great Plains and as far away as the Mississippi Valley are evidence of the importance of the obsidian trade.

John Colter, a member of the Lewis and Clark expedition, was probably the first non-Indian to visit Yellowstone. He left the expedition on the return trip in 1806 to join a party of trappers. The following year, alone and on foot, he crossed the Wind River Mountains and the Teton Range and explored the Yellowstone region, making contact with Indian tribes. Although wounded in a battle between Crow and Blackfoot Indians, he managed to make his way to a trading post at the mouth of the Bighorn River (south central Montana). There his description of a place of "fire and brimstone" was dismissed as incredible and the result of delirium.

Little was known about the region until the famous mountain man, Jim Bridger, after returning from an 1857 expedition, began spreading tales of spouting water, boiling springs, yellow rock, and a mountain of glass. Since Bridger was known as a "spinner of yarns," his reports were laughed at and regarded as preposterous. Nevertheless, Bridger's descriptions aroused the interest of influential men, among them the geologist and western explorer, F. V. Hayden. In 1859, he and an Army surveyor, W. F. Raynolds, began a two-year reconnaissance of the Upper Missouri region with Bridger as the party guide. They reached the approaches to Yellowstone but were unable to get through the mountain passes because of very heavy snows. The outbreak of the Civil War put a stop to their explorations. Hayden, who was a medical doctor before becoming a geologist, served as a battlefield surgeon with Union troops. Eleven years passed before he was able to fulfill his dream of making a scientific study of Yellowstone.

Meanwhile, intrigued by the reports of miners who had traveled up the Yellowstone River in an unsuccessful search for gold, a group of Montanans organized the Washburn-Langford-Doane expedition. Headed by Henry Washburn, the surveyor-general of Montana Territory, the party included Cornelius Hedges, a judge, and Nathaniel P. Langford (who later became known as "National Park" Langford). An Army detachment, commanded by Lt. Gustavus Doane, accompanied them. They spent nearly a month exploring the Yellowstone region, naming features, and collecting specimens. As they were gathered around a campfire on their last night in the area, they talked about what should be done with this strange and marvelous region. Out of this discussion came the idea of making it a preserve under the protection of the government and not open to settlement or mining. All agreed to work toward getting the land set aside.

The next year, in 1871, F. V. Hayden led a large, government-sponsored expedition into Yellowstone. His comprehensive report, supplemented by the excellent photographs of W. H. Jackson and paintings by Thomas Moran, helped to convince the Congress to withdraw this unique area from public sale. On March 1, 1872, President Grant signed a bill authorizing Yellowstone National Park. The new law provided that

> ". . . a tract of land in the territories of Montana and Wyoming, lying near the headwaters of the Yellowstone River is hereby reserved and withdrawn from settlement, occupancy, and sale under the laws of the United States, and dedicated and set apart as a public park or pleasuring ground for the benefit and enjoyment of the people. . . ."

N. P. Langford, the first superintendent of the park, served for five years without pay, as did several succeeding superintendents. Congress did not appropriate funds to implement what was a noble intention. In 1886, the responsibility for protecting and managing the park was given to the Army. After the National Park Service was

YELLOWSTONE NATIONAL PARK

Stream

Highway

Park Border

0   5   10 miles

N

Figure 39.2   Yellowstone National Park, Wyoming, Idaho, and Montana.

organized in 1916, a corps of rangers, under the leadership of a civilian superintendent, took over the park's care, continued improvements that Army engineers had begun, and supervised facilities operated by concessionaires.

The Geographic Setting

Yellowstone National Park is astride the Continental Divide, the topographic ridgepole of North America that separates Atlantic and Pacific drainages. The Divide enters the park near the southeast corner, crosses between Yellowstone Lake and Shoshone Lake, and then goes into Idaho south of the West Yellowstone entrance to the park (fig. 39.3).

The Yellowstone River, rises south of the park, collects water from tributaries as it flows northward across the park, and joins the Missouri River in Montana. The Missouri, in turn, joins the Mississippi, which empties into the Gulf of Mexico. The drainage from the southwestern third of the park finds its way into the Snake River, which passes through Grand Teton National Park (chapter 40) and then flows west to join the Columbia River on its way to the Pacific Ocean.

Yellowstone National Park is on a high plateau, averaging about 8000 feet in elevation and nearly surrounded by ranges of the Middle Rockies, 10,000 to nearly 14,000 feet high. To the southwest, the plateau slopes gently down to the Snake River Plains. The ranges that encircle the Yellowstone Plateau are (beginning on the west and going clockwise) the Madison Range; to the north, the Gallatin Range and the Beartooth Mountains; on the east, the Absaroka Mountains; and to the south, the Teton Range and Grand Teton National Park (chapter 40).

Geologic Features

Geologically, as well as topographically, the two national parks in the Middle Rockies—Yellowstone and Grand Teton (chapter 40)—are related to Rocky Mountain National Park (chapter 22) in the Southern Rockies and to Waterton-Glacier International Peace Park (chapter 23) in the Northern Rockies. The ranges of the Rocky Mountain chain evolved through the same time span of geologic history; were subjected to Laramide tectonic forces of great intensity; and have rocks and structures of similar age.

The Surrounding Ranges

Many features of the Yellowstone landscape are associated with volcanic events that took place in relatively recent geologic time; but long before Cenozoic volcanism began, the structures of the ranges were evolving. Thus the story begins not on the Yellowstone Plateau but in the mountains that rise above it (fig. 39.3).

Madison and Gallatin Ranges. These ranges, which extend into the northwest section of the park, are fault-block mountains with north-south river valleys following downfaulted blocks. The Gallatin River flows north between the two ranges. The Madison River crosses the park's western boundary and heads north along the western flank of the Madison Range. The Yellowstone River outlines the eastern margin of the Gallatin Range.

Like many other Rocky Mountain ranges, the Madison and the Gallatin have a core of Precambrian granites and gneisses with flanking Paleozoic and Mesozoic sedimentary rocks, some of which are overlain by volcanics. The northwest-southeast structural trend of the two ranges is characteristic of the general trend of the Rocky Mountains. Typical, also, is evidence of pulses of Laramide mountain-building and uplift. The present mountains are large blocks bounded by post-Laramide faults that in Cenozoic time cut obliquely across older structural trends in a nearly north-south direction.

The Beartooth Mountains. Uplift and vertical faulting have raised the Precambrian granites and gneisses of the Beartooth massif high into the air. From its southern tip in Yellowstone National Park, the massif rises to Montana's highest point, Granite Peak, elevation 12,799 feet. At the top of the range are a few outliers of a once extensive sedimentary cover, Beartooth Butte being an example. Other Paleozoic rocks, mainly limestones, lap up on the flanks. U-shaped valleys, 3000 to 4000 feet deep, dissect the intensely glaciated uplands. The bold eastern front of the Beartooths rises 4000 to 5000 feet above the Great Plains. U.S. Highway 212, which leaves the northeast corner of Yellowstone National Park and winds eastward and then north to Red Lodge, Montana, goes through spectacular alpine terrain in crossing the Beartooth Mountains.

Absaroka Mountains. This large range, trending northwest–southeast, occupies the eastern part of Yellowstone National Park and extends into western Wyoming. Absaroka peaks, rising to elevations of 10,000 to 12,000 feet, have been intensely eroded by streams and glaciers and form a jumble of steep ridges above fairly broad valleys. Unlike the other ranges of the Middle Rocky Mountains, the Absarokas are made up of over 10,000 feet of mainly andesitic, pyroclastic volcanics. Eruptions beginning in the Eocene Epoch came from numerous centers in a large volcanic field. The tuffs, breccias, and lavas of the Absaroka Volcanic Supergroup overlie older metamorphic and sedimentary rock sequences. The volcanic rocks have been somewhat uplifted and warped, but are only mildly deformed.

The Teton Range. On the south, Yellowstone National Park is joined to the northern boundary of Grand Teton National Park (chapter 40) by the John D. Rockefeller, Jr. Memorial Parkway. The Teton Range (described in the

Figure 39.3   Locations of major topographic features in Yellowstone National Park and vicinity.

chapter 40) is a tilted and uplifted fault block. Jackson Hole, the valley at the foot of the Tetons, is a down-dropped block.

Island Park Caldera and the Snake River Plains. The Island Park caldera, a few miles west of Yellowstone National Park, is a plateau ringed with low hills. Beyond the caldera, the Snake River Plains of southern Idaho slope gently to the southwest, with many flows of young basaltic lava covering the surface. In Craters of the Moon National Monument (about 120 air miles southwest of Yellowstone), some basalt flows erupted as recently as 2000 years ago. The volcanism of the Island Park caldera and the Snake River Plains is an integral part of the volcanic history of the Yellowstone Plateau.

The Yellowstone Plateau

A huge collapsed caldera, nearly filled with thousands of cubic miles of young volcanic rocks and volcanic debris, occupies the major part of the Yellowstone Plateau. Geyser basins, hot springs, and other thermal features demonstrate the caldera's lasting heat. Except for a few peaks of older rock sticking up through the plateau's relatively flat surface, the rocks are younger than those of the surrounding ranges. Mount Washburn (elevation 10,243 feet) is the most prominent summit. A depression in the caldera contains Yellowstone Lake, which has an elevation of 7331 feet. With an area of 139 square miles, it is the largest high-altitude lake in North America. Sporadic earthquake activity in the Yellowstone Plateau indicates that faults are responding to crustal stresses and fracturing associated with volcanic activity.

The voluminous Yellowstone eruptions, which may have lasted but a few days or months, were separated by long, quiet intervals that lasted hundreds of thousands of years. This long-period, episodic pattern of volcanic activity is characteristic of silicic magmas in continental settings and is in marked contrast to a more nearly continuous style of volcanism typical of basaltic systems, especially those in oceanic regions such as the Hawaiian volcanoes (chapters 37, 38).

In the volcanic history of the Yellowstone Plateau, three phases, or volcanic cycles, are recognized, all of which took place during the last two and a half million years. Rocks of the first volcanic cycle are exposed in Island Park caldera and southwestern Yellowstone. Identifiable outcrops of the second volcanic cycle have been found near the edge of the Island Park caldera, but second cycle rocks have been largely destroyed or buried by the output of the third volcanic cycle. Eruptive materials of the third cycle make up most of the Yellowstone Plateau.

The Yellowstone Caldera

Mazama's explosion-collapse caldera (chapter 33) has an area of barely 20 square miles. What exists at Yellowstone is a multiple-eruption caldera with an area of about 1500 square miles! The volume of magma erupted at Yellowstone probably amounts to 50 times the output of Mazama. By means of extensive field investigations and with the aid of satellite imagery, the outlines of Yellowstone caldera have been accurately determined. About 40 miles long and 30 miles wide, it reaches from the Madison River (near Madison Junction) to the eastern shore of Yellowstone Lake and from Lewis Falls north to the foot of Mount Washburn (fig. 39.4).

Studies of the predominantly rhyolitic rocks of the three volcanic cycles in the Island Park–Yellowstone area suggest that during each episode great volumes of lava rose to the surface and erupted, building up thousands of feet of ash, tuff, breccia, and layers of lava. Small lava eruptions—related to the third cycle, but occurring after the Yellowstone caldera was formed—persisted until about 70,000 years ago. (fig. 39.5).

Any explanation for such great volumes of extrusive material requires the presence of a fairly large reservoir of magma, presumably originating in the upper crust and

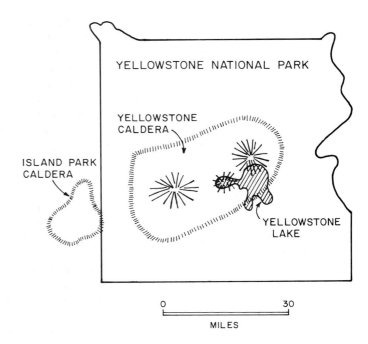

Figure 39.4 Location and extent of Island Park and Yellowstone calderas, both Pleistocene in age, but Island Park the older. Two currently expanding domes are within the Yellowstone caldera. The small crater in West Thumb of Yellowstone Lake is late Pleistocene.

Figure 39.5   Development of Yellowstone caldera during the most recent volcanic cycle, 600,000 years ago. *A.* Upwelling magma arches roof rocks, causing concentric ring fractures that extend down toward the magma chamber. *B.* When ring fractures tap the magma chamber, gases explode as pressure is released, triggering catastrophic pyroclastic flows. (Borne by prevailing winds, ash particles spread over thousands of square miles.) *C.* Roof rocks collapse, forming an immense caldera. *D.* Magma, depleted of gas, flows out from the ring fractures over the caldera floor. From W. H. Parsons, *Middle Rockies and Yellowstone.* © 1978 Kendall/Hunt Publishing Company. Used by permission.

extending fairly close to the surface. As magma was withdrawn by eruptions through fractures and weak zones, the roof rocks over the chamber stretched and thinned, finally collapsing of their own weight into the liquid magma below. Since no identifiable roof rocks or old volcanic cones have been found, we can assume that the missing surface material was either incorporated into the melt or ejected with ash. The cataclysmic explosions were of a magnitude not known in recorded history.

For thousands of years following the climactic eruption, lava continued to flow periodically until the tremendous depression was nearly filled. Then, about 150,000 years ago, the floor of the Yellowstone caldera began to bulge up again. This doming exposed rocks and structures that have enabled geologists to put together a sequence of events to account for this remarkable feature and also to develop hypotheses regarding possible future volcanic activities in Yellowstone. The significant lines of geological evidence are summarized below.

Yellowstone's Next Eruption?
Heat Flow. The earth's crust gets warmer with depth. The rate of heat increase with depth is called the *geothermal gradient.* Heat energy tends to move from hotter areas to

cooler areas. Despite the fact that rocks are poor conductors, the earth's internal heat moves from great depths toward the surface. Amounts of heat flow can be measured both at the surface and in drill holes. Heat flow values taken from a number of points on the Yellowstone Plateau are abnormally high, more than 30 times the average global heat flow. In the bottom of wells drilled by the U.S. Geological Survey, the average temperature is over 400° F at depths of 500 to 600 feet below the plateau surface.

Curie Depths. Most rocks lose their magnetism when they are heated beyond the *Curie point,* a threshold temperature of about 560° C. Thus information about the temperature of rocks below the surface can be inferred from the magnetic properties of rock. Rock under Yellowstone reaches the Curie point at a remarkably shallow depth. Using geophysical techniques, investigators have determined that beneath the surface of the Yellowstone Plateau, Curie depth is about 6 miles down (10 kilometers). Furthermore, in some spots under the caldera, Curie depth is less than 4 miles (6 kilometers) (Smith and Christiansen 1980). By contrast, Curie depths under other parts of the continent are between 10 and 20 miles (15 to 30 kilometers) below the surface.

Electrical Conductivity. At higher temperatures, earth materials tend to be good conductors of electricity. Since data obtained by geophysical techniques show that electrical conductivity at Curie depth under Yellowstone is good, the patterns of magnetic and electrical data are in agreement.

Seismic Wave Velocities. Seismologists have found that earthquake waves increase or decrease in speed as they travel through the rocks of the crust, depending upon the density of the materials they encounter. Seismic *P* waves, for example, travel faster through the high-density basalts of oceanic crust and then slow down when passing through granitic continental crust, which has lower density. Seismograph monitoring in the Yellowstone caldera and vicinity by the U.S. Geological Survey has recorded velocities of *P* waves arriving from distant earthquakes. Mathematical modeling of the travel-time delays suggests the presence of a body of low-density material close to the surface and extending downward to the upper mantle. (*P* waves are the fastest seismic waves that travel through the earth's interior.)

Gravity Data. A *gravity survey* consists of measuring the intensity of the gravitational field at a number of different locations and then plotting the results on a map. The object is to associate gravity variations with differences in the distribution of rock densities. If the crust consists of homogeneous materials, the variations are due solely to differences in surface elevation. Thus interpretation of gravity data can yield two sorts of evidence. First, the location of rock masses of different types can be determined. Then, after a survey has been repeated at intervals, perhaps five or ten years, any changes in elevation that have occurred can be calculated and the results compared.

Gravity measurements derived from 900 locations on the Yellowstone caldera and the surrounding region have indicated the presence of low-density material, both under the caldera and also in the subvolcanic basement, with the latter showing even lower readings than the material nearer the surface. What this implies is that the presumed shallow chamber contains magma that is high in silica, and that the silica content in rocks *above* the chamber is less high, perhaps because the silica has been somewhat dispersed. Rhyolites, which are plentiful in Yellowstone, are high in silica and have a density of about 2.5 to 2.6. Mafic rocks, such as basalts, which are less common in Yellowstone, have somewhat higher densities of around 3.0. Gravity data have also confirmed geodetic surveys that indicate the caldera area is rising at the geologically rapid rate of a little less than half an inch (12.5 mm) per year.

Depth of Foci and Earthquake Frequency. The *focus* (pl. *foci*) is the point within the earth that is the center of an earthquake. It is the initial rupture point of the earthquake from which seismic waves emanate. The *epicenter* is a point on the earth's surface *directly above* the focus of an earthquake. A seismologist can pinpoint the location of an earthquake and the depth of focus by comparing the seismic wave data recorded on the seismograph at his station with similar information from other stations and then analyzing the results.

With one earthquake of over magnitude 7 and seven magnitude 6 earthquakes, Yellowstone National Park and adjacent parts of Idaho have been have been the most seismically active region in the Northern and Middle Rocky Mountains in historic time. The strongest earthquake was the Hebgen Lake quake (7.1 Richter scale) in August 1959, which caused extensive landsliding. The epicenter was just outside the northwest corner of the park. Monitoring of many earthquakes in the Yellowstone area (most of which have registered between 1 and 6 on the Richter scale) has revealed that the average depth of focus is around 9 miles, which is considered shallow. This means that the rocks are solid to that depth because only solid rocks can fracture and produce earthquake waves. *Earthquake swarms* (clusters of minor earthquakes) occurring in Yellowstone suggest movement of magma below the surface.

Ages of the Caldera Rocks. U.S. Geological Survey scientists have ascertained absolute ages for volcanic rocks from the three volcanic cycles using the potassium-argon method of radiometric dating. Combining the dates with information from stratigraphic studies and geologic mapping produced the following sequence. The first volcanic cycle began about 2.2 million years ago in the Island Park area and lasted until about 1.6 million years ago. The climactic eruptions and caldera collapse occurred about 2 million years ago. Climactic eruptions of the second volcanic cycle, which also began in the Island Park area, were about 1.3 million years ago. Overall, a smaller volume of material was erupted. The third volcanic cycle, which was marked by a shift in volcanic activity from the Island Park area to the center of the Yellowstone Plateau, began about 1.2 million years ago. Climactic explosions about 630,000 years ago produced the present Yellowstone caldera.

Composition of the Lavas. While most of the volcanic rocks produced by eruptions from the Yellowstone caldera are rhyolitic, some basalt flows were extruded late in the third volcanic cycle. Basaltic lava flows also followed the rhyolitic flows of the first and second volcanic cycles. The pattern seems to be one of extensive rhyolitic volcanism, with the center slowly moving from southwest to northeast, followed periodically by episodes of basaltic volcanism along the same general trend.

The eruption of rhyolitic lavas and then basaltic lavas from the same vents with no intermediate lavas (e.g., andesites) appearing in the sequence calls for a special explanation. The composite volcanoes described in chapters

Figure 39.6   Simplified southwest-northeast cross section through the Yellowstone caldera and the Lamar valley, based on seismic and other geophysical interpretations. (Modified after Smith and Braile 1984).

32 through 36 changed the composition of their lavas as eruptions ran their course, but the transition from one type of lava to another was usually more gradual.

Chemically and mineralogically, the two rock types are quite different. Rhyolite is high in silica and low in iron and magnesium, while basalt is high in iron and magnesium and low in silica. The predominant minerals in rhyolite are potassium feldspar and quartz; in basalt, they are usually calcium plagioclase feldspar, olivine, augite, and hornblende. Minerals that form basalt crystallize out from a melt or magma at considerably higher temperatures than do the minerals that form rhyolite.

It seems logical to theorize that at Yellowstone two different source materials melted in order to form two distinct liquids, one rhyolitic and the other basaltic. Basaltic magma may be formed by the partial melting of the upper mantle, while rhyolitic magma may be derived by the partial melting of metamorphic rocks (such as gneisses and schists that are high in silica) in the earth's lower crust. Long-continued uplift and extension, or stretching, of the crust might cause upward displacement of the mantle. This, in turn, causes reduction of confining pressure that allows materials with low melting points to be liquified. As heat radiates upward in the crust, additional high-silica rocks are incorporated into the melt because they tend to melt first. In this way, a high-silica magma, having lower density and greater viscosity and being somewhat more buoyant, begins to accumulate in the upper crust. In the late phase of a volcanic cycle, after rhyolitic magmas have been ejected, molten basalt from the upper mantle begins to rise under conditions of reduced pressure.

Expanding Domes. According to geophysical and geochemical evidence, a body of very hot, low-density, high-viscosity material, capable of generating and extruding tremendous volumes of volcanic material, has existed beneath the Yellowstone region for the past several million years. Within the Yellowstone caldera, two small resurgent domes (Sour Creek dome and Mallard Lake dome) are bulging up the plateau center at a faster rate—a little more than one-half inch (15 mm) per year—than the general uplift going on in the Yellowstone caldera.

Scientists do not yet have enough information to make specific predictions, but past volcanism, contemporary uplift with high uplift rates, length of time between volcanic eruptions, earthquake swarms, recent volcanism in the caldera, and the presence of a shallow heat source in the upper crust all point to a future of continuing volcanic activity in the Yellowstone Plateau (fig. 39.6). The very high heat flow along with the anomalous behavior of seismic waves imply that shallow hot bodies, probably containing a large volume of fluid in the form of steam, water, and partially molten rock, are present underneath the plateau (Smith and Braile 1984, p. 108).

What form might future volcanic activity take? At present, there is no way of knowing whether small eruptions of lava may occur around the margins of the Yellowstone Plateau, or medium-size eruptions within the caldera, or perhaps even a major explosion and flow of ash. However, any renewal of volcanism will undoubtedly be preceded by such warning signs as localized earthquakes, variations in the velocity of seismic waves, increasing gas emissions and doming or other deformation. All these phenomena can be monitored and continuously evaluated.

The Hot Spot Theory

What is the source of the heat that drives the Yellowstone thermal and volcanic activity? And why have the eruptive centers shifted over time along a trend from southwest to northeast? Geologists theorize that a hot spot in the upper mantle, comparable to the one that produced the Hawaiian Island chain (chapter 37), may be the cause of volcanic episodes in the Yellowstone region. However, the Pacific hot spot lies under thin oceanic crust, while thick continental crust overlies the Yellowstone hot spot. As explained earlier, eruptive patterns in the two regions have developed along profoundly different lines. Although high heat flow keeps the geyser basins and hot springs active between eruptive cycles, major eruptions in Yellowstone have been more episodic than eruptions in the Hawaiian chain, which have tended to be more or less continuous.

At present, the North American tectonic plate is moving about 4.5 centimeters per year in a southwesterly direction over the hot spot. Since the assumption is that the hot spot has remained stationary, the centers of volcanic activity have shifted northeastward as the crust drifts toward the southwest. In general, the ages of the volcanic rocks tend to fit this model. The older rocks are in the vicinity of Island Park and younger rocks are in the central part of the Yellowstone Plateau. The young basaltic rocks of the Snake River Plains, which largely covered earlier rhyolitic eruptions, may represent partial melting of upper mantle material that occurred after the emptying of rhyolitic magma from an older chamber. Geologists have speculated that within a few hundred thousand years, if the trend continues, a fourth caldera may form in the northeast corner of the Yellowstone National Park.

Hot Springs and Geysers

For observing hydrothermal features, no place in the world can compare to Yellowstone, which has about 62 percent of all known geysers, plus thousands of hot springs, mud pots, and fumaroles. A *hot spring* is, by definition, any thermal spring whose temperature is above that of the human body. A *geyser* is an intermittent hot spring that regularly or irregularly erupts jets of hot water and steam, as a result of ground water being heated by hot rock, creating steam under conditions preventing free circulation. Confined steam builds up pressure underground until it attains enough force to be forcibly ejected at the surface.

Hot springs, which occur in many parts of the world, were well known to Europeans who settled North America. But none of the early western explorers had ever seen a geyser before coming to Yellowstone and had doubtless never heard of Geysir in Iceland, the namesake for all the world's geysers. Geyser fields are rare, and most are located in remote areas, such as Iceland, New Zealand, and Indonesia, places visited by few nineteenth-century Americans and Europeans. To come upon an area in the wilderness where great jets of steam and scalding water suddenly spouted from the ground, shooting 100—even 200—feet into the air, only to stop and then, inexplicably, begin again, must have been unnerving, as well as awesome to the mountain men. That people in the more settled parts of the country did not believe the "tall tales" is not surprising.

The source of the geysers' heat and the origin and operation of their "plumbing systems" were subjects of scientific speculation for more than a hundred years, but a generally accepted theory of how they work has been developed. For geysers to function, three conditions must be present: (1) a powerful heat source, usually volcanic; (2) a plentiful supply of water; (3) a fairly watertight plumbing system, capable of holding a large volume of water and withstanding the tremendous pressure of steam and superheated water (fig. 39.7).

Geysers do not have large open underground chambers for accumulating water. The tube or vent at the surface leads down to a shallow plumbing system consisting of intricate conduits that connect small openings in the rock and lenses of porous sediment capable of storing water. Normally, several geysers in a geyser basin are supplied by an interconnecting fracture system. The passageways in the system are lined with *sinter,* which is a chemical sedimentary rock, mainly siliceous, deposited as a hard incrustation by precipitation. Very hot water, deep underground, dissolves silica from the silica-rich rhyolitic rocks and carries it through the system and to the surface, depositing precipitated silica along the way. The sinter deposits help to keep the system nearly watertight and also tend to produce constrictions that prevent free circulation of water. Most of the world's geyser fields are in areas where rhyolite is unusually abundant.

Rainwater and meltwater from snow infiltrate the ground, percolating slowly down through cracks and pores in the rock. Some water seeps into the geyser plumbing systems; but much of it goes several thousand feet lower, where it is heated by contact with the hot rocks over the magma chamber to a temperature of more than 400° F (fig. 39.8). Due to hydrostatic pressure, the water cannot boil; but it begins to expand and rise toward the surface. Steam bubbles, rapidly forming in the rising water, heat the cooler water that is filling the lower part of the plumbing system. This filling and mixing process continues until the system is full of water ready to boil over. The average boiling point at the surface of the geyser basins in Yellowstone (elevation about 7500 feet) is about 199° F (93° C).

Steaming water flows, instead of spouting, from the hot springs in the geyser basins, because the hot springs

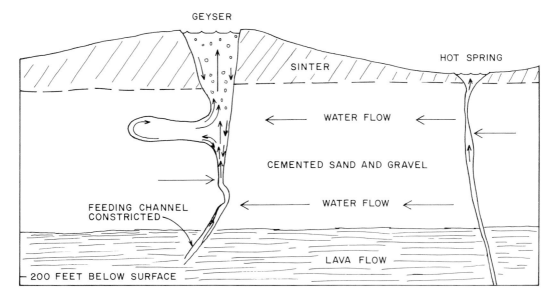

Figure 39.7   How geysers and hot springs work. Circulating ground water is heated at depth under pressure. Water boiling in the geyser system produces steam bubbles, causing expansion and overflow at the surface. With pressure suddenly reduced, water trapped in side chambers flashes into steam. The resulting "explosion" clears the upper part of the geyser tubes and shoots water and steam into the air. An eruption is not repeated until enough water seeps into the system and becomes again superheated, filling all tubes and side chambers. In hot springs, where boiling water is not constricted, it rises through the tubes and flows out continuously.

Figure 39.8   Relationship between a geyser or hot springs basin and a magma chamber at depth. Ground water, infiltrating from the surface, may percolate down to a depth of more than a mile before it becomes very hot and then begins to rise. From W. H. Parsons, *Middle Rockies and Yellowstone.* © 1978 Kendall/Hunt Publishing Company. Used by permission.

have larger openings and less restricted underground circulation. In the geysers, rapidly rising steam bubbles choke or clog constricted passageways, and the weight of the water above creates a pressure-cooker effect. As the water temperature rises, it reaches a critical point, which varies from geyser to geyser. Upward pressure of steam bubbles forces out some of the water; this lowers the pressure enough so that the water below flashes to vapor. The vapor then blasts out of the vent with such force that the whole system is nearly emptied as boiling water and steam shoot into the air. After that, the whole cycle begins again.

Because the silica in solution precipitates both above and below ground as the water temperature drops, sinter may build up enough to seal off diversionary passages in the plumbing system, thus changing a geyser's eruptive behavior. Some geysers build large cones or mounds of sinter around the vents. Hot springs may have fantastically shaped mounds of sinter over which water flows out. Most of the Yellowstone sinter is light colored, but some is tinted various hues.

The Geyser Basins. Upper Basin has the highest concentration of geysers and hot springs, the most famous being Old Faithful. Its eruptions are watched, photographed, and cheered by the thousands of visitors who come to view the spectacle. Old Faithful overflows for a few minutes at the beginning of the eruption, as if to clear its throat. Then with a roar, the mighty stream shoots up 100 to 180 feet into the air. (Average height of its eruptions is about 130 feet.) After a few minutes, the top of the jet slowly drops and then sinks down. Some puffs of steam drift out, and Old Faithful settles back to fill again.

Old Faithful plays as high as it did when first described, but average intervals between eruptions have lengthened after each of the last three major regional earthquakes. This is probably due to shifts in circulation of hot water away from the geyser. Nevertheless, Old Faithful's predictability remains unique, even though its timing is not so precise as many visitors think. Park rangers are able to set the "Old Faithful clock" (that gives the estimated time of the next eruption) by timing the length of the previous eruption. If an eruption is short, perhaps around 2 minutes, the interval before the next one will be about 55 minutes. If the play of water lasts 3 to 5 minutes, the next eruption may take place within 75 to 80 minutes. During an eruption, Old Faithful ejects from 3700 to 8400 gallons of water, most of which drains into the Firehole River.

In Lower Basin, south of Upper Basin, the groups of geysers and hot springs are more scattered. The Fountain Paint Pots, a cluster of multicolored mud pots, are in this basin amid a large geyser complex. *Mud pots* are hot springs that contain boiling mud and dissolved compounds that hot water brings up from the rocks below. Mud pots

do not have enough water to flush out sediment as it accumulates (fig. 39.9).

Fumaroles, or steam vents, puff in both Upper and Lower Basins. Some fumaroles become spouters or even small geysers when the water table is high.

Norris Basin, near the northwest edge of the Yellowstone caldera, is the hottest geyser basin in the park, presumably because magma is closest to the surface here. Another difference in Norris Basin is that the water is acidic rather than alkaline, as it is in the other geyser basins. The largest geyser in the world, Steamboat, is located in Norris, but its eruptions are few and far between. Six other geyser basins are in Yellowstone, but several are in less accessible areas of the park.

A quite different, spectacular thermal feature is in the northwest corner of the park, but not in a geyser basin. This is Mammoth Hot Springs, a mountain of travertine terraces covering nearly a square mile. *Travertine* is a dense, massive or concretionary limestone formed by rapid chemical precipitation of calcium carbonate. At Mammoth Hot Springs, the build-up of travertine deposits by precipitation is much more rapid than the accumulation of sinter deposits around the hot springs and geysers in the geyser basins (fig. 39.10).

Thick beds of marine limestone underlie Mammoth Hot Springs. Since limestone is highly soluble, the rising hot water dissolves and brings to the surface enormous quantities of calcium carbonate—over 2 tons per day in solution. As warm, mineralized water flows from the springs and down over the terraces, particles of calcium carbonate precipitate out as growing icicles and spreading sheets of travertine. The millions of algae that live in the warm pools tint the rock shades of brown, red, orange, and green. Springs and terraces change rapidly. New outlets form as old ones close over; active springs may become inactive in a year's time. Terrace Mountain, still under construction by Mammoth Hot Springs, continues to grow, change, and enlarge.

Seismic Activity and the Thermal Features

Earthquakes disrupt and change hydrothermal systems in Yellowstone. Immediately after the main shock of the Hegben Lake quake in 1959, hundreds of springs in Yellowstone began to spout jets of water. Clepsydra Geyser erupted wildly for three years after the quake and then became dormant. Sapphire Pool, which used to be a placid spring with occasional eruptions a few feet high, began erupting violently and voluminously. Hundreds of tons of sinter broke up and washed away before Sapphire's eruptive power gradually declined. Fumaroles opened near Firehole River, and mud pots spouted. Seismic Geyser in Upper Basin is the direct result of a crack in the earth

Figure 39.9   Black Dragons Cauldron, a constantly boiling mud pot in Lower Geyser Basin. National Park Service photo.

Figure 39.10   New Highland Spring, behind the main terraces at Mammoth Hot Springs, became active in 1952, depositing travertine in a forested valley. Trees are killed by hot water and then buried by encroaching rock. Photo by W. H. Parsons*

*The Parsons photographs in this chapter are from his field guide, *Middle Rockies and Yellowstone*, © 1978 Kendall/Hunt Publishing Company. Used by permission.

that opened during the 1959 quake. Old Faithful's eruptive intervals have lengthened and been less regular since the Borah Peak earthquake (magnitude 7.3) in 1983.

Glacial Features

Because violent eruptive events and later glaciations buried or destroyed most of the older glacial features in Yellowstone National Park, those that remain in the area belong to the Pinedale substage of Wisconsinan time, which ended in Yellowstone about 8000 years ago. Although winters on the high Yellowstone Plateau are severe, and snowfalls are heavy, no glaciers are presently on the plateau. At higher elevations in the encircling mountains, small glaciers assigned to Neoglacial time (post-Wisconsinan) are moderately active.

In the colder and wetter Pleistocene climates, valley glaciers started high in the mountains. As they grew and flowed down to lower elevations, they coalesced to form piedmont glaciers. The Lamar Glacier, which originated in the Beartooth Mountains, flowed down from the northeast; the Upper Yellowstone glacier, advancing from the southeast, was fed by valley glaciers in the southern part of the Absaroka Range. Small icecaps covered the plateau and the foothills during prolonged episodes. Mount Washburn and other high peaks stuck up above the ice as nunataks most of the time. At its greatest extent, the Pinedale icecap flowed radially outward from about the present location of Yellowstone Lake. To the south and southwest, ice flowed over the Continental Divide and down into Jackson Hole (chapter 40).

A large glacial erratic of Precambrian gneiss, left by a Pinedale glacier, gives some idea of its transporting power. The erratic, weighing more than 500 tons and measuring about 24 × 20 × 18 feet, stands near Inspiration Point, a short distance from the western rim of the Grand Canyon of the Yellowstone River. To get from its source to its present location, ice must have carried the erratic at least 15 miles.

The Grand Canyon of the Yellowstone River

The Yellowstone River, which rises in the Absaroka Range southeast of the park, carries runoff from the eastern side of the Continental Divide. The river flows into the southern end of Yellowstone Lake, out through the northern end, and then northward into Montana. A principal tributary is the Lamar River that drains the northeastern part of the park.

The Grand Canyon of the Yellowstone is a narrow, steep-walled canyon that evolved through several stages of cutting and filling during the past half a million years. Ash and volcanic debris from eruptions repeatedly filled the river valley, but the early Yellowstone River cut down

its channel again after each episode. Then advances of Bull Lake glaciers (early Wisconsinan time) blocked the Lamar-Yellowstone drainage system. Water backed up, creating a long lake in the canyon, and layers of lakebed sediment accumulated. After glacial retreat, rapid erosion cleared the canyon, but the Pinedale icecap overrode the canyon and again blocked the system. About 12,000 years ago, the last ice in the canyon melted away. Since that time, the Yellowstone River has eroded most of the volcanic debris and the glacial and lakebed sediment from the canyon.

The canyon walls are cut mainly in rhyolitic lava and tuff. Intense chemical weathering from hot ground water and hot springs has changed the color of the rock from an unweathered grayish black to a startling array of reds, browns, and yellows, producing the strange "yellow rocks" for which Yellowstone was named. Near the bottom of the canyon, where hot springs bubble out and fumaroles emit steam, the processes of hydrothermal alteration soften and dissolve the rock, changing its composition and physical properties.

At the head of the deepest and most spectacular part of the gorge is Lower Falls, 308 feet high (fig. 39.11). Here the river flows over the edge of an unaltered, resistant rhyolite flow that erodes more slowly than the tuff and softened lava below. Above Lower Falls is Upper Falls, which is 109 feet high. The Grand Canyon of the Yellowstone ends near Tower Falls, where Tower Creek flows over the rim of a hanging valley and drops 132 feet into the Yellowstone River below.

Columnar Jointing and Obsidian

Prominent columnar jointing is displayed in basaltic lava flows in Yellowstone National Park, especially in the walls of the Grand Canyon of the Yellowstone and at Sheepeaters Cliff along the Gardner River. The characteristic parallel columns of *columnar jointing,* polygonal in cross section, form by contraction as a lava flow cools. The rock tends to divide into columnar segments approximately perpendicular to the cooling surface (fig. 39.12).

The Yellowstone obsidian, which was of such value to the Indians, is a dark volcanic glass. Obsidian Cliff, along the road from Norris to Mammoth Hot Springs, is presumed to be the "Glass Mountain" that Jim Bridger talked about. *Obsidian,* or natural glass, forms when lava (usually rhyolitic) cools and hardens before mineral grains have had time to crystallize.

Buried Forests

At Specimen Ridge, on the south side of the Lamar River valley in the northeast corner of the park, petrified tree trunks (some with root systems preserved) stick up through

Figure 39.11   Lower Falls, 308 feet high, and the Grand Canyon of the Yellowstone River. At Lower Falls, the river drops from a ledge of resistant rhyolite. The canyon is cut in the young lavas and pyroclastics of the central part of the Yellowstone Plateau. Hot water, rising through fractures, accelerates chemical weathering processes, softening rock in the canyon walls and changing the color to shades of yellow and orange. Photo by W. H. Parsons.

Figure 39.12   Sheepeaters Cliff, a basalt flow near the Gardner River, displays prominent columnar jointing. As the lava cooled and contracted, intersecting joints produced the hexagonal columns. Photo by W. H. Parsons.

volcanic material in lifelike positions. Erosion of volcanic material on the slopes has exposed these fossil forests that were buried long ago by Eocene ash flows and lahars (mudflows). The trees were lithified as silica-enriched ground water infiltrated and preserved the woody structure (fig. 39.13). Another, smaller buried forest is at the tip of the Gallatin Range where it extends into the northwest corner of Yellowstone.

At Specimen Ridge, 20 or more successive forests (the exact number is uncertain) were buried, one above the other from bottom to top, during Eocene time. Perhaps as much as 2000 years passed between burials, as again and again violent eruptions from the Absaroka volcanic field sent ash flows and mud rolling down slopes, smashing and devastating the forests. The tree varieties most commonly identified are sycamore, walnut, magnolia, chestnut, oak, redwood, maple, and dogwood. Eocene climate throughout this time remained mild and humid, probably like the climate of southeastern United States today.

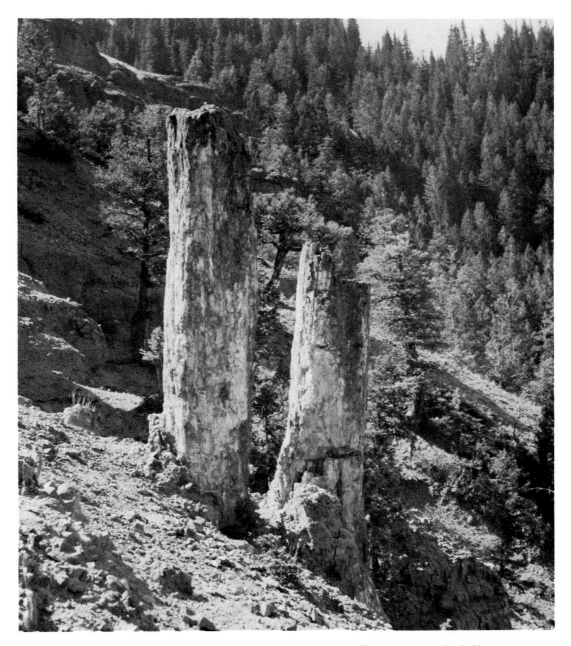

Figure 39.13   Petrified stumps on Specimen Ridge. Trees that grew in Eocene time were buried by volcanic breccia during an eruption and subsequently lithified by silica-enriched, percolating ground water. The silica that filled the wood cells is more resistant to weathering and erosion than the breccia that originally completely covered the stumps. Photo by W. H. Parsons.

Geologic History

1. Precambrian erosion, sedimentation, and tectonism.

In the northern part of the park, in the Gallatin and Beartooth Ranges, outcrops of contorted schists and gneisses represent fragments of the geologic history of Precambrian time. Some have been dated as over 2.7 billion years old. The schists were probably volcanics, sandstones, and shales deposited in basins in primitive crust.

The gneisses may have been derived from granitic intrusions. Presumably, the rock materials have been "recycled" several times by the processes of erosion, deposition, metamorphism, and tectonism.

2. Paleozoic and Mesozoic deposition.

Over 40 separate rock formations of Paleozoic and Mesozoic age have been identified in Yellowstone National Park. They crop out across the northern section of

Table 39.1 Generalized Geologic Column, Yellowstone National Park.

Era	Period	Epoch	Rock Units	Geologic Events
Cenozoic	Quaternary	Holocene	Alluvial, landslide, glacial, and hot spring deposits	Earthquakes; landsliding / Geothermal activity, hot springs, geysers / Neoglaciation / Regional uplift; doming in center of Yellowstone Plateau / Erosion of Grand Canyon of Yellowstone River / Pinedale glaciation
Cenozoic	Quaternary	Pleistocene	*Yellowstone Group (about 20 formations and members): (Plateau rhyolites and basalts), (Lava Creek Tuff), (Mesa Falls Tuff not exposed in park), (Huckleberry Ridge Tuff)	Post-caldera lava flows; Obsidian Cliff / Bull Lake glaciation / Pre-Bull Lake glaciation / Climactic eruptions and caldera collapse of third volcanic cycle; rhyolitic ash falls, mudflows, welded tuff, breccias, etc. / Second volcanic cycle; Island Park caldera and western part of Yellowstone Plateau / First volcanic cycle; caldera collapse in Island Park area; ash falls, pyroclastics, breccias, etc.; rhyolitic eruptions
Cenozoic	Tertiary — Neogene	Pliocene / Miocene	(hatched)	Erosion of ranges, canyon cutting, sedimentation in basins
Cenozoic	Tertiary — Paleogene	Oligocene		Regional uplift; block-faulting
Cenozoic	Tertiary — Paleogene	Eocene	*Absaroka Volcanic Supergroup (9 formations)	Repeated eruptions of andesitic pyroclastics and lavas; burial of fossil forests / Basaltic breccias and flows / Post-orogenic faulting, uplift and erosion
Cenozoic	Tertiary — Paleogene	Paleocene		
Mesozoic	(Cretaceous)		*Over 40 formations, some in northern and some in southern sections of park	Laramide orogeny; intense folding and thrust-faulting; uplifting of Rocky Mountains
Paleozoic / Mesozoic				Extensive deposition of marine and nonmarine sedimentary rocks as seas advanced and retreated over downwarped area
			(hatched)	Major angular unconformity
Precambrian			Gneisses and schists	Orogenic cycles; intrusions, metamorphism; sedimentation and erosion

*For details of the stratigraphic record, see Keefer (1971, pp. 9–11), Parsons (1978, p. 214), Christiansen and Blank (1972, p. B6), Love and Keefer (1975, p. D9), Ruppel (1972, p. A8), Smedes and Prostka (1972, p. C7).

the park in the Gallatin and Beartooth Ranges, as well as in the uplands in the southern margin of the park. Many of the Mesozoic and Paleozoic beds were stripped off during uplift or lie buried under many feet of volcanic rock. Little correlation exists between the northern and southern sedimentary rock units in the Yellowstone region.

Predominantly marine, the rocks are an assortment of conglomerates, sandstones, shales, and limestones. Approximately 3000 feet of Paleozoic rocks and 6000 feet of Mesozoic rocks were deposited, for the most part in shallow seas that repeatedly advanced and withdrew over what is now western North America. Beds of terrestrial deposits on beaches and flood plains are interspersed in the Mesozoic formations. For several hundred million years, this was a relatively quiet region of low elevation and low relief.

3. Laramide orogeny in late Cretaceous and early Tertiary time; faulting and erosion.

The mountain-building, compression, folding, and rapid increase in sedimentation that began in late Cretaceous and Paleocene time was the result of plate convergence to the west. The subducted Pacific plate underthrust the North American plate for hundreds of miles, shortening the crust and raising up the Rocky Mountain chain. Intense tectonic activity continued for some 30 million years.

In south central Yellowstone, Paleozoic and Mesozoic beds were tightly folded into anticlines (trending northwest–southeast), separated by synclines and faults. Movement along one reverse fault in this area was locally more than 10,000 feet. Reverse faulting of this magnitude also occurred in the northern part of Yellowstone. Fault blocks, created by the tremendous stresses, were raised, dropped, or tilted differentially. Intensified erosion was an immediate result of the uplift. Sediment stripped from the heights was deposited in nearby downwarped areas and basins. The Harebell Formation, for example, in south central Yellowstone, of latest Cretaceous age, received more than 8000 feet of sediment.

4. Eocene volcanism; the Absaroka volcanic field.

Tremendous outpourings and explosions of andesitic lava and pyroclastics built up the Absaroka and Washburn Ranges, as well as part of the Gallatins, during the Eocene Epoch. Most of the Yellowstone area was at least partially covered by volcanic debris. The eruptions of viscous lava, ash, and breccia came from large vents, one of which was in Mount Washburn. Composite thickness of the Absaroka Volcanic Supergroup is estimated to be over 12,000 feet. Intrusive equivalents (having the same age and composition) of the Absaroka volcanics were emplaced as stocks, dikes, and sills. Interbedded with the volcanics are sedimentary beds of reworked ash and volcanic

debris. Many deposits show the effects of extensive mass wasting, such as landsliding and slump. As eruptions subsided, large volumes of volcanic sediment were eroded and transported out of the Yellowstone area.

5. Late Tertiary (Pliocene) uplift, faulting, and erosion.

Regional uplift raised Yellowstone and the Middle Rockies several thousand feet. Mountain blocks were further elevated by faulting, while downthrown blocks became basins. The Gallatin Range was uplifted to its present height. To the south, the Teton Range was raised and Jackson Hole dropped down (chapter 40). With these tectonic events, the landscape of the Yellowstone region began to assume its present framework.

Uplift rejuvenated the streams, which began cutting deep canyons and removing more of the Absaroka volcanic rocks. Remnants exposed in the park, such as the Mount Washburn eruptive center, represent only a small amount of the great volume of Absaroka lavas that buried the Yellowstone surface in Eocene time. About 40 million years elapsed before volcanic activity—unrelated to the Absaroka episodes—resumed in Yellowstone.

6. Pleistocene rhyolitic eruptions; volcanic cycles and caldera collapse.

The largely rhyolitic rocks that built up the Yellowstone Plateau during the three Pleistocene volcanic cycles are designated the Yellowstone Group (fig. 39.14). The first volcanic cycle produced a large dome, quantities of welded tuff, and eventually a collapsed caldera, approximately 30 by 50 miles in size and several thousand feet deep. The Huckleberry Ridge Tuff is associated with the explosive eruptions that occurred as the caldera collapsed, about 2 million years ago (fig. 39.15).

The second volcanic cycle, similar in pattern but shorter in duration, was not so widespread or productive of lava as the first cycle. The second cycle reached its climax about 1.3 million years ago with the explosive eruptions that formed the Mesa Falls Tuff. Again, the caldera collapsed.

About 1.2 million years ago, the third, and most recent, volcanic cycle began with a definite shift in activity from the Island Park area to the Yellowstone Plateau. Ring fractures opened around the rising bulge. Through these fissures, gas, viscous rhyolitic lava, and ash were ejected until the roof rocks thinned and became seriously weakened. Very likely, ground water, plentifully supplied by rain and melting snow, played a role in triggering the catastrophic explosions of about 630,000 years ago, during which more than 1000 cubic kilometers of magma were expelled. Gaseous steam, suddenly released, propelled enormous fiery clouds of ash high into the air. Foaming lava poured from the vents, devastating the surroundings. Borne on prevailing winds, ash blanketed the

EXPLANATION

Third volcanic cycle
- PRM Roaring Mountain Member
- PCP Central Plateau Member
- PSL Shoshone Lake Tuff Member
- POC Obsidian Creek Member
- PUB Upper Basin Member
- PML Mallard Lake Member
- OS   Osprey Basalt
- MR   Madison River Basalt
- SLF Swan Lake Flat Basalt
- FR   Falls River Basalt
- ML   Basalt of Mariposa Lake
- LAC Lava Creek Tuff
- UF   Undine Falls Basalt
- MJ   Mount Jackson Rhyolite

Plateau Rhyolite

Second volcanic cycle
- MF   Mesa Falls Tuff

First volcanic cycle
- TN   Sediments and basalts of The Narrows
- LEC Lewis Canyon Rhyolite
- HR   Huckleberry Ridge Tuff
- BC   Rhyolite of Broad Creek
- JB   Junction Butte Basalt

Figure 39.14   Names and stratigraphic relationships of volcanic rock units recognized in the Yellowstone caldera area. (From Christiansen and Blank 1972, p. B7)

Figure 39.15   Cliffs of welded Huckleberry Ridge Tuff at Golden Gate, northwest section of Yellowstone National Park. A great pyroclastic flow over 2,000,000 years ago, during the first volcanic cycle, produced this dense, resistant welded tuff. Photo by W. H. Parsons.

Table 39.2 Comparison of estimated volumes of recent explosive eruptions with estimated volumes of Yellowstone volcanic explosions. Note that the Yellowstone eruptions were much greater in volume (20 times or more) than any modern volcanic explosive eruptions.

Volcano	Date	Volume of ejecta (cubic kilometers)
St. Helens	1980	1.2 km³
Lassen	1914–21	1.3
Katmai	1912	7
Krakatoa	1885	18
Mazama	6845 B.P.	75
Tambora	1815	150
Yellowstone/Island Park		
First cycle	2.2 m.y.a.	2500
Second cycle	1.2 m.y.a.	280
Third cycle	0.6 m.y.a.	1000
Yellowstone total		3780

Sources: Williams and McBirney 1979; Decker and Decker 1981, in Smith and Braile 1984.

midcontinent area. Collapse of the chamber roof produced a caldera 45 by 25 miles in extent and a half a mile deep. Among the features of the Yellowstone landscape formed at this time were the welded tuff deposits of the Lava Creek Tuff, the depression holding Yellowstone Lake, and the blurred but still visible caldera rim.

A resurgence of fairly strong volcanic activity took place about 160,000 years ago in the depression now occupied by the West Thumb of Yellowstone Lake. After dome expansion, explosive eruptions and collapse left a small caldera.

For thousands of years after the collapse of the main caldera, viscous lava periodically poured out over the caldera floor, sometimes in flows 1000 feet thick. Most flows stayed within the caldera, but a few spilled over the rim. Lava occasionally came up through the ring fissures. A prominent, small flow of this last type is the one exposed at Obsidian Cliff. Minor basalt flows also oozed out.

7. Glaciation and canyon-cutting.

At various times, glacial ice blocked major river valleys, diverted streams, and dammed Yellowstone Lake. Toward the end of Wisconsinan glaciation, Yellowstone Lake was approximately 140 feet higher than it is now. In some places in the park, glacial drift and alluvium are interbedded with volcanic debris, indicating that eruptions occurred when glaciers were in existence.

After the disappearance of Pinedale glaciers during the postglacial optimum (between 6000 and 2000 B.C.), the Yellowstone River cleared out its Grand Canyon. The processes of weathering, mass wasting, and stream erosion continue to modify this remarkable feature (fig. 39.16).

9. Post-Pleistocene uplift and erosion; geothermal activity.

The Yellowstone Plateau and the caldera have undergone uplifting in Holocene time. The caldera's more rapid rise is continuing at the present time. Two small resurgent domes in the caldera may presage a resumption of volcanic activity. Geysers and hot springs have probably been active on the Yellowstone Plateau ever since the end of Pinedale glaciation, and they may have been intermittently active during interglacial periods. Location of the hydrothermal features is believed to be related to faults and fractures in the bedrock.

Rates of heat flow on the Yellowstone Plateau, the frequency of shallow earthquakes, and the inferred presence of a body of hot, low-density material below the surface all point toward the possibility of eventual new eruptions. Investigators have noted that a major caldera has formed in the Yellowstone region about every 600,000 to 900,000 years in the past two million years. Inasmuch as the present Yellowstone caldera collapsed about 630,000 years ago, we can speculate that a new one may be in preparation.

Figure 39.16   Two resistant lava flows with columnar jointing jut out in the Narrows of Yellowstone Canyon, near Tower Falls. Between the two flows are numerous layers of stream gravel. A blanket of glacial till rests on the upper flow. Photo by W. H. Parsons.

Geologic Maps and Cross Sections

Christiansen, R. D. et al. 1986. Wyoming geologic highway map. Western Geographics and Geological Survey of Wyoming.

Taylor, R. L., Ashley, J. M. et al. n.d. Geological map of Montana and Yellowstone National Park. Department of Earth Sciences, Montana State University.

U.S. Geological Survey. 1972. Geologic map of Yellowstone National Park. Miscellaneous Geological Investigations, Map I-711.

U.S. Geological Survey. 1972. Surficial geologic map of Yellowstone National Park. Miscellaneous Geological Investigations, Map I-710.

Bibliography

Baker, R. D. 1976. Late Quaternary vegetation history of the Yellowstone Lake basin, Wyoming. U.S. Geological Survey Professional Paper 729E.

Bartlett, R. D. 1962. *Great surveys of the American West.* Norman, Oklahoma: University of Oklahoma Press.

Boyd, R. D. 1961. Welded tuffs and flows in the rhyolite plateau of Yellowstone Park, Wyoming. *Geological Society of America bulletin* 72:387–436.

Bryan, T. S. 1979. *The geysers of Yellowstone.* Boulder, Colorado: Colorado Associated University Press.

Christiansen, R. L. 1984. Yellowstone magmatic evolution: its bearing on understanding large-volume explosive volcanism, in National Research Council 1984. *Explosive volcanism: inception, evolution, and hazards,* Studies in geophysics, series II. Washington, D.C.: National Academy Press. p. 85–95.

———, and Blank, H. R. 1972. Volcanic stratigraphy of the Quaternary rhyolite plateau in Yellowstone National Park. U.S. Geological Survey Professional Paper 729-B.

Crandall, H. 1977. *Yellowstone, the story behind the scenery.* Las Vegas, Nevada: KC Publications.

Dorf, E. 1981. *Petrified forests of Yellowstone.* National Park Service pamphlet.

Hammond, A. L. 1980. The Yellowstone bulge. *Science 80* 1:68–73.

Keefer, W. R. 1971. *The geologic story of Yellowstone National Park.* U.S. Geological Survey Bulletin 1347.

Love, J. D., and Keefer, W. R. 1975. Geology of sedimentary rocks in southern Yellowstone National Park. U.S. Geological Survey Professional Paper 729D.

Marler, G. D. 1964. *Studies of geysers and hot springs along Firehole River, Yellowstone National Park.* Yellowstone Library and Museum Association.

Parsons, W. H. 1978. *Middle Rockies and Yellowstone,* K/H Geology Field Guide Series. Dubuque, Iowa: Kendall/Hunt Publishing Company.

Pierce, K. L. 1979. History and dynamics of glaciation in the northern Yellowstone Park area. U.S. Geological Survey Professional Paper 729-F.

Ruppel, E. T. 1972. Geology of the pre-Tertiary rocks in the northern part of Yellowstone National Park. U.S. Geological Survey Professional Paper 729-A.

Smedes, H. W., and Prostka, H. J. 1972. Stratigraphic framework of the Absaroka Volcanic Supergroup in the Yellowstone National Park region. U.S. Geological Survey Professional Paper 729-C.

Smith, R. B., and Braile, L. W. 1984. Crustal structure and evolution of an explosive silicic volcanic system at Yellowstone National Park, in National Research Council 1984. *Explosive volcanism: inception, evolution, and hazards,* Studies in geophysics, series II. Washington, D.C.: National Academy Press. p. 96–109.

———, and Christiansen, R. L. 1980. Yellowstone Park as a window on the earth's interior. *Scientific American* 242:104–117.

U.S. Geological Survey. 1977. *Geysers.* 23 p.

White, D. E.; Fournier, R. O.; Muttler, L. J. P.; and Truesdell, A. H. 1975. Physical results of research drilling in thermal areas of Yellowstone National Park, Wyoming. U.S. Geological Survey Professional Paper 892.

Note: The Yellowstone Institute, formed in 1976 in response to public interest in the Yellowstone ecosystem, offers summer seminars on the natural history of the park. College credit is available for courses in wildlife management, geology, botany, and the like. For information, write Yellowstone Institute, Box 515, Yellowstone National Park, Wyoming 82190.

Yellowstone National Park

Area: 2,219,8232 acres; 3468 square miles
Established: March 1, 1872
Designated a Biosphere Reserve 1976
Designated a World Heritage Site: September 6, 1978
Address: P.O. Box 168, Yellowstone National Park, Wyoming 82190

PART V

Landscapes and Structures in Areas of Complex Mountains

© by G. Shimer

Figure V.1 In Great Smoky Mountains National Park, in the high ranges of the Southern Appalachians, ancient and complex rocks and structures have created a mountain landscape of surpassing grandeur. Drawing by G. Shimer from J. A. Shimer, *Field Guide to Landforms in the United States;* The Macmillan Company. © 1972 by J. A. Shimer. Used by permission.

Mountains are challenging to climbers and to geologists. The rocks exposed, the shapes and elevations of summits, and the character of the land surrounding mountain ranges all give clues about the earth's internal processes—crustal movements and tectonic forces—that created mountains and placed them where they are today. The results of external processes show up on mountains, too. How have weathering, mass wasting, and the agents of erosion altered the rocks and the scenery?

Geologists like to do field work in mountains. They find more different kinds of rocks and geologic structures than are ordinarily visible in flatlands. But deciphering the nature and processes of mountain-building can be difficult and sometimes hazardous. A vital outcrop may be covered by glacial snow and ice or may be inaccessible to even skilled rock-climbers. Avalanches may have buried geologic structures; or erosion may have removed them.

As you have seen in earlier chapters, mountains are seldom isolated. They are generally components of a mountain mass, or *mountain range,* consisting of a succession of mountains or mountain ridges. *Major mountain belts,* which are chains or cordilleras consisting of many mountain ranges, tend to be more than a thousand miles

in length, but are usually only a few hundred miles wide. The Appalachians, which extend from Newfoundland south into Georgia, are an example. Acadia National Park (chapter 21) along with Shenandoah and Great Smoky Mountains National Parks (chapters 47 and 48 in this section) are in the Appalachians. Rimming the western side of the continent—from the Aleutian Islands in the Bering Sea southward into Mexico—are the multiple ranges of the North American Cordillera. Within its domain are 34 American national parks and seven Canadian national parks. Of the ten American national parks described in Part V, seven (Grand Teton, Great Basin, Sequoia, Kings Canyon, Channel Islands, Redwood, and Big Bend) are in the North American Cordillera.

Characteristics of Major Mountain Belts

What we know about the origin and development of mountain ranges and major mountain belts is a compilation of decades of study by many scientists. Mountain ranges are the result of severe deformation of crustal material plus igneous activity in a particular area. Mountain belts are produced along plate margins by plate convergence or collision with accompanying tectonic activities

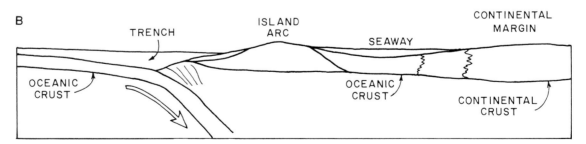

Figure V.2 A,B.   Schematic representations of two ways in which subduction may occur along a plate boundary. *A.* In this convergence of an oceanic plate and a continental plate, hundreds of miles of oceanic crust are disappearing beneath an on-thrusting subduction complex. Sediments accumulate in the shallow trench and the fore-arc basin; hot magma rises from the subduction zone; young mountain ranges, including a volcanic chain, build up parallel to the continental margin. *B.* Here a deep trench has formed between converging oceanic plates. Oceanic crust is subducted beneath an island arc, which is separated from the continental margin by a seaway. Sediments accumulate in the trench and the seaway. Volcanoes erupt along the island arc. Eventually, the island arc may become a tectonostratigraphic terrane, jammed against the continental margin, with crushed and severely deformed oceanic crust and sediments in the seaway slivered between.

Figure V.3   West side of the snow-capped Snake Range, Great Basin National Park Nevada, overlooking arid Spring Valley. The Snake Range is an upthrown, tilted fault block. Spring Valley is a basin formed by a down-dropped block. National Park Service photo.

on a greater and more prolonged scale. In general, three stages occur in the evolution of mountain belts: 1) an accumulation stage; 2) an orogenic stage; and 3) a block-faulting and uplift stage with prolonged erosion.

In an *accumulation stage,* great thicknesses of sedimentary and volcanic material are deposited in a marine environment. Sediments that accumulate, for example, between converging oceanic and continental plates are compressed and pulled down or scraped off when the oceanic plate is subducted. Over time, the squeezed and folded sedimentary rocks may be elevated and eventually exposed by erosion in mountainsides. For example, in Redwood National Park (chapter 44), which is located in the Northern California Coast Ranges, a greenish rock that is a contorted mixture (mélange) of deep-water, shallow-water, land-derived, and volcanic sediments is conspicuous in road cuts and cliffs along the shore.

The *orogenic* ("mountain-building") *stage* begins while accumulation of sediments is still going on. The convergence or collision of plates intensifies compression and causes the layered rocks to be tightly folded and then broken by thrust faults. Deeply buried rock layers are metamorphosed by heat and pressure; still deeper in the

crust or upper mantle, magma forms batholiths that may supply volcanoes or eventually cool below the surface as plutons. The plunging folds in Hot Springs National Park, Arkansas (chapter 45) are evidence of orogenic activity involving compression and deformation due to plate collision (fig. V.2 A,B).

Block-faulted mountain ranges are predominant features in several national parks in Part V. Two dramatic and well known examples of tilted block-fault ranges are Wyoming's Teton Range in Grand Teton National Park (chapter 40) and California's Sierra Nevada, displayed in Sequoia and Kings Canyon National Parks (chapter 42), as well as Yosemite National Park (chapter 25). Great Basin National Park, Nevada (chapter 41) and Big Bend National Park, Texas (chapter 46) are in the vast Basin and Range physiographic province of western United States where hundreds of block-fault ranges jut up, separated from each other by down-faulted valleys. Two national parks in Part II are also in the Basin and Range province; namely, Carlsbad Caverns National Park, New Mexico (chapter 14), and Guadalupe Mountains National Park, Texas (chapter 15).

The *block-faulting* and *uplift stage* is probably a reflection of isostatic adjustments following the orogenic stage. Continental crust, being less dense, tends to ride higher on the mantle than oceanic crust. Mountain-building causes a mountain belt to be out of equilibrium in relation to the *craton* (the continent's stable interior). The continental crust of the mountain belt, greatly thickened by accumulation of sedimentary rocks and intrusions of igneous rock, begins a long slow rise as erosion strips rock from summits and slopes. During uplift, normal faults may develop due to tensional (horizontal) forces pulling apart the rising crust so that fault-block mountains are created. Block-faulting may also result when vertical movements in the crust are unequal. When a mountain mass "floats" upward in order to regain isostatic balance, tilting may occur, or some blocks may rise faster than others.

Continents enlarge as major mountain belts evolve along continental margins. Some of the growth may be due to *tectonic accretion,* or addition of tectonostratigraphic terranes to a continental margin. In chapters about the national parks in Alaska, this process is described in more detail, especially in chapters 28 through 31 (Part III) and 35 and 36 (Part IV). The terranes (also called "suspect" or "exotic," depending on how much is known about their origin) may be island arcs or microcontinents that have been rafted on an oceanic plate for great distances before colliding with a continental margin. While most of the terranes that have been welded to western North America are relatively young in terms of geologic time, terranes added to North America's eastern margin are older and are associated with earlier (Paleozoic and Precambrian) orogenies. Chapters 47 and 48, describing Shenandoah and Great Smoky Mountains National Parks, deal with the evolution of the Appalachians as a major mountain belt.

National Parks in Part V

40

Grand Teton National Park

Northwest Wyoming

by Sherwood D. Tuttle
University of Iowa

Silvery gray, ice-sculptured peaks—
distinctive landmarks for Indians, pioneers,
and pilots—rise more than a mile above
sagebrush flats.

Tectonic forces are still raising the
upthrown Teton Range and tilting the
downdropped Jackson Hole block.

The Tetons, youngest mountains in the
Rocky Mountain system, are made up of
some of the continent's oldest rocks. Ice,
frost, running water, and mass wasting
chisel the tough, ancient rocks of the range
and deliver great volumes of sediment to
valley floor.

A few modern glaciers cling to cirque
basins and precipitous slopes.

Tiny, rock-basin lakes and patchwork
alpine meadows nestle among the peaks;
moraine-dammed lakes stretch along the
fault trace at the foot of the range.

The Snake River, wandering over the floor
of Jackson Hole, has cut steep terraces in
outwash deposits.

Nearly 3000 elk summer in the park and
migrate to the adjoining National Elk Refuge
in winter. The Teton Range and Jackson
Hole are part of the Greater Yellowstone
Ecosystem.

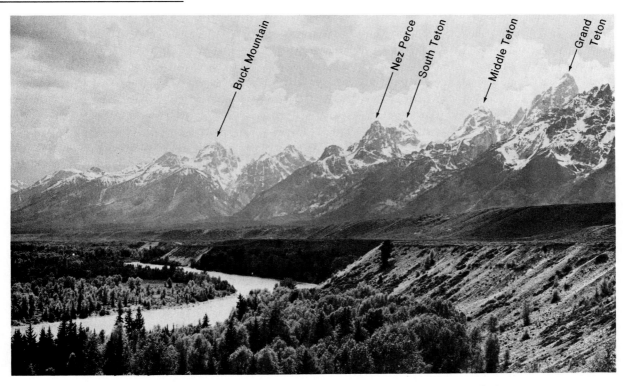

Figure 40.1 The Teton Range, the Snake River, and part of Jackson Hole in Grand Teton National Park. The Snake River, flowing south away from the viewer, has cut several terrace levels in the thick outwash deposits that retreating Pleistocene glaciers left in Jackson Hole. National Park Service photo.

Figure 40.2 Grand Teton National Park, Wyoming

Local History

Small bands of early hunters, drifting down from the Northern Rockies, camped beside Jackson Lake some 12,000 years ago. While roaming over the land, following the seasons and the migrations of animals, hunters and gatherers looked for the distinctive skyline of the Teton Range, just as we today scan the horizon, waiting for those dramatic peaks to come into view, whatever our mode of travel. Grand Teton and its companion peaks, rising without foothills from the flats, are like a beacon. Once seen, their outline against the sky is never forgotten. Who has stood on the floor of Jackson Hole, face tilted upward, and regarded the formidable east face of the Teton Range without a sense of awe?

Winters in Jackson Hole were too severe for habitation, but for thousands of years many tribes hunted, fished, and gathered plants in the valley in the warm months. Several routes of travel and trade passed through Jackson Hole. The Indians could follow the Snake River north into Yellowstone for obsidian, or go south to the Colorado Plateau or east to the Great Plains. Teton Pass, at the southern end of the range, was a shortcut to the Pacific Northwest. The Blackfoot, Crow, Flathead, Gros Ventre, Nez Perce, Bannock, Shoshone, and Ute Indians knew the area well and most of the time regarded the valley as neutral ground.

To the trappers and traders of the eighteenth and early nineteenth centuries, a deep valley ringed by high mountains was a "hole," likely to be a safe place where they could "hole up." Jackson Hole, the favorite valley of trapper David Jackson, was named for him in 1829 by his fur trade partner. Early French voyageurs gave the name *Les Trois Tetons* to the three peaks now called Grand Teton, Middle Teton, and South Teton. The common translation used by the American mountain men was "The Three Paps." These easily distinguished peaks, which can be seen from both western and eastern approaches, were the landmarks that guided the trappers, who followed trails earlier established by Indians. At the height of the fur trade in the 1830s, Jackson Hole was a crossroads through which major fur expeditions passed. By cutting through Teton Pass, trappers could reach mountain streams rich in beavers—streams draining the western slopes of the Tetons—as well as the wild rivers to the north. Around 1840, the fur trade ended abruptly as men of fashion began wearing tall silk hats instead of stovepipe hats made of beaver pelts. Jackson Hole became an isolated and nearly forgotten valley for some twenty years.

F. V. Hayden, the geologist who led the first government-sponsored, surveying and scientific expedition into Yellowstone in the summer of 1871, decided to also map the Teton valley that he had visited in 1860 with the Raynolds expedition. In the spring of 1872, Hayden split his group into two parties. He led one team into Yellowstone from the north. Geologist James Stevenson led the second party from Utah north to Fort Hall in Idaho and then east across Teton Pass into Jackson Hole. The two parties met late in the summer in Yellowstone. With Stevenson was N. P. Langford, who had been appointed superintendent of the newly established Yellowstone National Park; the photographer, William Henry Jackson; the artist-geologist, William Henry Holmes; several other geologists and topographers, packers, and a cook; and a mountain man as guide. Stevenson and Langford climbed to the top of a summit that they thought was Grand Teton, but apparently they were mistaken. A later surveyor, William O. Owen, made the first officially affirmed ascent of Grand Teton (elevation 13,770 feet) 26 years later in 1898.

Among the lasting accomplishments of that summer and fall of 1872 were the remarkable photographs of Teton country taken by William Henry Jackson; the superb drawings of William Henry Holmes, including his famous panorama showing the Teton Range; topographic maps; scientific reports; fossil and rock collections; and plant and animal specimens, including birds and insects. Two more expeditions in 1877 and 1878 completed the mapping of most of Wyoming, Idaho, and Montana and furnished material for the voluminous Hayden Survey reports covering those years. By the 1880s, this part of the American West was no longer unknown territory to the civilized world.

When Jackson Hole was opened to homesteaders after the completion of the surveys, a few settlers established homes and ranches. However, the growing season was too short for most crops, and the valley was isolated for weeks each winter when snow blocked the passes. The population remained sparse well into the twentieth century, even after Teton County was organized in 1921.

Meanwhile, a movement was begun to protect the timber south of Yellowstone. Pierce Cunningham, a rancher whose cabin is now a historic site, circulated a petition asking that the valley be set aside "for the education and enjoyment of the Nation as a whole." In 1929, part of Teton National Forest was designated Grand Teton National Park. This act by Congress protected the Teton Range and some of the lakes, but most of the floor of Jackson Hole, including Jackson Lake, remained in private hands. Cattlemen strongly opposed proposals to incorporate grazing acreage into the national park.

Nevertheless, interested citizens throughout the country realized that historically, culturally, and geologically, the Teton Range and the valley floor below were two halves of a natural environment that needed to be protected and managed as an integrated unit. John D. Rockefeller, Jr., who believed in this concept, formed a land company in 1927 and began buying up land to be held until such time as it could be administered by the National Park Service. President Franklin Roosevelt took the next step in 1943 by proclaiming an area of 223,000 acres (much of which was already public land) as Jackson Hole National Monument. The Rockefeller land was accepted as an addition to the monument in 1949. Finally, after considerable political maneuvering, Congress established the enlarged area as Grand Teton National Park in 1950, with part of the former Jackson Hole National Monument being absorbed by the National Elk Refuge south of the park. In recognition of Rockefeller's vision and generosity, the scenic highway that goes through Grand Teton National Park and into Yellowstone National Park (chapter 39) was named John D. Rockefeller, Jr. Memorial Parkway.

Geologic Features

Photographs cannot capture the grand scale of the landforms that the human eye sees in Grand Teton National Park. When you drive along the highway on the floor of Jackson Hole, the massive, gray mountain front looms up almost within reach. Snow clings to the roughhewn rock heights, and small glaciers hang in high, narrow, steep-walled valleys. In the canyons incising the slopes, waterfalls drop in silvery white lines, easily visible from many points along the highway.

Grand Teton National Park includes the major portion of two great landforms: (1) the Teton Range, an elongate, upfaulted block tilted to the west and about 45 miles long; and (2) Jackson Hole, a narrow, downdropped fault block, about the same length and 6 to 12 miles wide. The valley's remarkably flat floor varies in elevation from 6000 to 7000 feet. At least seven of the Teton peaks exceed 12,000 feet in elevation; Grand Teton is over 13,000 feet. These tall peaks, concentrated in the 4-mile center of the range, make up the "Cathedral Group," between Avalanche and Cascade Canyons.

The precipitous drop from peaks to valley floor, might be even greater if not for intense erosion that has crenulated and dissected the mountain range and deposited thousands of feet of sediment that cover the bedrock of the valley floor (fig. 40.1). An overriding question is, what tremendous force split the ancient crust, swinging the mountain block skyward and dropping the valley block down? In searching for answers to this question, we begin with a look at the ancient rocks.

Rocks and Structures

The sedimentary rocks that once covered the mountain block have been stripped from the top of the range, exposing Precambrian crystalline rock. Remnants of sedimentary rock flank the western slopes of the range. Similar units beneath Jackson Hole are buried by alluvial and glacial deposits. In geomorphic terms, the mountain features are erosional landforms and the valley features are depositional landforms.

The Crystalline Rocks. The Precambrian rocks that dominate the Teton Range consist of layered gneisses, granitic gneisses, granites, pegmatites, and diabase dikes. The banded nature of some of the gneisses suggests that they were sedimentary and volcanic rocks that accumulated in a subsiding basin or trench and were subsequently metamorphosed during orogenic activities. The massive gneisses were once granitic intrusions that were later metamorphosed. The gneisses are the oldest of the Precambrian rocks in the park. Strontium-rubidium dating places the age of the granites at about 2.5 billion years, and the black diabase dikes are 1.3 billion years old.

Extending vertically up the face of Mount Moran from base to summit is a black band that is the exposed edge of one of the diabase dikes that cut through the older rocks. The Mount Moran dike is about 150 feet thick and can be traced westward for about 7 miles. *Diabase* is a dense, black, igneous rock studded with light-colored plagioclase crystals (resembling small laths) incorporated with pyroxene crystals.

Some of the gneisses contain dark crystals of garnet set in white "halos," which look like eyes peering out from the rock face. Geologists call these rocks "bright-eyed gneiss." Pods of heavy dark green or black serpentine (soapstone) are scattered throughout the gneisses. Indians carved bowls from this smooth, relatively soft rock. Pebbles of serpentine—if found in stream gravels outside the park—can be polished and used as gems. This so-called "Teton jade" is softer and less lustrous than true jade.

Paleozoic Sedimentary Rocks. On the western flanks of the Teton Range and at the northern and southern ends of the range, gently dipping sequences of Paleozoic marine rocks top ridge crests and crop out on valley walls. The beds lie unconformably over the Precambrian rocks. Projections updip of the rock layers indicate that they formerly extended eastward clear across the Teton Range but were eroded as the mountain block was uplifted and tilted. Breaks in the rock record, shown by erosional unconformities, indicate many interruptions of deposition. Within the sequence are limestones, dolomites, shales, and sandstones. Careful study of the sedimentary units yields details of source areas, directions of sediment movement, positions of the Paleozoic seas, and evolving marine life forms. Drilling and subsurface geophysical exploration have made possible the identification of the same Paleozoic sequences deeply buried beneath Jackson Hole.

Mesozoic Sedimentary Rocks. Marine and nonmarine Mesozoic rocks that were deposited on top of the Paleozoic sequence were stripped off the Teton Range, exposing the Precambrian and Paleozoic rocks. Mesozoic rocks crop out in the northern, eastern, and southern parts of the Teton region, outside the park, and are also buried below the Jackson Hole sediments.

Figure 40.3   High altitude view, looking north, at Grand Teton (13,770 feet), highest peak of the range. The peak on the right, Mount Owen (12,928 feet), is second highest. The glacial trough behind the peaks is Cascade Canyon. The Precambrian crystalline rocks exposed in the peaks are gneisses. National Park Service photo.

Cenozoic Rocks and Sediments. Tertiary rocks are exposed along the eastern margin of Jackson Hole and at the north end of the Teton Range, but most of the Tertiary units are buried beneath younger deposits that cover the floor of Jackson Hole. The Tertiary sequence of conglomerates, sandstones, shales, volcanic breccias, and tuffs have an aggregate total thickness of about 6 miles! No other region in the United States contains a more complete nonmarine Tertiary section than the Teton region. The rock units reflect the tectonic activity of the Laramide orogeny, Eocene volcanism that produced the Absaroka lavas and ashfalls, and the movement of the fault blocks.

Uplift of the mountains accelerated erosion, producing vast amounts of debris; while the downfaulted basin provided places for sediments to accumulate. A resumption of volcanic activity in Oligocene and Miocene time contributed additional material. Then in late Pliocene and Pleistocene time, the great volcanic events occurring in the Yellowstone area (chapter 39) spread rhyolitic debris and tuff over the northern end of the Teton Range and over Jackson Hole.

All surface materials in the Jackson Hole section of the park, with the exception of outliers of older rock in Signal Mountain and Blacktail Butte, are Pleistocene or Holocene in age. These young deposits, which consist of volcanics, lakebed sediment, glacial drift, and alluvial materials, have blanketed the thick sequences of Paleozoic, Mesozoic, and Tertiary rocks that fill the Jackson Hole basin. Scattered bones, pollens, and snail shells in the lakebed sediment document the presence of Cenozoic plants and animals that lived in the Teton region prior to the present.

Faults and Fault Blocks

Tensional forces produced the Teton fault system, split apart the two large crustal blocks, and tilted them toward the west, with the mountain block hinging up and the east-side block sliding down. Separating the huge blocks is the 40-mile-long trace of the fault, which runs along the base of the Teton Range, approximately where the slope breaks, between the mountain front and the western side of Jackson Hole. In structural terms, the Teton fault is a

Figure 40.4   High altitude view looking west across Grand Teton, Teton Canyon, and, in the far distance, the Teton Basin. Most of the area shown lies outside the park's western boundary. The layered rocks in the middle distance are Paleozoic beds dipping gently to the west on the back of the Teton Range. In the right foreground, beds of upturned Paleozoic rock slope away from the Buck Mountain fault, which trends north-south, parallel to the range crest. National Park Service photo.

steeply dipping *normal fault*. By definition, the hanging wall of a normal fault has moved down in relation to the footwall. The Teton fault dips to the east. Therefore, the hanging wall side, the Jackson Hole block, moved downward in relation to the Teton Range block, which moved up (fig. 40.5).

How did geologists know the fault was there? How did they determine when the movement occurred? How much did the blocks move and in what direction? What do geologists know about the rate of movement, and what do they think caused this displacement of crustal units? Most of the geologic evidence was found in the rocks that were briefly described above.

1. Geologists suspect faulting when rocks of quite different types and ages are next to each other and the stratigraphic and geomorphic conditions make an unconformity unlikely. Beneath the floor of Jackson Hole, the Precambrian crystalline rocks of the Teton Range butt up against much younger rocks. Although the fault contact is obscured by glacial and stream deposits, geologists can locate the fault system by drilling and seismic prospecting, and by projecting data collected from isolated outcrops.

2. The absence of foothills is significant. This indicates that the mountains were not the result of erosion.

3. The relatively straight and smooth eastern front of the Teton Range is suggestive of faulting. Not many geologic processes other than faulting produce linear features. Because the mountain front has been eroded and cut by canyons, it does not represent the original

Figure 40.5   Simplified cross section of the Teton fault system, showing displacement of Paleozoic beds, the amount of ''throw,'' and a projection of volume of rock removed by erosion. (Buck Mountain fault not shown.)

fault scarp, but does suggest the approximate slope and location of the scarp. A few small, triangular facets, remnants of the original fault scarp, are spaced along the base of the range.

4. The most likely cause of the asymmetry of the range is faulting and tilting.

5. Small, fresh fault scarps in the vicinity of the Teton fault system are regarded as evidence of postglacial faulting and tilting.

With the presence of the Teton fault inductively confirmed, the next step is to determine which side moved up and how much movement occurred by matching rock types (stratigraphic correlation) across the fault. In matching one side with the other, the best marker is the contact between the Paleozoic rocks and the underlying Precambrian rocks. On the Teton block, we project the westward-dipping Precambrian-Paleozoic contact toward the east until it intersects the steeply dipping plane of the fault. On the Jackson Hole block, the Paleozoic-Precambrian contact can be located far below the surface by deep drilling and by geophysical techniques. The matching of contacts shows that the western block appears to have moved up relative to the eastern block a distance of roughly 30,000 feet (fig. 40.5).

The shape of the Teton block suggests that the Teton fault dips steeply eastward. Data from geophysical studies support this conclusion. Therefore, the Jackson Hole block (the hanging wall) moved downward in relation to the Teton block (the footwall), making the Teton fault a normal fault.

According to the principle of cross-cutting relationships, a fault is younger than the youngest rock that it cuts and older than the oldest rocks that are *not* displaced or cut by fault movement. Applying this rule to the Teton fault, we find that it truncated the youngest sedimentary rocks in the basin, that it began about 9 million years ago, and that it is still active. Most of the movement, which was spasmodic rather than continuous, occurred late in Tertiary time before Pleistocene glaciation. The average rate of movement (derived by dividing the amount of "throw" by the number of years) was about one foot for every 300 years. The continued tilting of the Jackson Hole block affects stream drainage and causes some slopes to be susceptible to landsliding.

Several other faults have significantly affected the Teton landscape. The highest mountains in the central part of the Teton Range were elevated by a reverse fault that is older than the Teton fault. The Buck Mountain fault, a reverse fault of moderate displacement, can be traced for a distance of about 10 miles in a north–south direction on the western side of the Teton summit area (figs. 40.6, 40.7). (A *reverse fault* is one where the hanging wall moves up in relation to the footwall.) West of the Buck Mountain fault trace, the Paleozoic sedimentary rocks are bent sharply upward, and on the narrow, uplifted block on the

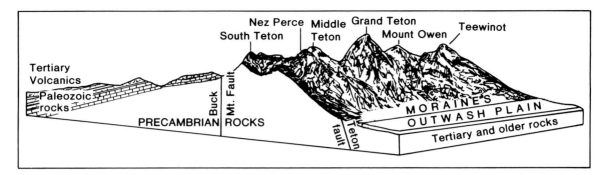

Figure 40.6   Block diagram of the crest of the Teton Range, bounded on the east by the Teton fault and by the Buck Mountain fault on the west. Outwash and moraines obscure the base of the fault scarp. On the western slopes are remnants of Paleozoic rocks that once lay unconformably over the Precambrian crystalline rocks. (After Horberg 1938; reproduced by permission of Augustana Library Publications.)

east side of the fault are the dominant peaks of the range—the Three Tetons and several others—in a fairly straight alignment. A logical assumption is that the Buck Mountain reverse fault was a result of orogenic activity that preceded the normal Teton fault of late Tertiary time. Thus the crest area that had been raised by reverse faulting was later elevated by normal faulting to its present height.

On the east side of the Jackson Hole block is a nearly vertical normal fault with the west side upthrown and the east side dropped down. Other faults have been mapped at both the northern and southern ends of the valley.

The ultimate cause of the faulting in the Teton region is related to the geologic history of the surrounding region. Fractures, or faults, in the earth's crust are responses of rock to stresses that may originate in the subcrustal zone of the earth. Normal faulting, looked at mechanically and geometrically, can be explained by stretching or extension of the crust. Such forces are described as tensional. Reverse faulting, on the other hand, is usually the result of compression of the crust. At different times, movements of tectonic plates may cause crustal compression or crustal extension.

During the Laramide orogeny, which began at the end of Cretaceous time, whole regions of western North America were folded, uplifted, and subjected to reverse faulting and thrust faulting on a grand scale. After compressive forces subsided, stretching and uplift produced many fault-block ranges and basins. However, the relationship of regional tectonic events to the unique geologic history of the Tetons is neither obvious nor simple. Perhaps the key to the puzzle may come eventually from a better understanding of the causes of the volcanism that has played an important role in the development of the Yellowstone-Teton landscape.

How Surface Processes Modified the Teton Landscape

Faulting has been the most significant geologic process at work in developing the landscape in Grand Teton National Park. Nevertheless, as the structurally controlled mountain framework rose, moisture-bearing clouds were forced to release rain in order to surmount the heights. Greater precipitation at the high elevations fed snowfields and increased the volume and velocity of streams. Running water and glacial ice, with the help of weathering and mass wasting processes, set about sharpening peaks, excavating canyons, and sculpturing the range. Sediments soon covered the floor of Jackson Hole. The most profound surficial changes were those wrought by major Pleistocene glaciers.

The Teton Range is a world-famous example of *alpine topography;* it is certainly as spectacular as the European Alps and the New Zealand Alps. The Teton skyline, as seen from below, is a series of concave curves passing from peak to peak with lower passes in between. The serrated ridge, conical horns, and great cirques are evidence of severe glacial erosion. The higher peaks in the center of the range are the focus of this grand scenic display.

Despite the fact that the Precambrian rocks are extremely hard and resistant to erosion, the high peaks, which remained above the ice as glaciers grew and shrank, were badly shattered and cracked by frost action on all exposed surfaces. Below the rough, jagged rock of the peaks are smooth ledges that have been scoured and scraped by ice. Shallow, ice-eroded depressions in the bedrock are occupied by small lakes. Cirques and rock basins in the glacial troughs also contain lakes.

The large cirques and groups of cirques that head the major canyons are like amphitheaters with sheer walls hundreds of feet high. The canyons, or glacial troughs,

Figure 40.7 Location map showing major faults and geologic features. The moraines shown are Pinedale in age. The Jackson Lake recessional moraine holds in Jackson Lake. Burned Ridge is a terminal moraine. The Potholes are in glacial outwash that extends to the south over Jackson Hole.

Location key
Mountain peaks (from north to south): MO = Mount Owen, GT = Grand Teton, MT = Middle Teton, ST = South Teton, BM = Buck Mountain.
Small fault blocks on the floor of Jackson Hole (from north to south): SM = Signal Mountain, BB = Blacktail Butte, EGVB = East Gros Ventre Butte, WGVB = West Gros Ventre Butte.
Moraine-dammed lakes at the base of the Teton Range (from north to south): L = Leigh, S = String, J = Jenny, B = Bradley, T = Taggart, P = Phelps.
Pre-Pinedale morainal lakes: TO = Two Ocean, EM = Emma Matilda.
Landslide topography: GVS = Gros Ventre Slide.

dissect the eastern face of the range. Excavated several thousand feet deep by ice, they have broad floors and sheer, oversteepened walls, forming the U-shape typical of glacial valleys. Tributary glaciers carved hanging valleys, now occupied by waterfalls. Some canyons have glacial staircases with falls and cascades rushing down the slopes.

Several times during the Pleistocene, valley glaciers became so large that they coalesced and spilled out over the floor of Jackson Hole. Here they joined great tongues of ice from the north and east that filled the basin. Moraines built at the mouths of the canyons mark the extent of the last Pleistocene glaciation. These moraines now hold a series of lakes (Leigh, String, Jenny, Bradley, Taggart, and Phelps) close against the foot of the mountains. Several of the lakes are surprisingly deep, such as Jenny Lake (226 feet) and Leigh Lake (250 feet). Meadows in the moraines have replaced former, more shallow lakes that have been filled in or have drained away. Large glacial boulders, carried down by the ice, dot the moraines and rim the lakes.

Jackson Lake, the largest body of water in the valley, occupies the trough of the last major glacier from the north that combined with ice from the northern canyons and covered part of Jackson Hole. A large recessional moraine that was built across the valley floor holds in Jackson Lake. South of this moraine is a pitted outwash plain, called the Potholes. The small depressions, some filled with water, are *kettleholes* that were left when stagnant, debris-covered ice blocks melted. Farther down the valley, the Burned Ridge moraine marks the terminus of the last glacial advance. Beyond this moraine, a thick outwash plain, built of glacial gravel, extends for some 10 miles down the valley, its surface sloping gently southward (fig. 40.7).

During times of glacial advance, more debris is put into the meltwater streams than they can carry, and they rapidly aggrade their beds by building up thick accumulations of sand and gravel. When glaciers retreat, streams are swollen with meltwater but carry less debris. They begin cutting trenches into the accumulated outwash, thereby deepening their channels and producing terraces. Terrace tops are built up during glacial advances; terrace faces are cut during times of glacial retreat.

Taming the Snake River. In Jackson Hole, the Snake River has cut several levels of prominent terraces (fig. 40.1) that can be related to moraines of various ages. The amount of work the Snake has done resorting and redepositing unconsolidated material is impressive. With its rapid changes in volume, high velocity, and shifting channels, the Snake deserved the name "Mad River" given it by the mountain men. Even today, despite the damming of the Jackson Lake outlet and the building of levees, the Snake is occasionally difficult to control. Most of the time, however, the Snake River twists around gravelly bars and quietly reworks loose material in its braided channel.

From its source in the Teton Wilderness, the Snake River flows into Jackson Lake, goes out the southeast end, and then runs on south through Jackson Hole, collecting runoff from the front slopes of the Teton Range and from the highlands on the east side of the valley. The tributaries deliver sediment to the valley floor, and much of it eventually reaches the Snake. The streams from the less steep mountains on the east tend to build alluvial fans where they enter the valley and then flow directly west, some of them joining the Snake almost at right angles. By contrast, the tributaries on the west side of the valley have much steeper gradients but do not develop extensive alluvial fans. Instead, the streams leave the canyons, flow east a short distance, and then bend south, flowing along the mountain front and joining the Snake at the south end of Jackson Hole. Several of the streams flow into interconnected morainal lakes at the base of the Teton Range on their way to the Snake River. This "lopsided drainage," which is due to the tilting of the Jackson Hole block, is another indication of tectonic control of topography.

Mass Wasting. Great quantities of broken rock and sediment were carried down the mountain sides by glaciers, meltwater, and landslides during Pleistocene time, and the dumping of rock debris in Jackson Hole did not stop with the retreat of the large-scale glaciers. The work of water and ice goes on, aided by the force of gravity, although only small glaciers in shaded locations now occupy cirques and the upper parts of glacial troughs. Talus accumulates at the bottom of cliffs and slopes, but rests there only temporarily. The journey of rock debris downslope can be fast or slow or by fits and starts. Some talus, mixed with water, snow, and ice, flows down slowly in rock glaciers. Active rock glaciers can be seen in Granite Canyon (southern part of the park) and in a north-trending valley near Eagles Rest (a peak in the northern part of the range). Some talus moves by the very slow process of creep. Some is brought down rapidly by snow avalanches, especially in late winter when avalanches are frequent.

Vegetation Patterns. In Jackson Hole, vegetation patterns reflect in an interesting way the general locations of glacial and alluvial deposits. The evergreen forests that grow on the mountain slopes below treeline extend out onto the valley floor, but only along the moraines (fig. 40.8). On the outwash plains and where outwash deposits have buried moraines, only sagebrush, coarse grasses, and a few other hardy plants can survive. Lodgepole pines grow well

Figure 40.8   Mount Moran and the outwash-covered floor of Jackson Hole, viewed from Snake River Overlook. A prominent diabase dike on Mount Moran extends from the summit to Falling Ice Glacier. The tree-covered moraine at the base of the range is Burned Ridge, terminus of the last major advance of Pinedale glaciation. The Snake River and several terrace levels are in the foreground. From W. H. Parsons, *Middle Rockies and Yellowstone.* © 1978 Kendall/Hunt Publishing Company. Used by permission.

on the moraines because the unsorted glacial debris is more compact and retains moisture better. Also, the soil has a higher clay content and is richer in minerals that plants need for growth. The coarse, quartzite gravels of the outwash plains, on the other hand, are dry and infertile. Moisture rapidly drains away to the nearest streambed. Along streambeds, below the parched and barren outwash plains, aspens, cottonwoods, and willows thrive on the available moisture and nutrients. Some irregular patches and lenses of loess, blown into the valley between glacial episodes, provide the only good soil in Jackson Hole (except for small amounts of silt). Badgers and coyotes dig their burrows where loess crops out on banks because loess is easy digging and does not collapse.

The Gros Ventre Landslide

In June 1925, after an unusually wet spring season, a large rockslide moved several cubic miles of rock and debris down Sheep Mountain and across the Gros Ventre River valley in about a minute's time. This occurred about three miles upstream from the southeastern boundary of Grand Teton National Park. The slide dammed the river, backing up a lake 200 feet deep and five miles long. What caused the slide was a combination of saturation of sedimentary rock layers on Sheep Mountain and the undercutting of the slope by the Gros Ventre River that left the base of the slope unsupported. Triggered by minor earth tremors, a rain-soaked slab of Tensleep Sandstone (of Pennsylvanian age) broke loose and slid downdip over soft shale underneath. Two years later the debris dam blocking the river gave way due to the pressure of high water from melting snow. The resulting flood washed out the town of Kelly five miles downstream, and six people drowned. Today, many years later, the jumbled landslide topography in the valley and a landslide scar 200 feet above the river on Sheep Mountain are still clearly visible.

Table 40.1 Generalized Geologic Column, Grand Teton National Park

Era	Period	Epoch	Rock Units	Geologic Events
Cenozoic	Quaternary	Holocene	Alluvium, talus, landslide deposits; minor glacial deposits	Weathering, erosion, mass wasting, Neoglaciation, Continued faulting
		Pleistocene	Moraine and outwash deposits	Pinedale glaciation (morainal lakes, outwash plain, terrace cutting, etc.)
			Loess deposits (interglacial)	Bull Lake glaciation
			Outwash deposits	Buffalo glaciation, extensive glacial erosion and deposition; Preglacial lakes in Jackson Hole; Volcanic eruptions in southern Jackson Hole
	Tertiary — Neogene	Pliocene	Bivouac Formation	Welded tuff eruptions from Yellowstone area
			Teewinot Formation	Teewinot Lake on subsiding Jackson Hole block; lakebed deposits, ashfalls
		Miocene	Conglomerates, tuff	Development of Teton fault system; uplift of Teton Range; increasing erosion
		Oligocene	Volcanic conglomerates, tuff, sandstones, etc.	Volcanic eruptions in Jackson Hole; basin filling
	Tertiary — Paleogene	Eocene	Volcanic tuffs, breccias, conglomerates, sandstones, claystone, shale, coalbeds	Extensive volcanism in the Absarokas
		Paleocene	Pinyon Conglomerate	Buck Mountain fault, uplift; Laramide orogeny; Sediment accumulation
Mesozoic	Cretaceous		Harebell Formation	Extensive nonmarine sedimentation
			About 15 nonmarine and marine formations in Jackson Hole region (for details, see p. 80, Love and Reed 1971)	Upwarps and subsidence, intermittent shallow sea; marine, transitional, and nonmarine sedimentation
	Jurassic			
	Triassic			
Paleozoic			About 9 major formations, limestones, dolomites, shales, sandstones (for details, see p. 68, Love and Reed 1971)	Intermittent deposition in shallow seas between episodes of upwarp and erosion
Precambrian time			Diabase dikes	Uplift and erosion
			Light-colored granites, pegmatites	Intrusions
			Dark granites	Orogenies
			Gneisses	Metamorphism
				Accumulation of marine and volcanic sediment in crustal trough

Source: modified after Love and Reed 1971, Behrendt et al. 1968

Geologic History

1. Accumulation of sedimentary and volcanic rocks in a Precambrian trough.

The oldest rocks were deposited about 2.8 billion years ago as sandstones, limestones, and shales in a marine environment. Volcanic rocks, probably from an island arc, were interbedded with sedimentary rocks.

2. Precambrian orogenies and intrusions.

The sedimentary and volcanic rocks were intensely folded and metamorphosed, possibly more than once. The gneisses that resulted were intruded by granites that were also metamorphosed. Subsequently, lighter-colored granites and pegmatite dikes were emplaced.

3. Intrusion of diabase dikes late in Precambrian time.

About 1.3 billion years ago, black diabase dikes, ranging from 5 to 200 feet thick, cut through the older Precambrian rocks. Long erosion ensued.

4. Deposition of Paleozoic sedimentary rocks.

About nine major formations, representing all periods of the Paleozoic Era except the Silurian, were deposited in Paleozoic oceans that covered the Teton region. The sedimentary units have a composite thickness of about 4000 feet and include sandstones, shales, limestones, and dolomites. Deposits are discontinuous, but no significant deformation occurred. The best fossil record is in the Paleozoic rocks of the Alaska Basin, just outside the park's western boundary. The limestones and dolomites contain brachiopods, trilobites, bryozoans, and corals that once grew in Paleozoic seas. Remnants of the Paleozoic rocks are exposed in the park, but more complete sequences are found outside park boundaries on the north, south, and west (fig. 40.4).

5. Mesozoic deposition.

Although Mesozoic rocks lie conformably over the Paleozoic rock units, the depositional environments of the Mesozoic Era were more varied, shifting from marine to transitional to continental, and alternating back and forth as crustal conditions changed in the Teton region. In general, sedimentation became more rapid throughout the era, especially toward the close. More than 15 formations with a composite thickness of between 10,000 and 15,000 feet accumulated during Mesozoic time. The Cretaceous units, which are the most numerous and have the greatest thickness, record the last transition from marine to continental rocks. Much of the sediment came from the mountains rising to the west that preceded the mountain-building that was soon to take place in the Teton region. Beds of bentonite between marine sands and shales represent ashfalls from western volcanoes that drifted down over a shallow sea. Landslides caused by the swelling of wet bentonite clay sometimes block access roads in the northern part of the park.

Coal accumulated in swamps as the last Cretaceous sea withdrew. Coal beds crop out in abandoned mine workings near the park's eastern margin. A broad, general uplift and the beginning of folding brought the Mesozoic to a close.

6. Laramide orogeny, beginning late in the Cretaceous Period.

The compressive forces of the Laramide orogeny initiated crustal buckling along a northwest-southeast trend and broad uplift throughout the region. From the Targhee uplift, northwest of the Teton region, rapidly eroding rivers brought large quantities of sand, gravel, and quartzite cobbles, which were spread south and east over the Teton region. These sediments made up the 5000 feet of conglomerates and sandstones of the Harebell Formation (northern and northeastern parts of the park).

Mountains in the region were raised higher as the orogeny intensified with more folding and faulting. The climax came early in Eocene time as great thrust faults and reverse faults produced mountain blocks and basins. The Buck Mountain fault probably formed at this time.

7. Tertiary nonmarine sedimentation.

In Jackson Hole and the immediate vicinity, vast quantities of clastic sediment that came from uplifted adjacent areas spread over the land. The Pinyon Conglomerate, which overlies the Harebell Formation, is of Paleocene age. On top of coal beds and claystone at the base of the Pinyon are thick zones of well rounded quartzite pebbles derived from the Targhee uplift, like those in the Harebell Formation. The quartzite fragments increase in size toward the northwest and many of them show percussion scars from being pounded during transport by swift mountain streams with steep gradients (Love and Reed 1971). The nature of the upwarp and subsidence of that time can be inferred from the rock record, which demonstrates how tectonic activities controlled sedimentation. Uplift was rapid in the mountain areas that supplied the coarse sediment. At the same time, Jackson Hole was subsiding more and more, thus providing a site for thick accumulations of eroded sediment.

8. Volcanic activity in Tertiary time.

As the intensity of Laramide mountain-building diminished in Eocene time, volcanoes began erupting in the Yellowstone-Absaroka area, producing the tremendous volume of volcanic rocks that make up the Absaroka Volcanic Supergroup. Tuff from these eruptions accumulated in Jackson Hole. In Oligocene and Miocene time, volcanoes on the eastern margin of Jackson Hole produced additional volcanic debris that filled basins and buried some of the mountains. The Colter Formation, for example, filled a downwarped trough in Jackson Hole with about 7000 feet of volcanic breccia, tuff, and other debris.

9. The Pliocene Teewinot Lake in Jackson Hole.

Movement on a fault that trended east-west across the floor of Jackson Hole, near the present south boundary of the park, impounded the first large freshwater lake in the valley. The water in the Teewinot Lake, dammed by the fault scarp, backed up toward the north. For about 5 million years, sediment (including tuff from nearby volcanoes) accumulated in the lake, which persisted because the rate of subsidence kept up with the rate of sedimentation. The presence of fossil snails, clams, and small mammal bones in the sediment indicates that the lake remained shallow.

10. Pliocene and Pleistocene volcanics.

Rhyolitic lava, spewing from volcanoes in Yellowstone, flowed down into the northern end of Teewinot Lake and solidified as obsidian, which has been dated by the potassium-argon method as 9 million years old. (Indians found the obsidian many years later and used it for cutting tools, spears, and arrowpoints.) After the Teewinot Lake dried up, glowing avalanches (nuées ardentes) poured down from Yellowstone, spreading welded tuff over the old lake bed. The Bivouac Formation, exposed on the flanks of Signal Mountain, was the result of one of these catastrophic eruptions. The welded tuff juts out as a ledge on the mountainside with conglomerates above and below. A series of volcanic eruptions also occurred in Pleistocene time, near the southern boundary of the park. Remnants of the volcanic rocks cap East and West Gros Ventre Buttes.

11. Teton block-faulting over the last 9 million years.

The long-continued upfaulting that elevated the Teton Range and tilted it westward initiated the erosion of the Mesozoic and Paleozoic rocks that once covered the summits. The range still stands high, however, because of ongoing faulting and the resistance of the Precambrian rocks. In Jackson Hole, sedimentation keeps up, more or less, with subsidence. Local faulting, upwarping, and downwarping caused the impounding of two successive preglacial lakes in the south and southwestern part of the valley. East and West Gros Ventre Buttes, as well as Blacktail Butte and Signal Mountain, are small fault blocks that stick up above the flat surface of Jackson Hole.

12. Pleistocene glaciation.

Cenozoic climates, which were initially subtropical and humid, gradually cooled. By Pliocene time, the Teton region had about the same climate as today. Pollen preserved in the Teewinot lakebed sediment came from plants and trees such as fir, spruce, pine, and sage, not unlike those that now grow in the park. With the onset of much colder Pleistocene climates, glacial ice accumulated in the Teton Range, in the Absaroka Mountains to the north, and in the Wind River Range and the Gros Ventre Mountains to the east. At least three times during the Pleistocene, glaciers from these highlands flowed down over Jackson Hole.

During Buffalo glaciation, the first well documented invasion of ice, glaciers from the north converged and flowed south along the eastern front of the Teton Range. Glaciers from the east flowed down the Gros Ventre River valley and joined the main ice stream near the Gros Ventre Campground. Signal Mountain and the other three buttes in southern Jackson Hole were overridden and severely abraded by a 2000-foot-thick mass of ice. At its maximum extent, the ice flowed down into eastern Idaho. Neither of the two later glaciations reached that far. After the ice melted, Jackson Hole was a wasteland. All accumulated soil had been scraped or washed away, and the floor of the valley was covered with quartzite boulders and cobbles. The parts of the valley not affected by the later glaciations are still in this barren state; that is, lacking in soil and too stony for all but the hardiest plant varieties to grow.

Bull Lake, the second glaciation, was less extensive. Some moraines and outwash deposits of Bull Lake glaciation were covered by wind-deposited loess after the glaciers melted back and before vegetation was reestablished. Later glaciation scraped off or covered much of the loess.

The glacial features associated with Pinedale glaciation, especially those in Jackson Hole, are familiar elements of the park landscape because this was the last major glacial episode. Beginning about 20,000 years ago and lasting until about 9000 years ago, Pinedale glaciers left behind most of the lake basins, moraines, and gravelly outwash plains that we see today, although Pinedale glaciation had less volume and erosive power than its predecessors. The terminal moraine of Pinedale ice is the Burned Ridge moraine, a series of low, hummocky, tree-covered ridges that slant southward across the middle part of Jackson Hole. The Potholes area, described earlier, is between Burned Ridge moraine and the recessional moraine that dams Jackson Lake. In the Teton Range, it was the valley glaciers of Pinedale time that put the finishing touches on the sculpturing of the horns, cols, and cirques that we see along the skyline. The string of morainal lakes at the foot of the range are also the result of Pinedale valley glaciers. The basins of Two-Ocean Lake and Emma Matilda Lake, however, were left by Bull Lake glaciation.

13. Holocene and historic time.

Pinedale glaciers probably melted completely during the warm period at the end of the Pleistocene, but small Neoglaciers formed in the Teton Range during the colder

Figure 40.9   Roadcut in morainal till (east of Jackson Hole) showing a wide range of sizes, shapes, and rock types, from boulders to fine clay, dumped by a large and powerful Pleistocene glacier. From W. H. Parsons, *Middle Rockies and Yellowstone.* © 1978 Kendall/Hunt Publishing Company. Used by permission.

climate of the Little Ice Age. These glaciers have not done much eroding, but they are picturesque and are regarded with caution and respect by mountain-climbers.

By building a dam at the south end of Jackson Lake, engineers have continued the work begun by Pinedale glaciers. The purpose of the dam is to prevent excessive raising and lowering of the lake level and to keep the Snake River from going on a rampage every spring. Sometimes in dry seasons, the needs of the Idaho ranchers downstream for more water from the Snake conflicts with the needs of people in the park who want the lake level kept high for recreation purposes.

South of Jackson Lake the Snake River keeps trying to cut a more westerly channel because of continued tilting of the Jackson Hole block. Engineers keep reinforcing levees in order to force the river to remain where it is.

Although severe earthquakes have not occurred in the valley since it was settled, one could happen at any time. Geologists who monitor the Teton fault system think that in the not-too-distant future, a major shock is possible, and perhaps probable. Meanwhile, the usual rockslides,

earthflows, and avalanches interfere from time to time with human activities.

The continual battle on the earth's surface goes on between the "wearing-down forces" and the "resisting framework." In Grand Teton National Park, this age-old conflict produces an interesting, diverse, and ever-changing landscape.

Geologic Maps and Cross Sections

Christiansen, R. D. et al. 1986. Wyoming geologic highway map. Western Geographics and Geological Survey of Wyoming.

Love, J. D.; Reed, J. C., Jr.; Christiansen, R. L.; and Stacy, J. R. 1972. Geologic block diagram and tectonic history of the Teton region, Wyoming-Idaho. U.S. Geological Survey miscellaneous geologic investigations, Map I-730.

Bibliography

Bartlett, R. A. 1962. *Great surveys of the American West.* Norman, Oklahoma: University of Oklahoma Press.

Behrendt, J. C.; Tibbetts, B. L.; Bonini, W. E.; Laven, P. M.; Love, J. D.; and Reed, J. C. 1968. A geophysical study in Grand Teton National Park and vicinity, Teton County, Wyoming. U.S. Geological Survey Professional Paper 516E.

Crandall, Hugh. 1978. *Grand Teton, the story behind the scenery.* Las Vegas, Nevada: KC Publications.

Fryxell, F. 1946. *The Tetons, interpretations of a mountain landscape.* Berkeley, California: University of California Press.

Horberg, L. 1938. *The structural geology and physiography of the Teton Pass area, Wyoming.* Publication No. 16. Rock Island, Illinois: Augustana Library Publications.

Love, J. D.; and Reed, J. C., Jr. 1971. *Creation of the Teton landscape* (with geologic map). Grand Teton Natural History Association.

———, and Love, J. M. 1983. Road log, Jackson to Dinwoody and return. Public Information Circular 20, Geological Survey of Wyoming. 32 p.

National Park Service. 1984. *Grand Teton, a guide to Grand Teton National Park, Wyoming.* Handbook 122. 96 p.

Parsons, W. H. 1978. *Middle Rockies and Yellowstone,* K/H Geology Field Guide Series. Dubuque, Iowa: Kendall/Hunt Publishing Company.

Note: Love and Reed (1971) is a well illustrated, nontechnical book about Grand Teton geology that explains features visitors can see in the park. The book is a handy size for tucking into a backpack.

Grand Teton National Park

Area: 310,521 acres; 485 square miles
Established: February 26, 1929
Part of former Jackson Hole National Monument incorporated: March 15, 1943 (with portions of monument absorbed by National Elk Refuge)
Address: Box 170, Moose, Wyoming 83012

41

Water is present in the semiarid Great Basin country, but it never reaches an ocean. Great Basin National Park is the only national park in this region of interior drainage.

High above the desert floor are snowy peaks, cirques, alpine meadows, and cirque lakes.

Five ecologic zones exist between the crest of the block-faulted range and the desert floor.

In Lehman Caves, remarkable rooms and passages, with beautiful dripstone features, extend deep into the mountain flank within a wedge of Cambrian limestone.

A 75-foot-high natural arch stands like a sentinel in a creek valley.

Great Basin National Park

Eastern Nevada

by Sherwood D. Tuttle
University of Iowa

Figure 41.1   A small glacier clings to Wheeler Peak (elevation 13,063 feet), high above the desert floor. The bedrock is resistant quartzite of Cambrian age. National Park Service photo by W. F. Heald.

GREAT BASIN
NATIONAL PARK

Figure 41.2   Great Basin National Park, Nevada.

Geographic Setting

Lying on the Southern Snake Range in eastern Nevada is recently established Great Basin National Park. It is located in a sparsely populated area near the Utah border, just south of U.S. Highways 6 and 50. The small village of Baker is the closest community.

As a descriptive term, "Great Basin" applies to a large area encompassing most of Nevada, a sizable portion of western Utah, and small slices of Wyoming, Idaho, Oregon, and California. *Drainage divides* (the high areas separating drainage basins) delineate the boundaries of this geographical region characterized by internal drainage. No streams flow outward from the Great Basin. All surface runoff and streams flow into saline lakes and marshes or gradually sink into sand and gravel in the beds of disappearing streams. Great Salt Lake in Utah, the Humboldt River, and the Carson Sink in Nevada are well known examples of the hydrologic concept of interior drainage.

The Great Basin region is not to be confused with the physiographic province called Basin and Range, which includes most of the Great Basin but is delineated by geologic structure rather than by drainage divides. Block-faulted mountains and sediment-filled grabens characterize the Basin and Range province, which extends farther south than the Great Basin, to include more of southern California, the southern half of Arizona and New Mexico, the western corner of Texas, and adjoining areas of Mexico. (A *graben* is a down-faulted trough or depression.)

Great Basin National Park, its boundaries well within both these overlapping regions, displays the notable characteristics of both areas as well as some distinctive and unusual features of its own. Recent past and present climatic conditions in the park region have been responsible for many of the landscape features.

The present climate of the Great Basin is arid to semiarid, but local precipitation is strongly influenced by topography. Whereas 4 inches of rain per year may fall in the lower basins, more than 30 inches may drop annually on the higher mountain summits, and much of it is in the form of snow in winter. In summer, rain falls as infrequent but torrential showers of short duration. These conditions have prevailed in historic time, but during the pluvial episodes of the Pleistocene Epoch, (not long ago in terms of geologic time), the climate was much wetter.

Temperatures averaged 8° to 15° F cooler during Pleistocene time. Glaciers accumulated on Wheeler Peak and the other high summits of the Snake Range. Late Pleistocene snowlines are estimated to have been at about the 10,500- to 11,800-foot level. *Pluvial lakes* (lakes formed during cooler, wetter glacial climates) occupied Spring Valley and Snake Valley, the basins on either side of the Snake Range. During some of that time, the lake in Snake Valley was the westernmost arm of Lake Bonneville, a large pluvial lake that spread over eastern Nevada, Utah, and southern Idaho. (Today, Great Salt Lake represents what's left of Lake Bonneville.) Dry saltpans on the flats of Spring Valley and Snake Valley remind us of the pluvial lakes that sparkled there before the waters dried up.

In Great Basin National Park, between the desert floor and the high peaks of the Snake Range, displayed in about 8000 feet of relief, are five remarkable ecosystems existing in a desert region in temperate latitudes. Controlled at their respective levels by temperature and moisture, the five zones are, from bottom to top: (1) sage and pinyon-juniper of the desert; (2) manzanita and mountain mahogany; (3) aspen and Douglas fir; (4) Engelmann spruce and pine; and (5) hardy alpine plants above the treeline. In the highest zone, at the edge of the treeline, are the centuries-old, gnarled and twisted bristlecone pines surviving in an ancient forest, now under strict protection (fig. 41.3). On Wheeler Peak, enough precipitation falls to support two small glacial lakes, an alpine meadow, small snowfields and glacial ice that lasts through the summer.

The desert climate of the Great Basin is under the dominance of a rain shadow created by the Sierra Nevada, some 350 miles to the west in California. Hence the Great Basin is called a *topographic desert*. Moisture-laden air masses coming in from the Pacific Ocean are forced to rise into much cooler air in order to pass over the high ridges of the Sierra. This reduces the ability of the clouds to hold moisture, causing rain and snow to fall on the windward side of the ranges. Descending on the lee side, the air warms and picks up moisture. This deprives the land below of rain and creates a desert climate where surface moisture quickly evaporates. The water table is discontinuous and usually far below the surface.

Local History

At 13,063 feet, Wheeler Peak is the second highest mountain in Nevada. It was named for Lieutenant George Montague Wheeler, a topographer in the U.S. Army Corps of Engineers, who led expeditions to the the Western Territories after the Civil War, from about 1869 to 1879. G.K. Gilbert, one of the giants of American geology, was a member of Wheeler's party and wrote the first geologic report on the Great Basin region.

Figure 41.3   Rare, ancient bristlecone pines at about 11,000-foot elevation on the east side of Mount Washington in the Snake Range. National Park Service photo by E. B. Young Jr.

Early inhabitants of the region came and went as climates ranged from moister to drier. The people of the Fremont culture farmed on the lower slopes during times of higher precipitation before A.D. 1300, but were forced to move on by a prolonged drought that began around that time. Nomadic tribes, mainly Shoshones and Paiutes, later roamed the area, hunting and gathering. They were here when the explorers and pioneers arrived. Some of the early routes to California crossed this part of Nevada. The journey was difficult because of the ups and downs of the ranges and basins as well as the scarcity of water and grasslands for sustaining draft animals and stock. The Pony Express trail, the Overland Telegraph, and the Overland Stage route all passed along the northern edge of the Snake Range. This was a direct way for those who could travel fast.

During the latter part of the nineteenth century, mining camps and widely scattered ranches were started in the region, with boom-and-bust activities in mining continuing to the present time. An enterprising miner, Absalom S. Lehman, who came to the area in 1867, failed to strike it rich on his mining claims but was alert to the heavy demand for food in the mining camps. He cleared a site on the lower slopes of the east face of Wheeler Peak and began to farm. A stream coming down the mountain and a spring supplied water for crops, stock, and an orchard. Soon he was supplying meat, dairy products, eggs, and garden produce to the mining camps and the venture became ever more successful as time went on.

In 1885 Lehman discovered an opening in the rocks above his ranch that, to his amazement, led into a sizable underground chamber. He lowered himself by rope into the cave and began exploring the passages with a lantern. After announcing his discovery through the local newspaper, he guided over 800 people through the passages in the first year. Lehman continued to escort visitors through the caves that bear his name for the rest of his life. In 1922, the caves and a square mile of land above ground became Lehman Caves National Monument. The protected area was greatly enlarged and established as Great Basin National Park in 1986. Included in the park are

Figure 41.4   The adit (opening) of this beryllium mine is just outside the park's west boundary, but the tunnel goes into Mount Washington, beneath the surface of park land. National Park Service photo by E. B. Young Jr.

Figure 41.5   Abandoned placer mining equipment on Snake Creek, which drains to the east from Wheeler Peak and Mount Washington. Large masses of quartz monzonite crop out in the creek's headwaters area. National Park Service photo by E. B. Young Jr.

Wheeler Peak, the neighboring peaks of the Southern Snake Range, and Lexington Arch, a 75-foot-high natural arch near the southern boundary. The National Park Service maintains a visitor center and campgrounds in the area. A scenic drive winds up to the 10,000-foot elevation on Wheeler Peak. Cattle grazing is permitted in some areas because this activity is considered an integral part of the Great Basin scene. Prospecting is not permitted in the park, but unpatented mining claims may be worked, provided a plan of operation has been approved by the National Park Service. None of the claims are producing at the present time (figs. 41.4, 41.5).

Geologic Features

The Rocks

At least 14,000 feet of marine sedimentary rocks, ranging in age from Cambrian through Pennsylvanian, are exposed in Great Basin National Park. Wheeler Peak owes its topographic dominance to the superior resistance of the thick Prospect Mountain Quartzite of Cambrian age from which it has been eroded.

Carbonates are the dominant rock type, with older rock units thicker than those near the top of the section. Because of complex faulting, stratigraphic sections of the Paleozoic rocks are incomplete, and precise correlations are difficult. Cutting through the sedimentary rocks are much younger *stocks* (small plutons) that form resistant peaks and ridges. Lapping up against the Paleozoic rocks are younger nonmarine units—coarse volcanic clastics, lavas, ash, and widespread alluvial deposits.

The two stocks in the range have been identified as coarse-grained quartz monzonite and coarse-grained, porphyritic quartz monzonite. They are close in age and probably came from the same magma chamber. The stock with the porphyritic texture is unusual in that it contains potassium feldspar phenocrysts, 1/4 to 1/2 inch long, set in a coarse-grained matrix of plagioclase, quartz, biotite and muscovite. (The more common porphyritic texture in igneous rocks occurs with *phenocrysts,* the large crystals, set in a fine-grained groundmass.) The crystal sizes in the porphyritic stock probably represent unusual cooling conditions at the time the plutons were emplaced, presumably during late Cretaceous and early Tertiary time.

Ore bodies. For over a century, mining has been a principal industry in Nevada. Hillsides everywhere are pockmarked with prospect pits. A goodly number of these sites have yielded commercial amounts of metallic ores. Copper, lead, gold, and silver have been mined in eastern Nevada; beryllium and tungsten have also been reported. Most of the ore bodies were deposited in the interval from late Cretaceous time to late in the Tertiary and are presumed to be related to the emplacement of plutons.

Fault-Block Mountains

Great Basin National Park extends for some 20 miles along the crest of the southern end of the Snake Range, a north–south-trending *fault-block mountain range* about 150 miles in length. This type of feature is a crustal unit formed by uplift along normal (or sometimes vertical) faults. Tensional forces, pulling apart the uplifting crust, cause normal faults to divide the crust into structural blocks of different elevations and orientations. As is typical of fault blocks, the Snake Range is asymmetrical and, in this case, steeper on the western side. The block has been tilted down toward the east and southeast. The amount of vertical displacement is difficult to estimate because the faults bounding the range are buried by gravels. However, the

mountain block has probably moved up more than 9000 feet. Numerous high-angle normal faults within the range have produced a series of slices and smaller blocks of rock that are unevenly tilted relative to each other. On the eastern side, fault movement was about half as much as on the western flanks.

Deep erosion has gullied the mountain slopes and de-roofed the plutons, exposing igneous rock on summits and slopes. Several thousand feet of roof rocks have been removed and washed downslope, blanketing the basins on either side with great quantities of fill (fig. 41.6 A,B).

Thrust Faults and Folds

Complicating the geology of the Snake Range are numerous nearly horizontal thrust faults that moved upper slabs eastward and northeastward relative to lower slabs. Detailed mapping in the southern part of the range has identified at least six discrete thrust-fault surfaces where younger beds have been pushed over older beds. This type of bedding-plane faulting is called *décollement*. In one area, for example, formations younger than middle Cambrian lie on the upper surface of the décollement and are themselves broken by numerous younger (Mesozoic) thrust faults, as well as by even younger (Tertiary) block faulting.

Figure 41.6   Simplified cross sections of the Snake Range. *A.* Structural relationships when thrust faulting and normal faulting occurred late in Mesozoic time and early Tertiary, prior to major episodes of block-faulting. Cambrian sedimentary rocks are below the thrust fault, with younger Cambrian and Ordivician strata above. Thrust faulting generally followed bedding plane traces. The pluton is Cretaceous in age. *B.* Block-faulting, from late Tertiary time through the Quaternary, uplifted and tilted the Snake Range toward the east. Erosion has stripped rock material from the upthrown block and deposited fill in the downdropped basin. (Modified after Drewes 1958 and Whitebread 1969.)

The rock layers and fault slices have also been bent into broad folds that trend in several directions. These structures may be associated with uplift or plutonic intrusions, or both, rather than with compression.

Some geologists have suggested that the décollement thrust faulting in the Snake Range may be better visualized as beds sliding off an unmetamorphosed cover of Paleozoic rocks overlying an uplifted core of low-grade metamorphic rocks of Precambrian and Cambrian age. This type of structure is called a *metamorphic core complex.*

Late Mesozoic or early Tertiary crustal extension in the Snake Range has been estimated as causing over 30 miles of displacement, measured by use of a separation of the Prospect Mountain Quartzite. This low-angle normal faulting in the region preceded the high-angle normal faulting that produced the typical basin-and-range structures. The low-angle normal faulting moved the upper rocks here in a northeasterly direction. If the fault plane dipped westward, it could be called a thrust fault. However, if the plane dipped eastward, a normal fault would have resulted. With possible upward movement of the core of the Snake Range in Tertiary time, the dip of the low-

angle fault plane may have been changed. To picture the relationship as "extensional," rather than "compressional" may be clearer than describing it in terms of "thrust faulting vs. normal faulting" (Bartley and Wernicke, 1984).

Glaciation

The erosive action of successive Pleistocene cirque glaciers gave Wheeler Peak its distinctive, hornlike profile against the sky. Neighboring peaks also display cirques, and the highest summits of the range have snowfields year-round, a remarkable sight when one is traveling across the hot desert floor in midsummer. Two episodes of glaciation left moraines on the upper slopes at about the 8100-foot elevation for the older Lehman Creek moraine and at 9500 feet for the Dead Lake moraine.

On Wheeler Peak, deep snow banks and ice fringe the cirque walls, forming a small cliff or cirque glacier. An active rock glacier nestles in a draw on the mountain's northeast flank. A *rock glacier* (a type of mass wasting) is a slowly moving accumulation of talus, ice, snow, and water that creeps and slides down a mountain slope, usually only a few inches per day (fig. 41.7). Lovely alpine

Figure 41.7 From this north-facing cirque on Wheeler Peak, a rock glacier descends in the foreground. An ice glacier is in the shadow of the rock wall, on which beds of Cambrian Prospect Mountain Quartzite are exposed. National Park Service photo by E. B. Young Jr.

GYPSUM
ANNEX

TALUS
ROOM

GRAND
PALACE

WEST
ROOM

NATURAL
ENTRANCE

GOTHIC
PALACE

LODGE
ROOM

0   100   200   300
SCALE IN FEET

Figure 41.8   Underground passages of Lehman Caves. (Adapted from National
Park Service.)

meadows and several small cirque lakes in the summit
areas of the Snake Range remind us of the Ice Age by
preserving a habitat for alpine wildflowers and other cold-
acclimated life forms.

Lehman Caves

Part way up on Wheeler Peak's eastern flank, behind an
attractive National Park Visitor Center, is the entrance to
Lehman Caves (fig. 41.8), a marbly limestone solution

cavern known for the diversity of its cave decorations. The
cave passages are in a thick, fairly uniform formation, the
Pole Canyon Limestone, Middle Cambrian in age, that
has been faulted upward as a large, wedge-shaped unit of
bedrock. Although the beds dip steeply, the 10,000 feet of
cave passages are nearly horizontal. This indicates that
the cave passages were formed late in the history of the
range, *after* most of the tilting and uplift had occurred
and most likely within the last three or four million years.

Hence base level and the water table had more control over the attitude of the cave system than the bedding planes in the limestone. However, the size and location of chambers and passages were influenced by fractures and bedding planes in the limestone.

The Pole Canyon Limestone is alternately pale-gray and white and (for the most part) massive. When exposed at the surface in this area, the rock is a cliff-former. The Pole Canyon Limestone underwent low-grade metamorphism during an earlier time, which made it purer, denser, and whiter in color than it was originally. During the process, fossils were mostly obliterated, and carbonate grains were dissolved and redeposited on adjacent grains, thus compacting most of the intergranular spaces. The Pole Canyon Limestone is considered a low-grade marble.

Many beautiful and varied dripstone features (speleothems) are displayed in the underground rooms and passages (fig. 41.9). Besides the usual stalactites, stalagmites, and columns, visitors can see "soda straws," scallops on sidewalls, shields (called "parachutes"), draperies, cave popcorn, and forms called *helictites,* which are twiglike decorations that grow in curlicues and spirals out from the walls. Aragonite crystals and delicate gypsum deposits form lacy decorations in a few chambers. (How speleothems form is explained in Part II, in chapters 12, 13, and 14.)

The Talus Room, a large chamber partly filled with broken rock and rubble, reveals a record of earthquakes, possibly 10,000 to 30,000 years ago, that brought down the entire ceiling of this 280-foot room. Earthquakes in Nevada, which occur from time to time, indicate that the region is tectonically active.

Some of the processes that formed Lehman Caves are still going on today although probably at a slower rate. Most of the hollowing out of the passages took place during times of higher precipitation, presumably during the pluvial episodes of Pleistocene time. Ground water (and meltwater) percolated downward through the fractured rocks of Wheeler Peak. At times, all the rock openings and cracks were filled above the present level of the cavern floor. Flowing and eddying water dissolved, rounded, and smoothed the walls, ceilings, and floors of the passages. The fastest moving and largest volume of ground water was probably dissolving rock just below the water table.

When the climate became drier, the water table dropped. Then air entered the cave passages, evaporation took place, and dripstone began to precipitate. Some of the cave features may have undergone several cycles of solution-deposition-solution-redeposition as the water table rose and fell with changes in climate.

Figure 41.9   Helictites and soda straws in Lehman Caves. National Park Service photo.

At present the year-round temperature inside the caves is about 50° F and the humidity is in the high 90s, making the environment exceedingly pleasant for visitors who come in from the desert on a hot day.

Lexington Arch

This imposing natural arch, 75 feet high, is the result of weathering and erosion of limestone (fig. 41.10). The arch stands near the base of Granite Peak in the southern part of Great Basin National Park. A trail from Lexington Creek leads up to Lexington Arch. Geologists surmise that the arch is a remnant of a cave passage that became exposed by erosion.

Figure 41.10   Lexington Arch, in the southwest corner of the park, is sculptured from Cambrian Notch Peak Limestone and may be a remnant of an exhumed cave system. National Park Service photo.

Figure 41.11   Looking southeast toward Wheeler Peak and the Snake Range, an uplifted fault block, from Spring Valley, a downthrown block. The scenery is typical of the Basin and Range. Photo by S. D. Tuttle.

Table 41.1 Simplified Geologic Column, Great Basin National Park

Time Units			Rock Units	Geologic Events
Era	Period	Epoch		
Cenzoic	Quaternary	Holocene	Basin fill, talus slopes, sand dunes Speleothems Rock glaciers	Increasing aridity Erosion/deposition of speleothems Mass wasting
Cenzoic	Quaternary	Pleistocene	Glaciofluvial and lakebed deposits	Two episodes of Wisconsinan glaciation Expanding and shrinking pluvial lakes Glacial cirque erosion Morainal deposition Solution erosion of cave passages Rising and falling water table Forming of speleothems
Cenzoic	Tertiary	Oligocene	Basaltic lavas, ash Tuffs, conglomerates, andesitic lavas	Development of basin-and-range topography by block faulting and uplift Volcanism (two phases)
Cenzoic	Tertiary		Older conglomerates Rhyolite porphyry dikes, sills; mafic sills	Dikes and sills intruded
Mesozoic	——?—— Cretaceous		Quartz monzonite, quartz monzonite porphyry	Emplacement of stocks East/west compression, complex thrust faulting, décollement Uplift
Paleozoic	Ordovician to Pennsylvanian (?)		10 or more formations (limestones, dolomites, chert, shales, sandstones, quartzites)	Long-continued deposition of marine sediments in continental shelf, near-shore, and shore environments
Paleozoic	——?——— ? ———— Upper Cambrian		Pogonip Group Corset Spring Shale Johns Wash Limestone Lincoln Peak Formation	
Paleozoic	Middle Cambrian		Pole Canyon Limestone Pioche Shale	
Paleozoic	Lower Cambrian		Prospect Mt. Quartzite	

Source: Drewes, 1958; Whitebread, 1969

Geologic History

1. Accumulation of marine sedimentary rocks, Cambrian to Pennsylvanian Periods.

Throughout most of Paleozoic time, this region lay along the edge of a continental landmass that eventually became western North America. The older rocks, including the Cambrian strata, accumulated in a continental shelf environment. Seas transgressed and receded numerous times across the region. Later in the era, from middle to upper Paleozoic time, complex geologic events took place with increasing frequency. Subduction occurred during orogenies (Antler and Sonoma).

2. Late Mesozoic and early Tertiary thrust faulting.

Evidence of geologic events in early and middle Mesozoic time is missing from the geologic record; but late in Mesozoic time, the region was affected by repeated episodes of thrust faulting caused by the pressure of eastward-moving tectonic plates. This tectonic activity, which was related to the Laramide orogeny and accompanied by plutonism, continued through Tertiary time. In the Snake Range, as described earlier, stocks were emplaced and tectonic pulses moved thrust blocks eastward. Although the stocks cut the thrust faults, the intrusions are quite close in age to the faulting. Some low-angle faults slid along bedding planes in the shales.

One might logically think that the low-grade metamorphism evident in the Pole Canyon Limestone of the Snake Range would have been caused by contact metamorphism from the plutons. However, field evidence indicates that the metamorphism preceded the thrust faulting, inasmuch as unaltered rocks are thrust over metamorphosed rocks and the plutons are younger than the thrust faults. Perhaps the rocks were metamorphosed during a Paleozoic orogeny. Mafic dikes and rhyolitic dikes and sills intruded the Snake Range following thrust faulting and emplacement of the stocks.

Some geologists have described the Snake Range as a metamorphic core complex, similar to others that have been identified in the Great Basin region. According to this hypothesis, the lightly metamorphosed Paleozoic rocks make up the core of the Snake Range. Nonmetamorphosed Paleozoic rocks form an overlying cover. Separating the core from the cover is the slide zone, or décollement. The low-angle faults may reflect eastward gravity sliding, perhaps in Tertiary time, from the up-arched highland.

3. Middle Tertiary rocks.

Andesitic volcanism was extensive in the Great Basin area during this time. Conglomerates, ash flows, and tuffs accumulated in the Snake Range. A unit of tuff has been dated by the potassium-argon method as 31 million years ± 1.6 million (Oligocene). The volcanic activity was accompanied by irregular uplifts and deep erosion.

4. Block faulting, Pliocene to present.

The entire region was uplifted and broken by many high-angle normal faults that produced the fault-block mountain ranges and down-faulted basins (fig. 41.11). A second phase of volcanism, typically basaltic in composition, accompanied the block faulting. The geometry of normal faulting, which is produced by tensional stresses, indicates crustal extension; that is, a thinning and stretching of the crust. Fault movements were both continuous and intermittent, with many individual fault surfaces on both sides of the range.

5. Effects of Quaternary climatic changes.

The summits of the Snake Range were eroded by alpine glaciers during two episodes of Wisconsinan glaciation. Pluvial lakes occupied the basins on either side of the range. A valuable legacy of Pleistocene time are the reservoirs of ground water stored in the basin fill beneath the lakebeds. Lehman Caves was formed largely during Pleistocene time, and dripstone was deposited in the chambers and passages. Extensive weathering, downslope movement, and erosion have continued throughout Quaternary time and vast quantities of fill have accumulated around the base of the range. Snow collects in the sheltered hollows at the summit. Sand dunes pile up on the flats below.

Bibliography

Bartley, J. M., and Wernicke, B. P. 1984. The Snake Range décollement interpreted as a major extensional shear zone. *Tectonics,* 3:(6) 647–657, November.

Chronic, Halka. 1986. *Pages of stone; Geology of western national parks and monuments.* v. 3 Desert southwest. Seattle: The Mountaineers.

Drewes, H. 1958. Structural geology of the Southern Snake Range, Nevada. *Geological Society of America Bulletin,* 69:221–240.

———, and Palmer, A. R. 1957. Cambrian rocks of the Southern Snake Range, Nevada. *American Association Petroleum Geologists Bulletin,* 41:104–120.

Fiero, Bill. 1986. *Geology of the Great Basin.* Reno: University of Nevada Press

Halladay, O. J., and Peacock, V. L. 1972. *The Lehman Caves Story.* Baker, Nevada: Lehman Caves Natural History Association.

Mackin, J. H. 1960. Structural significance of Tertiary volcanic rocks in southwestern Utah. *American Journal of Science,* 258:81–131.

McKee, E. H. 1971. Tertiary igneous chronology of the Great Basin of western United States: Implications for tectonic models. *Geological Society of America Bulletin,* 82:3492–3502.

McPhee, John. 1981. *Basin and range.* New York: Farrar, Straus, Giroux.

Misch, P., and Hazzard, J. C. 1962. Stratigraphy and metamorphism of late Precambrian rocks in central northeastern Nevada and adjacent Utah. *American Association Petroleum Geologists Bulletin,* 46:289–343.

Nolan, T. B. 1943. The Basin and Range province in Utah, Nevada, and California. U.S. Geological Survey Professional Paper 197D, p. 141–196.

Stewart, J. H. 1971. Basin and Range structure: A system of horsts and grabens produced by deep-seated extension. *Geological Society of America Bulletin,* 82:1019–1044.

Box 41.1.
Crustal Extension: What's Next?

An intriguing aspect of plate tectonics deals with possible mechanisms whereby plates composed of continental crust split or open to form a new ocean, creating new islands or differently shaped continents. Geologists speculate that in western North America, possibly in Nevada, the block faulting and crustal extension going on in the Basin and Range province may lead to large-scale rifting, resulting in even larger grabens similar to those in the Middle East and Africa. Such a depression may develop into an ocean arm, such as the Red Sea or the Gulf of California, both of which are examples of crustal rifting where new seaways formed.

For just about a century, geologists have theorized and argued over the nature of the tectonics that produced the structures of the Basin and Range. The hypothesis of fault-block mountain ranges and grabens bounded by normal faults has been generally accepted, but consensus has not been reached regarding the fundamental cause of the tensional stresses and the crustal extension.

The following critical factors make the Basin and Range different from neighboring regions: (1) numerous normal faults; (2) relatively thin crust (with low-velocity seismic waves indicating shallow depth to the upper mantle); (3) high regional heat flow; (4) "rift faulting"; (5) basaltic volcanism; (6) and high seismicity. Data collected show that the upper mantle here has an anomalously low P wave velocity through its upper 20 kilometers. This low-density, low-velocity upper mantle material must supply isostatic support for the uplifted Great Basin region.

Two episodes of Tertiary volcanism followed by vertical faulting in the Basin and Range have been explained as follows. Between 40 and 19 million years ago, eastward-moving plates underthrust themselves to sufficient depth to bring about the eruption of volcanic rocks of andesitic to rhyolitic composition in the central part of the Great Basin. Great sheets of ash and tuff were extruded during this time and the hot and glowing pyroclastics hardened into a dense rock called *ignimbrite*. This was the first phase of Tertiary volcanism. When the former spreading center (an oceanic ridge) being subducted reached a point under the Great Basin, about 16 million years ago, olivine basalt (from the old ocean floor) began to erupt. This was the second phase of volcanism that accompanied widespread block faulting.

Several tectonic models have been proposed to explain the extension tectonics: (1) The East Pacific Rise extends northward beneath the Basin and Range; (2) A converging zone with a Cenozoic oceanic plate was subducted beneath the region; (3) After subduction ceased, a bulge (diapir) or hot spot below the Great Basin rose from the mantle and spread out beneath the crust during late Cenozoic time. Which model is the one most likely to fit the situation? Take your pick!

Wallace, R. E. 1975. Basin and Range province, in Fairbridge, R. W., editor. *The encyclopedia of world regional geology, Part 1: Western hemisphere,* p. 541–548. Stroudsburg, Pennsylvania: Dowden, Hutchinson and Ross.

Whitebread, D. H. 1969. Geologic map of the Wheeler Peak and Garrison quadrangles, Nevada and Utah. U.S. Geological Survey Miscellaneous Geological Investigations, Map I 578, with accompanying text.

Great Basin National Park
Area: 77,109 acres; 120 square miles
Proclaimed Lehman Caves National Monument: January 24, 1922
Enlarged and established as a national park: 1986
Address: Baker, Nevada 89311

42

Sequoia and Kings Canyon National Parks together cover over 1300 square miles, preserving a major section of the Sierra Nevada wilderness and the largest groves of giant sequoias.

The bedrock of the parks is primarily late Cretaceous granitic batholiths with thick sections of early Mesozoic meta-sedimentary and metavolcanic rocks preserved as roof pendants. The latter rocks record the active subduction system at the western margin of North America from the end of the Paleozoic through the Mesozoic.

Uplift of the range occurred episodically with the most recent events beginning in the Pliocene along the Sierra Nevada fault system.

Glaciation at high elevations led to the formation of major glacier streams extending to low elevations in both parks.

Sequoia and Kings Canyon National Parks

East Central California

by Donald F. Palmer
Kent State University

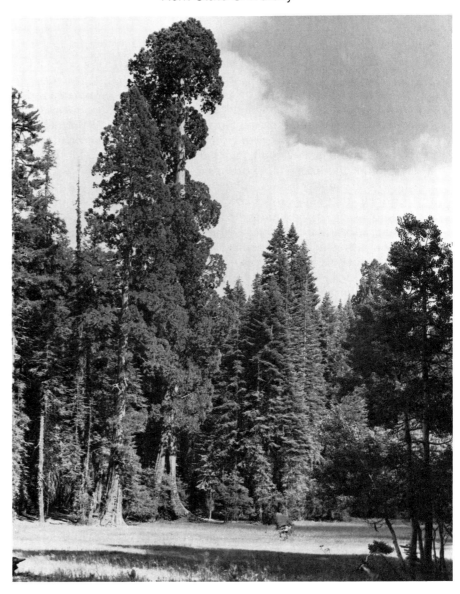

Figure 42.1   Giant sequoias growing at the edge of Crescent Meadow in Sequoia National Park. National Park Service photo.

Local History

Sequoia and Kings Canyon National Parks have grown significantly over a period of 75 years. They began in 1890 when Sequoia and the General Grant Grove were established as national parks, just in time to save their giant sequoia trees from destruction by the lumbering industry. Sequoia National Park was more than doubled in size in 1926 to include the upper drainage basins of the Kaweah and Kern Rivers and the Sierran crest, including Mt. Whitney. In 1940, a large tract of land extending northward from Sequoia was combined with General Grant Grove to form Kings Canyon National Park. Redwood Canyon, between General Grant Grove and Sequoia National Park, has a large grove of sequoias on the flanks of Redwood Mountain and was also included within Kings Canyon National Park. In 1965, the areas of Cedar Grove on the south fork of the Kings River and the Tehipite valley were added to Kings Canyon. The Middle Fork and the South Fork of the Kings River and the upper reaches of the North Fork and South Fork of the Kern River were added to the Wild and Scenic Rivers System by Act of Congress in 1987. The two parks, Kings Canyon and Sequoia, are administered as a single unit.

The major impetus for the founding of the parks was the preservation of the giant sequoias. About 13,000 giant sequoias (*Sequoiadendron gigantea*) grow in scattered patches on the moist western slopes of the Sierra Nevada at elevations between 4000 and 8000 feet, from Yosemite National Park (chapter 25) in the north to Sequoia in the south. Giant Forest in Sequoia National Park is the largest single grove and has about 3500 Big Trees, among them General Sherman, the largest living tree in the world. The circumference of its trunk is 102.6 feet, and it is almost 275 feet tall. The mature giant sequoias in the parks are between 3000 and 4000 years old.

The giant sequoias and their close relatives, the coast redwoods (*Sequoia sempervirens*), which grow in a narrow belt along the Pacific coast, are the surviving members of a once large family of huge conifers that spread throughout the temperate regions of North America during the Mesozoic Era. Fossilized stumps of trees of the sequoia family have been identified in Yellowstone National Park (chapter 39) and in Petrified Forest National Park (chapter 8). Most of the sequoia species became extinct during the Ice Age, which began about 2 million years ago and ended about 10,000 years ago. Only the two Pacific coast species survived Pleistocene glaciations and were able to reestablish themselves in the limited environments in which they now grow (see chapter 44, Redwood National Park). The mountain ranges paralleling the Pacific coast provide an ideal environment for the two sequoia species for two reasons. (1) The elevation of the mountains causes the moisture-bearing clouds from the Pacific Ocean to precipitate large amounts of rain and snow. (2) The fresh, sandy soils required by sequoias are derived from glacial erosion and the weathering of the crystalline rocks in the mountains.

Although hunters and explorers had found giant sequoias in the Sierra region as early as 1833, A. T. Dowd is credited with the first well publicized discovery of sequoias in 1852 while hunting north of the Yosemite region. The first of Sequoia's Big Trees were found by Hale Tharp in 1858, and those in the General Grant Grove of Kings Canyon National Park were discovered in 1862 by Joseph Thomas. News spread rapidly about the "mammoth trees," and in the 1860s and 1870s claims for land were filed and sawmills were erected. A giant sequoia was felled, cut into segments, and shipped to Philadelphia for the 1876 Centennial Exposition, where it was reassembled and exhibited.

The loggers of the day regarded felling the Big Trees as the ultimate challenge to their skill and bravery. The wood from sequoias is of exceptionally high quality, straight-grained, and naturally resistant to decay. A single tree might yield 300,000 to 400,000 board feet of lumber. Reminders of these early logging days can be seen in Big Stump Basin in Kings Canyon National Park near Generals Highway; where a grove of giant sequoias once flourished, only huge stumps remain.

The exploitation of the area was halted by the action of many far-sighted individuals who worked for the preservation of the Big Trees and the wilderness of the Sierra Nevada. Notable among these people is Colonel George Stewart whose practical efforts resulted in the preservation of Sequoia and led him to be called of the "father" of Sequoia National Park. But surely the most important person in helping to preserve the Sierran wilderness was the naturalist John Muir. It was Muir who founded the preservationist movement in the United States, and it was his life and writings that formed the basis of argument for the establishment of parks and wilderness preserves in countries throughout the world.

John Muir came to California in 1868 as a young man and settled in the Yosemite Valley. After a thorough study of the Yosemite region, he began in 1872 to explore the area to the south. On foot and often alone, he walked and climbed the entire Sierra between the Tuolumne and Kern Rivers, observing and documenting its natural wonders and its plants and animals. In 1873, he met Hale Tharp near the place that Tharp had first seen the Big Trees; a place Muir named Giant Forest. From that time on, he dedicated his life to the preservation of the Big Trees and the wilderness of the High Sierra.

Appalled by the devastation of the natural environment by loggers and miners, Muir enlisted the aid of politicians, newspaper editors, and other citizens in the

Figure 42.2   General location map for Sequoia and Kings Canyon National Parks, California.

campaign to save the Sierra Nevada from further exploitation. His writings were instrumental in informing the public of the value of wilderness. He helped found the Sierra Club and became its first president. A tireless naturalist and conservationist, he spent many years of his long and remarkable life traveling throughout the world, studying nature and spreading his message of wilderness preservation for the well-being of all people. The significant expansion of the boundaries of Sequoia and Kings Canyon National Parks was the realization of Muir's dream for the High Sierra. The magnificent John Muir Trail, established in his memory, follows the crest of the Sierra from Tuolumne Meadows in Yosemite to Mount Whitney in Sequoia, winding through three national parks and three national forests on its 218-mile route.

The California Geological Survey, under the direction of geologist Josiah D. Whitney, undertook the mapping of the High Sierra in the 1860s. Their primary mission was to locate occurrences of gold and other minerals with economic value. Despite insufficient funds, this group of scientists produced outstanding reports and maps that aided the conservation movement and laid the groundwork for subsequent scientific investigations. The survey party included William Brewer, a botanist who played an important role in the development of soil science; Charles Hoffman, an engineer who became known as the father of modern topographic surveying; and Clarence King, who later became the first director of the U.S. Geological Survey.

Geologic Features

The bedrock of Sequoia and Kings Canyon National Parks is dominated by the late Jurassic and Cretaceous igneous rocks of the Sierra Nevada batholith, which is interspersed with steeply inclined and complexly folded layers of metamorphosed early Mesozoic (late Triassic to early Jurassic) volcanic and sedimentary rock. In a few sites, isolated outcrops of Cenozoic volcanic rock attest to continued igneous activity throughout the period of uplift of the range. In large areas at high elevations and in many of the deeper valleys, thick sections of glacial sediments overlie the bedrock.

The shape and appearance of the landforms in Sequoia and Kings Canyon are strongly influenced by the batholithic structure and by the weathering characteristics of the intrusions. Because many of the granites are massive and have few joints or fractures, exfoliation processes have shaped many of the exposed bedrock surfaces (fig. 42.3). Exfoliation domes, similar to those in Yosemite National Park (chapter 25), are prominent features in Sequoia and Kings Canyon National Parks. Near the Middle

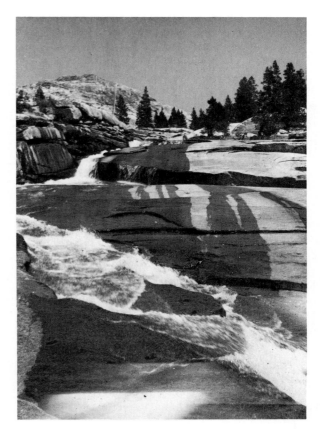

Figure 42.3   Exfoliation occurs as a series of horizontal joints in the upper Kaweah drainage basin of Sequoia National Park. Photo by P. M. Palmer.

Fork of the Kings River are a group of conspicuous exfoliation domes, including Tehipite and Kettle Domes and the Obelisk. Two large domes on opposite sides of Kings Canyon are Sentinel Dome and North Dome. These domes, over 8500 feet in elevation, are formed in massive granite. Examples of exfoliation in Sequoia National Park are Alla Peak, Beetle Rock, Sunset Rock, and Moro Rock (fig. 42.4).

In well-fractured granites, especially at high altitudes, frost action shatters rock into angular blocks and fragments of various sizes. Vertical cliffs with sharply incised gashes are formed by frost action and mass wasting. At moderate altitudes, where more moisture is available, cascading streamlets have cut slotlike gulches along fractures in canyon walls.

Avalanche chutes are numerous on steep slopes, cirque walls, and canyon walls in the High Sierra. An *avalanche chute* is the steep, smooth track, or trough, left on a slope after an avalanche—carrying ice, rocks, and trees—has descended. Most of the avalanche chutes probably formed during the severe conditions of the Pleistocene; some are

Figure 42.4   Moro Rock, an immense exfoliation dome of massive granite, rises to an elevation of 6725 feet above the canyon of the Kaweah River, Sequoia National Park. National Park Service photo.

still active. Along the crest of the Sierra where parallel, vertical fractures are closely spaced, the avalanche chutes that have developed on the fractures give the slope faces a fluted appearance. On the west flank of Mount Whitney, branching avalanche chutes follow fractures, and some of these chutes are more than 100 feet deep. In the massive granites (most notably on slope faces of the Great Western Divide), large, semicircular, almost perfectly smooth avalanche chutes have developed. They have been altered very little since they were repeatedly abraded and polished by sliding ice and rock during Pleistocene time.

At lower altitudes where the vegetation is heavy and moist soil was undisturbed by glaciation, intense chemical weathering has decomposed some of the older granites. This "rotten granite," which is loose and granular, can be seen in roadcuts along the Generals Highway. In some places, the granite is so soft that blasting was not required during highway construction and the roadcuts were excavated by power shovels.

Climatic Zones and Precipitation

In many respects, the weathering and erosional processes of the High Sierra are related to climatic zoning, which is fairly well defined on the slopes. The highest parts of

Sequoia and Kings Canyon National Parks are barren, windswept, rocky uplands in an Arctic-Alpine climatic (or life) zone. Rainfall is limited, and snow accumulates only in scattered patches because of the constant wind. The flat surfaces, such as the Boreal Plateau and the western slope of Mt. Whitney, were never glaciated because snow was blown away before it could accumulate. In general, relief is subdued in the high uplands. Only the hardiest plants survive in this environment.

In the transitional Subalpine zone, thin patches of soil support a few hardy shrubs and low plants. Peaks and basins show evidence of past glaciation. The sides of some of the flat uplands are scalloped by small cirques, and small lakes occupy some of the abraded rock basins. Near the treeline, the mountain landscape becomes more irregular. Ridges are serrated with horns and cols. A short distance below the main Sierran crest on the eastern side are the saw-toothed peaks that stand against the skyline so dramatically when viewed from the Owens Valley far below. Prominent cirques are carved out below peaks and arêtes. Many of the avalanche chutes, described above, extend down into the forested zones.

In the Canadian life zone, which reaches higher on south-facing slopes because plants receive more sunlight than on north-facing slopes, the firs and lodgepole pines resemble those of forests in more northerly latitudes, and heavy snows cover the ground from November to May. Some of the streams flow in U-shaped valleys that were formerly occupied by glaciers. Mountain meadows and marshes mark the locations of old lake basins that were eroded by glacial ice and are now filled in with sediment.

On the western slopes, precipitation is sufficient to support forests between elevations of roughly 2000 feet and 11,000 feet. The zone of maximum precipitation, approximately between the 7000-foot and the 8000-foot levels, is about where the Canadian zone blends into the Transition life zone. Here, the lodgepole pines give way to the taller ponderosa pines and the groves of giant sequoias.

Drainage Systems and Divides

The Great Western Divide is a north–south range of rugged mountains located about 10 miles west of the main crest of the Sierra. This high divide separates the westward-flowing Kaweah Basin drainage from the southward-flowing Kern River drainage (fig. 42.5). The Kaweah River is the southernmost of the major streams that flow generally southwest, following the regional slope of the Sierra Nevada. The western part of Sequoia, which is drained by the Kaweah River, is rugged mountain country, deeply dissected by branching canyons and ridges. The main canyon of the Kaweah at Ash Mountain Park Headquarters is more than 4000 feet deep. Several of the branch canyons, such as East Fork and Middle Fork, are even deeper. However, some plateaulike uplands are found in

Figure 42.5   The glacially eroded peaks of the Great Western Divide trend generally north-south across the boundary between Sequoia and Kings Canyon National parks. National Park Service photo.

the Kaweah Basin. Giant Forest occupies one of these gently sloping platforms. The relatively flat surface is apparently due to the resistant nature of the underlying massive granite bedrock that preserves the subdued topography developed prior to the most recent uplift of the range.

The Kern River rises in the upper Kern Basin in the eastern part of Sequoia National Park, between the Great Western Divide and the main crest of the Sierra. The Kern's glacially modified U-shaped valley is incised some 2000 feet in the high upland surface, between the Boreal Plateau on the east and Chagoopa Plateau on the west (fig. 42.6). Tributaries in hanging valleys at right angles to the main stream are notched in the canyon walls. The Kern River's southward course generally follows the trend of a north-south fault in the Sierra block. The fault zone provided a less erosion-resistant course for the stream as the crustal block was uplifted. The canyon becomes shallower as it approaches the southern boundary of the park, and thick glacial and alluvial deposits cover the valley floor.

Many of the topographic features of Kings Canyon National Park are grand in size and simple in form. The relief is very high, and drops of several thousand feet between ridge top and canyon bottom are characteristic of the coarse drainage texture in this part of the Sierra. *Coarse-textured drainage* means that the number of streams and valleys per square mile (or given area) is relatively low. On the massive granites of the tilted Sierra block, only a few major streams have developed. The reason is that as the streams cut deep valleys during the rapid uplift, the larger rivers collected (or captured) the waters of the smaller streams. Moreover, the drainage

networks were greatly modified by several cycles of glaciation during uplift. Like the Kaweah drainage in Sequoia, the main tributaries of the Kings drainage have a dendritic pattern and follow the trend of the regional slope.

Pleistocene glaciers scoured some of the domes and drainage divides (fig. 42.7) and excavated U-shaped valleys (fig. 42.8). The largest and longest of the valley glaciers was the one in Kings Canyon; it was a river of ice 44 miles in length that reached a relatively low altitude of 2500 feet on the western Sierra slope. Hanging valleys, some with beautiful waterfalls, such as Roaring River Falls and Mist Falls, once held tributary glaciers that fed the Kings Glacier. Sixty-Lake Basin, near the Sierra Crest, is a glacial staircase. Many small rock-basin lakes dot the broad Monarch Divide (elevation 10,000 to 12,000 feet) that separates the Middle Fork from the South Fork of the Kings River.

The floor of Kings Canyon (South Fork) is from one-fourth to one-third of a mile wide and is covered with outwash and glacial debris. Steep granite walls rise on each side, giving the canyon the typical U-shape of a glacial trough. By contrast, the lower canyon of the Middle Fork of the Kings River is V-shaped, indicating erosion by stream processes rather than by glacial ice (fig. 42.9). Only the upper reaches of the valley are U-shaped, giving the limit of glacial advance in Pleistocene time.

A few modern glaciers that formed in Neoglacial time, such as the Darwin Glacier, cling to cirques on the mountains along the northeast boundary of Kings Canyon National Park. Their activity is minor compared to the severe glacial erosion and extensive deposition of their Pleistocene predecessors.

Figure 42.6   View looking south down the U-shaped glacial valley of the Kern River from near the mouth of Whitney Creek. The broad uplands that slope gently to the top of the canyon walls are approximately 2000 feet above the river. The stream valley was incised into the high upland surface along the Kern River fault during the most recent uplift of the Sierra Nevada fault block. The valley was later modified by glacial erosion. U.S. Geological Survey low-altitude, oblique photo.

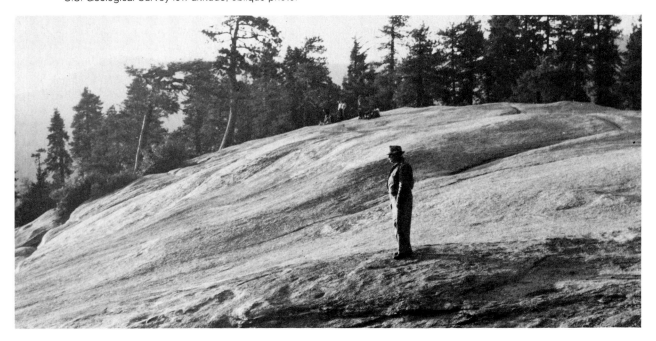

Figure 42.7   Glacially smoothed and polished granite along a bedrock ridge west of the Great Western Divide shows where Pleistocene glaciers topped the ridges and moved from one drainage basin to another. Photo by M. B. Palmer.

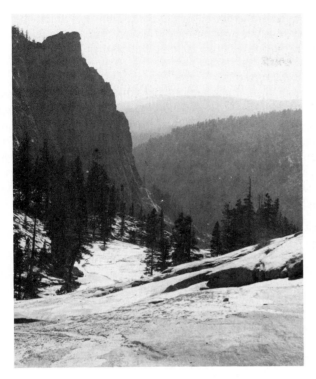

Figure 42.8 The typical U-shaped cross section of a glaciated valley is seen in the steep walls and the rounded and scoured bedrock bottom of a valley. Photo by M. B. Palmer.

The Sierra Nevada Uplift

In broad outline, the geologic story of Sequoia and Kings Canyon National parks is like that of Yosemite (chapter 25). All three parks lie along the crest and western slopes of the Sierra Nevada, a huge crustal block that has been lifted and tilted westward. Thus, the mountain range is asymmetrical; it has steep slopes on the fault-bounded eastern margin and a more gentle regional slope to the west. Along the eastern foot of the range is the Sierra Nevada normal fault system, which trends north–northwest and marks the break between the bold mountain front and Owens Valley. With an elevation of 14,494 feet, Mount Whitney is the highest summit in the United States outside Alaska. The floor of Owens Valley lies nearly 11,000 feet lower. Death Valley, some 80 miles to the east, contains the lowest point in North America, 282 feet below sea level. These figures give an indication of the magnitude of uplift and downfaulting that has occurred in this part of California. The vertical movement along the Sierra Nevada fault system has been at least 15,000 feet.

Numerous recent earthquakes on the Sierra fault system are evidence that the mountain-building process continues to the present day. The largest earthquake in historic time occurred on March 26, 1872, near Lone Pine on the floor of Owens Valley. (The epicenter was about 15

Figure 42.9 View up the Tehipite Valley on the Middle Fork of the Kings River from near Hume Lake, outside of Kings Canyon National Park. Glaciers did not extend into the lower part of the valley. The tops of Tehipite and Kettle Domes are visible in the far distance. Both domes were formed by the exfoliation of massive granite. National Park Service photo.

miles east of Mount Whitney.) The quake is estimated to have had a Richter magnitude of at least 8.0. Vertical movement on the fault was as much as 28 feet, and horizontal displacement was about 20 feet. The ground was broken for nearly 100 miles along the fault trace. Nearly all the buildings in Lone Pine were destroyed, and 27 people were killed (Hill 1975).

Bedrock Geology

The composite batholith now exposed in the High Sierra is made up of granitic rocks, largely quartz diorites, quartz monzonites, and granodiorites. While the Sierra Nevada batholith consists of intrusions of a wide range of Mesozoic ages, those in the area of Kings Canyon and Sequoia National Parks are predominantly late Cretaceous with very few from the late Jurassic. Single intrusions can cover as much as 200 square miles and are often elongated parallel to the trend of the Sierra Nevada range. Many of the individual intrusions have been separated on the basis of age by using the rules of intrusive cross-cutting relationships and the existence of inclusions or xenoliths of earlier rocks in younger ones. Within the parks, a number of intrusive sequences have been identified. Such sequences consist of an associated group of igneous rocks intruded over a relatively short period of time. One such sequence near Giant Forest is the concentrically zoned Sequoia intrusive sequence that was emplaced between 102 and 97 million years before present (mybp). The Sequoia sequence involves a set of four nested intrusions in which the outer zones are more mafic granodiorites surrounding younger intrusions of more silica-rich granite and alaskite (a granite rock having a very low percentage of dark minerals). The existence of these associated intrusions provides geologists with a natural laboratory that confirms the general processes of differentiation in granitic magmas and gives us an idea of the total time an individual magma system may operate and undergo differentiation. *Magmatic differentiation* involves the partial crystallization of an initially homogeneous magma and the development of a residual magma of different chemical composition. By this process, more than one type of igneous rock develops from a common magma. (For more on the intrusive suites of the Sierra Nevada batholith, see chapter 25, Yosemite National Park.)

Within Sequoia, the oldest intrusions appear as relatively small, sheared bodies of granodiorite on the margins of later plutons or within the fault zone of the Kern River drainage. These have been dated as approximately 160 mybp. Emplacement of the main mass of the batholith appears to have started nearly 120 mybp and to have ended at 80 mybp. The youngest large intrusions occur near the crest of the Sierra with the masses of Paradise and Whitney granodiorite in the south and the LeConte

Canyon granodiorite and the alaskite of Evolution Basin in the northern part of Kings Canyon National Park. These were intruded between 86 and 80 mybp.

Roof Pendants

The intrusion of the Sierra Nevada batholith obliterated most of the preexisting rocks in the area. Fortunately, small sections of the country rocks into which the batholiths were emplaced occur between separate granite intrusions as sheets or curtains of metamorphic rock called *roof pendants*. Both Sequoia and Kings Canyon National Parks have roof pendants of varying lithologies and ages (fig. 42.10).

CENOZOIC VOLCANIC ROCKS

LATE MESOZOIC GRANITIC INTRUSIONS

METASEDIMENTARY ROCKS } EARLY

METAVOLCANIC ROCKS } MESOZOIC

Figure 42.10   Simplified geologic map of Sequoia-Kings Canyon National parks (after Saleeby and others, 1978). The early Mesozoic (late Triassic to early Jurassic) metasedimentary rocks contain both deep-water oceanic and shallow-water continental components mixed with volcanic rocks. The Mesozoic volcanic rocks are largely rhyolitic and dacitic ash flows and tuffs with some sedimentary material. The age of the metasedimentary rocks in the roof pendants on the crest of the Sierra on the eastern edge of Kings Canyon National Park is uncertain. Some geologists have suggested that they are Paleozoic in age.

In the northern part of Kings Canyon National Park, the southern part of the Mt. Goddard pendant trends northwest–southeast and separates the youngest intrusions of the batholith on the northeast from slightly older intrusions on the southwest. The Mt. Goddard pendant is composed primarily of early Jurassic metavolcanic rocks, representing the vestiges of a thick volcanic pile of rhyolitic ash flows and tuffs mixed with more mafic lava flows and clastic sedimentary rocks and blocks of limestone. Intrusive breccias crop out along the margins of the pendant, and many dikes, both felsic and mafic, are emplaced in the metamorphic rock and the sheared granites associated with the pendant structure. Along the same trend, remnants of this volcanic-sedimentary sequence are seen in the Kaweah Peaks and Mineral King pendants.

The Sequoia Park pendants record a thick section of metamorphosed sedimentary rock, including limestones, quartzite, and calcareous sandstones with lesser amounts of volcanic rock. The relationship between the two sequences may be seen in the Boyden Cave pendant where the sedimentary sequence of shales, sandstones, and limestones is juxtaposed along the Kings River fault against the kinds of rhyolite ash flows and tuffs found in the eastern pendants.

Just to the west of the two parks, the lower Kings River and Kaweah River pendants record the emplacement of a complex ophiolite mass representing oceanic crust of late Paleozoic to early Triassic age. The roof pendants are all complexly deformed, and geologists do not agree completely upon the stratigraphic sequences within them or upon the original location of the sequences prior to the intrusion of the Sierra Nevada batholith. There is agreement, however, that the mixed volcanic and sedimentary sections do represent deposition at the edge of the continent where island-arc volcanics, derived in part from continental crust, were interleaved and mixed with sediments from both oceanic and continental sources.

Caves and Mineralization

In the pendants in the western parts of the parks, marbles have been partially dissolved by ground water to form underground caverns such as Crystal Cave, Clough Cave, Palmer Cave, and Paradise Cave. The metamorphism that changed limestone to marble destroyed any fossils that may have existed, but the rocks are assigned a late Triassic to early Jurassic age, based on fossils in other nearby pendants. As these carbonate rocks were being raised above sea level, ground water began percolating along fractures and bedding planes, enlarging openings by removing more of the material. When the openings were raised above the water table by continued uplift, ground water then began to deposit as well as to dissolve calcite. Typical dripstone features formed on the ceilings, walls, and floors of the underground chambers. (Cave features and solution processes are explained in chapters 12, 13, and 14.)

The mines in Redwood Canyon and along the North Fork of the Kaweah River are in *tactite,* a complex calcareous metamorphic rock formed as a result of hydrothermal activity and contact metamorphism. During the emplacement of the adjacent plutons, chemically active solutions from the hot magma penetrated the country rock and replaced some of its constituents with new minerals. Contained in the tactite are garnet, pyroxene, quartz, the tungsten ore scheelite, and the copper ores chalcopyrite and bornite.

Geologic History of Sequoia and Kings Canyon National Parks

1. Paleozoic deposition

The western coast of North America has been active over much of geologic time. During the early Paleozoic, the coast was part of a passive continental margin and not much is known about the marine sedimentary rocks that were deposited in Paleozoic seas that spread over the region where the Sierra Nevada now stands. Some Paleozoic rocks crop out east of the range, and a few scattered patches have been found in the mountains some distance north of Sequoia and Kings Canyon National Parks. In Devonian to Mississippian time, the plate convergence began that led to the Antler orogeny in Nevada. Subduction along the continental margin led to the emplacement of oceanic crust of late Paleozoic to Triassic age along the foothill faults of the Sierra Nevada.

2. Mesozoic deposition and mountain-building.

The Mesozoic history of the two parks may be divided into two parts. The first episode is recorded in the metamorphosed volcanic and marine sedimentary rocks of Triassic and early Jurassic age that are preserved as roof pendants and inclusions in the Sierra Nevada batholith. The second episode tells of the plutonic intrusion of the batholith largely in Cretaceous time.

The rocks of the roof pendants demonstrate that a subduction system existed off the western coast of North America early in Mesozoic time. A volcanic island-arc environment developed adjacent to the old continental shelf, and the subduction system very likely caused large-scale lateral transport and tectonic juxtaposition of the volcanic and sedimentary terranes. Much of the extreme deformation seen in the roof pendants comes from this period.

Toward the end of Jurassic and the beginning of Cretaceous time, the subduction system had moved farther to the west. Igneous activity, resulting in the emplacement of plutons in the Cretaceous batholith, occurred apparently as a series of pulses. The older, early Cretaceous plutons are found in the western parts of the range; and the

Table 42.1 Generalized Geologic Column, Sequoia and Kings Canyon National Parks

Time Units			Rock Units	Geologic Events
Era	Period	Epoch		
Cenozoic	Quaternary	Holocene	Alluvial, glacial, and mass wasting deposits	Stream erosion; Neoglacial activity; landslides, rockfalls Earthquakes, faulting, continuing uplift
		Pleistocene	Till, moraines, glacial outwash of various ages Avalanche deposits	At least three major glaciations, including substages; intense glacial erosion, avalanche chutes, ice-sculptured landscapes, glacial deposition Faulting, relatively rapid uplift
	Tertiary / Neogene	Pliocene		Probable onset of glaciation Southern Sierra volcanism (outside parks) Canyon-cutting, intense erosion Beginning of faulting and uplift Westward tilting of Sierra Nevada block
		Miocene		
	Paleogene	Oligocene Eocene Paleocene		Erosion Pause in tectonic activity Probable beginning of cave development
Mesozoic		Cretaceous Jurassic Triassic	Quartz diorites, granodiorites, quartz monzonites Limestones, shales, sandstones, interbedded with volcanics	Emplacement of plutons in composite batholith Metamorphism of country rock Nevadan orogeny Marine sedimentation and island arc volcanism Subduction and plate convergence
Paleozoic			(not exposed in parks)	Long accumulation of sediments in Paleozoic seas

Source: modified after Hill 1975; Bateman 1974; Bateman and Wahrhaftig 1966

younger, late Cretaceous granites are near the crest. In general, the older intrusions are more mafic in composition (quartz diorites, for example), and the younger granites are more felsic (granodiorites, quartz monzonites).

The Nevadan orogeny that uplifted the Sierra Nevada is regarded as the first wave of the Laramide mountain-building episodes that created the great North American Cordillera. The extensive erosion that accompanied the orogeny stripped off many thousands of feet of Paleozoic and Mesozoic rocks. Estimates are that a layer of material over seven miles thick was stripped from the top of the

Sierra Nevada. The uncovering of the batholith also exposed the roof pendants of metamorphosed volcanic and marine sedimentary rocks.

3. Cenozoic faulting and uplift.

Erosion continued during the beginning of the Cenozoic Era, but tectonic activity had slowed. Then the Sierra Nevada began to tilt westward. Faulting continued as the mountain block rose higher and higher, and the range began to assume its present position and form. The most recent faulting, uplift, and tilting probably began 2.5 million years ago. Sporadic earthquakes in the southern Sierra are an indication that movement and uplift are continuing. Late in the Cenozoic, volcanism occurred in the southern Sierra, but only a few remnants of those lava flows are found within park boundaries (fig. 42.10).

4. Pleistocene and Holocene glaciation.

Episodes of glaciation in the southern Sierra were less severe and less extensive than in the northern Sierra. Pleistocene climates were less harsh in the southern Sierra, and less precipitation was available. Overall, however, the patterns of advance and retreat of ice were similar. Alpine glaciation began early in the Sierra Nevada, probably late in the Pliocene and long before the large continental glaciers accumulated. In Sequoia and Kings Canyon, valley glaciers moved down major stream valleys, enlarging and deepening them. Glaciers were largest and most numerous in the zones of high precipitation on the western slopes. The valley glaciers on the east side of the crest strongly eroded the peaks but did not extend down into the Owens Valley.

At its maximum, the Kern Glacier, the most southerly of the Sierra Nevada trunk glaciers, nearly reached the southern boundary of Sequoia National Park. In the Kaweah Basin, glaciers partly filled many of the converging canyons, but few of them overtopped the divides except in the upper reaches of the basin. Even at their maxima, the largest glaciers in the northeast part of the basin did not extend much below the 5000-foot level.

Glacier systems were more extensive in Kings Canyon National Park, especially toward the north. Valley glaciers occupied the canyons of the Kings River and its tributaries, as well as the upper reaches of the San Joaquin River basin at the northern tip of the park. Glacial erosion was severe.

Three major episodes of glaciation occurred during the Pleistocene, with substages involving advances and retreats of ice during these generally colder climatic periods. Till patches and moraines of varying ages mark glacial recessions. The Pleistocene glaciers did not last through the post-Pleistocene climatic optimum.

In Kings Canyon National Park, a few glaciers began occupying the cirques of earlier glaciers during the Little Ice Age (A.D. 1450–1850). These small valley glaciers are in the vicinity of Mount Goddard in the northern part of the park; others are in the Palisades region, just east of the park's northeastern boundary.

Prospects for the Future

More than one hundred years have passed since John Muir began his fight to preserve the California wilderness. Yet today, the national parks and national forests of the High Sierra are seriously threatened by throngs of visitors. The crowds have created problems of overuse that tend to damage the fragile environment, especially in the Alpine and Subalpine life zones. Because Sequoia and Kings Canyon are the national parks closest to the Los Angeles metropolitan area, they are particularly endangered.

Ironically, a recent ten-year study by the National Park Service indicated that the giant sequoias may also be threatened by *overprotection*. Periodic thinning by fires is needed to remove undergrowth and allow seeds to reach the soil and root in the sunlight. A program of "controlled burning" has been started by the National Park Service to help ensure new generations of sequoias. How can we balance the needs of a large urban population that seeks enjoyment of the parklands and, at the same time, protect the park environment from the damaging effects of too many people and vehicles and yet avoid overprotection? For the National Park Service, the answer to these questions involve careful management in the present, long-range planning for the future, and, above all, cooperation and understanding on the part of park visitors.

Geologic Maps and Cross Sections

Bateman, P. C. 1965. Geologic map of the Black Cap Mountain quadrangle, Fresno County, California. U.S. Geological Survey GQ Map 428.

Bateman, P. C., and Moore, J. G. 1965. Geologic map of the Mt. Goddard quadrangle, Fresno and Inyo Counties, California. U.S. Geological Survey GQ Map 429.

Jennings, C. W. Geologic map of California. California Division of Mines and Geology. Sheets covering southern Sierra Nevada.

Moore, J. G. 1978. Geologic Map of the Marion Peak Quadrangle, Fresno County, California. U.S. Geological Survey GQ Map 1399.

Moore, J. G. 1981. Geologic Map of the Mount Whitney Quadrangle, Inyo and Tulare Counties, California. U.S. Geological Survey GQ Map 1545.

Ross, D. C. 1958. Igneous and metamorphic rocks of parts of Sequoia and Kings Canyon National Parks, California. California Division of Mines, Department of Natural Resources Special Report 53.

Bibliography

Bateman, P. C. 1974. Model for the origin of Sierra granites. *California geology* 27(1):3–5.

Bateman, P. C., and Wahrhaftig, C. 1966. Geology of the Sierra Nevada. In *Geology of Northern California*, ed. E. H. Bailey, p. 107–172. California Division of Mines and Geology Bulletin 190.

Bowen, E. 1976. *The High Sierra, American Wilderness Series.* Rev. ed. New York: Time-Life Books.

Colby, W. E., ed. 1960. *John Muir's studies in the Sierra.* San Francisco: Sierra Club.

Gray, W. R. 1975. *The Pacific crest trail.* Washington, D.C.: National Geographic Society.

Hamilton, W. 1969. Mesozoic California and the underflow of Pacific mantle. *Geological Society of America Bulletin* 80:2409–2430.

Hartesvelde, R. J. 1973. *Field guidebook for Sequoia and Kings Canyon National Parks; the natural history, ecology, and management of the giant Sequoias.* American Association of Stratigraphic Palynologists, 6th annual meeting, Anaheim, California, October 16–20.

Hill, M. 1975. *Geology of the Sierra Nevada.* Berkeley: University of California Press.

Matthes, F. E. (ed. F. Fryxell). 1950. *Sequoia National Park, a geological album.* Berkeley: University of California Press.

———. (ed. F. Fryxell). 1965. *Glacial reconnaissance of Sequoia National Park, California.* U.S. Geological Survey Professional Paper 504-A.

Oakeshott, G. B. 1978. *California's changing landscapes.* 2nd ed. New York: McGraw-Hill Book Company.

Oberhansley, F. R. 1974. *Crystal Cave in Sequoia National Park.* Sequoia Natural History Association.

Ross, D. C. 1958. Igneous and metamorphic rocks of parts of Sequoia and Kings Canyon National Parks, California. California Division of Mines, Department of Natural Resources Special Report 53.

Saleeby, J. B, Goodwin, S. E., Sharp, W. D., and Busby, C. J. 1978. Early Mesozoic paleotectonic-paleogeographic reconstruction of the southern Sierra Nevada, in *Mesozoic Paleogeography of the United States,* Society of Economic Paleontologists and Mineralogists, p. 311–336.

Sharp, R. P. 1976. *Geology field guide to Southern California,* K/H Geology Field Guide Series. Rev. ed. Dubuque, Iowa: Kendall/Hunt Publishing Company.

White, John R., and Pusateri, S. J. 1949. *Sequoia and Kings Canyon National Parks.* Stanford University Press, Stanford, 212 p.

Wheelock, W., and Condon T. 1974. *Climbing Mount Whitney.* 3rd ed. Glendale, California: La Siesta Press.

Sequoia National Park/Kings Canyon National Park

Areas: Sequoia: 402,488 acres; 629 square miles
Kings Canyon: 460,636 acres; 720 square miles

Established: September 25, 1890 for Sequoia
March 4, 1940 for Kings Canyon

Designated a Biosphere Reserve: 1976

Address: Sequoia National Park, Kings Canyon National Park, Three Rivers, California 93271

43

Channel Islands National Park has been designated a Biosphere Reserve to protect unique geoecologic systems above and below sea level.

The park includes all the Northern Channel Islands—San Miguel, Prince, Santa Rosa, Santa Cruz, and Anacapa Islands—and Santa Barbara Island, one of the Southern Channel Islands.

The Channel Islands have a complex tectonic history and are related geologically to the Transverse Ranges north of the Los Angeles Basin.

Marine terraces record sea level changes that attest to a recent history of subsidence and uplift producing impressive topographic features. The rise in sea level in Pleistocene time isolated the islands from the mainland and each other and led to their distinctive biological systems. Archeological sites that record early human occupation also denote changes in the character of plant and animal populations of the islands during the last 15,000 years.

Channel Islands National Park

Offshore Southern California

by Donald F. Palmer
Kent State University

Figure 43.1 Looking north toward barren Prince Island from Cuyler Harbor beach on San Miguel. The sand dunes on the backshore (foreground) are partially stabilized by sparse vegetation. Photo from Channel Islands National Park Collection.

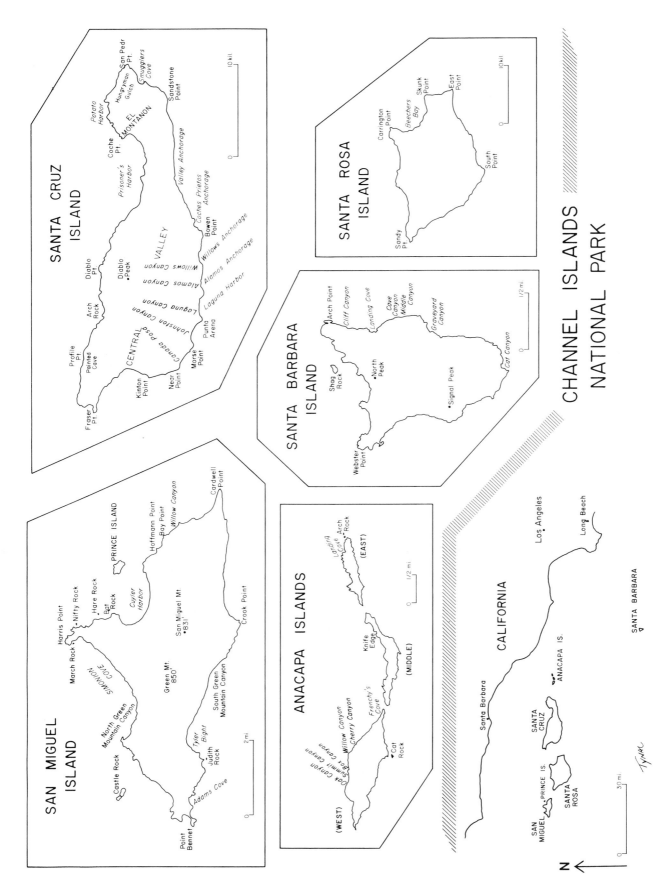

Figure 43.2 Channel Islands National Park, California.

Geographic Setting and Local History

The Northern Channel Islands, which lie across the Santa Barbara Channel from the California mainland, are geographically separate from the Southern Channel Islands, which are set off from the mainland by the San Pedro Channel and the Gulf of Catalina. (The Southern Channel Islands not in the national park are San Nicholas, Santa Catalina, and San Clemente.) Three Channel Islands, Santa Barbara (from the southern group), Anacapa and San Miguel (in the northern group), were set aside as a national monument in 1938. The addition of Santa Cruz, Santa Rosa, and Prince Islands in 1980 and the creation of the national park expanded this haven for endangered species of plants and wildlife, many of which are unique to the islands. Also protected are rookeries for sea birds, fur seals, elephant seals, sea lions, and other marine animals. Santa Cruz, for example, is the home of nine endangered plant species, as well as the endangered California brown pelican. San Miguel has the only marine mammal rookery in the world inhabited by six species of seals and sea lions. The endangered island fox, much smaller than mainland species, is found on four of the islands. Santa Barbara has a treelike sunflower of the genus *Coreopsis.*

Before the park was established, the fragile ecosystems of the islands and their surrounding waters were threatened by such human activities as ranching, trapping, oil spills, offshore oil exploration, a proposed pipeline compressor facility for crude oil, and residential development.

Under the present management plan for Channel Islands National Park, primitive camping is permitted only on Santa Barbara and Anacapa Islands. The remaining islands may be visited for one day only with permission from the National Park Service or property owners. Tourists on San Miguel must be accompanied at all times by Park Service personnel. Qualified scientists carry on environmental research projects on several islands, but sensitive areas, defined as habitats of endangered plant or animal species, are closed to the public.

The Channel Islands have generally cool, foggy summers and rainy winters (about 12 inches of rain annually). In winter, during the migration season, the islands are some of the best areas along the California coast for watching whales.

Early Inhabitants. The Channel Islands had long supported prehistoric cultures prior to the arrival of the Spanish explorers in the sixteenth century. Roasting pits with charred dwarf mammoth bones have been found on the islands. Definite human skeletal remains have been dated as 11,000 B.P. (Before Present), which was approximately the end of Pleistocene time. The climate then was more moist; the drier climate of today began around 7500 years ago.

When sea levels were lower during Pleistocene time, the Channel Islands occupied a larger land area as part of the Transverse Ranges of Southern California. Pleistocene mammoths that had migrated to the southwestern tip of the area were cut off from the mainland when sea level rose, creating the islands. Once isolated, the mammoths evolved into a pigmy variety, only four feet high at the shoulder. Indian hunters may have been responsible for their extinction.

By 3000 years ago, the seafaring Canalino Indians lived in the region. They used tar from oil seeps on San Miguel and the mainland coast as caulking material for their canoes, for mending bowls, and as adhesive for attaching arrow points and other weapon heads to shafts. It was the Canalino Indians who greeted Rodriguez Cabrillo, the discoverer of California, when he stopped at San Miguel in 1542. Cabrillo, a Portuguese mercenary in the service of Spain, led an expedition for the Spanish viceroy to find a sea route to China. After sailing north along the California coast beyond San Francisco, Cabrillo returned to San Miguel to spend the winter. He died on the island in January of 1543.

Diseases introduced by Europeans reduced the native population of the islands in the seventeenth and eighteenth centuries. The Canalino Indians abandoned Santa Rosa after a severe earthquake in 1812, and the last Indian was removed from Santa Cruz in 1813.

In the 1870s and 1880s, the city of Santa Barbara on the California coast opposite the Channel Islands was a popular health resort because of the numerous natural seeps of tar, oil, and gas along the shore. Natural oil slicks, spread over the coastal waters by winds and currents, were believed to purify the air, making infectious diseases and chronic illnesses less prevalent. Most of the seeps emanated from shale in the Monterey Formation, which crops out in some of the Channel Islands as well as on the mainland. San Miguel has two oil seeps from the Monterey Formation, one near Castle Rock, off the island's northwest shore, and the other near Tyler Bight, off the south shore. Another seep has been burning at the eastern end of Santa Cruz Island near Chinese Harbor since before written records existed.

Ever since the beginnings of European settlements, the Channel Islands, despite being relatively close together, have had separate histories based on the different ownerships and types of land use. These varying patterns of land use are still affecting the natural environments of the islands in the park.

Santa Rosa Island. In 1843 Santa Rosa was awarded to the Carrillo family as a land grant from the government of Mexico, and for more than fifty years, family members maintained a sheep ranch. W. L. Vail and J. V. Vickers bought Santa Rosa in 1902 and converted the ranch to a cattle operation. The Vail and Vickers Company still owns Santa Rosa, but eventually the island will be purchased by the federal government.

Santa Cruz Island. This island, largest in the group, supported a population of some 3000 Indians when the Spaniards first visited it. The island was named Santa Cruz ("holy cross") in 1769 by a Spanish naval officer after Indians found an iron crucifix that had been lost by the expedition's chaplain. The Mexican government had a penal colony on Santa Cruz for a short time in 1829, but long enough to give Prisoners Harbor its name. Sheep and other domestic animals were first brought over by an English company that acquired the island in 1859. Ten years later the island was sold to Justinian Caire who operated a sheep and cattle ranch and produced wine. His descendants still own a 6000-acre sheep ranch on the eastern end of Santa Cruz. The rest of the island (about nine-tenths of the total acreage) is owned by the Santa Cruz Island Company. Most of the sheep have been removed and those that remain are confined by fences. Cattle grazing is restricted to parts of the island that are less subject to damage by erosion.

A University of California research facility, the Channel Islands Field Station (with an office located on the University of California-Santa Barbara campus), was established in the 1960s in cooperation with the Santa Cruz Island Company. Eventually the entire island will be owned and managed by the Nature Conservancy in cooperation with the National Park Service.

San Miguel Island. The federal government acquired San Miguel when Mexico ceded California territory to the United States in 1848. George Nidever, who brought sheep, cattle, pigs, and horses to the island in the 1850s, was the first leaseholder. The adobe house that he constructed from local clay still stands. Prior to the introduction of domestic animals, the island had a luxuriant ground cover, but a combination of overgrazing and repeated droughts destroyed most of the vegetation. Sand dunes, formed by the prevailing northwest winds, prevented regrowth. The severe drought of 1863–64 killed 85 percent of Nidever's stock (5000 sheep, 180 cattle, and 30 horses). An 1870 drought caused the remaining sheep to dig as deeply as two or three feet into the ground for roots. What the sheep didn't kill, the drifting sand dunes did. Subsequent droughts in 1877, 1893–94, and 1923–24 changed the landscape to near-desert conditions supporting only a few grasses and wildflowers. The U.S. Navy owns the island and administers it in cooperation with the National Park Service.

Anacapa Island. Anacapa, which is actually three islets (East, Middle, and West) linked by sand bars, was acquired by the United States from Mexico at the same time as San Miguel. In the late nineteenth and early twentieth centuries, a few sheep were pastured on the island, but all were removed when Anacapa became part of the national monument in 1938.

The name Anacapa is from an Indian word, *eneepah,* which means "deception, or mirage"; from a distance the Indians' view of the three islets was distorted. Under certain conditions of weather and light, this distortion can still be seen—probably because, when it is viewed from the north, the marked difference in slope between the north and south sides of the islets causes Anacapa to appear to be a larger landmass than it actually is.

Arch Rock, located at the eastern tip of East Island, is a well known Anacapa landmark. James Whistler, the nineteenth-century American painter, made an etching of this 40-foot high sea arch when, as a young man, he worked as a map engraver for the U.S. Coast and Geodetic Survey.

An automated 600,000-candlepower beam and foghorn were installed in the lighthouse on East Island in 1966. Prior to that time, Coast Guard personnel manned the lighthouse, which was constructed in 1932. An acetylene beacon had given warning to coastal shipping for many years before that. In spite of the warning system, wrecks and groundings occur on Anacapa's rocky shores, usually because of dense fog. The most famous shipwreck was that of the paddle-wheel steamer "Winfield Scott," which sank in 1853. All 250 passengers got off safely, but spent several anxious days on Anacapa before being rescued. What remains of the wreck lies offshore in about 20 feet of water.

Santa Barbara Island. Sometimes referred to as "Santa Barbara Rock" because of its small size (approximately a square mile), this island is the only one of the Southern Channel Islands to be included in Channel Islands National Park. Archeological evidence indicates occasional extended visits here by prehistoric people but no permanent camps, probably because of lack of fresh water. A few seeps, like one in Indian Water Cave, are the only natural sources of fresh water.

Originally the island was well covered with vegetation and supported a number of bird and shore species and small animals. However, a century or more of farming and livestock-raising drastically changed the island's natural environment. Many of the changes are irreversible. What happened to the island's ecosystem can be summarized as follows. Most of the items also apply to other Channel Islands.

1. The introduction of cats eliminated several species of sea birds.
2. Domestic sheep, rabbits, goats, etc., reduced or eliminated many native plants.
3. Clearing of land for farming and introduction of grasses and mainland weeds removed or forced out

much of the original vegetation. Native species are left only in isolated, protected areas.

4. Fires, both deliberate and accidental, burned most of the vegetation to the water's edge and destroyed so much of the island's habitat that the Santa Barbara song sparrow has become extinct.

5. The planting of grains brought about overpopulation of native deer mice who, in turn, decimated land mollusks and sea bird eggs.

6. The introduction of the "ice plant" (*Mesembryanthemum crystallinum*), a fleshy-leaved plant with glistening encrustations, led to the crowding out of native species.

7. The combined effects of burning, exotic plants, and domestic animals caused considerable erosion.

Environment and Organisms

Clearly, Channel Islands National Park occupies an area much changed by human occupation. In this sense, the park is an example of how misuse and overuse of land can change an environment forever. Nevertheless, the park does provide one of the best marine wildlife sanctuaries in the United States, as well as preserving local habitats for a number of endangered species.

The Channel Islands are located in a transitional region between northern and southern marine fauna, with Point Conception (located on the mainland 23 miles due north of San Miguel) being the dividing point. The westernmost islands—San Miguel, Prince, and San Nicholas (not in the park)—support mainly northern marine fauna since they are in the path of the cold water of the south-flowing California Current. Because of the warm countercurrent from the south, the easternmost islands—San Clemente and Santa Catalina (outside the park) and Santa Barbara—support primarily southern species. Anacapa, Santa Cruz, and Santa Rosa, which lie between the two currents, have mixtures of southern and northern species. Waters surrounding all the islands are rich in marine algae and kelp, the food supply for marine organisms of many types. About 20 species of whales and porpoises abound in the area, along with six species of sea lions and harbor seals and many species of sea birds.

Although some islands lack trees, most of the islands have sage scrub and grasses along the shorelines, with oak woodlands mixed in on the larger islands. Islands that have (or have had) sheep are nearly stripped of vegetation, except where sheep have been restricted to fenced areas.

Each island has endemic species and subspecies of plants and animals that live only on one island or, in some cases, adjacent islands; examples are unique snails, deer mice, island foxes, and many kinds of plants. Both native and feral animals live on the islands. The feral animals—rabbits, cats, sheep, and pigs that have reverted to the wild—have altered the environment and caused the disappearance of some native species.

Among the threats still endangering the environments of these peaceful islands are: (1) overharvesting of marine resources, such as anchovies (main food of the California brown pelican) and kelp; (2) overgrazing; (3) oil spills from passing tankers; (4) development of offshore oil and gas; (5) sonic booms that disturb seal and bird rookeries; (6) and the introduction of exotic species.

Geologic and Other Natural Features of the Channel Islands

If the ocean water were to disappear so that only land could be seen, the islands would stand as high ridges on the continental shelf, separated by a series of basins. Marine terraces on the islands indicate that intermittent uplift of the islands or lowering of sea level, or both, has occurred in the geologic past. Although the islands are now in close proximity to one another, they do have distinct geologic characters and histories.

San Miguel Island

Structurally, San Miguel, 14 square miles in area, is the northern flank of a faulted anticline. Its axis trends northwest, which is also the strike of the beds that dip northeast. Ages of the bedrock range from Cretaceous to Miocene. The rocks are mostly sedimentary, and the beds form a homoclinal structure.

Several levels of beaches encircle the island. Due to overgrazing, slopes are severely gullied and the western two-thirds of the island is covered by sand dunes. Rainwater trapped by the dunes is the source of fresh water springs that make San Miguel one of the few Channel Islands with potable water (drinkable, although poor tasting). Some of the dunes are being stabilized by regrowth of vegetation and by cementation due to groundwater action.

Since most of San Miguel's land area is a plateau (400 to 500 feet above sea level) with two rounded hills, it is lower in relief and more subject to wind erosion than the other islands (fig. 43.3). Prevailing northwest winds have blown away topsoil, exposing caliche and dune rock at the surface. *Caliche*—composed of crusts of soluble calcium salts, mixed with sand and silt—is formed mainly by capillary action drawing mineral-rich ground water to the surface where it evaporates in the wind and sun, leaving the salts. Caliche also tends to form in a layer just below the surface when not enough moisture is available to flush the calcium salts from the soil. Both surface and subsurface caliche deposits occur over most of the island. *Dune rock,* an *eolianite* made up largely of dune sand, is formed

Figure 43.3   On the exposed western end of San Miguel Island, on land formerly covered by vegetation, sand dunes pile up in long rows, parallel to the direction of prevailing northwest winds. Photo from Channel Islands National Park Collection.

from wind-deposited clastic sediment that has been cemented by calcite deposited by ground water. Gradually dunes were consolidated into a dune rock layer.

On San Miguel the dune rock is interbedded with layers of soil that represent periods when conditions were stable long enough to permit weathering to form soil. Unstable and drier periods brought about resumption of dune growth, and dunes buried vegetation. Concretions of caliche formed as nodules, tubes, and casts around the buried roots and tree trunks. A caliche-encased log, 30 feet long and over 2 feet in diameter, was found on the island. Erosion has also uncovered trunk casts of other logs.

Case-hardened caliche is common on the northern and western parts of the island. Here the caliche deposits were exposed to rain and sea spray so that calcium salts on the surface were dissolved and reprecipitated many times, making the crusts thick and hard. Slabs and large chunks of caliche are scattered over the gound, and the sand dunes have become semi-indurated.

Of interest to geologists are the "elevator rocks" of San Miguel. They are found as a pebble-and-cobble pavement on old marine terraces that have been raised above

sea level and then buried by sediment washed down from still higher terraces or slopes. When enough moisture is present, soil churning forces the coarser gravel and cobbles to the surface. The finer grains are blown away by wind or washed away by water.

Many paleontological and archeological sites are on the island, including over 500 Indian sites. A major paleontological find was the remains of the pigmy mammoth (*Mammuthus exilis*), which was only four feet high at the shoulder. A large accumulation of pigmy mammoth bones was found at Running Springs (near Dry Lake on the west side of the island). Another pigmy mammoth site is about three miles west of Cardwell Point on the southeast shore. The bones at Running Springs have been well preserved by calcareous tufa deposited around the springs, which have a high dissolved load of calcite.

Carbon-14 dating of mammoth remains yields ages such as 15,630 ± 460 and 16,520 ± 150 years. Burned and calcined mammoth bones, charcoal made of cypress, mammoth tusks and teeth all suggest that prehistoric people were present at the sites and that they hunted the pigmy mammoth. There is geologic evidence on San

Miguel of overgrazing by the pigmy mammoths, periodic forest fires set by hunters or by lightning, and a 20-year drought cycle.

San Miguel is noted for its unique plant and animal species. The island is on the migration pathway of the humpback, blue, finback, and sei whales. Six species of seal (five of which breed here) frequent the seal rookery at Point Bennett on the western tip of San Miguel. These species are the harbor seal, California sea lion, northern elephant seal, Steller sea lion, northern fur seal, and the Guadalupe fur seal (nonbreeding at this rookery). Eight kinds of plants are found only on San Miguel.

Prince Island

Tiny Prince Island (also called Gull Island) lies just off the northeast shore of San Miguel (fig. 43.1). Prince Island's highest elevation is only 296 feet, and its dimensions are about 1000 by 1500 feet. Thin, shallow soils mantle the rocky slopes of this foggy, windswept piece of land. The bedrock, consisting of massive dacite intrusives, flows, and clastics, is Oligocene in age (table 43.1). The island is a major sea bird rookery and is also in the whale migration pathway.

Santa Rosa Island

Santa Rosa, 84 square miles in extent, is the second largest island in Channel Islands National Park, has about 45 miles of shoreline, and is approximately 15 miles long and 10 miles wide. Geologically, the island is a series of anticlines and synclines that have been faulted. A major fault, the Santa Rosa fault, trends east–west across the island. Movement along this fault, producing both vertical and horizontal displacement, has caused distinct differences in topography between the northern and southern parts of the island. The northern half of the island, the downthrown side, has broad, plateaulike terraces into which streams have cut steep canyons. High bluffs face the northern shore. White sandy beaches stretch along the northern and western coast. Most of the beaches are flanked by dunes up to 400 feet high. South of the Santa Rosa fault, the land is higher and more rugged. The highest point (1589 feet) is on a mountainous ridge that follows the east–west trend of the fault. The shales, sandstones, and volcanics on the south side of the island are more resistant to erosion than the shales (Monterey Formation) exposed on the north side.

Many archeological and paleontological sites have been located on Santa Rosa, including some with remains of pigmy mammoths. A few sites have been studied, but much archeological work still needs to be done. A rare subspecies of Channel Islands fox is found only on this island, as well as an endemic spotted skunk. The Torrey pine grows only here and in Torrey Pines State Park in San Diego County on the mainland. Twenty-six other plants that do not grow on the mainland have been identified on Santa Rosa. The profuse kelp beds along the shore are a refuge for sea otters and a breeding ground for the California sea lion. Humpback, blue, and sei whales swim by the island during their migrations.

Santa Cruz Island

With 96 square miles of land surface, Santa Cruz is the largest Channel Island. It has 65 miles of shoreline, is 24 miles long, and between 1 to 6.5 miles in width. With its steep shoreline, pocket beaches, and mountains, Santa Cruz is the most rugged of the islands. Sea caves have been eroded in the cliffs along the shore by wave action. Unlike the other Northern Channel Islands, which have relatively few safe anchorages, Santa Cruz's irregular shoreline has a number of protected coves, especially on the south side.

A major east–west-trending fault cuts across Santa Cruz, separating northern and southern mountain ranges, the principal rock types, and the island's drainage networks. The floor of the intervening valley, eroded along the fault zone by streams, is about the only flat land on the island. Most of the Central Valley is drained by a stream that flows into Prisoner Harbor on the northeastern side of the island. A smaller stream drains to the west. The fault zone is the result of strike-slip movement along the East Santa Cruz Basin fault that brought together the northern and southern halves of the island during Miocene time.

In the northern mountain range, Diablo Peak, elevation over 2400 feet, is the highest point. Sharp ridges and steep canyons leading down to the coast dissect the northern range. The less rugged southern mountain range rises 1523 feet above sea level. Broader canyons on the south side, with gentler slopes, provide access to the island's interior.

Some of the sedimentary rocks on Santa Cruz are a continuation of those in neighboring Santa Rosa; but the geology on Santa Cruz is more complex, with exposures of numerous rock types, including volcanic and metamorphic rocks of various ages. Prehistoric inhabitants of the Channel Islands found excellent raw materials for stone tools among this rich variety. Some of the volcanic rocks were useful for chipping and grinding weapons and tools. Especially prized was a chert of high quality in beds of the Monterey Formation cropping out on the eastern end of the island. Points and tools chipped from this chert were bartered for goods from mainland Indians.

Among the endemic plants and animals on Santa Cruz are the great island pine, the Santa Cruz pine, spotted skunk, a species of salamander, and the dwarf fox. About 500 native plant species and subspecies grow on the island;

Figure 43.4   Low tide at Anacapa Island exposes abalone clinging to the base of the cliff. Blocks of rock, loosened along joint planes, are strewn on the wave-cut platform. A piece of heavy anchor chain (foreground) is washed up by the surf. Photo from Channel Islands National Park Collection.

7 of which are endemic, and 31 of which are not found on the mainland. Resident bird species number 44 and include the endangered California brown pelican. Kelp beds surrounding the island provide food for seal and bird rookeries, as well as for fish, shellfish, and other organisms.

Anacapa Island

The three Anacapa islets are segments of a volcanic ridge made up of lava flows, breccias, ash, and cinders erupted from the ocean floor in Miocene time. Many of the lava flows have amygdules of olivine and chalcedony, and a large vein of chalcedony is exposed near Frenchys Cove on West Island. Cobbles and pebbles weathered from chalcedony are scattered over the cove beach, one of the few places on Anacapa where boats can land. Sheer cliffs and rocks make up most of the shoreline. The cliffs are particularly high and almost vertical from the ridge crest

to the water along the island's south side. Slopes on the north side of the ridge are somewhat less steep.

The highest point on the volcanic ridge, 930 feet, is Vela Peak on West Island. The ridge gradually lowers to 250 feet above sea level, which is the highest point on East Island. From west to east, Anacapa is 5 miles long, but only about one-half mile wide. The largest islet, West Island, has about half of Anacapa's total acreage of a little over one square mile.

Wave-erosional features, such as sea arches, sea caves, stacks, black sand and cobbles, wave-cut platforms, surge channels, and blowholes, are prominent on Anacapa's shore (fig. 43.4). The sea caves, arches, and surge channels tend to form in bedrock fracture zones . A well developed surge channel and blowhole is near Cat Rock on the south side of West Island. When a wave swell moves into the narrow channel, air that is trapped and compressed within the fissures forces out sea water explosively, spraying the surrounding area. Raised beaches and marine terraces on parts of Anacapa are indications of changing sea levels in the fairly recent geologic past.

Santa Barbara Island

Santa Barbara is geologically related to Anacapa, since both islands are almost entirely volcanic material, primarily basalt. Geologic features of Santa Barbara include cinders, pillow basalts mixed with sedimentary marine beds, marine and nonmarine terraces, eolianites, sea caves, landslides, and evidence of churning and slickensides in the soil, caused by differential expansion and contraction of soil materials.

Overgrazing and burning on Santa Barbara have caused severe erosional problems. Cliffs and slopes are unstable and subject to landsliding, especially on the island's north side. Another area of severe erosion is near the base of Signal Peak on its eastern side. Signal Peak (elevation 635 feet), the highest point on the island, overlooks the southwest coast.

Marine terraces, cut by canyons, are prominent on Santa Barbara's northern and eastern sides. The highest Pleistocene sea stands probably submerged the entire island. Faunal remains and numerous archeological sites have been found in the terrace gravels. Prehistoric people must have camped often on the island to fish and hunt when canoeing between the Southern and Northern Channel Islands.

Although Santa Barbara is a small island, it is important for the protection of certain native plants and animals: the giant *coreopsis,* which attains a height of eight feet and has bright yellow, daisylike flowers in the spring; the live-forever (*Crassulaceae*) and other plants not found on the mainland; an endemic night lizard and certain species of land snails; rookeries of the California sea lion, elephant seal, harbor seal, and species of sea birds; and the habitat of a few endemic mammals, such as a small bat and deer mice.

Geologic History

Channel Islands National Park is located on the west side of the boundary zone between the North American plate and the Pacific plate.* The lateral movement along the San Andreas fault system that has been occurring for several million years in this boundary zone has acted to: 1) open the Gulf of California by separating the Baja peninsula from the mainland of Mexico; 2) move Baja California and that part of California west of the San Andreas fault to the northwest; and 3) cause large-scale displacements and rotations of crustal blocks near the California borderland. Tectonic changes of such magnitude, when superimposed upon an already complex geologic history, have made interpretation of the geologic history of the Channel Islands and adjacent mainland very difficult. Continued movement along the San Andreas fault system will lead to greater displacement and perhaps further rotation of the Channel Islands as time goes on.

The Channel Islands are geologically associated with the Transverse Ranges section of the California Coast Ranges, which are included in the Pacific Border Province. In the Transverse Ranges, the basins, ranges, and most other structures trend east–west or "transverse" to the dominant northwest–southeast trend of the adjoining

*The national parks in the Sierra Nevada—Yosemite (chapter 25), Sequoia and Kings Canyon (chapter 42)—are in the part of central California that is *east* of the San Andreas fault system and the plate boundary.

Figure 43.5   The Northern Channel Islands shown in relation to the east-west trending Santa Monica Mountains and other Transverse Ranges. Modified from R. P. Sharp, *Southern California.* © 1976 Kendall/Hunt Publishing Company. Used by permission.

Table 43.1 Geologic Column, Channel Islands National Park

Era	Period	Epoch	Formation	Geologic Events
Cenozoic	Quaternary	Holocene	Alluvium, stream and landslide deposits	Continued uplift and sea level changes
		Pleistocene	Terrace gravels	Repeated uplift and development of marine terraces
			Middle Anchorage	Sedimentation of conglomerate
			Alluvium	
			Potato Harbor Formation	Marine transgression, late volcanic activity
	Tertiary (Neogene)	Miocene	Monterey and Blanca Formations	Uplift, faulting, and exposure of the basement complex associated with development of the Transverse ranges, movement and rotation due to plate tectonics.
			Santa Cruz Island Volcanics	Rapid accumulation of volcanic lavas, breccias, and pyroclastic debris. Erosion and reworking of older sediments and deposition under increasing depths of water.
	Tertiary (Paleogene)		San Miguel clastic and volcanic units	
			San Onofre Breccia	
		Oligocene	Rincon Formation	Deposition of marine and nonmarine clastic sediments mixed with volcanics. Uplift in source areas to the north and east.
			Vaqueros Formation	
			Sespe Formation	
		Eocene	Cozy Dell Formation	Inception of coarse clastic sedimentation due to uplifts in surrounding area. Deep water sedimentation
			South Point and Jolla Vieja Formations	
			Cañada Formation	Continuous deposition in a continental slope environment with a gradually transgressive sea.
		Paleocene	Pozo Formation	
Mesozoic	Cretaceous (Upper)		Jalama Formation	Marine sedimentation and the Laramide orogeny
	Jurassic		Alamos Pluton Willows Plutonic Complex Santa Cruz Schist	Volcanic activity and sedimentation, intrusion of plutons and regional metamorphism.

Source: modified after Weaver 1969; Howell 1976.

Note: ～～～ erosional events.

areas (the California Coast Ranges, the Sierra Nevada, and the Great Valley). The aligned peaks of the Northern Channel Islands are, in fact, the tops of a partially submerged, westward extension of the Santa Monica Mountains on the mainland. The Santa Barbara Channel (north and east of the islands) is the submerged extension of the Ventura Basin, a broad valley north of and approximately parallel to the Santa Monica Mountains (fig. 43.5).

The geologic history of the Channel Islands is told by four kinds of evidence: 1) igneous rocks and metamorphism and deformation of preexisting rocks, showing the effects of tectonic processes; 2) the sedimentary rocks of the area that reveal the depositional and uplift history of the islands and the surrounding region; 3) paleomagnetic data that demonstrate wholesale clockwise rotation of the Channel Islands and adjacent Transverse Ranges relative to the rest of California, which occurred in post-Miocene time; and 4) Pleistocene fossils and topographic features that help to explain recent history.

1. Mesozoic volcanic activity, intrusion, and metamorphism.

The earliest rocks in Channel Islands National Park are a mixture of metamorphosed volcanic and sedimentary rocks called the Santa Cruz Schist. They formed over 160 million years ago as a volcanic pile made up of basaltic and rhyolitic lavas interlayered with clastic sediments of volcanic origin. Later intrusions by the Willows Plutonic Complex and the Alamos Pluton occurred at about 160 and 140 million years ago, respectively. Then in Jurassic and Cretaceous time, the area underwent regional (dynamic) metamorphism that recrystallized and deformed the rocks.

2. Deposition and deformation during the Cretaceous and early Tertiary.

The Cretaceous ushered in a marine environment with a tropic ocean that extended north of the Channel Islands area. Over 6300 feet of shales, siltstones, and sandstones were deposited as the Jalama Formation. Deposition continued into the early Tertiary and is marked by coarsening of some sediments as a function of uplift of the adjacent continent and the nearness of the source terrain. During early Tertiary time, 5000 feet of clastic sediments consisting of cobbles, sands, and shales were deposited to form the Pozo, Cañada, Jolla Vieja, South Point, and Cozy Dell Formations. These formations were deposited as turbidites, the sediment coming from a continental mass to the south and southwest of the present islands. (*Turbidites*, which tend to be graded and well sorted, are deposited by a dense, swift, bottom-flowing ocean current; i.e., a *turbidity current*.)

During the Oligocene Epoch, terrestrial sediments of the Sespe, Alegria, Vaqueros, and Rincon Formations were deposited to a thickness of 3400 feet over a wide area from Santa Cruz Island across the Santa Barbara Channel to the Santa Monica Mountains.

Sedimentation in early Tertiary time was controlled by extensive mountain-building and a general uplift throughout western North America, which affected specific ranges such as the Santa Ynez Mountains (one of the Transverse Ranges) and other ranges in southern California.

3. Late Oligocene-Early Miocene deformation and continued sedimentation.

Deformation at the end of the Oligocene involved folding and faulting and was followed by renewed volcanic activity in the early Miocene. Volcanic activity was initiated by mafic flows and intrusions and was followed by more felsic flows, breccias, and pyroclastic debris. This 1400 feet of volcanic rock is known as the Blanca Formation, which, in turn, was eroded along with earlier sedimentary rocks and the basement complex. The resulting coarse clastics were deposited as the Beecher Member of the Monterey Formation. Most of the Monterey Formation is a siliceous deposit formed on the continental shelf in warm seas.

4. Clockwise rotation of the Channel Islands and adjacent Transverse Ranges.

Paleomagnetic evidence on Miocene igneous rocks shows that the Channel Islands were rotated between 70° and 80° clockwise since Miocene time. This rotation, which affected much of the Transverse Ranges, appears to be due to "rolling" of the Transverse Range block between the moving north Pacific and North American plates during late Tertiary time. This shearing also created the major offsets along the San Andreas and parallel faults in southern California. The sequence of these events as reconstructed from the paleomagnetic evidence is shown in figure 43.6. The rotation does much to explain the sources of sediment and the location of the seas described earlier. The ocean to the north of the Channel Islands and the south and southwest sources of Tertiary sediments are rotated, upon reconstruction to pre-Miocene time, to a western ocean and eastern sources of sediment, such as we find today.

5. Emergence of the Channel Islands during early Pleistocene time.

Uplift of the Transverse Ranges and Channel Islands formed an area of high relief. Initial uplift formed a major land extension called the Cabrillo Peninsula that stretched from the Santa Monica Mountains out along the present

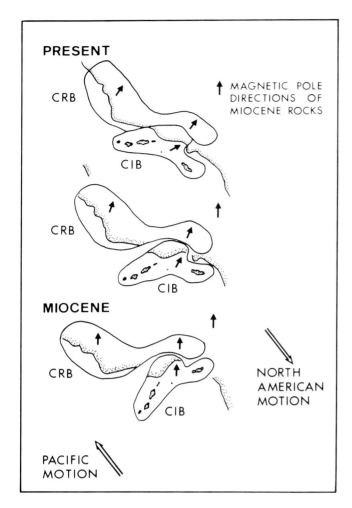

Figure 43.6 Rotation of the Channel Islands from Miocene to the present. The diagram shows a schematic representation of the rotations of the Channel Islands Block (CIB) and the Coast Range Block (CRB) relative to the rest of North America. Evidence for rotation is found in the paleomagnetic directions to the pole found in Miocene igneous rocks (shown by black arrows on the different blocks) which point in different directions at the present time. These directions were parallel in Miocene time when the igneous rocks cooled in the Miocene magnetic field (bottom of figure). Subsequent rotation in a clockwise direction led to the situation seen at present (top of figure). The rotation was caused by the relative motion of the Pacific and North American plates (long open arrows), which effectively "rolled" both the Channel Islands and Coast Range Blocks as small fragments caught between the two larger plates. In the figure, the borders of the blocks are approximate, and the land areas (stippled line) are given as a reference only.

Channel Islands. It was along this peninsula that Pleistocene land mammals migrated, only to become isolated when the land bridge between the Channel Islands and the mainland subsided. Marine terraces found at various levels on San Miguel Island, and as high as 1800 feet on Santa Cruz Island, attest to the dynamic uplift of parts of the island chain during the Pleistocene. Such uplifts

and areas of subsidence have caused the isolation of the Channel Islands, led to the distinctive biological heritage of many of the islands, and produced the impressive topographic features of today.

Geologic Maps and Cross Sections
See Howell, D. G., ed. 1976. and Weaver, D. W. 1969.

Bibliography

Ernst, W. G., ed. 1981. *The geotectonic development of California,* Rubey volume 1. Englewood Cliffs, New Jersey: Prentice-Hall.

Glassow, M. 1977. An archeological overview of the Northern Channel Islands, California, including Santa Barbara Island. Prepared for the Western Archeological Center, National Park Service, Tucson, Arizona. National Technical Information Service No. 298855.

Hall, C. A., Jr. 1981. San Luis Obispo transform fault and Middle Miocene rotation of the western transverse ranges, California. *Journal of geophysical research* 86:1015–1031.

Howell, D. G., ed. 1976. *Aspects of the geologic history of the California continental borderland.* Pacific Section, American Association of Petroleum Geologists, Miscellaneous Publication 24. (Reports by various investigators on the basement rocks, stratigraphy and paleontology, sedimentology, and tectonics presented at the California Borderland Symposium, Santa Cruz Island, 1975.)

Howorth, P. C. 1982. *Channel Islands, the story behind the scenery.* Las Vegas, Nevada: KC Publications.

Johnson, D. L. 1967. *Caliche on the Channel Islands.* Mineral Information Service, Natural Resources Program, Naval Undersea Center, San Diego, California.

Kamerling, M. J., and Luyenduk, B. P.. 1979. Tectonic rotations of the Santa Monica Mountains region, western transverse ranges, California, suggested by paleomagnetic vectors, *Geological Society of America bulletin* 90: 331–337.

Luyenduk, B. P., Kamerling, M. J., and Terres, R. 1980. Geometric model for Neogene crustal rotations in Southern California, *Geological Society of America bulletin* 91:211–217.

Power, D. M., ed. 1980. The California Islands: Proceedings of a multidisciplinary symposium. Santa Barbara, California: Santa Barbara Museum of Natural History.

Resource Management Workshop, Channel Islands, November 1–3, 1978. Mimeographed sheets.

Sharp, R. P. 1976, *Field guide to southern California,* K/H Geology Field Guide Series. Dubuque, Iowa: Kendall/Hunt Publishing Company.

U.S. Dept. of the Interior. April 1980. Channel Islands National Monument, California legislative background.

Weaver, D. W. 1969. *Geology of the Northern Channel Islands.* Special Publication. Pacific Sections, American Association of Petroleum Geologists; Society of Economic Paleontologists and Mineralogists.

Channel Islands National Park

Area: 246,870 acres; 386 square miles (includes 122,870 acres of submerged land; 124,115 acres above sea level)

Proclaimed a National Monument: April 26, 1938

Designated a Biosphere Reserve: 1976

Established as a National Park: March 5, 1980

Address: 1699 Anchors Way Drive, Ventura, California 93003

44

Coastal redwood forests along 40 miles of the northern California shore.

The world's tallest trees, hundreds of years old.

Designated a World Heritage Site and Biosphere Reserve.

Sea cliffs, marine terraces, and lagoons, along a dynamic and scenic coast.

Striking exposures of the Franciscan rock assemblage.

Landslides on the unstable slopes of the Coast Ranges; earthquakes due to continuing tectonic activity.

Redwood National Park

Northern California

by Lisa A. Rossbacher
California State Polytechnic University,
Pomona

Figure 44.1   The tallest trees on earth are the coast redwoods, which may reach over 300 feet in height. They grow only in a narrow belt along the northern California coast. National Park Service photo.

Local History

The region around Redwood National Park in northern California is rich in legends of the Klamath Indians. They made dugout canoes from redwood trees, as did more recent tribes, including the Yurok, Tolawa, and Chilula. These later Indians also constructed homes and sweathouses out of redwood planks.

Portola, a Spanish explorer, led an expedition through this area in 1769. In a notebook, Father Juan Crespi, a Franciscan missionary on the expedition, referred to the *palo colorado* or "red trees." The redwoods were officially named in 1823, when A. B. Lambert designated them *Taxodium sempervirens,* or evergreen cypress. Later, a German botanist realized that these trees did not belong to the bald cypress group, and he renamed them *Sequoia sempervirens.* Both the coast redwood and the giant sequoia of the Sierra belong to the family Taxodiaceae, or swamp cypress family.

In a sense, the stately grandeur of the redwoods has been a factor in their destruction. The attributes that make them so impressive and beautiful in the forest are the same qualities that make them attractive for construction uses; their bark and wood are resistant to decay, fire, and insects. Humans are the trees' only real enemy. Cutting the trees is the ultimate destruction, but even the stress of people walking on their shallow root systems can bruise them.

This part of northern California was once dense with redwoods. When the explorer Jedediah Smith led a group into the forests in 1828, he and his company were able to travel only about a mile per day, while winding their way through the huge trees. The pioneers and early settlers cut many trees in the 1880s for buildings in San Francisco and northern California towns. Much of this building was the result of the discovery of gold in northern and central California. Later the advent of motorized equipment made more remote areas accessible to loggers, and more trees were felled. By 1964, only 15 percent of the original stand of redwoods remained.

In order to avert the total destruction of the trees, the Save-the-Redwoods League was established in 1918 to promote the creation of Redwood National Park; it took 50 years to accomplish the League's goal. Feasibility studies were made in the 1920s and 1930s, but no legislative action was taken. In 1946, the Douglas Bill proposed the establishment of a national memorial forest to include 2.5 million acres, about 180,000 of which would be administered according to National Park Service regulations. The Douglas Bill never passed.

Aided by funds from the National Geographic Society, the National Park Service surveyed redwood areas in 1963 and 1964. This study revealed some dreadful statistics. Of the original forest of nearly 2 million acres, only 15 percent remained—and only 50,000 acres (2.5 percent) were protected in state parks! If the logging rate continued unchecked, all unprotected trees would disappear before the year 2000. The study also found a grove of the earth's tallest trees, including one 367 feet high, on private—and unprotected—land.

Redwood National Park was finally established by Congress in 1968, guaranteeing protection of these endangered trees. The park links three California state parks; from north to south, these are Jedediah Smith, Del Norte, and Prairie Creek State Parks. In addition to virgin redwood groves, the park includes stands of maple, cedar, and oak, private homes and small businesses, some cutover land, and more than 40 miles of coastline. Some of the most impressive redwood stands are located in the three adjacent state parks. The state parks and the national park have some cooperative programs but are administered as separate units.

Establishment of Redwood National Park assured protection of the surviving trees from logging, but other problems remained. Logging operations outside park boundaries threatened to cause serious erosion of Redwood Creek watershed, which includes the tallest trees. Topsoil eroded from drainage basins at higher elevations outside the park was being deposited in stream channels within the park, and this was causing the streams to rise and literally drown the redwood trees. Other effects of logging include soil compaction, loss of topsoil, elimination of shade and ground cover, and changes in small drainage basins. To help solve these problems, Congress extended the park boundaries in 1978 by increasing the total area by over 80 percent in order to protect a total of 106,000 acres.

The added park area is primarily cutover lands, including 300 miles of logging roads, 3000 miles of skid trails, and countless charred stumps marking former locations of stately redwoods. The Park Service is involved in a watershed rehabilitation program with a long-range goal of speeding recovery of the natural systems in the park area. Since restoration began in 1978, efforts have been concentrated on the 30,000 acres that were logged before the park's establishment. Erosion control priorities are aimed at logging haul roads, tractor-logged hillslopes, active landslides, and natural prairies that were gullied because of road construction. Techniques of slope stabilization and revegetation being developed in the cutover parts of Redwood National Park may aid in reclamation of damaged land in other parts of the country.

The Tall Trees

The redwoods include the tallest trees on earth; many of them are over 200 feet tall, and some stand more than 300 feet high. The tallest coast redwood was discovered by the National Geographic Society in 1963, near the southern

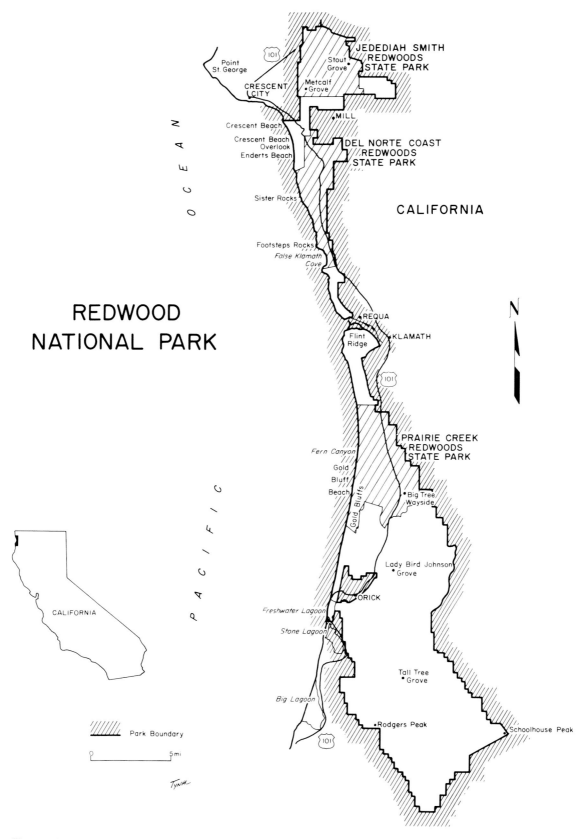

Figure 44.2   Redwood National Park, California.

end of Redwood National Park in what is now Tall Trees Grove. In 1965, this huge tree measured 367.8 feet high, with a trunk 14 feet in diameter near its base. The ages of the tallest trees average only 400 to 800 years, although a tree that was cut in 1934 had reached an age of 2000 years. This is only half the age of the giant sequoias.

The trees are called redwood because the color of their bark is a dull reddish brown. The bark is dense and fibrous, and may be as much as a foot thick on a very old tree. The leaves are usually less than an inch long, sharply pointed, and evergreen. Redwood cones are egg-shaped and about 1.5 inches in diameter. A coast redwood tree must be at least 20 years old before it produces cones.

The coast redwood, like its cousin the giant sequoia (chapter 42), grew over much of North America when dinosaurs were alive. Pleistocene glaciation, beginning about 2 million years ago and ending only 10,000 years ago, caused major changes in soil and climate conditions, resulting in a much narrower distribution of the Sequoia family. The coast redwoods are now restricted to a band approximately 35 miles wide and 500 miles in length along the Pacific coast. Fossils of redwood trees have been found in the Dakotas, Montana, and Yellowstone National Park in Wyoming (chapter 39).

Favorable conditions for redwood trees include moderate temperatures, rainy winters (25 to 122 inches per year), dry summers, and year-round fog. An old saying is that the redwoods grow "as far inland as the fog flows," and that is about 30 miles here. Fog limits the trees' loss of water by evaporation and transpiration, as well as adding moisture to the soil.

Geologic History and Features

Redwood National Park is entirely within the Coast Ranges of California, which extend from the Oregon border south to Santa Barbara. This region has provided some important insights into the theory of plate tectonics. The northern Coast Ranges lie east of the San Andreas fault zone, which crosses the coast and goes offshore about 50 miles south of the park. The moist climate that makes this region so favorable for redwoods and other vegetation also makes the geology very difficult to see; thick soils and dense forests result in few exposures. However, the sea cliffs provide an excellent view of the geology for anyone willing to brave the rocky beach.

1. **Deep-water, continental, and volcanic deposits in a subducting trough: accumulation of Jurassic-Cretaceous Franciscan rocks.**

    The oldest rocks found in the northern Coast Ranges are part of the Franciscan rock assemblage, which is described as a *mélange,* a term reflecting the mixture of diverse materials it contains. These rocks are the product of deep-water sedimentation on the continental slope and on oceanic crust in deeper trenches offshore. The sands and muds included submarine volcanics, radiolarian oozes (mud composed of tiny siliceous animal shells), and sediments eroded from land. This mixture of sediments—carried along on an oceanic plate—was subducted beneath the continental margin about 100 million years ago. Tectonic activity probably occurred at the same time as deposition, so the unconsolidated sediments were severely disturbed. Some outcrops of the Franciscan have rocks so contorted that they look as if they had been stirred with a spoon. To complicate the situation further, ultramafic rocks intruded the Franciscan beds during subduction. Some geologists believe that fragments of tectonostratigraphic terranes have also been mixed into Franciscan rocks.

    The subduction ended relatively suddenly, and the complex Franciscan assemblage was uplifted and exposed at the surface, probably about 70 million years ago. Its characteristic minerals reflect the low-temperature and high-pressure conditions of their origin.

    The Franciscan assemblage is composed primarily of grayish green graywackes. These sandstones contain relatively high proportions of plagioclase feldspar, as well as quartz; a chlorite matrix gives the Franciscan its distinctive greenish color. The volcanics were altered to serpentinite by metamorphism, and these metavolcanics are also green.

    The Franciscan is a thick and widespread rock unit in the Coast Ranges. A conservative estimate of its volume is 350,000 cubic miles—which could cover all of the lower 48 states to a depth of 600 feet!

2. **Basin-forming and sedimentation in the Miocene.**

    The sea advanced over the Coast Range area in the early Miocene, creating a number of small, nearly isolated marine basins. Volcanism, predominantly submarine, also contributed to the accumulation of deposits during this time.

    In the northern Coast Ranges, the late Miocene brought further tectonism. Downwarped areas formed basins, defined by the intervening anticlinal arches trending to the northwest. The area where Redwood National Park is now located was surrounded by two of these depressed and flooded embayments, one at Crescent City to the north and the other along the Mad River near Eureka. A few downfaulted remnants of the basin deposits are all that have survived late Cenozoic erosion to show that marine rocks were formerly widespread here.

3. **Thick stream gravels in the Pliocene.**

    The orogenic activity that began in the northern Coast Ranges during the late Miocene extended southward throughout the Coast Ranges in the Pliocene. By late Pli-

Table 44.1 Generalized Geologic Column, Redwood National Park

Era	Period	Epoch	Rock Units	Geologic Events
Cenozoic	Quaternary	Holocene	Alluvial and mass-wasting deposits	Coastal erosion and spit deposition; accelerated erosion on land; landsliding; continued faulting and coastal subsidence
Cenozoic	Quaternary	Pleistocene		Continuing uplift and erosion; formation of marine terraces in middle to late Pleistocene
Cenozoic	Tertiary (Neogene)	Pliocene	Stream gravels	Coast Range orogeny extending to the south; thick stream gravels deposited in valleys (especially south of park)
Cenozoic	Tertiary (Neogene)	Miocene	Marine sediments, gravels	Basin forming and sedimentation in Coast Ranges; some volcanism, predominantly submarine
Cenozoic	Tertiary (Paleogene)	Oligocene / Eocene / Paleocene	Marine sediments, some quartzose sandstone	Decreasing depositional rates as seaways narrow and become smaller; weathered clays and clean sandstones indicate nearly tropical climate in Eocene. Thrusting over Franciscan rocks
Mesozoic	Cretaceous / Jurassic		Franciscan rock assemblage (sandstone with quartz, plagioclase feldspar, and chlorite mica), shale, limestone, radiolarian cherts, ultrabasic intrusions	Extensive deposition of marine sediments (over 50,000 feet), followed and accompanied by subduction beneath continental margin, tectonic mixing, and intrusion of ultrabasic peridotite. Metamorphism of sediments and older rocks
Mesozoic	Triassic		(Not exposed in park)	
Paleozoic			(Not exposed in park)	Long accumulation of sediments in Paleozoic seas (No Precambrian or early Paleozoic rocks have been identified in the Coast Ranges; if they exist, they have been severely metamorphosed.)

Source: modified after Norris and Webb 1976

Figure 44.3   This generalized block diagram shows the coast and offshore area of northern California. The San Andreas fault swings west and merges with the Mendocino ridge and fracture zone south of Eureka. Key: CC = Crescent City, T = Trinidad, E = Eureka.

ocene time, most of the ranges that exist today were exposed as dry land. Thick layers of stream gravels were deposited in valleys, sometimes nearly burying the surrounding hills. Deposition was most extensive in the southern Coast Ranges, but a few patches of gravel have been found north of Eureka. One place where Pliocene-Pleistocene deposits are exposed is along the beach at Gold Bluff, where the sand does contain small flecks of gold.

4. Quaternary erosion and sea-level changes.

Because most of the land surfaces in the northern Coast Ranges have been exposed since the end of the Pliocene (about the last 2 million years), the geologic record has been slowly destroyed. Only a few coastal Quaternary deposits cover the Franciscan in this area. The only major Quaternary activity recorded in the northern Coast Ranges is the development of raised marine terraces. Some terrace formation was caused by worldwide sea-level changes due to glaciation. However, most of the terraces in northern California are wave-cut platforms that have been eroded in bedrock and then raised by localized uplift. Marine ter-

races are visible in the Crescent City area at the northern end of the park.

Tectonic activity has been going on in the Pacific, off the northern California coast, at least since the early Cenozoic. At Cape Mendocino, about 50 miles south of the park, the San Andreas fault zone goes offshore and merges with the east-west trending Mendocino ridge and fracture zone. About 50 miles offshore, the Gorda plate is being subducted under the North American plate (fig. 44.3). This subduction zone was responsible for the November 8, 1980, earthquake in this area; essentially the quake was caused by friction between the two plates. The magnitude of the tremor was 7.0, which made it the largest earthquake in California in 28 years! Northern coastal California has had at least 20 earthquakes with magnitudes of 6.0 or greater in this century.

5. Landslides and accelerated erosion today.

Franciscan rocks are extremely fractured and sheared, which makes them highly susceptible to mass wasting. Landslides are so common in this area that they cause more

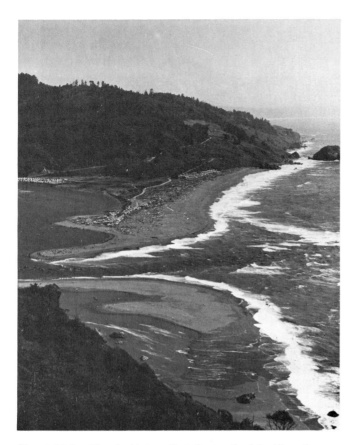

Figure 44.4 View looking south at the mouth of the Klamath River, drowned by coastal subsidence. River and tidal currents keep open an outlet through the bar, built by longshore drift of sand. The lagoon behind the bar is filling with sediment. National Park Service photo.

downslope movement of material than any other geologic process—including stream action! Water is often an important agent in initiating landslides, and recent studies have shown that rainfall amounts are useful in predicting landslides. Rock fracturing in fault zones also decreases slope stability. Construction on steep slopes is another significant cause of landslides. Each year, millions of dollars worth of property damage is done by landslides in the northern Coast Ranges.

Stream activity is important in the park area today. Stream deposits support some of the largest stands of tall trees. Both the Tall Trees Grove on Redwood Creek and Stout Grove on the Smith River grow on alluvial flats. Logging activities in the upstream areas have had numerous effects on downstream reaches. These impacts include accelerated bank erosion, elevated and wider streambeds, higher winter stream discharge, and lower summer discharge. An elevated water table directly threatens redwoods growing on alluvial terraces adjacent to Redwood Creek.

The coastal region of Redwood National Park has a number of lagoons. These are the drowned mouths of stream valleys that have sand spits separating them from the sea. The lagoons are evidence that this part of the California coastline is sinking. The reason for the subsidence is not known, but it may be related to subduction offshore.

A recent noteworthy geologic event in the Crescent City area, at the northern end of the park, was the *tsunami* (seismic sea wave) caused by the 1964 Good Friday earthquake in Alaska (chapters 28, 30). Sudden changes in the sea floor, caused by earthquakes or large submarine landslides, can start ocean swells that move across the ocean and hit coastlines with destructive violence. Tsunamis occur with some frequency in Alaska and in the Hawaiian Islands (chapters 37, 38), but Crescent City is the only place in conterminous United States to have recorded serious damage in historic time from such an event.

Erosion in the area around Redwood National Park has been accelerated by human activities. Extensive lumbering and clear-cutting have nearly eliminated the soil protection in some areas. Gully formation, the loss of rich topsoil, and streams choked with debris are all consequences of thoughtless exploitation of the redwoods as a natural resource. Redwood National Park was established to save the area from further damage, but it will take many years and considerable work to repair the existing damage.

An innovative approach to restoring the whole area—beyond the official park boundaries—involves the establishment of the Park Protection Zone. This 30,000-acre tract, upstream from the park, along Redwood Creek, includes the upper two-thirds of the drainage basin. Because activities upstream have such a powerful impact on downstream reaches within the park, National Park Service resource managers are working with private companies to limit the effects of land use in upstream areas. This philosophy, by recognizing the ways in which land areas within a watershed are connected, is an important step in reclaiming and restoring Redwood National Park.

Bibliography

Alt, D.; and Hyndman, D. W. 1975. *Roadside geology of northern California.* Missoula, Montana: Mountain Press Publishing Company.

Bailey, Edgar, ed. 1966. *Geology of northern California.* California Division of Mines and Geology Bulletin 190.

Dewitt, J. B. 1982. *California redwood parks and preserves.* San Francisco: Save-the-Redwoods League.

Norris, R. M.; and Webb, R. W. 1976. *Geology of California.* New York: John Wiley & Sons.

Oakeshott, G. B. 1978. *California's changing landscapes.* New York: McGraw-Hill Book Company.

Savina, M. E. ed. 1985. *Redwood country: Field trip guidebook.* Northfield, Minnesota: Carleton College and the American Geomorphological Field Group.

Schrepfer, S. R. 1983. *The fight to save the redwoods: A history of environmental reform, 1917–1978.* Madison: University of Wisconsin Press.

Sunset Books editorial staff. 1969. *Redwood country and the Big Trees of the Sierra.* Menlo Park, California: Lane Books.

William, R. L. 1976. *The loggers,* The Old West Series. New York: Time-Life Books.

Redwood National Park

Area: 109,027 acres; 107 square miles

Established: October 2, 1968

Designated: World Heritage Site, 1980; Biosphere Reserve, 1983

Address: 1111 Second Street, Crescent City, California 95531

(707) 464-6101 (24-hour information telephone)

45

Hot Springs National Park

West Central Arkansas

From 750,000 to 950,000 gallons a day flow from 47 thermal springs. Year-round temperature of the water remains near 143° F.

Tufa deposits; novaculite, used for whetstones; and large, tinted quartz crystals are displayed in the park.

The hydrologic system of the springs developed in intensely metamorphosed, folded, and faulted Paleozoic rocks.

Figure 45.1 Display Springs in Hot Springs National Park. Photo courtesy Hot Springs Chamber of Commerce.

Local History

The Indian name for the hot springs was *Tah-ne-co,* meaning "the place of the hot waters," where the Great Spirit's warm breath heated healing waters and vapors. Battles were fought for possession of the sacred springs. Finally the tribes decided that the the springs were meant to be for the benefit of all who were sick. From that time on, the area became neutral ground. Braves from enemy tribes could lie side by side in the hot mud to ease the pain from rheumatism and old wounds. When the foundations for the Quapaw Bathhouse in the park were dug, a narrow winding cave used by Indians that led to the springs was discovered. A number of Indian campgrounds and burial mounds have been found in the region.

Hernando de Soto, in his search for treasure in the New World, visited the hot springs in 1541 shortly before his death. Chronicles of his ill-fated expedition describe his discovery of "large streams of hot water and a small lake" near an Indian village, when he was journeying between the Mississippi River and what is now Oklahoma. Since there are no other hot springs near his route along the Arkansas River, the presumption is that he was in the area of Hot Springs National Park.

In 1804, after the Louisiana Purchase, President Jefferson sent William Dunbar and Dr. George Hunter, a chemist, to explore the Ouachita Mountains and to investigate the hot springs, which Jefferson had heard about. Spending a month in the area, they counted some 70 hot springs and analyzed the water issuing from the rocks. Theirs was the first scientific report to be written about the springs. Other explorers who visited the hot springs were U.S. Army officers, Zebulon M. Pike in 1806 and Stephen H. Long in 1818. Two members of Long's expedition, Kearney Swift and Edwin James, the latter a botanist and geologist, studied the springs. They suggested that build-up of *tufa* (calcareous-siliceous deposits around spring openings) blocked the flow from some springs, causing variations from time to time in the total number of springs. In their natural state, the springs flowed from a number of openings in a belt about a quarter of a mile long and several hundred feet wide along the southwest slope of Hot Springs Mountain.

As time went on, local residents began putting up cabins, bathhouses, and even hotels to accommodate visitors who sought the benefits of thermal bathing. The natural environment of the hot springs became completely altered. Springs were excavated and lined with masonry to increase and control the flow and then covered to protect the water from contamination. To prevent commercial exploitation and ensure that all people would have access to the springs, President Andrew Jackson persuaded Congress to withdraw from public sale four sections of land (four square miles) around the hot springs. This was in 1832, four years before Arkansas became a state.

In 1876, a year after a rail line was extended to the area, the city of Hot Springs was incorporated. Part of the city is now surrounded by the park. The preserve around the springs was dedicated as a permanent reservation for public use in 1880, and in 1921 it was redesignated a national park. Today in this famous health resort, there are 12 concessionaire bathhouses (five in the city, seven in the park), all under federal regulations. The former Army and Navy Hospital is now the Arkansas Rehabilitation Center, operated by state and federal authorities. A private facility, the Libbey Memorial Center for Physical Medicine, operates under federal lease.

Of the 47 hot springs in the park, two Display Springs have been left in a natural state so that visitors can see what the springs looked like originally (fig. 45.1). At Tufa Terrace on Arlington Lawn, a hot water display has been developed, using the runoff from the upper springs that created the terrace. Where the thermal water drains down the slope, blue-green algae accumulate and vapor rises from the cascade in cool weather. The remaining springs are sealed to prevent contamination and loss of gas. The combined flow from the springs ranges from a high of around 950,000 gallons a day in winter to 750,000 gallons in summer and fall. The water temperature remains year-round near 143° F. The water is carried by insulated pipes from the springs to large underground reservoirs. Hot spring water is mixed with cooled spring water from the reservoirs to reach desired temperatures. Cooled water is also pumped to drinking fountains and jug fountains. The spring water in Hot Springs National Park is of exceptional purity, with no unpleasant taste or odor.

Through the years, the National Park Service has acquired additional land in the region of the springs for the purpose of protecting the infiltration and recharge areas from pollution. The mountains in the park (between 1000 and 1400 feet in elevation) are part of the ZigZag Mountains, a range in the Ouachitas. Music Mountain, in the southwest corner of the park, is the highest point (1405 feet) and is at the center of a great horseshoe-shaped ridge, with West Mountain and Sugarloaf Mountain forming the roughly parallel extended "arms." A few roads and many miles of bridle and hiking trails wind through the woodlands and rugged hill country.

Geologic Features

How Did the Hot Springs Originate?

Hot Springs National Park is in a unique geological setting. The hot springs have aroused scientific curiosity since Dunbar and Hunter first made their report to President

Figure 45.2 Hot Springs National Park, Arkansas.

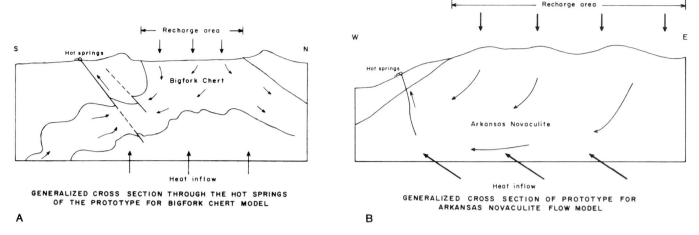

Figure 45.3 A,B   Cross-sectional models illustrating directions of ground-water flow in Hot Springs National Park (from Bedinger et al. 1979). A. North-south cross section approximately through the center of Bigfork Chert recharge area, about two miles from the hot springs. Details of folds and faults not shown. B. East-west cross section of the Arkansas Novaculite recharge area, less than a mile from the hot springs.

Jefferson. Scientists (mainly geologists) have argued over the origin of the water and the heat source. Is rain the source of the water, or is it newly risen from deep in the earth? If the former is the case, how did the water get so hot? Has the water been stored for a long period of time at considerable depth? How and from where has it moved to its point of issue? Since the mid-1800s, numerous investigations have been made, and hypotheses both simple and complex have been proposed.

The modern explanation, which appears fairly conclusive, draws on several lines of evidence. The seasonal variations in flow—higher in wet seasons and lower in dry periods—suggest rainwater as the most likely source of supply for the hot springs. Reinforcing this idea is the fact that the chemical composition of the hot springs water is similar to that of the local ground water, cold spring water, and well water—after allowances are made for the higher content of dissolved silica contained in the hot water. (Hot water can hold greater concentrations of dissoved compounds than cold water.)

Bedinger and his co-investigators (1979) used mathematical models to test several conceptual models of the hot springs flow system. They concluded that the "geochemical data, flow measurements, and geologic structure of the region support the concept that virtually all the hot springs water is of local, meteoric [i.e., from rain] origin." They further deduced that:

> Recharge to the hot springs artesian flow system is by infiltration in the outcrop areas of the Bigfork Chert and the Arkansas Novaculite. The water moves slowly to depth where it is heated by contact with rocks of high temperature. Highly permeable zones, related to jointing or faulting, collect the heated water in the aquifer and provide avenues for the water to travel rapidly to the surface (fig. 45.3 A,B).

The Paleozoic sedimentary rocks of the park, which collect the rainwater and from which the hot springs emerge, have been deformed and metamorphosed by episodes of intense folding, faulting, and thrusting. Many complex geologic structures have resulted. The ZigZag Mountains basically owe their topographic expression to the resistant ridges of the Arkansas Novaculite Formation, which crop out in a zigzag pattern because of the erosion of tightly compressed plunging folds. Weaker, less resistant rocks underlie the valley floors. The landscape in Hot Springs National Park is a good example of control by geologic structure (fig. 45.4).

The main collectors of rainwater are the permeable ridge crests of novaculite (locally intensely fractured) and the Bigfork Chert, which forms lower ridges and rounded knobs. The shale beds below the permeable rocks tend to be impermeable. On Hot Springs Mountain, between the novaculite on the ridge and the Stanley Shale beds in the valley, the Hot Springs Sandstone crops out. This massive, quartzitic sandstone (which structurally is near the axis of a plunging anticline) contains large joints and fractures. The water of the hot springs emerges from the sandstone between the traces of two thrust faults that parallel the anticlinal axis. This structural situation accounts for the hot springs' alignment in a belt along the mountain slope. We can visualize the general hydrologic and geothermal flow system as follows:

1. In the park's higher elevations among the ridges and higher valleys, rainwater infiltrates recharge areas on the Arkansas Novaculite and the Bigfork Chert.
2. Cold ground water works its way very slowly to depths more than a mile below the surface where, with increasing geothermal gradient, the rocks are very hot.

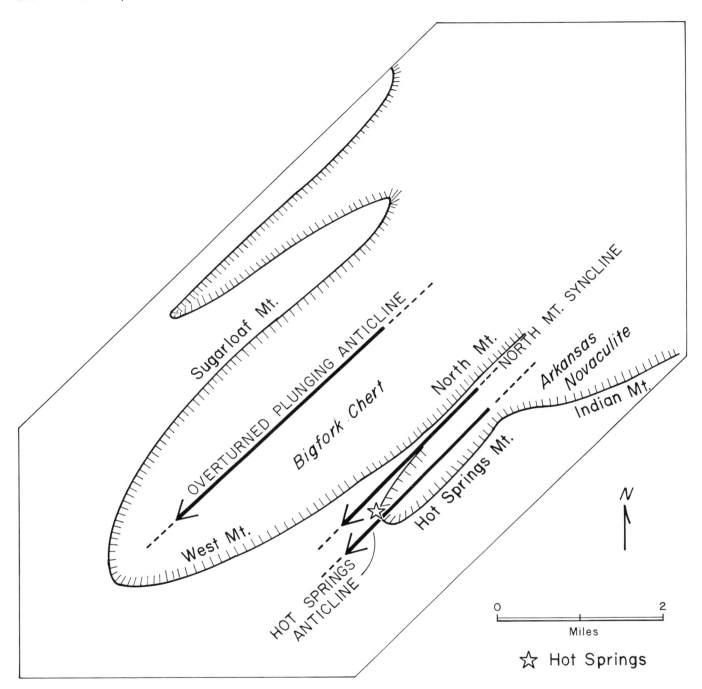

Figure 45.4 Sugarloaf Mountain, West Mountain, North Mountain, Hot Springs Mountain, and Indian Mountain illustrate zigzag topography produced by erosion of plunging folds. Arrows show the northeast-southwest-trending folds plunging to the southwest. All the folds shown have overturned southeast limbs, which means that both limbs dip northwest. Reverse faults (not shown) that cut the folds trend parallel to them and dip steeply to the northwest. Water that supplies the hot springs infiltrates recharge areas of Bigfork Chert and Arkansas Novaculite, both of which crop out in the middle of anticlines. (Simplified from geologic map in Bedinger et al. 1979).

3. As cold water sinks, heated water rises rapidly, following paths of least resistance—in this case, the permeable fault zones that bring the water close to the surface.

4. From the fault zones, water moves along joints in the Hot Springs Sandstone aquifer; while still very hot, it flows out from openings at the surface. Since the recharge area is at a higher elevation than the discharge area, the springs are artesian; that is, the water has built up a hydrostatic head due to artesian pressure and it flows without pumping.

Estimates as to the age of the hot springs water and the travel time involved in the journey from the surface to depth and back to the surface have been derived from tritium and carbon-14 analyses. *Tritium* (H_3), a naturally occurring, radioactive isotope of hydrogen, is formed by cosmic rays in the upper atmosphere and enters the earth's hydrologic cycle via precipitation. Tritium's half-life is about 12¼ years. *Tritium dating,* involving measuring the rate of radioactive decay in water samples, is useful for calculating the age of a sample and also for tracing subsurface movements and velocities of ground water. The carbon-14 dating of ground water is calculated from dissolved soil gas (CO_2) with corrective adjustments made for carbonates dissolved from rock through which the water travels. The results from many carefully collected and analyzed water samples indicate that water in the hot springs is a mixture of water that is about 4400 years old with a small amount of water less than 20 years old (Bedinger et al. 1979). The inference is that movement of water down through the rocks takes much longer than its rise through the fault zones after being heated.

As noted earlier, except for the higher temperature and hence higher content of dissolved silica, the hot springs water and the local cold ground water are very similar. The Arkansas hot springs are less mineralized than the water in most hot springs. What this means is that the rocks through which the Arkansas hot springs water travels have fewer mineral ions available for solution. The dissolved constituents are listed in table 45.1.

Minute amounts of radium and its radioactive by-product, radon gas, are present in the hot springs water and have been studied by a number of investigators. The source of the radium is not known, but its presence in trace amounts in the ground water of the region is not unusual. Most of the gas in the hot springs is carbon dioxide (CO_2) with smaller amounts of dissolved oxygen and nitrogen. Analyses of water samples from the different springs and from samples collected over a period of time have shown slight variations in dissolved constituents; but on the whole, the water has remained fairly uniform and consistent. Besides being odorless, the water is remarkably free of bacteria, viruses, or other impurities.

Table 45.1
Chemical Analysis of Hot Springs Water*

(milligrams per liter)	
Silica (SiO_2)	42.0
Calcium (Ca)	45.0
Magnesium (Mg)	4.8
Sodium (Na)	4.0
Potassium (K)	1.5
Bicarbonate (HCO_3)	165.0
Sulfate (SO_4)	8.0
Chloride (Cl)	1.8
Fluoride (F)	.2
Dissolved oxygen (DO)	3.0
Free carbon dioxide (CO_2)	10.0
Radioactivity through radon gas emanation** 0.81 millimicrocurie per liter	

*Analysis by U.S. Geological Survey, 1972
**Analysis by University of Arkansas

Rocks of Special Interest in the Park

The novaculite found in Hot Springs National Park and vicinity is a relatively uncommon rock. *Novaculite* is a dense, hard, even-textured, light-colored, siliceous sedimentary rock that resembles chert. Novaculite is *cryptocrystalline* (i.e., having crystals too small to be distinguished under a microscope); but unlike chert, true novaculite is made up of more microcrystalline quartz than chalcedony. Novaculite was formerly thought to be the result of primary deposition of silica; but in its type occurrence in the Lower Paleozoic rocks of the Ouachita Mountains (the region in which the park is located), true novaculite appears to be a thermally metamorphosed, bedded chert (Keller et al. 1977). The parent rock, a bedded chert, was recrystallized by heat (and to a lesser extent by confining pressure) and thus altered to novaculite.

In some areas outside the Hot Springs National Park, novaculite beds are quarried for use as whetstones, and some specimens that are richly colored are suitable for lapidary work. The Indians prized the rock for making arrowheads and stone tools. Old quarries, opened by Indians, are in the northeast corner of the park on Indian Mountain. The largest quarry is 150 feet in diameter and 25 feet deep. The Indians' method of breaking up and excavating this hard, dense rock was to build a fire to get the rock as hot as possible and then throw cold water onto it. The process fractured the rock into small pieces; but it was dangerous. Sometimes cold dousing of hot rock caused an explosion that filled the air with sharp projectiles of novaculite.

The Hot Springs region is also noted for beautiful quartz crystals of many shapes and sizes. Some crystals are 2 to 3 feet long. Specimens must not be collected within

Table 45.2 Geologic Column, Hot Springs National Park

Time Units			Rock Units		Geologic Events
Era	Period	Epoch			
Cenozoic	Quaternary	Holocene	Alluvium		Erosion and alluviation
		Pleistocene	Terrace deposits		Development of hot springs hydrologic/geothermal system Pluvial climate
	Tertiary		/////////////		Long erosion Major unconformity End of uplift
Mesozoic	Cretaceous		/////////////		Syenite stocks intruded close to park; small dikes in park
	Jurassic		/////////////		Long erosion, major unconformity
	Triassic		/////////////		Uplift, faulting
Paleozoic	Permian		/////////////		Ouachita orogeny (plate collision, thrust faulting, intense deformation, uplift)
	Pennsylvanian		(not exposed in park)		Shales, sandstones deposited in closing seaway.
	Mississippian		(Formation) Stanley Shale	(Members) Chickasaw Crk. Shale Hatton Tuff Hot Springs Sandstone	Flysch deposits in deep trough between closing plates Ash falls Beginning of flysch facies: turbidites, conglomerates, etc. Unconformity
	—?—		Arkansas Novaculite	Upper	Siliceous shales, ash falls
	Devonian			Lower	"Starved basin" deposition
				Middle	
	Silurian		Missouri Mountain Shale		
			Blaylock Sandstone		
	—?—	(late)	Polk Creek Shale		Slow accumulation in deep-water, stable basin
	Ordovician	(middle)	Bigfork Chert		
			Womble Shale		

Source: modified after Bedinger et al. 1979; Viele 1979

park boundaries, but may be purchased at mines and rockshops in the vicinity.

Tufa deposits surround the Display Springs and can be seen at other locations where spring water flowed out naturally in the past. Some of the tufa was precipitated by blue-green algae, the only form of life that can exist in water that hot.

Geologic History

1. Ordovician to Mississippian deposition.

Sediments accumulated very slowly in deep, quiet ocean water. (This ancient ocean basin lay between tectonic plates that eventually became North and South America.) Fine silts, organic material, siliceous particles, and precipitated silica made up the sediment that eventually became shales, chert, and novaculite in a sedimentary sequence characteristic of "starved basin" deposition. Such a sequence implies that the ocean basin remained stable for a long period of time with the volume of sediment gradually decreasing.

Beds in the sequence now exposed in or close to Hot Springs National Park are the Womble Shale, Bigfork Chert, and Polk Creek Shale of Ordovician age; the Silurian Blaylock Sandstone and Missouri Mountain Shale; and the Arkansas Novaculite Formation, which is Devonian and Lower Mississippian in age. The rocks are not highly fossiliferous, although a few graptolites (extinct colonial marine organisms) have been found in the Polk Creek Shale.

2. Plate convergence in Mississippian time.

The appearance of interbedded siliceous shales in the uppermost part of the Arkansas Novaculite Formation is an indication that the ocean was beginning to close. Some geologists believe that siliceous ash, blown from volcanoes on the south side of the narrowing ocean and deposited on the sea floor, was the principal source of sediment for these beds.

As continental plates converged, the sea-floor margin of the southward-moving North American plate plunged under the leading edge of the north-moving southern plate and was subducted. The Stanley Shale, a flysch formation, marks the beginning of this tectonic event. A *flysch* is a marine sedimentary facies that is the result of accelerated accumulation of sediment and rapidly changing conditions of deposition. Typically a flysch is deposited in a deep-water trough between approaching plates and consists of thinly bedded, graded deposits of marls and sandy shales interbedded with conglomerates, graywackes, and the like. Beds that show evidence of deposition by fast-moving turbidity currents (i.e., turbidites) are characteristic.

The Hot Springs Sandstone, the lower member of the Stanley Shale Formation, contains novaculite pebbles in conglomerate layers and shows some interbedding of thin black shale. It is from the Hot Springs Sandstone outcrops on Hot Springs Mountain that the springs emerge.

The Hatton Tuff (of volcanic origin) and the dark shales and graywackes of the upper part of the Stanley Shale are less resistant beds that underlie the wide valleys in the Hot Springs region.

3. The Ouachita orogeny in late Pennsylvanian and Permian time.

Caught between colliding continental plates, the Paleozoic marine beds were severely deformed by intense folding, thrust-faulting, and uplifting. As the plates locked, some thrusting was to the north, and some was to the south. Plunging anticlines and synclines, some of them overturned, produced complex mountain structures as the whole region buckled and was pushed upward. It was this collision of plates that formed the Ouachita Mountains.

4. Continuing uplift and erosion during Mesozoic time.

The geologic history of the Ouachita Mountains in postorogenic time is somewhat obscure. Most probably, isostatic adjustments were made as crustal blocks rose differentially in order to attain equilibrium. A gradual uplift continued in the Ouachitas until early in the Tertiary. The local faults that control the flow of the hot springs water may have developed during uplift. In Cretaceous time, intrusions of syenite (an intermediate igneous rock) were emplaced near the park, with a few dikes extending into the park.

Erosion was rapid as streams stripped off weaker beds, exposing more resistant rock layers and revealing fold patterns, as in the ZigZag Mountains.

5. Development of the hot springs hydrologic system.

The hot springs hydrologic system is probably older than the oldest water now carried in the system. Pluvial climates of the Pleistocene, when moisture was plentiful, may have brought about the development of the hot springs. Flow will probably continue as long as the regional climate is temperate and humid. Only a slight lessening of the average flow has been observed since monitoring of the system began. The main concern of National Park Service officials is to protect the recharge areas, maintain the quality of the local ground water, and prevent contamination of the hot springs so that people can continue to enjoy and benefit from these unique thermal waters and the beautiful surroundings.

Geologic Maps and Cross Sections

Bedinger, M. S.; Pearson, F. J., Jr.; Reed, J. E.; Sniegocki, R. T.; and Stone, C. G. 1979. *The waters of Hot Springs National Park, Arkansas—their nature and origin.* U.S. Geological Survey Professional Paper 1044-C.

Viele, G. W. 1979. Geologic map and cross section, Eastern Ouachita Mountains, Arkansas. Map and Chart Series MC-28F. Plate Margins Working Group, U.S. Geodynamics Committee. Geological Society of America. (Cross section is drawn a short distance east of Hot Springs National Park and is helpful in showing ancient plate margins and explaining the regional tectonic history.)

Bibliography

Bedinger, M. S., et al. 1979. *The waters of Hot Springs National Park, Arkansas—their nature and origin.* (see above).

————. 1974. *Valley of the vapors: Hot Springs National Park.* Eastern National Park and Monument Association.

Hanor, J. S. 1980. *Fire in the folded rocks: geology of Hot Springs National Park.* Eastern National Park and Monument Association.

Keller, W. D., Viele, G. W., and Johnson, C. H. 1977. Texture of Arkansas novaculite indicates thermally induced metamorphism. *Journal of sedimentary petrology* 47:834–843.

Stone, C. G.; Hale, B. R.; and Viele, G. W. 1973. *Guidebook to the geology of the Ouachita Mountains, Arkansas.* Arkansas Geological Commission.

Stone, C. G., ed. 1977. *Symposium on the geology of the Ouachita Mountains,* vols. 1, 2. Arkansas Geological Commission.

Hot Springs National Park

Size: 5,824 acres; 9 square miles

Hot Springs Reservation set aside, April 20, 1832

Dedicated to public use as a park: June 16, 1880

Redesignated Hot Springs National Park, March 4, 1921

Address: P.O. Box 1219, Hot Springs National Park, Arkansas 71901

46

One of the great rivers of the continent, the Rio Grande, loops deeply to the south, forming the "Big Bend" in the border between Texas and Mexico.

Three spectacular canyons, cut in uplifted Cretaceous strata by the Rio Grande, reveal much of the region's geologic history.

Rising abruptly from the Chihuahuan Desert floor, the Chisos Mountains display Tertiary lava flows and intrusions amid thick sequences of Cretaceous sedimentary rocks.

Superb desert scenery, fantastically eroded rocks, high-angle faults, a breached anticline, unusual fossils, hot springs, old mines, archeological sites, and a high, dry wilderness ecosystem are some of Big Bend's wonders.

Big Bend National Park

Southwest Texas

by Sherwood D. Tuttle
University of Iowa

Figure 46.1 From Big Bend National Park on the Texas side, visitors can look at dipping beds of Cretaceous Santa Elena Limestone displayed in the cliffs across the Rio Grande on the Mexican side. Here the river has carved Boquillas Canyon through Sierra del Carmen, a range that extends from Texas southward into Mexico for many miles. National Park Service photo.

BIG BEND NATIONAL PARK

Figure 46.2   Big Bend National Park, Texas.

Local History and Geographic Setting

The first superintendent of Big Bend National Park was Ross Maxwell, a geologist who worked in Big Bend country for many years. In his guidebook (Maxwell 1968), he relates an Indian legend to show that prehistoric people recognized the rock formations and landforms as being strange, unusual, and puzzling. According to the Indian story, when the Great Spirit had finished making the earth, the stars, and all living things, he couldn't decide what to do with the scraps and stony trash left over. Finally, he piled it all up in a heap and left it. That was how Big Bend country came to be.

Big Bend country is a triangular-shaped region formed where the Rio Grande runs southeast and then northeast in a great loop. At the elbow of the bulge is Big Bend National Park. Mexico lies across the river. Although Big

Bend National Park occupies only about a third of what is thought of as Big Bend country in west Texas, it is still one of the larger national parks in the United States. Within its bounds are spectacular canyons of the Rio Grande, the jagged igneous peaks of the Chisos Mountains, barren desert highlands, rare plants and animals.

The Chihuahuan Desert, which is in the Basin and Range Province, extends for many miles both north and south of the international boundary. Because of the generally sparse vegetation, angular landforms and rock units exposed in all their starkness dominate the landscape of this geologically young desert. In 1963 and 1964, in preparation for moon landings, astronauts studied rocks and practiced moving about on the rocky terrain in Big Bend National Park.

Big Bend National Park is far from a settlement of any size. The town of Marathon, 70 miles north of the park, is on the Southern Pacific Railroad as well as the nearest east–west highway. The U.S. Customs facility for the area is also in Marathon. Interstate 10, the main highway across southern Texas, is an additional 50 miles north of Marathon. Big Bend is not the isolated region it once was because a modern highway (US 385) goes from Marathon to Persimmon Gap (the north entrance of the park), and a second highway goes north from the western part of the park to Alpine, Texas.

Today's climate is harsh, but 10,000 years ago, nomadic bands hunted elephants, camels, bison, horses, and other big game in Big Bend country. Animals and hunters had drifted south, well beyond the limits of the last great continental glacier to what was then a temperate, well watered region of forests and grasslands. Then, in a relatively brief period, as the climate warmed, the ice sheet melted and dryness spread over the land. Moisture-loving plants died out, except for a few that survived around springs and at high elevations. The Ice Age mammals and the hunters disappeared.

In a later time, Indians who had adapted to desert life camped in sheltered caves along the river while they gathered plants and hunted desert animals. They left hearths, grinding pits, burial sites, petroglyphs, and artifacts in the park area.

Chisos Indians from Mexico, a more recent group, were in Big Bend country when the Spaniards came in the sixteenth century, but were driven out by marauding Mescalero Apaches. During the next three centuries, the Spaniards, later the Mexicans, tried to control the Indians along the Rio Grande, but neither missionaries nor soldiers had much success in subduing or taming the tribes. The fierce Apaches forced the Spaniards to abandon their fort on the river and flee back to Mexico. The Comanches, a nomadic group of raiders, began appearing in the region late in the 1700s. The Comanche Trail goes through Big Bend. Every autumn, the ruthless Comanches rode from Oklahoma across Texas and into Mexico to raid ranches and steal horses and cattle. Both Apaches and Comanches resisted for many years the incursions of settlers and other intruders into their territory.

After the Mexican War in the 1840s, Major W. H. Emory surveyed the Rio Grande and the Big Bend area for the International Boundary Commission. His 1852 report was the first scientific description of the region. Emory Peak (7835 feet), the highest point in the park, is named for him.

A few years later, the United States War Department sent an expedition to Big Bend to locate the Comanche Trail and try to devise ways of protecting the Texas settlements from Indians. The expedition was also for the purpose of determining the capabilities of camels imported for military transport in the arid southwest. A young Army lieutenant named Echols led the camel car-

avan along the base of the Santiago Mountains, entering territory that is now Big Bend National Park near Persimmon Gap and Dog Canyon. They crossed the difficult desert terrain and followed Terlingua Creek to the mouth of Santa Elena Canyon on the Rio Grande, a trip that took more than a month (fig. 46.3). For several days in the desert, death from thirst seemed imminent, but finally enough water was found to enable them to complete their journey to the border. Lt. Echols reported that the twenty camels stood the trip better than the mules and even better than the men. However, with the beginning of the Civil War, the camel experiment ended. The camels were turned loose in the desert and disappeared.

Mining activities began in Big Bend in the 1890s with the opening of quicksilver (mercury) mines near Terlingua, now a ghost town outside the park's western boundary. Some quicksilver was also produced at the Mariscal mine in the southern part of the park. Maxwell (1968) reports that the quicksilver mined and smelted in the Terlingua district from 1896 until the 1940s amounted to one-fourth the total mercury production in the United States. The quicksilver was extracted from the mineral cinnabar (red mecuric sulfide), which the Indians had been using for war paint and for pictographs for thousands of years. Except for the quicksilver, a hundred years of prospecting in and around the mountains of the park failed to turn up significant deposits of other ore metals, despite legends of gold and silver from "lost mines" supposedly worked by the Spaniards. However, across the Rio Grande from Boquillas, fluorspar (fluorite) ore, lead, and zinc have been mined in Mexico.

A few hardy ranchers and homesteaders settled in wider parts of the canyon bottoms along the Rio Grande in the early 1900s. The settlers were able to grow cotton and livestock feed by drawing irrigation water from the river. At the site of some warm springs near the river, a small village and a bathhouse were erected. Invalids came there to soak in tubs of warm water. Nevertheless, Big Bend was still the "Wild West" and the last frontier. Raiders, gangs of outlaws, and Mexican revolutionaries crossed the border to plunder ranches and settlements. U.S. Army outposts were established to protect the settlers and patrol the border, but raids persisted until about the time of World War I.

The movement to establish a park in the Big Bend began in 1933 when Texas Canyons State Park was set aside by an act of the Texas legislature, and the process of acquiring additional land was started. Two years later the United States Congress authorized the inclusion of the area in the National Park System, pending completion of land acquisition. Through donations and state appropriations during the next ten years, the work went on. Finally in 1944, a large tract of mountains, desert, and 107 miles of international boundary was transferred to the federal government, thus establishing Big Bend National Park.

Figure 46.3   Air view, looking southwest toward the mouth of Santa Elena Canyon, one of the most spectacular landscapes in Big Bend National Park. Leaving the canyon, the Rio Grande flows to the left along the base of the escarpment. A large slump block juts out on the left side of the canyon mouth. The broad wash (right foreground), filled with alluvium, is Terlingua Creek, which joins the Rio Grande from the north. The cliffs on the right of the canyon and the land in the foreground are in the United States. The cliffs to the left are in Mexico. Santa Elena Canyon extends upstream for about 18 miles, cutting across the Mesa de Anguila, a large fault block. The canyon's maximum depth is over 1500 feet. The imposing fault scarp of the uplifted Mesa de Anguila block stretches across the photograph, with the trace of the Terlingua fault running along the base. The foreground area is part of the downdropped block. Faulting with a throw of about 3000 feet began late in Cretaceous time, with renewed movement occurring near the end of the Tertiary. Nearly 1600 feet of Lower Cretaceous formations are exposed at the mouth of the canyon. The prominent cliff-forming rock unit near the top is the Santa Elena Limestone; below is shaly Sue Peaks Formation, with talus-covered slopes; below that another limestone, the Del Carmen. Two more formations crop out at the base of the escarpment. National Park Service photo.

Because of its location and unique ecosystem, Big Bend National Park was designated a Biosphere Reserve in 1976. Rare species of plants, birds, and animals live in the park, especially in the Chisos Mountains, where there is a little more moisture at higher elevations. A few species of colder-climate trees, such as ponderosa pine, cypress, and aspen, have survived among the peaks since the Ice Age. When the climate was cooler and moister, these species grew over a wider area, but desert vegetation replaced them as conditions became drier and hotter.

Throughout Big Bend country, rainfall averages less than 10 inches per year. When rain does come, usually during late summer or early fall, precipitation is local, fre-

quently in a cloudburst, perhaps as much as an inch in an hour. At these times, erosion is intense. Temperatures are warm the year round; cooler in winter and at higher elevations, and very hot in summer. NEVER hike or drive in Big Bend National Park without taking along drinking water.

Geologic Features

The Landforms

The locations and shapes of the infinite variety of landforms in Big Bend National Park can be explained in terms of geologic structure and the degree of resistance to erosion of specific rock units. For example, the three great canyons of Big Bend, through which the Rio Grande flows, are located at places where geologic structures have produced mountains. In the southeastern part of the park, Boquillas Canyon (longest of the three) cuts across the Sierra del Carmen, a range of mountains that trends north–south along the park's eastern boundary. Upstream, at the park's southern tip, is Mariscal Canyon, where the Rio Grande crosses the structure of Mariscal Mountain. At the western end of the park, Santa Elena Canyon, the most spectacular of the gorges, was cut by the river during uplift of Mesa De Anguila, a fault-block mountain (fig. 46.3). The principal bedrock of all these mountains is folded and faulted Cretaceous limestone.

Seemingly unrelated are the volcanic and plutonic rocks that make up the Chisos Mountains in the center of Big Bend National Park. The rugged Chisos, which are several thousand feet higher in elevation than the features along the Rio Grande, were built by igneous emplacements and extrusions, Oligocene in age or younger, that are the youngest bedrock in Big Bend.

The oldest rocks in the park are Paleozoic units exposed in Persimmon Gap at the northern entrance. Persimmon Gap, an opening in the Santiago Mountains, is a wind gap left by stream capture that probably took place early in Cenozoic time.

People have given vivid and colorful names to oddly shaped natural landmarks in Big Bend National Park. Most of these features were produced by the processes of weathering and erosion on rocks of differing resistance. Cowboy Boot, for example, which looks like a giant inverted boot, is a rock pillar eroded along joints in volcanic rock. Two resistant dikes, side by side and looking like a mule's ears sticking up on a mountaintop, are the result of differential erosion. The landform, of course, is named Mule Ear Peaks. Elephant Tusk, a steep-sided peak with a pointed top, was eroded from a resistant igneous intrusion. Alsate's Face is an eroded ridgetop that resembles the head in profile of a recumbent Indian gazing at the sky. (Alsate was an Apache chieftain of local renown.) Casa Grande ("big house") is a lava-capped mesa with nearly vertical sides and flat top. A small cave eroded in a cliff near the top of Lost Mine Peak is named the Watchman's House. The cave, according to legend, is occupied by the ghost of an Indian who was left to guard a mine once worked by the Spaniards.

A large open valley (elevation 5401 feet), nestled among the peaks of the Chisos Mountains, was thought by some early investigators to be either a large cirque or a caldera; but further study showed that the feature was the result of erosion by running water. (No glaciers formed in the Chisos Mountains in Pleistocene time, but erosion was intensified.) This landform, called the Basin, is the site of the Chisos Mountains Lodge and Campgrounds. The Window, a narrow gorge cut in the Basin's western rim, opens onto a 75-foot dry fall. The Window is the outlet for all the drainage from the Basin.

Evidence of deposition is as widespread throughout Big Bend as erosional features, although the sheetwash gravels, filled basins, terrace gravels, and gravel-strewn pediments lack the curious shapes of the erosional landforms. Great quantities of sand and gravel have been washed down from the mountains and deposited in valleys and basins. Deposition was especially heavy in Pleistocene and post-Pleistocene time.

Because of the scarcity of vegetation, Big Bend is a place where many of the features plotted on geologic maps (anticlines, fold axes, fault traces, fault blocks, etc.) can be readily seen. Noteworthy examples are Cow Heaven anticline and the Mariscal anticline (fig. 46.4).

In learning about such a multiplicity of structural features and landforms, it is helpful to remember that the limestones and other sedimentary units are mainly Cretaceous; the igneous rocks are early Cenozoic in age; and the major structures are related to Laramide tectonic activity in latest Cretaceous and early Cenozoic time. Block-faulting and mountain uplift, characteristic of the Basin and Range province, occurred in Tertiary time along with intense erosion and deposition.

The Rio Grande

Throughout most of its 1885-mile length, from southern Colorado to the Gulf of Mexico, the Rio Grande is younger than the geologic structures over which it flows. It developed its course by flowing across basin fills and fault blocks. This is clearly evident in the Big Bend international boundary area, where the Rio Grande is a *transverse stream;* that is, it flows in a crosswise direction over the faults and fault blocks. The river is also *superimposed,* which means that it was let down from higher levels by erosion, cutting through the formations on which the river originally developed and through unconformities below. During the downcutting, the Rio Grande and its tributaries removed quantities of sand and gravel (that once filled the basins) and exposed the bedrock of the upfaulted blocks. A major tributary is the Rio Concho, which crosses northeastern Mexico and joins the Rio Grande at Presidio upstream from Big Bend National Park. The Rio Concho had a significant role in the cutting of the canyons.

Detailed studies of river gravels, terraces, and pediments in the Rio Grande valley (mainly in New Mexico) show the occurrence of alternating periods of *aggradation* (the building up of the land surface by deposition) and *degradation* (the lowering of the land surface and general reduction of relief by stream action). This sort of cut-and-fill also went on in the Big Bend canyons. In some places on less steep sidewalls, pediments cutting across both bedrock and gravel were left high above the present valley floor.

Figure 46.4　Air view, looking south toward Mexico in the far distance, of Mariscal Mountain, a plunging anticlinal fold. An anticlinal mountain eroded from a plunging fold is a relatively rare landform. The axis of the fold runs from the lower right to upper left, along the mountain crest. Hogback ridges, formed by differential erosion, run along the base of the mountain on both sides, and converge in the foreground. Elevation of the mountain crest (the trace of the fold axis) decreases toward the foreground. Thus the anticline plunges toward lower right in this view. Beds in the hogbacks on the right dip about 40°, and those on the left, 15° to 20°, indicating that the fold is asymmetrical with a steeper right limb. The notch in the far distance on the mountain's right slope is the entrance to Mariscal Canyon, cut through Mariscal Mountain by the Rio Grande. The rocks exposed include almost the entire Cretaceous section in Big Bend National Park. The oldest rock, cropping out in the center of the fold is the Santa Elena Limestone. A small quicksilver mine, now abandoned, is visible (center foreground) between the road junction and the foot of the mountain. The cinnabar ore was recovered from the Boquillas Formation. National Park Service photo.

The volume of the river has fluctuated considerably in late Cenozoic time. During the Pleistocene Epoch, the Rio Grande was fed by valley glaciers in the Southern Rockies. Their meltwater greatly increased the river's volume and erosive power. Periods of pluvial climate brought greater precipitation to the whole southwest region; but at other times, the regional climate was even drier than at present. Climatic fluctuations profoundly affect the river's regimen, especially its cycles of flooding, and also cause changes in the processes of weathering, mass wasting, and runoff in the Big Bend region.

In 1978, a 191-mile strip on the American shore of the Rio Grande was placed under the protection of the Wild and Scenic Rivers system. The protected area begins in Big Bend National Park and continues downstream to the Terrell-Val Verde County line more than 75 miles from the park's eastern boundary. Except during flood stage, which is hazardous, float trips by permit are allowed.

Hot Springs

The Hot Springs area, which has been developed for tourists in southeastern Big Bend National Park, is the best known of many hot springs along the course of the Rio

Grande, on the Mexican as well as the Texas side. Three other hot springs, all downstream from Hot Springs, are located in the park. One is in a cane break next to the river, nearly a half-mile downriver; another is adjacent to Boquillas Canyon; the third serves as the water supply for Rio Grande Village. At Hot Springs, which has the highest temperature (106° F), the water is collected in a rectangular pool beside the river. The other springs have lower temperatures—105°, 95°, and 96° F, respectively. The springs maintain fairly constant temperatures and amounts of discharge, and are seemingly unaffected by seasonal changes in weather and rainfall. Discharge rates for the Big Bend hot springs are ordinarily less than 52 gallons per minute, which is not especially high.

Geologists think that the hot springs are remnants of what was a highly active hydrothermal system during late Tertiary time when crustal thinning (extension), involving normal faulting and numerous eruptions, was occurring. The hot springs are related to ground-water circulation. In this region of high heat flow, hot rocks are comparatively near the surface. Ground water percolates down through the many faults and fractures in the limestone, is heated at relatively shallow depths, and then rises to the surface.

Vertebrate Fossils

Near Park Headquarters at Panther Junction, an exhibit shelter displays mammal and dinosaur bones that were discovered in sedimentary rocks in the park. The remains of very ancient, primitive mammals were found in Paleocene beds near the roadside exhibit. These extinct mammals, the oldest that have been found in Texas, were small, forest-dwelling carnivores and herbivores having no direct relationship to modern species. (Such mammals may have coexisted earlier with late Mesozoic dinosaurs, whose fossilized remains have also been found in Big Bend.)

Eocene beds overlying the Paleocene rocks near the exhibit contain the bones of somewhat more advanced mammals. These animals were not direct ancestors of living mammals, but had an anatomical similarity to later species. The remains include the earliest "horse" (*Hyracotherium,* or eohippus), an early cat, piglike and hippopotamuslike animals, and a rhinoceroslike species. In younger rocks elsewhere in Big Bend, doglike mammals, a rabbit, camels, a small deer, and sheeplike animals have been identified.

Dinosaur fossils in Big Bend have been found in shallow marine and nonmarine deposits, late Cretaceous in age, that accumulated in a warm, swampy, coastal environment. In addition to dinosaur bones, fossils in these rock units include petrified wood, turtle shells, fish scales, shark teeth, and crocodile bones.

By far the most remarkable finds have been wing bones and other fragments from *pterosaurs* (flying reptiles). In 1971, a student at the University of Texas, Douglas Lawson, discovered wing bones (including a large forearm bone) of a giant pterosaur, which has become known as the "Big Bend pterodactyl." Diligent later searches turned up fossil fragments of other individuals of the same species, probably the largest creatures ever to fly. The Big Bend pterodactyl had an estimated wing span of 38 feet! The mechanics of pterosaur flight are puzzling. How were such enormous wings controlled and maneuvered? How did the flying reptiles take off and how did they land? Did they "ride" updrafts and air currents as hang gliders do? Did the huge creatures hunt over land as well as water? How did they get around when they were on the ground? Recently, paleontologists and aerodynamicists have been constructing working models of giant pterodactyls in an effort to answer some of these questions. (Langston 1981, 1986).

At the end of the Cretaceous Period, the pterosaurs, who had dominated the skies for millions of years, died out—along with all other dinosaurs on earth. Modern birds, who took over the realm of the sky, developed from another form of reptilian life, unrelated to the pterosaurs.

Mining in the Big Bend Area

The quicksilver that was mined for many years in Big Bend country came from cinnabar ore hydrothermally deposited, probably during Tertiary igneous activity. The mineral *cinnabar* (mercuric sulfide) was brought upward in hydrothermal solutions that were then caught in a stratigraphic trap under the impermeable Del Rio Clay and deposited along with calcite in the Santa Elena limestone, just below the clay. The stratigraphic trap was formed by a faulted anticline. Most of the ore was found within 50 feet of the bottom of the Del Rio Clay, but in some places the ore extended downward as much as 200 feet into the Santa Elena beds. Some of the richest ore was discovered in collapsed sinks and pipes.

In the Terlingua district (outside the park's western boundary), the cinnabar ore occurs in an east–west belt about 16 miles long and 6 miles wide, extending from Study Butte to Lajitas on the Rio Grande. Scattered occurrences of cinnabar have also been reported at sites in the lower Big Bend area. A mine was operated for a few years in the Mariscal Mountain area (fig. 46.4). Here the ore is associated with hydrothermal activity during the emplacement of sills in the Boquillas Formation.

Cinnabar ore from the Big Bend mines was smelted locally and the recovered quicksilver was shipped out in iron flasks. Obtaining fuel for the furnaces was always a problem. Most of the forests on the lower slopes of the Chisos Mountains were cut down for fuel. Because of the soil erosion that resulted and the arid climate, regrowth of trees was not possible and the slopes have remained barren. After deposits of subbituminous coal were located northeast of the mines, coal was used in the smelters.

Table 46.1 Geologic Column, Big Bend National Park

Era	Periods	Epochs		Rock Units	Geologic Events
Cenozoic	Quaternary		Holocene / Pleistocene	Younger gravels and terraces	Erosion and deposition Erosion, deposition, major canyon-cutting Uplift, renewed faulting
	Tertiary	Neogene	Pliocene / Miocene	Older gravels, pediment gravels, older terraces	Basin filling
				/////////////////	Block faulting
		Paleogene	Oligocene ----?----	South Rim Formation	Volcanic eruptions Hydrothermal mineralization Igneous intrusions
			Eocene	Chisos Formation	
				Canoe Formation	Block faulting Mixed clastics and ash
			Paleocene	Hannold Hill Formation Black Peak Formation	Nonvolcanic deposits (mammal bones)
Mesozoic	Cretaceous			/////////////////	Laramide orogeny, folding, thrust-faulting
			(Upper)	Javelina Formation Aguja Formation	Nonmarine deposition (dinosaur bones, silicified wood)
				Pen Formation	
				Boquillas Formation	
				Buda Limestone	
			(Lower)	Del Rio Clay	
				Santa Elena Limestone	Transitional and shallow marine deposition in transgressing and withdrawing Cretaceous seas
				Sue Peaks Formation	
				Del Carmen Limestone	
				Telephone Canyon Formation	
				Glen Rose Limestone	
Paleozoic	(Ordovician to Pennsylvanian)			/////////////////	
					Ouachita orogeny
				Deformed shales, sandstones, novaculite, conglomerates, probably Tesnus, Caballos, and Maravillas Formations	Shelf deposits; deep-ocean deposits

Source: modified after Maxwell 1968

Geologic History

1. Sedimentation and mountain-building in the Paleozoic Era.

The oldest rocks in Big Bend National Park are deformed Paleozoic units exposed in thrust slices at Persimmon Gap. Here in the northernmost section of the park are the Santiago Mountains, which trend northwestward into the Marathon Basin and the Glass Mountains of west Texas. The rocks at Persimmon Gap, which include shales, limestones, chert, and novaculite, are related on the basis of age and lithology to Paleozoic sequences of the Ouachita fold belt in Oklahoma and Arkansas. The novaculite, a dense, light-colored, siliceous rock, is similar to that found in Hot Springs National Park (chapter 45).

The Paleozoic rocks were deposited in an ancient ocean between a continental plate that eventually became the North American landmass and another continent to the south. In late Paleozoic time, the ocean closed and the continental plates collided. During folding and faulting, some of the rock units from the deep part of the ocean basin were thrust over units along the continental margin.

Except for the outcrops at Persimmon Gap, the Paleozoic rocks of the Big Bend area have been stripped off or buried by much younger rocks. However, in Mexico, across the Rio Grande from Boquillas, some older metamorphic Paleozoic rocks are exposed in the Sierra del Carmen escarpment.

2. Accumulation of Cretaceous sediments.

The absence of Triassic and Jurassic rocks in the Big Bend sequence probably indicates a long period of erosion. Throughout Cretaceous time, the Big Bend region was part of a coastal plain and a shallow marine shelf along the margin of a continental landmass that sloped toward an ancestral Gulf of Mexico. Unconformities in the sedimentary sequence mark several advances and withdrawals of the ocean during the 90 million years of the Cretaceous Period. The final draining away of seawater occurred when the region was uplifted late in Cretaceous time.

Massive, limey units like the Glen Rose Formation, the Del Carmen Limestone, and the Santa Elena Limestone accumulated during higher sea stands (fig. 46.5). Clastic sediment from the north and northwest spread over the ocean floor when sea level was lower or the land higher. Units interbedded with the limestones include claystone and mudstone (some of it bentonitic), shales, calcareous sandstones, thin sandstone beds, stringers of silt and coal, gypsiferous marls, chalk, cherty limestone, and basal conglomerates. Careful study of these sequences allows geologists to document sedimentary facies changes that show transgressions and withdrawals of the sea, episodes of erosion, and a wide variety of sedimentary environments.

Most of the rocks are fossiliferous and contain molluscan remains typical of marine Cretaceous sequences.

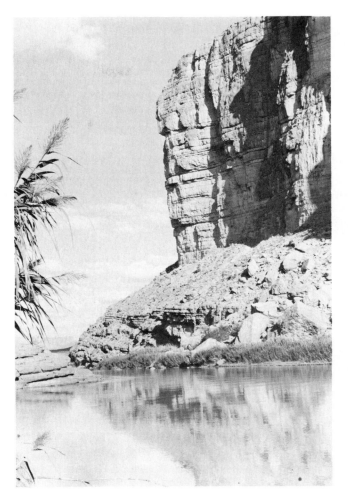

Figure 46.5   Vertical cliff in Santa Elena Canyon. Exposed here from bottom to top are the Glen Rose Formation (beds just above the waterline), the Telegraph Canyon Formation (covered with talus), and the Del Carmen Limestone, a cliff-former. Rock units are Lower Cretaceous in age. National Park Service photo.

In the transitional and nonmarine units (the Aguja and Javelina Formations) that accumulated late in Cretaceous time are fossilized logs and various remains, including dinosaur bones.

3. Laramide orogeny, late Cretaceous-early Tertiary time.

The Big Bend region is close to the eastern edge of the vast area of western North America that was profoundly affected by the complex and long-lasting tectonic events associated with the Laramide orogeny. Deformation was moderate in Big Bend in comparison to that occurring in regions farther north and west. Compressive forces produced tight folds and overthrusting along the eastern and western margins of the Big Bend and broader anticlinal folds in between. Some of the older folds and structures, such as those at Persimmon Gap, were realigned and refolded with the overlying Cretaceous beds.

Figure 46.6   South side of Vernon Bailey Peak, on the northwest side of the Basin, in the Chisos Mountains. The peak consists of resistant igneous rock (a felsic quartz-riebeckite) emplaced in a small stock and subsequently exposed by erosion. The rock is believed to be the intrusive equivalent of the Burro Mesa Riebeckite Rhyolite, which is the youngest eruptive rock in Big Bend National Park and has been dated as Oligocene and younger. National Park Service photo.

If you draw a line on a geologic map from Santa Elena Canyon to the northeast corner of Big Bend National Park, you will note that the line crosses more than a dozen folds and several major faults.

On the mountainous topography that developed as a result of tectonic activity, the processes of erosion went to work. Structural trends controlled the patterns of valleys and ridges that evolved. Not much evidence of these landforms can be found in the present landscape of Big Bend because they were destroyed or buried by subsequent geologic events.

4. Paleocene and Eocene nonvolcanic deposition.

Erosion of the Cretaceous rocks and structures produced clastic sediments that accumulated between folds and fault blocks, forming nonvolcanic rock units, Lower Tertiary in age and approximately 3000 feet in thickness. These units lie unconformably over the older Cretaceous rocks. Lithologically the rock units are composed of con-

glomerates, sandstones (including channel sandstones), and shales. In the Canoe Formation nonvolcanic beds grade into mixtures of clastics, tuff, and ash, interbedded with lava flows, signifying the onset of volcanic activity. Because fairly intense erosion continued on the mountain slopes and on the nonvolcanic sequences that had accumulated in the valleys, local unconformities cut across all three early Tertiary formations and some of the uppermost Cretaceous units as well.

5. Igneous extrusive and intrusive activity.

In early Tertiary time, after the slackening of Laramide compressive forces, magma began rising into the crust. Molten rock reached the surface and was extruded as lava or pyroclastics. Magma below the surface cooled in dikes, sills, sheets, laccoliths, plugs, and other discrete bodies of irregular shape. Many were covered by lavas and pyroclastics and some may have been feeders for volcanic eruptions. Because most of the intrusions are fine grained, they probably cooled at depths of a few thousand feet or less. Subsequent erosion in Big Bend National Park and adjacent areas has exposed many of the intrusions, which are largest and most abundant in a broad, northwest-trending belt centering in the Chisos Mountains. In terms of composition, the intrusive rocks range from felsic rhyolites (the riebeckites) through intermediate rocks, such as trachytes and andesites, to mafic basalts. They range in age from latest Cretaceous to late Oligocene (fig. 46.6).

Volcanic centers opened in the Big Bend area and from these vents, lavas and pyroclastics were erupted. In basins and valleys, volcanic rocks are interbedded with sandstone and other clastics. Some beds, formerly continuous, have been interrupted by faulting and erosion. The sequence shows that between eruptions, streams reworked and redeposited loose debris that was later buried by new volcanic material.

The Chisos and South Rim Formations are made up of 3000 to 5000 feet of massive lava flows, flow breccias, conglomerates, tuffaceous sandstones, mudstones, tuffaceous clays, and ash beds. Chisos rocks have been dated as mainly late Eocene in age, and the South Rim Formation as probably Oligocene or younger. Potassium-argon dates taken from basalts in the lower part of the Chisos Formation yield dates of 38 to 42 million years before the present. Dates from riebeckite rhyolite near the top of the South Rim Formation record eruptions that occurred about 30 million years ago.

6. Tertiary block-faulting.

Big Bend country, which is in the easternmost part of the Basin and Range province, has a landscape characteristic of the province; that is, shaped by the erosion of upfaulted blocks and deposition in downfaulted basins. Toward the close of Laramide tectonism, segments of crustal material that had been contorted and squeezed together by compression began to readjust isostatically. As the crust sagged, thinned, and stretched, tensional forces broke it into blocks that moved up or down. In the Big Bend area, evidence based on cross-cutting relationships indicates that throughout Tertiary time high-angle faulting occurred intermittently, producing a series of northwest-trending fault blocks. Faulting was most intensive, however, during early and middle Tertiary time. Most geologists believe that the igneous activity was associated with the faulting, although the causal relationship (if it exists) is not entirely clear.

7. Extensive late Tertiary erosion and deposition.

Unconsolidated sediments, shown as "older gravels" on geologic maps of the Big Bend, are the remnants of a sedimentary cover that was once much more widespread. Indirect evidence suggests that eroded clastic and volcanic debris filled the fault basins and probably covered some of the fault-block ranges. It was also during this time that the present drainage system, graded to the Rio Grande, was organized.

8. Pleistocene and Holocene faulting and erosion.

Due to complex changes in rainfall and runoff patterns, a large part of the basin fill and other Tertiary deposits were eroded and carried away by swollen Pleistocene streams. Renewed uplift on some of the Tertiary faults elevated the highlands, at the same time increasing the erosive power of the streams. The Rio Grande and its larger tributaries cut canyons and widened and deepened valleys. Landforms referred to as "older terraces" developed at this time, and cliff recession produced pediments that truncated bedrock and older gravels.

In post-Pleistocene time, as the regional climate became increasingly arid, streams continued to erode and deposit but at a slower rate. Young gravels have accumulated in valley bottoms and on narrow flood plains. Pediments, fans, and terraces are being dissected by ongoing erosion.

Geologic Map and Cross Section

Maxwell, R. A., and Lonsdale, J. T. 1966. Geologic map of the Big Bend National Park, Brewster County, Texas. University of Texas Bureau of Economic Geology

Bibliography

Chesser, K. and Estepp, J. D. 1986. Hot springs of Big Bend National Park and Trans-Pecos Texas. in Pause, P. H. and R. G. Spears, editors. 1986 (citation below). p. 97–104.

Langston, W., Jr. 1981. Pterosaurs. *Scientific American* 244 (2):122–136.

—— 1986. Rebuilding the world's biggest flying creature: the second coming of *Quetzalcoatlus northropi*. in Pause, P. H. and R. G. Spears, editors. 1986 (citation below). p. 125–128.

Maxwell, R. A. 1968. *The Big Bend of the Rio Grande,* a guide to the rocks, landscape, geologic history, and settlers of the area of Big Bend National Park. Guidebook no. 7 (geologic maps in pocket). Austin, Texas: University of Texas Bureau of Economic Geology.

——; and Dietrich, J. W. 1965. *Geology of the Big Bend area, Texas;* Field trip guidebook with road log and papers on the natural history of the area. West Texas Geological Society Publication 65–51.

——; Lonsdale, J. T.; Hazzard, R. T.; and Wilson, J. A. 1967. *Geology of the Big Bend National Park, Texas.* Publication no. 6711. Austin, Texas: University of Texas Bureau of Economic Geology.

National Park Service. 1983. *Big Bend.* Handbook 119. 128 p.

Pause, P. H. and Spears, R. G., editors. 1986. *Geology of the Big Bend area and Solitario Dome, Texas.* West Texas Geological Society 1986 Field Trip Guidebook, publication 86–82 (geologic maps in pocket). 277 p.

Sheldon, R. A. 1979. *Roadside geology of Texas.* Missoula, Montana: Mountain Press Publishing Company. p. 83–91.

Schiebout, J. 1986. Big Bend: a crossroads in the beginning of the Age of Mammals. in Pause, P. H. and R. G. Spears, editors. 1986. (citation above). p. 129–134.

Tyler, R. C. 1984. *The Big Bend: a history of the last Texas frontier.* National Park Service. 288 p.

Big Bend National Park

Area: 741,118 acres; 1158 square miles
Authorized: June 20, 1935
Established: June 12, 1944
Designated a Biosphere Reserve: 1976
Address: Big Bend National Park, Texas 79834

47

From Skyline Drive, along the crest of the Blue Ridge Mountains, spectacular vistas overlook Shenandoah Valley to the west and the rolling Piedmont to the east.

Deformed, crystalline Precambrian and early Paleozoic rocks and complex structures record ancient orogenies.

Post-Paleozoic uplift and erosion created beautiful Appalachian scenery. Waterfalls, racing streams, high meadows, ridges, and hidden hollows can be seen along 500 miles of trails, including a 95-mile segment of the Appalachian Trail.

Shenandoah National Park

Northwest Virginia

Figure 47.1   Climbers stand on a ledge of massive Catoctin lava near the summit of Little Stony Man in the north central section of Shenandoah National Park. They are looking to the west over a winter landscape of ridges and foothills descending to the Shenandoah Valley. The summit profiles of the rounded mountains, eroded from Precambrian bedrock, are characteristic of unglaciated ranges formed from massive, resistant rock. National Park Service photo.

SHENANDOAH NATIONAL PARK

Figure 47.2   Shenandoah National Park, Virginia.

Geographic Setting

South of New England, the Appalachian Mountain belt divides into four geologic provinces that trend south to Georgia in roughly parallel bands. From east to west, these divisions are the Piedmont (the foothills), the Blue Ridge, the Valley and Ridge, and the Appalachian Plateau (fig. 47.3). In the Blue Ridge province, which has the highest elevations in the Appalachian system, are two national parks—Shenandoah in Virginia and Great Smoky Mountains (chapter 48) in Tennessee and North Carolina.

The Blue Ridge Mountains, which are the easternmost high range of the Appalachians, begin in southern Pennsylvania as a single narrow ridge only about 10 miles wide. Trending southwestward, the mountains become closely spaced ridges of greater height and width, ending finally in a broad bulge of complex ridges that cross North and South Carolina and Tennessee and dip into Georgia. Great Smoky Mountains National Park (chapter 48) occupies the highest part of the southern end of the Blue Ridge province.

Shenandoah National Park, near the northern end of the Blue Ridge, is long and narrow. The park follows the crest of the Blue Ridge for about 80 miles from Front Royal, Virginia (about 72 miles west of Washington, D.C.), to the southern entrance of the park near Waynesboro. Elevations in Shenandoah National Park range from around 600 feet at Front Royal to 4051 feet at the top of Hawksbill Mountain.

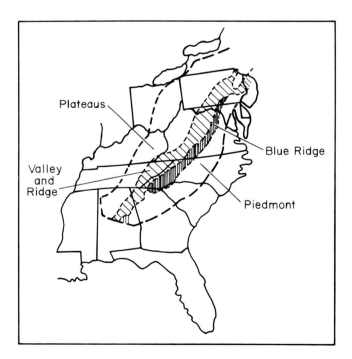

Figure 47.3   Physiographic provinces of the Southern Appalachians.

Skyline Drive, a scenic highway famous for its overlooks, runs the length of Shenandoah National Park and connects with the Blue Ridge Parkway, which continues southward to Great Smoky Mountains National Park. The Appalachian Trail goes through both parks as it follows the crest of the Appalachian Mountains from Maine to Georgia.

The Blue Ridge Mountains and other ranges of the Southern Appalachians are more rounded and less angular than the mountains of western United States. The greater age of the Appalachians and a more humid climate are reasons for this aspect of the landscape. Much of the bedrock of the Appalachians is mantled by *residual soil* that accumulated in place through millions of years of weathering. This soil and ample precipitation distributed evenly throughout the year support dense hardwood forests on the mountain slopes. Evergreens that have persisted at higher elevations are probably remnants of coniferous forests that migrated from the north during cold Pleistocene climates. Glaciers did not reach Virginia, but periglacial conditions prevailed at high altitudes and must have intensified weathering and erosional processes. The large boulder fields on many slopes of the Blue Ridge are doubtless the result of periglacial activity. Even in today's climate, snow blankets higher levels of the Blue Ridge for at least part of the winter season. Frost-wedging is a significant factor in the weathering of rocks.

Major streams are the Potomac River, which cuts through the Blue Ridge at Harpers Ferry (north of Shenandoah National Park) and the James River, which has carved scenic gorges for its course through the Blue Ridge south of the park. No streams cross the Blue Ridge within the park. To the west in the Shenandoah Valley, the Shenandoah River flows north—paralleling the Blue Ridge—and joins the Potomac at Harpers Ferry. The Blue Ridge itself is dissected by tributary streams, many with waterfalls, that drain the mountain slopes. Streams have been the dominant agent of erosion in the region since the mountains were uplifted in Paleozoic time. Erosion is fairly intense, despite the protective cover of vegetation. During heavy rains, quantities of loose sediment muddy the rivers. When storms are prolonged or extremely heavy, mudslides occur on steep or denuded slopes.

Local History

Primitive food gatherers, followed later by Indian hunters, frequented the Blue Ridge for thousands of years before Europeans came to Virginia. Indians left stone burial mounds, arrowheads, spear points, and other artifacts in Shenandoah National Park. They camped in the uplands during summer and fall to gather berries, hickory nuts, and chestnuts, and hunt game. Elk, deer, and bison grazed in the high open meadows and clearings in the forest. Fires

Figure 47.4   The Virginia Piedmont stretches away to the east below Old Rag summit (elevation 3268 feet) on the eastern edge of the Blue Ridge. National Park Service photo.

that kept the meadows free of trees were started by lightning, or sometimes kindled by the Indians, as this was a method of driving game into the range of spears and arrows.

The English who settled in the Virginia Tidewater did not expand westward for about a hundred years. The Piedmont (the foothills of the Blue Ridge), the Blue Ridge, and the Shenandoah Valley on the far side remained unsettled and largely unexplored. In the early 1700s, Alexander Spotswood, an able and energetic royal governor of the Virginia colony, saw a need for expansion as the colony's population grew. He encouraged settlement of the outlying frontiers by exempting settlers from quitrents (taxation). In 1716, he led an expedition to the Blue Ridge and the Shenandoah Valley and a few years later negotiated a treaty by which the Iroquois Indians agreed to stay north of the Potomac River. From that time on, settlers began to establish farms and villages on both sides of the Blue Ridge and in the Shenandoah Valley.

Settlements in the Shenandoah Valley and the Piedmont became increasingly prosperous as time went on, but the mountain people who farmed the upland valleys of the Blue Ridge became ever poorer and more isolated. They were limited in their choice of favorable sites not only by

the rugged topography but also by the nature of the bedrock. Although they may not have understood the reason, it was obvious that the well watered, fertile soils of the broad summit meadows and the gentler eastern slopes—underlain mostly by granites and basaltic greenstones—were more suitable for homesteads than the rocky, steep western ridges that developed thin, infertile soils on resistant quartzites and phyllites. Moreover, on the west side, springs were fewer and intermittent. What good soils there were in the uplands soon became depleted as crops were planted year after year in the same small fields. Erosion was severe, especially after timber cutting removed protective vegetation from the slopes. A disastrous blight early in this century exterminated the chestnut trees, which had been one of the finest hardwoods of the region.

The discovery of small deposits of iron, manganese, and copper in the Blue Ridge helped the region for a time during the nineteenth century; but as ores were worked out, mines were abandoned. Some manganese mining was carried on as late as World War II. In the western foothills of the park and outside the western boundary are more than 40 old mine sites.

As times became ever harder, the mountain people and their unique way of life seemed doomed. But toward the end of the nineteenth century, the development of mountain resorts high in the Blue Ridge brought money, new people, and a revival of interest in the region's potential. The beauty of the scenery and the benefits of pure mountain air were now accessible to the large populations of the nation's capital and nearby cities, who sweltered in summer heat and humidity. The idea of establishing a national park in the Virginia Blue Ridge country began to grow. George Freeman Pollock, who operated Skyland Resort on the mountaintop and controlled some 5000 acres along the crest, became enthusiastic about the project. After Congress authorized the park's acquisition in 1926, he helped to raise over a million dollars for land purchases. With no public lands in the area, all ground for the park had to be purchased from private owners or acquired by donation. President Herbert Hoover gave his land and fishing camp on the Rapidan River. The Virginia legislature appropriated a million dollars for compensation and relocation of landowners. In 1935, a large tract along the mountain crest was turned over to the federal government for the establishment of Shenandoah National Park. Later additions brought the total area to over 300 square miles.

Skyline Drive, with its beautiful panoramas to the east and west, goes through the park along the spine of the Blue Ridge. Most of the land has been allowed to revert to a wilderness state. Several ancient clearings, including Big Meadows, are kept open, but new forests have reclaimed the mountain farms, orchards, and logged-over areas. More than three million people visit this exceedingly popular national park each year.

Geologic Features

Bedrock and Geologic Structures

Fresh exposures of bedrock are rare in Shenandoah National Park, and outcrops show evidence of deep weathering. Ledges may be shattered or largely covered by broken and crumbling talus. Weathering and vegetation may mask bedrock and structures, but their controlling influence on landscape development is plain to see.

Consider Big Meadows, which is part of a broad summit area on the crest of the Blue Ridge. Here is a plateau of low relief, about 3550 feet in elevation and roughly 5 square miles in extent. Elsewhere along the northern Blue Ridge, the crest is relatively narrow and deeply dissected. Two intermittent swamps fed by springs and runoff from higher slopes have long existed in shallow basins in Big Meadows. What lies beneath Big Meadows are at least

Box 47.1
The Evolution of the Appalachian Highlands

For nearly two centuries, the Appalachian Mountains have been the inspiration and testing ground for many of the grand ideas and concepts of American geology. Up until the twentieth century, American geologists were trained on Appalachian rocks and structures because most centers of learning were in the eastern part of the country. It was in dealing with the complexities of these ancient mountains that American geologic sciences came of age. From observations in the field, geologists learned the nature of folds and thrust faults in mountain belts and worked out the anticlinal theory for the accumulation of petroleum. They recognized an association of sea-floor rocks with initial stages of mountain-building, developed hypotheses to account for orogenies, and postulated principles of landscape development. For many years, the processes of mountain-building that created the Appalachian belt were conceived of as one major "revolution" that brought the Paleozoic Era to a close.

The related concept of the geosyncline, based on studies of Appalachian geology and first put forth in 1857, was elaborated on for more than a hundred years. The theory postulated an extensive belt of down-sinking crust as an essential prologue to orogeny. Thousands of feet of sediment, which accumulated in a linear trough during prolonged subsidence, gradually became consolidated as layered rock and finally buckled due to compressive force. The height of the mountains that resulted was thought to be directly related to the amount of accumulated sediment.

But many questions regarding the origin and development of the Appalachians remained unanswered. One puzzling fact was that masses of rock exposed in the higher Appalachians appeared to be geologically incongruent (unrelated) to rock units in the foothills and lowlands on either side. Also, Paleozoic deposits thickened and became coarser toward the east, but where was the great landmass that had supplied the sediments? Had a "lost continent" foundered beneath the surface of the broad Atlantic? Furthermore, the model did not account satisfactorily for the underlying cause of mountain-building. Great lateral contractions—due to immense horizontal compression—produce folded mountain belts like the Appalachians. How could compressive stresses be caused by presumably vertical forces associated with instability in the root of a geosyncline? And what was the nature and origin of this vague instability?

Pre-plate tectonics concepts have been expanded by evidence indicating that tectonic pulses in a series of episodes throughout Paleozoic time were involved in the geologic history of the Appalachian belt region. Plate tectonic theory, with its envisioning of moving crustal plates, removed the need for having fixed landmasses as sediment sources. Instead, continents and continental fragments (microplates) were visualized as coming together and breaking apart, with intervening oceans closing or opening during the process. When crustal plates come together in such a way that structural downbuckling or subduction occurs, an area of subsidence forms, which may fill with sediment. Colliding landmasses may crumple and pile up. Microplates become tectonostratigraphic (suspect) terranes by being sutured to continental plates during an orogeny. The suspect terranes, with their sharply defined boundaries, usually contain rocks that are completely unrelated to the rocks of neighboring provinces in terms of age and stratigraphy, fossil forms, plutonic or metamorphic history, paleomagnetic signatures, and the like.

Present approaches toward gaining an understanding of Appalachian belt evolution are based on reconstructions of former positions and movements of landmasses during Paleozoic and Mesozoic time. In the Paleozoic Era, landmasses were loosely bonded together while drifting through low latitudes in a westerly direction. Episodes of crustal compression and rifting built mountains and opened seaways. Mesozoic time saw supercontinents form and break up with landmasses drifting in different directions. Present-day continents and modern oceans came into being in the Mesozoic. Data derived primarily from studies of magnetic anomalies on the ocean floors have enabled geologists to reconstruct with reasonable accuracy the positions of continents since the beginning of the Mesozoic Era.

Deductions and conclusions regarding paleolatitude positions of landmasses are drawn from investigations of ancient magnetism in continental rocks, from fossil assemblages, and depositional features, such as evaporites and coalbeds.

Modeling of tectonic movements and orogenies associated with old mountain belts involves paleomagnetic data, orientation of fold belts, the distribution of fossil forms and sedimentary facies, and geophysical techniques, such as deep seismic reflection profiling—which was originally developed for oil exploration by the petroleum industry. The outline that follows summarizes tectonic mechanisms that most geologists feel help to explain the development of major geologic features of the Appalachian highlands.

Although the Appalachian Mountains belt divides into three major tectonic segments along its course, the overall structural features exhibit a remarkable parallelism of strike, from Newfoundland to northern Mexico. The three segments, which have related but somewhat different orogenic histories, are 1) the

continued

Box 47.1 continued

Newfoundland-New England segment; 2) the Pennsylvania-Carolina segment, which includes the Blue Ridge; and 3) the Ouachita-Marathon segment. Recesses, or unexposed portions of the mountain belt, such as those in New York and Mississippi, separate the segments. Emphasis in this discussion is on the second segment.

1. Precambrian events.

The granites, gneisses, and schists exposed in the Appalachian highlands, some more than a billion years old, were originally sedimentary, volcanic, and plutonic rocks. They underwent intense deformation and metamorphism, presumably more than once, during the latter part of Precambrian time. Ancient Appalachian rocks are distinctly different from North American cratonic rocks. Appalachian crystalline rocks probably belonged to primitive microplates or tectonostratigraphic terranes that were deformed in continental collisions somewhere in the Southern Hemisphere. Parts of what eventually became North America, Africa, and Europe were involved in this coming together of tectonic plates that formed a Precambrian supercontinent.

2. Opening of the Proto-Atlantic (Iapetus) Ocean.

Rifting of the supercontinent very late in Precambrian time brought about the opening of an ocean basin, the Proto-Atlantic Ocean (called Iapetus, after the father of Atlas, for whom the Atlantic Ocean was named). Rocks on the west side of the ocean ultimately became the North American plate, while those on the east side remained with an Afro-European plate. (For convenience, in this outline we use the modern orientations of the landmasses. Because tectonic plates tend to rotate as they drift, their relationships to each other in terms of compass directions were different in Precambrian and Paleozoic time than now.)

3. Cambrian and early Ordovician erosion and deposition.

As the Precambrian highlands (that had been elevated by orogenic episodes) were eroded, clastic sediments that washed into the ocean

throughout the Cambrian were deposited on a shallow continental shelf. Carbonates, mostly Ordovician, covered the Cambrian sandstones. Fossil remains of early Paleozoic life forms, such as brachiopods and stromatolites, indicate that the climate was tropical or subtropical.

4. Oceanic-continental plate convergence.

Oceanic crust, pushed westward from a spreading center in the Proto-Atlantic Ocean, began to sink into an oceanic trench and was subducted beneath the North American plate. A volcanic island arc, parallel to the continental margin, developed above the trench. As plate convergence continued, the island arc with its accumulated volcanic debris, plus slices of sea-floor rocks, were jammed together and thrust westward onto the margin of eastern North America. Some of these rocks, metamorphosed and deformed, crop out in the Piedmont province east of the Blue Ridge.

5. Taconic orogeny in late Ordovician time.

Compression and subduction on both sides of the Proto-Atlantic Ocean brought about an intense tectonic episode, called the Taconic orogeny, that created the Appalachian mountain belt. The Taconic was the first of the Appalachian orogenies that took place in Paleozoic time. Slivers of continental crust, in suspect terranes, were pushed up against the already deformed Precambrian and early Paleozoic rocks. Granitic plutons in the terranes, dated as Precambrian, were emplaced before they docked at the continental margin.

6. Prolonged erosion and deposition.

Great volumes of eroded sediment washed down from the uplifted mountains. Some sediment accumulated in the ocean basin to the east. Lavas and pyroclastics, generated by subduction, were interbedded with the clastics. A great deal of the sediment was transported down the west side of the ranges and deposited in a vast, shallow continental sea that spread over the central part of the North American craton, beginning in Cambrian time.

Figure 47.5   Approximate locations of major suspect terranes in the Southern Appalachians. The western edge of the Piedmont terrane was pushed up against the margin of the old North American continent during the Taconic orogeny. Avalon and other terranes docked during later Paleozoic mountain-building episodes. (The phrase "continental sediments" in the diagram indicates source areas—not sites of deposition.)

continued

Box 47.1 continued

7. Acadian orogeny in Devonian time.

In late Silurian and early Devonian time, the Proto-Atlantic Ocean was closing as the Afro-European and North American plates moved nearer together. The Acadian orogeny (chapter 21) in middle and late Devonian time resulted from the collision of these landmasses. Most of the Acadian tectonic activity took place in the Northern Appalachian region because this was where the ocean closed first. The high Acadian ranges, produced by this intense and long-lasting episode of mountain-building, furnished great volumes of sediment that accumulated in swamps (that became coalbeds) and on coastal plains.

8. Final docking of terranes and a culminating orogeny late in Paleozoic time.

The last slices of suspect terranes making up the Appalachian Highlands arrived near the end of Paleozoic time. These younger terranes now lie outboard of the older terranes that docked earlier along the eastern edge of North America. Some younger terrane rocks lie under the Atlantic Coastal Plain while others crop out in the Piedmont.

As Europe and the northern part of North America were being fused together, the African plate moved toward what is now southeastern United States. The ensuing continental collision and orogeny—the Alleghenian orogeny—took place over a 50-million-year interval in Pennsylvanian and Permian time and finally destroyed the Proto-Atlantic Ocean. The force of the Alleghenian orogeny, which mainly affected the Southern Appalachians and the Ouachita-Marathon segment of the Appalachian belt, caused intense folding, severe metamorphism, and major thrusting. Included in the series of tectonic events was the northward movement of the South American plate against what is now southern United States. The creation of the supercontinent Pangea was the eventual result of the Paleozoic continental collisions and orogenies.

In the Southern Appalachians, thrust sheets are the dominant structural features. The ancient crystalline rocks of the Blue Ridge province were thrust westward over younger sedimentary rocks of the former continental margin. New data derived from deep seismic reflection profiling suggest that the thrusting (especially at the southern end of the Blue Ridge) was on a much greater scale than had been previously estimated. The thrust sheets may have been pushed over the continental margin in a series of pulses for as much as 150 miles (in some places), burying younger rocks thousands of feet below the surface!

9. Sedimentation, isostatic adjustments, and the break-up of Pangea in Mesozoic time.

At the beginning of the Mesozoic Era, the Southern Appalachians, which stretched across the interior of Pangea in a humid tropical zone south of the equator, probably looked more like the modern Himalayas than the worn-down summits that we see in Appalachia today. Making up these grand Pangean mountains was a great mass of rocks, estimated to be in excess of 30,000 to 50,000 feet thick—the result of millions of years of plate convergence, sedimentation, volcanism, and plutonism.

As the compressive force of the orogeny declined, the relentless processes of erosion pared down the mountain peaks. The mountain mass, made up of less dense continental crust (specific gravity around 2.5), "floated" upward as material was stripped off, so as to regain its isostatic equilibrium. These vertical adjustments caused extensional stresses in the crust, producing structural basins (down-faulted blocks) and block-faulted mountains. Shales and sandstones accumulated in the basins.

Pangea began to split up along a new rift east of the Appalachians. Small segments of crustal material that had formerly been part of European, African, and South American plates remained attached to what is now northeastern and southeastern United States. A spreading center opened in Jurassic time, and the Atlantic Ocean began to fill the widening basin between the drifting North American and European plates. (The South Atlantic Ocean opened between the African and South American plates.)

10. A quieter Cenozoic time.

Since Cretaceous time, sediments from the Appalachian Mountains have accumulated along the trailing edge of the North American continent, spreading over the Atlantic Coastal Plain and the continental shelf. Isostatic uplift of the Appalachians continues at a subdued rate. Fluvial erosion digs ever deeper, uncovering more of the ancient rocks and structures as they slowly rise.

The most intense activity of the Appalachian orogenic belt took place along the eastern side of the United States in the northern and middle segments. The northern segment, which reaches as far as Newfoundland in Canada, includes the coastal mountains in Acadia National Park (chapter 21). As noted earlier, Shenandoah and Great Smoky Mountains National Parks (chapters 47, 48) are in the middle segment. The two national parks in Florida, Everglades (chapter 17) and Biscayne (chapter 18), expose only very young rocks of the coastal plain and continental shelf, but older rocks are buried many feet below the surface. Deep drilling has revealed the presence of Paleozoic sedimentary rocks in northern Florida that appear to have been originally part of West Africa. To the southeast, probably extending beneath the Everglades, ancient crystalline rocks are buried. Geologists surmise that when Pangea broke up, Florida remained attached to the North American plate.

In the third segment, the Appalachian orogenic belt extends west beneath the Mississippi embayment and the Gulf Coastal Plain and rises again in the Ouachita Mountains of Arkansas and Oklahoma. Hot Springs National Park (chapter 45) is in the Ouachita Mountains, while Big Bend National Park (chapter 46) is at the extreme western end of this trend. The Marathon region of west Texas was also involved in the late Paleozoic orogenic events that produced the Ouachitas. The sedimentary accumulations that formed the rocks in Guadalupe National Park (chapter 15) and Carlsbad Caverns National Park (chapter 14) were largely controlled by the tectonic events of that time.

1800 feet of massive Precambrian lava flows (Catoctin Formation) that were originally basaltic but later metamorphosed to greenstones. These highly resistant greenstones cap the highest ridges and peaks of the park (Hawksbill Mountain, for example). In the Big Meadows area the top of the formation and overlying sedimentary rocks have been removed by erosion, producing the gently rolling topography of this part of the Blue Ridge crest.

Prominent cliffs of columnar-jointed lava, such as Franklin Cliffs and Crescent Rocks, can be seen from Skyline Drive (fig. 47.6). The columnar jointing in the Catoctin lavas indicates that they were erupted on land. Most of the waterfalls in the park cascade over cliffs and benches of resistant greenstone. Dark Hollow Falls, which drains much of the Big Meadows area, is an example.

Some of the Precambrian plutonic rocks that are older than the Catoctin Formation have been cut by greenstone feeder dikes that represent intrusions of molten basalt that did not reach the surface. Because the ancient granitic rocks are even more resistant than the greenstones, differential erosion has produced deep notches of varying widths where the dikes were emplaced. Along the Ridge Trail on Old Rag Mountain, a flight of natural stairs has been weathered out between vertical granite walls. The stairs on the dike surface weathered from columnar joint blocks.

The Old Rag Granite, exposed at the summit of Old Rag Mountain and elsewhere in the park, is easily recognized by its very light color and coarse grain. Because this metamorphosed granitic bedrock is very resistant, weathering is mostly along the joints (fractures) where water and expanding ice can attack the rock. Between the joints, large spheroidal boulders (some as much as 10 feet in diameter) have weathered in place (fig. 47.7).

The resistance of the gneissic Pedlar Formation relates to the character of the bedrock exposures; that is, whether they are massive or strongly foliated. Excellent outcrops of strongly foliated Pedlar gneisses can be seen along Skyline Drive, especially at the Tunnel Parking and Buck Hollow Overlooks. The Pedlar and Old Rag rock units are the oldest rocks in Shenandoah National Park.

In general, the ancient greenstones and granites are the rocks that form the crest and gentler eastern slopes of the Virginia Blue Ridge. These areas were favored by the Indians and later by the settlers. On the western side are the Cambrian clastic rocks of the Chilhowee Group that have been metamorphosed to quartzites and phyllites. Steep, sharp ridges, V-shaped hollows, and slopes covered by coarse rock debris characterize the topography formed on these rocks (fig. 47.8).

Figure 47.6   Columnar jointing in lava of the Catoctin Formation is displayed at Crescent Rock, beside Skyline Drive. National Park Service photo.

Figure 47.7   This huge block of Old Rag Granite, perched on the crest of Old Rag Mountain, was weathered in place from highly resistant, massive, granitic bedrock. Over time, frost action and spheroidal weathering, attacking the rock along joint planes, produced the unusual feature. National Park Service photo.

Figure 47.8   Erosional processes gave this outcrop of Hampton Formation phyllite (Chilhowee Group) the shape of a crude kiln. The feature is near Bull Run Overlook in the southern part of the park. National Park Service photo.

The Cambrian rocks vary in resistance, with the quartzites, like those in the Weverton Formation, being more resistant to erosion. The thin, sandy, and shaly soils that developed on the rocks support forests, but the trees tend to be less diverse, smaller, and less luxuriant than those that grow in more fertile soil on the plutonic and volcanic rocks. Part of the difference may be due to lower moisture content in the soils weathered from the sedimentary rocks.

Cleavage and Joints

Except for the Triassic diabase dikes (emplaced long after the Paleozoic orogenies), the ancient bedrock in the park is cut by thousands of cleavage planes, which are the dominant and most obvious feature on outcrops. *Rock cleavage,* which is the tendency of rocks to split along closely spaced, parallel planes, is independent of bedding and is the result of regional metamorphism and deformation. (Rock cleavage is not the same as *mineral cleavage,* which is a physical property of certain minerals that break along planes related to molecular structure and crystal faces.) In the Shenandoah region, metamorphism and compressive deformation were so severe that cleavage

completely obscures the original bedding of some fine-grained metasedimentary rocks. Cleavage is pervasive even in massive plutonic rocks and quartzites.

Joints (bedrock fractures without displacement) are conspicuous in the Blue Ridge, with most outcrops displaying fracture trends (joint sets) in several directions. Joint planes, which are ordinarily less closely spaced than cleavage planes, are caused by stress due to deformation and also by pressure release or unloading.

Weathering operates on both cleavage and joints, breaking up the bedrock into boulders, blocks, and rubble. Talus slopes tend to reflect the particular characteristics of the cleavage and joint systems of the bedrock source. White angular blocks, for example, accumulate below ledges of massive, resistant quartzite. From a distance, the talus slopes look like fresh rockslides because vegetation cannot grow over them. Soluble minerals available for soil enrichment that may have been present in the original sandstone have long since been removed, and the quartzite is extremely resistant to chemical weathering. Talus slopes of basaltic blocks, on the other hand, are difficult to see from a distance because trees have grown over the rocks in soil developed by weathering. Sizes of granite blocks in talus slopes range from small to enormous, depending upon

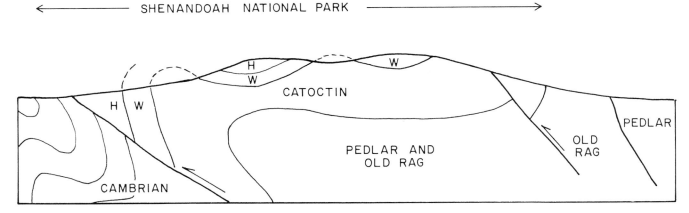

Figure 47.9 Simplified geologic cross section of the northern Blue Ridge in the area of Shenandoah National Park. Complex folding and thrust faulting are the result of intense tectonic compression that engulfed the region during much of the Paleozoic Era. Key: W=Weverton Formation; H=Hampton Formation. Both are Cambrian in age. (Adapted from Gathright, 1976.)

the fractures in the bedrock source. Below ledges and cliffs of layered rocks, the talus is usually made up of many small blocks of near-uniform size because cleavage planes, bedding, and joints are normally closely spaced.

Folds and Faults

Large and small asymmetric folds are visible on some Chilhowee outcrops, especially in the southern section of the park. Folds in light-colored, resistant beds of quartzite are easy to trace along hillside slopes. Most folds trend northeast–southwest, parallel to the long axis of the Blue Ridge.

Faults have strongly influenced landform development in Shenandoah National Park. The low- to high-angle reverse and thrust faults, which dip generally to the southeast, are the most important structural features in the park, although fault traces are apt to be obscured by talus, gravel, or alluvium. Displacements produced by faulting moved the rock mass of the Blue Ridge over younger Paleozoic rocks (fig. 47.9).

Also significant are high-angle transverse faults that trend northwestward across the Blue Ridge. Movement was mainly vertical, generally less than 500 feet. Uneven movement on the two types of faults produced uneven summits on the Blue Ridge and an irregular lobe and embayment pattern of rock distribution along the Shenandoah Valley's eastern edge.

In the massive crystalline rocks, linear shear zones indicate the presence of faults. More spectacular are the zones of tectonic breccia that mark reverse faults in the Erwin quartzites. Red iron oxide cement has filled voids in the fault zone, binding the brecciated quartzite into very resistant rock.

Water Gaps and Wind Gaps

A *water gap*—like the James River gorge, south of Shenandoah National Park—is a deep pass in a mountain ridge through which a stream flows. A *wind gap* is a former water gap that has been abandoned by the stream that eroded it. Some of the wind gaps in Shenandoah National Park that were carved into the resistant rocks of the Blue Ridge by rivers of an earlier time are more than a thousand feet deep. Tranverse valleys of this type are called wind gaps because strong winds blow through them. Their shape tends to funnel the summit winds. Typical examples are Compton Gap, Thornton Gap, Swift Run Gap, Powell Gap, Jarman Gap, and Rockfish Gap.

What happened to the streams that eroded the wind gaps? And why, for such a long time, did the rivers flow *across* the Blue Ridge, a formidable topographic and structural barrier, instead of going around it? A plausible explanation for evolution of drainage on the northern Blue Ridge might be as follows:

1. Streams flowed southeastward to the Atlantic Ocean from a drainage divide in the Appalachian Plateau province, which lay to the west. As the streams crossed the Blue Ridge, which was being uplifted, their course and direction were controlled by the regional slope of the land surface.
2. As streams cut down, erosion removed a considerable volume of rock material, creating water gaps that were transverse to the trend of the Blue Ridge structures. In this way, the original dendritic stream pattern was *superimposed* on structures and rock units.
3. Because of lithologic and structural variations in the bedrock, some streams had an advantage in downcutting and maintaining their courses. The Potomac

Table 47.1 Geologic Column, Shenandoah National Park

Time Units			Rock Units		Geologic Events
Eras	Periods	Epochs	Group	Formation	
Cenozoic	Quaternary	Holocene Pleistocene	Alluvium Periglacial talus		Stream erosion, mass wasting Intensive frost action and stream erosion
	Tertiary				Stream erosion, evolution of modern drainage Isostatic upwarp
Mesozoic	Cretaceous Jurassic Triassic		Diabase dikes		Uplift and erosion Tensional rifting and block faulting, intrusion of diabase
Paleozoic	Permian Pennsylvanian				Alleghenian orogeny, thrust faulting, folding, metamorphism, uplift and erosion
	Mississippian ––––– Devonian Silurian –––––				Acadian orogeny, uplift and erosion
	Ordovician				Arrival of suspect terranes Taconic orogeny, uplift and erosion Plate rifting—opening of Proto-Atlantic Ocean, sedimentation
	Cambrian		Chilhowee Group	Erwin (Antietam)	Quartzites (with Skolithos tubes), interbedded shales
				Hampton (Harpers)	Shales, siltstones, interbedded quartzites
				Weverton	Conglomeratic quartzites, shales
Proterozoic Z (Late Precambrian)			Catoctin		Erosion –––––––––– Dark greenstones (originally basaltic lavas, dike intrusions, ash, and tuff)
			Swift Run		Volcano-clastic deposits
Proterozoic Y			Old Rag Granite		Orogeny and erosion –––––––– Metamorphosed light-gray plutonic rock
			Pedlar		Metamorphosed greenish gray gneisses

Sources: Gathright, 1976; Reed, 1969; Rodgers, 1987.

River, for example, found a course through the Blue Ridge, north of the park, where the resistant layers of Catoctin lava thin to less than 50 feet. Favored streams and their tributaries were able to establish lower local base levels and erode headward faster than the streams without such advantages.

4. The Shenandoah River, a tributary of the Potomac, cut headward up (south) the Shenandoah Valley, which is underlain by limestone, much more erodible in a humid climate than the crystalline rocks of the Blue Ridge. With its development aided by the underlying bedrock structure—a belt of nonresistant limestone—the Shenandoah River was able to *capture* the headwaters of streams flowing across the Blue

Ridge between the Potomac and the James Rivers. The water gaps then became wind gaps, usually with two small streams flowing away from the crest of the gaps, one to the east and the other to the west. By the capture of the Blue Ridge tributaries, the drainage in the Shenandoah Valley became *adjusted* to the structure. In other words, *adjustment of drainage* had taken place.

5. Differential resistance of bedrock and structures continues to control the development of the topography. During isostatic upwarping in Cenozoic time, streams have maintained their courses by downcutting and adjusting to rocks and structures.

Geologic History

1. The oldest Precambrian rocks in Shenandoah National Park (Proterozoic Y).

Radiometric dating procedures have yielded an approximate age of 1100 million years for the crystallization and metamorphism of the Pedlar Formation and the Old Rag Granite, which make up the resistant core of the Blue Ridge. The intensely foliated gneisses of the Pedlar Formation were probably intruded by the Old Rag Granite. Both were metamorphosed at depth under conditions of high temperatures and pressures, presumably during Precambrian orogenies before the ancient rocks were joined to the North American plate (box 47.1).

2. Deposition of the Swift Run Formation in latest Precambrian time (Proterozoic Z).

A discontinuous sequence of conglomerates, sandstones, and pyroclastics was deposited by streams eroding Precambrian highlands and building up valley floors. An orogeny accompanying plate convergence and collision elevated the highlands and initiated the sedimentation.

3. Eruption of the Catoctin Formation lavas.

Plateau basalts, pouring out over the ancient landscape, filled the valleys and eventually covered the eroded mountains. The rifting that allowed the basaltic magma to well up preceded the initial opening of the Proto-Atlantic Ocean early in Cambrian time. Ash falls interlayered with the lava flows indicate that some of the eruptions were explosive. Stream-deposited clastic materials are interbedded with the ash and lava. The uppermost layer in the volcanic sequence is a purplish slate (representing an extensive ash fall) that marks the boundary between the Catoctin Formation and the overlying Cambrian beds.

4. Deposition of the Chilhowee Group, Cambrian marine sedimentary rocks.

Great quantities of clastic (detrital) sediments were washed down from the highlands and deposited on a shallow continental shelf along North America's eastern margin. The Weverton Formation, which lies unconformably on Catoctin rocks and, in some places, on the Pedlar

Figure 47.10 The "face" of Stony Man, seen here from Stony Mountain Overlook on Skyline Drive, is eroded from lavas of the Catoctin Formation. National Park Service photo.

Formation, is made up of 100 to 500 feet of conglomerates, sandstones, and shales metamorphosed to quartzites and phyllites.

The earliest traces of life forms appear in the quartzites of the Hampton Formation. In these beds are tracks, trails, and *Skolithos* tubes, or burrows, left by a worm-like, marine organism. The Hampton Formation is composed of 1800 to 2200 feet of clastic rocks—thin-bedded sandstones and siltstones (now quartzites and phyllites). In the overlying Erwin Formation, *Skolithos* tubes are more abundant and varied, especially in the quartzites. The resistant, massive beds of Erwin quartzite form sharp peaks, flatirons, and ridges along the western flanks of the Blue Ridge (fig. 47.11). Most of the manganese and iron deposits in the area have been found in the topmost Erwin beds.

5. Paleozoic orogenies and docking of tectonostratigraphic terranes.

A long series of tectonic episodes, beginning with the Taconic orogeny in Ordovician time and culminating in Permian time with the Alleghenian orogeny, greatly altered rocks and structures of the Blue Ridge (box 47.1). During the Taconic orogeny, suspect terranes moved in from the southeast, oceanic crust was subducted under the continental margin, and the Proto-Atlantic Ocean was compressed and partly closed. The Devonian Acadian orogeny had less effect on the Blue Ridge since its severity was greatest in the Northern Appalachians. Suspect terranes from the Afro-European plate that were accreted at the time of the Alleghenian orogeny were crumpled against the Blue Ridge and some were thrust over younger rocks that had accumulated along the continental margin. Folds that are oversteepened to the northwest, thrust faults pushed from southeast to northwest, and the cleavage and joint systems oriented to compressive stresses from the southeast were all produced by repeated convergence and collision of tectonic plates.

6. Intrusion of Triassic dikes, erosion, and uplift.

Long, slow uplift during the Mesozoic Era was accompanied by tensional faulting. Active erosion removed great volumes of rock from the uplands. The largely vertical crustal movements were the result of isostatic adjustments that followed the mountain-building of the last Paleozoic orogeny. The diabase dikes that rose through fractures in the Blue Ridge bedrock in Triassic time are associated with block-faulting in the mountain belt. Only a few of the dikes are exposed in Shenandoah National Park, most of them in the central and southern sections. McCormick Gap, near the southern entrance, is eroded along the northwest–southeast trend of an extensive diabase dike. The dike rock is less resistant than the Catoctin lavas that it intruded.

Figure 47.11   Chimney Rock, an outcrop of light-colored, resistant quartzite of the Erwin Formation (Chilhowee Group), is a prominent landmark along a trail near the park's southern boundary. Weathering processes, such as "root pry," plus mass wasting, break up the rock along fracture planes. Tree roots enlarge cracks, allowing plant acids to penetrate ever deeper into the rock. National Park Service photo.

7. Cenozoic upwarping and development of the modern landscape.

The upwarping of the Appalachian region in Cenozoic time (which continues at a very slow rate) was due to isostatic compensation of the crust caused by the gradual widening of the Atlantic Ocean.

During Pleistocene time, erosion intensified because of increased frost action and precipitation. The wide alluvial fans along the foothills, sloping westward to the Shenandoah Valley, were built up in the Pleistocene. The fans are now being eroded by the same streams that deposited them. The talus slopes described earlier are also mainly the result of intensified Pleistocene weathering.

Bibliography

Ben-Avraham, Zvi. 1981. The movement of continents, *American scientist* 69(3): 291–299, May–June.

Bird, J. M. and Dewey, J. F. 1970. Lithosphere plate-continental margin tectonics and the evolution of the Appalachian orogen. *Geological Society of America Bulletin* 81:1031–1060.

Bobyarchich, A. R. 1981. The Eastern Piedmont fault system and its relationship to Alleghenian tectonics in the Southern Appalachians, *Journal of geology* 89(3): 335–345, May.

Crandall, H. 1975. *Shenandoah, the story behind the scenery.* Las Vegas, Nevada: KC Publications.

Espenshade, G. H. 1970. Geology of the northern part of the Blue Ridge anticlinorium. In *Appalachian geology,* ed. G. W. Fisher et al., p. 199–211. New York: Interscience Publishers.

Frye, K. 1986. *Roadside geology of Virginia.* Missoula, Montana: Mountain Press. p. 97–116.

Gathright, T. M., II. 1976. *Geology of the Shenandoah National Park, Virginia* (geologic maps in pocket). Virginia Division of Mineral Resources Bulletin 86.

Heatwole, H. T. 1981, 2nd edition. *Guide to Skyline Drive and Shenandoah National Park.* Shenandoah Natural History Association 226 p.

Oliver, J. 1980. Exploring the basement of the North American continent, *American scientist* 68(6): 676–683, November–December.

Pique, Alain. 1981. Northwestern Africa and the Avalonian plate relations during late Precambrian and late Paleozoic time, *Geology* 9(7): 319–321.

Reed, J. C., Jr. 1955. Catoctin Formation near Luray, Virginia. *Geological Society of America Bulletin* 66: 871–896.

———. 1969. Ancient lavas in Shenandoah National Park near Luray, Virginia (includes geologic map). U.S. Geological Survey Bulletin 1265.

Rodgers, J. 1970. *The tectonics of the Appalachians.* New York: John Wiley & Sons, Inc.

———. 1987. The Appalachian-Ouachita orogenic belt. *Episodes* 10(4):259–270, December.

Williams, H. and Hatcher, R. P. 1983. Appalachian suspect terranes. In *Contributions to tectonics and geophysics of mountain chains.* Geological Society of America Memoir 158:33–53.

Shenandoah National Park

Area: 195,072 acres; 305 square miles
Wilderness area: 79,579 acres
Authorized: March 2, 1926
Established: December 26, 1935
Wilderness designated: October 20, 1976;
*   September 1, 1978*
Address: Route 4, Box 292, Luray, Virginia 22835

48

Loftiest range east of the Black Hills and one of the oldest uplands on earth.
Intricately folded, metamorphosed Precambrian bedrock forms rounded, resistant summits.
Major faults join ancient suspect terranes to the North American continent.
Year-round moist air from the Gulf of Mexico waters dense forests and produces smoky haze, for which the range is named.

Great Smoky Mountains National Park

Tennessee-North Carolina Border Region

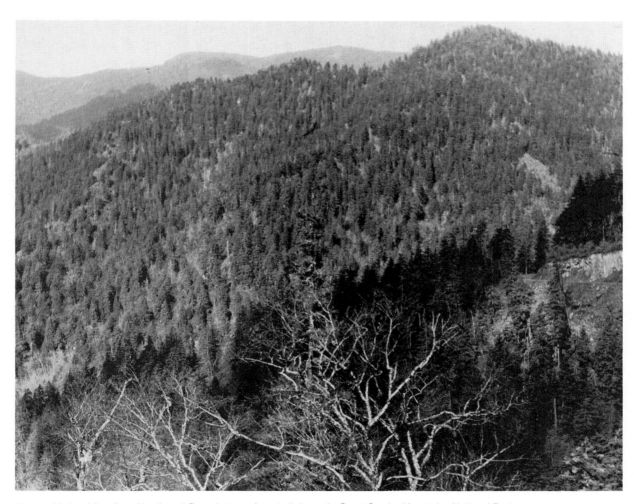

Figure 48.1   View from Newfound Gap of steep, forested slopes in Great Smoky Mountains National Park. Mount LeConte is in the far distance. National Park Service photo.

GREAT SMOKY MOUNTAINS

NATIONAL PARK

Figure 48.2 Great Smoky Mountains National Park, North Carolina and Tennessee.

Geographic Setting

Great Smoky Mountains National Park, which straddles for some 60 miles the boundary between Tennessee and North Carolina, is in the highest part of the Blue Ridge and the entire Appalachian Mountains belt. Only Mount Mitchell (6684 feet in elevation) in the Black Mountains northeast of the park is higher than the major peaks of the Great Smokies. Clingmans Dome, highest point in Great Smoky Mountains National Park, is 6642 feet in elevation; the second highest peak, Mount Guyot, is 6621 feet. More than 20 other summits in the Great Smokies are over 6000 feet in height. The two highest mountains in the park were named for T. L. Clingman and Arnold Guyot, the geologists who made the first scientific reconnaissance of the range in the late 1860s.

The crest of the southern Blue Ridge in Great Smoky Mountains National Park is more than 2000 feet higher than the Blue Ridge crest in Shenandoah National Park (chapter 47), some 400 miles to the north. In Shenandoah, the spine of the Blue Ridge is relatively narrow, with spurs jutting out on either side; while in Tennessee, North Carolina, and Georgia—at the southern end of the Blue Ridge physiographic province—the range becomes a sprawling complex of ridges, lacking the distinctive lineation of the mountains to the north. What is locally called "the Blue Ridge" in North Carolina is a prominent frontal scarp that rises from the Piedmont along the eastern edge of the Blue Ridge province.

No rivers cross the Blue Ridge upland south of the Roanoke River in Virginia. The drainage divide follows the crest of the Blue Ridge province. Streams on the eastern slopes and the Piedmont drain to the Atlantic. Most of the water from the southern Blue Ridge flows west or northwest and then south to the Gulf of Mexico.

The southern highlands are more humid because warm, wet winds, sweeping northeastward from the Gulf of Mexico, discharge quantities of moisture on the mountain slopes. Clingmans Dome and the other high summits in the Great Smokies usually receive more than 85 inches of precipitation annually, making these mountains almost as wet as those of the Pacific Northwest. Elevation determines temperature as well as precipitation. Average summer temperatures on Clingmans Dome range from 51° F to 64° F. By contrast, at Park Headquarters in Gatlinburg, Tennessee, elevation 1500 feet, summer temperatures average 59° F to 87° F. Gatlinburg and the valleys in the southern Blue Ridge receive 50 or more inches of rainfall per year.

An abundance of moisture and a thick mantle of residual soil in valleys and foothills has supported for millions of years a luxuriant and diversified plant life, with many species attaining extraordinary size. Some tree and plant species that became established in the region during the colder, wetter climates of the Pleistocene have continued to thrive in the mountain environment.

Droplets of mist in the air, vapors (bearing resins and plant oils) rising from the lush vegetation, especially in spring and fall, and wisps of fog cause the natural "smoke" that gave the Great Smokies their name. The gently swirling rise and fall of the haze under varying conditions of temperature and light produces an effect of bluish gray smoke around the summits and in the valleys (fig. 48.3).

Because of its unique ecological resources, Great Smoky Mountains National Park, designated a Biosphere Reserve in 1976, is being carefully monitored to sense the impact of pollution. An ecological problem causing concern at the present time is the increasing damage to the *biota* (the region's plant and animal life) due to acid rain, resulting from air pollution. Trees at higher elevations are especially vulnerable, and a large number are dying or have died. Increased emissions of sulfur dioxide particles from coal- and oil-burning power plants are the main cause of acid rain. The particles are carried into the park region in droplets of water. Unfortunately, haze resulting from pollution has begun to outweigh and mask the natural haze that has characterized the Great Smokies atmosphere. Since the 1950s, visibility in the mountains has declined 30 percent.

Great Smoky Mountains National Park is visited by more than 10 million persons annually, partly because a major highway (US 441) crosses the mountains through the park. Here, also, is the southern terminus of the scenic Blue Ridge Parkway (chapter 47). Yet despite the fact that this park has more visitors than any other in the National Park system, about 75 percent of its 800 square miles is kept in a wilderness state. Although two-thirds of the total area had been logged over or burned before coming under national park protection, establishment of Great Smoky National Park saved the largest remaining virgin forest in eastern United States.

Local History

The Cherokee Indians, primarily an agricultural people, had developed a settled and advanced culture in the Southern Appalachians, long before the coming of Europeans. The Great Smoky Mountains were the sacred ancestral home of the Cherokee nation. Their early contact with white settlers was friendly, although the Indians fought when they were provoked. After a war with Carolina settlers ended in the 1760s, the Cherokee withdrew to their lands in the mountains and foothills of the Blue Ridge. Their way of life ended after gold was discovered on Cherokee land. In 1838, the Cherokee were forcibly deported from their ancient homeland and escorted by U.S. Army troops to Oklahoma Territory. Thousands died from

Figure 48.3 "Smoke" (haze) hangs over valleys and wreathes summits in the Great Smokies when certain atmospheric conditions prevail. This view, from Cave Creek Gap, looks south over North Carolina. National Park Service photo.

starvation and exposure during the march, called the "Trail of Tears." Some Cherokee escaped deportation by hiding in the mountains. Eventually the federal government permitted Cherokee who wished to do so to return from Oklahoma. About 4500 descendants of the Cherokee now live in the Qualla Indian Reservation in North Carolina on land adjoining Great Smoky Mountains National Park.

In the late eighteenth and early nineteenth centuries, white settlers, mainly emigrants from England, Scotland, and Ireland, began to occupy Cherokee land (legally and illegally) in the foothills and valleys on both the Tennessee and North Carolina sides of the Great Smokies. After the fertile bottom lands in the valleys were taken, families sought farm sites higher on the slopes. A fortunate few found hidden "coves" and "balds." Cove is the local term for an open, flat, or gently sloping, valley underlain by limestone bedrock and protected from wind by high ridges. (Their geologic origin is discussed later in the chapter.) The balds are treeless openings in the forest on ridges at higher altitudes. Some are grassy, but most are covered by a nearly impenetrable maze of closely intertwined shrubs, like rhododendron and mountain laurel. The balds

may be long-weathered scars of old mudslides or tracts that Indians burned over to encourage growth of berry bushes.

Cades Cove, first settled in 1818, is maintained as a historical and cultural preserve by the National Park Service. In this isolated valley, deep in the Great Smoky Mountains, descendants of early pioneers work their farms, keeping up the traditions of their forebears. Exhibits, museums, a blacksmith shop, and a working grist mill tell the story of life as it was in the southern highlands for well over a century—self-sufficient, independent, and shut off from the outside world. The overshot water wheel that powers the grist mill is an example of the dozens of small water wheels that settlers built on mountain creeks for grinding grain. With high gradients and high velocities, the streams were a reliable and convenient energy source.

In the early 1900s, large lumber companies acquired vast tracts of land in the Great Smoky Mountains. Railroads and towns were built and sawmills set up. Forests centuries old were leveled, and mountain streams were used as sluiceways for logs. Splash dams were constructed on creeks so that the water could be released in a flood that washed logs downstream. The Sinks, a series of cascades on the Little River, carried logs by the thousands from the slopes of Clingmans Dome.

The economic boon was short-lived. Logging, as practiced at that time, was the most destructive form of land use the region was ever subjected to. The skidding of logs downslope, as forest cover was removed, caused rapid erosion. Landslides occurred with every heavy rain. Streams were so clogged with sediment that they were unable to sustain aquatic life. Disastrous forest fires, spreading from heaps of dry branches, roared up the slopes. Charlies Bunion, a bare slate knob on the crest of the range, was completely denuded of soil and vegetation by a fire and subsequent flash flood in the 1920s. Only a few hardy plants clinging to crevices in the jagged bedrock have been able to grow back.

People in Tennessee and North Carolina who deplored the destruction of the forests and the pollution of the streams began to organize a conservation movement for the purpose of establishing a national park in the Smokies. Other Americans, realizing that the mountains were a botanical treasure house, joined in the effort. After Congress authorized the park in 1926, the difficult task of raising money to buy land from thousands of individual owners began. Citizen groups gathered contributions, and the Tennessee and North Carolina state legislatures each appropriated two million dollars. John D. Rockefeller, Jr.'s donation of five million dollars, plus a million and a half from the federal government, finally brought the undertaking to successful completion. Congress established Great Smoky Mountains National Park for full development in 1934, and it was dedicated in 1940.

Geologic Features

"*What* geology?" the ranger naturalist said with a chuckle to an enquiring tourist. "We don't have any geology in this park!" What he meant is that the bedrock geology in Great Smoky Mountains National Park is not obvious. Outcrops are scanty, and geologic structures are obscured by the forest cover. These difficulties, field geologists agree, are "compounded by the enigmatic nature of the rocks themselves" (King et al. 1968, p. 1).

Some geomorphic aspects of the mountain landscape can be related to geologic processes and climatic factors. Summits and ridges in the park are more rounded than sharp. The outward appearance of these magnificent and imposing mountains is the result of millions of years of weathering and erosion in a humid, predominantly warm to temperate climate.

Glaciers did not sculpture the Southern Appalachian skyline. The climatic fluctuations of Pleistocene time mainly affected the plant and animal life of the region, but intense frost action attacked the mountain slopes and rocks

when great ice sheets were advancing over what is now northern United States. Although they are fairly stable and masked by tree growth, the bouldery talus slopes in the steep draws and gullies at higher altitudes probably developed during the Pleistocene. At lower altitudes, patches of deeply weathered soils may reflect the increased warmth and humidity of interglacial episodes.

The results of mass wasting processes are evident, especially on steep, high slopes where soil cover on bedrock is thin and fragile, or regrowth, since lumbering ceased, has not become well established. In September 1951, when a thunderstorm dumped four inches of rain in one hour on Mount LeConte, more than 40 slides and avalanches, carrying rocks, trees, and debris, roared down the mountainside. Landslide scars are still clearly visible from Newfound Gap Road (fig. 48.4).

Throughout the park, large and small streams, fed by high rainfall, race down slopes through stone-filled channels and plunge over waterfalls. Scenic highways, such as Little River and Laurel Creek Roads, follow winding streams along valleys at lower elevations. Hiking trails, from easy to strenuous, lead to waterfalls that plunge over bedrock ledges (fig. 48.5).

Bedrock and Geologic Structures

The rocks and structures in Great Smoky Mountains National Park are related to those in Shenandoah National Park (chapter 47). In general, the large-scale tectonic events outlined in box 47.1, "The Evolution of the Appalachian Highlands," apply to both parks. Both regions were profoundly affected by geologic conditions that acted on the Appalachian Mountains belt. Tectonostratigraphic (suspect) terranes, which may have originated on an early Afro-European continental plate, make up the great bulk of the rocks in Shenandoah and Great Smoky Mountains National Parks. The terrane units were lodged against the North American continent in Paleozoic time during a succession of plate collisions and orogenies. A significant difference between the two areas is that deformation was more severe and the extent of thrust faulting was considerably greater in the southern section of the Blue Ridge province than in the northern part.

In gross terms, three groups of rocks make up the Great Smokies and surrounding foothills. In chronological and roughly geographical order from southeast to northwest across the mountains, the groups are as follows:

1. The basement complex consists of rocks, middle Precambrian in age, that have been deeply buried, have undergone repeated episodes of metamorphism, and have been uparched, uplifted, and exposed by erosion.
2. The Ocoee Supergroup (also called Ocoee Series) is a thick sequence of metamorphosed sedimentary rocks of late Precambrian age that form the main bulk of the Great Smokies.

Figure 48.4   Landslide scars on Mount LeConte (ridge in background) are the result of slides that stripped trees and soil from the steep slope, exposing bedrock. Four inches of rain, falling in less than an hour, in September, 1951, triggered the slides. View is from Chimney Tops, looking northwest over Anakeesta Ridge (foreground). National Park Service photo.

3. The Chilhowee Group and other early Paleozoic rock units crop out on lower ridges, in the coves, and in the Appalachian Valley, mostly outside the park.

The Basement Complex

In the southeastern foothills on the North Carolina side of Great Smoky Mountains National Park are exposures of ancient, deformed crystalline rocks. Deep Creek Campground and the Oconaluftee Visitor Center in the southeast corner of the park are underlain by Precambrian basement rocks. Some of the layered gneisses and schists were probably formed from sand, silt, and volcanic sediments. The granitic gneisses may originally have been plutonic rocks. The origin of the basement rocks is obscure; they may be part of the North American craton (Rodgers 1987).

The Ocoee Supergroup

Great thicknesses of Ocoee rocks make up most of the mountain mass in the park and extend for many miles along the mountain trend, both northeast and southwest of the Great Smokies. These are the rocks of the great thrust sheets that were pushed northwestward over younger rocks by a series of plate collisions in Paleozoic time. Ocoee rocks, which were to some degree metamorphosed and deformed before reaching North America, are part of a large suspect terrane (called the Piedmont terrane, but not the equivalent of the Piedmont province) that was subjected to the Taconic (Ordovician) orogeny and subsequent Paleozoic orogenies.

Included in the Ocoee Supergroup are the Snowbird Group, which lies unconformably over the basement complex; the Great Smoky Group, by far the thickest and most massive; some unclassified sandstone formations; and the Walden Creek Group, which crops out mostly in the

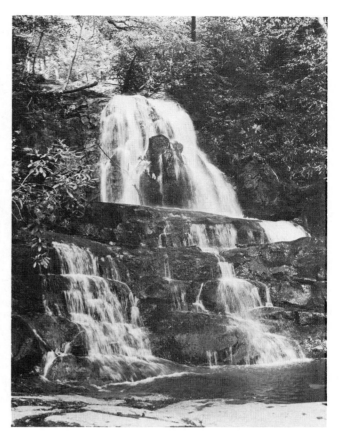

Figure 48.5 Waterfalls tumble over resistant ledges of Precambrian Thunderhead Sandstone. *Left:* Laurel Falls, accessible by trail from Little River Road. *Right:* Rainbow Falls on LeConte Creek, reached by trail from Cherokee Orchard Road. The blocks in the foreground broke off from the ledge along fracture planes. National Park Service photos.

foothills on the western and northern sides of the Great Smokies. All are nonfossiliferous, clastic marine sedimentary rocks that have been metamorphosed in varying degrees. Walden Creek rocks contain shaly and silty clastics and lenses of conglomerate, but the other Ocoee rock units consist mainly of many layers of fine- to coarse-grained sandstones.

The most prominent Ocoee units are the formations of the Great Smoky Group, which are exposed on the summits, in numerous roadcuts along the transmountain highway (US 441), and in waterfalls. The fine-grained Elkmont Sandstone is at the base of the Great Smoky Group. The coarse-grained Thunderhead Sandstone, thick-bedded and resistant, forms ledges and cliffs and the high summits of Clingmans Dome and Mount Guyot. Overlying the Thunderhead Sandstone on Mount LeConte, Chimney Tops, Sawteeth, and other parts of the crest, are the dark, slaty rocks of the Anakeesta Forma-

tion. Craggy pinnacles and steep-sided ridges, eroded from Anakeesta rocks, form prominent landmarks along the Appalachian Trail.

The massive rock units of the Great Smoky Group were initially remarkably homogeneous. Variations in their lithologies are due to the several metamorphic facies that reflect differing degrees of heat and pressure when the rocks were undergoing progressive regional metamorphism. The more intensely metamorphosed rocks (amphibolite facies) are in the southeastern part of the mountains, while rocks of the same formations to the northwest show decreasing grades of metamorphism (greenschist facies). The chemical and mineralogical variations in the rocks of the thrust sheet indicate that on the southeast side metamorphism was probably more prolonged and increasingly intense.

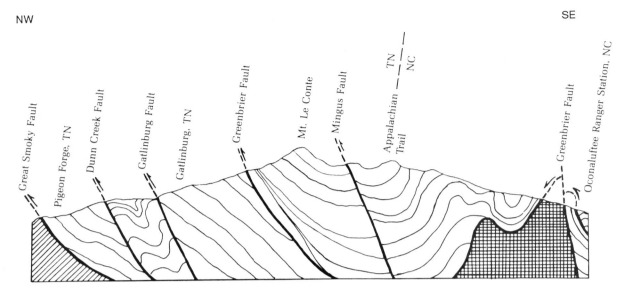

Figure 48.6   Northwest-southeast cross section through Great Smoky Mountains National Park. Rocks beneath the Great Smoky fault (*far left*) are Paleozoic. Rocks between Great Smoky fault and Greenbrier fault are Precambrian (Ocoee Supergroup). Rocks beneath the Greenbrier fault (*lower right*) are Precambrian basement complex. The region has undergone intense compression, which is shown by the relative movement along the faults (directions indicated by arrows). From H. L. Moore, *A Roadside Guide to the Geology of the Great Smoky Mountains National Park.* © 1988 The University of Tennessee Press. Used by permission.

The Paleozoic Rocks

A few beds of the Cambrian Chilhowee Group are exposed in roadcuts along the Foothills Parkway that skirts the western edge of Great Smoky Mountains National Park. Primitive marine fossils (trilobites, brachiopods, etc.) have been found in these rocks. Lower Ordovician limestones of the Knox Group underlie the coves nestled among the western foothills.

Igneous Rocks

A geologic map of the Great Smokies region shows extensive sequences of the dominant sedimentary and metamorphic rocks, but only a very few minor bodies of igneous rocks. (Volcanic and plutonic components in the basement complex rocks have been too greatly altered to be identifiable bodies.) Narrow sills of metamorphosed diorite, Paleozoic in age, crop out between beds of the Thunderhead Sandstone near the top of Clingmans Dome. Since the sills extend southwestward down the mountain, they were probably the source of ore in old copper mines along Hazel Creek in the southern part of the park.

Folds and Faults

Folds in the layered rocks of Great Smoky Mountains National Park range from broad and open, with long stretches of uniform dip, to small and sharp. In the southeastern area, where metamorphism was more intense, folds of several ages intersect, producing complex relationships. The majority of folds align with the dominant regional trend of northeast–southwest. They are asymmetric in that the northwest limbs dip more steeply than the southeast limbs.

Three major thrust fault systems dominate the mountain structures of the region: the Greenbrier faults, the Great Smoky faults, and the Gatlinburg faults. All three fault systems trend generally northeast-southwest and are the result of strong compressive forces from the southeast that thrust late Precambrian Ocoee rocks as much as a hundred miles northwestward over younger rocks. The major fault systems initially may have been boundaries or sutures between suspect terranes. Their present complexity reflects their involvement in several episodes of crustal rifting and plate convergence throughout the Paleozoic Era.

Oldest of the three are the faults of the Greenbrier system. Prominent exposures of the faults are in the southeast boundary area of the park between the Ocon-aluftee section and the Cherokee Indian Reservation. Greenbrier faults also loop around the northeast corner of the park and extend westward, where they are broken up and obscured by younger faults. The low-angle Green-brier thrust faults carried rocks of the Great Smoky Group over the underlying rocks of the Snowbird Group, perhaps for a distance of 15 miles. Basement complex rocks in the southeast were also involved in Greenbrier faulting. In the Straight Fork area they overrode younger Snowbird rocks that are now exposed in a window, or fenster. (*Window*, when used as a tectonic term, refers to an eroded area of a thrust sheet that displays the rocks beneath the thrust sheet. The German term *fenster* is synonymous.)

Unlike the Greenbrier fault system, which is within the Great Smokies district and mainly local in scope, the Great Smoky fault system is only one segment of an extensive displacement involving Southern Appalachian structures from Virginia to Alabama. The Great Smoky fault system and associated faults, all low-angle thrusts, moved Precambrian Ocoee rocks in the Great Smokies area for many miles over Ordovician and occasionally younger Paleozoic rocks. Most exposures of the Great Smoky fault system are between the foothills and the Appalachian Valley, west and north of the park, but some are within the park, as, for example, fault segments along the coves. The main thrust sheet of the Great Smoky fault is split on the western side of the mountains into several fault blocks. The coves—the windows in the thrust sheet—are located in these breaks and openings. From the leading edge of the thrust fault to where the fault surface is exposed in windows is a distance of nine miles, but this represents only a small part of the overall movement of the thrust sheet. Movement along the Great Smoky fault system occurred after the Ocoee rocks had undergone regional metamorphism.

The Gatlinburg faults, believed to be younger than the Great Smoky faults, dip more steeply and cut both Great Smoky and Greenbrier faults in some places. Although most of the Gatlinburg fault system trends east-northeast and lies outside the park's western boundary, the Oconaluftee fault, which is related to the Gatlinburg system, branches off toward the southeast from the vicinity of Meigs Mountain, crosses the crest of the Great Smokies, and goes down the valley of the Oconaluftee River. Gatlinburg faults have had a controlling effect on stream drainages where streams have eroded shattered rock along the fault traces. Some of these faults cross from one stream valley to the next through notches in intervening ridges.

How the Coves Developed

Cades Cove, Crib Gap, and White Sink in the western part of the park are windows in the thrust sheet of Precambrian Ocoee rocks where younger Paleozoic rocks of the Knox Group (Ordovician) are exposed. Tuckaleechee and Wear Coves, just west of the park boundary, are windows of larger size. The coves have been described as miniature Appalachian Valleys because they are fertile, slightly rolling patches of lowland underlain by the same sedimentary, unmetamorphosed carbonate rocks as the long, narrow lowland stretches of farming country in eastern Tennessee's Appalachian Valley.

The coves developed among the western foothills of the Great Smokies where the thrust sheet becomes thinner and broken up along its leading edge. Because the limestones beneath the thrust sheet were much less resistant to erosion than the metamorphosed Ocoee rocks, breaks in the thrust sheet became enlarged. Runoff from plentiful rains was channeled into the openings as the mountains were uplifted. Quantities of sediment built alluvial fans and spread over the bottoms. This eventually produced the rounded and roughly oval coves surrounded by high ridges. A few very small, remote coves that were not cleared by settlers or lumbermen remain in their natural state. The rich soils of these protected, well watered coves support a unique community of plants, the "cove hardwoods" (yellow birches, basswoods, buckeyes, sugar maples, etc.), and a profusion of wild flowers.

Geologic History

1. The Precambrian basement complex.

The oldest isotopic dates derived from minerals in these rocks indicate that plutonic and metamorphic episodes occurred about 1.1 billion years ago, or Proterozoic Y time. A scattering of later dates in these rocks suggests that radioactive clocks were reset by recrystallization that took place during Paleozoic orogenies and metamorphism.

2. Deposition of the Ocoee Supergroup late in Precambrian time (Proterozoic Z).

For millions of years ancient oceans received the clastic sediments that make up Ocoee rocks. Mostly sand from a cratonic source, the sediments were deposited on broad continental shelves. Paleogeographers speculate that the ocean and its adjacent lands lay well south of the equator during that time. Although fossils have not been found in the rocks, changes in grain size and bedding are indications of changing environments of deposition over

Table 48.1 Geologic Column, Great Smoky Mountains National Park

Time Units			Rock Units		Geological Events
Eras	Periods	Epochs			
Cenozoic	Quaternary	Holocene Pleistocene	Alluvium Talus slopes		Stream erosion, mass wasting Increased frost action, weathering, stream erosion
	Tertiary				Erosion; establishment of modern drainage systems, sporadic uplift
Mesozoic					Rifting; initial opening of Atlantic Ocean, prolonged erosion. Uplift, block faulting
Paleozoic	Permian Pennsylvanian				Proto-Atlantic Ocean destroyed Alleghenian orogeny Thrust faulting Regional metamorphism
	Mississippian Devonian		(Outside park)		Deposition Ocean closing Acadian orogeny
					Erosion ————
	Silurian				Arrival of Ocoee Supergroup as suspect terrane
	Ordovician	(Middle)	(Outside park)		Taconic orogeny. Folding and faulting, sedimentation
		(Lower)	Knox Group		Uplift and erosion ———— Carbonates (exposed in coves)
	Cambrian (Lower)		Chilhowee Group		Plate rifting; opening of Proto-Atlantic Ocean, thick sedimentation on continental shelf
	—?——				Uplift and erosion ————
Proterozoic Z (Late Precambrian)			Ocoee Supergroup	Walden Creek Group Several formations Great Smoky Group Snowbird Group	Deformation, metamorphism (before arrival at North American plate) Thick sedimentary accumulations
					Uplift, erosion, orogeny
Proterozoic Y			Basement complex		Highly metamorphosed and deformed igneous and sedimentary rocks

Source: King, Neuman, and Hadley, 1968; Rodgers, 1987

time; i.e., varying depths of water, ocean currents, types of sediment (pebbly, sandy, muddy), and the like. Approximately 50,000 feet of sediment accumulated in the Ocoee Supergroup. Of this thickness, the Snowbird Group made up about 13,000 feet; the Great Smoky Group, 25,000 (in some places); unclassified formations, 3000 to 4000; and the Walden Creek Group, 8000. The sedimentary processes by which Ocoee units accumulated were under tectonic control. Ultimately, tectonic pulses brought about uplifting, deformation, and metamorphism of the rocks.

3. Crustal rifting.

Presumably, the continental plate bearing Ocoee rock units drifted for millions of years, possibly northwestward, as sediments accumulated along its margin. Perhaps the plate wandered above a zone deep in the crust from which heat flowed and magma rose. The crustal rifting that ensued led eventually to the opening of the Proto-Atlantic (Iapetus) Ocean, which may have been, at maximum extent, 1000 miles long and several hundred miles wide. In this northeast–southwest-trending ocean, a microcontinent made up largely of Ocoee rocks drifted toward North America.

Figure 48.7   Schematic representation of geologic structures, northeast section of the park and vicinity. Here the Great Smoky fault, once nearly horizontal, has been folded. Rocks on top of the fault have moved from right to left. Complex faulting and folding produced a syncline under English Mountain and an anticline under the valley between English Mountain and Green Mountain. The Wilhite Formation under Green Mountain is part of the Walden Creek Group of the Ocoee Supergroup. The Sevier Shale and the Knox Group are Paleozoic. From H. L. Moore, *A Roadside Guide to the Geology of the Great Smoky Mountains National Park.* © 1988 The University of Tennessee Press. Used by permission.

4. Deposition of Paleozoic rocks.

Marine sedimentary rocks, ranging in age from Cambrian to Mississippian, were deposited on continental shelves in the Proto-Atlantic Ocean. The Cambrian through Lower Ordovician rock units accumulated in relatively shallow water; but as sea levels rose, sedimentation slowed, and the quartz sands gradually changed to carbonates. Complex deformation operated throughout part of the time that the rocks were being deposited. Some rock units became part of the thrust sheets associated with the Great Smoky fault system, as, for example, Chilhowee rocks exposed on the northwest face of Chilhowee Mountain (outside the park). Other units were overridden by the thrusting and now crop out in valleys and coves, some inside the park and some outside (fig. 48.7).

5. Paleozoic tectonic episodes.

Plate convergence ultimately followed the rifting that had begun late in the Precambrian. An unconformity between Lower and Middle Ordovician stratigraphic units marks the beginning of the Taconic orogeny in the Southern Appalachian region. Subduction and compression resulting from plate convergence during the orogeny initiated folding and faulting and caused uplifted land to shed clastic sediments. The Piedmont suspect terrane arrived, bearing its largely Precambrian rocks of the Ocoee Supergroup, which were pushed up onto the North American continental margin (fig. 47.5, box 47.1).

In the Devonian Period, the Acadian orogeny coincided with the arrival of additional suspect terranes. Although they lie to the east of the Blue Ridge, their docking caused intense compression and a renewal of thrust faulting. Plutons, generated by plate convergence, were emplaced in the foothills east of the Blue Ridge.

Late in Paleozoic time, Mississippian rocks northwest of the park were deformed by compressive forces of the Alleghenian orogeny. Thrust sheets, by a series of movements, were pushed farther to the northwest. The collision of plates destroyed the Proto-Atlantic Ocean and produced the supercontinent Pangea. The Great Smokies were on the crest of a major mountain belt that stretched across Pangea (box 47.1).

6. Mesozoic isostatic adjustments, erosion, and renewed rifting.

Although tectonic activity in the Appalachian Mountains belt subsided, isostatic adjustments caused continued uplift of the Great Smoky Mountains. Erosion shaped the ridges and valleys and uncovered many of the older rocks. To the south and east, new rifts opened and a spreading center developed. The basin of the present Atlantic Ocean began to form, and Africa and Europe drifted away from North America. The climate of the Southern Appalachian region became less tropical and more temperate as the North American plate moved westward and northward.

7. Continued erosion and sculpturing of the Great Smoky Mountains throughout Cenozoic time.

Sporadic uplift continued into Cenozoic time. Modern drainage patterns developed partly in the direction of regional slopes and partly by differential erosion of rock units. In the Pleistocene Epoch, periglacial processes affected rocks and slopes at higher elevations, although the mountains were not glaciated. Climatic oscillations not only caused increased erosion, but also brought about great diversity in the plant and animal life of the region. Some of the unusual plant species found in the Great Smoky Mountains National Park vicinity may have had a much wider range throughout North America prior to the Ice Ages, which put great stress on both plant and animal life. In the Southern Appalachian region fossil remains have been found of extinct giant mammals, such as mammoths, saber-toothed tigers, and giant sloths. Their bones have yielded carbon-14 dates of 11,000 to 12,000 years ago, which means that these creatures managed to survive episodes of cold, harsh weather, probably by sheltering in protected valleys and caves. Why the giant mammals became extinct after the Pleistocene ended is a scientific mystery.

Bibliography

Allen, G. C., and Ragland, P. C. 1972. Chemical and mineralogical variations during prograde metamorphism, Great Smoky Mountains, North Carolina and Tennessee. *Geological Society of America bulletin* 83:1285–1298.

Bird, J. M., and Dewey, J. F. 1970. Lithosphere plate-continental margin tectonics and the evolution of the Appalachian orogen. *Geological Society of America bulletin* 81:1031–1060.

Cantu, R. 1979. *Great Smoky Mountains, the story behind the scenery.* Las Vegas, Nevada: KC Publications.

Catlin, D. T. 1984. *A naturalist's Blue Ridge Parkway, a guidebook to the rocks, flora, and fauna of the Blue Ridge.* Knoxville: University of Tennessee Press. 208 p.

King, P. B. 1977 (2nd edition). *The evolution of North America.* Princeton: Princeton University Press. p. 52–54.

————, Neuman, R. B., and Hadley, J. B. 1968. Geology of the Great Smoky Mountains National Park, Tennessee and North Carolina (geologic map and cross section in pocket). U.S. Geological Survey Professional Paper 587. 23 p.

Moore, H. L. 1988. *A roadside guide to the geology of the Great Smoky Mountains National Park.* Knoxville: University of Tennessee Press. 192 p.

Reed, J. C., Jr. 1987. Precambrian geology of the U.S.A. *Episodes* 10(4):243–247.

Rives, M. R. 1982. *Blue Ridge Parkway, the story behind the scenery.* Las Vegas, Nevada: KC Publications. 48 p.

Rodgers, J. 1987. The Appalachian-Ouachita orogenic belt. *Episodes* 10(4):259–266.

Shelton, N. 1981. *Great Smoky Mountains.* Handbook 112, National Park Service. 128 p.

Williams, H. and Hatcher, R. D., Jr. 1983. Appalachian suspect terranes. In *Contributions to tectonics and geophysics of mountain chains.* Geological Society of America Memoir 158:33–53.

Great Smoky Mountains National Park

Area: 520,269 acres; 813 square miles
Authorized: May 22, 1926
Established for full development: June 15, 1934
Designated a Biosphere Reserve: 1976
Designated a World Heritage Site: December 6, 1983
Address: Gatlinburg, Tennessee 37738

Appendix

U.S. Geological Survey Topographic Maps of National Parks plus selected USGS Topographic Maps (United States Series and others), showing National Park areas.

Index maps covering individual states list additional USGS topographic maps of various scales that are available. Because the U.S. Geological Survey is shifting to the metric system, maps for some states are now published in metric scales (e.g., 1:100,000, 1:50,000, etc.). In some states metrication has been delayed in order to maintain continuity—for example, until complete coverage by 1:24,000-scale maps is available. Other states use a combination of systems with partial metrication.

The U.S. Geological Survey publishes some of the National Park topographic maps in both contour (C) and shaded relief (SR) editions. On the shaded relief maps, the shading accentuates the physical features by simulating in color the appearance of sunlight and shadows on the terrain, thereby creating the illusion of a three-dimensional land surface. On contour maps, the configurations of the earth's surface are depicted by contour lines. National Park topographic maps and other U.S. Geological Survey maps and references can be ordered as follows:

Map Publications
Map Distribution Section
U.S. Geological Survey
Box 25286, Federal Center
Denver, Colorado 80255

Book Publications
Book and Open-File Reports Section
U.S. Geological Survey
Box 25286, Federal Center
Denver, Colorado 80255

Many of the 1:250,000-scale maps are available in raised-relief, plastic editions. Publication and distribution of USGS topographic maps in this format are handled by Geoscience Resources, 2990 Anthony Road, Burlington, North Carolina 27215.

USGS Topographic Maps of National Parks (C = Contour; SR = Shaded Relief)	United States Series, Scale 1:250,000, Covering National Park Areas
Acadia National Park and vicinity, ME 1:62,500 (C)	Bangor
Alaska 1980 (shows Alaska Parklands) 1:2,500,000 and 1:5,000,000 (C)	
Arches National Park, UT 1:50,000 (C)	Moab
Badlands National Park, North/South Units, SD 1:50,000 (C, SR)	Hot Springs, Martin
Big Bend National Park, TX 1:100,000 (C, SR)	Emory Peak
Bryce Canyon National Park, UT 1:31,680 (C, SR)	Cedar City
Canyonlands National Park and vicinity, UT 1:62,500 (C, SR)	Salina, Moab, Cortez
Channel Islands National Park (Anacapa and Santa Barbara Islands only), CA, 1:24,000 (C)	Long Beach, Los Angeles
Crater Lake National Park and vicinity, OR, 1:62,500 (C, SR)	Roseberg, Crescent, Medford, Klamath Falls
Denali National Park and Preserve, Mount McKinley section, AK 1:250,000 (C, SR)	Healy, Talkeetna, Tanana
Everglades National Park, FL 1:250,000 (C)	Miami, Key West
Gates of the Arctic National Park and Preserve, AK	Hughes, Survey Pass, Killik River, Chandler, Phillip Smith Mtns., Wiseman, Ambler River, Chandler Lake

Glacier Bay National Park and Preserve, AK	Juneau, Mt. Fairweather, Skagway, Yakutat
Glacier National Park, MT, 1:125,000 (C)	Kalispell, Cutbank
Grand Canyon National Park and vicinity, AZ, 1:62,500 (C, SR); former National Monument Section 1:48,000 (C)	Grand Canyon, Marble Canyon, Williams, Flagstaff
Grand Teton National Park, WY 1:62,000 (C, SR)	Ashton, Driggs
Great Smoky Mountains National Park (east half), NC, TN, 1:62,500 (C); Great Smoky Mountains National Park (west half), NC, TN, 1:62,500 (C); Great Smoky Mountains National Park, NC, TN, 1:125,000 (C, SR)	Knoxville
Hot Springs National Park (north), Hot Springs National Park (south), AR 1:24,000 (C)	Little Rock
Isle Royale National Park, MI 1:62,500 (C, SR)	Thunder Bay, Hancock
Katmai National Park and Preserve, AK	Afognak, Iliamna, Mt. Katmai, Karluk, Naknek
Kenai Fjords National Park, AK	Seldovia, Kenai, Seward, Blying Sound
Kobuk Valley National Park, AK	Misheguk, Howard Pass, Baird Mtn., Ambler River
Lake Clark National Park and Preserve, AK	Lime Hill, Lake Clark, Tyoney, Kenai, Iliamna, Seldovia
Lassen Volcanic National Park and vicinity, CA, 1:62,500 (C, SR)	Susanville
Mammoth Cave National Park and vicinity, KY 1:24,000 (C)	Evansville
Mesa Verde National Park, CO, 1:24,000 (C, SR)	Cortez
Mount Rainier National Park, WA, 1:50,000 (C)	Yakima
North Cascades National Park, WA, 1:250,000 (C)	Concrete
Olympic National Park and vicinity, WA, 1:125,000 (C, SR)	Cape Flattery, Victoria, Copalis Beach, Seattle
Petrified Forest National Park, AZ, 1:62,500 (C)	Gallup, St. Johns
Rocky Mountain National Park, CO, 1:62,500 (C, SR)	Greeley
Sequoia and Kings Canyon National Parks, CA, 1:125,000 (C, SR)	Fresno, Mariposa
Shenandoah National Park, VA (one sheet) (C); Shenandoah National Park (north section), 1:62,000 (C); Shenandoah National Park (central section) 1:62,500 (C); Shenandoah National Park (south section), 1:62,500 (C)	Charlottesville
Theodore Roosevelt National Park (north unit), ND 1:24,000 (C); Theodore Roosevelt National Park (south unit) ND, 1:24,000 (C)	Dickinson, Watford City
Voyageurs National Park, MN, 1:50,000 (C)	International Falls
Wind Cave National Park and vicinity, SD, 1:24,000 (C, SR)	Hot Springs
Wrangell-St. Elias National Park and Preserve AK	Mt. St. Elias, Yakutat, Gulkana, Bering Glacier, Cordova, Nabesna, McCarthy, Valdez
Yellowstone National Park and vicinity, WY, MT, ID, 1:125,000 (C, SR)	Ashton, Cody, WY; Bozeman, Billings, MT
Yosemite National Park and vicinity, CA, 1:125,000 (C, SR); Yosemite Valley, CA, 1:24,000 (C, SR)	Walker Lake, Mariposa
Zion National Park (Kolob section), UT, 1:31,360 (C, SR); Zion National Park (Zion Canyon section) UT, 1:31,360 (C, SR)	Cedar City

Additional USGS Topographic Maps Covering National Park Areas Not Included in the Preceding List
Biscayne National Park, FL, Miami, 1:250,000
Capitol Reef National Park, UT, Salina, 1:250,000
Carlsbad Caverns National Park, NM, 1:250,000
Great Basin National Park, NV, Ely, Lund, 1:250,000
Guadalupe Mountains National Park, TX, Van Horn, 1:250,000
Haleakala National Park, HI, Maui, 1:250,000 and 1:62,500
Hawaii Volcanoes National Park and vicinity, HI, Hawaii, 1:100,000
Redwood National Park, CA, Eureka, Weed, 1:250,000
Virgin Islands National Park, VI, Puerto Rico and Virgin Islands, 1:250,000; Western St. John, 1:24,000

American Association of Petroleum Geologists, Geologic Highway maps; P.O. Box 979, Tulsa, Oklahoma 74101

Map # 1   1966, MidContinent Region, AR, KS, MO, OK.
Map # 2   1967, Southern Rocky Mountain Region, AZ, CO, NM, UT.
Map # 3   1968, Pacific Southwest Region, CA, NV,
Map # 4   1970, MidAtlantic Region, DE, KY, MD, NC, SC, TN, VA, WV
Map # 5   1972, Northern Rocky Region, ID, MT, WY
Map # 6   1973, Pacific Northwest Region, OR, WA
Map # 7   1973, Texas
Map # 8   1974, Alaska and Hawaii
Map # 9   1975, Southeastern Region, AL, FL, GA, LA, MS
Map #10   1976, Northeastern Region, CT, ME, MA, NH, NJ, NY, PA, RI, VT
Map #11   1976 Great Lakes Region, IL, IN, MI, OH, WI
Map #12   1984, Northern Plains Region, IA, MN, NE, ND, SD

Selected Videotapes

Acadia National Park and Cape Cod National Seashore (30 minutes) Finley-Holiday Films
Anatomy of a Volcano, Mt. St. Helens (59 minutes) Time-Life Video
Big Bend National Park (30 minutes) Finley-Holiday Films
Big Bend Encounter (12 minutes) National Park Service
Bryce, Zion and the North Rim of the Grand Canyon (25 minutes) Finley-Holiday Films
Carlsbad Caverns and Guadalupe Mountains (20 minutes) Finley-Holiday Films
Crater Lake (18 minutes) National Park Service
Denali Wilderness, Alaska (25 minutes) Finley-Holiday Films
Eruption at the Sea (30 minutes) Ka'Io Productions
Eruption of Kilauea Volcano (includes 1955 and 1959–60 eruptions) (42 minutes) Crystal Productions
Eruptive Phenomena of Kilauea's East Rift Zone (includes Pu'u O'o 1980s eruptions) (43 minutes) Ka'Io Productions
Fire and Ice (Mt. Rainier) (20 minutes) National Park Service
Geology of Capitol Reef (18 minutes) Gould Media
Giant Sequoias of Sequoia, Kings Canyon and Yosemite (28 minutes) Finley-Holiday Films
Glacier Bay National Park & Preserve (20 minutes) National Park Service
Glacier Bay (15 minutes) National Park Service
Glacier National Park (30 minutes) Finley-Holiday Films
Grand Canyon and the Petrified Forest (45 minutes) Finley-Holiday Films
Grand Canyon (55 minutes) International Video Works/Readers Digest
Grand Canyon, Zion & Bryce (22 minutes) Finley-Holiday Films
John Wesley Powell, Canyon Geologist (22 minutes) Modern Talking Picture Service; U.S. Geological Survey
Katmai (15 minutes) Media Design Associates, Inc.
Mammoth Cave National Park, Finley-Holiday Films
Mesa Verde (23 minutes) Finley-Holiday Films
Mount Rainier (28 minutes) Finley-Holiday Films
Mount Rushmore and the Black Hills of South Dakota (30 minutes) Finley-Holiday Films
Rocky Mountain National Park (30 minutes) Finley-Holiday Films
Rivers of Fire—An Eruption of Hawaii's Mauna Loa Volcano (25 minutes) Hawaii Natural History Association
River Song: A Natural History of the Colorado River (40 minutes)
Shenandoah—The Gift (20 minutes) Finley-Holiday Films
Spectacular Canyons (Bryce and Zion Canyons, Grand Canyon of the Yellowstone) (17 minutes) Journal Films
Stone Forest (17 minutes) National Park Service
Ten Trails of Yosemite (45 minutes) Where To Productions
Teton Country (30 minutes) Finley-Holiday films
The Denali Wilderness (30 minutes) Finley-Holiday Films
The Grand Tetons (30 minutes) Finley-Holiday Films
The Land that Came in from the Cold (13 minutes) Journal Films
Touring Alaska (52 minutes) Encounter Productions
Touring America's National Parks (60 minutes) (covers 17 National Parks) Encounter Productions
Video Field Trip to the Keweenaw Rift: Lava Flows and Copper (42 minutes) Michigan Technological University
Volcanoes of the United States (20 minutes) Scott Resources

Voyageurs National Park (16 minutes) National Park Service
Wild Alaska (60 minutes) Encounter Productions
Yellowstone High Country Treasure (78 minutes) The Yellowstone Association
Yellowstone in Winter (60 minutes) The Yellowstone Association
Yellowstone/Teton (42 minutes) Finley-Holiday Films
Yellowstone (55 minutes) International Video Works/Readers Digest
Yosemite (55 minutes) International Video Works/Readers Digest
Yosemite National Park—Season and Splendor (40 minutes) Finley-Holiday Films

Videos can be obtained in national parks, from commercial outlets, and from
the address below:
> Park Education Materials Center
> National Parks and Conservation Association
> 1015 31st Street, NW
> Washington, D.C. 20007

Slides with Narration Tapes (available in national parks and from Carolina Biological Supply Company)
Carlsbad Caverns National Park (15 minutes)
Glacier National Park (15 minutes)
Grand Canyon National Park (15 minutes)
Mammoth Cave National Park (15 minutes)
Mesa Verde National Park (15 minutes)
Mount St. Helens (15 minutes)
Olympic National Park (15 minutes)
Painted Desert and Petrified Forest (15 minutes)
Rocky Mountain National Park (15 minutes)
Yellowstone National Park (15 minutes)
Yosemite National Park (15 minutes)

Note: Slide sets without tapes are obtainable from national parks and from commercial outlets, such as Geoscience Resources and Wards National Scientific Establishment.

Selected 16 MM Movie Films
Coulter's Hell (13 minutes) Phoenix Films
Eruption of Kilauea Iki (40 minutes) United States Geological Survey
Geology of Yellowstone National Park (17 minutes) McGraw-Hill
John Muir: Father of Our National Park (15 minutes) Disney
National Parks: Our American Heritage (20 minutes) Bailey Films
National Parks: Promise and Challenge (23 minutes) National Geographic
No Solid Ground to Stand On (Everglades) (28 minutes) Macmillan
The Redwoods (21 minutes) Macmillan
Yellowstone—Our First National Park (15 minutes) Bailey Films
Yosemite National Park (19 minutes) Encyclopedia Britannica Educational Corporation
Yosemite National Park—Its History and Beauty (19 minutes) Bailey Films

Glossary

aa
Lava flows characterized by a spiny, rubbly surface.

abrasion
Mechanical wearing, grinding, scraping, or rubbing away of rock by friction or impact. Solid rock particles, transported by wind, ice, waves, running water, or gravity, are tools of abrasion.

absolute age
Age given in years or some other unit of time.

accreted terrane
See *tectonostratigraphic terrane*.

adjustment of drainage, adjusted drainage
Condition whereby streams are fitted to structure by headward erosion along outcrops of less resistant rocks.

agate
Banded or variegated chalcedony, characterized by colors arranged in alternating strips.

aggradation
Building up of a land surface by deposition.

alkalic basalt
A group (or suite) of basalts, undersaturated in silica, containing one variety of pyroxene and a considerable amount of olivine.

alluvial fan
Large, fan-shaped pile of sediment that forms where a stream's velocity decreases as it emerges from a steep canyon onto a plain or valley floor.

alluvial terrace
Landform, made up of alluvium, produced by dissection or downcutting of a flood plain (see *terrace*).

alluvium
Deposits of gravel, sand, silt, and clay laid down by streams.

alpine glaciers, mountain glaciers, valley glaciers
Glaciers that originate in high cirques or icefields and flow in confined paths from higher elevations to lower elevations.

alpine topography
Mountainous, glacially eroded topography resembling the European Alps.

amphiboles
Group of common, rock-forming mafic minerals (e.g. *hornblende*).

amygdule
Deposits of a secondary mineral, such as calcite or quartz, precipitated by ground water in a gas cavity or vesicle in an igneous rock.

anastomoses (in caves)
Nets of braided, interconnected tubes, commonly confined to a discrete layer of rock.

andesite
Fine-grained, extrusive rock of intermediate composition.

angular unconformity
An unconformity in which younger rock strata overlie an erosional surface on tilted or folded layered rock; bedding planes of the rock units above and below the erosion surface are not parallel.

anhydrite
Mineral composed of anhydrous calcium sulfate, common in evaporite deposits.

anticline
An upfold opening downward; a fold with older beds in the center (see *syncline*).

aphanitic texture
Describing the texture of an igneous rock in which grains are too small to be seen with the unaided eye.

aquifer
Body of rock, of moderate to high permeability, from which ground water can be pumped.

aragonite
Mineral with the same composition as *calcite* but belonging to a different crystal system.

arch
Archlike landform produced by weathering and erosion.

arête
Sharp, serrated ridge that separates abutting cirques or glacial valleys.

arkose
Variety of sandstone containing quartz grains and at least 25 percent feldspar fragments.

atoll
Roughly circular coral reef, composed of low islets that enclose or nearly enclose a lagoon.

aureole
Zone surrounding an igneous intrusion where the country rock has been altered by contact metamorphism.

avalanche chute
Track or trough left after ice, rocks, soil, and trees have swept down a steep slope in a slide.

back reef
Landward side of a reef, including deposits between reef and mainland.

badlands, badlands topography
Intricately dissected topography characterized by a fine drainage network, high drainage density, short, steep slopes, and narrow interfluves. Badlands develop on a surface with little or no vegetative cover in unconsolidated or poorly cemented clays and silts.

bajada
Broad, gently sloping, depositional surface (landform) that develops in a semiarid or arid basin sometimes by coalescence of alluvial fans.

balanced rock
Large rock resting more or less precariously on its base, formed in place by weathering and erosion.

bar (stream)
Ridgelike accumulation of sand and gravel in a stream channel, along banks, or wherever a drop in velocity induces deposition.

barchan (dune)
Crescent-shaped dune with convex windward side and horns pointing downwind.

barrier reef
Long, narrow reef, parallel to a shoreline and separated from the shore by a lagoon.

basal conglomerate
Layer of pebbles or coarse rock fragments that forms the bottom stratigraphic unit of a sedimentary series and rests on an erosion surface, thereby marking an unconformity.

basalt
Dark, fine-grained mafic igneous rock composed of calcium-rich feldspar and ferromagnesian minerals.

base level
The theoretical limiting surface below which a stream (or network of streams) cannot erode.

batholith
Large, discordant, plutonic mass with no known floor that has more than 40 square miles in surface exposure.

beach
Strip of sediment (commonly sand, but sometimes shell fragments, cobbles, or mud) that extends from low-water line inland to a cliff or zone of permanent vegetation.

beach drifting
Lateral movement of sand and pebbles along the foreshore caused by oblique arrival of waves against a coast.

bedding plane
Nearly flat surface separating each layer in a sedimentary sequence from the beds above and below.

bentonite
Very fine-grained sedimentary rock (composed mainly of montmorillonite and colloidal silica) derived from volcanic debris, tuff, and ash. Greasy and soaplike to the touch, it absorbs water quickly and swells when wet. When saturated, it slumps or may even flow as a fine muck. After it dries, it hardens again.

biota
All living organisms (plants, animals) of an area.

blanket sand
Sedimentary deposit of fairly uniform thickness and unusually wide distribution, typically a quartz sandstone deposited by a transgressing sea.

blowout
Hollow formed by wind sweeping away sand or fine soil; the accumulating material may form a *blowout dune*.

bomb (volcanic)
Fragment of lava, or ejecta, rounded, ropy, or spiral in shape, that cooled and hardened while spinning in the air.

boulder
A detached rock mass having a diameter greater than 10 inches (at least the size of a volleyball); the largest size of rock fragments regarded as sediment.

boxwork (cave)
After a network of intersecting blades of calcite was deposited in cracks in highly fractured limestone, the host rock was removed by solution, leaving the intricately crisscrossed fins of the boxwork on cave walls.

braided stream
Stream that flows in an interlacing, tangled network of rivulets and branches separated from each other by islands and bars.

breakdown (cave)
Debris from collapse of the ceiling or walls of a cave.

breaker
Wave that has become so steep that the crest topples forward, moving faster than the main body of the wave.

breccia
Coarse-grained clastic rock composed of large, angular, broken rock fragments cemented together in a finer-grained matrix; it can be of any composition, origin, or mode of accumulation.

brecciated texture
Describing rocks formed from an accumulation of angular fragments or broken and crushed particles.

butte
An isolated, steep-sided hill, with or without a flat top, but usually smaller than a mesa.

calcite
A rock-forming mineral composed of calcium carbonate ($CaCO_3$); the principal constituent of limestone.

caldera
Large, bowl-shaped volcanic depression, more or less circular in form and having a diameter many times greater than that of the original crater. Most calderas result from the explosion and collapse of volcanic cones due to emptying of a magma reservoir by voluminous pyroclastic eruptions.

caliche
A rock layer, common in arid regions, composed of crusts of soluble calcium salts mixed with sand and silt; formed mainly by capillary action drawing mineral-rich ground water toward the surface where it evaporates, precipitating the salts.

calving
Breaking away of blocks of ice from a floating glacier front, forming icebergs.

canyon passage (caves)
Underground stream bed that resembles a chasm.

capture (stream)
The natural diversion of a stream's headwaters into the channel of another stream having greater erosional activity and flowing at a lower level.

carbonates
Sedimentary rocks that contain more than 50 percent by weight of calcite or other carbonate minerals.

carbonation
Type of chemical weathering in which carbon dioxide in air and soil combines with water to form carbonic acid, which reacts with minerals or rocks.

carnelian
Translucent red or orange-red variety of chalcedony.

cave (cavern)
Naturally formed underground chamber, large enough for a person to enter and usually developed in carbonate rock by solution processes.

cave earth
Residue of clay particles and other insoluble materials found in all caves; accumulates as insoluble residue from the solution of limestone or as fine sediment washed into a cave from outside sources.

cave-in lake
A shallow body of water whose basin is produced by collapse of the ground following contraction due to unequal thawing of ground ice in regions underlain by permafrost.

cave lining
Coating of calcite or silica deposited on cave walls during an interval when cave was filled with water.

cave pearls
Concretions of calcite formed in caves when small particles or sand grains become coated with calcium carbonate.

cave rafts
Thin, rectangular pieces of calcite that form in bodies of standing water in caves.

cave popcorn
Nodular, grapelike protrusions formed where water seeps out of pores in the rock walls of a cave.

cementation
Process by which clastic sediments become consolidated into compact rock, usually through precipitation of minerals in spaces between sediment grains.

chalcedony
A cryptocrystalline (having no visible crystalline structure) variety of quartz, sometimes found in chert or as concretions in sedimentary rocks; possesses *conchoidal fracture.*

chattermarks
Small, close, curved cracks made by vibratory chipping of ice-embedded rock fragments against a bedrock surface.

chemical weathering
See *decomposition.*

chert
Hard, fine-grained sedimentary rock formed almost entirely of silica.

cinder cone
Relatively small volcano constructed of loose fragments (*tephra*) ejected from a central vent.

cinnabar
A heavy, red mineral, mercuric sulfide (HgS), the principal ore of mercury.

circum-Pacific belt
The great volcanic belt that rings the Pacific Ocean and marks the location of plate boundaries. Earthquakes and major tectonic activity occur along the belt.

cirque
Steep-walled, amphitheaterlike hollow carved into a high mountain slope by an alpine glacier.

cirque glacier
Glacier occupying a cirque basin at the head of a glacial valley; usually the remnant of a much larger glacier.

cirque lake
Small, almost circular lake occupying a cirque; has no prominent inlet or outlet, is fed by runoff from the cirque walls, and is dammed by a lip of bedrock or a small moraine.

clastic (detrital)
Describing fragments derived from pre-existing rocks or minerals, transported as particles to depositional sites by water, ice, wind, or gravity. When consolidated, such sediment becomes *clastic sedimentary rock.*

clastic dike
Body of clastic materials, derived from underlying or overlying beds, that cuts across bedding of a sedimentary formation.

clastic texture
Rock texture produced by the lithification of an accumulation of fragments.

clay
Sediment made up of particles less than 4 microns in diameter (i.e., smaller than very fine silt grains).

clay minerals
Group of finely crystalline, hydrous aluminum silicates, formed chiefly by weathering of silicate minerals (feldspars, pyroxenes, etc.); able to absorb substantial amounts of water.

cleavage (mineral)
Property of some minerals to break along planes related to a mineral's crystal structure.

cleavage (rock)
Characteristic or tendency of a rock to split along parallel, closely spaced planar surfaces.

closed drainage basin
A drainage basin lacking sufficient streamflow to discharge into another basin, river, or the sea; an area of interior drainage.

coarse-textured drainage
Descriptive of an area where the number of streams and valleys per unit of area is relatively low.

coast
Strip of land of indefinite width that extends from low-tide line inland to the first major change in landform features.

cobble
Rock fragment, somewhat rounded by abrasion, larger than a pebble and smaller than a boulder (i.e., between 2.5 to 10 inches in diameter).

col
Saddle-shaped depression or pass in a ridge, carved by glacial ice breaking down an arête between cirques.

collapse block (in a cave)
Segment of a cave ceiling or wall that has fallen to the cave floor.

collapse breccia
Breccia resulting from the collapse of rock overlying an opening, as in a cave.

colloidal dispersion
Condition existing when submicroscopic particles carry a slight charge of electricity that causes them to repel each other, keeping the particles in suspension.

column (dripstone)
A cave feature formed as a stalactite growing downward meets and joins a stalagmite growing upward.

columnar structure, columnar jointing
Volcanic rock solidified from lava in parallel, vertical columns, polygonal in cross section; formed as a result of contraction during cooling.

composite volcano
Volcano (or volcanic cone) constructed of alternating layers of pyroclastics and rock solidified from lava flows; also called *stratovolcano*.

compression, compressive force
A force that tends to shorten a body (see *tensional force*).

conchoidal fracture
The characteristic of certain minerals or rocks to break along smoothly curved surfaces.

concordant
Having boundaries parallel to bedding or earlier developed planar structures; structurally conformable.

concretions
Hard, compact, rounded masses, or nodules, of mineral matter, of varying size and composition, that have been precipitated out from solution, usually around nucleii of some type (such as fossils).

conglomerate
Coarse-grained clastic sedimentary rock formed by cementation of rounded gravel in a matrix of sand.

contact
Interface between two types or ages of rock.

contact metamorphism
A local process of thermal metamorphism related to the intrusion of magma; affects country rock at or near its contact with the intruded igneous body.

continental crust
Granitic crust that underlies continents and continental shelves.

continental glacier
Ice sheet of considerable thickness covering a substantial part of a continent.

continental shelf
Submarine platform at the edge of a continent, inclined gently seaward.

continuous permafrost
A permafrost zone uninterrupted by pockets or patches of unfrozen ground.

converging plate boundary
Boundary between two tectonic plates that are moving toward one another.

coquina
Soft, porous, coarse-grained, off-white limestone made up of loosely aggregated shells and shell fragments.

coral
Cylindrical-shaped marine invertebrate that lives either singly or bound together structurally with other corals in a colony attached to the sea floor; a reef-forming organism.

correlation
Determination of age relationships between rock units (or geologic events) in separate areas.

country rock
Any rock older than and intruded by an igneous body.

crag and tail
Glacially carved, streamlined ridge with a knob (the "crag") of resistant bedrock and an elongate body (the "tail") of less resistant bedrock or till on the lee side.

crater
Saucerlike depression over a vent on a volcanic cone.

craton
A portion of continental crust that has been structurally stable for a prolonged period of time.

creep
Very slow, continuous downslope movement of soil or unconsolidated debris; type of mass wasting.

crevasse
Open fissure in a glacier caused by fracturing of ice during movement.

cross-bedding
A sequence of crisscross, dipping beds deposited by water or wind currents of variable direction.

cross fold
Complex structure produced when a fold intersects a pre-existing fold of different orientation.

crust
The thin, outermost layer of rock that makes up the earth's surface.

crustal rebound
See *glacial rebound*.

cryptocrystalline
Having an internal crystalline structure so fine that it cannot be detected by a petrographic microscope.

crystalline (rock)
Referring to a rock consisting wholly of crystals or crystal fragments, such as an igneous rock that cooled from a melt or a metamorphic rock that has undergone recrystallization.

cuesta
An asymmetrical ridge that slopes gently on one side, conforming to the gentle dip of resistant beds, and has on the opposite side a steep scarp or cliff formed by the outcropping of resistant beds (cf. *hogback*).

Curie point
The temperature above which thermal agitation prevents spontaneous magnetic ordering. Specifically, the temperature at which there is a transition from ferromagnetism to paramagnetism. Above the Curie point, rocks do not retain their magnetic properties.

cutbank
Develops on the outside of a stream meander where the bank is being undercut by the stream's lateral migration.

dacite
Fine-grained, extrusive igneous rock, intermediate in composition between rhyolite and andesite.

daughter product
Isotope of an element produced by radioactive decay.

debris flow
Mass wasting involving rapid flowage of debris of many sizes.

décollement
Detachment structure of beds showing disharmonic folding and characterized by independent styles of deformation in the rocks above and below the décollement; associated with overthrusting.

decomposition
Weathering processes by which rock is changed due to chemical activity, mainly involving water and atmospheric gases; also called *chemical weathering*.

deflation
Removal by wind of fine sediment particles from a land surface; form of wind erosion.

degradation
The lowering of a land surface and general reduction of relief by erosion.

dendritic drainage pattern
Drainage pattern of a stream and tributaries that resembles a branching tree or veins in a leaf.

desert pavement
Closely packed surface of wind-polished pebbles and rock fragments, somewhat cemented together by calcium carbonate, which tends to form on flat surfaces in arid regions after wind (or water) has removed soil and fine particles. Once formed, desert pavement tends to protect underlying sediment from further deflation (wind erosion).

desert varnish
Thin, dark, brownish red or black patina that develops on the surface of pebbles, boulders, or bedrock in a desert region; consists of iron and manganese oxides with some silica; probably caused by repeated evaporation of water drops from the rock surface.

detrital (detritus)
See *clastic*.

diabase
Dense, black, intrusive igneous rock (frequently found in dikes) studded with light-colored plagioclase crystals (resembling small laths) incorporated with pyroxene crystals; similar in composition to *diorite*.

differential erosion (or weathering)
Erosion (or weathering) that occurs at irregular or varying rates, caused by differences in the resistance and hardness of rocks, minerals, or surface materials.

differentiation
Process by which more than one rock type develops from a common magma.

dike
A tabular, discordant body of intrusive igneous rock (cf. *sill*).

diorite
Coarse-grained, intrusive igneous rock, intermediate in composition; composed of approximately equal amounts of plagioclase feldspar and ferromagnesian minerals.

dip
The vertical angle and direction at which a bedding plane or fault plane is inclined from the horizontal.

disconformity
An unconformity that indicates missing rock layers, although beds above are parallel to those below.

discontinuous permafrost
An intermediate permafrost zone containing patches of unfrozen ground.

discordant
Cutting across bedding or earlier developed planar structures.

disintegration
The breaking up of rock into smaller pieces by frost action, abrasion, etc. Also called *physical weathering* or *mechanical weathering*.

displacement
Refers to the relative movement of two sides of a fault, as measured in a specific direction.

diverging boundary
Boundary separating two tectonic plates that are moving away from each other; a *spreading center*.

dogtooth spar
Variety of calcite with sharply pointed crystals resembling eyeteeth.

dolomite
Sedimentary rock composed of the mineral dolomite (calcium-magnesium carbonate), commonly formed by replacement of limestone.

dome pit (cave)
Underground vertical passage or high chamber formed by solution; has domed ceiling and vertical walls; sometimes called a *vertical shaft*.

drainage basin
Total area drained by a stream and its tributaries.

drainage divide
Boundary between adjacent drainage basins.

driblet cone, driblet spire
Spire-shaped mound resulting from the spattering of lava ejected from an opening in a lava tube; also called *hornito*.

dripstone
General term for calcite or other deposits formed in caves by dripping water.

drumlin
Smooth-surfaced, elongate hill or ridge (the steeper slope being at the up-ice end), made up of compacted glacial till.

dune rock
Consolidated dune sand (see *eolianite*).

earthquake swarm
Series of minor earthquakes occurring in a limited area and time.

ejecta
Material, both solid and liquid, thrown out by a volcano. Sizes may range from ash particles to large blocks of rock and volcanic bombs.

endellite
Clay mineral formed by alteration of other clay minerals in an acidic, sulfate-rich environment.

end moraine
Ridgelike accumulation of till deposited at the lower or outer end of a glacier.

entrainment
The process of water currents picking up and transporting sediments.

eolian
Referring to wind action.

eolianite
Sedimentary rock consisting of clastic material deposited by wind.

eon
The longest geologic-time unit, next in magnitude above an *era;* e.g., the Archean Eon.

ephemeral stream
Flows briefly only in direct response to precipitation in the immediate locality (cf. *intermittent stream*).

epicenter
The point on the earth's surface directly above the *focus* of an earthquake.

epoch
Unit of geologic time shorter than a *period;* e.g., the Holocene Epoch of the Quaternary Period.

epsomite
Mineral consisting of native Epsom salts; occurs in colorless, prismatic crystals, masses, and encrustations in gypsum mines, limestone caves, or in mineral waters in solution.

era
Geologic-time unit next in order of magnitude below an eon and including two or more *periods;* e.g., the Paleozoic Era.

erosion
The general process (or group of processes) whereby the materials of the earth's crust are loosened, dissolved, or worn away and removed by running water, glacial ice, wind, or other geologic agents.

erosion surface
A lowered land surface, smoothed by prolonged erosion.

erratic
Large rock fragment that has been transported by ice from its place of origin; commonly different in composition from underlying bedrock.

esker
Long, narrow, sinuous ridge of stratified, glacial meltwater deposits (sand, gravel) that accumulated in channels in or beneath stagnant ice.

evaporite
Sediment that precipitated from a saline solution; usually marine but sometimes nonmarine.

exfoliation, sheet jointing
Weathering process by which concentric scales, plates, or shells of rock are successively spalled or stripped from the surface of a large rock mass; caused by physical and chemical forces producing differential stresses within the rock, *or* by release of confining pressure (i.e., *pressure release jointing*).

exfoliation dome
Large, rounded landform developed in massive rock, such as granite, by the process of exfoliation (sheet jointing).

extrusive rock
Igneous rock that has cooled or solidified at the earth's surface, whether from a lava flow or from fragments of volcanic debris.

facies, facies change
The term *facies* refers to the characteristics of an assemblage of rock, mineral, or fossil features that reflect a rock's conditions of origin. *Facies change* refers to gradual lateral or vertical changes in sediments or fossils that indicate an evolving depositional environment (see *metamorphic facies, sedimentary facies*).

fault
Fracture in bedrock along which movement has taken place.

fault-block mountains
Uplift along normal or vertical faults forms structural, isolated mountain blocks or ranges that are products of crustal extension.

feldspar group
Includes the earth's most common minerals that make up about 60 percent of the crust. All feldspars contain silicon, aluminum, and oxygen and may contain potassium, sodium, calcium, or other elements in varying combinations. Feldspars are components of many types of rock.

felsenmeer, block field
Frost-riven blocks and coarse, angular fragments that have accumulated on a summit area.

fenster
See *window.*

felsic
Describing silica-rich igneous rocks with a relatively high content of potassium and sodium.

ferromagnesian minerals
Generally dark, iron-magnesium-bearing minerals, such as olivine, hornblende, biotite, augite; the *mafic* minerals.

fetch
The extent of open ocean over the surface of which the wind blows with constant speed and direction, thereby creating a wave system.

fin (ridge)
Local term (Colorado Plateaus) for a sharp, narrow ridge, usually eroded from sandstone.

fine-textured drainage
Descriptive of an area where the number of streams and valleys per unit of area is relatively high.

fiord
Glacially carved coastal inlet occupied by an arm of the ocean.

firn
Compacted mass of granular snow, transitional between snow and glacial ice.

firn line
Irregular, shifting line marking the highest level at which a glacier's winter snow cover is lost during a melting season.

flood basalt
Fluid basaltic lava that erupts repeatedly from fissures and spreads over a vast area.

flood plain
That part of a valley floor, periodically covered by flood waters, which leave a blanket of sediment on either side of the stream channel.

flowstone
General term for any accumulation in caves of calcium carbonate or other mineral matter formed by flowing water on the walls or floor of a cave.

flute (cave)
An incised, vertical or nearly vertical groove developed in a cave wall by dripping water.

flysch
Marine sedimentary facies that is the result of accelerated accumulation of sediment and rapidly changing conditions of deposition (e.g., alternating beds of shales and sandstones).

focus
The point within the earth from which seismic waves originate in an earthquake (cf. *epicenter*).

fold
A bend or warping in bedding or foliation in rock, the result of deformation.

foliated, foliation
Planar arrangement of textural or structural features in any type of rock, especially in metamorphic rock.

footwall
The underlying side of a fault, especially the wall rock beneath an inclined fault (see *hanging wall*).

fore reef
Seaward-facing section of a reef; normally covered with reef talus.

foreshore
Area between mean low water level and mean high water level; the zone regularly covered and uncovered by rise and fall of tides, and upon which waves usually break.

formation
Mappable, lithologically distinct rock unit, usually made up of one or more beds of sedimentary rock, but may be composed of metamorphic or igneous rock units. Contacts separating a formation from adjacent rock units must be recognizable.

fossil
Any evidence of ancient organisms naturally buried and preserved in rocks or sediments.

fringing reef
Relatively small, linear reef formed parallel to a shoreline and directly attached to or bordering the shore, normally without a lagoon.

frost action
Mechanical weathering processes caused by cycles of freezing and thawing water in pores, cracks, and other openings at or near the surface.

frost heaving
Uneven lifting of rock or soil by expansion of freezing water.

frost splitting, frost wedging
Breaking up of rock or soil due to pressure exerted by freezing of water contained in cracks or fissures.

frostwork (cave)
Bristlelike cluster of aragonite needles deposited on cave popcorn.

fumarole
Volcanic vent from which gases and vapors are emitted.

gabbro
Coarse-grained, mafic igneous rock composed mainly of ferromagnesian minerals and calcium-rich plagioclase feldspar.

geode
Hollow, or partly hollow, globular-shaped body occurring in sedimentary rock that weathers out as a discrete nodule or concretion; type of concretion.

geologic agent, geologic process
Running water, wind, ice, and the like, are *geologic agents* that do geologic work by means of natural *geologic processes,* such as erosion and deposition.

geothermal gradient
Rate of temperature increase with depth of crust. Average geothermal gradient in the crust is about 25° C/km of depth.

geyser
Hot spring that intermittently ejects jets of hot water and steam.

glacial drift
Unconsolidated rock debris, stratified and unstratified, that has been transported by glaciers or meltwater streams and deposited directly on land or in the sea.

glacial pavement
A bedrock surface that has been planed down, striated, smoothed, and polished by glacial abrasion.

glacial rebound
Upward movement of the earth's crust after the weight of glacial ice that had depressed the crust is removed; crustal rebound.

glacier
Large, long-lasting mass of ice, formed on land by compaction and recrystallization of snow, which moves because of its own weight; forms where winter snowfall exceeds summer melting.

glacieret
Miniature alpine glacier or cirque glacier; commonly a remnant of a once much larger glacier.

glacier outburst, glacier outburst flood, glacier burst
Sudden release of meltwater from a glacier or glacially dammed lake; may be caused by break out, buoyant lifting of ice, or subglacial thermal or volcanic activity.

glassy (vitreous) texture
Extrusive igneous rock texture similar to that of broken glass; develops as a result of rapid cooling of lava without visible crystallization into individual mineral grains.

glauconite
Dull-green mineral, iron-potassium silicate of the mica group; its presence in a sediment or sedimentary rock indicates very slow accumulation.

glowing avalanche, glowing cloud
Incandescent, turbulent, swiftly flowing cloud of hot ash, dust, and gas erupted from a volcano. The denser, lower part of the cloud may form an ash flow (also called *nuée ardente*).

gneiss
Coarse-textured metamorphic rock comprised of bands or lenses of light and dark minerals.

graben
Crustal block between faults that has been dropped relative to adjacent blocks (see *horst*).

gradient
Degree of inclination or steepness of a surface or slope; ratio of vertical to horizontal, expressed in feet/mile, meters/kilometer, degrees, etc.

granite, granitic rock
Any felsic, coarse-grained, light-colored, intrusive igneous rock containing quartz as an essential component, along with feldspar and mafic minerals.

granular disintegration
Type of mechanical weathering consisting of grain-by-grain breakdown of rock masses. Mineral grains tend to separate from one another along their natural contacts to produce coarse sand or gravel.

gravel
Rounded sediment particles coarser than sand.

gravity survey
Measuring direction and intensity of the earth's gravitational field at different locations, using a gravimeter, and plotting the measurements on a map.

ground moraine
Blanket of till deposited by a glacier or released as glacial ice melts.

ground water, underground water
The water beneath the ground surface that fills the cracks, crevices, and pore space of rocks.

group (stratigraphy)
Two or more geographically associated formations with notable features in common.

gypsum
A mineral (hydrous calcium sulfate) which, when lithified, becomes a sedimentary rock, called *rock gypsum*.

gypsum flower, cave flower
Gypsum that occurs as a curved, elongate cave deposit resembling the shape of a flower. (Epsomite also can occur as cave flowers.)

halite
Mineral composed of sodium chloride; referred to as *rock salt* when lithified.

hanging valley
Smaller valley that terminates abruptly above a main valley or shoreline. A waterfall may develop because the floors of the small valley and the main valley (or shoreline) are discordant in level.

hanging wall
Overlying side of, or surface above, an inclined fault plane (see *footwall*).

harmonic tremors
Earthquake vibrations, associated with volcanic activity, which are harmonious or regular in displacement and arrival time.

headward erosion
Uphill growth of a valley above its original course by gullying, mass wasting, and sheet erosion.

helictite (cave)
A contorted cave deposit (calcite speleothem) that angles or twists erratically.

high-energy coast
Exposed to ocean swell and stormy seas; characterized by average breaker heights of greater than 50 cm.

hogback
Long, narrow, sharp-crested ridge formed by differential erosion on the outcropping edges of steeply inclined or highly tilted resistant rocks (cf. *cuesta*).

hoodoo, earth pillar, rock pillar, toadstool
Column or pillar of bizarre shape, produced by rainwash and differential weathering of horizontal beds; may be small and delicate or large and massive.

horn, matterhorn
Sharp peak formed when cirques cut back into a mountaintop from three or four sides.

hornblende
Commonest mineral of the amphibole group; found in both igneous and metamorphic rocks.

hornfels
Fine-grained, unfoliated metamorphic rock.

horst
Crustal block between faults that has been raised relative to adjacent blocks (see *graben*).

hot spot
Deep, persistent heat source in the earth's mantle that causes magma to rise in a plume toward the surface; associated with Hawaiian-type volcanoes but not island-arc volcanoes.

hot spring, thermal spring
Natural spring with water temperature warmer than human body temperature.

hydration
Type of chemical weathering in which water combining with minerals in rock causes mineral grains to expand, weakening the rock.

hydromagnesite
White, earthy magnesium mineral, sometimes called "moon milk."

hydrothermal metamorphism
Alteration of rock by passage through it of hot water bearing mineral ions in solution.

iceberg
Block of glacier-derived ice floating in water.

icecap
Large glacier, not restricted to a valley, extending in all directions.

ice cave
An artificial or natural cave in which ice forms and persists throughout all or most of the year.

ice sheet
See *continental glacier*.

ice wedge
Large, wedge-shaped, vertical, or inclined sheet, dike, or vein of ice that tapers downward into permafrost; originates by freezing of water in a narrow crack produced by thermal contraction of the permafrost. An ice wedge enlarges with subsequent freezing and thawing.

igneous rock
Rock formed from solidification of magma on the earth's surface or within the crust.

ignimbrite
Rock formed by consolidation of pyroclastic debris, welded tuff, recrystallized ash flows, etc.

impermeable
Rock, soil, or sediment that cannot transmit fluid, such as water or petroleum, under pressure.

incised meander
Meander that retains its sinuous curves as it cuts downward below the level at which it originally formed. Environmental change is the likely cause of renewed downcutting.

inclusion
Fragment or enclave of rock distinctly different in composition from the enclosing igneous rock.

intermediate rocks
Transitional or intermediate in composition between felsic and mafic igneous rocks.

intermittent stream
Flows only at certain times of the year (cf. *ephemeral stream*).

intrusion
Upward movement of a magma body into the crust; also, an igneous rock mass so formed within surrounding rock.

intrusive rock
Igneous rock that has cooled within the earth's crust.

intrusive sheet
Tabular igneous intrusion, usually concordant.

island arc, volcanic arc
Curved chain of volcanic islands above a subduction zone and on the landward side of an ocean trench. Not all volcanic arcs are island arcs. Some volcanic arcs (not island arcs) develop above a subduction zone that dips down beneath a continental plate.

isoclinal fold, isocline
Fold having parallel limbs, with both limbs having the same amount and direction of dip.

isostasy
State of balance or equilibrium, comparable to floating, between adjacent blocks of crust resting on the mantle's plastic layer.

isostatic adjustment
Concept of vertical movement of blocks or sections of the earth's crust in order to attain equilibrium.

jasper
Variety of chert, characteristically reddish in color.

joints, joint sets
Surface fractures, or partings, in rock without displacement. A joint surface is usually a plane. A group of parallel joints forms a *joint set*.

karst topography
Topography that develops due to erosion of limestone or dolomite (occasionally gypsum) by solution of rock; characterized by closed depressions, sinkholes, caves, underground drainage, etc.

keratophyre
Light-colored volcanic rock, composed predominantly of plagioclase feldspar and quartz; extruded from submarine vents.

kettle, kettlehole
Depression in glacial outwash caused by melting of a buried block of stagnant ice.

kettle lake
Body of water occupying a kettle in a pitted outwash plain or moraine.

kipuka
Islandlike surface area surrounded by a lava flow; highly variable in size (Hawaiian word meaning "opening").

klippe
An outlier or isolated, erosional remnant of a thrust sheet.

knob (topographic)
Protruding mass of resistant rock; any rounded mound or hillock.

lahar
Mudflow on a volcano flank; debris carried in the flow consists of pyroclastics, mud, vegetation, etc.; sometimes triggered by glacial meltwater.

lamina
The thinnest recognizable unit or layer of deposition in a sediment or sedimentary rock.

landform
Any of the natural features (canyon, stream, arch, mountain, etc.) that make up the earth's surface character.

landscape
An assemblage of landforms.

landslide, landsliding
General inclusive term for slope failure or downslope movement of rock and debris; may be sudden, fairly rapid, or slow.

lapilli
Pyroclastic fragments, with diameters ranging from 2 to 64 mm, blown out in a solid state during an eruption.

lateral moraine
Low, ridgelike accumulation of till carried along the side of a valley glacier.

lava
Magma that reaches the surface, either through fissures or volcanic vents.

lava ball
Accretionary feature that forms when a fragment of solidified lava rolls along, growing in size as it picks up sticky lava.

lava fountain
Jet of incandescent lava that shoots into the air as lava is forced out of a vent by hydrostatic pressure and expansion of gas bubbles.

lava lake
Lake of lava (usually basaltic) in a volcanic crater or depression; may be solidified or partly solidified.

lava stalactite
Iciclelike pendant of lava that hardened while dripping from the roof of a lava tube or lava tunnel.

lava tree
Mold of lava projecting above the surface (see *tree mold*).

lava tube, lava tunnel
Hollow space, cave, or tunnel beneath the surface of a solidified lava flow; formed by withdrawal of molten lava after a crust forms over the flow.

lee (glacier)
Indicates the side of a hill, knob, or prominent rock located away from an advancing glacier or ice sheet; facing the downstream side of a glacier and sheltered from severe abrasive action (see *stoss*).

lignite
Brownish black coal, intermediate in coalification between peat and subbituminous coal.

limb
One side of a fold; each limb is usually part of two adjacent folds.

limestone
Sedimentary rock, organic or inorganic in origin and composed mainly of calcite; a carbonate rock.

lithification
Consolidation and conversion of sediment to form sedimentary rock.

loess
Fine-grained deposit of windblown dust, composed of angular, unweathered grains of quartz, feldspar, and other minerals; unconsolidated, but may be held together by calcareous cement.

longshore current
Moving mass of water that develops parallel to a shoreline, caused by oblique impingement of waves against a coast. This piles up water, some of which then flows along the shore.

longshore drift
Movement of sediment parallel to the shore carried by longshore currents.

low-energy coast
A sheltered coast protected from strong wave action by headlands, islands, or reefs and characterized by average breaker heights of less than 10 cm.

mafic
Describing silica-poor igneous rocks having a relatively high content of magnesium, iron, and calcium.

magma
Naturally occurring, mobile, melted rock, containing water and gases under pressure; generated at high temperatures within the lower crust and upper mantle.

magma chamber
Reservoir in the earth's crust occupied by magma.

magnetic field (of the earth)
Region of magnetic forces surrounding the earth.

magnetic poles (of the earth)
The points of emergence at the earth's surface where the strength of the magnetic field is greatest. Magnetic lines of force appear to enter at the negative (presently south) pole and leave at the positive (presently north) pole; also called *geomagnetic poles*.

magnetic reversal
A change in the earth's magnetic field between *normal polarity* and *reverse polarity*.

mantle
The thick rock shell of the earth's interior below the crust and above the core. The upper mantle is believed to be partially plastic in character.

marble
Coarse-grained metamorphic rock made up of interlocking calcite crystals, or recrystallized calcite. Some marble contains dolomite crystals.

marker flow
Lava flow with distinctive characteristics that allow it to serve as a stratigraphic reference or be traced over a distance.

massive (rock)
Describes stratified rock that appears to be without internal structures, such as minor joints and lamination; also applies to igneous rocks (granite, diorite) that may have fairly homogeneous texture over a wide area.

mass wasting
Loosening and downslope movement of soil and rock in response to gravitational stress; rates of movement range from creep (very slow) to very rapid falls and slides.

matrix
Finer-grained material (groundmass) enclosing larger grains or particles of a rock or filling spaces between grains.

meander
A sinuous curve or loop in a stream's course; produced by lateral erosion.

meander scar
Meander that has dried up or filled with sediment and vegetation after abandonment by the stream that formed it.

mechanical (physical) weathering
See *disintegration*.

medial moraine
A single, long ridge of till riding on top of a glacier; formed by joining of adjacent lateral moraines of merging glaciers moving downslope.

mélange (tectonic)
Body of rock, large enough to be mapped, that includes many fragments and blocks of various sizes and kinds of rock embedded in a sheared matrix of malleable rock, such as claystone or shale.

member (stratigraphy)
Rock unit comprising some portion of a formation.

mesa
Broad, flat-topped hill bounded by cliffs and capped by a resistant rock layer.

metamorphic core complex
Domal uplift of metamorphic and igneous rock overlain by a technically detached, unmetamorphosed cover of sedimentary rock.

metamorphic facies
Group of metamorphic rocks having similar mineral compositions and physical characteristics and hence assumed to have formed under a particular range of pressure and temperature conditions.

metamorphic rock
Rock that has been transformed from pre-existing rock into texturally or mineralogically distinct new rock as a result of high temperature, high pressure, or both, but without the rock melting in the process.

mica
Group of rock-forming minerals (muscovite, biotite) that commonly occur in flakes, sheets, shreds, or scales. Micas, which are complex silicates, are characterized by low hardness and perfect basal cleavage.

microplate
Small tectonic plate.

microseisms
Very small earth tremors.

mid-oceanic ridge
A continuous, mostly submarine mountain range, with a central rift valley, extending through the Arctic, Atlantic, Indian, Pacific, and Antarctic Oceans; see *spreading center*.

migmatite
Complex rock composed of mixed igneous and metamorphic material; has a banded and veined appearance.

mineral
A naturally occurring, inorganic, usually crystalline solid with a definite chemical composition and characteristic physical properties that are either uniform or variable within definite limits.

mirabilite
White or yellow monoclinic mineral occurring as a residue from saline lakes, playas, and springs, and as an efflorescence; also called Glauber's salt.

monadnock
Upstanding hill or mountain, rising conspicuously above the general level of an erosion surface; remnant of a former upland.

monocline, monoclinal fold
An open fold (usually in relatively flat-lying beds) that bends, or flexes, from the horizontal in only one direction and has only one limb.

montmorillonite
Group of clay minerals derived from the alteration or weathering of calcic feldspars, ferromagnesian minerals, and volcanic glass; common in soils, a typical insoluble residue in cave earth, and the main constituent of bentonite.

moraine
Body of till being carried on or by a glacier or left behind after a glacier has receded.

mountain belt, cordillera, mountain chain
Complex, interconnected series of mountain ranges having a well defined, longitudinal trend; commonly several thousand miles in length and several hundred miles wide.

mountain range
A single mass of mountain ridges, closely related in age and origin.

mudball
Spherical mass of mud or mudstone (may be as large as 8 inches in diameter) in a sedimentary rock; developed by weathering and breakup of clay deposits.

mud cracks
Crudely polygonal pattern formed in drying clay or silt by shrinkage.

mudflow
Debris flow consisting of soil, sand, silt, ash, clay, etc., mixed with enough water so that it moves downslope as a viscous mass.

mud pot
Hot spring choked by dissolved compounds in boiling mud.

natural bridge
Arch-shaped landform, produced by weathering and erosion, that spans a water course or a dry stream valley.

natural levees
Low ridges of flood-deposited sediment, formed on either side of a stream channel, that thin away from the channel.

needle
Narrow, pointed spire of resistant rock formed by weathering and erosion.

nonclastic
Describes a sediment or rock formed by chemical precipitation or organic secretion (see *clastic*).

nonfoliated
Pertaining to metamorphic rocks (or massive igneous rocks) lacking *foliation*.

normal fault
Caused by tensional force, which moves the hanging wall down in relation to the footwall (cf. *reverse fault*).

normal polarity
Natural remanent magnetization closely parallel to the present ambient magnetic field direction; that is, the direction of magnetic north in a rock set when it was formed and approximately the same as modern polarity.

novaculite
Dense, hard, even-textured, light-colored, siliceous sedimentary rock resembling chert.

nuée ardente
See *glowing avalanche*.

nunatak
Isolated bedrock peak, projecting above an icefield or glacier.

obsidian
Volcanic glass, usually black or reddish brown and rhyolitic in composition; characterized by conchoidal fracture.

oceanic crust
That part of the earth's crust that underlies the ocean basins; predominantly basaltic.

oceanic trench
Deep, narrow submarine trough, parallel to an island arc or continental margin; associated with a subduction zone.

offset
Horizontal displacement along a fault, measured perpendicular to the interrupted horizon.

olivine
Common, dark-colored, rock-forming ferromagnesian mineral.

ooids, ooliths
Tiny, spherical grains about the size of fish roe, usually composed of calcium carbonate, that form by precipitation around a nucleus (shell fragment, sand grain, etc.) in shallow, wave-agitated water. The grains accumulate as sedimentary layers.

oolite
Sedimentary rock, usually limestone, made up of ooliths, cemented together.

ophitic texture
Igneous rock texture (especially in diabase) in which lath-shaped plagioclase crystals are partially or completely included in pyroxene crystals (typically augite).

ore
Naturally occurring earth material from which metal(s) can be profitably extracted.

original horizontality, principle of
Most water-laid sediment is deposited in horizontal or near-horizontal layers closely parallel to the earth's surface.

orogeny
Forming of mountain systems by deformational (compressive) crustal processes, accompanied by metamorphism and igneous activity.

orthoclase feldspar
Light-colored, rock-forming mineral of the alkali feldspar group.

outwash (glacial)
Sand and gravel deposited by meltwater streams in front of or beyond the terminal moraine or margin of a glacier. Coarser material is dropped nearer to the ice.

outwash plain
Smooth, broad, gently sloping plain formed by deposits of heavily loaded meltwater streams draining a glacier margin.

overbank flow
That part of high water or flood water that rises out of the stream channel, depositing sediment over the flood plain surface.

oxbow lake
Crescent-shaped body of standing water occupying a cutoff or abandoned meander.

oxidation
Uniting of oxygen with another element or compound; a process of chemical weathering.

pahoehoe
Basaltic lava flow having a ropy or billowy surface texture (see *aa*).

paleomagnetism
The record of the earth's ancient magnetic fields.

paleosol
Soil layers of past geologic time buried by younger sediments; usually recognizable in a soil profile.

parabolic dune
Deeply curved dune in a region of abundant sand; horns (arms) point upwind and may be anchored by vegetation.

patch reef
Small, moundlike organic reef, usually part of a larger reef complex.

patterned ground
More or less symmetrical forms—circles, polygons, nets, steps, and stripes—characteristic of, but not necessarily confined to, surficial material subject to intensive frost action—as in polar and subpolar regions.

pebbles
Small, rounded, waterworn stones, ranging in size between a small pea and a tennis ball (1/6 to 2.5 inches).

pediment
Broad, flat (or gently sloping) bedrock erosional surface that develops at the base of a mountain front, usually in a dry region; typically covered with a thin veneer of gravel.

pedestal, mushroom, rock pedestal
Slender rock column or neck topped by a bulge of more resistant rock; formed by undercutting due to weathering and erosion.

pegmatite
Extremely coarse-grained igneous rock, usually granitic in composition and made up mostly of large quartz and feldspar crystals; commonly found around the margins of deep-seated plutons.

Pele's hair, Pele's tears
The natural spun glass of "Pele's hair" tends to stream out as threads of lava from the pear-shaped "tears," which are ejected from a vent in a liquid state and solidify in flight. Lava threads break off and may be borne by wind for some distance.

periglacial
Refers to processes, conditions, areas, climates, and topographic features at the margins of glaciers and ice sheets, where freezing and thawing occur seasonally.

period
Geologic time unit longer than an *epoch;* subdivision of an *era;* e.g., the Cambrian Period.

permafrost
Ground that remains permanently frozen over a period of many years.

permeability
Capacity of a rock to transmit a fluid (water, petroleum, etc.) (see *porosity*).

permineralization
Process of fossilization whereby the original hard parts of an organism have had additional mineral material deposited in their pore spaces.

petroglyphs
Prehistoric carvings on rock.

phaneritic texture
Describing the texture of an igneous rock in which individual components are visible to the unaided eye.

phenocryst
Relatively large, conspicuous crystal in the groundmass of a porphyritic igneous rock.

phreatic eruption
Explosive, steam-propelled ejections of ash and mud caused by heating and expansion of ground water.

phreatic explosion
Explosion of steam, mud, and nonincandescent material; caused by heating and consequent expansion of ground water due to an underlying heat source; more violent than a *phreatic eruption.*

phreatic zone
Subsurface zone (below the water table) in which all rock openings are filled with water; also called the *saturated zone.*

phreatomagmatic explosion
Violent volcanic explosion that extrudes magmatic gases and steam along with pyroclastics; caused by mixing of magma with ground water or ocean water.

phyllite
Fine-grained metamorphic rock in which clay minerals have recrystallized into microscopic micas. A silky sheen on corrugated cleavage surfaces is a characteristic.

physiographic province
Geomorphic region showing in all its parts similar climate and geologic structures and having relief features readily differentiated from those of neighboring regions.

pictograph
Picture-writings or signs carved or painted on rock, hides, etc.

piedmont
General term specifying a feature at the base of a mountain; a foothill.

piedmont glacier
Thick, continuous sheet of ice resting on land at the base of a mountain range; formed by the spreading out and coalescing of valley glaciers from higher elevations.

piercement structure, diapir
Piercing or rupturing of domed or uplifted rocks by mobile material pushed up by tectonic stresses.

pillar (cave)
Bedrock support in a cavern remaining after removal of surrounding rock by solution.

pillow lava, pillow structure
Submarine extrusions of basaltic lavas that solidify as pillow-shaped masses, closely fitted together. "Pillows" range in size from a few inches to a few feet in diameter.

pingo
Large, blisterlike mound, pushed up by pressure of freezing water in permafrost.

piping
Erosion by percolating water in a layer of subsoil, resulting in caving and in the formation of narrow conduits, tunnels, or "pipes" through which soluble or granular soil material is moved.

pisolite
Sedimentary rock, usually limestone, made up chiefly of small, round accretionary bodies cemented together; similar to *oolite.*

placer
Surficial deposit, usually on a beach or in a stream bed, containing particles of valuable minerals or native metal (e.g., gold) in unusual concentrations due to their higher specific gravities.

plagioclase feldspar
Abundant group of rock-forming minerals containing sodium and/or calcium in addition to aluminum, silicon, and oxygen.

plastic
Describes a substance capable of being molded, bent, or deformed under strain without rupturing (up to the limit of its malleability).

plastic flow zone (glacier)
Bottom part, or sole, of a glacier where ice is molded or bent under stress so that it flows without fracturing.

plateau
Broad, comparatively flat landform of considerable extent, elevated above adjacent countryside; usually bounded by cliffs or an escarpment on at least one side.

plate, tectonic plate
Large, moving slab of rock making up part of the earth's crust.

plate tectonic theory
The concept that intense geologic interactions occur along boundaries between rigid, slowly moving plates that make up the earth's surface.

plucking, quarrying
Glacial erosional process by which freezing of water in cracks and joints loosens and detaches large rock fragments and blocks from bedrock, which are then torn out by advancing ice.

pluton
Any massive body of igneous rock formed at depth in the earth's crust by consolidation of magma.

pluvial climate
Climatic interval characterized by abundant rainfall, as during a transitional or interglacial period. A *pluvial lake* is one that formed during such an interval.

polyp
An individual living coral.

porosity
The percentage of openings (interstices) in the total volume of a rock.

porphyritic texture
Igneous rock texture in which larger crystals (phenocrysts) are set in a finer groundmass, which may be crystalline or glassy.

pothole
A general term for any naturally formed circular pit or depression in soil or rock; sometimes used as a synonym for *kettle* or *kettlehole;* in desert regions, may refer to a "rock tank," in which water collects. Most commonly, a *pothole* designates a cylindrical hollow eroded into the rock of a stream bed by abrasive action of the stream's sediment load.

Precambrian shield
Extensive area of ancient Precambrian plutonic and metamorphic rocks at or near the surface; part of a *craton.*

pressure-release jointing
Exfoliation that occurs in once deeply buried rocks that have been unloaded by erosion, thus releasing their confining pressure.

primary structures
Features formed during the time of a rock's origin but before its final consolidation.

protalus rampart
Ridge of loose, angular rock fragments that accumulated along the toe of a large perennial snowbank lying at the base of a cliff. When the climate warmed, the snowbank melted, leaving an arcuate ridge of boulders and other coarse debris.

pseudomorph
Mineral whose outward crystal form is that of another mineral species.

pumice
Light-colored, vesicular, glassy rock, commonly of rhyolitic composition and often sufficiently bouyant to float on water; formed during explosive eruptions from lava containing gas and water vapor in large amounts.

pyroclasts, pyroclastic debris or material
Rock fragments formed by volcanic explosion or ejection from a volcanic vent. Rock consolidated from such material is said to have *pyroclastic texture.*

pyroxene group
Dark, rock-forming silicate minerals (e.g., augite) similar in chemical composition to the amphiboles.

quartz
Crystalline silica (SiO_2), an important rock-forming mineral.

quartzite
a) Metamorphic rock formed by recrystallization of sandstone or chert during regional metamorphism. *b*) Very hard but unmetamorphosed sandstone, consisting of quartz grains solidly cemented with secondary silica.

quicksand
Bed of fine, smooth sand, saturated with water flowing upward through voids, forming a soft, shifting, semiliquid mobile mass that yields easily to pressure and will not support much weight.

radioactive decay
Spontaneous disintegration or transformation of certain unstable atoms, called the parent material, to daughter products through gain or loss of nuclear particles.

radiometric dating
Methods for determining absolute age in time units of geologic materials, based on nuclear decay of naturally occurring radioactive isotopes (e.g., carbon-14, uranium-lead ratio, etc.).

rain-shadow zone
Region on the downwind (lee) side of a mountain range that receives little precipitation because clouds drop much more rain and snow on the upwind side of the range.

recharge
Addition of water to an aquifer or to the phreatic (saturated) zone.

recrystallization
Development in a rock of new crystalline grains that may or may not have the same composition but are usually larger than the original grains.

reef
Resistant mound or ridge formed in the ocean by accumulation of calcareous remains of plants and animals and other debris.

reef crest
Upper surface of a reef.

reef terrace
Shelflike, eroded surface that may develop at several levels on a reef because of successive episodes of submergence and emergence.

regional metamorphism
Metamorphism affecting an extensive area and involving relatively high temperatures and pressures; associated with orogenic belts.

regression (marine)
Retreat or withdrawal of a sea from land areas, and the consequent evidence of such withdrawal.

relative time, relative age
Refers to a sequence of geologic time determined by organic evolution or superposition (see *absolute age*).

relict
Describes a topographic feature developed by erosional or depositional processes no longer operating because of a pronounced change in environmental conditions.

relief
Difference in elevation between highest and lowest points on an area of the earth's surface.

remanent magnetism
The permanent magnetism induced in *a*) an igneous rock when it cooled from a liquid to a solid state (i.e., past the *Curie point*), *or* induced in *b*) a sedimentary rock by settling of magnetic grains during deposition so that they became aligned with the earth's magnetic field.

reservoir rock
Any rock sufficiently porous and permeable to store and transmit petroleum.

residual soil
Soil cover formed in place by weathering of underlying bedrock (cf. *transported soil*).

reversed polarity
Natural remanent magnetization opposite to the present ambient magnetic field direction. When the rock was formed, the direction of magnetic north was approximately 180° different from what it is now (see *normal polarity*).

reverse fault
Fault, caused by compression, in which the hanging-wall block moved up relative to the footwall block (see *normal fault*).

rhyodacite
Extrusive porphyritic rock, intermediate in composition between dacite and rhyolite.

rhyolite
Fine-grained, felsic extrusive rock composed chiefly of feldspar and quartz.

rift zone
a) System of crustal fractures (e.g., the rift valley along the crest of the mid-oceanic ridge) where tensional forces pull plates apart. *b*) On Hawaiian volcanoes, a linear zone of volcanic features associated with underlying lava conduits.

rigid zone (glacier)
Upper, brittle part of a glacier; does not flow plastically.

rimstone (cave)
Thin, crustlike deposit of calcite forming a ring around an overflowing basin or pool of water in a cave.

ring dikes, ring complex
Arcuate or roughly circular igneous intrusions, vertical or inclined away from the area in the center, which may have collapsed.

ripple marks
Small-scale troughs and ridges produced by wind or water currents sweeping over loose sediments.

roche moutonnée
Elongate knob or hillock of bedrock that has been smoothed and scoured by moving ice on the upglacier (stoss) side so that the rock is gently inclined and rounded. On the downglacier (lee) side the rock is steep and hackly from glacial plucking or quarrying.

rock
Naturally formed, solid aggregate composed of one or more minerals or a mass of mineral matter, sediments, precipitates, or organic matter.

rock-basin lake
Small body of water occupying a depression scoured or excavated in bedrock by glacial action; also called *tarn*.

rock cycle
Theoretical, graphical concept for relating tectonism, erosion, and rock-forming processes to common rock types.

rockfall
Sudden free fall of a bedrock segment from a cliff or steep slope; the fastest form of mass wasting.

rock flour
Fine, powdery glacial sediment consisting of ground-up rock produced by abrasion underneath a moving glacier.

rock glacier
Mass of poorly sorted, angular boulders and fine materials, cemented by interstitial ice and flowing downslope by means of glacier motion. An inactive or relict rock glacier may descend by *creep* (mass wasting).

rock salt
Sedimentary (evaporite) rock composed chiefly of the mineral *halite*.

rockslide
Rapid sliding of a block or mass of bedrock that has become detached from an inclined surface of weakness (such as a bedding plane).

roof pendant
Downward projection of *country rock* into an igneous intrusion (batholith or pluton).

salt anticline
Anticlinelike structure formed by the bulging upward of a linear salt core into overlying sedimentary beds.

saltation
Sediment transport by which stream currents move sand grains along a channel in short, abrupt leaps.

salt dome
Circular piercement structure produced by upward movement of a pipelike salt plug.

sandblasting
Abrasion by wind-blown sand.

sandstone
Sedimentary rock formed by cementation of sand-sized grains, usually 85 to 90 percent quartz; may be deposited by water or wind.

savanna
Tropical or subtropical grassland, characterized by scattered shrubs, sparse trees, and a pronounced dry season.

scallop (cave)
Asymmetrical solution hollow formed by turbulent water flow.

schist
Strongly foliated, crystalline metamorphic rock with excellent cleavage; characterized by coarse-grained minerals oriented approximately parallel to each other.

scoria
Frothy, highly vesicular basalt.

sea wall (beach)
Long, steep-faced, natural embankment or berm of shingle or boulders, built by powerful storm waves along the high-water mark.

sediment
Unconsolidated, loose, solid fragments or particles that can originate by 1) weathering and erosion of pre-existing rocks; 2) chemical precipitation from solution, usually in water; and 3) secretion by organisms. Sediment particles may be transported and deposited by water, wind, or ice.

sedimentary facies
Significantly different rock types occupying laterally distinct parts of a layered rock unit (formation).

sedimentary rock
Rock formed from 1) lithification of any type of sediment; 2) precipitation from solution; 3) consolidation of remains of plants and/or animals; or 4) combinations of processes.

seismic waves
Oscillations of energy that produce earth vibrations during an earthquake.

selenite
Clear, colorless, cleavable variety of gypsum.

shale
Fine-grained, clastic sedimentary rock, formed by the compaction mainly of clay; splits easily into thin slabs parallel to bedding planes.

shatter zone
Area of randomly fissured or cracked rock; may have a network of veins filled with mineral deposits.

sheet jointing
See *exfoliation*.

shield volcano
Broadly convex, gently sloping, volcanic cone built by flows of low-viscosity lava.

shingle
Coarse, loose, well rounded waterworn detritus of various sizes.

sill
Tabular, concordant body of intrusive igneous rock (cf. *dike*).

silt
Fine-grained sediment intermediate in size between clay particles and sand particles.

siltstone
Fine-grained clastic sedimentary rock consisting of compacted silt and mud; does not split like shale.

sinkhole (karst)
Closed depression found on land surfaces underlain by carbonate rock; formed by solution or by collapse of a cavern roof.

sinking stream
Surface stream that disappears into an underground channel; may not reappear in the same or even an adjacent drainage basin.

sinter
Chemical sedimentary rock, mainly siliceous, deposited as a hard incrustation by precipitation.

slate
Fine-grained metamorphic rock that splits easily along smooth, parallel cleavage planes.

slough
Wet, marshy area; may be a sluggish channel or a backwater.

slump block
Mass of rock material that slides downward (slumps), usually with a backward tilting relative to the slope from which it descends.

slope wash, rain wash
Movement of sediment downslope by mass wasting assisted by trickling or running water not confined to a channel.

solfatara
Fumarole that emits sulfurous gases.

solifluction
Downslope flow of water-saturated debris over impermeable material; a type of mass wasting common in permafrost areas.

solution (weathering)
Process of chemical weathering by which soluble mineral and rock constituents pass into solution and become *dissolved load*.

source rock (petroleum)
Sedimentary rock containing organic matter that is converted to petroleum by burial and other postdepositional changes over a period of time.

sparry crystal growth
See *dogtooth spar*.

spatter cone
Low, steep-sided mound formed of globs of lava piled up around a volcanic vent.

speleothem
Any secondary mineral deposit formed in a cave by the action of water.

spheroidal weathering
Process of chemical weathering that tends to round off a block of rock by loosening and detachment of successive, concentric shells of decayed rock from the block; similar to *exfoliation*.

spilite
Green, mafic igneous rock composed predominantly of plagioclase feldspar and chlorite; an altered basalt.

spit
Fingerlike extension of a sandy beach, with the far end terminating in open water.

spongework (cave)
An entangled complex of irregular, interconnecting, tubular channels or cavities of various sizes in the walls of limestone caves; separated by amazingly intricate and perforated partitions and remnants of partitions.

spreading center
The crest of the mid-oceanic ridge, where the ocean floor is being pulled apart and new crust is forming; a crustal rift where tensional forces separate tectonic plates.

stack, sea stack
Pillarlike erosional remnant left detached from a headland as a wave-eroded coast retreats inland.

stalactite
Icicle-shaped, dripstone pendant formed on a cave ceiling.

stalagmite
Cone-shaped dripstone mass formed on cave floors, generally directly below a stalactite.

steppe
Generally flat, treeless, semiarid region, with short grasses and scattered bushes.

stock
Small pluton having an area of surface exposure of less than 40 square miles.

stoss
Describes the side of a hill, knob, or prominent rock facing the upstream side of an advancing glacier; the side most exposed to abrasive action (see *lee*).

striations
Furrows or lines, generally parallel, inscribed on rock surfaces as rock fragments embedded in ice are dragged along by a moving glacier.

strike
Compass direction or trend of a line formed by the intersection of an inclined plane (such as a fault plane) with a horizontal plane.

strike-slip fault
Fault in which movement is parallel to the strike of the fault surface. Two blocks slide past each other, resulting in lateral displacement.

stromatolite
Laminated, calcareous sedimentary structure consisting of the remains of blue-green algae; occurs in various shapes and forms—horizontal, columnar, branching, etc.

subduction
The sliding down of the leading edge of an oceanic plate beneath a continental margin or an island arc.

suite
Set of geographically related crystalline rocks (igneous and/or metamorphic).

supergroup (stratigraphy)
Assemblage of related formations and groups in a geographic locality.

superimposed stream
Stream (one established on a new surface) that maintained its course while cutting across buried rock structures as it eroded downward into underlying rocks.

surf zone
Strip of shore where the waves break and roll up the beach or beat against a cliff face (see *foreshore*).

suspect terrane
See *tectonostratigraphic terrane*.

suspension
Process of sediment transport in which fine particles are held indefinitely in eddying, turbulent water.

syenite
Coarse-grained intrusive rock containing one or more feldspars, less than 50% dark minerals, and little or no quartz (see *trachyte*).

syncline
Downfold opening upward; fold with younger beds in the center.

tactite
Complex, calcareous metamorphic rock formed by contact metamorphism.

talus
Broken rock accumulating at the base of a cliff or slope, most commonly as a result of frost action and mass wasting.

tectonostratigraphic terrane, accreted terrane, suspect terrane, exotic terrane
Fault-bounded region made up of rock units having a geologic history markedly different from that of contiguous terranes; considered to be a discrete portion of oceanic or continental material not formed in place.

tension, tensional force
Stress that tends to elongate or pull apart a solid body.

tephra (pyroclastics)
General term for fragmental material ejected from a volcanic vent.

terminal moraine
End moraine marking the farthest advance of a glacier.

terrace
General term for a landform that resembles a stairstep with a flat tread (the top) and a steep riser (the face).

terrain
Physical features characterizing a region of the earth's surface; an ecologic environment.

texture (rock)
General physical appearance of a rock, including size, shape, arrangement of grains, etc.

thaw lake
Pool of water formed on a glacier surface by meltwater accumulation; also, cave-in lake in permafrost.

tholeiitic basalt
Suite of basalts that contain several varieties of pyroxenes and little or no olivine; oversaturated in silica.

thread-lace scoria, reticulite
Formed by bursting of lava bubbles, creating a delicate, three-dimensional network of glass threads.

thrust fault
Reverse fault in which the fault plane has a low angle of dip.

tidewater glacier
Glacier that reaches the sea; usually terminates in an ice cliff, from which icebergs are discharged.

till
Glacial drift consisting of unsorted and unstratified rock debris carried or deposited by a glacier.

time-transgressive rock unit
Sedimentary formation that becomes younger (or older) when traced across country; usually the result of laterally shifting shorelines when a sea is advancing or retreating.

topographic desert
Region of aridity caused by location in a continental interior or on the lee side of high mountain ranges (cf. *rain-shadow zone*).

trachyte
Fine-grained, alkaline extrusive rock (extrusive equivalent of *syenite*).

traction
Process of sediment transport by which water currents or wind roll or slide particles and fragments along a stream bed (in water), or (in air) over a land surface.

transform fault
Variety of strike-slip fault along which displacement abruptly stops or changes form; may be associated with a mid-oceanic ridge.

transform plate boundary
Boundary where two plates are sliding past each other.

transgression (marine)
Spread or extension of the sea over land areas, and the consequent evidence of such advance.

transported soil
Weathered surficial material unrelated to bedrock or older soil on which it rests (see *residual soil*).

transverse stream
Flows across bedrock structures, such as folds, faults, and fault blocks.

travertine
Any porous deposit of calcite formed by rapid precipitation of calcium carbonate from surface water or ground water *or* by evaporation around the mouth of a spring, especially a hot spring.

tree mold
Cylindrical hollow in a lava flow formed by envelopment of a tree by the flow, solidification of lava in contact with the tree, and disappearance of the tree by burning and subsequent removal of charcoal and ash. The mold preserves surficial features of the tree.

tritium
Naturally occurring, radioactive hydrogen isotope, formed by cosmic rays in the upper atmosphere; enters the earth's hydrologic cycle via precipitation.

tritium dating
Calculation of age in years by measuring the concentration of tritium in water. The method provides a means of tracing subsurface movement of water and determining its velocity.

truncated spur
Faceted lower end of a ridge projecting into a glacial valley that has been eroded by glacial ice.

tsunami
Seismic sea wave produced by a large-scale, short-duration disturbance of the ocean floor, principally by a submarine earthquake.

tubular passage (cave)
Cavern or tunnel with an elliptical or lenticular cross section formed in the phreatic zone.

tufa
Chemical sedimentary rock composed of calcium carbonate; formed by evaporation as an incrustation around the mouth of a hot or cold calcareous spring or seep.

tuff
Rock formed by solidification of volcanic dust, ash, or pumice.

tumescence
Swelling of a volcano edifice due to rising magma.

tumuli
Mounds or hillocks on the surface of pahoehoe lava (sing. *tumulus*).

tundra
Treeless, gently undulating plain, commonly underlain by muck and permafrost; characteristic of arctic and subarctic regions.

turbidites
Rocks or sediments deposited in a marine environment by settling out from muddy (turbid) water flowing along a sloping ocean bottom.

turbidity current
Flowing mass of sediment-laden water (heavier than clear water) that slides downslope along a lake bottom or ocean floor.

type locality
Geographic location where a rock unit is situated and from which it takes its name.

ultramafic rock
Dark plutonic rock (such as peridotite) composed almost entirely of mafic minerals.

unconformity
Planar surface between rock units that represents a time break in the geologic record, with the rock unit immediately above being considerably younger than the rock beneath.

uniformitarianism, principle of
The assumption that geological processes operating today are the same processes that operated in the geologic past.

U-shaped valley
Characteristic cross profile of a glacially excavated trough (i.e., flat-floored with nearly perpendicular side walls).

vadose zone
Subsurface zone located above the *phreatic zone* and between the water table and the surface; zone in which rock openings are filled partly with water and partly with air; also called *zone of aeration*.

valley glacier
See *alpine glacier*.

valley train
Outwash dropped or spread within the valley of a meltwater stream that flows out from a glacier (see *outwash plain*).

ventifact
A boulder, cobble, or pebble with flat surfaces worn or polished by abrasion of wind-blown sand.

vesicle
Small cavity or void in volcanic rock caused by gas bubbles trapped in solidifying lava.

vesicular texture
Igneous rock texture characterized by abundant cavities formed by entrapment of gas bubbles during solidification of lava.

viscosity
Resistance to flow. Thin, runny lava has low viscosity; thick, sticky lava has high viscosity.

volcanic dome
Steep-sided, pluglike, or spinelike mass of volcanic rock solidified from viscous lava in or immediately above a volcanic vent; also called *plug dome, lava dome*.

volcanic neck, volcanic plug
Vertical, pipelike body of rock that represents the conduit to a former volcanic vent; an erosion remnant of a volcanic cone.

water gap
Deep, narrow, low-level pass, through which a stream flows, that penetrates to the base of and across a mountain ridge; a narrow gorge or ravine cut through resistant rocks by a stream.

water table
The surface within the ground below which rocks are saturated; separates the *phreatic zone* (below) from the *vadose zone* (above).

wave-cut platform, wave-cut terrace
Horizontal bench in the surf zone cut by wave erosion on a coast; usually overspread by beach sediment.

weathering
Natural processes that alter or break down rock at or near the earth's surface (see *decomposition, disintegration*).

weather(ing) pits
Shallow depressions on flat or gently sloping exposures of granitic rocks; attributed to strongly localized solvent action of impounded water.

welded tuff
Glass-rich volcanic rock composed of silicic pyroclasts welded together.

wind gap
Shallow notch in the crest or upper part of a mountain ridge; former *water gap,* abandoned by the stream that formed it.

window (landform)
Hole that appears in a fin (or narrow ridge) as a result of differential weathering; also, an opening under an arch or natural bridge.

window (structure)
Eroded area within a thrust sheet that displays rocks beneath the thrust sheet; also called *fenster.*

xenocryst
Crystal foreign to the igneous rock body in which it occurs; resembles a phenocryst.

zone of accumulation
The portion of a glacier with a perennial snow cover, where more snow accumulates than melts each year.

zone of wastage
The portion of a glacier where melting or loss of ice and snow exceeds the accumulation.

References

Bates, R. L. and Jackson, J. A., editors. 1987. *Glossary of geology,* 3rd edition. Alexandria, Virginia: American Geological Institute. 788 p.

Lapidus, D. F. with D. Coates. 1987. *The facts on file dictionary of geology and geophysics.* New York: Facts on File Publications. 347 p.

Index

Alaskan National Parks and Preserves.